AF324723

# Numerical Simulations of Incompressible Flows

# Numerical Simulations of Incompressible Flows

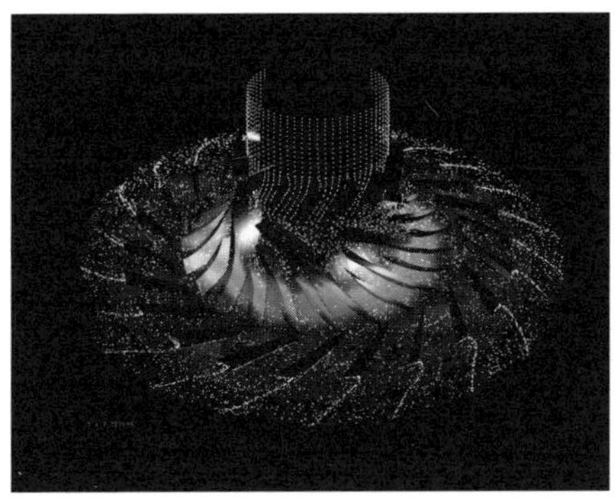

edited by

## M. M. Hafez

University of California, Davis

*Published by*

World Scientific Publishing Co. Pte. Ltd.

5 Toh Tuck Link, Singapore 596224

*USA office:* Suite 202, 1060 Main Street, River Edge, NJ 07661

*UK office:* 57 Shelton Street, Covent Garden, London WC2H 9HE

**British Library Cataloguing-in-Publication Data**
A catalogue record for this book is available from the British Library.

**NUMERICAL SIMULATIONS OF INCOMPRESSIBLE FLOWS**



ISBN 981-238-317-4

Printed in Singapore by World Scientific Printers (S) Pte Ltd

# CONTENTS


**Dedication**                                                          ix

**Biographical Sketch and Publications of Dr. Dochan Kwak**             xiii

**I    Navier-Stokes Solvers**

Incompressible Navier-Stokes Solvers in Primitive Variables
and Their Applications to Steady and Unsteady Flow
Simulations                                                             3
  *Cetin C. Kiris, Dochan Kwak, and Stuart Rogers*

Evolution of Artificial Compressibility Methods in CFD                  35
  *Sankaran Venkateswaran and Charles L. Merkle*

On Incompressible Flow Solvers                                         50
  *Rainald Lohner, Chi Yang, Juan Cebral, Orlando Soto, and
  Fernando Camelli*

Direct Primitive Variable Solution Techniques for
Incompressible Flows                                                   72
  *Prem K. Khosla and Stanley G. Rubin*

Mass Conservation and the Accuracy of Non-staggered Grid
Incompressible Flow Schemes                                            86
  *Michael C. Wendl and Ramesh K. Agarawal*

**II    Projection Methods**

Accuracy of Projection Methods for the Incompressible
Navier-Stokes Equations                                               101
  *David L. Brown*

A Split-Step Scheme for the Incompressible Navier-Stokes
Equations                                                             108
  *William D. Henshaw and N. Anders Petersson*

Higher-Order Semi-Implicit Projection Methods                         126
  *Michael L. Minion*

Numerical Solutions of the Incompressible Navier-Stokes
Equations Using Helmholz Velocity Decomposition                       141
  *Koorosh Nikfetrat and Mohamed Hafez*

Three Dimensional Viscous Incompressible Flow Simulations
Using Helmholz Velocity Decomposition    165
*Koorosh Nikfetrat and Mohamed Hafez*

## III    Finite Element Methods

On the Numerical Simulation of Incompressible Viscous Fluid
Flow around Moving Rigid Bodies of Elliptical Shape    179
*Roland Glowinski, Hector L. Juarez, and Tsorng-Whay Pan*

Steady Incompressible Inviscid and Viscous Flow Simulation
Using Unstructured Tetrahedral Meshes    203
*Kenneth Morgan, Dhemi Harlan, Oubay Hassan,
Kaare Sorensen, and Nigel Weatherill*

Stabilized Finite Element Formulations and Interface-Tracking
and Interface-Capturing Techniques for Incompressible Flows    221
*Tayfun E. Tezduyar*

Computation of Moving Boundaries and Interfaces and
Stabilization Parameters    240
*Tayfun E. Tezduyar*

## IV    Higher-Order Methods

A Fourth-Order Difference Method for the Incompressible
Navier-Stokes Equations    263
*Bertil Gustafsson, Per Lotstedt, and Anders Goran*

Solving the Navier-Stokes Equations Using $B$-spline Numerical
Methods    277
*Olivier Botella and Karim Shariff*

Viscous Flow Simulation Using Compact Differencing and
Filtering Schemes    288
*Miguel R. Visbal*

Incompressible Flow Simulation by Using Multidirectional
Finite Difference Scheme    311
*Kunio Kuwahara, Satoko Komurasaki, and
Angel Bethancourt*

## V  Innovative Methods

Use of Lattice Boltzman Method for Computing
Incompressible Flows  329
*Nobuyuki Satofuka, Mitsuru Ishikura, and Ishikawa Yutaka*

Simulation of Structure-Fluid Interaction by Universal Solver
CIP for Solid, Liquid, and Gas in Cartesian Grid  340
*Takashi Yabe, Youichi Ogata, and Takao Kawai*

Vorticity Confinement — Recent Results: Turbulent Wave
Simulations and a New, Conservative Formulation  350
*J. Steinhoff, M. Fan, and L. Wang*

## VI  Applications in Aeronautics

Optimization of Propellers Using a Vortex Model  375
*Jean-Jacques Chattot*

A Noniterative Method for Boundary-Layer Equations I.
Two-Dimensional Laminar Flows  385
*Tuncer Cebeci and Jian Ping Shao*

A Noniterative Method for Boundary-Layer Equations II.
Two-Dimensional Laminar and Turbulent Flows  397
*Tuncer Cebeci and Jian Ping Shao*

Parallelized Flow Analysis over High-Lift Airfoils Using
Incompressible/Compressible Navier-Stokes Equations  415
*Chang Sung Kim, Chongam Kim, and Oh Hyun Rho*

Ice Effects on Aerodynamic Performance of Airfoils  450
*K. D. Lee, J. Shim, and J. Chung*

## VII  Applications Beyond Aeronautics

Flow in the Stenotic Carotid Bifurcation  463
*Stanley A. Berger and Vitaliy L. Rayz*

Direct Simulation of the Vaporization of Droplet Arrays  474
*M. T. Parra, H. A. Dwyer, and F. Castro*

Incompressible Simulations on the Flowfield around a Square
Cylinder at Ground Proximity  484
*Dong-Ho Lee and Jaeho Hwang*

Numerical Simulation of an Underwater Hot Launch Exhaust
Plume — 501
*W. G. Szymczak, T. Hsieh, W. P. Chepren, A. E. Berger,
and T. F. Zien*

## VIII Multiphase and Cavitating Flows

AUSM-family Schemes for Multiphase Flows at All Speeds — 517
*Meng-Sing Liou and Jack R. Edwards*

Modeling and Computation of Unsteady Cavitating Flows
Based on Bubble Dynamics — 544
*G. H. Schnerr*

## IX Special Topics

Advances in Hydrodynamics Stability Theory: the
Wave/Vortex Eigenmodes' Interaction — 577
*Oleg S. Ryzhov*

Evaluation of a Preconditioned Flow Solver for a Broad Range
of Mach Number and Temperature Ratio — 605
*B. K. Lambert, L. K. Taylor, and W. R. Briley*

Evolution of Topology in Axi-symmetric and 3-D Viscous
Flows — 622
*G. Scheuermann, W. Kollman, X. Tricoche,
and T. Wischgoll*

Data Compression for Incompressible Flow Solutions — 644
*Dohyung Lee*

The Historic Remains 'Drifting Cup in a Meandering Stream'
in China, Korea and Japan — 673
*Keun-Shik Chang*

# Dedication

This volume consists mainly of papers presented at a workshop on Numerical Simulations of Incompressible Flows, held in Half Moon Bay, California, June 19-21, 2001. The purpose of the workshop was twofold: to discuss the state-of-the-art in this field and to honor Dr. Dochan Kwak on the occasion of his $60^{th}$ birthday and in recognition of his contributions over the past twenty years. The workshop consisted of twelve sessions with speakers from U.S., Britain, France, Germany, Sweden, Japan, and Korea. The workshop was co-organized by M. Hafez of University of California Davis and Dong Ho Lee of Seoul National University with partial support from Brain Korea 21, Mechanical Engineering Research Division.

The book covers many topics including theory and applications, algorithm developments and modern techniques for incompressible flow simulations. A few of the authors were not able to attend the workshop but their papers are included in this collection. There are 37 articles in the nine sections of the book dealing with different aspects of the field.

Dr. Dochan Kwak obtained his BS degree from Seoul National University, Department of Mechanical Engineering in 1964. His Ph.D. was from Stanford University, Department of Aeronautics and Astronautics / Mechanical Engineering in 1975. He was one of the first Ph.D. students of Stanford – NASA LES research program, using ILLIAC IV.

As a staff-scientist in Los Alamos Laboratory in 1979 he developed a hydro code using Particle-In-Cell method. From 1979 to 1992 he was a research-scientist at NASA Ames. From 1993 to present he has been a chief of Advanced Computational Methods, Computational Physics and Simulation, and currently NAS (NASA Advanced Supercomputing) Applications branch.

He has served in many professional organizations, wrote 90 technical papers and holds two patents including NASA / DeBakey Ventricular Assist Device. He has lectured in the States and abroad including The University of Tennessee Space Institute in 1985, IBM CFD short course in 1985, VKI Lecture series in Belgium in 1989, and AGARD CFD short courses in Portugal in 1996.

Dr. Kwak's major technical contributions have been in the development of incompressible flow simulation procedures and in the application of his codes for solving practical problems. Dr. Kwak has been instrumental in developing the INS3D family of codes. The original version (1982-1987) was based on the Artificial Compressibility Method using ADI with application to the Space Shuttle main engine. During the period from 1988-1997 other versions emerged: INS3D_UP (upwind / line relaxation), INS3D_LU (LU-SGS Scheme), INS3D-FS (pressure projection / fractional step) with applications to advance engines for RLV (NASA Marshall Space Flight Center's Pump Consortium) as well as to Penn State Artificial Heart. The latter is an example where CFD plus rocket propulsion technology contributed to developing a biofluid device which has saved many human lives. In November 1994, he received the first NASA Software of the Year Award for developing INS3D codes.

Dr. Kwak has received the H. Julian Allen Award, NASA Ames Research Center, December 1985 for Best Scientific Research Paper. Between 1983 and 1994 he has received numerous NASA awards. In April 1999, Dr. Kwak was inducted into the Space Technology Hall of Fame, together with Dr. Michael DeBakey. Currently Dr. Kwak is working on developing the next generation RLV as well as advanced bio-computing capability, with high-fidelity unsteady flow simulations.

Dr. Kwak has been a leader in the field of CFD. His contributions have literally impacted the lives of many people. Dr. Kwak has been a very successful scientist and a very effective manager at the same time. His friends and colleagues wish him many productive and pleasant years to come.

Dr Dochan Kwak

# Biographical Sketch of Dr. Dochan Kwak

## Education

BS Seoul National University, Mechanical Engineering, 1964

Ph.D. Stanford University, Aeronautics and Astronautics / M.E., 1975

Stanford-NASA started LES research using ILLIAC IV. He was the first Ph.D. under this program.

## Professional Experience

Staff Scientist, Los Alamos National Laboratory, 1979

During this period, he developed a CFD code for simulating stress-supporting multi-material problem by devising a 2nd order Particle-In-Cell scheme for elastic-plastic flow.

Research Scientist, NASA Ames research center, 1979 – 1992

Chief of Advanced Computational Method Branch, Computational Physics and Simulation Branch, and NAS (NASA Advanced Supercomputing) Applications Branch (current position), 1993 – Present

## Services to Professional Organizations

AIAA Fluid Dynamics Technical Committee member, 1990 - 1993

AIAA 10th Computational Fluid Dynamics Conference Chairman, June 1991

NASA Ames Basic Research Council, 1993- 1998

Stanford-NASA Center for Turbulence - Research Steering Committee, 1993-1996

International Conference in CFD (ICCFD) Science Committee member, 2000- (Exec committee member since July 2002)

## Publications, Patents, and Lectures

Publication list attached

Holds 2 patents including NASA/DeBakey Ventricular Assist Device

Lectured nationally and internationally including:

The University of Tennessee Space Institute, 1985

VKI (Von Karman Institute) Lecture in Belgium, 1989

IBM CFD Short Course, 1990

AGARD CFD Short course in Portugal, 1996.

## Major Technical Contributions

During his career with NASA, he has made significant contributions toward advancing the state of the art in computational physics, especially in incompressible flow CFD area, and has authored and co-authored 90 technical publications, and generated two patentable ideas, one as a sole inventor. He has made many significant contributions to the technology of CFD and has demonstrated an outstanding ability to conduct and direct complex research programs. He has received many honors and awards for his contributions in aerospace related activities as well as for his technology transfer efforts most notably to bio-medical research. A list of his personal contributions is given below:

**Developed incompressible flow simulation procedures for solving real-world problems (has been instrumental in developing and applying INS3D family of codes)**

Developed incompressible Navier-Stokes solution procedures for obtaining

- Steady-state solutions based on pseudo-compressibility approach, and
- Time-accurate solutions based on pseudo-compressibility approach and a fractional-step method

These procedures are primarily designed for three-dimensional applications, and resulting publications have been widely cited in the CFD community.

**INS3D Code development: 1982-1987 INS3D-Original version was developed by Kwak et al. Artificial compressibility method, ADI**

Developed the first version of incompressible Navier-Stokes codes, INS3D family of codes. This is one of the pioneering efforts that enabled three-dimensional incompressible Navier-Stokes simulations for real-world applications. The INS3D code has made a far-reaching impact on aerospace problems and beyond, and has been widely disseminated to US aerospace, automobile, chemical, mechanical and bio-medical industries, and to other government agencies. The first version alone has been disseminated to 105 research organizations, including 28 universities, 18 US government agencies, and 59 industry laboratories. Subsequent versions of the code are routinely requested by users throughout the nation. The impact of INS3D code was recognized by being selected as the first recipient of NASA Software of the Year Award in 1994.

With NASA Administrator, Dr. Dan Goldin, in November 1994, during **the First NASA Software of the Year Award** ceremony

**Major Applications: Space Shuttle Main Engine Enhancements**

The Space Shuttle Main Engine (SSME) developed in the 1970s by NASA's Marshal Space Flight Center is the world's most sophisticated reusable rocket engine. During a Shuttle lift off and ascent, three main engines burn more than 500,00 gallons of liquid hydrogen and liquid oxygen. NASA continues to increase the reliability and safety of the Shuttle flights through a series of enhancements to the SSME. One such effort was power head redesign during the early 80s.

The powerhead, considered the backbone of the engine, consists of the main injector assembly and pre-burners. Partially burned hot gas pass through Hot Gas Manifold (HGM) to main injector assembly.

*Phase II + Hot Gas Manifold (HGM) redesign work*

During 1983-84, the powerhead redesign was undertaken, focusing on two-duct HGM. Kwak and several Rocketdyne engineers, J. Chang, S. Barson and G. Belie, collaborated in applying INS3D code for this task. The INS3D code was still under development. Kwak et al. developed and successfully applied CFD simulation procedure to this task of enhancing the SSME powerhead. The new design was developed based on two-duct hot gas manifold. The two-duct design replaced the previous three-duct engine, smoothing the fuel flow, reducing pressure and turbulence, and lowering temperatures in the engine during operation. This two-duct design first flew on the shuttle in July 1995. It significantly improved fluid flow in the system, thus reducing maintenance and enhancing the overall performance of the engine. This pioneering work was probably the first major application of CFD to rocket propulsion system. Besides developing flow solvers for applications, this work defined requirements on flow visualization. Kwak was instrumental in developing visualization software as a part of the CFD processes. His group purchased the first work station at NASA Ames to initiate this task, resulting in Producing widely disseminated graphics packages later on.

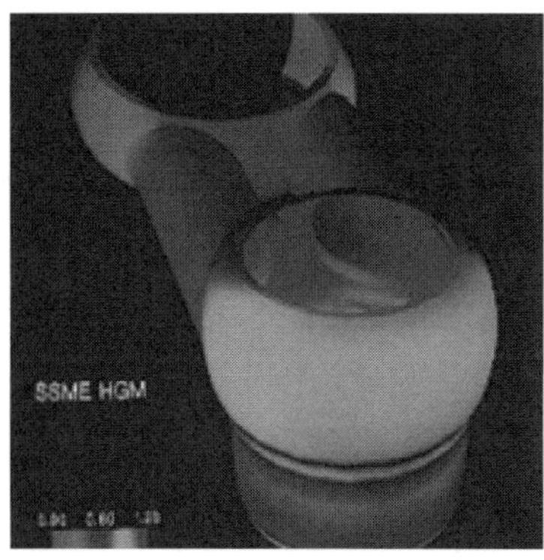


SSME Phase II+ new 2-duct HGM design

Flight engines in 1995

1983-84 In front of the new two-duct power head with Rocketdyne engineers (Steve Barson, Jim Chang and Gary Belie)

## INS3D Code development

Starting in 1988, he embarked on enhancing the capability of INS3D to improve some of the limitations inherent to ADI scheme by incorporating numerous numerical schemes into the code. At this point he formed INS Group, most notably Stuart rogers and Cetin Kiris joined the Group. Together they produced INS3D family of codes and application procedures.

1988-1997     INS3D-UP an upwinding version (Rogers, Kirs and Kwak)

INS3D-LU artificial compressibility with LU scheme (Yoon and Kwak)

INS3D-FS using pressure projection and fractional step (Rosenfeld and Kwak)

1998-Present   INS3D parallel production version(Kiris, Rogers and Kwak)

Current activity: Developing high-fidelity unsteady flow simulation capability.

## Applications to RLV Turbopump

During this period, Kwak and Kiris started developing simulation technology for turbopump flow in future Reusable Launch Vehicle (RLV). Pump technology is playing a key role in developing liquid rocket engine which is the basis of the Shuttle and also for the next generation RLV after Shuttle. Kwak and Kiris participated in Pump Technology Consortium organized by NASA's Marshal Space Flight Center.

## Application to an artificial heart

INS3D was initially developed for obtaining steady state solutions. For that artificial compressibility method was very effective. To obtain time-accurate solutions, sub-iteration scheme was adopted into the code. Resulting unsteady solver was applied to a piston type artificial heart, developed by Penn State University, with moving valves and pusher plate. To simulate this flow, moving geometry capability was developed using overset technology.

Computational simulation of the flow in the Penn State artificial heart set a new milestone in biofluid computation, and was nominated for the Core Award in medicine category for the

1993 Computerworld Smithsonian Award. Together with Kiris and Rogers, they advanced the state-of-the-art in artificial heart flow analysis.

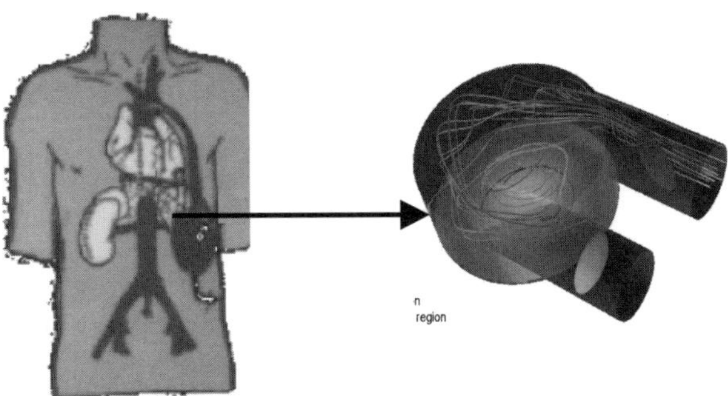


Penn State Artificial Heart

## Application to development of a Ventricular Assist Device (VAD)

The pump simulation technology and moving boundary procedure enabled Kiris and Kwak to participate in the development of DeBakey VAD. The DeBakey VAD is based on fast rotating axial flow pump with much similarity in flow characteristics with the SSME inducer. In addition flow features are directly tied to physiology of blood flow. They, through the application of CFD tools, made the device usable for human implantation.

Next two charts show the technical highlights. As a result of this contribution, they were inducted into 'space Technology Hall of Fame' in 1999.

The DeBakey VAD was chosen as the Commercial Invention of the Year in 2001.

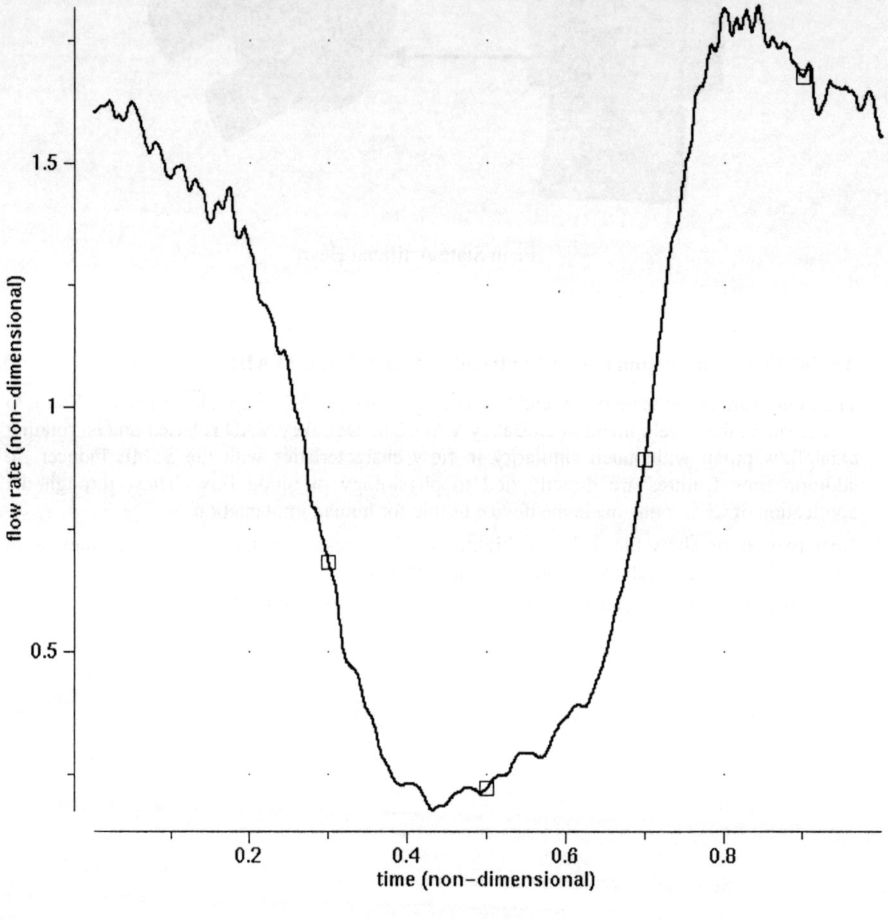


Unsteady Flow

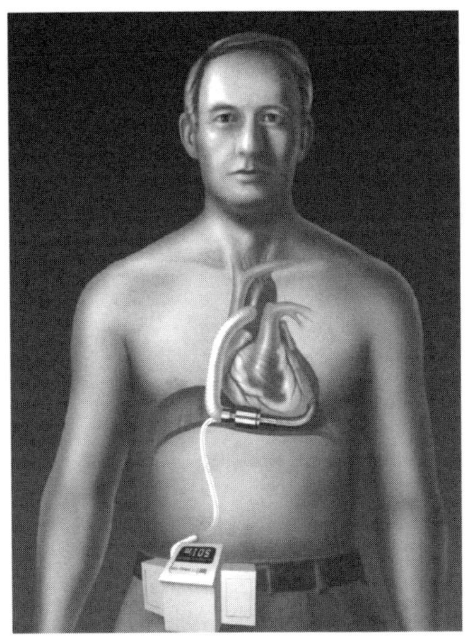

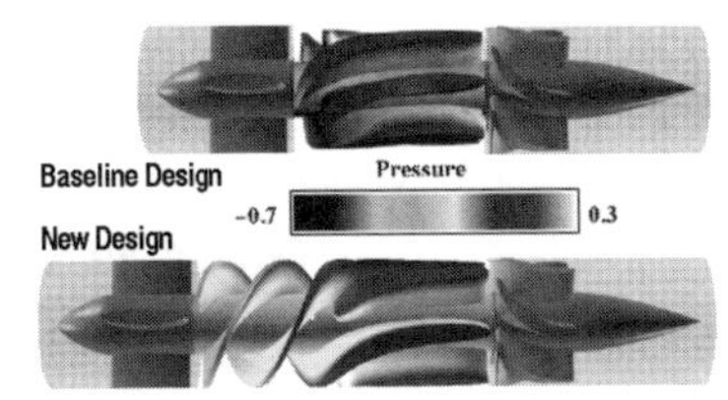


|  | Baseline Design | New Design |
|---|---|---|
| Hemolysis Index | 0.02 | 0.002 |
| Thrombus Formation | Yes | no |
| Test Run Time | 2 days | 30+ days |
| Human Implantation |  | 1 year |

**First Human Implantation in Europe**

On November 13, 1998, the first six DeBakey VADs are implanted in European patients by Drs. Roland Hetzer and DeBakey at the German Heart Institute of Berlin.

One of the patients, fifty six year old Josef Pristov, is able to return home and spend Christmas with his wife after a month's stay for recovery and monitoring at the clinic.

**April 1999 - Inducted into "Space Technology Hall of Fame"** together with Cetin Kiris of NASA Ames, Dr. Michael DeBakey (above) of Baylor Colleger of Medicine and a group of NASA/JSC engineers

With NASA Administrator, Sean O'Keefe, June 25, 2002, during **2001 NASA Commercial Invention of the Year Award**

## Additional Awards

- Special Achievement Award, March, 1983 (National Full Scale Wind Tunnel Simulation)
- H. Julian Allen Award, NASA Ames Research Center, December 1985 For the best scientific research paper.

- Engineer of the Year Award, June 1986 (AIAA SF Section)
- **NASA** Exceptional Scientific Achievement Medal, October 23, 1986 (with Dr. James Fletcher, then NASA Administrator, and Dr. W. Balhaus, NASA Ames director)

- **Engineering Project of the Year Award**, February, 1987 The Institute for the Advancement of Engineering
- **Distinguished Science and Technology Award**, June 20, 1987 National Society of Professional Engineers
- **NASA Space Act Award**, August 13, 1992 (Artificial heart flow simulation)
- **NASA Manned Flight Awareness Award**, June 1994 NASA Special Act Award, August 1, 1994

# Publications of Dr. Dochan Kwak


1. Kwak, D., 'Nonreflecting Far-Field Boundary Conditions for Unsteady Transonic Flow Computation', *AIAA J.*, vol 19, No. 11, pp 1401-1407, Nov 1981. Also, *AIAA Paper 80-1393*, AIAA 13th Fluid and Plasma Dynamics Conference, Snowmass, Colorado, July 14-16, 1980.
2. Kwak, D., 'A Comparative Study of Nonreflecting Far-Field Boundary Condition Procedures for Unsteady Transonic Flow Computation', *NASA CP-2201*, Symposium on Numerical Boundary Condition Procedures, NASA Ames Research Center, October 19-20, 1981.
3. Kwak, D., 'An Implicit Transonic Full Potential Code for Cascade Flow on H-Grid Topology,' *AIAA Paper* 83-0506, AIAA 21st Aerospace Sciences Meeting, Reno, Nevada, January 10-13, 1983.
4. Flores, J., Holst, T. L., Kwak, D., and Batiste, D., 'A New Spatial Differencing Scheme for the Transonic Full Potential Flow Calculation,' *AIAA J.*, vol 22, No. 8, August 1984. Also, *AIAA Paper 83-0373*, AIAA 21st Aerospace Sciences Meeting, Reno, Nevada,, January 10-13, 1983.
5. Kwak, D., Chang, J. L. C., Shanks, S. P., and Chakravarthy, S., 'An Incompressible Navier-Stokes Flow Solver in Three-Dimensional Curvilinear Coordinate Systems Using Primitive Variables,' AIAA *J.*, vol 24, No. 3, 390-396, Mar. 1986. Also, *AIAA Paper 84-0253*, AIAA 22nd Aerospace Sciences Meeting, Reno, Nevada, January 9-12, 1984.
6. Chang, J. L. C. and Kwak, D., 'On the Method of Pseudo Compressibility for Numerically Solving Incompressible Flows,' *AIAA Paper 84-0252*, AIAA 22nd Aerospace Sciences Meeting, Reno, Nevada, January 9-12, 1984.
7. Kwak, D. and Chang, J. L. C., 'A Computational Method for Viscous Incompressible Flows,' *Fifth IMACS International Symposium on Computer Methods for PartialDifferential Equations*, Lehigh University, Bethlehem, Pa. 18015, June 19-22, 1984.
8. Kwak, D., Chang, J. L. C., and Shanks, S. P., 'A Solution Procedure for Three-Dimensional Incompressible Navier-Stokes Equation and Its Application,' *Ninth International Conference on Numerical Methods in Fluid Dynamics*, CEN-Saclay, France, June 25-29, 1984.
9. Chang, J. L. C., Kwak, D., Dao, S. C., and Rosen, R., 'A Three-Dimensional Incompressible Flow Simulation Method and Its Application to the Space Shuttle Main Engine, Part I - Laminar Flow,' *AIAA Paper 85-0175*, AIAA 23rd Aerospace Sciences Meeting, Reno, Nevada, January 14-17, 1985.
10. Kaul, U. K., Kwak, D., and Wagner, C., 'A Computational Study of Saddle Point Separation and Horseshoe Vortex System,' *AIAA Paper 85-182*, AIAA 23rd Aerospace Sciences Meeting, Reno, Nevada, January 14-17, 1985.
11. Kwak, D. and Chang, J. L. C., 'A Three-Dimensional Incompressible Navier-Stokes Flow Solver, Part I - INS3D Code,' *Lecture note for Workshop on CFD in Aerospace Design*, The University of Tennessee Space Institute, Tullahoma, Tennessee, June 4-6, 1985.
12. Chang, J. L. C., and Kwak, D., 'A Three-Dimensional Incompressible Navier-Stokes Flow Solver, Part II - Space Shuttle Main Engine Power Head Flow Simulation,' *Lecture note for Workshop on CFD in Aerospace Design*, The University of Tennessee Space Institute, Tullahoma, Tennessee, June 4-6, 1985.
13. Kaul, U., and Kwak, D., 'Computations of Internal Turbulent Flow with Large Separated Flow Regions,' *International Journal for Numerical Method in Fluids,* vol. 6, 927-937, December 1986. Also, *AIAA Paper 85-1687*, AIAA 18th Fluid Dynamics and Plasmadynamics and Laser Conference, Cincinnati, Ohio, July 16-18, 1985.
14. 14. Rogers, S. E., Kwak, D., and Kaul, U., 'On the Accuracy of the Pseudo- compressibility Method in Solving the Incompressible Navier-Stokes Equations,' *Applied Mathematical*

*Modeling*, vol.11, February 1987. Also, *AIAA Paper 85-1689*, AIAA 18th Fluid Dynamics and Plasma- dynamics and Laser Conference, Cincinnati, Ohio, July 16-18, 1985.

15. Chang, J. L. C., Kwak, D., Dao, S. C., and Rosen, R., 'A Three-Dimensional Incompressible Flow Simulation Method and Its Application to the Space Shuttle Main Engine, Part II - Turbulent Flow,' *AIAA Paper 85-1670*, AIAA 18th Fluid Dynamics and Plasmadynamics and Laser Conference, Cincinnati, Ohio, July 16-18, 1985.

16. Rogers, S. E., Kaul, U., and Kwak, D., 'A Numerical Study of Single and Multiple LOX Posts and Its Application to the Space Shuttle Main Engine,' *AIAA Paper 86-0353*, AIAA 24th Aerospace Sciences Meeting, Reno, Nevada, January 6-8, 1986.

17. Kwak, D., Chang, J. L. C., Rogers, S. E., and Kaul, U., 'On the Grid-Induced Error in Numerical Simulation of Viscous Incompressible Flows,' *AIAA Open Forum 86-0586*, AIAA 24th Aerospace Sciences Meeting, Reno, Nevada, January 6-8, 1986.

18. Rogers, S. E., Kwak, D., and Chang, J. L. C., 'Numerical Solution of the Incompressible Navier-Stokes Equations in Three-Dimensional Generalized Curvilinear Coordinates,' *NASA TM 86840*, Jan. 1986.

19. Lee, W. H. and Kwak, D., 'On the PIC Method for Modeling the Shape-Charged Problems,' *The Proceedings of the 9th Int'l Symposium on Ballistics*, Shrivenham, ed, Royal Military College of Science, Apr. 29-May 1, 1986.

20. 20. Rogers, S. E., Chang, J. L. C., and Kwak, D., 'A Diagonal Algorithm for the    Method of Pseudocompressibility,' *J. Comp. Phys.* vol. 73, No. 2, December 1987,pp364-379. Also, *AIAA Paper 86-1060*, AIAA/ASME 4th Fluid Mechanics, Plasma Dynamics and Lasers Conference, Atlanta, GA, May 12-14, 1986.

21. Kwak, D., Rogers, S. E., Kaul, U. K., and Chang, J. L. C., 'A Numerical Study of Incompressible Juncture Flows,' Tenth International Conference on Numerical Methods in Fluid Dynamics, Beijing, Peoples Republic of China, June 23-27, 1986.

22. Chang, J. L. C., Yang, R-J, and Kwak, D., 'A Full Navier-Stokes Simulation of Complex Internal Flows,' *Tenth International Conference on Numerical Methods in Fluid Dynamics*, Beijing, Peoples Republic of China, June 23-27, 1986.

23. Lee, W. H. and Kwak, D., 'On the PIC Method for Elastic-Plastic Flow,' *Tenth International Conference on Numerical Methods in Fluid Dynamics*, Beijing, Peoples Republic of China, June 23-27, 1986.

24. Yang, R-J, Chang, J. L. C., and Kwak, D., 'Navier-Stokes Flow Simulation of the Space Shuttle Main Engine Hot Gas Manifold,' *J. of Spacecraft and Rockets*, vol. 29, Number 2, March-April, 1992, pp 253-259. Also, *AIAA Paper 87-0368*, 1987.

25. Kwak, D. and Lee, W. H., 'PIC Numerical Scheme for a Two-Dimensional Eulerian Code,' *AIAA 8th Computational Fluid Dynamics Conference*, Honolulu, Hawaii, June 9-11, 1987.

26. Lin, S.J., Yang, R-J, Chang, J. L. C., and Kwak, D., 'Numerical Simulation of Flow Path in the Oxidizer Side of Hot Gas Manifold of the Space Shuttle Main Engine,' *AIAA Paper 87-1800*, AIAA/SAE/ASME/ASEE 23rd Joint Propulsion Conference, San Diego, CA, June 29 - July 2, 1987.

27. Rogers, S. E., Kwak, D., and J.L.C. Chang, 'INS3D - An Incompressible Navier-Stokes Code in Generalized Three-Dimensional Coordinates, User's Guide,' *NASA TM - 100012*, November 1987.

28. Rosenfeld, M., Kwak, D., and Vinokur, M., 'A Solution Method for Unsteady, Incompressible Navier-Stokes Equations in Generalized Curvilinear Coordinate Systems,' *J. Comp. Phys.* Vol. 94, No. 1, May 1991, pp102-137. Also, *AIAA Paper 88-0718* AIAA 26th Aerospace Sciences Meeting, Reno, Nevada, January 11-14, 1988.

29. Chang, J. L. C. and Kwak, D., 'Numerical Study of Turbulent Internal Shear Layer Flow in an Axisymmetric U-Duct,' *AIAA Paper 88-0596*, AIAA 26th Aerospace Sciences Meeting, Reno, Nevada, January 11-14, 1988.

30. Lee, W.H. and Kwak, D., 'PIC Method for a Two-Dimensional Elastic-Plastic-Hydro Code,' Computer *Physics Communications*, vol. 48, 1988, pp11-16.
31. Lee, W.H. and Kwak, D., 'Elastic-Plastic Flow using Operator Splitting and Particle-in-Cell Methods in a Two-Dimensional Eulerian Hydrodynamic Code,' Los Alamos National Laboratory Report LA-10857, Los Alamos, New Mexico, March 1988.
32. Chang, J.L.C., Kwak, D., Rogers, S. E. and Yang, R-J, 'Numerical Simulation Methods of Incompressible Flows and an Application to the Space Shuttle Maine Engine,' *International Journal for Numerical Methods in Fluids*, vol.8, 1241-1268, 1988.
33. 33. Kwak, D., Chang, J.L.C., Chang, Rogers, S. E. and Rosenfeld, M., 'Potential Applications of Computational Fluid Dynamics to Biofluid analysis,' *International Symposium on Biofluid Mechanics*, Palm Springs, CA, April 27-29, 1988.
34. Rogers, S. E. and Kwak, D., 'An Upwind Differencing Scheme for the Time-Accurate Incompressible Navier-Stokes Equations,' *AIAA J.* vol. 28 No. 2, pp 253-262, February 1990. Also, *AIAA Paper 88-2583*, AIAA 6th Applied Aerodynamics Conference, June 6-8, 1988.
35. Rosenfeld, M., and Kwak, D., 'Numerical Simulation of Unsteady Incompressible Viscous Flows in Generalized Coordinate Systems,' *11th International Conference on Numerical Methods in Fluid Dynamics*, Williamsburg, Virginia, June 27- July 1, 1988.
36. Yoon, S. and Kwak, D., 'Artificial Dissipation Models for Hypersonic Internal Flow,' *AIAA Paper 88-3277*, AIAA/ASME/SAE/ASEE 24th Joint Propulsion Conference, July 11-13, 1988.
37. Yoon, S. and Kwak, D., 'Artificial Dissipation Models for Hypersonic External Flow,' *AIAA Paper 88-3708*, AIAA/ASME/ASCE/SIAM/APS 1st National Fluid Dynamics Congress, July 25-28, 1988.
38. Kwak, D., Chang, J.L.C., Chang, Rogers, S. E. and Rosenfeld, M., 'Three-Dimensional Incompressible Navier-Stokes Computations of Internal Flows,' *Numerical and Applied Mathematics*, W.F. Ames (editor), Scientific Publishing Co., 1989, pp181-186. Also, *12th IMACS World Congress on Scientific Computation*, Paris, France, July 18-22, 1988.
39. Rogers, S. E. and Kwak, D., 'An Upwind Differencing Scheme for the Steady-state Incompressible Navier-Stokes Equations,' J. *Applied Numerical Mathematics*, vol 8, pp 43-64, 1991.
40. Rogers, S. E. and Kwak, D., 'Numerical Solution of the Incompressible Navier-Stokes Equations for Steady and Time-Dependent Problems,' *AIAA Paper 89-0463* AIAA 27th Aerospace Sciences Meeting, Reno, Nevada, January 9-12, 1989.
41. Rosenfeld, M., and Kwak, D., 'Numerical Solution of Unsteady Incompressible Flows in Generalized Moving Coordinate Systems,' *AIAA Paper 89-0466* AIAA 27th Aerospace Sciences Meeting, Reno, Nevada, January 9-12, 1989. Revised Version: 'Time-Dependent Solutions of Viscous Incompressible Flows in Moving Coordinates,' *International Journal for Numerical Methods in Fluids,* vol.13, 1311-1328, Dec. 1991.
42. Kwak, D., 'Computation of Viscous Incompressible Flows,' ***von Karman Institute for Fluid Dynamics, Lecture Series*** 1989-04, Mar. 6-10,1989.
43. Rogers, S. E., Kutler, P., Kwak, D. and Kiris, C., 'Numerical Simulation of Flow Through an Artificial Heart,' *Fourth International Conference on Supercomputing*, Santa Clara, CA, April 30 - May 5, 1989.
44. Yoon, S., Kwak, D. and Chang L., 'LU-SGS Implicit Algorithm for Three-Dimensional Incompressible Navier-Stokes Equations with Source Term,' *AIAA J.* vol. 29, N0. 6, pp 874-875, June 1991. Also, *AIAA Paper 89-1964*, AIAA 9th Computational Fluid Dynamics Conference, Buffalo, New York, June 13-15, 1989.
45. Kwak, D., Rogers, S., Yoon, S., Rosenfeld, M. and Chang, J.L.C., 'Numerical Simulation of Viscous Incompressible Flows,' Fluids Engineering Seminars, Seoul, Korea, September 3-8, 1989. *Fluids Engineering*, Kim, Hyun and Lee, ed., Hemisphere Publishing Corp, 1991

46. Rogers, S. E., Kwak, D. and Kiris, C., 'Numerical Solution of the Incompressible Navier-Stokes Equations for Steady and Time-Dependent Problems,' *Tenth Australasian Fluid Mechanics Conference*, Melbourne, Australia, December 1989.

47. Kiris, C., Chang, I., D. Kwak and Rogers, S. E., 'Numerical Simulation of the Incompressible Internal Flow Through a Tilting Disk Valve,' *AIAA Paper 90-0682*, AIAA 28th Aerospace Sciences Meeting, January, 1990.

48. Rogers, S. E., Kwak, D., Kiris, C. and Chang, I-D, 'Numerical Simulation of Flow Through Two Biofluid Devices,' NASA TM 102270, March 1990. Published in special video edition of the *Int. J. of Supercomputer Applications* 4(2), 1990, pp96-106.

49. Kwak, D., Rogers, S. E., Kiris, C. and Yoon, S., 'Numerical Methods for Viscous Incompressible Flows,' *12th International Conference on Numerical Methods in Fluid Dynamics*, Oxford, England, July 9-13, 1990.

50. Yoon, S. and Kwak, D., 'Implicit Methods for the Navier-Stokes Equations,' *Computer and Structures*, vol. 1, No. 2-4, pp. 535-537, November 1990.

51. Rogers, S. E., Kwak, D. and Kiris, C., 'Steady and Unsteady Solutions of the Incompressible Navier-Stokes Equations,' *AIAA J*. vol. 29, N0. 4, pp 603-610, April 1991.

52. Yoon, S. and Kwak, D., 'Implicit Three-Dimensional Navier-Stokes Solver for Compressible Flows,' *AIAA J*. vol. 30, No. 11, pp 2653-2659, November 1992. Also, *AIAA Paper 91-1555-CP*, AIAA 10th Computational Fluid Dynamics Conference, Honolulu, Hawaii, June 24-27, 1991.

53. Kiris, C., Chang, I-D, Kwak, D. and Rogers, S., 'Computation of Incompressible Viscous Flows Through Artificial Heart Devices with Moving Boundaries,' *Contemporary Mathematics*, vol. 141, 1993. Also, Joint AMS-IMS-Siam Summer Research Conferences in Biofluiddynamics, Seattle, WA, July 6-12, 1991.

54. Kwak, D., Kiris, C., Rogers, S. E., Kiris, Yoon, S., and Rosenfeld, M., 'Several Computational Aspects of Three-Dimensional Viscous Incompressible Flow Simulations,' *Proc. 4th International Symposium on Computational Fluid Dynamics*, Davis, CA, September 9-12, 1991.

55. Kiris, C., Rogers, S. and Kwak, D. 'Computational Approach for Probing the Flow Through Artificial Heart Devices,' *J. Biomechanical Engineering, Transaction of ASME*, 1997. Also, 112th ASME Winter Annual Meeting, Atlanta, Georgia, Dec. 1-6, 1991.

56. Rogers, S. E., Wiltberger, N.L., and Kwak, D., 'Efficient Simulation of Incompressible Viscous Flow Over Single and Multi-Element Airfoils,' *J. of Aircraft*, vol. 30, Number 5, Sep-Oct, 1993, pp 736-743. Also, *AIAA Paper 92-0405*, AIAA 30th Aerospace Sciences Meeting, Reno, Nevada, Jan. 6-9, 1992.

57. Rosenfeld, M. and Kwak, D., 'Multi-Grid Acceleration of a Fractional-Step Solver in Generalized Curvilinear Coordinate Systems,' *AIAA J*. vol. 31, No. 10, October 1993, pp 1792-1800. Also, *AIAA Paper 92-0185*, AIAA 30th Aerospace Sciences Meeting, Reno, Nevada, Jan. 6-9, 1992.

58. Kiris, C., Kwak, D. and Rogers, S., 'Computational Viscous Incompressible Flows Through Turbopump Components,' *Fourth International Symposium on Transport Phenomena and Dynamics of Rotatin Machinery (ISROMAC-4)*, Honolulu, Hawaii, April 5-8, 1992.

59. Kwak, D., Kiris, C., Rogers, S. and Rosenfeld, M., 'Numerical Methods for Simulating Unsteady Incompressible Flows,' *13th International Conference on Numerical Methods in Fluid Dynamics*, Rome, Italy, July 6-10, 1992.

60. Rosenfeld, M., Kwak, D. and Vinokur, M., 'Development of A Fractional-Step Method for the Unsteady Incompressible Navier-Stokes Equations in Generalized Coordinate Systems,' *NASA TM 103912*, November 1992.

61. Kiris, C., Chang, L., Kwak, D. and Rogers, S., 'Incompressible Navier-Stokes Computations of Rotating Flows,' *AIAA Paper 93-0678*, AIAA 31th Aerospace Sciences Meeting, Reno, Nevada, Jan. 11-14, 1993.

62. Dacles-Mariani, J., Rogers, S., Kwak, D., Zilliac, G. and Chow, J. 'A Computational Study of Wingtip Vortex Flowfield,' *AIAA Paper 93-3010*, AIAA 24th Fluid Dynamics Conference, Orlando, FL, July 6-9, 1993.

63. Yoon, S., Chang, L. and Kwak, D.,'Multigrid Convergence of an Implicit Symmetric Relaxation Scheme,' *AIAA J.* Vol. 32, No. 5, May 1994, pp950-955. Also, *AIAA Paper 93-3357*, AIAA 11th Computational Fluid Dynamics Conference, Orlando, FL, July 6-9, 1993.

64. Wiltberger, N.L., Rogers, S. and Kwak, D., 'A Comparison of Two Incompressible Navier-Stokes Algorithms For Unsteady Internal Flows,' *NASA TM 108794*, November 1993.

65. Kiris, C., Rogers, S. E., Kwak, D. and Lee, Y-T, 'Time -Accurate Incompressible Navier-Stokes Computations with Overlapped Moving Grids,' ASME *Fluids Engineering Summer Meeting*, Lake Tahoe, NV, June 19-23, 1994.

66. Eyi, S., Lee, K.D., Rogers, S. E. and Kwak, D., 'High-Lift Design Optimization Using Navier-Stokes Equations,' *J. of Aircraft*, vol. 33, No. 3, May-June 1996, pp 499-504. Also, *AIAA Paper 95-0477*, AIAA 33rd Aerospace Sciences Meeting, Reno, Nevada, Jan. 9-12, 1995.

67. Kiris, C., Kwak, D. and Rogers, S. E., 'Dual-Use Applications A Computational Fluid Dynamics Code for Viscous Incompressible Flow,' *32nd Space Congress*, Cocoa Beach, FL, April 25-28, 1995.

68. Kiris, C., Kwak, D. and Benkowski, R., 'Computational Flow Analysis of a Left Ventricular Assist Device,' Sixth *International Symposium on Computational Fluid Dynamics*, Lake Tahoe, Nevada, September 4-8, 1995.

69. Dacles-Mariani, J., Kwak, D., and Zilliac, G., 'Incompressible Navier-Stokes Simulation Procedure for a Wingtip Vortex Flow Analysis,' *Sixth International Symposium on Computational Fluid Dynamics*, Lake Tahoe, Nevada, September 4-8, 1995.

70. Kwak, D., Rogers, S. E., Yoon, S. and Chang, J.L.C., 'Numerical Solution of Incompressible Navier-Stokes-Equations,' **Computational Fluid Dynamic Technique**, ed. W.G. Habashi and M. Hafez, Chapter V.1, pp 367-396, 1995.

71. Dacles-Mariani, J., Kwak, D., and Zilliac, G., 'Accuracy Assessment of a Wingtip Vortex Flowfield in the Near-Field Region,' *AIAA Paper 96-0280*, AIAA 34th Aerospace Sciences Meeting, Reno, Nevada, Jan. 15-18, 1996.

72. Kiris, C. and Kwak, D., 'Numerical Solution of Incompressible Navier-Stokes Equations Using a Fractional-Step Approach,' *AIAA Paper 96-2089*, AIAA 27th Fluid Dynamics Conference, New Orleans, LA, June 17-20, 1996.

73. Dacles-Mariani, J., Hafez, M., and Kwak, D., 'Prediction of Wake-Vortex Flow in the Near- and Intermediate-Fields Behind Wings, *AIAA Paper 97-0040*, AIAA 35th Aerospace Sciences Meeting, Reno, Nevada, Jan. 6-10, 1997.

74. Sa, J-Y and Kwak, D., 'A Numerical Method for Incompressible Flow with Heat Transfer,' *NASA TM 110444*, April 1997.

75. Kwak, D., Kiris, C., Dacles-Mariani, J., Rogers, S. and Yoon, S., "Incompressible Navier-Stokes Computations in Aerospace Applications and Beyond," *Frontiers of Computational Fluid Dynamics 1998*, Caughey, D.A. and Hafez, M., ed. World Scientific, Singapore, May 1998.

76. Kiris, C., D, Kwak, Rogers, S. and I-D Chang 'Computational Approach for Probing the Flow Through Artificial Heart Devices,' *J. Biomechanical Engineering, Transaction of ASME*, vol. 119, pp 452-460, November 1997.

77. Kwak, D.Kiris, C., Dacles-Mariani, J., Rogers, S., and Yoon, S., 'Incompressible Navier-Stokes Solvers for Three-Dimensional Steady and Unsteady Flow

Simulations,'*Computational Fluid Dynamics 1997*, Hafez, M. and Oshima, K., ed. World Scientific, Singapore, May 1998.

78. Kwak, D. and Kiris, C., 'Computing the Flow in Mechanical Heart Devices,' *IMACS98 International Conference on Scientific Computing and Mathematical Modeling*, Alicante, Spain, June 25-27, 1998.

79. Kwak, D. and Kiris, C., 'An Assessment of Artificial Compressibility and Pressure Projection Method for Incompressible Flow Simulations,' *16th International Conference on Numerical Methods in Fluid Dynamics*, Arcachon, France, July 6-10, 1998.

80. Kiris, C., Kwak, D., and Benkowski, R., 'Incompressible Navier Stokes Calculations for the Development of a Ventricular Assist Device,' *Computers & Fluids*, Vol. 27, No. 5-6, pp709-719, 1998.

81. Dacles-Mariani, J., Kwak, D., and Zilliac, G., 'On Numerical Errors and Turbulence Modeling in Tip Vortex Flow Prediction,' *International Journal for Numerical Methods in Fluids*, 30,65-82, 1999.

82. Kiris, C., Kwak, D., and Chan, W., 'Parallel Unsteady Turbopump Simulations Using Overset Grid System,' *5$^{th}$ Symposium on Overset Grids and Solution Technology*, University of California at Davis, Davis, CA, September 18-20, 2000.

83. Kiris, C., and Kwak, D., 'Parallel Unsteady Turbopump Simulations for Liquid Rocket Engines,' *Super Computing 2000*, November 2000.

84. Kiris, C., and Kwak, D., 'Numerical Solution of Incompressible Navier-Stokes Equations Using a Fractional-Step Approach,' *Computers & Fluids*, vol 30, 2001, pp829-851.

85. Kwak, D., and Kiris, C., 'Blood Pump Development Using Rocket Engine Flow Simulation Technology,' *The 9$^{th}$ International Symp. On Transport Phenomena and Dynamics of Rotating Machinery, Honolulu*, Hawaii, February 10-14, 2002.

86. Kiris C., and Kwak D., 'Progress in Unsteady Turbopump Flow Simulations,' JANNAF 2002 Meeting, Destin FL, Aprill 8-12, 2002.

87. Kiris C., Kwak D., and Chan W., ' A Three-Dimensional Parallel Time-Accurate Turbopump Simulation Procedure Using Overset grid Systems,' The *Second International Conference on Computational Fluid Dynamics*, Sydney, July 15-19, 2002.

88. Kiris C., and Kwak D., "Aspects of Unsteady Incompressible Flow Simulations," **Computers and Fluids**, Vol 31, No: 4-7, 2002, pp 627-638.

89. Kiris, C., and Kwak, D., 'Parallel Unsteady Turbopump Simulations for Reusable Launch vehicle,' *Frontiers of Computational Fluid Dynamics 2002*, Caughey, D.A. and Hafez, M., ed, World Scientific, 2002.

# Navier-Stokes Solvers

# Incompressible Navier-Stokes Solvers in Primitive Variables and Their Applications to Steady and Unsteady Flow Simulations

Cetin C. Kiris,  Dochan Kwak, and Stuart E. Rogers

NAS Applications Branch, M.S, T27B-1
NASA-Ames Research Center, Moffett Field, CA 94035

**Abstract.** This paper reviews recent progress made in incompressible Navier-Stokes simulation procedures and  their application to problems of engineering interest.  Discussions are focused on the methods designed for complex geometry applications in three dimensions, and thus  are limited to primitive variable formulation. A summary of efforts in flow solver development is given followed by numerical studies of a few example problems of current interest.  Both steady and unsteady solution algorithms and their salient features are discussed.  Solvers discussed here are based on a structured-grid approach using either a finite-difference or a finite-volume frame work. As a grand- challenge application of these solvers, an unsteady turbopump flow simulation procedure has been developed which utilizes high performance computing platforms. In the paper, the progress toward the  complete simulation capability of the turbo-pump for a liquid rocket engine is reported. The Space Shuttle Main Engine (SSME) turbo-pump is used as a test case for evaluation of two parallel computing algorithms that have been implemented in the INS3D code. The relative motion of the grid systems for the rotor-stator interaction was obtained using overset grid techniques. Unsteady computations for the SSME turbo-pump, which contains 114 zones with 34.5 million grid points, are carried out on SGI Origin 3000 systems at NASA Ames Research Center.  The same procedure has been extended to the development of NASA-DeBakey Ventricular Assist Device (VAD) that is based on an axial blood pump. Computational, and clinical analysis of this device are presented.

# 1   Introduction

Incompressible flow can be considered as a limiting case of compressible flow as the flow speed approaches to a significantly low value compared to the speed of sound. There are a large number of flow problems of practical importance in aerospace and other fields  which belong in this category. The incompressible Navier-Stokes equations, which govern these flows, pose a special problem of satisfying the mass conservation equation because it  is not coupled to the momentum equations. Physically,  these equations are characterized by the elliptic behavior of the pressure waves, the speed of which are infinite.

Various methods have been developed, which can be classified in numerous ways depending on the choice of formulations, variables, or algorithms. Since three-dimensional applications involving complex geometries are of our primary interest, the primitive variable formulation is chosen in the present study. The primitive variables, namely the pressure and the velocities, can easily be defined in real geometry compared to derived quantities like stream function or vorticity. Therefore, for convenience and flexibility, primitive variable formulations were used for developing incompressible Navier-Stokes codes (INS3D family of codes) at NASA Ames Research Center. The present article is intended to present our progress made since the review given by the second author in 1989 as a VKI Lecture note (Kwak, 1989). The solution procedures presented here are mainly within a structured-grid framework. During the last several years, a large number of review articles and books on CFD discussed incompressible flow methods. For a more comprehensive review of computational methods for incompressible flow in general, readers are referred to these materials, i.e. Hirsch (1988), Hafez and Oshima (1995).

## 2 Solution Methods

In this section, two solution methods used in the development of INS3D, namely, an artificial compressibility method and a pressure projection method, are reviewed. The governing equations will be given first, followed by a discussion on the computational procedure related to the two methods. Three-dimensional incompressible flow with constant density is governed by the following Navier-Stokes equations:

$$\frac{\partial u_i}{\partial x_i} = 0 \tag{2.1}$$

$$\frac{\partial u_i}{\partial t} + \frac{\partial u_i u_j}{\partial x_j} = -\frac{\partial p}{\partial x_i} + \frac{\partial \tau_{ij}}{\partial x_j} \tag{2.2}$$

where $t$ is the time, $x_i$ the Cartesian coordinates, $u_i$ the corresponding velocity components, $p$ the pressure, and $\tau_{ij}$ the viscous-stress tensor. All the variables have been non-dimensionalized by a reference velocity and length scale. The viscous stress tensor can be written as

$$\tau_{ij} = 2\upsilon S_{ij} - R_{ij} \tag{2.3}$$

$$S_{ij} = \frac{1}{2}\left(\frac{\partial u_i}{\partial x_j} + \frac{\partial u_j}{\partial x_i}\right) \tag{2.4}$$

where, $\upsilon$ is the kinematic viscosity, $S_{ij}$ is the strain-rate tensor, and $R_{ij}$ are the Reynolds stresses. Various levels of closure models for $R_{ij}$ are possible. In the present article, turbulence is simulated by an eddy viscosity model using a constitutive equation of the following form:

$$R_{ij} = \frac{1}{3}R_{kk}\delta_{ij} - 2\upsilon_t S_{ij} \tag{2.5}$$

where $\upsilon_t$ is the turbulent eddy viscosity. By including the normal stress, $R_{kk}$, in the pressure, $\upsilon$ in equation (2.3) can be replaced by $(\upsilon + \upsilon_t)$ as follows.

$$\tau_{ij} = 2(\upsilon + \upsilon_t)S_{ij} = 2\upsilon_T S_{ij} \qquad (2.6)$$

In the remainder of this note the total viscosity, $(\upsilon + \upsilon_t)$, will be represented simply by $\upsilon$. The present formulations allow for spatially varying viscosity.

In general curvilinear coordinates, $(\xi, \eta, \zeta)$, the governing equations can be written as

$$\frac{\partial}{\partial t}\hat{u} = -\frac{\partial}{\partial \xi_i}(\hat{e}_i - \hat{e}_{vi}) + \hat{s} = -\hat{r} \qquad (2.7)$$

$$\frac{\partial}{\partial \xi_i}\left(\frac{U - (\xi)_t}{J}\right) = 0 \qquad (2.8)$$

where $\hat{r}$ is the right-hand side of the momentum equation, and

$$\xi_i = \xi, \eta \text{ or } \zeta \text{ , for } i=1, 2, \text{ or } 3$$

$$\hat{u} = \frac{1}{J}\begin{bmatrix} u \\ v \\ w \end{bmatrix}$$

$$\hat{e}_i = \frac{1}{J}\begin{bmatrix} (\xi_i)_x p + uU_i \\ (\xi_i)_y p + vU_i \\ (\xi_i)_z p + wU_i \end{bmatrix},$$

$$\hat{e}_{vi} = \frac{\upsilon}{J}\nabla\xi_i \cdot \left(\nabla\xi_l \frac{\partial}{\partial \xi_l}\right)\begin{bmatrix} u \\ v \\ w \end{bmatrix} \qquad (2.9)$$

$$U_i = (\xi_i)_t + (\xi_i)_x u + (\xi_i)_y v + (\xi_i)_z w$$

$$J = \text{Jacobian of the transformation}$$

$$\hat{s} = \text{source term}$$

The source term $\hat{s}$ is used to represent centrifugal and Coriolis forces in a steady rotating reference frame, and will be discussed later in this section. For most flow applications, this term is set to zero.

## 2.1  Artificial Compressibility Method

Major advances in the state of the art in CFD have been made in conjunction with compressible flow computations. Therefore, it is of significant interest to be able to use some of these compressible flow algorithms for incompressible flows. To do this, the artificial compressibility method of Chorin (1967) can be used. In this formulation, the continuity equation is modified by adding a pseudo-time derivative of the pressure, resulting in

$$\frac{1}{\beta}\frac{\partial p}{\partial \tau} + \frac{\partial u_i}{\partial x_i} = 0 \qquad (2.10)$$

where $\beta$ is an artificial compressibility parameter and $\tau$ is a pseudo-time parameter. This forms a hyperbolic-parabolic type of pseudo-time dependent system of equations. Thus, implicit schemes developed for compressible flows can be implemented to solve for steady-state solution. In the steady-state formulation the equations are to be marched in a time-like fashion until the divergence of velocity in equation (2.10) converges to a specified tolerance. The time variable for this process no longer represents physical time, so in the momentum equations t is replaced with $\tau$, which can be thought of as a pseudo-time or iteration parameter.

Physically, this means that waves of finite speed are introduced into the incompressible flow field as a medium to distribute the pressure. For a truly incompressible flow, the wave speed is infinite, whereas the speed of propagation of these pseudo waves depend on the magnitude of the artificial compressibility parameter. In a truly incompressible flow, the pressure field is affected instantaneously by a disturbance in the flow, but with artificial compressibility, there is a time lag between the flow disturbance and its effect on the pressure field. Ideally, the value of the artificial compressibility parameter is to be chosen as high as the particular choice of algorithm will allow so that the incompressibility is recovered quickly. This has to be done without lessening the accuracy and the stability property of the numerical method implemented. On the other hand, if the artificial compressibility parameter is chosen such that these waves travel too slowly, then the variation of the pressure field accompanying these waves is very slow. This will interfere with the proper development of the viscous boundary layer. In viscous flows, the behavior of the boundary layer is very sensitive to the streamwise pressure gradient, especially when the boundary layer is separated. If separation is present, a pressure wave traveling with finite speed will cause a change in the local pressure gradient which will affect the location of the flow separation. This change in separated flow will feed back to the pressure field, possibly preventing convergence to a steady state. When the viscous effect is important for the entire flow field as in most internal flow problems, the interaction between the pseudo-pressure waves and the viscous flow field is especially  important.

Artificial compressibility relaxes the strict requirement of satisfying mass conservation in each step.  However, to utilize this convenient feature, it is essential to understand the nature of the artificial compressibility both physically and mathematically. Chang and Kwak (1984) reported details of the artificial compressibility, and suggested some guidelines for choosing the artificial compressibility parameter. Various applications which evolved from this concept have been reported for obtaining steady-state solutions (e.g., Steger and Kutler, 1977; Kwak et al. 1986; Chang et al., 1988; Choi and Merkle, 1985). To obtain time-dependent solutions using this method, an iterative procedure can be applied in each physical time step such that the continuity equation is satisfied (see, Merkle and Athavale, 1987, Rogers and Kwak ,1988, Rogers, Kwak, and Kiris, 1991, Belov et. al., 1995). Further discussions on the artificial compressibility approach can be found in the literature (see, Temam, 1979, Rizzi and Eriksson, 1985).

Combining equation (2.10) and the momentum equations gives the following system of equations:

$$\frac{\partial}{\partial \tau}\hat{D} = -\frac{\partial}{\partial \xi_i}\left(\hat{E}_i - \hat{E}_{vi}\right) + \hat{S} = -\hat{R} \qquad (2.11)$$

where $\hat{R}$ is the right-hand-side of the momentum equation and can be defined as the residual  for steady-state computations, and where

$$\hat{D} = \frac{D}{J} = \frac{1}{J}\begin{bmatrix} p \\ u \\ v \\ w \end{bmatrix}, \qquad \hat{E}_i = \begin{bmatrix} \beta U_i / J \\ \hat{e}_i \end{bmatrix}, \qquad \hat{E}_{vi} = \begin{bmatrix} 0 \\ \hat{e}_{vi} \end{bmatrix} \qquad (2.12)$$

When the governing equations are solved in a steadily rotating reference frames, the source term, $\hat{S}$, represents centrifugal and Coriolis terms. If the relative reference frame is rotating around the x-axis, the source term $\hat{S}$ is given by

8

$$\hat{S} = \begin{bmatrix} 0 \\ 0 \\ \Omega(\Omega y + 2w) \\ \Omega(\Omega z - 2v) \end{bmatrix}$$

where $\Omega$ is the rotational speed. In this report, the source term, $\hat{S}$, is set to zero other than for rotational steady solutions. Relative velocity components are written in terms of absolute velocity components $u_a$, $v_a$, and $w_a$ as

$$u = u_a$$
$$v = v_a + \Omega z$$
$$w = w_a - \Omega y$$

Time-dependent calculation of incompressible flows are especially time consuming due to the elliptic nature of the governing equations. This means that any local change in the flow has to be propagated throughout the entire flow field. Numerically, this means that in each time step, the pressure field has to go through one complete steady-state iteration cycle, for example, by Poisson-solver-type pressure iteration or artificial compressibility iteration method. In transient flow, the physical time step has to be small and consequently the change in the flow field may be small. In this situation, the number of iterations in each time step for getting a divergence-free flow field may not be as high as regular steady-state computations. However, the time-accurate computations are generally an order of magnitude more time-consuming than steady-state computation. Therefore, it is particularly desirable to develop computationally efficient methods either by implementing a fast algorithm and by utilizing computer characteristics such as vectorization and parallel processing.

A time-accurate method using artificial compressibility developed by Rogers, Kwak, and Kiris (1991) is summarized next. In this formulation the time derivatives in the momentum equations are differenced using a second-order, three-point, backward-difference formula

$$\frac{3\hat{u}^{n+1} - 4\hat{u}^n + \hat{u}^{n-1}}{2\Delta t} = -\hat{r}^{n+1} \tag{2.13}$$

where the superscript $n$ denotes the quantities at time $t = n\Delta t$ and $\hat{r}$ is the right-hand side given in equation (2.7). To solve equation (2.13) for a divergence free velocity field at the $(n+1)$ time level, a pseudo-time level is introduced and is denoted by a superscript m. The equations are iteratively solved such that $\hat{u}^{n+1,m+1}$ approaches the new velocity $\hat{u}^{n+1}$ as the divergence of $\hat{u}^{n+1,m+1}$ approaches zero. To drive the divergence of this velocity to zero, the following artificial compressibility relation is introduced:

$$\frac{p^{n+1,m+1} - p^{n+1,m}}{\Delta\tau} = -\beta\nabla\cdot\hat{u}^{n+1,m+1} \qquad (2.14)$$

where $\tau$ denotes pseudo-time and $\beta$ is an artificial compressibility parameter. Combining equation (2.14) with the momentum equations gives

$$I_{tr}\left(\hat{D}^{n+1,m+1} - \hat{D}^{n+1,m}\right) = -\hat{R}^{n+1,m+1} - \frac{I_m}{\Delta t}\left(1.5\hat{D}^{n+1,m} - 2\hat{D}^n + 0.5\hat{D}^{n-1}\right) \qquad (2.15)$$

where $\hat{D}$ is the same vector defined in equation (2.13), $\hat{R}$ is the same residual vector defined in equation (2.11), and $I_{tr}$ is a diagonal matrix given by

$$I_{tr} = diag\left[\frac{1}{\Delta t}, \frac{1.5}{\Delta t}, \frac{1.5}{\Delta t}, \frac{1.5}{\Delta t}\right]$$

Finally, the residual term at the $m+1$ pseudo-time level is linearized giving the following equation in delta form

$$\left[\frac{I_{tr}}{J} + \left(\frac{\partial\hat{R}}{\partial D}\right)^{n+1,m}\right]\left(\hat{D}^{n+1,m+1} - \hat{D}^{n+1,m}\right)$$

$$= -\hat{R}^{n+1,m} - \frac{I_m}{\Delta t}\left(1.5\hat{D}^{n+1,m} - 2\hat{D}^n + 0.5\hat{D}^{n-1}\right) \qquad (2.16)$$

As can be seen, this equation is very similar to the steady-state formulation which can be rewritten for the Euler implicit case as

$$\left[\frac{1}{J\Delta\tau}I + \left(\frac{\partial\hat{R}}{\partial D}\right)^n\right]\left(\hat{D}^{n+1} - \hat{D}^n\right) = -\hat{R}^n \qquad (2.17)$$

Both systems of equations will require the discretization of the same residual vector $\hat{R}$. The matrix equation is solved iteratively by using a non-factored Gauss-Seidel type line-relaxation scheme employed by MacCormack (1985), which maintains stability and allows a large pseudo-time step to be taken. Details of the numerical method can be found in paper by Rogers, Kwak, and Kiris (1991). The GMRES scheme has also been utilized for the solution of the resulting matrix equation (Rogers, 1995). Computer memory requirement for the flow solver (INS3D-UP code) with line-relaxation is 35 times the number of grid points in words, and with GMRES-ILU(0) scheme is 220 times the number of grid points in words.

## 2.2  Pressure Projection Method

In 1965, Harlow and Welch published the first primitive variable method using a Poisson equation for pressure.  In this method, called the marker-and-cell (MAC) method, the pressure is used as a mapping parameter to satisfy the continuity equation. By taking the divergence of the momentum equation, the Poisson equation for pressure is obtained:

$$\nabla^2 p = \frac{\partial h_i}{\partial x_i} - \frac{\partial}{\partial t}\frac{\partial u_i}{\partial x_i}$$

$$(2.18)$$

where

$$h_i = -\frac{\partial u_i u_j}{\partial x_j} + \frac{\partial \tau_{ij}}{\partial x_j}$$

The usual computational procedure involves choosing the pressure field at the current time step such that continuity is satisfied at the next time step.  The original MAC method is based on a staggered arrangement on a 2-D Cartesian grid. The staggered grid conserves mass, momentum, and kinetic energy in a natural way and avoids odd-even point decoupling of the pressure encountered in a regular grid (Gresho and Sani, 1987). Even though the original method used an explicit Euler solver, various time advancing schemes can be implemented with this formulation.  Ever since its introduction, numerous variations of the MAC method  have been devised and successful computations have been made.

The MAC method can be viewed as a special case of  the projection method (i.e. Chorin, 1968). In this method the strict requirement of obtaining the correct pressure for a divergence-free velocity field in each step  may significantly slow down the overall computational efficiency. To satisfy the mass conservation in grid space, the difference form of the second derivative in the Poisson equation has to be constructed consistent with the discretized momentum equation (see Kwak, 1989).

To solve for a steady-state solution, the correct pressure field is desired only when the solution is converged. In this case, the iteration procedure for the pressure can be simplified such that it requires only a few iteration at each time step. The best known method using this approach is the Semi-Implicit Method for Pressure-Linked Equations (SIMPLE) (Patankar, 1980 ; Chen et al., 1995). The unique feature of this method is the simple way of estimating the velocity and the pressure correction. This feature simplifies the computation but introduces empiricism into the method.  Despite its empiricism, the method has been used successfully for many steady-state computations. It is not the intention of the present paper to evaluate this method, and readers interested in this approach are referred to the above cited references.

The time-integration scheme is based on operator splitting, which can be accomplished in several ways by combining the pressure, convective, and viscous terms in the momentum equations. The fractional-step method is based on the decomposition of vector field into a divergence free component and a gradient of a scalar field. Since its inception, this approach is perhaps the most widely used method in computing incompressible flow. Variations of this idea are too numerous to list here.

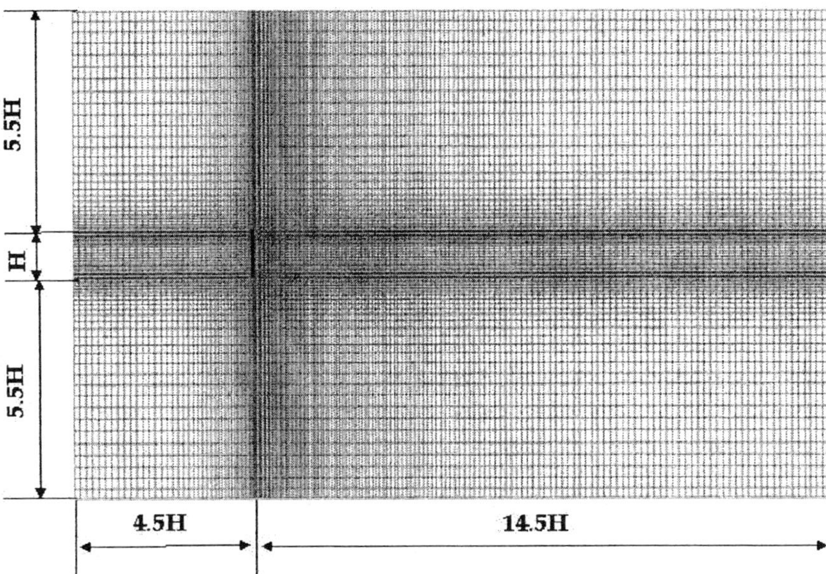

**Fig. 1.** Computational grid for the flow past a 90-degree float plate (plate thickness = 0.03 H).

The common application of fractional-step method is done in two steps. The first step is to solve for an auxiliary velocity field using the momentum equations. In the second step, the velocity field is corrected by using the pressure, which can map the auxiliary velocity onto a divergence free velocity field. In the first step, the momentum equations are discretized in time using a second-order implicit Runga-Kutta method. The Poisson equation for pressure is obtained by taking the divergence of the momentum equations and by using the continuity equation. For the spatial discretization, a finite-volume formulation is used where pressure is defined at the cell center and the mass fluxes at the faces of each cell. The mass-conservation equation is evaluated by computing the mass flux across faces of a computational cell. When the mass fluxes are chosen as unknowns, the continuity equation is satisfied automatically in generalized coordinate systems. The continuity equation with this choice of the dependent variables takes a form identical to the Cartesian case. Therefore, the mass fluxes are considered as the *natural* dependent variables for projection methods in curvilinear coordinates. Treating the mass fluxes as dependent variables in a finite-volume formulation is equivalent to using contravariant velocity components, scaled by the inverse of the transformation Jacobian, in a finite-difference formulation. This choice of mass fluxes as dependent variables complicates the discretization of the momentum equations. In order to replace Cartesian velocity components by the new dependent variables, namely, the contravariant velocity components, the corresponding area vectors are dotted with the momentum equations. Then the integral momentum equation is evaluated on different computational cells for each unknown. For the definition of variables, a staggered grid orientation was selected to eliminate checker-board-like oscillations in pressure and provides more compact stencils. Full details on the derivation of momentum equations and the solution procedure is outlined in references by Rosenfeld, Kwak, and Vinukur (1991) and by Kiris and Kwak (1996). A flow solver using the above procedure is designated as INS3D-FS. Since each equation is solved in a segregated fashion, the memory requirement for GMRES solver in INS3D-FS is only 70 times the number of grid points in words.

# 3  Computed Methods

## 3.1  Flow Past $90^0$ Flat Plate

Numerical results for the time evolution of twin vortices behind a two-dimensional flat plate are presented. Several cases were run to with various algorithmic parameters. To expedite the process, a two-dimensional test case is selected here. It should be noted that the associated flow solvers, INS3D-UP for the artificial compressibility method and INS3D-FS for the pressure projection method, are written for three-dimensional applications. This numerical experiment is studied to help select a method for large three-dimensional unsteady applications where computing resources become a critical issue.

Computed results from both methods are compared with the experimental data by Taneda and Honji (1971). The experiment was carried out in a water tank 40 cm wide. A thin, 3 cm high plate was immersed in water. The flow was started from rest impulsively at the velocity u=0.495 cm/s. The Reynolds number for this case is 126 based on the plate height. The computational grid size is 181x81 in flow and vertical directions, and 3 layers of this grid are used to obtain two-dimensional results (figure 1). Since INS3D-FS is written in a finite-volume staggered-grid formulation, it requires one additional ghost cell in each direction. Figure 2 shows calculated velocity vectors obtained from INS3D-FS at various times. The flow separates at the edges of the plate and forms a vortex pair. The twin vortices become longer in the flow direction with time. The time history of the stagnation point is compared with experimental results and other numerical results in figure 3. Symbols represent experimental measurements, and the solid line and the dashed line represent results from INS3D-UP and INS3D-FS, respectively. In addition the dotted line shows the numerical results from finite element formulations of Yoshida and Nomura (1985).

In figure 3, the interval for time integration was 0.5 sec, which corresponds to nondimensional value of 0.0825. Even though the plate started impulsively in the experiment, the computations presented in figure 3 have a slow start procedure. In figure 4, two different ways of starting the flow are prescribed, namely, an impulsive start as in figure 4a and a slow start as prescribed in figure 4b. Yoshida and Nomura (1985) used the same slow start procedure in their calculations. For the slow start case, the velocity profile shown in figure 4b is prescribed and the starting time of calculation is appropriately shifted from that of experiment.

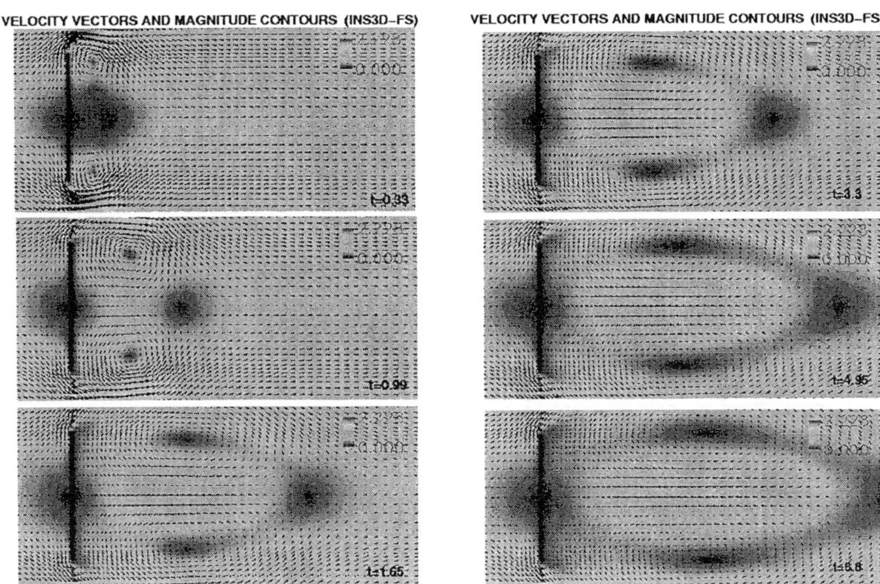


**Fig. 2.** Velocity vectors at various non-dimensional times (INS3D-FS).

First, INS3D-FS results are presented in figures 5 and 6. In figure 5, the effect of starting procedure on the development of the flow is shown. Here, a non-dimensional time step of 0.0825 was used. There are measurable differences in the resulting flows. Increasing the spatial resolution does not change the results significantly while decreasing the time step improves the agreement with experiments, as shown in Fig. 6. These unsteady computations using a time step size of 0.0825 was completed in two hours of CPU time on single processor Cray-J90. Computational results using the artificial compressibility code, INS3D-UP, are presented in figure 7. Results using two different artificial compressibility parameters, BETA, are compared. For time accurate solutions, sub-iterations should be terminated when the divergence of the velocity reached a specified error limit.

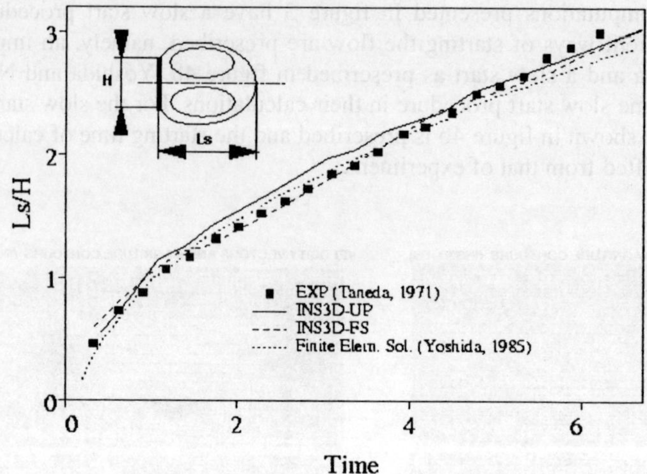



**Fig. 3.** Calculated time history of the stagnation point.

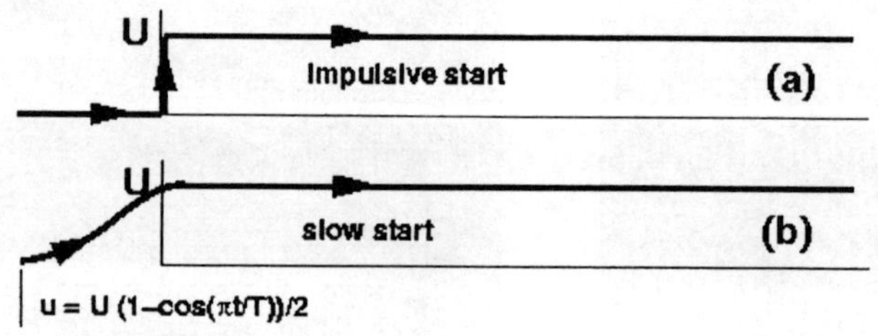



**Fig. 4.** Prescribed velocity for an impulsive start (a) and for a slow start (b).

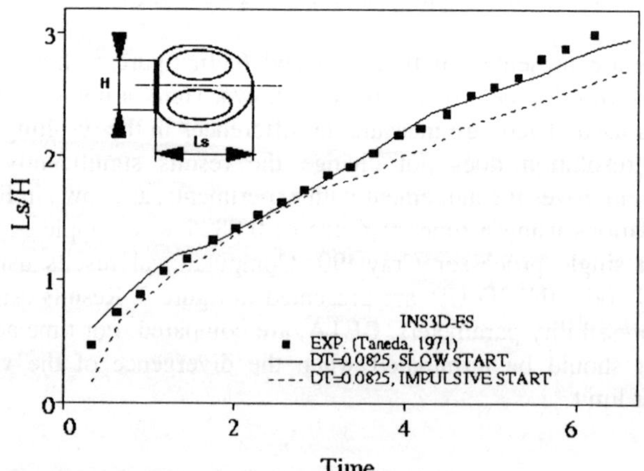



**Fig. 5.** Effects of starting procedure.

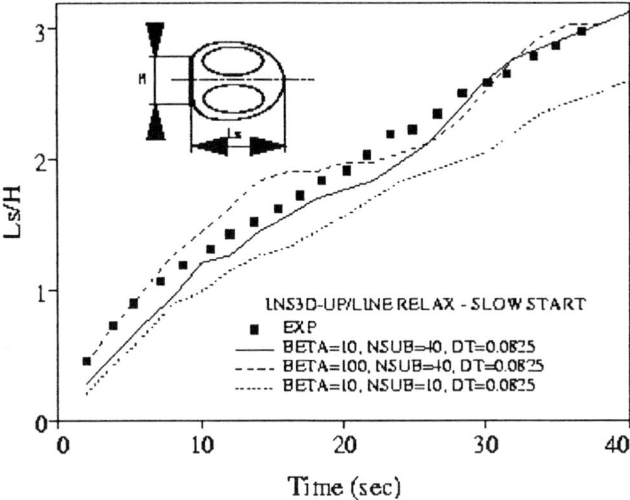



**Fig. 6.** Evaluation movement of stagnation point from INS3D-UP calculation with line-relaxation scheme.

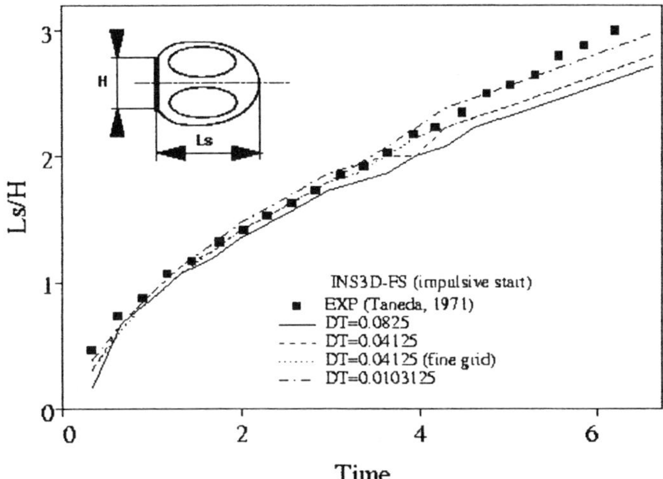



**Fig. 7.** Effects of time-step size for impulsive start.

In reality this will impose heavy burden on available computing resources. Therefore, the maximum number of sub-iterations, NSUB, is artificially fixed at 10 and 40 for the present experiment. With 10 subiterations, the incompressibility conditions is not fully satisfied at each physical time step resulting in large error as time progresses. Computing time requirement using line relaxation scheme is large ranging 4 and 10 hours of CPU time for 10 and 40 subiterations respectively on a single processor Cray-J90 computer. It is observed that for engineering applications, a fast convergence scheme is necessary at each physical time step in order to meet incompressibility condition within reasonable accuracy. Otherwise, artificial compressibility method with line relaxation scheme can be expensive for 3D time-accurate computations. In order to alleviate this difficulty, GMRES-ILU(0) solver is implemented in INS3D-UP at the expense of increasing memory requirement. The results shown in figure 8 were obtained with less than 4 hours of Cray-J90 computer. The agreement between the computed results and experimental data is better. With GMRES-ILU(0) solver, the mass flow ratio between inflow and exit is always satisfied. In addition, the discrepancies between numerical results are very small when two different values of artificial compressibility parameter were used.

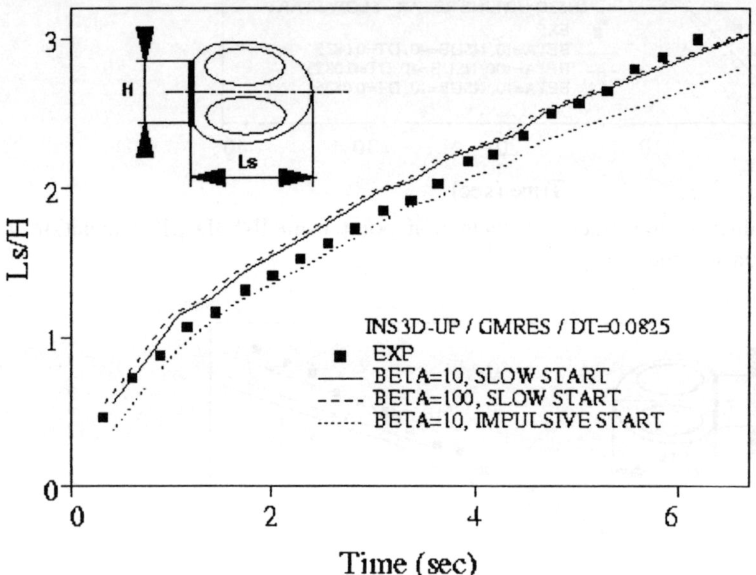


**Fig. 8.** Evaluation movement of stagnation point from INS3D-UP calculation with GMRES-ILU(0) scheme.

When a fast converging scheme, such as a GMRES-ILU(0) solver, was implemented into artificial compressibility method, reasonable agreement was obtained between computed results and experimental data. Memory requirement of this scheme is the major drawback for three-dimensional large-scale applications. However, using parallel computing platforms, such as SGI Origin systems, memory requirement may not be a significant issue. The line-relaxation scheme in artificial compressibility method becomes very expensive for time accurate computations and could lead to erroneous solutions if incompressibility is not enforced in each time step. The pressure projection method is usually more expensive for steady state solutions due to the time required for the Poisson equation for pressure. For cases where very small physical time step is required, the pressure projection method was found to be computationally efficient since it does not require subiterations procedure. However, the governing equations are not fully coupled as in the artificial compressibility approach, and this may affect the robustness and limit the maximum allowable time step size for complicated geometries in engineering applications.

## 3.2 Pump Technology For Liquid Rocket Engine

Until recently, the high performance-pump design process was not significantly different from that of 30 years ago. During the past 30 years a vast amount of experimental and operational experience has demonstrated that there are many important features of pump flows which are not accounted for in the current semi-empirical design process. Pumps being designed today are no more technologically advanced than those designed for the Space Shuttle Main Engine (SSME). During that same time span huge strides have been made in computers, in numerical algorithms, and in physical modeling. The major accomplishment of this work is to extend the CFD technology  to validate advanced CFD codes on pump flows and to demonstrate their value to the pump designer.  Rocket pumps involve full and partial blades, tip leakage, and an exit boundary to diffuser. In addition to these geometric complexities, a variety of flow phenomena  are encountered in turbopump flows. These  include turbulent boundary layer separation, wakes, transition, tip vortices, three-dimensional effects, and Reynolds number effects. In order to increase the role of Computational

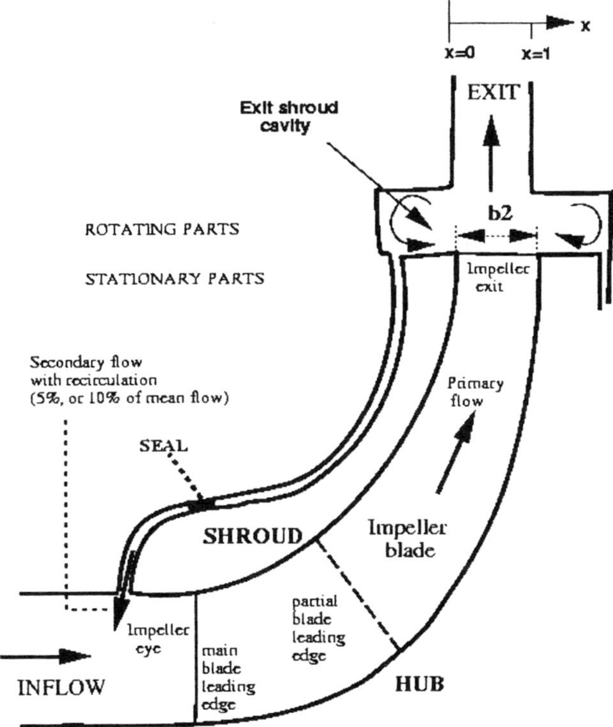


**Fig. 9.** Schematic view of an advanced pump impeller cross-section.

Fluid Dynamics (CFD) in the design process, the CFD analysis tools   must be evaluated and validated  so that designers gain confidence in their use.

The incompressible Navier-Stokes flow solver, INS3D-UP,  has been validated for pump component analysis. In this validation effort, computed results obtained from a rocket-pump inducer simulation were compared with experimental data. Further details can be found in the paper by Kiris at al (1993). The resulting computational procedure was applied to the flow through the SSME High Pressure Fuel Turbopump impeller and to the development of an advanced pump impeller (Kiris and Kwak 1994). The results from the advanced-pump impeller-flow analysis are presented next.

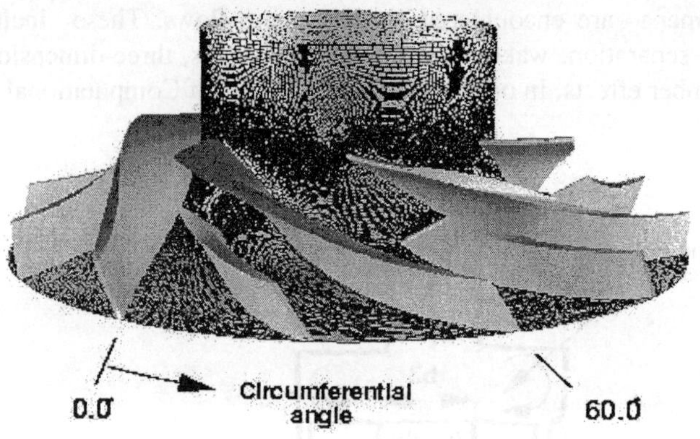


**Fig. 10.** Advanced pump impeller computational grid on the hub surface.

In Figure 9, a cross-sctional view of the advanced-pump impeller is shown schematically. The computational model of this pump includes the impeller and the exit cavity region. Figure10 shows the computational grid near the hub region of the impeller. The impeller design flow rate is 1,205 gal/min with a design speed of 6,322 rpm. The Reynolds Number for this calculation was 181,283 per inch. In Figure 11, the meridional velocity obtained from steady-state calculations in the rotating referance frame is shown at the impeller discharge. A relative x-distance is measured from the shroud to hub, where x=1.0 is the hub. The meridional velocities, $Cm$, were integrated along a radial strip for each constant x position and they were non-dimensionalized by the wheel speed of 249.5 ft/sec. The meridional velocity distribution for 5% and 10% recirculation from the exit shroud cavity were also plotted.  When the exit shroud cavity has leakage to the impeller eye, the velocity peak at the impeller exit moves toward to the center of the b2 width, where b2 is defined as the blade height at the impeller exit (see figure 9). However, the shroud leakage has only minor effects on the solution at impeller exit (Figure 11).

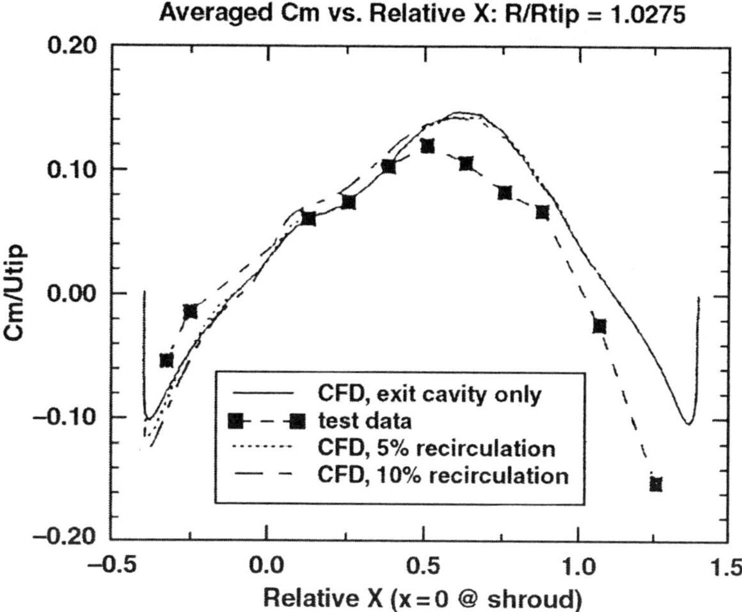



**Fig. 11.** Comparison of circumferentially averaged meridional velocity at the impeller exit.

In Figure 11, the symbols represent experimental data, and the lines represent *Cm* distributions for the flow with vaneless space at the exit of the impeller. The test data shows that the peak is closer to the center of the b2 width. The discrepancy between the computed results and experimental data is partially due to the recirculation flow in the hub cavity. The leakage at the hub cavity leads to a stronger recirculation region which shifts the velocity peak to the center of the b2 width. Since the CFD analysis did not include the leakage at the hub cavity, the predicted recirculation region in the vaneless space is not as strong as in the experimental study.

Figure 12 shows blade-to-blade velocity distributions at the impeller exit. The blade-to-blade velocity distribution illustrates the impeller-exit flow distortion. Symbols represent the experimental data and the lines represent computed results. The jet-wake like pattern, which produces and unsteady load in the diffuser vanes, was captured at both meridional locations. Overall, the numerical results compare reasonably well with the experimental data.

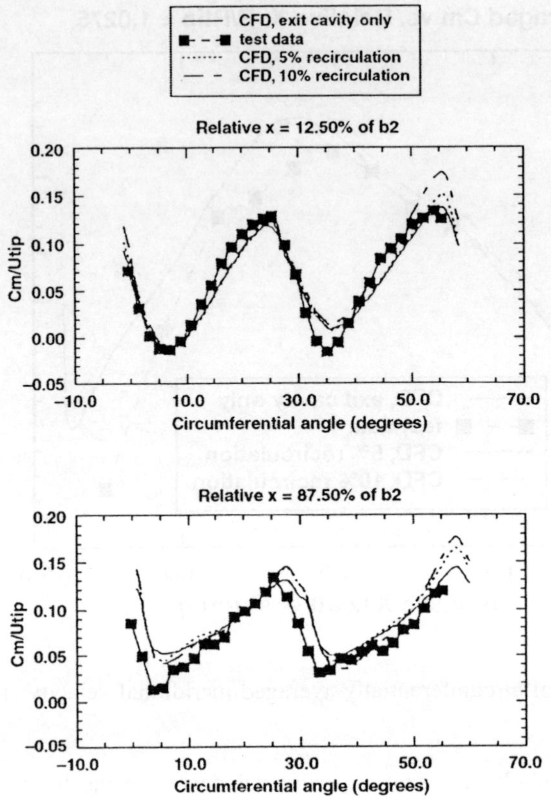


**Fig. 12.** Comparison of blade-to-blade meridional velocity at the impeller exit.

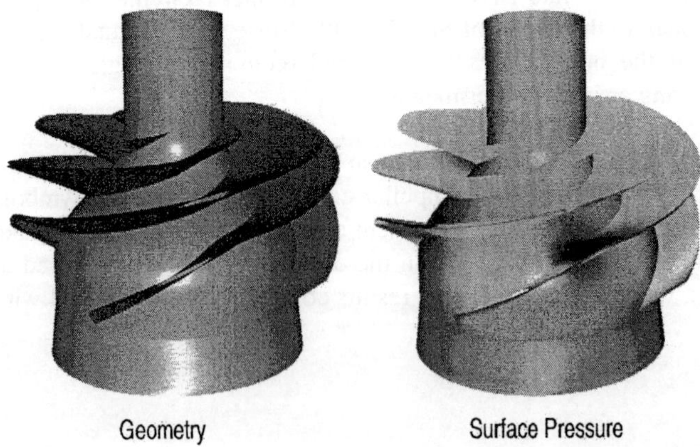


**Fig. 13.** Geometry and surface pressure for a pump inducer.

More recently, an unsteady-flow simulation capability utilizing overset grid approach for a multi-component turbopump geometry was developed at NASA-Ames Research Center. The motivation of this effort was primarly was based on two elements. First, the entire turbo pump simulation is intended to provide a computational framework for the design and analysis of an entire liquid rocket engine fuel supply system. The second motivation for this research was to support the design of liquid rocket systems for development of space transportation systems. Since the space launch systems in the near future are likely to rely on liquid rocket engines, increasing the efficiency and reliability of the engine components is an important task. A substantial computational time reduction for these 3D unsteady flow simulations is required to reduce the design-cycle time of the pumps. Part of this speed up will be due to enhancements in computer hardware. The remaining portion of the speed-up must be contributed by advances in algorithms and by efficient parallel implementations. The following section outlines the initial effort and steps taken in order to reach this speed-up.

The geometry for a typical liquid oxygen pump has various rotating and stationary components, such as flow-straightener, inducer, impeller, diffuser, where the flow is extremely unsteady. Figure 13 shows the geometry and computed surface pressure of the inducer from steady-state components analysis. When rotating and stationary parts are included, time-dependent simulations need to be carried out due to relative motion of the components. To handle the geometric complexity, an overset grid approach is used.

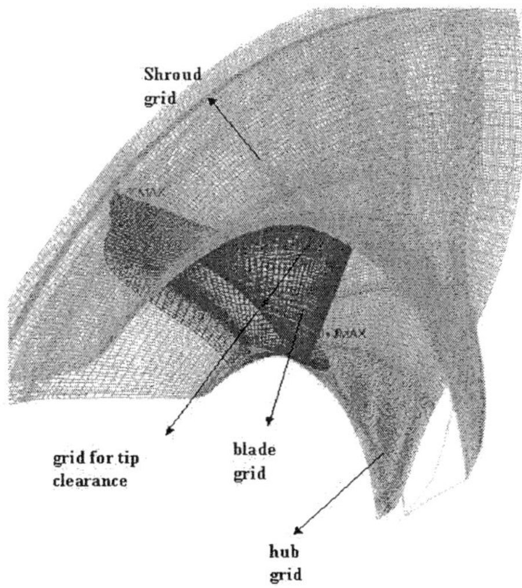


**Fig. 14.** Overset grid system for the impeller long blade section with tip clearance.

**Fig. 15.** Geometry of SSME-rig1 shuttle upgrade pump impeller

The overset structured grid approach to flow simulation has been utilized to solve a variety of problems in aerospace, marine, biomedical and meteorological applications (Chan 2002). Flow regimes can range from simple steady flows as that of a commercial aircraft, to unsteady three-dimensional flows with bodies in relative motion, as in the case of turbopump configurations. A geometrically complex body is decomposed into a number of simple grid components, as shown in figure 14. In Figure 14, only long-blade impeller section is shown. For the entire configuration including inlet guide vanes, impeller blades and diffuser blades as shown in Figure 15, the computational grid has been generated by using 34.3 Million grid points with 114 zones. The freedom to allow neighboring grids to overlap arbitrarily implies that these grids can be created independently from each other and each grid is typically of high quality and nearly orthogonal. Connectivity between neighboring grids is established by interpolation at the grid outer boundaries (Meakin 2001). Addition of new components to the system and simulating arbitrary relative motion between multiple bodies are achieved by establishing new connectivity without disturbing the existing grids. Scalability on parallel compute platforms is naturally accomplished by the already decomposed grid system. For certain problems, it is more efficient to gather the grids into groups of approximately equal sizes for parallel processing.

The performance of two different approaches in implementing multi-level parallelism of the INS3D code is reported in this section. The first approach is a hybrid MPI/OpenMP and the second approach is Multi Level Parallelism (MLP) developed at NASA-Ames Research Center (Taft 2000). The first approach is obtained by using message-passing interface (MPI) for inter-zone parallelism, and by using OpenMP directives for intra-zone parallelism. INS3D-MPI is based on the explicit message-passing interface across MPI groups and is designed for coarse grain parallelism. The primary strategy is to distribute the zones across a set of processors. During the iteration, all the processors would exchange boundary condition data between processors whose zones shared interfaces with zones on other processors. A simple master-worker architecture was selected because it is relatively simple to implement and it is a common architecture for parallel CFD applications. All I/O was performed by the master MPI process and data was distributed to the workers. After the initialization phase is complete, the program begins its main iteration loop.

The MLP approach differs from the MPI/OpenMP approach in a fundamental way in that it does not use messaging at all. All data communication at the coarsest and finest level is accomplished via direct memory referencing instructions, however, this can only executed on shared-memory computers. The coarsest level parallelism is implemented by spawning independent processes via the standard UNIX fork. The advantage of this approach over the MPI procedure is that the user does not have to change the initialization section of the large production code. Library of routines are used to initiate forks, to establish shared memory arenas, and to provide synchronization primitives. The boundary data for the overset-grid system is updated in the shared memory arena by each process. Other processes access the data from the arena as needed. Figure 16 and figure 17 show the speed-up for the SSME impeller computations using 19.2 million grid points by using MPI/OpenMP and MLP strategies, respectively.

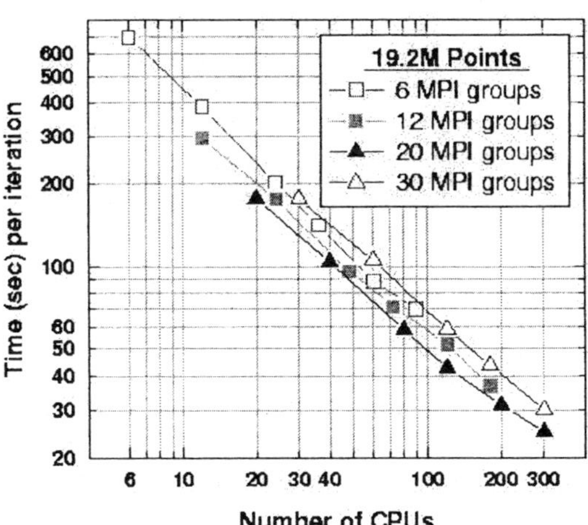


**Fig. 16.** Time (sec) per iteration for SSME impeller computations using INS3D-MPI/OpenMP.

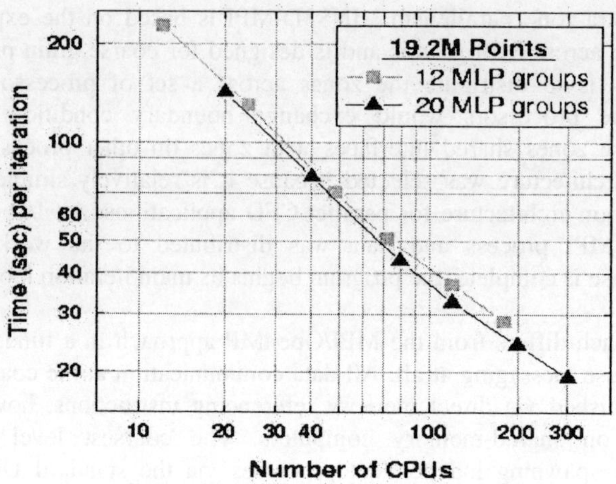



**Fig. 16**. Time (sec) per iteration for SSME impeller computations using INS3D-MLP.

**Fig. 17.** Snapshots of particle traces and pressure surfaces from unsteady turbopump computations.

Using the MLP parallel implementation, time-accurate computations for the SSME-rig1 configuration have been carried out on SGI Origin 2000 and 3000 platforms. Instantaneous snapshots of particle traces and pressure surfaces from these computations are shown in Figure 17. The initial conditions for these simulations used flow at rest, and then the impeller started to rotate impulsively. Three full impeller rotations were completed in the simulations using 34.3 million grid points. Using 128 SGI Origin 3000 CPUs, one impeller rotation was competed in less then 3.5 days. This capability is needed to support the design of pump sub-systems for advanced space transportation vehicles that are likely to involve liquid propulsion systems. To date, computational tools for design/analysis of turbopump flows are based on relatively lower fidelity methods. An unsteady, three-dimensional viscous flow analysis tool involving stationary and rotational components for the entire turbopump assembly has not been available for real-world engineering applications. The present effort provides developers with information such as transient flow phenomena at start up, and non-uniform inflows, and will eventually impact on system vibration and structures.

## 3.3 Ventricular Assist Device

Approximately 20 million people worldwide suffer annually from congestive heart failure (CHF), a quarter of them in America alone. In the United States, an alarmingly low 2,000 to 2,500 donor hearts are available each year. One potential approach to improve this situation is to use a mechanical device to boost or to create blood flow in patients suffering from hemodynamic deterioration; that is, loss of blood pressure and lowered cardiac output. The goal of this device can be to replace the natural heart, i.e. total artificial heart, or to assist an ailing heart, i.e. ventricular assist device (VAD). In either approach, the device can be used to bridge the gap while waiting for a matching donor heart for transplantation. However, to ease the shortage of donor hearts, making these devices suitable for long-term or permanent use would be an ultimate goal.

Another benefit of an assist device is the potential for providing time for the natural heart to recover. In some patients, it has been observed that the natural heart can recover by unloading the pumping requirement through the use of a VAD. In what conditions this might happen is not very well quantified at this time and should involve physiological particulars of patients among other factors. From pump technology point of view, the challenge is to design a device which can deliver the required blood circulation while not adversely impacting human physiological conditions.

Requirements for a VAD related to fluid dynamics are demanding such as: simplicity and reliability; small size for ease of implantation; pumping capacity to supply 5 liter/min of blood against 100 mmHg pressure; high pumping efficiency to minimize power requirements; and minimum hemolysis and thrombus formation. In addition to fluid dynamic issues, there are many other important aspects to be taken care of such as material compatibility with the human body, controls and implantation procedures. Due to the complexity of the flow physics and the delicate operating conditions, an empirical approach to quantify the flow phenomena in a VAD is very time consuming and expensive, especially to study many design variations. CFD simulation tools hold the potential to be invaluable for the development of these devices. In this section, the discussion is focused on how fluid dynamic issues of VAD can be resolved a computational analysis, which is extremely challenging. Flow is unsteady and involves moving parts. For a complete analysis of a VAD, a simulation of the human circulatory system has to be coupled to the device in use. However, for the purpose of developing mechanical components, a truncated circulation system can be modeled. For example, empirical inflow condition can be specified at the inlet of a VAD. Even with this type of simplifications, computational approach can produce flow field data in great detail, thus shedding lights to obtain a better understanding of the dominant flow physics produced by an artificial device. Especially, computational analysis can be utilized to optimize the design of mechanical devices at a significantly lower cost and time than required by an empirical approach.

In 1989, NASA Johnson Space Center (JSC) began a joint project with the DeBakey Heart Center of the Baylor College of Medicine (BCM) in Houston to develop a new implantable prototype LVAD system.  This LVAD is based on a fast rotating axial pump requiring a minimum number of moving parts.  To make it implantable, the device has been made as small as possible, requiring a very high rotational speed. The computational procedure described in the pump section has been used to provide the designers with a view of the complicated fluid dynamic processes inside this device.

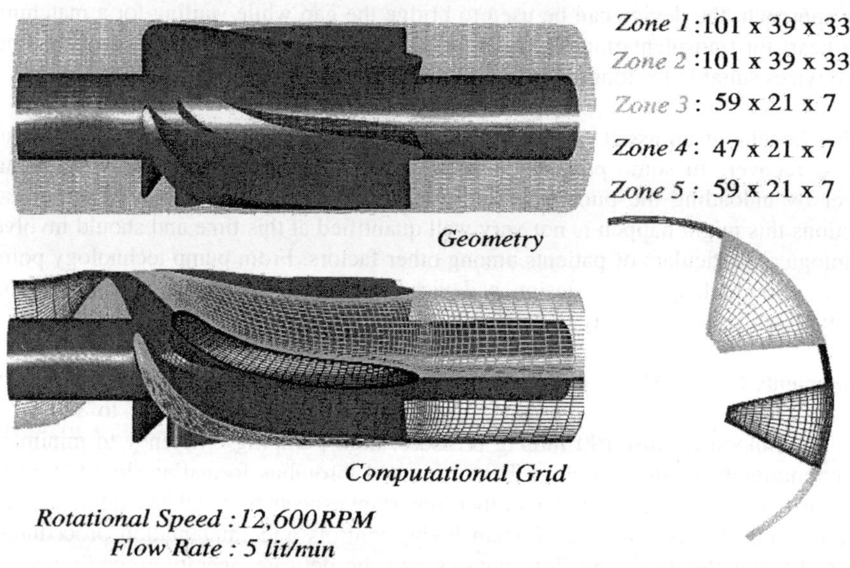


**Fig. 18.** Computational grid for the baseline model of DeBakey VAD.

The flow through the baseline design of the VAD impeller was numerically simulated by using the INS3D-UP flow solver and a steady rotating frame of reference. Zonal multiblock grids were used in this component analysis. The surfaces of the computational grids for the VAD baseline  impeller are shown in Figure 18. The domain is divided into five zones with dimensions of  127 x 39 x 33,  127 x 39 x 33, 59 x 21 x 7, 47 x 21 x 5, and 59 x 21 x 7, respectively. Zone 1 is the region between the suction side of the partial blade and the pressure side of the full blade;  the region between the pressure side of the partial blade and the suction side of the full blade is filled by zone 2; and zones 3 through 5 allow tip-leakage effects to be included in the computational study  and occupy the regions between the impeller blade tip and the casing. At the zonal interfaces, grid points were matched one-to-one. For all zones, an H-H type grid topology was used. An H-type surface grid was generated for each surface using an elliptic grid generator. The interior region of the three-dimensional grid was filled using an algebraic-grid generator coupled with an elliptic smoother. Periodic boundary conditions were used at the end points in the rotational direction. The design flow of this impeller is 5 liters per minute and the design speed is 12,600 revolutions per minute (rpm). The problem was non-dimensionalized by the tube diameter (0.472 inches) and the impeller tip-velocity. The solution was considered converged when the maximum residual had dropped  at least five orders of magnitude. Figure 19 shows the flow pattern near the suction side and pressure side of the baseline impeller blades.

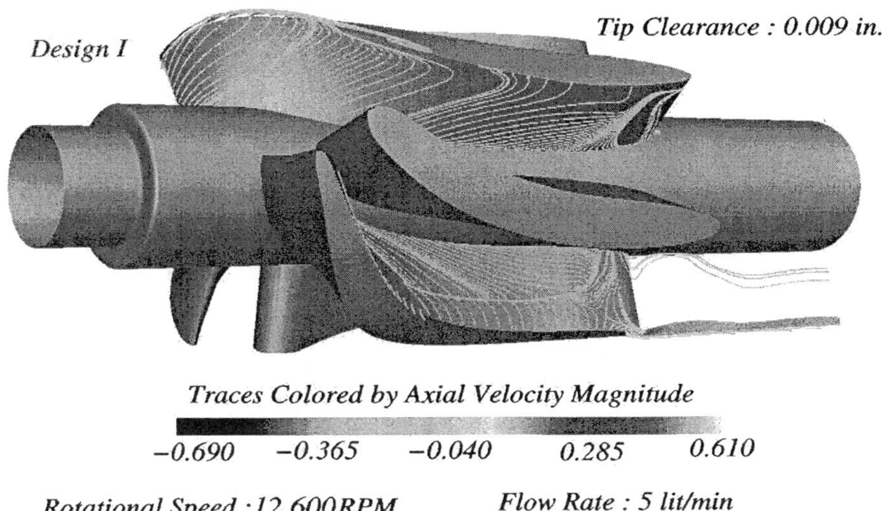


**Fig. 19.** Flow pattern near the suction and pressure sides of VAD baseline impeller full blade.

A parametric study was performed to optimize the impeller blade shape and the tip clearance. Initially, three different impeller-blade designs with a tip clearance of 0.009 inches were analyzed. Then baseline blade shape  was analyzed with two tip clearances; the tip clearance of 0.0045 inches shows better hydrodynamic  performance in terms of efficiency and head coefficient than with a tip clearance of 0.009 inches.

Using this design with a tip clearance of 0.0045 inches as  the baseline impeller design, ideas from rocket propulsion were introduced to develop a new implantable VAD.  In collaboration with Micromed Technologies and NASA-JSC engineering team and BCM researchers, a new design consisting of the baseline impeller plus an inducer was investigated. The hub and blade surfaces of the baseline impeller and the new impeller, colored by nondimensionalized pressure, are shown in Figure 20. The pressure gradient across the blades, due to the action of centrifugal force, and the pressure rise from inflow to outflow are shown. The inducer provides a sufficient pressure rise to the flow in order to prevent the cavitation on the impeller blades. Figure 21 shows the particle traces through the new impeller design.

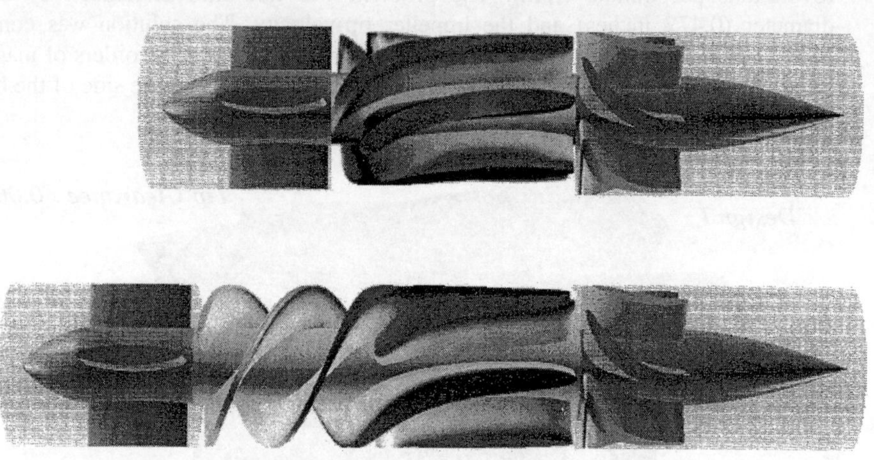

**Fig. 20**.  Pressure surfaces of the baseline design  (top) and new impeller design.

**Fig. 21.**  Pressure surfaces of the baseline design  (top) and new impeller design.

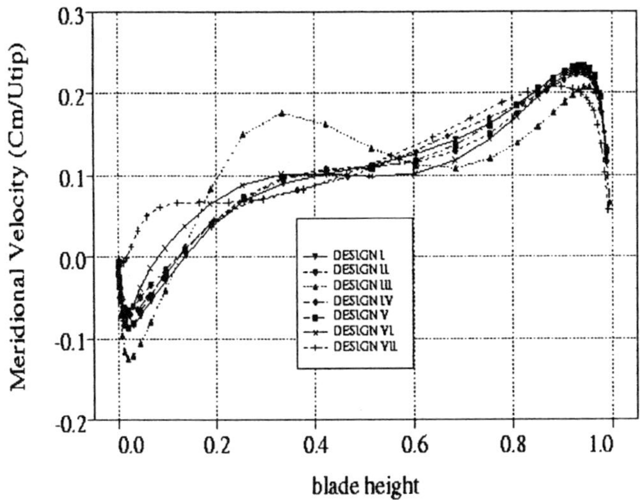

**Fig. 22.** Meridional velocity distribution along impeller blade height of various designs.

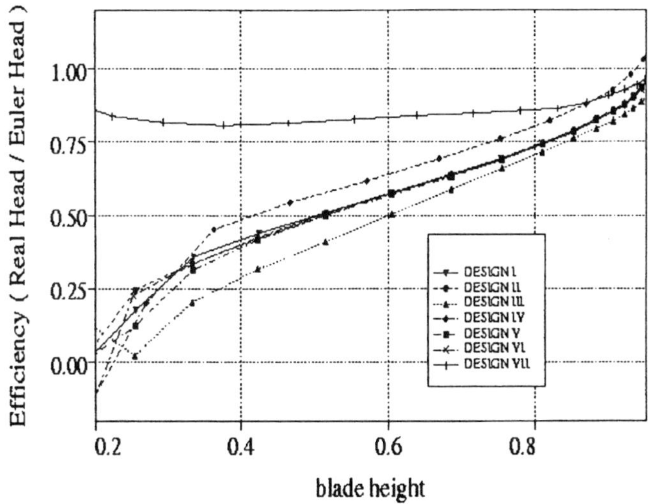

**Fig. 23.** Hydrodynamic efficiency distribution along impeller blade height of various designs.

The traces are colored by the relative total-velocity magnitude. The particles were released near the inducer leading edge, the hub, the  inducer blade pressure side, and the tip regions. The swirling motion of the particles indicates a secondary flow  region between the partial and the full blades.The particles released near the pressure side of the blade indicate  a radial velocity component inside the blade boundary layer. The particles tend to flow from the hub to the tip of the blade. The particles near the inducer leading edge and full blade trailing edge  indicate the presence of back flow.

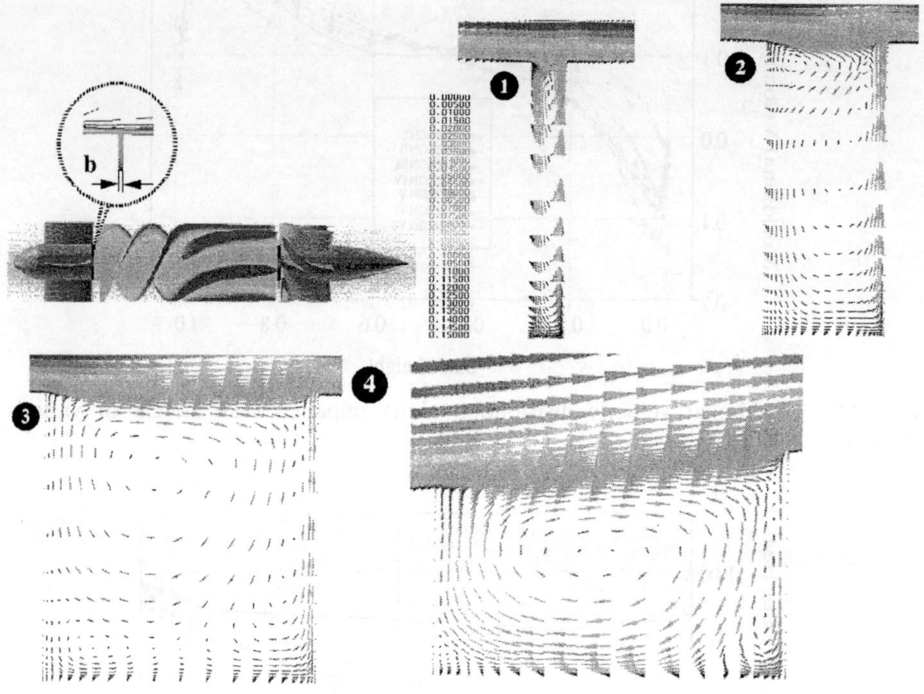

**Fig. 24.** Velocity vectors inside various bearing geometries.

Parametric studies to eliminate the back flow near the hub region by tapering the hub surface have also performed for this configuration. Figure 22 shows the circumferentially averaged meridional velocity distribution along the blade height for various designs. The original blade design is referred  to as Design I. Design II has less blade curvature than Design I in the trailing edge region, and Design III has more blade curvature than Design I. In Design IV, the blade shape for Design I is kept and the tip clearance is reduced from 0.009 inches to 0.004 inches. In Design V, the hub region has the blade shape for Design I and the tip region has the blade shape for Design II. In this design, the impeller blades have backward lean near the trailing edge region. In Design VI, the blades have forward lean which includes Design III in the hub region and Design I in the tip region. Design VII has small tip clearance gap, Design I blade shape and an inducer geometry upstream of impeller blades. In Figure 22, all designs except Design VII showed back flow near the hub region. The back flow has been reduced with forward blade lean which is suggested as a design change. Figure 23 shows the efficiency curves for these design variations. The inducer addition clearly shows substantial improvement in the hydrodynamic efficiency.

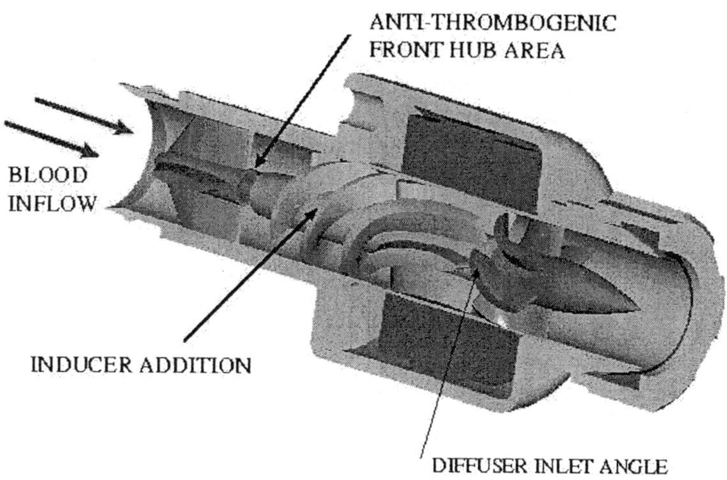



**Fig. 25.** Contribution of CFD analysis to VAD design.

Besides improving the pumping efficiency, the design of the VAD requires good wall washing near the solid walls and reducing the stagnation regions. One of the critical regions for potential blood clotting is near the bearing area between rotating and non-rotating components. Clotting can be caused in the hub area due to either high shear or stagnation, depending on the gap and configuration of the area. Figure 24 shows velocity vectors colored by velocity magnitude for four different bearing designs. Design 1 is the original baseline design with the cavity width of b. This design showed very high shear stresses near the rotating hub face and very stagnant fluid region in the lower portion of the cavity. Increasing the cavity width to 3.5b (Design 2), and to 8b (Design 3 and 4) showed that the recirculation was increased in the cavity. In order to eliminate stagnant areas in the lower portion of the cavity, the hub surface was tapered. Tapering the hub surface reduced the cavity height, accelerated the flow near the hub region, and resulted in stronger recirculation in the cavity (Design 4). A modified version of Design 4 has been adopted in the current DeBakey VAD design. Figure 25 shows the areas that the VAD design is improved by using the present CFD analysis tool. This unique insight into the internal fluid structures led to an improved heart-assist device which enabled human implantation of the device. As of June, 2002, over 160 patients have successfully received this VAD. Thus, improved designs made possible because of the current work is making a far-reaching impact on human health.

## 4 Summary

In this paper, incompressible Navier-Stokes solvers designed for three-dimensional flow simulations have been discussed. The discussion has been limited to the primitive-variable formulation as it causes fewer complications in setting the boundary conditions. Numerous computed results have been presented to illustrate the numerical procedures. Even though computer speed and memory have been increased substantially in the recent past, the speed and the memory requirements of a flow solver are still major factors affecting the turnaround time. INS3D-UP, which is an upwind finite-difference code based on an artificial compressibility approach, has been being applied to a wide variety of applications for steady-state, time-accurate and rotational-steady solutions. INS3D-FS, which is based on a pressure projection method using a finite volume discretization on staggered grids, was written solely for solving time-dependent problems. These solvers have been utilized in many applications of major engineering significance. As an example, an efficient and robust solution procedure for 3-D turbopump analyses and its spin-off application to VAD impeller has been presented. The flow through an advanced turbopump impeller and SSME rig-1 configuration have been successfully simulated. The validated solution procedure was then applied to the development of the DeBakey VAD. Various design improvements were made through the use of this computational tool. For example, the addition of an inducer dramatically increased pumping efficiency, thereby reducing the hemolysis to an acceptable level for human use, and an optimum cavity redesign practically removed thrombus formation in the bearing area. Overall, the VAD development was expedited by extending the incompressible-flow simulation procedure originally developed for a rocket pump, thus enabling human implantation. The final measure of success has been demonstrated through successful human implantations.

## Acknowledgements

The authors would like to thank Micromed Technologies, NASA Johnson Space Center and Baylor College of Medicine VAD Team for providing the VAD geometry, and the clinical results.

## References

**Belov, A. Martinelli, L. and Jameson, A.,** "A New Implicit Algorithm with Multigrid for Unsteady Incompressible Flow Calculations," AIAA Paper 95-0049, 1995.

**Chan, W. M., Gomez, R. J., Rogers, S. E. and Buning, P. G.,** "Best Practices in Overset Grid Generation, " AIAA Paper 2002-3191, 32nd AIAA Fluid Dynamics Conference, St. Louis, Missouri, June, 2002.

**Chang, J. L. C., and Kwak, D.,** "On the Method of Pseudo Compressibility for Numerically Solving Incompressible Flows," AIAA Paper 84-0252, AIAA 22nd Aerospace Sciences Meeting, Reno, NV, January 9-12, 1984.

**Chang, J.L.C., Kwak, D., Rogers, S. E. and Yang, R-J,** "Numerical Simulation Methods of Incompressible Flows and an Application to the Space Shuttle Main Engine," Int. J. Numerical Method in Fluids, Vol. 8, pp. 1241-1268, 1988.

**Chen, Y.S., Shang, H.M and Chen, C.P.,** "Unified CFD Algorithm with a Pressure Based Method," 6th Int'l. Symposium on Comp. Fluid Dyn., Sept 4-8, 1995, Lake Tahoe, NV.

**Choi, D. and Merkle, C.L.,** "Application of Time-Iterative Schemes to Incompressible Flow," AIAA J., Vol. 23, No. 10, 1518-1524, 1985.

**Chorin, A. J.,** "A Numerical Method for Solving Incompressible Viscous Flow Problems," J. Comp. Phys., Vol. 2, pp.12-26, 1967.

**Chorin, A.J.,** "Numerical Solution of Navier-Stokes equations," Mathematics of Computation, Vol. 22, No. 104, 745-762, 1968.

**Gresho, M. P. and Sani, R. L.,** "On Pressure Boundary Conditions for the Incompressible Navier-Stokes Equations," Int. J. Numerical Methods in Fluids, Vol. 7, pp. 1111--1145, 1987.

**Hafez, M. and Oshima, K.,** ed. <u>Computational Fluid Dynamics 1985</u>, John Wiley and Sons, 1995.

**Harlow, F. H. and Welch, J. E.,** "Numerical Calculation of Time-Dependent Viscous Incompressible Flow with Free Surface," Phys. Fluids, Vol. 8, No. 12, pp. 2182-2189,1965.

**Hirsch, C.,** "Numerical Computation of Internal and External Flows," John Wiley & Sons, 1988.

**Kiris, C., Chang, L., Kwak, D., and Rogers, S. E.,** "Incompressible Navier-Stokes Computations of Rotating Flows," AIAA Paper No. 93-0678, 1993.

**Kiris, C. and Kwak, D.,** "Progress in Incompressible Navier-Stokes Computations for the Analysis of Propulsion Flows,'" NASA CP 3282, Vol II, Advanced Earth-to-Orbit Propulsion Technology, 1994.

**Kiris, C. and Kwak, D.,** "Numerical Solution of Incompressible Navier-Stokes Equations Using a Fractional-Step Approach," AIAA Paper 96-2089, AIAA 27th Fluid Dynamics New Orleans, LA, June 17-20, 1996.

**Kiris, C., Kwak, D., and Chan, W.,** "A Three-Dimensional Parallel Time-Accurate Turbopump Simulation Procedure Using Overset Grid Systems," $2^{nd}$ International Conference on Computational Fluid Dynamics, Sydney, July 15-19, 2002.

**Kwak, D., Chang, J. L. C., Shanks, S. P., and Chakravarthy, S.,** "A Three-Dimensional Incompressible Navier-Stokes Flow Solver Using Primitive Variables," AIAA J, Vol. 24, No. 3, pp 390-396, Mar. 1986.

**Kwak, D.,** "Computation of Viscous Incompressible Flows," von Karman Institute for Fluid Dynamics, Lecture Series 1989-04. Also NASA TM 101090, March 1989.

**MacCormack, R. W.,** "Current Status of Numerical Solutions of the Navier-Stokes Equations," AIAA Paper 85-0032, 1985.

**Meakin, R. L.,** "Object X-rays for Cutting Holes in Composite Overset Structured Grids, " AIAA Paper 2001-2537, 15th AIAA Computational Fluid Dynamics Conference, Anaheim, California, June, 2001.

**Merkle, C. L. and Athavale, M.,** "Time-Accurate Unsteady Incompressible Flow Algorithms Based on Artificial Compressibility," AIAA Paper 87-1137, 1987.

**Patankar, S.V.,** <u>Numerical Heat Transfer and Fluid Flow,</u> Hemisphere Publishing Co., New York, 1980.

**Peyret, R. and Taylor, T.,** <u>*Computational Methods for Fluids*</u>, Springer-Verlag, Nyew York, 1983.

**Rizzi, A. and Eriksson, L.-E.,** "Computation of Inviscid Incompressible Flow with Rotation," J. Comp. Phys., Vol. 153, pp 275-312 , 1985.

**Rogers, S. E. and Kwak, D.,** "An Upwind Differncing Scheme for the Time-Accurate Incompressible Navier-Stokes Equations," AIAA Paper 88-2583, AIAA 6th Applied Aerodynamics Conference, Williamsburg, VA, June 6-8, 1988.

**Rogers, S. E., Kwak, D. and Kiris, C.,** "Steady and Unsteady Solutions of the Incompressible Navier-Stokes Equations," AIAA J. Vol. 29, No. 4, 603-610, April 1991.

**Rogers, S. E.,** "A Comparison of Implicit Schemes for the Incompressible Navier-Stokes Equations and Artificial Compressibility," AIAA J., Vol. 33, No. 10, Oct. 1995.

**Rosenfeld, M., Kwak, D. and Vinokur, M.,** "A Fractional Step Solution Method for the Unsteady Incompressible Navier-Stokes Equations in Generalized Coordinate Systems," J. Comp. Phys., Vol. 94, No.1, pp 102-137, May, 1991.

**Steger, J. L. and Kutler, P.,** "Implicit Finite-Difference Procedures for the Computation of Vortex Wakes," AIAA J., Vol. 15, No. 4, pp. 581-590, Apr. 1977.

**Taft, J. R.,** "Performance of the OVERFLOW-MLP and LAURA-MLP CFD Codes and the NASA Ames 512 CPU Origin Systems," HPCC/CAS 2000 Workshop, NASA Ames Research Center, 2000.

**Taneda, S., and Honji, H.,** "Unsteady Flow Past a Flat Plate Normal to the Direction of Motion," J. Phys. Soc. Japan Vol. 30, pp. 262-273, 1971.

**Temam, R.,** <u>Navier Stokes Equations,</u> Revised Edition., North Holland, 1979.

**Yoshida, Y., and Nomura, T.,** "A Transient Solution Method for the Finite Element Incompressible Navier-Stokes Equations," Int. J. Num. Methods in Fluids, Vol. 5, pp. 873-890, 1985.

# Evolution of Artificial Compressibility Methods in CFD

Sankaran Venkateswaran and Charles L. Merkle

University of Tennessee, Tullahoma, TN 37388, USA

**Abstract.** We trace the evolution of artificial compressibility or preconditioning methods, originally introduced by Chorin to enable time-marching solutions of the incompressible equations and later adapted and generalized to a wide range of flow regimes. In particular, we use asymptotic expansions to investigate the theoretical underpinnings of the approach and to elucidate how it is instrumental in improving the accuracy and efficiency of a wide range of flow computations. These include low Mach compressible flows, viscous-dominated flows, unsteady flows and multi-phase flows. Representative computations are provided throughout to illustrate the efficacy of the techniques.

## 1  Introduction

Computational fluid dynamics methods for solving incompressible flowfields have followed two distinct evolutionary paths. The first set of incompressible flow algorithms were the MAC method due to Harlow and Welch [1], the projection method due to Chorin [2] and the popular SIMPLE family of schemes due to Patankar and Spalding [3]. These methods, collectively referred to as pressure-based methods, usually solve the governing equations in uncoupled fashion and are characterized by the introduction of a Poisson equation for the pressure field, in lieu of the standard continuity equation. The second group of methods for the solution of the incompressible equations is the artificial compressibility approach, originally proposed by Chorin [4] and widely applied to practical problems by Kwak [5,6] and others [7,8]. The foundations and continued development of this class of methods is the subject of the present article.

The artificial compressibility method evolved from density-based methods, that were developed for transonic and supersonic compressible flows [9]. These methods solve the governing equations in coupled fashion, typically retaining the time-derivatives, so that the steady state solution is obtained by marching the equations in time. Further, the continuity equation is solved in its standard form, with the density variable (which appears in the time-derivative of continuity) being used as a primary dependent variable. This latter aspect stymies the natural extension of these methods to incompressible flow because the density is constant (or nearly so) for these flows. Chorin's formulation circumvents this problem by replacing the physical time-derivative of density in the continuity equation with an artificial time-derivative of pressure, thereby enabling a well-posed time-marching solution procedure to be formulated for incompressible computations.

The notion of introducing artificial time-derivatives instead of the physical ones for the purpose of optimizing the time-iterative process proved to be a fundamentally revolutionary idea that led to the development of the so-called preconditioning methods [10–13]. The term preconditioning here refers to the introduction of a preconditioning matrix that premultiplies the standard time-derivative terms, thereby offering a systematic means of controlling the definition of these terms. In particular, the careful scaling of the time derivatives provides a means of controlling the characteristic eigenvalues of the system and, thereby, of enhancing the convergence process [12,13]. Perhaps more importantly, the artificial time-derivatives also influence the selection of the artificial dissipation terms in the formulation of the discretized system and, consequently, can impact the overall accuracy of the formulation [10,11]. These ideas have led to the generalization of the artificial compressibility or preconditioning methods to both incompressible and compressible flows. In particular, recent research has demonstrated that it is possible to optimize the scaling of the artificial time-derivative terms for a variety of flow regimes such as low-speed flows governed by arbitrary equations of state [13,14], viscous-dominated flows [15], flows with high aspect ratio cells [16,17], unsteady flows [18] and multi-species and multi-phase flows [19].

The theoretical basis for the above developments is illustrated by the use of asymptotic expansions to examine the limiting forms of the governing equations in various flow regimes [20,21]. Accordingly, we begin our discussion by applying asymptotic theory to the low-speed (or incompressible) limit of the fluid equations with an arbitrary equation of state. The analysis points to the underlying reasons for the difficulty in applying traditional time-marching procedures and suggests the appropriate scaling or preconditioning of the time-derivatives to rectify the situation. Moreover, by applying the perturbation analysis to the discrete form of the equations, it is also possible to verify the accuracy enhancements enabled by the preconditioned formulation. Following this, we discuss in turn the proper definition of the preconditioned time-derivative terms for the limits of viscous-dominated flows, flows with high aspect ratio grid cells and unsteady flows. Finally, we briefly discuss the extensions of the method to multiple species and multiple phases. Representative computational results are presented throughout the paper to demonstrate the accuracy and efficiency improvements enabled by the preconditioning formulation.

## 2   Euler Equations Analysis

### 2.1   Asymptotic Analysis of the Low Speed Limit

To examine the low speed asymptotic limit of the standard compressible Euler equations, we start with the one-dimensional equations,

$$\Gamma_p \frac{\partial Q_p}{\partial \tau} + \frac{\partial E}{\partial x} = 0$$

where the flux vectors are defined as:

$$Q_p = \begin{pmatrix} p \\ u \\ T \end{pmatrix} \qquad E = \begin{pmatrix} \rho u \\ \rho u^2 + p \\ (e + p)u \end{pmatrix}$$

To begin with, we take $\Gamma_p = \partial Q / \partial Q_p$, where $Q$ is the conservative variables' vector. In other words, the time-derivative terms take the same form as the physical time derivatives. These terms are defined as,

$$Q = \begin{pmatrix} \rho \\ \rho u \\ e \end{pmatrix} \qquad \frac{\partial Q}{\partial Q_p} = \begin{pmatrix} \rho_p & 0 & \rho_T \\ u\rho_p & 1 & u\rho_T \\ h_o\rho_p - (1 - \rho h_p) & \rho u & h_o\rho_T + \rho h_T \end{pmatrix}$$

The four variables $\rho_p$, $\rho_T$, $h_p$ and $h_T$ that appear in this matrix are thermodynamic quantities that describe the properties of the working fluid. The subscripts denote partial derivatives, i.e., $\rho_p = (\partial\rho/\partial p)_T$, etc. These relations can be obtained directly from the equations of state, $\rho = \rho(p, T)$ and $h = h(p, T)$. For a perfect gas, $\rho_p = (1/RT)$, while $\rho_T = -(\rho/T)$, $h_p = 0$ and $h_T$ is the specific heat. On the other hand, for incompressible flow, $\rho_p = \rho_T = 0$.

To simplify the algebra, it is easier to work with the non-conservative equations, which can be obtained from the conservative form by standard techniques. Further, we non-dimensionalize the equations of motion by introducing reference scales for all the variables,

$$L, \; p_r, \; \rho_r, \; T_r, \; h_r, \; u_r, \; \tau_r$$

Here, $L$ represents a length scale that is characteristic of the problem, while the reference pressure, $p_r$, density, $\rho_r$, temperature, $T_r$, and enthalpy, $h_r$ are taken as representative thermodynamic quantities. The reference velocity, $u_r$, is chosen as the oncoming or free-stream velocity. Finally, $\tau_r$ is the characteristic time scale, which will be defined later.

The resulting non-dimensional equations are,

$$\left(\frac{L}{\tau_r u_r}\right)\left(\tilde{\rho}_p \frac{\partial p}{\partial \tau} + \tilde{\rho}_T \frac{\partial T}{\partial \tau}\right) + \frac{\partial \rho u}{\partial x} = 0$$

$$\left(\frac{L}{\tau_r u_r}\right)\rho\frac{\partial u}{\partial \tau} + \rho u\frac{\partial u}{\partial x} + \left(\frac{p_r}{\rho_r u_r^2}\right)\frac{\partial p}{\partial x} = 0 \qquad (2)$$

$$\left(\frac{L}{\tau_r u_r}\right)\left[-\left(\frac{p_r}{\rho_r h_r} - \rho\tilde{h}_p\right)\frac{\partial p}{\partial \tau} + \rho\tilde{h}_T\frac{\partial T}{\partial \tau}\right] + \rho u\frac{\partial h}{\partial x} - \left(\frac{p_r}{\rho_r h_r}\right)u\frac{\partial p}{\partial x} = 0$$

For clarity, we have used tildes on the four properties, $\tilde{\rho}_p$, $\tilde{\rho}_T$, $\tilde{h}_p$ and $\tilde{h}_T$, to indicate that they are non-dimensional. The non-dimensionalization is of the form, $\tilde{\rho}_p = p_r\rho_p/\rho_r$, etc. We note all the other terms (except the reference quantities) are also non-dimensional, but the tildes have been dropped to avoid clutter.

The non-dimensional equations introduce several dimensionless quantities. Two of them are dimensionless pressures, the ratio of the pressure to the dynamic pressure, $(p_r/\rho_r u_r^2)$, and the ratio of the pressure to the other thermodynamic properties, $(p_r/\rho_r h_r)$. In addition, we have the dimensionless time scale given by $(\tau_r u_r/L)$.

The above non-dimensional equations provide a vantage point from which we can assess the order of magnitude of the various terms in the equations. Accordingly, we expand all the variables in a power series of the form,

$$p = p_0 + \epsilon p_1 + \dots \tag{3}$$

The definition of the small parameter, $\epsilon$, will depend upon the conditions as discussed later.

Substituting the expansions into Eqns. 2, we obtain the perturbation equations. We first consider the momentum equation, which takes the form:

$$\left(\frac{L}{\tau_r u_r}\right)\rho\frac{\partial u}{\partial \tau} + +\rho u \frac{\partial u}{\partial x} + \left(\frac{p_r}{\rho_r u_r^2}\right)\epsilon\frac{\partial p_1}{\partial x} = 0 \tag{4}$$

where the zeroth order pressure drops out from order of magnitude considerations. For all other variables, only the zeroth order quantities appear and, to avoid clutter, we drop the subscripts.

The continuity equation is the other equation of interest and it takes the form,

$$\left(\frac{L}{\tau_r u_r}\right)\left(\tilde{\rho}_p\epsilon\frac{\partial p_1}{\partial \tau} + \tilde{\rho}_T\frac{\partial T}{\partial \tau}\right) + \frac{\partial \rho u}{\partial x} = 0 \tag{5}$$

The energy equation does not add anything new to the analysis and, to conserve space, is not considered further.

We require that all the dominant terms in the perturbation equations balance in order for the system to be well-behaved. Specifically, in the momentum equation (Eqn. 4), we require that the pressure gradient term balances the dominant term in the momentum equation. The dominant term in the momentum equation is the convective term, which is order unity. At low speeds, the reference velocity, $u_r$, approaches zero so that the ratio of the dynamic pressure to the thermodynamic pressure, $(\rho_r u_r^2/p_r)$, is small. Accordingly, we specify the small parameter,

$$\epsilon = \rho_r u_r^2/p_r$$

This causes the pressure gradient to be of order unity and, thereby, balance the convective term.

We also require that the non-dimensional time term be of order unity. This condition requires that the reference pseudo-time scale,

$$\tau_r = L/u_r$$

which is a convective time scale. We note that this selection is appropriate for low-speed flows in the steady, inviscid limit.

Going back to the continuity equation (Eqn. 5), we now observe that the first-order pressure time-derivative is of order $\epsilon$. In the limit that $\epsilon \to 0$ (or at low speeds), the term becomes singular. There is then no means of updating the first-order pressure in the system. Note that in our analysis, we have not altered the equations in any way. We have simply analyzed the proper low Mach number limiting form and determined the existence of the well-known singularity problem in the incompressible limit.

## 2.2 Artificial Compressibility or Preconditioning

The advantage of deriving the above perturbation equations lies in the potential for altering the time-derivatives in such a manner as to allow updating of the pressure field. Since the difficulty arises in the time-derivative of $p_1$ in th continuity equation, we re-write the time derivative in the continuity equation by replacing the property $\tilde{\rho}_p$, by $\tilde{\rho}'_p$,

$$(\tilde{\rho}'_p \epsilon \frac{\partial p_1}{\partial \tau} + \tilde{\rho}_T \frac{\partial T}{\partial \tau}) + \frac{\partial \rho u}{\partial x} = 0 \tag{6}$$

To obtain an update relation for $p_1$ from this equation, we must require that the time term be order one or the coefficient of the time term be $1/\epsilon$. We enforce this condition by setting,

$$\tilde{\rho}'_p = \frac{k_i}{\epsilon} = \frac{k_i p_r}{\rho_r u_r^2}$$

or,

$$\rho'_p = \frac{k_i}{u_r^2} \tag{7}$$

where $k_i$ is a constant of order one.

We note that the physical property, $\rho_p$, represents the reciprocal of the isothermal sound speed squared. For an ideal gas, the artificial property $\rho'_p$ may equivalently be expressed as $\rho'_p = k_i \rho_p / \gamma M_r^2$, where $\rho_p = 1/RT$ and $M_r$ is the reference Mach number. This form is particularly well-suited for compressible codes since it approaches the standard equations near the transonic limit. Further, for incompressible flows, we note that $\rho'_p$ is analogous to the reciprocal of Chorin's artificial compressibility parameter, $\beta$ [4]. Thus, with the introduction of the artificial time derivative, the system is now well-posed and well-conditioned for time marching solutions.

The final preconditioning matrix for the Euler equations given in Eqn. 1 takes the following form,

$$\Gamma_p = \begin{pmatrix} \rho'_p & 0 & \rho_T \\ u\rho'_p & 1 & u\rho_T \\ h_o\rho'_p - (1 - \rho h_p) & \rho u & h_o\rho_T + \rho h_T \end{pmatrix} \tag{8}$$

where $\rho'_p$ is given by,

$$\rho'_p = Max[\frac{k_i}{u^2}, \rho_p]$$

The above definition automatically switches between the appropriate definition for the low-speed limit and the physical time-derivatives for transonic/supersonic flows. Note that, in actual implementation, the above definition is usually modified to account for local effects, particularly in flows involving stagnation regions (eg., see [10] and [22]).

## 2.3 Perturbation Analysis of the Discrete Form

The formulation of the discrete system is responsible for the overall accuracy of the numerical computation. In all CFD algorithms, the discrete system is obtained by appropriate flux reconstruction at the faces of the control volume. It is convenient to interpret the inviscid flux representations as being composed of a central difference part (which is non-dissipative) and an artificial dissipation contribution. We note that, in general, we may view artificial dissipation as being necessary for two reasons—one, for smoothing out high-frequency errors in the solution (conventionally done by adding higher order derivatives), and two, for obtaining monotonic solution behavior in the vicinity of steep gradients (usually achieved by adding second-order derivatives). The differences between schemes may then be related to the precise form of the artificial dissipation terms. In this article, we consider the flux-difference upwind scheme due to Roe [23], which may be viewed as adding a particular form of "matrix" dissipation. Similar interpretations are possible for other schemes and have been considered elsewhere [21].

An upwind flux-difference scheme applied to Eqn. 1 would correspond to central differencing coupled with the addition of matrix dissipation terms. We consider only the first-order representation here,

$$\Gamma_p \frac{\partial Q_p}{\partial \tau} + \frac{\partial E}{\partial x} = \frac{\Delta x}{2} \Gamma_p |\Gamma_p^{-1} A| \frac{\partial^2 Q_p}{\partial x^2} \tag{9}$$

We now consider applying the perturbation analysis procedure to above equation and assess the behavior of the terms in the matrix dissipation in the low-speed limit. The matrix coefficient is observed to take the form,

$$\tilde{\Gamma}_p |\tilde{\Gamma}_p^{-1} \tilde{A}| = \begin{pmatrix} \frac{2}{S} & \frac{\rho u}{S} & u\tilde{\rho}_T \\ \frac{u}{S} & \frac{\rho}{2S}(u^2 + S^2) & 0 \\ 0 & 0 & \rho u \tilde{h}_T \end{pmatrix} \tag{10}$$

where the tildes refer to the non-dimensional perturbation form of the terms and $S = \sqrt{u^2 + \frac{4}{\tilde{\rho}'_p \epsilon}}$.

In the low Mach limit, we observe that $S$ tends to $1/\epsilon$ or $1/M$ when the standard physical-time system is used. All other terms in Eqn. 10 are order unity. Examining the matrix dissipation coefficient, we can see that the $(2,2)$ entry scales as $1/M$, signifying excess dissipation in the momentum equation. On the other hand, the $(1,1)$ term scales as $M$, suggesting that the pressure field does not possess adequate dissipation. When the proper artificial compressibility or preconditioning system (Eqn. 7) is used, $S$ is order unity and, therefore, all the terms in the dissipation matrix are well-proportioned and good accuracy is obtained at all Mach numbers. Similar analyses of the discrete form have been reported by several other researchers as well [10,24].

## 2.4  Computational Results

In this section, we examine convergence efficiency and accuracy issues for inviscid flow in a channel over a circular bump. All computational examples are two-dimensional and employ first-order Roe differencing to highlight the issues. The alternating direction implicit (ADI) algorithm is used for inverting the implicit operator.

Figure 1 shows the results for two different Mach numbers with and without preconditioning. The properly scaled algorithm in converges to machine zero in about 2000 iterations and the convergence rate is independent of the Mach number. On the other hand, with the wrong scaling, the algorithm takes over 10,000 iterations to converge to similar levels and there is a marked slowdown as the Mach number becomes lower.

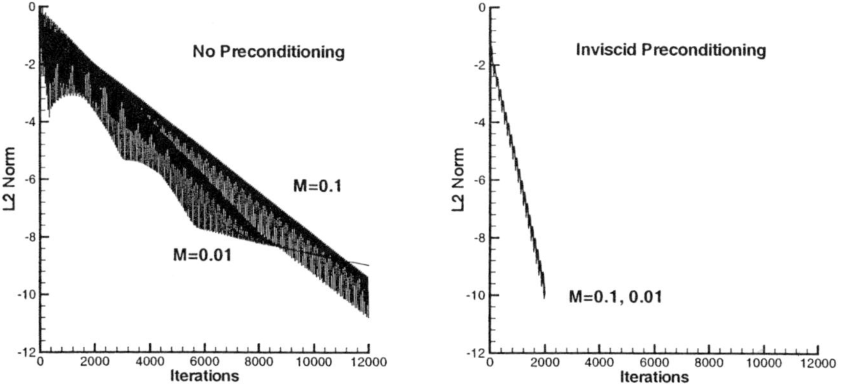


**Fig. 1.** Convergence for inviscid flow without and with preconditioned pseudo-time derivatives

Figures 2 and 3 show solution contours for the two Mach number cases without and with proper scaling of the time-derivative system. It should be recalled that the properly scaled system possesses well-conditioned dissipation terms as well. Indeed, both sets of results show that solution fidelity is poor when the non-preconditioned system is used. The solution at the lower Mach number is, in fact, completely wrong. On the other hand, when the proper preconditioned time-derivatives are used, the correct solution is obtained at both Mach numbers shown.

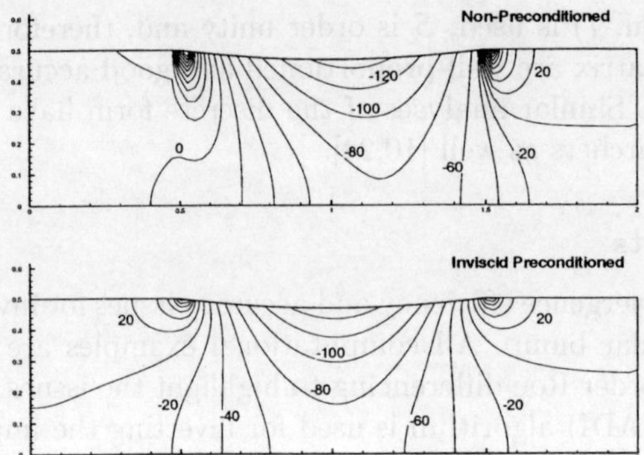

**Fig. 2.** Pressure contours for M=0.1 for inviscid flow over circular bump.

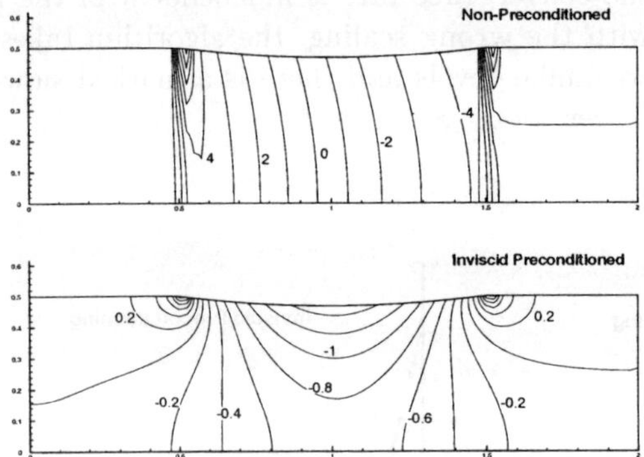

**Fig. 3.** Pressure contours for M=0.01 for inviscid flow over circular bump.

## 3 Extension to More Complex Equation Systems

### 3.1 Viscous Flows

We now consider the regime in which the local Reynolds number is small or, in other words, the viscous effects dominate the convective effects. Specifically, we consider the limit where the Reynolds number is of order unity or smaller in all of the coordinate directions. Analogous to the low Mach number perturbation procedure, we predicate our viscous analysis on ensuring that the equations remain well-balanced in the viscous limit. In the momentum equations, for example, the required balance is no longer between the convective terms and the pressure gradient, but is between the diffusive terms and the pressure gradient. Further, we again require that the time derivative of pressure in the continuity equation be of the same order as the dominant physical terms to be effective in removing errors from the flow-field. Although we do not show the details here, we note that these requirements are met by specifying,

$$\rho_p' = Max[Min(\frac{k_i}{u^2}, \frac{k_v Re^2}{u^2}), \rho_p] \tag{11}$$

This then is the appropriate definition of the time derivative in the viscous limit. Please see Ref. [20,21] for details.

We consider computational solutions in the same two-dimensional channel geometry as before. Figure 4 shows convergence results for three different channel-Reynolds numbers. Figure 4a shows the convergence when the wrong pseudo-time scaling (based on the inviscid definition) is employed, while Fig. 4b shows results when the proper scaling is used. Again, it is clearly apparent that the correct scaling influences convergence efficiency dramatically. For all Reynolds numbers shown, the viscous preconditioned case converges to machine zero in about 2000 iterations. On the other hand, the inviscid preconditioning choice performs well for the highest Reynolds number shown (Re=200), but its performance is observed to deteriorate significantly for the lower Reynolds numbers. These results thus verify the theoretical findings obtained from the asymptotic analysis.

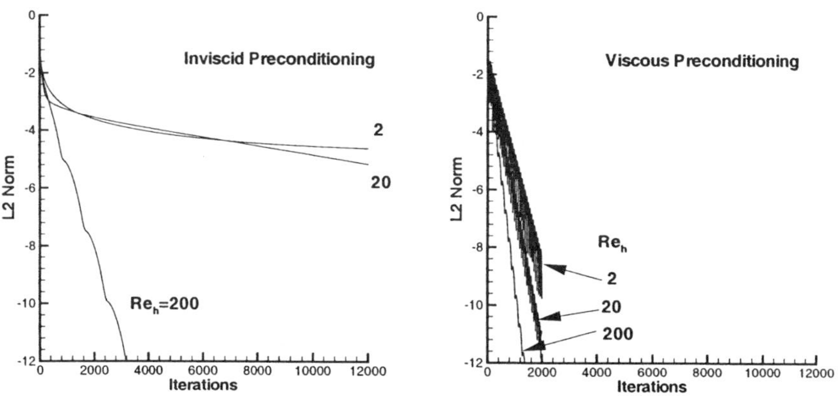


**Fig. 4.** Convergence for viscous flow with inviscid and viscous preconditioned pseudo-time derivatives.

## 3.2 Viscous Flows with High Aspect Ratio Grids

Practical high Reynolds number problems are characterized by highly stretched grids to resolve boundary layers or shear layers. Such high grid aspect ratio problems are generally dominated by viscous processes in the cross-stream direction, while inviscid wave propagation dominates in the streamwise direction. The

44

above viscous preconditioning procedure is therefore not valid and a new procedure needs to be devised. Applying asymptotic analysis to the two-dimensional
Navier-Stokes equations yields the following definition for the time-derivative,

$$\rho_p' = Max[Min(\frac{k_i}{V^2}, \frac{k_v Re_{\Delta y}^2}{V^2 AR^2}, \frac{k_v AR^2 Re_{\Delta x}^2}{V^2}), \rho_p]  \tag{12}$$

where $V^2 = u^2 + v^2$ and $AR = \frac{\Delta x}{\Delta y}$.

Note that the above definition essentially optimizes the viscous processes in
the cross-stream direction along with acoustic wave propagation in the streamwise direction. It should, however, be pointed out that high grid aspect ratios
cause convergence and stability difficulties for many CFD algorithms and the
above preconditioning definition cannot ameliorate these fundamental problems.
Detailed discussion of aspect ratio effects is given in Ref. [16,17]. Here, we simply note that when line implicit procedures are employed in two dimensions,
the above preconditioning scaling combined with proper definition of the local
time-step yields convergence rates that are independent of the aspect ratio.

Representative convergence results for a flat plate boundary layer are given
in Fig. 5. Results are shown for four different Reynolds numbers using grids
with correspondingly different extents of wall grid stretching. The three lower
Reynolds numbers (4000, $4 \times 10^4$ and $4 \times 10^5$) are laminar computations and the
corresponding maximum grid aspect ratios (located adjacent to the wall) are 10,
30 and 200. The highest Reynolds number ($4 \times 10^6$) is a turbulent computation
and the maximum grid aspect ratio for this case is 8000 ($y^+$ is less than one).
In all cases, the flow Mach number is 0.1, and the grid size is $61 \times 61$. For the
turbulent boundary layer computation, the algebraic Baldwin-Lomax turbulence
model is employed.

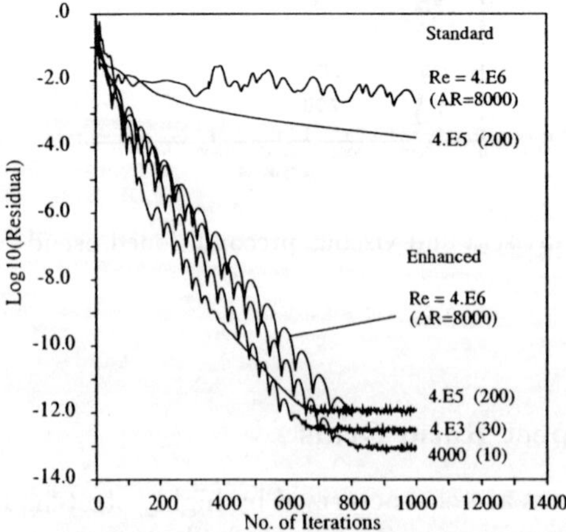


**Fig. 5.** Convergence for flat plate boundary layer for standard and viscous preconditioned algorithms

The convergence in Fig. 5 using the properly scaled algorithm is extremely good for all four Reynolds numbers. Despite the different aspect ratios of the grid cells for each of the cases, the convergence rates are almost identical reaching machine zero is about 800 iterations. For comparison, the convergence of one of the laminar cases ($Re = 4 \times 10^5$) and of the turbulent case ($Re = 4 \times 10^6$) using the standard algorithm (max-CFL/max-VNN) are also shown in Fig. 5. For the laminar case, the residuals drop just over an order of magnitude for every 1000 iterations requiring about 15,000 iterations to reach machine zero. The corresponding convergence speed-up of the enhanced algorithm is a factor of about 20. The turbulent computation in Fig. 5 reflects a convergence speed-up of about 100 times.

## 3.3 Unsteady Flows

As the final limiting condition, we consider the regime in which the unsteady effects are important and dominate the convective and diffusive effects. Unsteady effects are characterized by the Strouhal number. For acoustic wave problems, the physical time scales are controlled by the physical acoustic speeds and $St_r$ approaches $1/M_r$. On the other hand, for vortex propagation problems, the physical time scales are defined by the particle speed and $St_r$ is close to unity. Finally, for unsteady viscous-dominated problems, the characteristic physical time scale would be the diffusion time-scale, then $St_r \rightarrow 1/Re_r$. Thus, we are specifically interested in cases where the physical time ratio, $St_r \geq 1$.

For unsteady computations, we cast the equations in a dual-time framework, which means that distinct physical and pseudo-time derivatives will be present. The actual transient solution is advanced in physical time while, at each physical time-step, pseudo-time iterations are performed until the non-linear equations are satisfied. This strategy is commonly referred to as dual-time stepping [18]. We further note that since the pseudo-time step is used purely for iterative purposes, it typically does not need to be introduced until the system is discretized and/or linearized. Moreover, when iterations are performed at multiple levels— for instance, at the non-linear and linear levels—more than one set of pseudo time-derivatives may be employed (i.e., $\tau_1$, $\tau_2$, etc.) and appropriately defined. For the purposes of the present analysis, however, we will consider a single set of pseudo-time derivatives, introduced at the differential equation level.

We now carry out the analogous perturbation analysis of the dual-time form of the unsteady Navier-Stokes equations. In the momentum equation, we require that the pressure gradient term balances the unsteady time-derivative, while in the continuity equation, we again require that the pressure time-derivative balance the divergence term. We then get,

$$\rho'_p = Max[Min(\frac{k_i}{u^2}, \frac{k_v Re^2}{u^2}, \frac{k_u}{u^2 St^2}), \rho_p]  \tag{13}$$

This is the appropriate definition of the pseudo-time derivative in the unsteady limit. Note that, for simplicity, we have only given the one-dimensional version of the definition.

The above preconditioning definition deserves further comment. We noted earlier that for acoustic wave problems, $St_r \to 1/M_r$. This essentially means that $\rho'_p \to 1/c^2$, which is essentially $\rho_p$. The analysis thus stipulates that for acoustic problems, the time-derivative must not be preconditioned, i.e., in other words, the physical time-derivatives must be used. On the other hand, for vortex propagation problems, $St_r \to 1$. In that case, the above preconditioning definition prescribes the use of the inviscid low-speed choice (see Eqn. 7). Finally, for viscous-dominated unsteady flows, we get $St_r \to 1/Re_r$ which gives us the viscous low-speed choice from Eqn. 11. The definition in Eqn. 13 thus automatically selects between these different values depending upon the nature of the problem.

Representative inner iteration convergence results for the propagation of a Lamb vortex in a two-dimensional channel are shown in Fig. 6. Results are shown for the standard approximate-Newton scheme as well as the appropriately preconditioned dual-time scheme and both line iterative (LGS) and point-iterative (LU) schemes are used. For both iterative methods, the preconditioned dual-time convergence rates are observed to be far superior to those obtained with standard iterative procedures. More detailed results and discussions are given in Refs. [18,20].

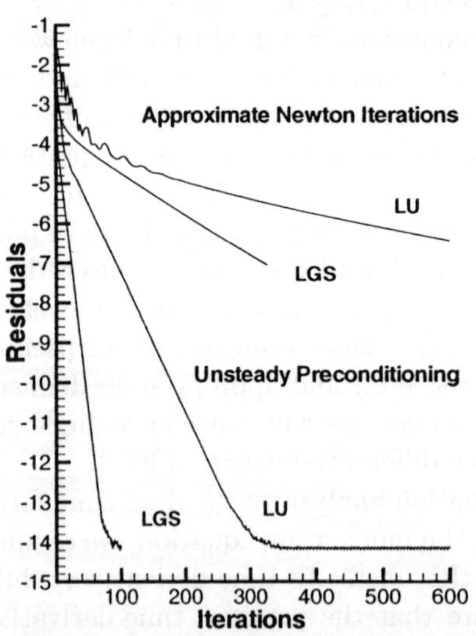


**Fig. 6.** Convergence of inner iterations for a single physical time-step for the computation of Lamb vortex propagation in a straight channel, M=0.001, $CFL_u = 1$.

## 3.4 Multi-Species and Multi-Phase Flows

Extension of the method to multiple species and phases is facilitated by the application of the perturbation analysis to the relevant set of governing equations. Of particular interest is the multiphase situation which involves the simultaneous presence of incompressible liquid and compressible vapor in the same flowfield. Moreover, the mixture region in such problems is characterized by very low sound speeds, which means that transonic and supersonic flow are frequently encountered in these regions. Details of the derivation and application of preconditioning methods to multi-phase flows are given in Ref. [19].

Here, we present two applications that involve two-phase effects. The first example (Fig. 7) is an underwater supersonic projectile with a flow Mach number is 1.03. The computational results show density contours. The liquid-to-gas density ratio is 1000. The flow adjacent to the body and in the wake section are observed to be fully vaporized. Further, both the experimental and computational results show evidence of a bowshock upstream of the nose. The second example (Fig. 8) is the supersonic plume flowfield of an underwater rocket. The plume is slightly under-expanded and, consequently, the computations reveal the classic expansion pattern in the shock function field. We also observe from the density contours that the interface between the liquid and gas phases is comprised of a two-phase mixture, which is supersonic because of the low mixture sound speed. These results showcase the capabilities of the present formulation to handle complex multi-phase mixture flows.

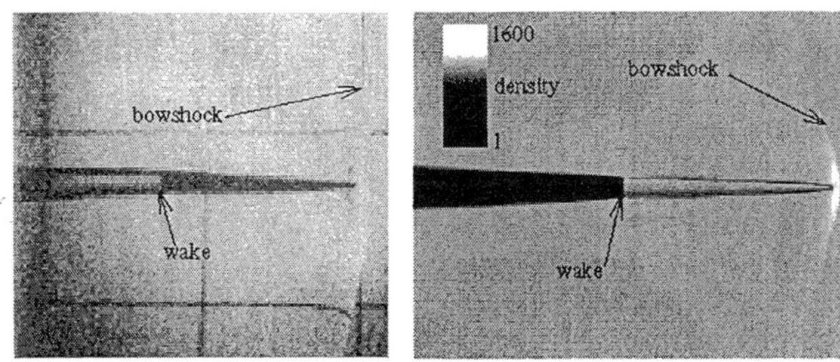


**Fig. 7.** Photograph [25] and multiphase model result of supersonic underwater projectile.

## 4 Summary

The artificial compressibility method originally suggested by Chorin and extensively applied by Kwak and others has served as the inspiration for the development of a larger class of methods, commonly referred to as preconditioning methods. These methods introduce appropriately scaled artificial time-derivatives to the governing equations and thereby enhance the accuracy of the discrete formulation and the convergence efficiency of the solution process. The derivation

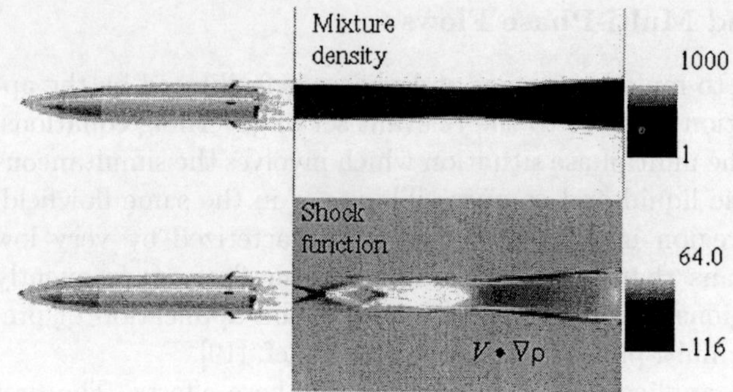


**Fig. 8.** Computation of supersonic gas plume in liquid water. Liquid to gas density ratio is 1000.

of proper form of the artificial time terms follows from a formal application of asymptotic theory to probe the behavior of the equations under various limiting circumstances. The method has been successfully implemented to problems involving low speeds, dominant viscous effects, high aspect ratio grid cells and unsteady flows. The formulation has been generalized for incompressible, compressible and arbitrary equations of state as well as to flows involving multiple species and phases. Future work will focus in part on extensions to even more complex systems such as multi-fluid and plasma systems.

## Acknowledgments

We would like to thank Dr. P. E. O. Buelow for providing Figure 5 and Dr. J. W. Lindau for providing Figures 7 and 8.

## References

1. Harlow, F.H. and Welch, J.E., "Numerical Calculation of Time-Dependent Viscous Incompressible Flow with Free Surface," *Physics of Fluids*, Vol. 8, pp. 2182-2189, (1965).
2. Chorin, A.J., "Numerical Solution of the Navier-Stokes Equations," *Math. comp.*, 22: 742-762, (1968).
3. Patankar, S.V. and Spalding, D.B., "A Calculation Procedure for Heat, Mass and Momentum Transfer," *Intl. Journal of Heat and Mass Transfer*, Vol. 15, pp. 1787-1806, (1972).
4. Chorin, A.J., "A Numerical Method for Solving Incompressible Viscous Flow Problems," *J. Comp. Phy.*, Vol. 2, pp. 12, (1967).
5. Kwak, D., Chang, J. L. C., Shanks, S. P., and Chakravarthy, S., "A Three-Dimensional Incompressible Navier-Stokes Flow Solver Using Primitive Variables," *AIAA Journal*, Vol. 24, No. 3, pp. 390-396, (1986).
6. Rogers, S. E., Kwak, D. and Kiris, C., "Numerical Solution of the Incompressible Navier-Stokes Equations for Steady and Time-Dependent Problems," *AIAA Journal*, Vol. 29, No. 4, pp. 603-610, (1991).

7. Choi, D. and Merkle, C.L., "Application of Time-Iterative Schemes to Incompressible Flow," *AIAA Journal*, Vol. 23, No. 10, pp. 1518, (1985).

8. Merkle, C. L. and Athavale, M., 'A Time Accurate Unsteady Incompressible Algorithm Based on Artificial Compressibility', AIAA Paper No. 87-1137, AIAA 8th Computational Fluid Dynamics Conference, Honolulu, HA, (1987).

9. Hirsch, C., *Numerical Computation of Internal and External Flows*, Vol. 1 and 2, John Wiley and Sons, 1991.

10. Turkel, E., "Preconditioning Techniques in Computational Fluid Dynamics," *Annual Review of Fluid Mechanics*, Vol. 31, (1999).

11. Van Leer, B., Lee, W. T. and Roe, P. L., "Characteristic Time-Stepping or Local Preconditioning of the Euler Equations," AIAA Paper 91-1552-CP, Computational Fluid Dynamics Conference, Honolulu, (1991).

12. Choi, Y.-H. and Merkle, C. L., "The Application of Preconditioning in Viscous Flows," *Journal of Computational Physics*, Vol. 105, pp. 207-223, (1993).

13. Weiss, J. M. and Smith, W. A., "Preconditioning Applied to Variable and Constant Density Flows," *AIAA Journal*, Vol. 33, pp. 2050-2057, (1995).

14. Merkle, C. L., Sullivan, J. A. Y., Buelow, P. E. O. and Venkateswaran, S., "Computation of Flows with Arbitrary Equations of State," *AIAA Journal*, Vol. 36, No. 4, pp. 515-521, (1998).

15. Venkateswaran, S. and Merkle, C. L., "Analysis of Time-Derivative Preconditioning for the Navier-Stokes Equations," 6th International Symposium on Computational Fluid Dynamics, pp. 1323-1328, (1995).

16. Buelow, P. E. O., Venkateswaran, S. and Merkle, C. L., "The Effect of Grid Aspect Ratio on Convergence," *AIAA Journal*, Vol. 32, No. 12, pp. 2401-2408, (1994).

17. Buelow, P. E. O., Venkateswaran, S. and Merkle, C. L., "Stability and Convergence Analysis of Implicit Upwind Schemes," *Computers and Fluids*, Vol. 30, pp. 961-988, (2001).

18. Venkateswaran, S. and Merkle, C. L., "Dual-Time Stepping and Preconditioning for Unsteady Computations," AIAA Paper 95-0078, 33rd Aerospace Sciences Meeting and Exhibit, (1995).

19. Venkateswaran S., Lindau J. W., Kunz R. F. and Merkle C. L., "Computation of Multiphase Mixture Flows with Compressibility Effects," *Journal of Computational Physics*, in press, 2002.

20. Venkateswaran, S. and Merkle, C. L., "Analysis of Preconditioning Methods for the Euler and Navier-Stokes Equations," 1999-03, *VKI Lecture Series on CFD*, March, 1999.

21. Venkateswaran, S. and Merkle, C. L., Efficiency and Accuracy Issues in Contemporary CFD Algorithms, AIAA 2000-2251, Denver, CO, (2000).

22. D.L. Darmofal and K. Siu., "A Robust, Multigrid Algorithm for the Euler Equations with Local Preconditioning and Semi-Coarsening," *Journal of Computational Physics*, Vol. 151, pp. 728-756, (1999).

23. Roe, P.L., "Approximate Riemann Solvers, Parameter Vectors, and Difference Schemes," *Journal of Computational Physics*, Vol. 43, pp. 357-372, (1981).

24. Guillard, H. and Viozat, C, "On the Behavior of Upwind Schemes in the Low Mach Number Limit," *Computers and Fluids*, 28, pp.63-86, (1999).

25. Kirschner, I.N., "Results of Selected Experiments Involving Supercavitating Flows," *RTO AVT/VKI Special Course of Supercavitating Flows*, von Karman Institute for Fluid Dynamics, Rhode Saint Genese, Belgium, (2001).

# On Incompressible Flow Solvers

Rainald Löhner[1], Chi Yang[1], Juan Cebral[1], Orlando Soto[1], and
Fernando Camelli[1]

School of Computational Sciences and Informatics
M.S. 4C7, George Mason University, Fairfax, VA 22030-4444, USA

**Abstract.** A family of low-order finite element solvers for incompressible flows is
described. Both the advection and divergence terms are treated using consistent numerical fluxes along edges. Several techniques to accelerate convergence to steady
state are explored and compared. The techniques are then used in a fully implicit
time-marching scheme that solves a steady problem at every timestep. Several examples demonstrate the usefulness of the developed scehmes.

## 1 Introduction

Among the flows that are of importance and interest to mankind, the category of low Mach-number or incompressible flows is by far the largest. Most
of the manufactured products we use on a daily basis will start their life
as an incompressible flow (polymer extrusion, melts, a large number of food
products, etc.). The air which surrounds us can be considered, in almost all
instances, as an incompressible fluid (airplanes flying at low Mach-numbers,
flows in and around cars, vans, buses, trains and buildings). The same applies
to water (ships, submarines, torpedoes, pipes, etc.) and most biomedical liquids (e.g. blood). Given this large number of possible applications, it is not
surprising that numerical methods to simulate incompressible flows have been
developed for many years, as evidenced by an abundance of literature [9], [48],
[20].

The equations describing incompressible, Newtonian flows may be written as

$$\mathbf{v}_{,t} + \mathbf{v}\nabla\mathbf{v} + \nabla p = \nabla\mu\nabla\mathbf{v} \ , \tag{1}$$

$$\nabla \cdot \mathbf{v} = 0 \ . \tag{2}$$

Here $p$ denotes the pressure, $\mathbf{v}$ the velocity vector and both the pressure $p$ and
the viscosity $\mu$ have been normalized by the (constant) density $\rho$. By taking
the divergence of Eqn.(1) and using Eqn.(2) we can immediately derive the
so-called pressure-Poisson equation

$$\nabla^2 p = -\nabla \cdot \mathbf{v}\nabla\mathbf{v} \ . \tag{3}$$

What sets incompressible flow solvers apart from compressible flow solvers
is the fact that the pressure is not obtained from an equation of state $p =$

$p(\rho, T)$, but from the divergence constraint. This implies that the pressure field establishes itself instantaneously (reflecting the infinite speed of sound assumption of incompressible fluids) and must therefore be integrated implicitly in time.

The remainder of the paper is organized as follows: Section 2 describes the basic elements of the solver, in particular the main design criteria. Section 3 treats the spatial discretization, with particular emphasis on advection and divergence terms. This is followed in Section 4 by the temporal discretization chosen. Section 5,6 are devoted to iterative solvers and the acceleration to steady state. Section 7 treats fully implicit timestepping. Several examples are given in Section 8. Finally, in Section 9 some conclusions are drawn and an outlook for future work is summarized.

## 2  Basic Elements of the Solver

The numerical schemes chosen to solve the incompressible Navier-Stokes equations given by Eqns.(1,2) were based on the following criteria:

- Spatial discretization using unstructured grids (in order to allow for arbitrary geometries and adaptive refinement);
- Spatial approximation of unknowns with simple finite elements (in order to have a simple input/output and code structure);
- Temporal approximation using implicit integration of viscous terms and pressure (the interesting scales are the ones associated with advection);
- Temporal approximation using explicit integration of advective terms;
- Low-storage, iterative solvers for the resulting systems of equations (in order to solve large 3-D problems);
- Steady results that are independent from the timestep chosen (in order to have confidence in convergence studies);

## 3  Spatial Discretization

As stated before, we desire a spatial discretization with unstructured grids in order to:

- Approximate arbitrary domains, and
- Perform adaptive refinement in a straightforward manner, i.e. without changes to the solver.

From a numerical point of view, the difficulties in solving Eqns.(11.1-3) are the usual ones. First-order derivatives are problematic, while second-order derivatives can be discretized by a straightforward Galerkin approximation. We will first treat the advection operator and then proceed to the divergence operator.

## 3.1  The Advection Operator

It is well known that a straightforward Galerkin approximation of the advection terms will lead to an unstable scheme (recall that on a 1-D mesh of elements with constant size, the Galerkin approximation is simply a central difference scheme). Three ways have emerged to modify (or stabilize) the Galerkin discretization of the advection terms:

- Integration along characteristics [15],
- Taylor-Galerkin (or streamline diffusion) [27], [6], [10], and
- Edge-based upwinding [30].

Of these, we only consider the third option here. The Galerkin approximation for the advection terms yields a right-hand side (RHS) of the form:

$$r^i = D^{ij}\mathcal{F}_{ij} = D^{ij}(\mathbf{f}_i + \mathbf{f}_j) \ , \tag{4}$$

where the $\mathbf{f}_i$ are the 'fluxes along edges'

$$\mathbf{f}_i = S^{ij}_k \mathbf{F}^k_i \ , \quad S^{ij}_k = \frac{d^{ij}_k}{D^{ij}} \ , \quad D^{ij} = \sqrt{d^{ij}_k d^{ij}_k} \tag{5}$$

$$\mathcal{F}_{ij} = \mathbf{f}_i + \mathbf{f}_j \tag{6}$$

$$\mathbf{f}_i = (S^{ij}_k v^k_i)\mathbf{v}_i \ , \quad \mathbf{f}_j = (S^{ij}_k v^k_j)\mathbf{v}_j \tag{7}$$

A consistent numerical flux is given by

$$\mathcal{F}_{ij} = \mathbf{f}_i + \mathbf{f}_j - |v^{ij}|(\mathbf{v}_i - \mathbf{v}_j) \ , \tag{8}$$

where

$$v^{ij} = \frac{1}{2}S^{ij}_k \left(v^k_i + v^k_j\right) \tag{9}$$

As with all other edge-based upwind fluxes, this first-order scheme can be improved by reducing the difference $\mathbf{v}_i - \mathbf{v}_j$ through (limited) extrapolation to the edge center [31].


## 3.2  The Divergence Operator

A persistent difficulty with incompressible flow solvers has been the derivation of a stable scheme for the divergence constraint (11.2). The stability criterion for the divergence constraint is also known as the Ladyzenskaya-Babuska-Brezzi or LBB condition [19]. The classic way to satisfy the LBB condition has been to use different functional spaces for the velocity and pressure discretization [12]. Typically, the velocity space has to be richer, containing more degrees of freedom than the pressure space. Elements belonging to this class are the p1/p1+bubble mini-element [45], the p1/iso-p1 element [48], and the p1/p2 element [46]. An alternative way to satisfy the LBB condition is through the use of artificial viscosities [29], 'stabilization' [13], [47], [14] or a 'consistent numerical flux' (more elegant terms for the same thing). The equivalency of these approaches has been repeatedly demonstrated (e.g., [45], [29], [31]. The approach taken here is based on consistent numerical fluxes. For the divergence constraint, the Galerkin approximation along edge $i, j$ is given by

$$\mathcal{F}_{ij} = \mathbf{f}_i + \mathbf{f}_j \quad , \quad \mathbf{f}_i = S_k^{ij} v_i^k \quad , \quad \mathbf{f}_j = S_k^{ij} v_j^k \quad . \tag{10}$$

A consistent numerical flux may be constructed by adding pressure terms of the form:

$$\mathcal{F}_{ij} = \mathbf{f}_i + \mathbf{f}_j - |\lambda^{ij}|(p_i - p_j) \tag{11}$$

where the eigenvalue $\lambda^{ij}$ is given by the ratio of the characteristic advective timestep of the edge $\Delta t$ and the characteristic advective length of the edge $l$:

$$\lambda^{ij} = \frac{\Delta t^{ij}}{l^{ij}} \quad . \tag{12}$$

Higher order schemes can be derived by reconstruction and limiting, or by substituting the first-order differences of the pressure with third-order differences:

$$\mathcal{F}_{ij} = \mathbf{f}_i + \mathbf{f}_j - |\lambda^{ij}|(p_i - p_j + \frac{l^{ij}}{2}(\nabla p_i + \nabla p_j)) \quad . \tag{13}$$

This results in a stable, low-diffusion, fourth-order damping for the divergence constraint.


## 4 Temporal Discretization

As stated before, one is usually interested in physical phenomena that propagate with the advective timescales. Diffusive phenomena typically occur at a much faster rate, and should therefore be integrated implicitly. Given that the pressure establishes itself immediately through the pressure-Poisson equation, an implicit integration of pressure seems appropriate. The hyperbolic character of the advection operator and the elliptic character of the pressure-Poisson equation have led to a number of so-called projection schemes. The key idea is to predict first a velocity field from the current flow variables without taking the divergence contraint into account. In a second step, the divergence constraint is enforced by solving a pressure-Poisson equation. The velocity increment can therefore be separated into an advective and pressure increment:

$$\mathbf{v}^{n+1} = \mathbf{v}^n + \Delta\mathbf{v}^a + \Delta\mathbf{v}^p = \mathbf{v}^* + \Delta\mathbf{v}^p \quad . \tag{14}$$

For an explicit integration of the advective terms, one complete timestep is given by:

- <u>Advective-Diffusive Prediction</u>: $\mathbf{v}^n \to \mathbf{v}^*$

$$\left[\frac{1}{\Delta t} - \nabla\mu\nabla\right](\mathbf{v}^* - \mathbf{v}^n) + \mathbf{v}^n \cdot \nabla\mathbf{v}^n + \nabla p^n = \nabla\mu\nabla\mathbf{v} \quad ; \tag{15}$$

- <u>Pressure Correction</u>: $p^n \to p^{n+1}$

$$\nabla \cdot \mathbf{v}^{n+1} = 0 \quad ; \tag{16}$$

$$\frac{\mathbf{v}^{n+1} - \mathbf{v}^*}{\Delta t} + \nabla(p^{n+1} - p^n) = 0 \quad ; \tag{17}$$

which results in

$$\nabla^2(p^{n+1} - p^n) = \frac{\nabla \cdot \mathbf{v}^*}{\Delta t} \quad ; \tag{18}$$

- <u>Velocity Correction</u>: $\mathbf{v}^* \to \mathbf{v}^{n+1}$

$$\mathbf{v}^{n+1} = \mathbf{v}^* - \Delta t\nabla(p^{n+1} - p^n) \quad . \tag{19}$$

At steady state, the residuals of the pressure correction vanish, implying that the result does not depend on the timestep $\Delta t$. This scheme has been widely used [4], [29], [35], [5], [1], [26], [38], [30].


## 5    Iterative Solvers

Both Eqn.(15) and Eqn.(18) lead to large (symmetric) systems of equations of the form:

$$\mathbf{K}\mathbf{u} = \mathbf{r} \quad . \tag{20}$$

Preconditioned conjugate gradient (PCG) solvers [41] are used to solve Eqn.(20). For isotropic grids, simple diagonal preconditioning has proven very effective. For highly stretched RANS grids, linelet preconditioning has proven more effective [35], [44]. We also remark that we have attempted repeatedly to use multigrid as a solver, but that for most cases to date the simpler, highly optimized PCG solvers have proven superior.

# 6    Acceleration to Steady-State

For steady flows, the use of a time-accurate scheme with uniform timestep $\Delta t$ in the domain will invariably lead to slow convergence. In order to obtain steady results faster, a number of possibilities can be explored. Among them, the following seem the most promising:

- Local timesteps;
- Reduced iteration for the pressure;
- Substepping for the advection terms;
- Implicit treatment of the advection terms; and
- Fully implicit treatment of advection, diffusion and pressure.

We review the main features of these in the sequel.

6.1 <u>Local Timestepping</u>: Faster convergence to steady-state may be achieved by employing local timesteps. Given that the results obtained by the schemes used do not depend on the timestep, this can be readily done. One simply defines a separate timestep for each gridpoint, and marches in time until a steady solution is reached.

6.2 <u>Reduced Pressure Iterations</u>: The most time-consuming part of projection schemes is the solution of the pressure-Poisson equation at every timestep. Recall that this is the equation that defines that pressure and establishes a divergence-free state at the next timestep. If we are only interested in the steady result, obtained after many timesteps, then the maintenance of an exact divergence-free state at every intermediate timestep is not required. Therefore, one can use a less stringent convergence criterion for the pressure-Poisson solver, saving a considerable amount of CPU-time. We have experimented extensively with this option. The results obtained have been mixed. For inviscid flows (Euler equations), this option works well. However, for the more demanding high Reynolds-number cases, it has proven difficult to define reliable convergence criteria.

6.3 <u>Substepping for the Advection Terms</u>: As stated before, the most time-consuming part of projection schemes is the solution of the pressure-Poisson equation at every timestep. The idea is then to march the (cheaper) advective-diffusive system for several steps, and then to apply the pressure-Poisson equation (with larger timestep). Another possibility is to perform a very small number of pressure-Poisson iterations for $m-1$-steps, and then apply a complete solution at step $m$. This latter procedure has proven quite reliable and robust, but may also fail for high Reynolds-number cases, particularly in the start-up phase of runs, when the initial solution is very far from the final solution.

6.4 <u>Implicit Treatment of the Advection Terms</u>: Any explicit integration of the advective terms implies that information can only travel at most one element per timestep. In order to allow for a faster transfer of information and larger timesteps, the advective terms have to be integrated implicitly. Eqn.(15) then becomes:

$$\left[\frac{1}{\Delta t} + \mathbf{v}^* \cdot \nabla - \nabla\mu\nabla\right](\mathbf{v}^* - \mathbf{v}^n) + \mathbf{v}^n \cdot \nabla\mathbf{v}^n + \nabla p^n = \nabla\mu\nabla\mathbf{v} \ , \qquad (21)$$

leading to a non-symmetric system of equations of the form:

$$\mathbf{A}\Delta\mathbf{v} = \mathbf{r} \ . \qquad (22)$$

This may be rewritten as

$$\mathbf{A} \cdot \Delta\mathbf{v} = (\mathbf{L} + \mathbf{D} + \mathbf{U}) \cdot \Delta\mathbf{v} = \mathbf{r} \ , \qquad (23)$$

where $\mathbf{L}, \mathbf{D}, \mathbf{U}$ denote the lower, diagonal and upper diagonal entries of $\mathbf{A}$. Classic relaxation schemes to solve this system of equations include:
a) Gauss-Seidel, given by:

$$(\mathbf{L} + \mathbf{D}) \cdot \Delta\mathbf{v}^1 = \mathbf{r} - \mathbf{U} \cdot \Delta\mathbf{v}^0 \ , \qquad (24)$$

$$(\mathbf{D} + \mathbf{U}) \cdot \Delta\mathbf{v} = \mathbf{r} - \mathbf{L} \cdot \Delta\mathbf{v}^1 \ ; \qquad (25)$$

b) Lower-Upper Symmetric Gauss-Seidel (LU-SGS), given by:

$$(\mathbf{L} + \mathbf{D}) \cdot \mathbf{D}^{-1} \cdot (\mathbf{D} + \mathbf{U}) \cdot \Delta\mathbf{v} = \mathbf{r} \ . \qquad (26)$$

These relaxation schemes have been optimized over the years, resulting in very efficient compressible flow solvers [32], [33], [34], [?]. Key ideas include:

- Using the spectral radius $\rho_A$ of $\mathbf{A}$ for the diagonal entries $\mathbf{D}$; for the advection case, $\rho_A = |\mathbf{v}|$, resulting in:

$$\mathbf{D} = \left[\frac{1}{\Delta t}\mathbf{M}_l^i - 0.5\sum \mathbf{C}^{ij}|\mathbf{v}|_{ij} + \sum k^{ij}\right]\mathbf{I} \ ; \qquad (27)$$

- Replacing:

$$\mathbf{A} \cdot \Delta\mathbf{v} \approx \Delta\mathbf{F} \ , \qquad (28)$$

resulting in:

$$\Delta\mathbf{F} = \mathbf{F}(\mathbf{v} + \Delta\mathbf{v}) - \mathbf{F}(\mathbf{v}) \ . \qquad (29)$$

The combined effect of these simplifications is a family of schemes that are matrix free, require no extra storage as compared to explicit schemes, and (due to lack of limiting) per relaxation sweep are faster than conventional explicit schemes. For the LU-SGS scheme, each pass over the mesh proceeds as follows:

- Forward Sweep:

$$\Delta\hat{\mathbf{v}}^i = \mathbf{D}^{-1}\left[\mathbf{r}^i - 0.5\sum_{j<i}\mathbf{C}^{ij}\cdot(\Delta\hat{\mathbf{F}}_{ij} - |\mathbf{v}|_{ij}\Delta\hat{\mathbf{v}}_j) + \sum_{j<i}\mathbf{k}^{ij}\Delta\hat{\mathbf{v}}_j\right] \quad (30)$$

- Backward Sweep:

$$\mathbf{r} = \mathbf{D}\cdot\Delta\hat{\mathbf{v}} \quad (31)$$

$$\Delta\mathbf{v}^i = \mathbf{D}^{-1}\left[\mathbf{r}^i - 0.5\sum_{j>i}\mathbf{C}^{ij}\cdot(\Delta\mathbf{F}_{ij} - |\mathbf{v}|_{ij}\Delta\mathbf{v}_j) + \sum_{j>i}\mathbf{k}^{ij}\Delta\hat{\mathbf{v}}_j\right] \quad (32)$$

Luo [32] has shown that no discernable difference could be observed when taking central or upwind discretizations for $\Delta\mathbf{F}$. As the CPU requirements of upwind discretizations are much higher, all relaxation passes are carried out using central schemes. Given that the same loop structure $(\mathbf{L}, \mathbf{D}, \mathbf{U})$ is required for both the Gauss-Seidel, the LU-SGS and the GMRES matrix-vector products, it is possible to write a single 'sweep' subroutine that encompasses all of these cases. The initialization of the Gauss-Seidel loop is accomplished with an LU-SGS pass.

## 7  Fully Implicit Integration

Using the notation

$$u^\Theta = (1 - \Theta)u^n + \Theta u^{n+1} \;, \quad (33)$$

which implies

$$u^{n+1} - u^n = \frac{u^\Theta - u^n}{\Theta} \;, \quad (34)$$

an implicit timestepping scheme may be written as follows:

$$\frac{\mathbf{v}^\theta - \mathbf{v}^n}{\theta\Delta t} + \mathbf{v}^\Theta\nabla\mathbf{v}^\Theta + \nabla p^\Theta = \nabla\mu\nabla\mathbf{v}^\Theta \;, \quad (35)$$

$$\nabla\cdot\mathbf{v}^\Theta = 0 \;, \quad (36)$$

Following similar approaches for compressible flow solvers [2], this system can be interpreted as the steady-state solution of the pseudo-time system:

$$\mathbf{v}^\theta_{,\tau} + \mathbf{v}^\Theta\nabla\mathbf{v}^\Theta + \nabla p^\Theta = \nabla\mu\nabla\mathbf{v}^\Theta - \frac{\mathbf{v}^\theta - \mathbf{v}^n}{\theta\Delta t} \;, \quad (37)$$

$$\nabla \cdot \mathbf{v}^{\Theta} = 0 \ , \tag{38}$$

Observe that the only difference between Eqns.(37,38) and the original in-compressible Navier-Stokes equations given by Eqns.(1,2) is the appearance of new source-terms. These source terms are pointwise dependent on the variables being integrated ($\mathbf{v}$), and can therefore be folded into the left hand side for explicit timestepping without any difficulty. The idea is then to march Eqns.(37,38) to steady state in the pseudo-time $\tau$ using either an explicit-advection projection scheme or an explicit artificial compressibility scheme using local timesteps.

## 8  Examples

1. <u>NACA0012</u>: The first example considered is the classic NACA0012 wing at $\alpha = 5^o$ angle of attack. This is a steady, inviscid case (Euler). Figures 1a,b show the surface mesh employed, as well as the surface pressures obtained. Although the mesh is rather coarse (`nelem=368,872, npoin=68,321`), it still allows for a meaningful comparison of the different solvers.

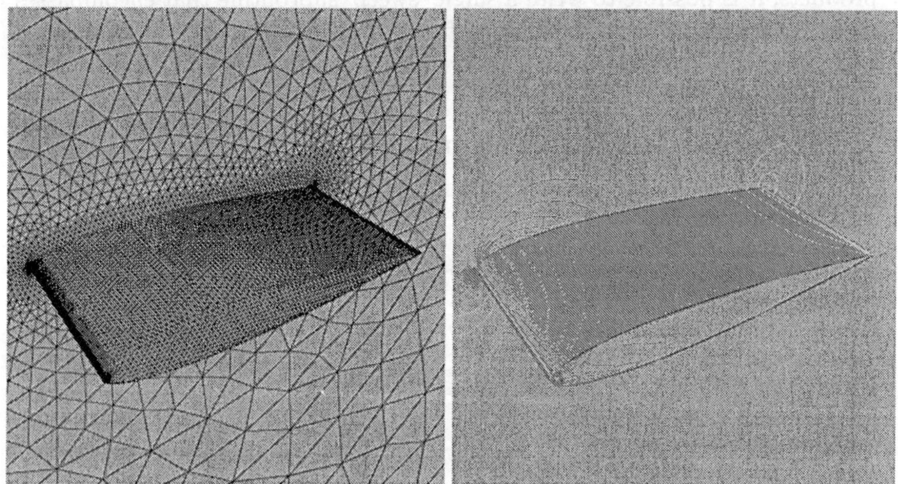

Figures 1a,b  NACA 0012: Surface Mesh and Pressure

This problem was solved using:

- The standard explicit-advection projection scheme ($C = 0.1$) with full pressure solution every 20 timesteps, and partial pressure solution in between (max 20 PCG iterations);
- The implicit-advection projection scheme ($C = 4.0$) with SGS relaxation;

- The implicit-advection projection scheme ($C = 4.0$) with LU-SGS-GMRES solver for the advection.

Each one of these runs was considered converged when the change in lift, normalized by the Courant-number, was below $t_l = 10^{-3}$ for five subsequent timesteps. We have found such a measure to be a better indicator of convergence than residuals, as this allows the user to state clearly what accuracy is desired.

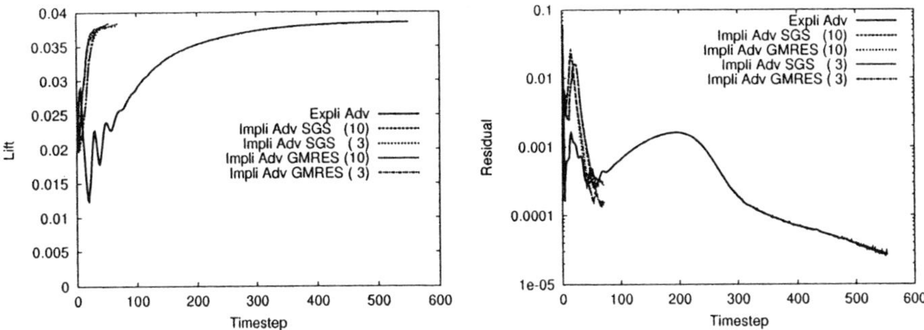

Figures 1c,d  NACA 0012: Convergence History for Lift and Residuals

Figures 1c,d show the convergence history for the lift and residuals respectively. One can see that the implicit-advection schemes converge significantly faster. The timings obtained are, of course, not as spectacular, as each implicit-advection step requires considerably more CPU than its explicit-advection counterpart. Nevertheless, the timings recorded, summarized in Table 1, indicate a considerable speedup.

| Scheme | `ntime` | CPU [sec] | Speedup |
|---|---|---|---|
| Expli | 550 | 2,316 | 1.00 |
| Impl-SGS (10) | 51 | 532 | 4.35 |
| Impl-SGS ( 3) | 70 | 188 | 12.30 |
| Impl-GMRES (10) | 55 | 772 | 3.00 |
| Impl-GMRES ( 2) | 70 | 216 | 10.70 |

**Table 1.** NACA0012

2. <u>High-Re Bump</u>: The second example considered is high Reynolds-number flow past a parabolic bump. This is essentially a 2-D example, but is run with the 3-D solver. The parameters were set to: $\rho = 1.0, \mathbf{v} = (1, 0, 0), \mu = 10^{-5}, L = 1$, which implies a Reynolds-number of $Re = 10^5$. The Baldwin-Lomax turbulence model is used. The resulting flow is steady and shows the

development of a boundary layer. Figures 2a,b show a zoom of the surface mesh employed, as well as the velocity field obtained. Although the mesh is rather coarse (`nelem=23,785`, `npoin=5,674`), it still allows for a meaningful comparison of the different solvers.

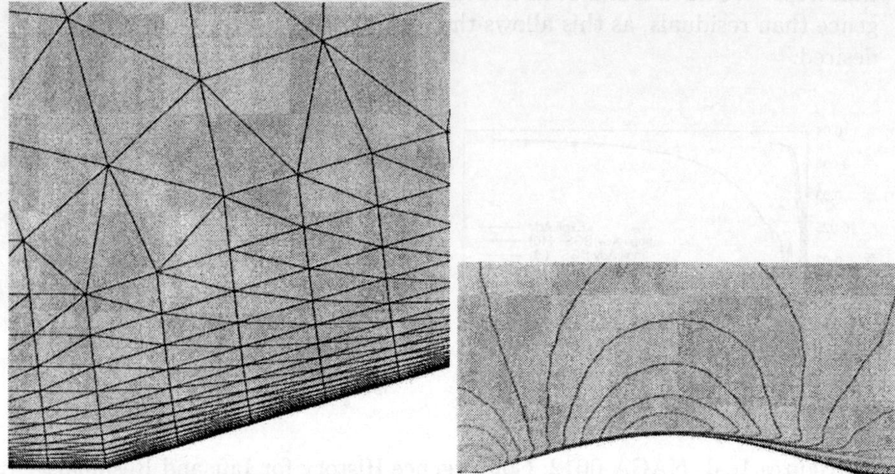

Figures 2a,b  Bump: Surface Mesh and Velocities

As before, this problem was solved using:

- The standard explicit-advection projection scheme ($C = 0.1$) with full pressure solution every 20 timesteps, and partial pressure solution in between (max 20 PCG iterations);
- The implicit-advection projection scheme ($C = 4.0$) with SGS relaxation;
- The implicit-advection projection scheme ($C = 4.0$) with LU-SGS-GMRES solver for the advection.

| Scheme | ntime | CPU [sec] | Speedup |
|---|---|---|---|
| Expli | 790 | 144 | 1.00 |
| Impl-SGS (10) | 100 | 31 | 4.65 |
| Impl-SGS ( 4) | 100 | 25 | 5.76 |
| Impl-GMRES (10) | 100 | 38 | 3.79 |
| Impl-GMRES ( 2) | 100 | 25 | 5.76 |

**Table 2.** High-Re Bump

For the pressure-Poisson equation (as well as the velocity PCG solver of the explicit-advection projection scheme), linelet preconditioning was used. Figures 2c,d show the convergence history for the lift and residuals respectively. As before, one can see that the implicit-advection schemes converge significantly faster. The timings recorded have been summarized in Table 2.

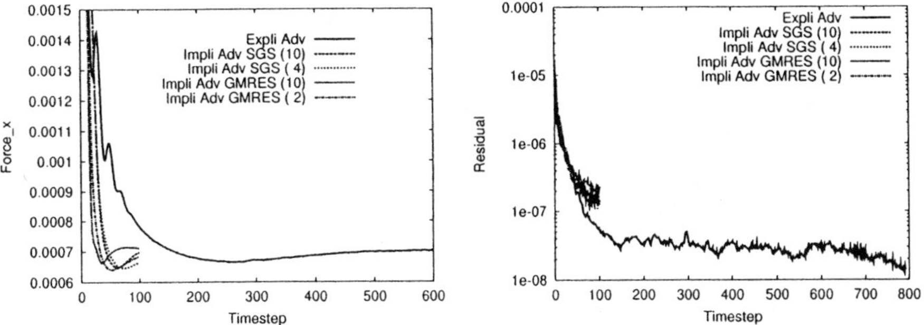


Figure 2c,d  BUMP: Convergence History for Lift and Residuals

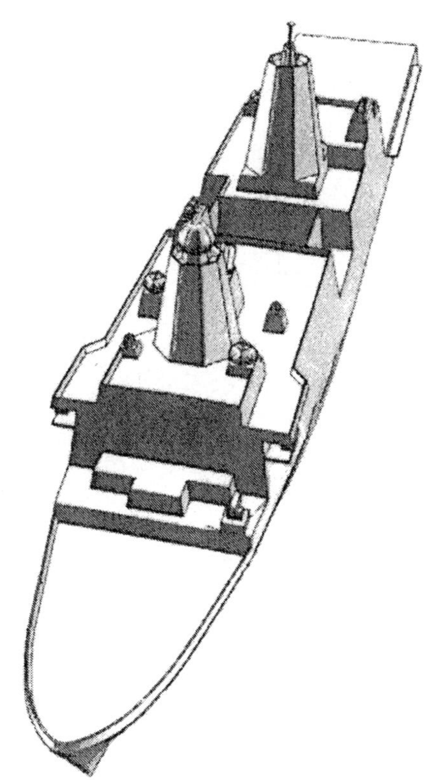

Figure 3a  LPD-17: Geometry

3. <u>LPD-17 Topside Flow Study</u>: This case considers the external flow (so-called airwake studies) past a typical ship, in this case the LPD-17. The geometry is illustrated in Figure 3a. The dimensions of the ship are approximately length: 200 $m$, width: 30 $m$ and height above the waterline: 50 $m$.

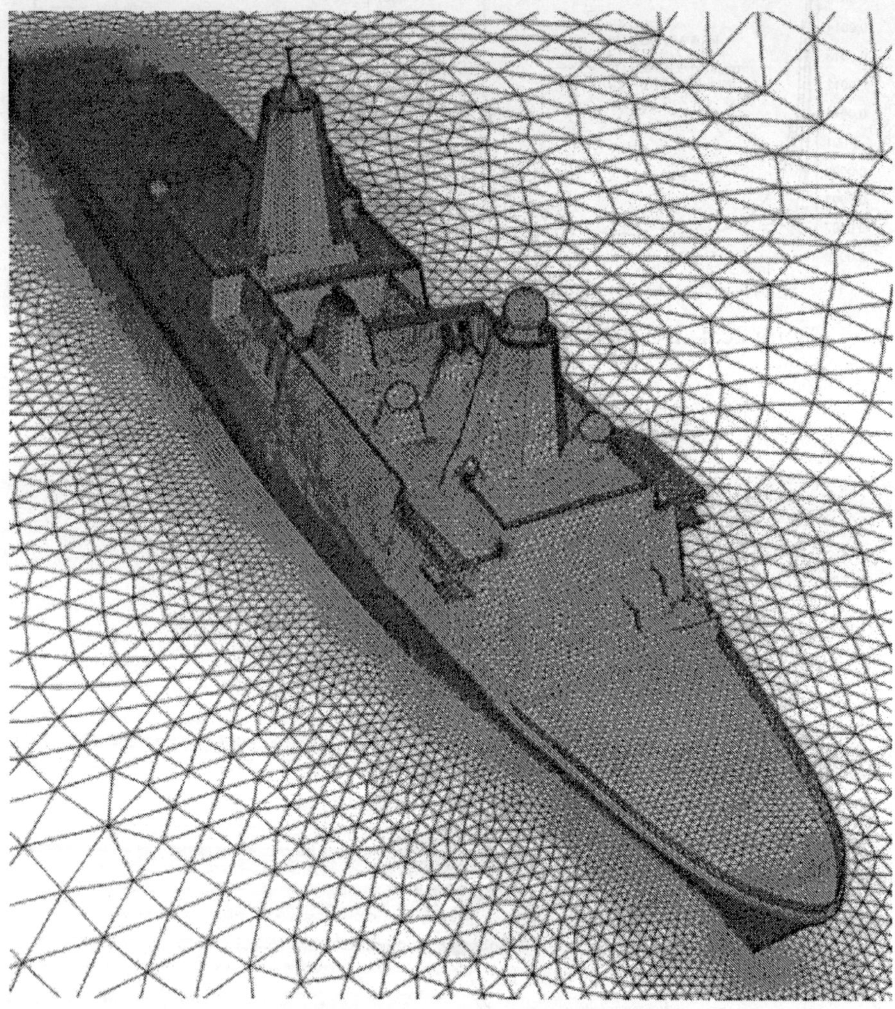

Figure 3b  LPD-17: Surface Mesh

The ship has two main masts. The standard configuration of the ship has 5 diesel generator engine stacks (D/G) and 4 main propulsion engine stacks (M/P). These stacks are grouped as follow: 2 D/G behind the front mast, 2 M/P and 1 D/G in the right side of the ship between the two masts, and 2 M/P and 2 D/G in the left side before the landing deck. The flow rates and gas exit temperatures used in the simulation corresponded to full power condition. The inflow condition is set to uniform flow with a velocity of 30 knots (15.43 m/s) assuming maximum ship speed. An initial step that must be taken before airwake computations can begin is the generation of a 3-D surface model that accurately represents the ship and is suitable for CFD computations. The process of converting the design IGES files to the grid generator file format is described in [39]. Since the primary objective of the LPD-17 experimental airwake measurement project was CFD valida- tion, great care was taken to ensure that the CFD geometry and wind tunnel model were in agreement to within construction accuracy. This highly accu- rate surface representation was the starting point for seveal studies of this type [39], [18]. Two different grid resolutions were used. The coarser mesh had approximately 490Kpts and 2.7Mtets. The finer mesh had 990Kpts and 5.6Mtets. A view of the surface mesh for the coarser mesh is shown in Fig- ure 3b. The numerical simulation was performed with the implicit monolithic scheme presented in [43]. The time step was set to 0.01 s. The density was 1.225 kg/m$^3$, and the viscosity was $1.789 \times 10^{-5}$ kg/m/s. The Reynolds num- ber based on the height of the ship is $5 \times 10^7$, implying that the flow is in the turbulent regime. Due to the unsteadiness and turbulent characteristics of the flow an LES simulation was performed with the Smagorinsky turbulence model [42]. In the first part of the run, a pseudo steady flow was established. After this state was reached, the flow was integrated for 90 s of real time. The whole run, including the initialization part, took approximately 2 weeks on a non-dedicated 16 processor SGI 3800 shared memory machine.

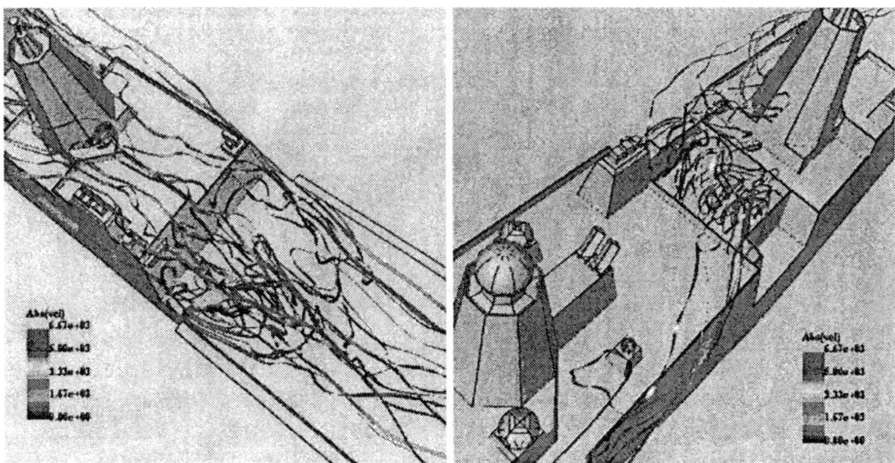

Figures 3c,d  LPD-17: Streamribbons

The pressure was set to hydrostatic. Buoyancy effects were considered in the simulation via Boussinesq's approximation. To indicate the complexity of the flows simulated, two groups of instantaneous streamribbons (coloured according to speed) are shown in Figures 3c,d. The ribbons show a large recirculation above the landing deck, as well as a pronounced crossflow from port to starboard near the middle of the ship. Experimental velocity data from wind tunnel LDV measurements was available at the positions shown in Figure 3e. The experimental velocities were averaged and compared with averaged numerical results. The numerical results were averaged over a period of 90 s.

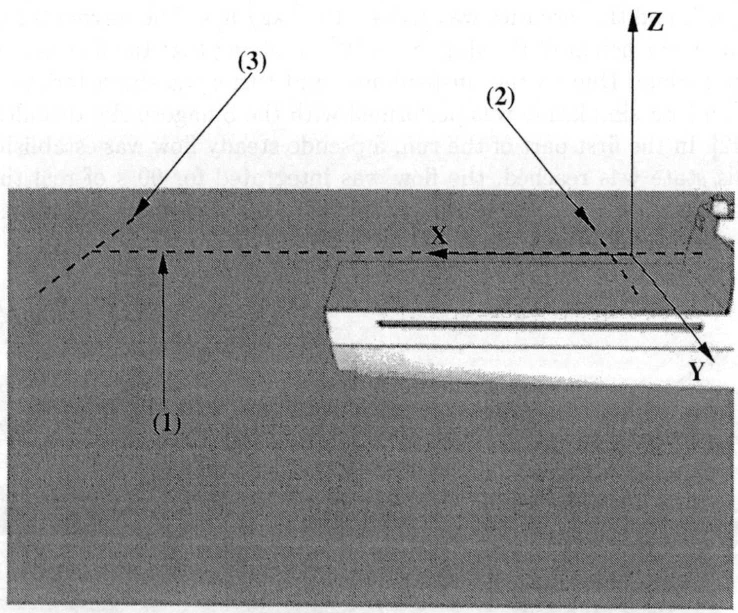


Figure 3e  LPD-17: Location of Experimental Measurements (Lines 1-3)

Figures 3d-f shows the comparison of computed and measured average velocities. The crosses indicate the experimental data, the stars the average of the numerical results, and the dash-dotted lines the standard deviation for the numerical results. The numerical results agree fairly well with the experimental data for most of the plots. The numerical velocity in the x-direction in line station (2) (see Figure (13) shows a significant deviation from the experimental data at the center position. One possible explanations is the lack of resolution of the experimental instruments. This was also seen in earlier computations. Experimental errors were not provided for these comparisons.

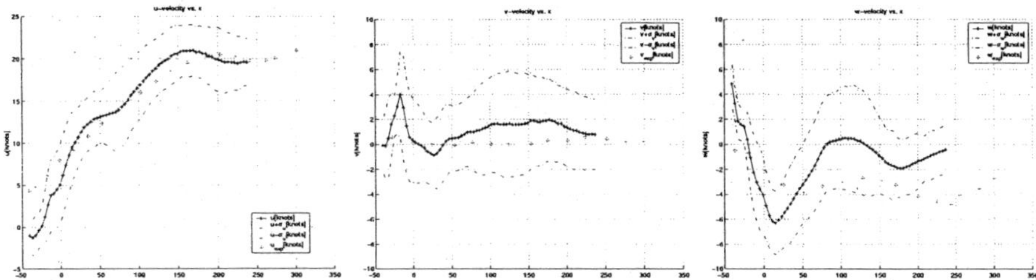

Figure 3d  LPD-17: Comparison of Velocities for Line 1

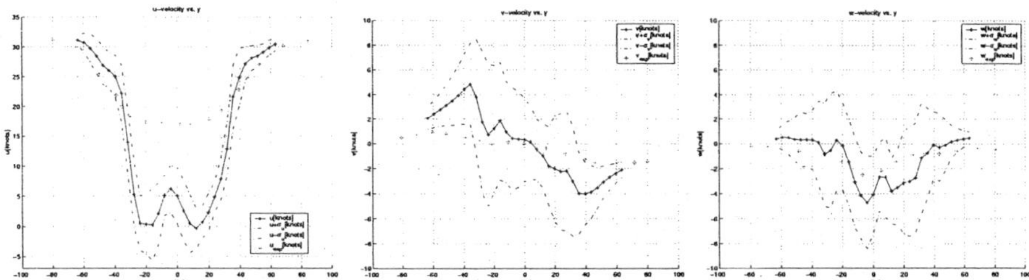

Figure 3e  LPD-17: Comparison of Velocities for Line 2

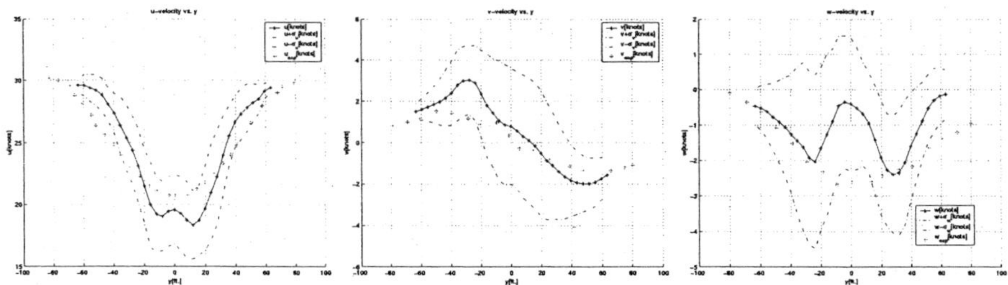

Figure 3f  LPD-17: Comparison of Velocities for Line 3

4. <u>Generic Submarine Forebody Vortex Flow Study</u>: This case considers the flow past a submarine with a 42-feet diameter. The speed is 7 knots and the Reynolds-number per foot is $Re = 1.1 \cdot 10^6$. Turbulence is simulated with

the Baldwin-Lomax model [3]. The front half of the subsmarine is modeled in order to study the vortex shed by the sail.

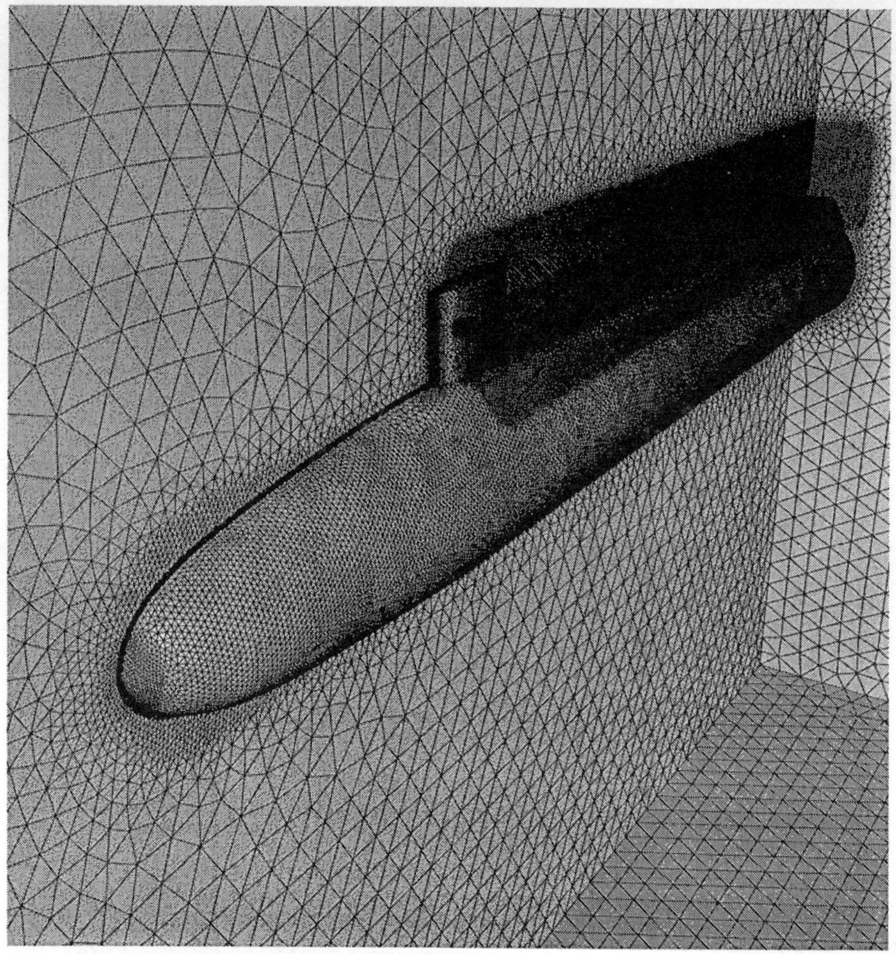

Figure 4a  Submarine Forebody: Surface Mesh

The RANS mesh had approximately `nelem=6Mtets` elements, and was suitably refined in boundary layer regions and the region behind the sail. Figures 4a-c show different views of the surface grid, as well as the surface pressure, together with contours of the absolute value of the velocities in 6 planes behind the sail. The development of a main tip vortex and several secondary vortices and boundary layer detachment is clearly visible.

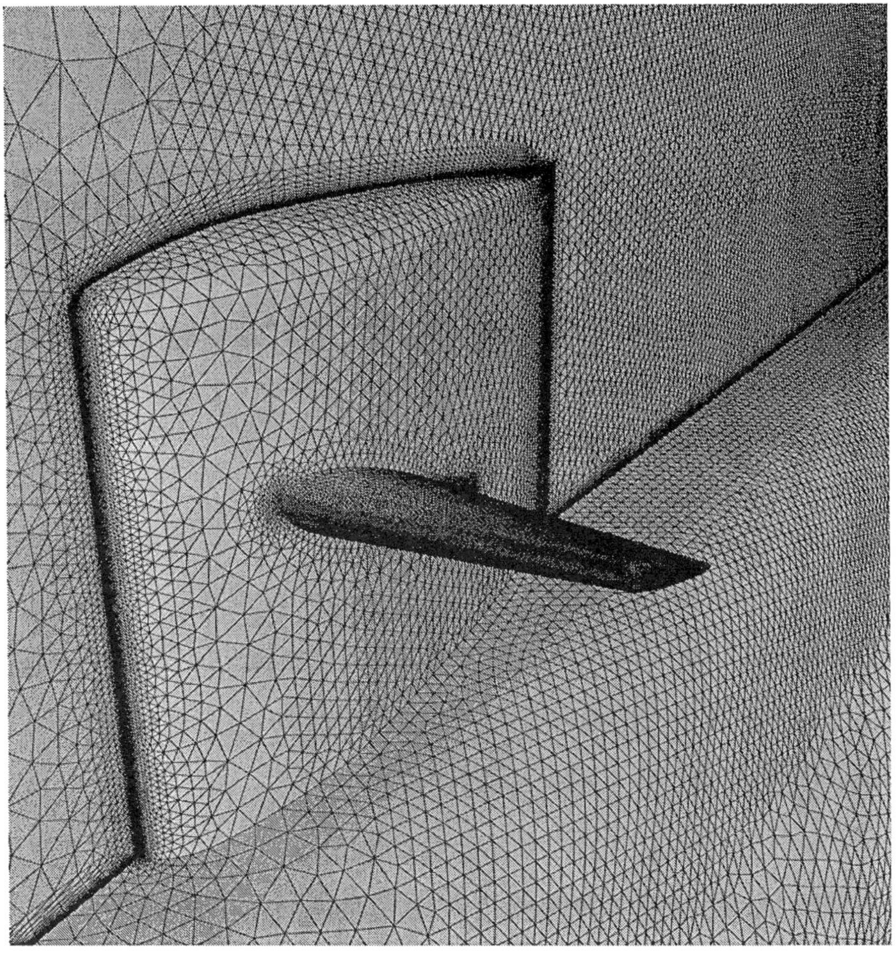

Figure 4b  Submarine Forebody: Surface Mesh (Detail)

## 9　Conclusions and Outlook

A family of low-order finite element solvers for incompressible flows has been described. Both the advection and divergence terms were treated using consistent numerical fluxes along edges. Several techniques to accelerate convergence to steady state were explored and compared. Paerticularly the SGS and LU-SGS-GMRES solvers for the advection step show considerable promise, reducing CPU times by 1:3-1:5. The schemes developed were applied to several large-scale problems, of which two were included in the present paper. Future work will consider:

- Further exploration of (parallel) acceleration techniques,
- Inclusion of vorticity confinement for implicit solvers,
- Adaptive mesh refinement for wakes and vortical structures.

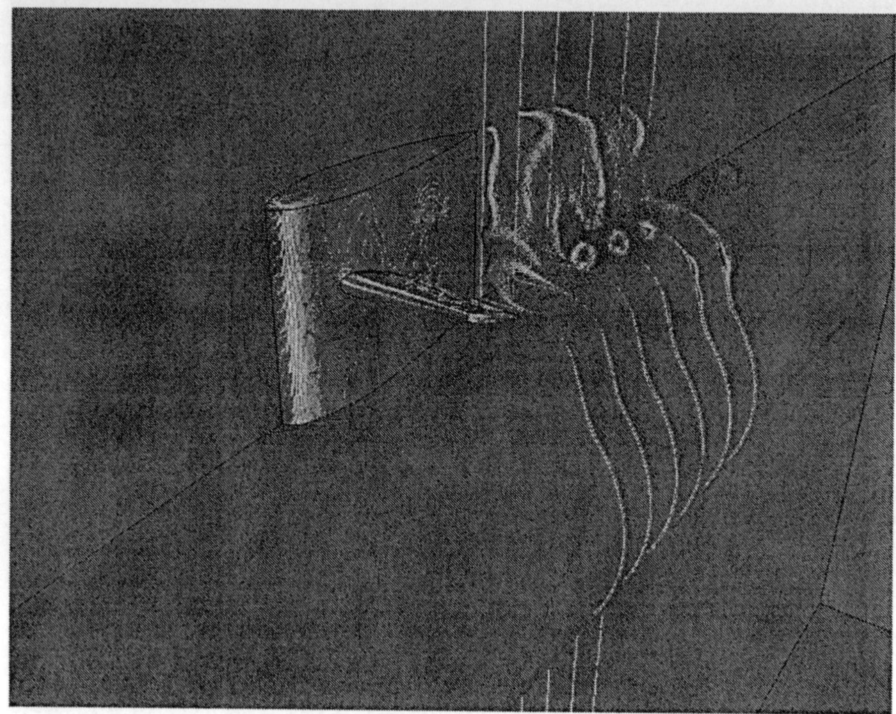

Figure 4c  Submarine Forebody: Surface Pressure and Velocity in Planes

## 10   Acknowledgements

This work was partially supported by ONR and AFOSR, with Drs. Patrick Purtell and Leonidas Sakell as the technical monitors.

## References

1. B. Alessandrini and G. Delhommeau - A Multigrid Velocity-Pressure- Free Surface Elevation Fully Coupled Solver for Calculation of Turbulent Incompressible Flow Around a Hull; *Proc. 21st Symp. on Naval Hydrodynamics*, Trondheim, Norway, June (1996).
2. J. Alonso, L. Martinelli and A. Jameson - Multigrid Unsteady Navier-Stokes Calculations with Aeroelastic Applications; *AIAA-95-0048* (1995).

3. B.S. Baldwin and H. Lomax - Thin-Layer Approximation and Algebraic Model for Separated Turbulent Flows; *AIAA-78-257* (1978).

4. J.B. Bell, P. Colella and H. Glaz - A Second-Order Projection Method for the Navier-Stokes Equations; *J. Comp. Phys.* 85, 257-283 (1989).

5. J.B. Bell and D.L. Marcus - A Second-Order Projection Method for Variable-Density Flows; *J. Comp. Phys.* 101, 2 (1992).

6. A.N. Brooks and T.J.R. Hughes - Streamline Upwind/Petrov Galerkin Formulations for Convection Dominated Flows with Particular Emphasis on the Incompressible Navier-Stokes Equations; *Comp. Meth. Appl. Mech. Eng.* 32, 199-259 (1982).

7. A.J. Chorin - A Numerical Solution for Solving Incompressible Viscous Flow Problems; *J. Comp. Phys.* 2, 12-26 (1967).

8. A.J. Chorin - Numerical Solution of the Navier-Stokes Equations; *Math. Comp.* 22, 745-762 (1968).

9. Conferences - See the following Conference Series: *Finite Elements in Fluis* I-IX, J. Wiley & Sons, *Int. Conf. Num. Meth. Fluid Dyn.* I-XII, Spinger Lecture Notes in Physics, *AIAA CFD Conf.* I-XV, AIAA CP, *Num. Meth. Laminar and Turbulent Flow*, Pinerige Press, and others.

10. J. Donea, S. Giuliani, H. Laval and L. Quartapelle - Solution of the Unsteady Navier-Stokes Equations by a Fractional Step Method; *Comp. Meth. Appl. Mech. Eng.* 30, 53-73 (1982).

11. J.R. Farmer, L. Martinelli and A. Jameson - A Fast Multigrid Method for Solving Incompressible Hydrodynamic Problems With Free Surfaces; *AIAA J.* 32, 6, 1175-1182 (1993).

12. M. Fortin and F. Thomasset - Mixed Finite Element Methods for Incompressible Flow Problems; *J. Comp. Phys.* 31, 113-145 (1979).

13. L.P. Franca, T.J.R. Hughes, A.F.D. Loula and I. Miranda - A New Family of Stable Elements for the Stokes Problem Based on a Mixed Galerkin/Least-Squares Finite Element Formulation; pp. 1067-1074 in *Proc. 7th Int. Conf. Finite Elements in Flow Problems* (T.J. Chung and G. Karr eds.), Huntsville, AL (1989).

14. L.P. Franca and S.L. Frey - Stabilized Finite Element Methods: II. The incompressible Navier-Stokes Equations; *Comp. Meth. Appl. Mech. Eng.* 99, 209-233 (1992).

15. J.P. Gregoire, J.P. Benque, P. Lasbleiz and J. Goussebaile - 3-D Industrial Flow Calculations by Finite Element Method; *Springer Lecture Notes in Physics* 218, 245-249 (1985).

16. P.M. Gresho, C.D. Upson, S.T. Chan and R.L. Lee - Recent Progress in the Solution of the Time-Dependent, Three-Dimensional, Incompressible Navier-Stokes Equations; pp. 153-162 in *Proc. 4th Int. Symp. Finite Element Methods in Flow Problems* (T. Kawai ed.), University of Tokio Press (1982).

17. P.M. Gresho and S.T. Chan - On the Theory of Semi-Implicit Projection Methods for Viscous Incompressible Flows and its Implementation via a Finite Element Method That Introduces a Nearly-Consistent Mass Matrix; *Int. J. Num. Meth. Fluids* to appear (1991).

18. M.J. Guillot and M.A. Walker - Unsteady Analysis of the Air Wake Over the LPD-17; *AIAA-00-4125* (2000).

19. M.D. Gunzburger - Mathematical Aspects of Finite Element Methods for Incompressible Viscous Flows; pp. 124-150 in *Finite Elements: Theory and Application* (Dwoyer, Hussaini and Voigt eds.), Springer Verlag (1987).

20. M.D. Gunzburger and R. Nicolaides (eds.) - *Incompressible Computational Fluid Dynamics: Trends and Advances*; Cambridge University Press (1993).
21. T. Hino - Computation of Free Surface Flow Around an Advancing Ship by the Navier-Stokes Equations; *Proc. 5th Int. Conf. on Numerical Ship Hydrodynamics*, Hiroshima, Japan (1989).
22. T. Hino, L. Martinelli and A. Jameson - A Finite-Volume Method with Unstructured Grid for Free Surface Flow; *Proc. 6th Int. Conf. on Numerical Ship Hydrodynamics*, Iowa City, August (1993).
23. T. Hino - An Unstructured Grid Method for Incompressible Viscous Flows with a Free Surface; *AIAA-97-0862* (1997).
24. J.D. Huffenus and D. Khaletzky - A Finite Element Method to Solve the Navier-Stokes Equations Using the Method of Characteristics; *Int. J. Num. Meth. Fluids* 4, 247-269 (1984).
25. T.C. Jue, B. Ramaswamy and J.E. Akin - Finite Element Simulation of 2-D Benard Convection with Gravity Modulation; pp. 87-101 in FED-Vol. 123, ASME (1991).
26. Y. Kallinderis and A. Chen - An Incompressible 3-D Navier-Stokes Method with Adaptive Hybrid Grids; *AIAA-96-0293* (1996).
27. D.W. Kelly, S. Nakazawa, O.C. Zienkiewicz and J.C. Heinrich - A Note on Anisotropic Balancing Dissipation in Finite Element Approximation to Convection Diffusion Problems. *Int. J. Num. Meth. Eng.* 15, 1705-1711 (1980).
28. J. Kim and P. Moin - Application of a Fractional-Step Method to Incompressible Navier-Stokes Equations. *J. Comp. Phys.* 59, 308-323 (1985).
29. R. Löhner - A Fast Finite Element Solver for Incompressible Flows; *AIAA-90-0398* (1990).
30. R. Löhner, C. Yang, E. Oñate and S. Idelssohn - An Unstructured Grid-Based, Parallel Free Surface Solver; *Appl. Num. Math.* 31, 271-293 (1999).
31. R. Löhner - *Applied CFD Techniques*; J. Wiley & Sons (2001).
32. H. Luo, J.D. Baum and R. Löhner - A Fast, Matrix-Free Implicit Method for Compressible Flows on Unstructured Grids; *J. Comp. Phys.* 146, 664-690 (1998).
33. H. Luo, J.D. Baum and R. Löhner - An Accurate, Fast, Matrix-Free Implicit Method for Computing Unsteady Flows on Unstructured Grids; *AIAA-99-0937* (1999).
34. H. Luo, D. Sharov, J.D. Baum and R. Löhner - A Class of Matrix-free Implicit Methods for Compressible Flows on Unstructured Grids; *First International Conference on Computational Fluid Dynamics*, Kyoto, Japan, July 10-14 (2000).
35. D. Martin and R. Löhner - An Implicit Linelet-Based Solver for Incompressible Flows; *AIAA-92-0668* (1992).
36. A.T. Patera - A Spectral Element Method for Fluid Dynamics: Laminar Flow in a Channel Expansion. *J. Comp. Phys.* 54, 468-488, (1984).
37. J. Peraire, K. Morgan and J. Peiro - The Simulation of 3D Incompressible Flows Using Unstructured Grids; Chapter 16 in *Frontiers of Computational Fluid Dynamics* (D.A. Caughey and M.M. Hafez eds.), J. Wiley (1994).
38. R. Ramamurti and R. Löhner - A Parallel Implicit Incompressible Flow Solver Using Unstructured Meshes; *Computers and Fluids* 5, 119-132 (1996).
39. R. Ramamurti and W.C. Sandberg - LPD-17 Topside Aerodynamic Study: FE-FLO; *NRL Memorandum Rep. NRL/MR/6410-00-8498*, LCP&FD, NRL (2000).
40. A. Rizzi and L. Eriksson - Computation of Inviscid Incompressible Flow with Rotation; *J. Fluid Mech.* 153, 275-312 (1985).

41. Y. Saad - *Iterative Methods for Sparse Linear Systems*; PWS Publishing, Boston (1996).

42. J. Smagorinsky - General Circulation Experiments with the Primitive Equations, I. The Basic Experiment; *Mon. Weather Rev.* , 91, 99-164 (1963).

43. O. Soto, R. Löhner and J.R. Cebral - An Implicit Monolithic Time Accurate Finite Element Scheme for Incompressible Flow Problems; *AIAA-01-2616-CP* (2001).

44. O. Soto, R. Löhner and F. Camelli - A Linelet Preconditioner for Incompressible Flow Solvers; *Int. J. Heat and Fluid Flow* to appear (2002).

45. A. Soulaimani, M. Fortin, Y. Ouellet, G. Dhatt and F. Bertrand - Simple Continuous Pressure Elements for Two- and Three- Dimensional Incompressible Flows; *Comp. Meth. Appl. Mech. Eng.* 62, 47-69 (1987).

46. C. Taylor and P. Hood - A Numerical Solution of the Navier-Stokes Equations Using the Finite Element Method. *Comp. Fluids* 1, 73-100 (1973).

47. T.E. Tezduyar, R. Shih, S. Mittal and S.E. Ray - Incompressible Flow Computations With Stabilized Bilinear and Linear Equal-Order Interpolation Velocity-Pressure Elements; *UMSI Rep.* 90 (1990).

48. F. Thomasset - *Implementation of Finite Element Methods for Navier-Stokes Equations.* Springer-Verlag (1981).

49. C. Yang and R. Löhner - Fully Nonlinear Ship Wave Calculation Using Unstructured Grids and Parallel Computing; pp. 125-150 in *Proc. 3rd Osaka Colloquium on Advanced CFD Applications to Ship Flow and Hull Form Design,* Osaka, Japan, May (1998).

# Direct Primitive Variable Solution Techniques for Incompressible Flows

Prem K. Khosla and Stanley G. Rubin
Department of Aerospace Engineering and Engineering Mechanics
University of Cincinnati, Cincinnati, Ohio 45221-0070, USA

## Abstract

A direct primitive variable solution technique, that does not require either artificial compressibility or a pressure Poisson solver, is discussed for the solution of the incompressible potential, Euler and reduced or full Navier Stokes equations. Both steady state and transient flows are evaluated. The procedure is equally applicable for the solution of compressible flows as given in a number of the references. With the compressible code, the incompressible limit of Mach number identically equal to zero provides results identical with those contained here for the incompressible form of the equations. The convergence properties for the two systems are essentially identical. No additional boundary conditions on pressure or any other flow variable are required in either formulation.

## 1    Introduction

Computational techniques for the steady state or transient solution of the primitive variable form of the incompressible flow equations; potential, full Euler or viscous Navier Stokes equations generally require an indirect and iterative approach for the solution of the first-order continuity equation. This is due to the lack of a time derivative term in the incompressible limit of this equation.

In order to circumvent this problem two methods have been popular with researchers. In the first method, artificial compressibility (a fictitious time derivative of the pressure) is added to the continuity equation and appropriate preconditioning is applied so that in the steady state this term approaches zero. For transient flows, dual time stepping is required. In the second method, the divergence of the momentum equations leads to a Poisson equation for the pressure. In this solution procedure the continuity equation is indirectly satisfied through the pressure Poisson equation and some form of velocity correction. This procedure requires the prescription of appropriate pressure boundary conditions that are not given by any specific physical boundary condition for the pressure. Related projection methods have similar difficulties with the numerical boundary conditions. Because of the differentiation of the momentum equations, lack of definitive numerical boundary conditions and indirect mass conservation condition, this procedure can lead to a lack of consistency, non-uniqueness and the development of spurious boundary layers [1,2].

In the present paper an alternate approach that allows for direct solution of the primitive variable form of the incompressible potential, Euler or Navier Stokes equations is considered. This technique is an outgrowth of the reduced compressible Navier Stokes methodology developed by the present authors. Numerous technical presentations and journal articles have resulted over a period of 20 years, see e.g.., the review article by Rubin and Tannehill [3] and the recent study by Srinivasan and Rubin [4,5]). In this approach, the primitive variable equations, including the discrete form of the first-order continuity equation are computed directly for inviscid, viscous, steady and transient flows. The technique does not require any artificial viscosity or numerical boundary conditions. Rather, the pressure on appropriate boundaries is computed as a part of the solution. At the workshop where this incompressible study was first presented, significant discussion was devoted to the pressure boundary condition. The role of this boundary condition will be clarified herein. In addition, the general methodology will be reviewed and the relationship to the other methods, as discussed previously, will be clarified

## 2   Difference Equations

An unconditionally stable flux-vector procedure that considers the acoustic and advective fluxes independently has previously been detailed; see, reference [3-5]. For incompressible flows, in the primary flow direction, this leads to upwinding of the advective terms, backward differencing or forward marching for the continuity equation, an effective forward differencing for the vorticity and forward differencing for the 'negative flux' axial pressure gradient. For large reverse flow velocities the directionality of the continuity and pressure gradient differencing can be reversed for improved accuracy. Both forms of differencing are unconditionally stable. This improvement has not been considered necessary for the present applications. All derivatives in direction perpendicular to the axial flow direction are two or three point central differenced.

For the two-dimensional case, where the flow variables are evaluated at the grid points, the difference scheme is illustrated below.

Continuity, centered at (i,j-1/2):

$$c_{i,j-\frac{1}{2}} \equiv \frac{u_{i,j} - u_{i-1,j} + u_{i,j-1} - u_{i-1,j-1}}{2\Delta x} + \frac{v_{i,j} - v_{i,j-1}}{\Delta y} = 0 \tag{1-a}$$

X-momentum, centered at (i,j):

$$\frac{\partial u_{i,j}}{\partial t} + u_{i,j}\frac{u_{i,j} - u_{i-1,j}}{\Delta x} + v_{i,j}\frac{u_{i,j+1} - u_{i,j-1}}{2\Delta y} + \frac{p_{i+1,j} - p_{i,j}}{\Delta x} = viscous\ terms \qquad (1\text{-}b)$$

Y-momentum, centered at (i,j-1/2):

$$\frac{\partial v_{i,j+\frac{1}{2}}}{\partial t} + u_{i,j+\frac{1}{2}}\frac{v_{i,j} - v_{i-1,j}}{\Delta x}$$

$$+ v_{i,j+\frac{1}{2}}\frac{v_{i,j+1} - v_{i,j}}{\Delta y} + \frac{p_{i,j+1} - p_{i,j}}{\Delta y} = viscous terms \qquad (1\text{-}c)$$

For convenience in the subsequent discussion of the linear system, the non-conservation Cartesian form of the equations is shown here. For computational purposes, conservation form and general non-orthogonal co-ordinates are applicable in a similar fashion. The accuracy of these difference equations is second order in y and somewhere between first and second order in x. An alternate and slightly more accurate form of the x-momentum is discussed in reference [6,7]. The question of the overall accuracy is discussed in a subsequent section. This form of the discretization is suggested by the nature of the prescribed physical boundary conditions. For example, at a free stream boundary, the pressure is prescribed and the normal component of velocity 'v' is predicted and is computed. At a solid surface 'v' is prescribed and the pressure is computed as a part of the solution. The advantage of two-point differencing is that it facilitates the application of the physical boundary conditions and allows for the direct computation of the unknown wall pressure.

## 2.1   Relationship to Central Difference Equations

The x-momentum equation is centered at the point (i,j) . The pointwise central difference form of the continuity and y-momentum equations can easily be recovered by adding the equations at (i,j-1/2) and (i,j+1/2) respectively. The overall accuracy of the resulting system remains unchanged. A somewhat higher order truncation error results from the averaging of the x-derivatives and the viscous terms. Instead of equations (1), the resulting pointwise equations are solved. However, at one or more of the boundaries, the two-point equations must be used as a connection condition to eliminate any oscillations. Typically, one would use the two-point y-momentum equation (1b) to calculate pressure at a solid boundary and the y-component of velocity 'v' at a free stream boundary when the conditions are imposed at a finite boundary. It should be noted that accurate solutions using the central difference form to approximate the y-derivatives in the equations (with flux splitting in axial direction) have previously been obtained in references [8,9] and in a number of other unpublished studies by the present authors.

## 2.2    Relationship to Poisson Equation for Pressure

A five point discretized version of the Poisson equation for pressure can easily be obtained from the two momentum equations by using a numerical operation for the divergence:

$$\frac{(X\,momentum)_i - (X\,momentum)_{i-1}}{\Delta x} + \frac{(Y\,momentum)_j - (Y\,momentum)_{j-1}}{\Delta y} = 0 \tag{2}$$

Although a divergence operation has been performed, the resulting discrete Poisson equation does not have a direct relation to discrete continuity equation. When this form of the Poisson equation is satisfied, it implies that the momentum equations are also satisfied; however, the conservation of mass cannot be inferred from this operation. The Poisson equation can be used as a replacement for one of the original momentum equations, provided that the momentum equation is also used as a connection condition at the appropriate boundary. The normal momentum equation at the boundary must be used to insure a well-posed condition. In this respect, the momentum equations provide the so-called missing "pressure boundary condition". The most important aspect of the above process is that in the present procedure, the resulting Poisson equation is compatible with the differencing of the momentum equations. For example, at a solid wall located at a fixed value of y, the y-momentum equation provides the appropriate boundary condition for pressure. If on the other hand, three-point central differencing is used to approximate the advection and the pressure gradient terms, five-point differencing for the Poisson equation is incompatible with the momentum equations. The Poisson equation will then represent a larger stencil and additional conditions are required at the boundary. An additional two-point version of the momentum equation must still be applied to provide the necessary connection conditions. It should be emphasized that this procedure does not provide any significant advantage, vis-à-vis, the mass conservation condition. Therefore, it is necessary that the original momentum and continuity equations, or the Poisson enhanced system of momentum equations together with the continuity equation, must be solved. When this consistent approach is applied, there is no problem in satisfying the discrete form of the continuity equation and there is no requirement for any numerical 'pressure boundary conditions'.

### 2.3 Relationship to Vorticity and Vorticity Transport

In a similar vein, the pressure can be eliminated by applying the discrete operation for the curl in order to obtain the discrete vorticity transport equation and discrete form of the vorticity.

$$\frac{\left(Y\,momentum\right)_{i+1,j} - \left(Y\,momentum\right)_{i,j}}{\Delta x}$$
$$-\frac{\left(X\,momentum\right)_{i,j} - \left(X\,momentum\right)_{i,j-1}}{\Delta y} = 0 \tag{3}$$

With constant nonlinear coefficients in the momentum equations equation (3), in the invscid limit, leads to the following form of the vorticity transport equation:

$$u\frac{\omega_{i,j} - \omega_{i-1,j}}{\Delta x} + v\frac{\left(\omega_{i,j+1} + \omega_{i-1,j+1} - \omega_{i,j-1} - \omega_{i-1,j-1}\right)}{4\Delta y} = 0\left(\Delta x^2, \Delta y^2\right) \tag{4-a}$$

where

$$\omega_{i,j} = \frac{v_{i+1,j} - v_{i,j} + v_{i+1,j-1} - v_{i,j-1}}{2\Delta x} - \frac{u_{i,j} - u_{i,j-1}}{\Delta y}. \tag{4-b}$$

Both of these equations are second order accurate. Additional discussion and details relating to the accuracy of these discrete equations is given in reference (7).

## 3 Semi-implicit relaxation

For incompressible flows, the speed of sound is infinite, and from the explicit CFL condition this leads to a limiting value of $\Delta t$ = zero. As such, a true time dependent marching technique in the incompressible limit is not possible. Time marching is achieved by introducing an artificial compressibility term to the continuity equation. This requires special handling and conditioning for both steady and unsteady computations.

An alternate semi-implicit relaxation procedure, which highlights the defined pressure-velocity coupling is considered herein (10). This procedure treats all advection terms explicitly; however, the continuity equation and all acoustic gradients are treated implicitly. In this manner, the CFL condition no longer is dependent upon the speed of sound and is determined solely by the flow velocity. The time step limitation is given as $\Delta t \leq \Delta x/u$. *Clearly, $\Delta t$ remains finite and does* *not vanish, as with conventional time dependent methods. For this reason, it is possible to solve the primitive variable system directly without the need for artificial compressibility.* All of the fully implicit terms are linear so that the coefficient matrix depends only upon the coordinates and time step. Therefore, for a fixed grid and time step, only a single matrix inversion is required for all time. For the two -dimensional cases, the single inversion is accomplished herein with the YSMP solver.

# 4  Boundary Conditions

At the workshop, considerable discussion was devoted to boundary conditions, in particular for the pressure. In the present flux-split procedure only physical boundary conditions are required to completely close the system of equations. The boundary conditions integrate into the differencing described previously. Any physical variable that is not prescribed at the boundary is automatically computed from the governing equations. It has been shown in earlier studies, references [3,4,5,7] that this self-consistent approach reproduces the asymptotic large Reynolds number limits of the Navier-Stokes equations, e.g., inviscid, boundary-layer and triple deck equations. Additional extrapolation or characteristic relations for the boundary conditions are not required. The boundary conditions for each system are now discussed.

## 4.1    Irrotational Inviscid Flow

The primitive variable form of the governing equations constitutes a system of linear first-order partial differential equations. An eigen-value analysis indicates that the system is elliptic and therefore requires the prescription of appropriate data on a closed boundary. For the combined continuity/vorticity system, with the differencing described previously, this suggests the following:

At an inflow boundary, either the primary (u) or secondary (v) flow velocity or a combination thereof should be prescribed. In each case, the prescribed boundary condition and the governing equations then automatically determine the other velocity component at the boundary. At an outflow boundary, a similar combination of conditions applies. On a solid or porous surface, the normal mass flux or streamline condition is prescribed and the tangential flow velocity is computed from the governing equations.

For external flows, at a far-field free surface, the free stream velocity is prescribed and the secondary flow velocities are determined. Since far field boundary conditions are typically applied at a finite distance, an acceptable small value of the normal velocity is computed. A sketch of the differencing and the boundary conditions is shown in figure (1-a,b). It must be emphasized that one can generate a Laplace equation for either of the velocities by cross differentiating the original equations. The above discussion of the boundary conditions and differencing is consistent with the Laplace equation. However, the present authors prefer to deal with the original primitive variable system.

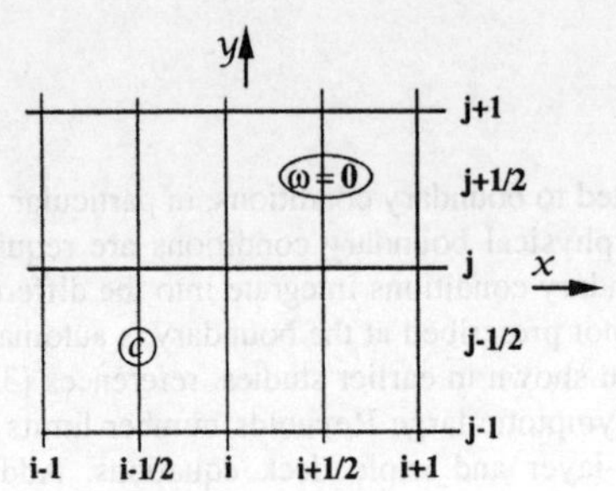


**Figure 1-a. Equation locations for potential flow discretization**

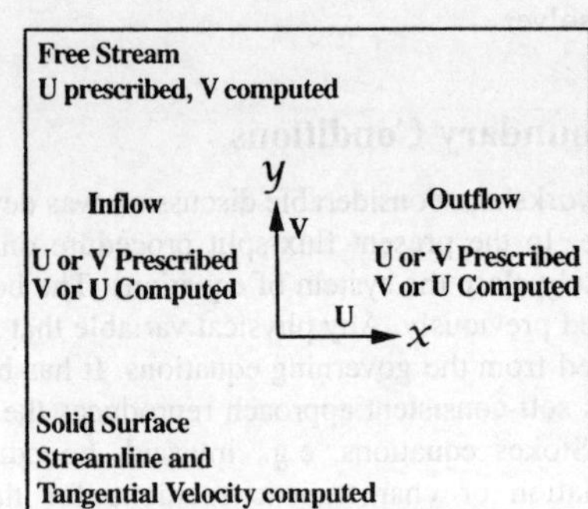


**Figure 1-b. Schematic of Boundary Conditions for Potential Flow Equations**

## 4.2   Euler Equations

Inviscid incompressible flows can be computed directly from the Euler equations. Once again, the governing equations constitute a system of first order elliptic partial differential equations. The difference equations and the physical boundary conditions form a complete system. As discussed previously, the present system is equivalent to the velocity-vorticity system; consistent inflow boundary conditions include one component of velocity and vorticity. This is comparable with the irrotational case. For most applications, uniform inflow conditions are prescribed; therefore, both the primary and secondary flow velocities are given at the inflow. The vorticity or pressure is computed at the inflow. At an outflow boundary, the pressure (for internal flows) or pressure gradient (for external flows) is prescribed and the velocities are then computed from the governing equations. At a solid surface, a normal mass flux or streamline condition is specified and the pressure and tangential velocities are computed from the governing equations. Alternately, a zero vorticity condition can be applied, in lieu of the tangential momentum equation. The difference in the two solutions is negligible. For external flows, in the free stream, once again the tangential velocity and pressure are prescribed and the secondary velocity is computed. A sketch of the differencing and the boundary conditions for the Euler equations is shown in figure (2-a,b). It should be emphasized again that differentiation of the momentum equations to generate either the vorticity transport or pressure Poisson equation is unnecessary. The first order system, (1-a,b,c) is unchanged.

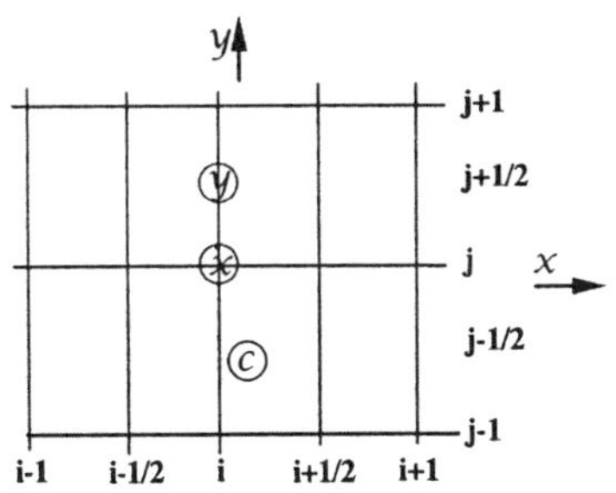

Figure 2-a. Equation locations for discretization of the $\xi$th 3-D $\eta$–$\zeta$ cross plan

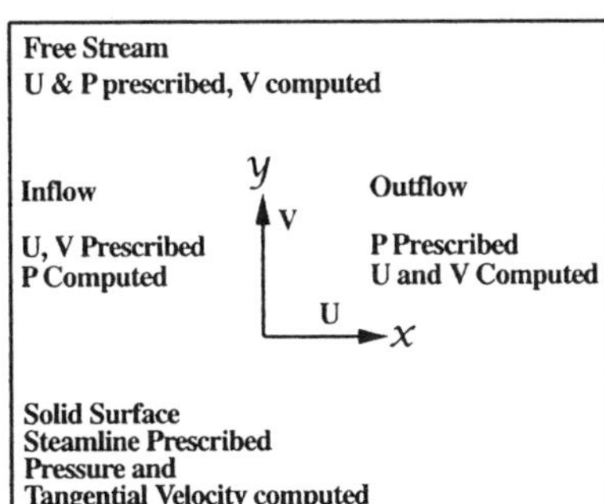


Figure 2-b. Schematic of Boundary Conditions for Euler Equations

## 4.3 Viscous Flows

In this case the boundary conditions remain unchanged, except that no-slip is applied at a solid surface; the surface pressure is again computed as a part of the solution. At an outflow boundary, the axial viscous terms are neglected or added as a deferred corrector.

## 5 Accuracy

The accuracy of the present procedure can be illustrated by the simple incompressible, irrotational system. The difference forms of the continuity equation and irrotationality condition are given as:

$$\frac{u_{i,j} - u_{i-1,j}}{\Delta x} + \frac{v_{i,j} - v_{i,j-1}}{\Delta y} = 0$$

$$\frac{v_{i+1,j} - v_{i,j}}{\Delta x} + \frac{u_{i,j+1} - u_{i,j}}{\Delta y} = 0$$

Since first order differencing has been used to approximate various derivatives, one would expect the solution to be also first order. However, elimination of either $u$ or $v$ between these equations leads to the well-known five-point second-order accurate difference approximation for the Laplace equation:

$$\frac{u_{i+1,j} - 2u_{,i,j} + u_{i-1,j}}{\Delta x^2} + \frac{u_{i,j+1} - 2u_{i,j} + u_{i,j-1}}{\Delta y^2} = 0$$

This difference equation is second order accurate, in spite of the fact that it was generated from two-point differencing of the first-order system. Higher order or three-point differencing of the continuity equation and irrotationality condition will lead to a more accurate discrete representation of the Laplace equation. This explains the fact that the two-point upwind/backward differencing in the full Euler system results in better than first order accuracy of the solution. For invscid Euler computations on successively finer grids, an accuracy of $O(\Delta x)^{1.97}$ has been observed [7]. This accuracy is reduced somewhat for viscous flow computations. For a flow over a finite flat plate with a discontinuity in pressure at the trailing edge, the order of accuracy reduces to $O(\Delta x)^{1.3}$. However, for most of the separated viscous flows where a dominant flow direction is present, an accuracy of $O(\Delta x)^{1.7}$ or better has been observed [7]. This suggests that the pressure velocity coupling for viscous flows is much more complex than for the invscid flows. It should be pointed out that in reverse flow region only axial convective flux is up-winded. The continuity and the acoustic fluxes remain unchanged. For most of the flows with a dominant direction, separated flow regions are comparatively small portions of the whole flow field, such an approach doesn't seriously affect accuracy. However, for flows where a large part of the flow is moving in an opposite direction to the 'main flow direction', in addition to convective up-winding, the acoustic gradient (flux) and the axial velocity gradient in the continuity should also be reversed. This will also result in revered (forward) differencing of the continuity equation (e.g. continuity at i+1/2,j-1/2). In the absence of such a switch, the numerical scheme remains stable, but can result in a further loss of accuracy. Flow in a driven cavity presents such a situation. In the lower part of the cavity the bulk of the flow is completely reversed and of $O(1)$. In such a situation the overall accuracy will be reduced. As noted previously the formulation remains stable even with large reverse flows; however, formal first-order accuracy $O(\Delta x)$ should be expected.

# 6    Results

Over the years, a large number of two and three-dimensional steady and unsteady solutions have been computed successfully using the present formulation. These are available in many of the references; see, specifically [3-5]. At the workshop some concerns were raised that the present formulation may not be valid in the absence of a preferred flow direction. In order to address this concern, the following problems have been chosen to illustrate the validity of the present formulation. All viscous flow computations have been carried out on four successively refined grids and for the steady state with $\Delta t = \infty$.

## 6.1    Flow in a driven cavity

In order to satisfy the kinematic boundary conditions on all the four solid surfaces, the continuity equation is now centered at (i-1/2,j-1/2). Solutions for Reynolds numbers from 100 to 1000 are obtained on four successively refined grids. Although the streamline pattern is accurately predicted, the solution in this case can be expected to be of lower order accuracy as the pressure flux and the continuity representation have not been reversed in regions of large negative u velocity. As discussed earlier however, there is no loss of stability. For the value of $u_{min}$ on four grids, first order accuracy of the solution is obtained. Richardson extrapolation between 81x81 and 161x161 grids provided an accurate second order solution.  On an 81x81 grid, the computed value of $u_{min}$ through the vertical centerline is .38297 that agrees quite well with .38289 of reference [11]. Similarly, the value of $v_{min} = -.51593$ agrees with -.51550 of reference [11]. The u-velocity profile along the centerline of the cavity and the associated streamlines are depicted in figures (3a,b). The agreement with the extrapolated solution on 81x81 and the solution of reference [11] is quite good.

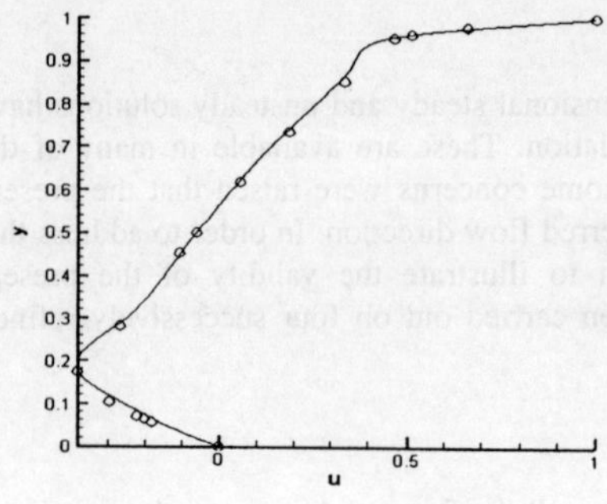

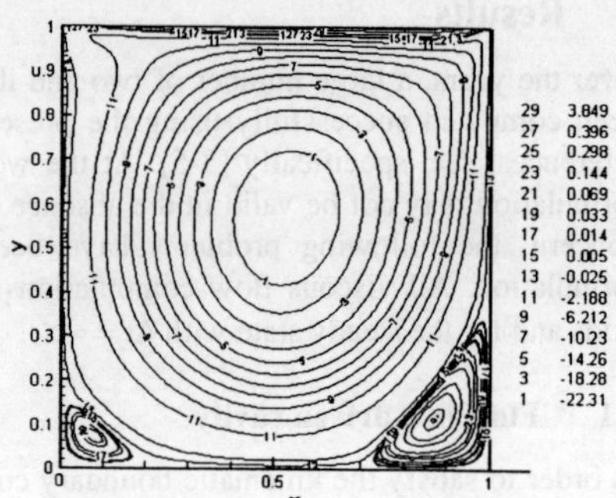

**Figure 3a Velocity profile through the center of the cavity, Re=10$^{13}$**

**Figure 3b Streamlines, Re=10$^3$**

## 6.2    Two dimensional flow past a cylinder of square cross-section

This problem is chosen because it has a preferred flow direction.  However, in the front bluff region, the dominant flow directionality is lost. Steady flow solutions or Re=25 have been obtained on successively refined grids. A mix of uniform and non-uniform grid is provided. On the top and bottom surfaces, uniform grid is used in the flow direction while over the front and rear, uniform grid is used in the direction normal to the flow. The grid is shown in figure (4-a). Streamlines lines are depicted in figure (4-b). The computations have been performed on two grids. The length of the wake is not influenced by the grid refinement and is about the same as for a circular cylinder of diameter equal to the side of the square block.

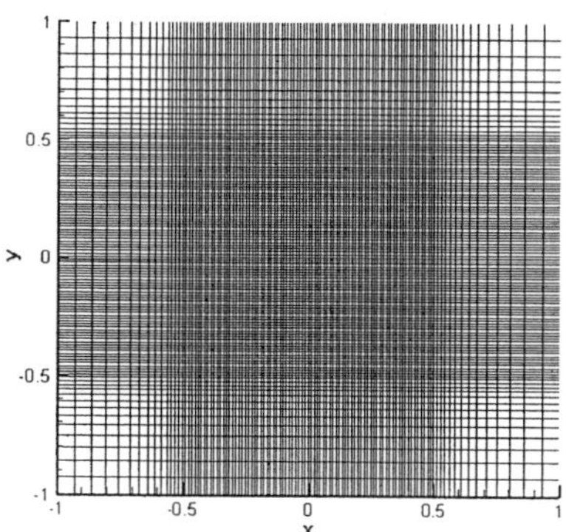

**Fig.4-a Grid**

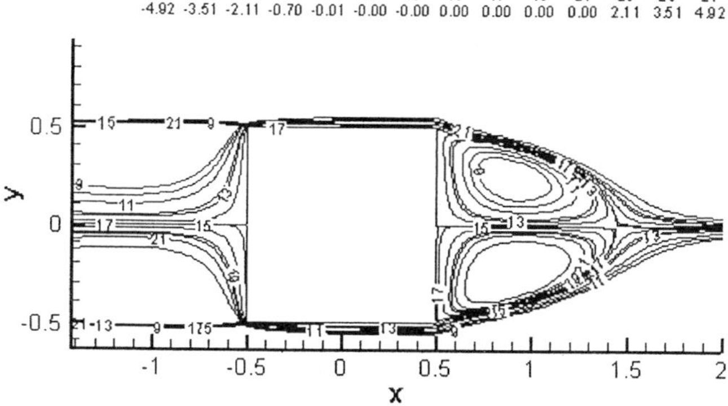

**Fig. 4-b Stream lines Re=25**

## 6.3    Flow in a channel with a backward facing step

This case has previously been considered in several of our previous papers. However, most of the solutions were carried out by reduced form of the Navier-Stokes equations with a deferred corrector. In the present case complete Navier-Stokes solution for Re=400 on a 161x81 uniform grid is presented in figure (5-a). The reattachment length of 8.6 times the step height agrees ver well with previous computations and experimental observation. On a stretched grid a secondary vortex in the corner is also observed. The streamlines in the corner region are shown in figure (5-b). The horizontal extent of this corner vortex is about the same as previously predicted by our deferred corrector procedure. However, the vertical extent is over predicted by the deferred corrected.

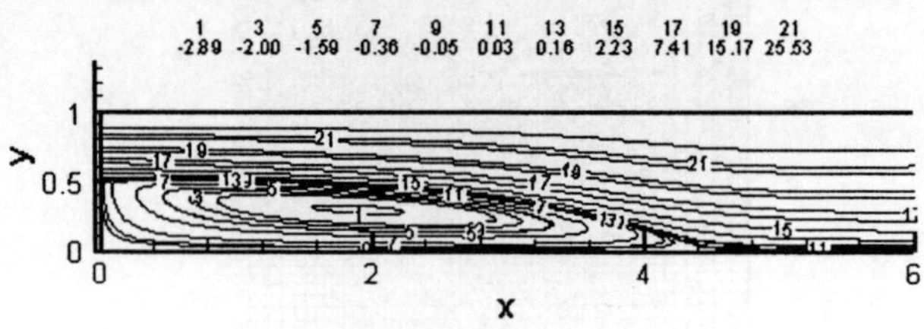

**Fig. 5-a Stream lines in the corner Re=400 161x81 non-uniform grid**

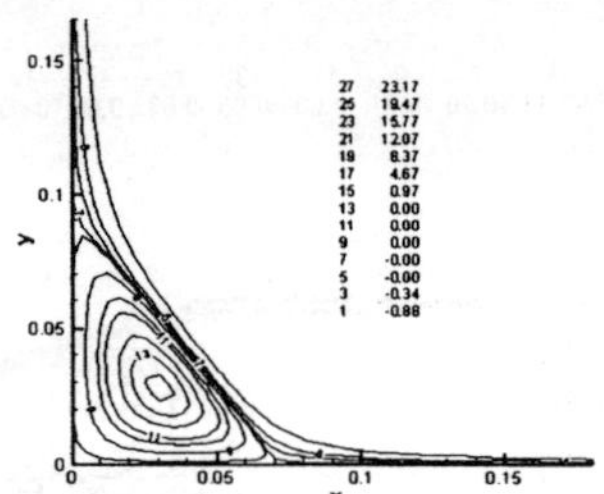

**Fig. 5-a Stream lines Re=400 161x81 uniform grid**

# 7 References


[1]  Orszag, S.A., Israeli, M. and Deville, M.O. Boundary conditions for incompressible flows. *J. Sci. Computing.* Vol.1, pp75-11 (1986)

[2]  Weinan, E. and Liu, J.G. Projection Method I :Convergence and numerical boundary-layers. *Siam j. Num. Anal.* Vol. 32 pp 1017-1057 (1995).

[3]  Rubin, S.G. and Tannehill, J.C. Parabolized/Reduced Navier-Stokes Computational Techniques. *Annual Review of Fluid Mechanics* Vol. 24, pp117-44 (1992).

[4]  Srinivasan, K and Rubin, S.G. Segmented multi-grid domain decomposition procedure for two and three-dimensional incompressible viscous flows. *J.Fluids Engineering* Vol.115, pp608-13 (1993).

[5]  Srinivasan, K. and Rubin, S.G. Solution based grid optimization through segmented multi-grid domain decomposition. *J. Computational Physics.* Vol. 136, pp467-93 (1997).

[6]  Israeli, M and Lin, A. Numerical solution and boundary conditions for boundary-layer like flows. *8$^{th}$ ICNMFD conference* pp 266-272 (1982).

[7]  Rubin, S.G. and Reddy, D.R. Analysis of global pressure relaxation for flows with strong interaction and separation. *Computers and Fluids* Vol. 11, pp281-306 (1983).

[8]  Lin, T.C. and Rubin, S.G. Viscous flow over a cone at moderate incidence J. Fluid Mechanics Vol. 59, pp593-620 (1973).

[9]  Zhu,Z. and Fletcher, C. A. J. Computation of the reduced Navier-Stokes Equations with Multigrid acceleration. Proceedings of the computational techniques and applications held at Griffith University, Australia. CTAC-89 pp373-382 (1989).

[10]  Kaushik, S. Optimized solution procedure for internal viscous flows with applications ranging from incompressible to supersonic flows. *Ph.D. Thesis, Department of Aerospace Engineering and Engineering Mechanics, university of Cincinnati (1992).*

[11]  Ghia, U., Ghia, K.N., Shin, C.T. High Re-solutions for incompressible flow using the Navier-Stokes equations and a multigrid method. J. Computational Physics Vol. 48 pp387-411 (1982).

# Mass conservation and the accuracy of non–staggered grid incompressible flow schemes

Michael C. Wendl and Ramesh K. Agarwal

Washington University

4444 Forest Park Blvd, Box 8501

Saint Louis, MO 63108

## 1  Introduction

Assessing solution accuracy continues to be an important issue in computational fluid dynamics [1, 2, 3, 4]. Given the complex interplay among such factors as numerical scheme, flow complexity, grid distribution, dissipation, and boundary conditions, this task is rarely trivial. When subjected to rigorous analysis, many well–developed codes have been shown to have difficulties in handling even fundamental flow configurations [5]. Recent work has thus focused on a more precise classification and explanation of uncertainties [6] and on recommendations for systematic error quantification [7].

Pressure–based non–staggered grid solvers have recently become more popular for simulating incompressible flows. Their origin lies in a set of schemes that eliminate the so–called "checkerboard" numerical instability via re–coupling of adjacent pressure and velocity nodes e.g. [8, 9, 10, 11, 12, 13]. In addition to the usual accuracy considerations mentioned above, there is another issue specific to these algorithms: they do not achieve divergence free solutions [14, 15, 16, 17, 18]. In other words, the pressure Poisson equation does not yield a pressure distribution compatible with satisfying the continuity equation, as is it does for a staggered grid arrangement.

This anomaly is often explained in the context of discrete control volumes in the computational grid [16]. Specifically, a staggered grid is naturally compatible with mass conservation in the finite–volume sense of individual nodes, since velocity components are defined at the same location where fluxes are computed. Converged pressure solutions imply both local and global mass conservation. Conversely, non–staggered grids require interpolation to compute fluxes since velocity components are defined at the center of each discrete volume. Converged pressure solutions do *not* imply local mass conservation in the finite–volume sense. This degree depends upon the type of interpolation chosen [15]. Some investigators argue that lack of mass conservation is a fundamental consequence of non–staggered grids which cannot be eliminated [18]. Most proposed solutions involve explicitly defining either extra fluxes or velocity components at cell faces, which are constrained to identically satisfy conservation of mass [19]. However, these arrangements necessitate extra storage and the evaluation of additional momentum equations. It is debatable whether such arrangements qualify as non–staggered grid schemes in a strict sense.

Lack of mass conservation causes other anomalies as well. For example, time marching schemes that use $\partial/\partial t$ for calculating steady flows yield solutions that depend upon the time step size via the non–vanishing residual dilation [18, 20]. No work has examined how accuracy is affected under these circumstances. Because such information is essential in further evaluating non–staggered grid solvers for large–scale simulations, we focus on this issue for a representative non–staggered scheme.

## 2   Analysis

The practical advantages of non–staggered grids over staggered grids are well known [13, 21, 17]. However, despite extensive application of non–staggered algorithms, the mass conservation problem appears to be neither widely known nor appreciated. In fact, to some degree, the whole issue has generated confusion among researchers. For example, though citing the earlier work of Abdallah [9, 10] Lee and Tzong [18] consider the divergence constraint on the boundary value problem for pressure $P$, given by

$$\iiint_{\Omega} \frac{\partial}{\partial x_k} \frac{\partial P}{\partial x_k} d\nu - \iint_{\partial\Omega} n_k \frac{\partial P}{\partial x_k} dA = 0 \tag{1}$$

equivalent to satisfying the continuity equation. Abdallah's scheme [9, 10] is, however, a clear contradiction of this assumption. Specifically, in spite of a differencing procedure which identically satisfies Eq. 1 over the grid, the Abdallah scheme has been shown to have relatively poor mass conservation properties [15]. Other algorithms [18, 22] have employed correction factors to enforce Eq. 1. For example Ghia et al. [22] simply distribute the amount by which Eq. 1 is violated uniformly over the grid via an extra source term in the pressure equation. While these algorithms concede the ability to conserve mass locally, they apply numerical compensation so that it is at least conserved globally.

Sotiropoulos and Abdallah [16] have argued that no equivalence between mass conservation and the divergence constraint can be assumed for non–staggered schemes. Eq. 1 should only be taken as a necessary condition for the well–posed pressure problem. The continuity equation is satisfied where discrete residuals of the momentum equations vanish, i.e. at nodes $(i, j)$, however, satisfying mass conservation in a finite volume sense requires residuals to also vanish when summed over the boundaries $(i \pm 1/2, j \pm 1/2)$. The conservation problem can therefore be viewed as an inconsistency between the discretized and continuum quantities. Thus, neither global nor local mass conservation can be identically satisfied to machine accuracy on a non–staggered grid (without numerical compensation), irrespective of the algorithm used to interpolate the discrete quantities. Numerical results reported in the literature support this view, although the choice of interpolation has a bearing on the degree to which continuity is violated [14, 15].

Sotiropoulos and Abdallah [16] used a Taylor series analysis to estimate dilation magnitude for individual control volumes. They show that the mass source is proportional to fourth derivatives of pressure and functions essentially as a dissipation factor. Other interpolations exhibit this behavior as well, for example Barton and Kirby [23] showed fourth–order dissipation in the Rhie–Chow interpolation scheme [8]. But is it possible to estimate dilation magnitude over the complete domain, i.e. in a global sense? Let us consider a Cartesian domain made up of discrete square control volumes. There are $M$ volumes in the $i$ direction and $N$ volumes in the $j$ direction. Resulting staggered and non–staggered grid arrangements are shown in Figure 1. In each case, the boundary is situated such that boundary value interpolations are minimized.

For a physically realistic calculation, global mass conservation must be satisfied on the boundary $\partial\Omega$ of the domain, that is

$$\iint_{\partial\Omega} u_k n_k \, dA = 0 \,. \tag{2}$$

Eq. 2 is independent of the algorithm and should be expected to be satisfied numerically in all cases. For example, according to Fig. 1, this equation can be expressed discretely on a two–dimensional staggered grid as

$$\sum_{j=1}^{N} (u_{M,j} - u_{0,j}) + \sum_{i=1}^{M} (v_{i,N} - v_{i,0}) = 0 \,, \tag{3}$$

where $u$ and $v$ are the velocity components in the $i$ and $j$ directions, respectively. Global mass conservation for the numerical scheme can be computed by integrating $\partial u_k / \partial x_k$ for each control volume of the grid and summing their contributions. Let $\Gamma$ denote the residual dilation, which measures deviation from identically satisfying the continuity equation. For the staggered grid,

$$\Gamma = \sum_{i=1}^{M} \sum_{j=1}^{N} (u_{i,j} - u_{i-1,j} + v_{i,j} - v_{i,j-1}) \,. \tag{4}$$

Eq. 4 is consistent with the standard compact Laplacian stencil that preserves proper numerical coupling between velocity and pressure. It is straightforward to show that Eq. 4 reduces to Eq. 3 confirming that $\Gamma = 0$. Thus, mass is identically conserved for a staggered grid variable arrangement if boundary conditions are appropriately prescribed such that Eq. 3 is satisfied.

Conversely, the analog of Eq. 2 for the non–staggered grid arrangement in Fig. 1 is

$$\sum_{j=2}^{N-1} \left( u_{M,j} - u_{1,j} \right) + \sum_{i=2}^{M-1} \left( v_{i,N} - v_{i,1} \right) = 0 \,. \tag{5}$$

The finite volume calculation of the continuity equation over the grid now requires a suitable interpolation scheme. For example, Abdallah [9, 10] uses standard linear interpolation, which for a uniform grid reduces to averaging. This yields

$$\Gamma = \sum_{i=2}^{M-1} \sum_{j=2}^{N-1} \left( \frac{u_{i,j} + u_{i+1,j}}{2} - \frac{u_{i,j} + u_{i-1,j}}{2} + \frac{v_{i,j} + v_{i,j+1}}{2} - \frac{v_{i,j} + v_{i,j-1}}{2} \right) \,. \tag{6}$$

Simplifying and making use of Eq. 5, Eq. 6 reduces to

$$\Gamma = \sum_{j=2}^{N-1} \left( \frac{u_{M-1,j} - u_{2,j}}{2} \right) + \sum_{i=2}^{M-1} \left( \frac{v_{i,N-1} - v_{i,2}}{2} \right) \,. \tag{7}$$

Therefore, $\Gamma$ is proportional to the integral form of the continuity equation about a closed surface displaced exactly one cell width inside of the physical boundary of the problem, i.e. $\Gamma \propto \iint_{\partial\Omega-\delta} u_k n_k dA$, where $\delta$ is the cell width. This anomaly arises because the physical boundary values of the problem naturally associate with cell centers, while finite volume calculations depend upon fluxes calculated at the faces. Of course for staggered grids, these locations coincide. Since local dilation does not vanish, we expect $\Gamma \neq 0$ in Eq. 7. The more interesting implication is, however, that

$$\lim_{\delta \to 0} \Gamma = 0 \,, \tag{8}$$

i.e. we should see increasingly improved global conservation as the grid becomes progressively finer at the boundaries. In the limit, we should obtain the exact conservation law given by Eq. 2.

# 3 Numerical Evaluation

The above analysis yields an estimate of $\Gamma$ in terms of the unknown dependent variables for Abdallah's scheme [9, 10]. However, the non–linear nature of the system prevents straightforward quantification of $\Gamma$, particularly its effect upon solution accuracy as a function of the time step. This phenomenon can be determined numerically. We therefore implement Abdallah's scheme with convective effects as the source term of the Poisson equation [24]. Characteristics are similar to many codes based on the Abdallah algorithm [25, 26, 27, 28], i.e. finite volume discretization, $\partial/\partial t$ used in the iterative role and modeled with Euler explicit time stepping, standard third order upwinding for convective effects, and the standard second–order explicit approximation for viscous terms. Spatial accuracy is formally second order. Successive Over Relaxation (SOR) is used to solve the pressure Poisson problem with a relaxation factor of 1.9 for all cases.

The planar shear driven cavity problem [29] is perhaps the most widely used incompressible flow benchmark. It is challenging from a numerical point of view because it exhibits non–linear effects and singularities, but provides a straightforward set of boundary conditions. Unfortunately, this configuration is largely unsuitable for evaluating numerical errors to very exact tolerances. No closed–form solution exists nor are rigorous experimental measurements available due to the inherent three–dimensional nature of the flow [30]. We therefore utilize a qualitatively similar configuration constructed by the so–called "method of manufactured solutions" [31, 32, 33]. Here, a closed form solution is obtained to the momentum equations for a given flow pattern of interest. This allows numerical quantities to be checked against an analytical component for every discrete control volume. The use of such solutions is highly recommended, provided sufficient structure exists so that the highest derivatives are non–trivial [6]. We utilize a flow pattern similar to the driven cavity (Fig. 2), where the stream function is given by $\psi = \left(y^3 - y^2\right)\left(x^2 - 2x^3 + x^4\right)$ [33]. Corresponding pressure distribution and forcing functions are given in Wendl [33].

Some investigators take $\Gamma$ (or a related quantity) as an accuracy indicator [15, 18], although these have not been proven to be rigorous accuracy metrics. We advocate a norm–based measure, which is more consistent with global accuracy evaluation: $Ac = -\log_e\left(\|\phi' - \phi\|_2 / \|\phi\|_2\right)$. This describes the relative accuracy of a computed scalar field, $\phi'$, when compared to the exact solution, $\phi$. It is essentially a non–dimensional "accuracy number", whose value increases with increasing solution fidelity.

All computations are performed at a Reynolds number of 100 and rest conditions are used as the initial state for each case. A uniform grid of $N \times N$ control volumes is employed, thus second–order accuracy is formally retained. Accuracy numbers for the horizontal component of velocity were computed every five iterations and convergence was assumed when this value was greatest. A more conventional convergence measure was also employed, the maximum $L_2/N^2$ change in $(u, v, P)$ from one iteration to the next. This value was of order $10^{-5}$ or lower at the termination of each case. Numerically, $\Gamma$ is measured as $L_\infty$ of the quantity $\iint u_k n_k dA$ computed for each computational cell. The size of the time step, $\delta t$, is varied by specifying the CFL number where $\delta t$ is CFL / $N$ divided by the magnitude of the largest velocity component in the problem domain.

In the first calculation, we evaluate the hypothesis that dilation can be taken as a direct indicator of solution accuracy [15, 18]. An example calculation using $N = 53$ shows that the two quantities, dilation and accuracy, are not linearly related (Fig. 3). In fact, the rate of gain in accuracy slows with decreasing dilation implying that other factors such as truncation error remain important even when dilation approaches zero. Therefore, without being able to individually characterize other types of error contributions, it is not possible to draw reliable conclusions about accuracy based only upon the dilation. It was difficult to obtain additional accuracy at this resolution because the time step becomes impractically low.

Dependencies of both dilation and accuracy upon $\delta t$ are studied for three different grid resolutions: $N = 29$, $N = 41$, and $N = 53$. The CFL number is varied in increments of 0.01 from a minimum of 0.01 (well below values used in actual practice) to a maximum dictated by the onset of numerical instability for each grid. No attempt was made to employ finer grids because of excessive computational time needed for low values of $\delta t$. Fig. 4 shows dilation, $\Gamma$, for the converged solutions. As our analysis predicts and as previous studies have found, $\Gamma$ decreases with increasing grid resolution. However, as a function of CFL number, dilation is maximum at the lowest time step, decreases almost linearly, and finally attains a relatively fixed value. For all grids, the linear region to the left of $\Gamma_{min}$ indicates cases where maximum dilation occurs in the upper left portion of the cavity where flow impinges on the wall. At $\Gamma_{min}$, locations of maximum dilation shift suddenly to the middle of the cavity and remain there for all higher CFL numbers. The numerical reasons for this behavior are unclear. The analysis of Sotiropoulos and Abdallah [16] and the numerical results of Aksoy and Chen [15] and Lee and Tzong [18] all suggest that dilation should always increase with CFL, so these results are somewhat unexpected. It is interesting that the CFL number at which $\Gamma_{min}$ occurs grows as the grid resolution increases, suggesting that low CFL numbers are not necessarily optimal for minimizing dilation in high–resolution calculations, at least for the standard Abdallah scheme. The fact that the percentage change in $\Gamma$ is also greatest for the fine grid also suggests that selecting an appropriate CFL is critical as $N$ becomes large.

Lee and Tzong [18] concluded that accuracy does not necessarily increase as a function of the grid resolution, based upon an interpretation of the discrete parameters. According to their conjecture, $Ac$ should also be a strong function of $\delta t$. In other words, for two grids of differing resolution, $\delta t$ will strongly influence which grid will yield the more accurate solution. This suggests that overlapping accuracy curves should be observed. However, Fig. 5, which shows the corresponding accuracy curves for the cases in Fig. 4, contradicts this hypothesis. A monotonic increase in $Ac$ is observed and no evidence of overlap is present. While $\delta t$ clearly plays a role in determining solution accuracy, it does not appear to be as significant as suggested by Lee and Tzong. Instead, our findings agree with those of Sotiropoulos and Abdallah [16] who concluded that accuracy increases with better resolution.

Trends in Fig. 5 again confirm that proper selection of CFL becomes important for finer grids. For example, the percentage change from the least accurate to most accurate solution is greatest for $N = 53$. Moreover, the range of CFL numbers for which solution accuracy is highest narrows as the resolution is increased. Small departures from the optimal CFL number may have appreciable impact on the solution accuracy for large $N$. In spite of these difficulties, users of codes based upon Abdallah's scheme claim acceptable results, usually within the prescribed experimental error of corresponding measurements [29]. Typically, a single $CFL$ value is used on a relatively fine grid. However, solution accuracy would possibly have been improved by determining the optimal $CFL$.

# 4   Conclusion

Mass conservation properties of non–staggered grid incompressible Navier–Stokes schemes have stimulated considerable discussion and investigation. It is clear from the literature that misconceptions still remain with respect to the validity of the standard pressure Poisson formulation for non–staggered grids. Moreover, most of the basic non–staggered schemes, including the one examined here, have fundamental numerical problems arising from this phenomenon. Hybrid schemes [19] may provide the best compromise between strictly staggered and non–staggered schemes for the near future. However, there is anticipation on the part of many investigators that tomorrow's super–dense grids enabled by increases in computing capacity will marginalize many of the numerical pitfalls we see on today's "coarse grids", e.g. pressure–velocity coupling and convective flux limiting [23]. Non–staggered grid mass conservation may be among the issues resolved by this brute force means.

# References


[1] Mehta, U. B. Some aspects of uncertainty in computational fluid dynamics results. *Journal of Fluids Engineering*, 113: 538–543, 1991.

[2] Douglass, R. W. & Ramshaw, J. D. Perspective: Future research directions in computational fluid dynamics. *Journal of Fluids Engineering*, 116: 212–215, 1994.

[3] Marvin, J. G. Perspective on computational fluid dynamics validation. *AIAA Journal*, 33: 1778–1787, 1995.

[4] Johnson, C., Rannacher, R., & Boman, M. Numerics and hydrodynamic stability: Toward error control in computational fluid dynamics. *SIAM Journal of Numerical Analysis*, 32: 1058–1079, 1995.

[5] Freitas, C. J. Perspective: Selected benchmarks from commercial CFD codes. *Journal of Fluids Engineering*, 117: 208–218, 1995.

[6] Roache, P. J. Quantification of uncertainty in computational fluid dynamics. *Annual Review of Fluid Mechanics*, 29: 123–160, 1997.

[7] American Institute of Aeronautics and Astronautics. Guide for the verification and validation of computational fluid dynamics solutions. Technical Report G–077–1998, 1998.

[8] Rhie, C. M. & Chow, W. L. Numerical study of the turbulent flow past an airfoil with trailing edge separation. *AIAA Journal*, 21: 1525–1532, 1983.

[9] Abdallah, S. Numerical solutions for the pressure Poisson equation with Neumann boundary conditions using a non–staggered grid, I. *Journal of Computational Physics*, 70: 182–192, 1987.

[10] Abdallah, S. Numerical solutions for the incompressible Navier–Stokes equations in primitive variables using a non–staggered grid, II. *Journal of Computational Physics*, 70: 193–202, 1987.

[11] Reggio, M. & Camarero, R. A calculation scheme for three–dimensional viscous incompressible flows. *Journal of Fluids Engineering*, 109: 345–352, 1987.

[12] Strikwerda, J. C. & Nagel, Y. M. A numerical method for the incompressible Navier–Stokes equations in three–dimensional cylindrical geometry. *Journal of Computational Physics*, 78: 64–78, 1988.

[13] Miller, T. F. & Schmidt, F. W. Use of a pressure–weighted interpolation method for the solution of the incompressible Navier–Stokes equations on a nonstaggered grid system. *Numerical Heat Transfer*, 14: 213–233, 1988.

[14] Acharya, S. & Moukalled, F. H. Improvements to incompressible flow calculation on a nonstaggered curvilinear grid. *Numerical Heat Transfer B*, 15: 131–152, 1989.

[15] Aksoy, H. & Chen, C. J. Numerical solution of Navier–Stokes equations with non–staggered grids using finite analytic method. *Numerical Heat Transfer B*, 21: 287–306, 1992.

[16] Sotiropoulos, F. & Abdallah, S. The discrete continuity equation in primitive variable solutions of incompressible flow. *Journal of Computational Physics*, 95: 212–227, 1991.

[17] Tafti, D. Alternate formulations for the pressure equation Laplacian on a collocated grid for solving unsteady incompressible Navier–Stokes equations. *Journal of Computational Physics*, 116: 143–153, 1995.

[18] Lee, S. L. & Tzong, R. Y. Artificial pressure for pressure linked equation. *International Journal of Heat and Mass Transfer*, 35: 2705–2716, 1992.

[19] Zang, Y., Street, R. L., & Koseff, J. R. A non–staggered grid fractional step method for time–dependent incompressible Navier–Stokes equations in curvilinear coordinates. *Journal of Computational Physics*, 114: 18–33, 1994.

[20] Majumdar, S. Role of underrelaxation in momentum interpolation for calculation of flow with nonstaggered grids. *Numerical Heat Transfer*, 13: 125–132, 1988.

[21] Rodi, W., Majumdar, S., & Schönung, B. Finite volume methods for two dimensional incompressible flows with complex boundaries. *Computer Methods in Applied Mechanics and Engineering*, 75: 369–392, 1989.

[22] Ghia, K. N., Hankey, W. L., & Hodge, J. K. Use of primitive variables in the solution of incompressible Navier–Stokes equations. *AIAA Journal*, 17: 298–301, 1979.

[23] Barton, I. E. & Kirby, R. Finite difference scheme for the solution of fluid flow problems on non–staggered grids. *International Journal for Numerical Methods in Fluids*, 33: 939–959, 2000.

[24] Gresho, P. M. & Sani, R. L. On pressure boundary conditions for the incompressible navier stokes equations. *International Journal for Numerical Methods in Fluids*, 7: 1111–1145, 1987.

[25] Lee, Y. H. & Chou, K. S. Modelling of growth rate uniformity of CdTe epilayer by MO–CVD. *Journal of Electronic Materials*, 21: 87–92, 1992.

[26] Babu, V. & Korpela, S. A. Numerical solution of the incompressible three–dimensional Navier–Stokes equations. *Computers and Fluids*, 23: 675–691, 1994.

[27] Cao, Z. & Esmail, M. N. Numerical study on hydrodynamics of short–dwell paper coaters. *AIChE Journal*, 41: 1833–1842, 1995.

[28] Pritchard, W. F., Davies, P. F., Derafshi, Z., Polacek, D. C., Tsao, R., Dull, R. O., Jones, S. A., & Giddens, D. P. Effects of wall shear stress and fluid recirculation on the localization of circulating monocytes in a three–dimensional flow model. *Journal of Biomechanics*, 28: 1459–1469, 1995.

[29] Burggraf, O. R. Analytical and numerical studies of the structure of steady separated flow. *Journal of Fluid Mechanics*, 24: 113–151, 1966.

[30] Koseff, J. R. & Street, R. L. On end wall effects in a lid–driven cavity flow. *Journal of Fluids Engineering*, 106: 385–389, 1984.

[31] Shih, T. M. A procedure to debug computer programs. *International Journal for Numerical Methods in Engineering*, 21: 1027–1037, 1985.

[32] Shih, T. M., Tan, C. H., & Hwang, B. C. Effects of grid staggering on numerical schemes. *International Journal for Numerical Methods in Fluids*, 9: 193–212, 1989.

[33] Wendl, M. C.  A method for obtaining benchmark Navier–Stokes solutions.  *International Journal for Numerical Methods in Fluids*, 31: 657–660, 1999.

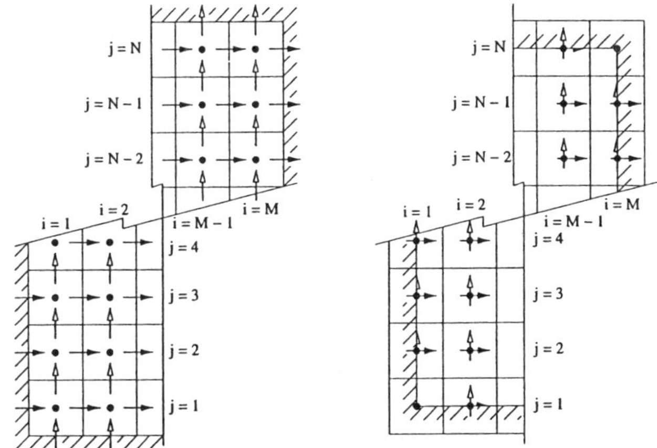

Figure 1: Staggered (left) and non–staggered (right) grid arrangements.

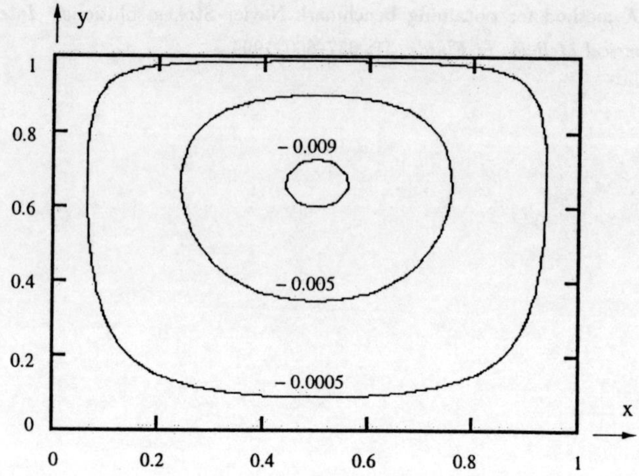

Figure 2: Driven cavity analog showing three stream function contours.

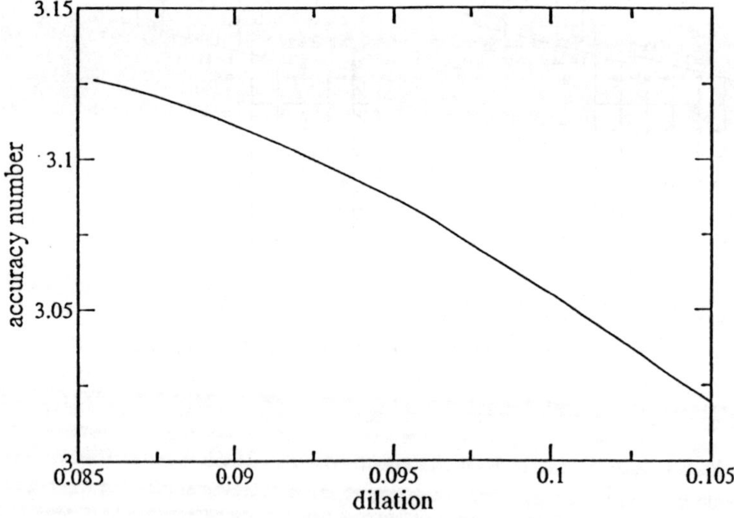

Figure 3: Solution accuracy versus dilation calculated for a square grid of $N = 53$.

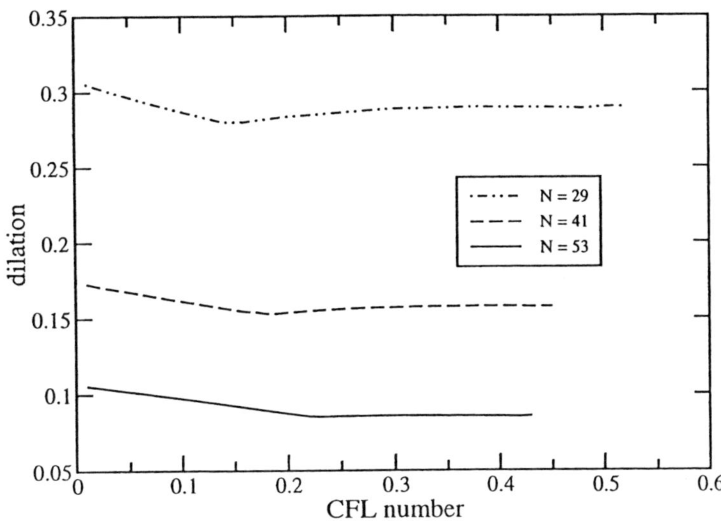

Figure 4: Dilation versus time step (CFL number) for coarse, medium, and fine grids.

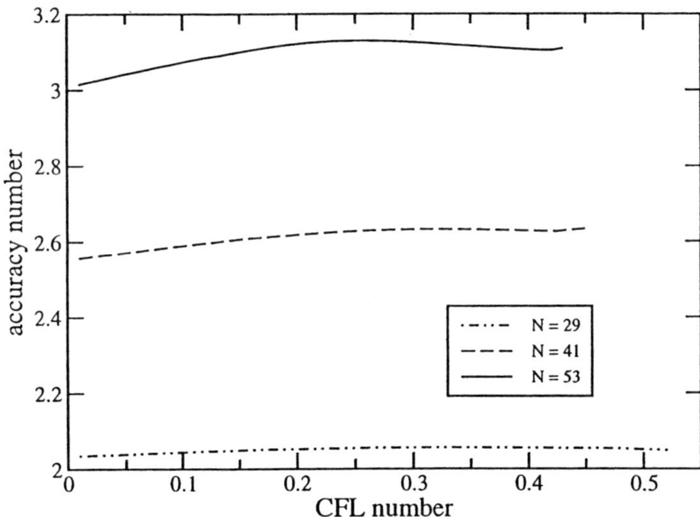

Figure 5: Solution accuracy versus time step (CFL number) for coarse, medium, and fine grids.

# Projection Methods

# Accuracy of Projection Methods for the Incompressible Navier-Stokes Equations

David L. Brown [*]

Center for Applied Scientific Computing, Lawrence Livermore National Laboratory, Livermore, CA 94551, USA; dlb@llnl.gov
UCRL-JC-144037

**Abstract.** Numerous papers have appeared in the literature over the past thirty years discussing projection-type methods for solving the incompressible Navier-Stokes equations. A recurring difficulty encountered is the proper choice of boundary conditions for the auxiliary variables in order to obtain at least second order accuracy in the computed solution. A further issue is the formula for the pressure correction at each timestep. An overview of boundary condition choices that give second-order convergence for all solution variables is presented here based on recently published results by Brown, Cortez and Minion [2].

## 1   Introduction

Denoting by $\mathbf{u}$, the velocity, $p$, the pressure, and $\nu$, the viscosity of the fluid, the incompressible Navier-Stokes equations

$$\mathbf{u}_t + (\mathbf{u} \cdot \nabla)\mathbf{u} + \nabla p = \nu \nabla^2 \mathbf{u} \tag{1}$$

$$\nabla \cdot \mathbf{u} = 0 \tag{2}$$

are considered in a region $\Omega \in \mathbb{R}^n$, with boundary conditions

$$B(\mathbf{u}, p) = 0 \qquad \text{on} \quad \partial \Omega. \tag{3}$$

Typical boundary conditions might be those for a solid wall:

$$\mathbf{u} \cdot \hat{\mathbf{n}} = 0 \quad \text{``no flow''} \tag{4a}$$

$$\mathbf{u} \cdot \hat{\tau} = 0 \quad \text{``no slip''}. \tag{4b}$$

where local normal and tangential vectors to the wall are given by $\hat{\mathbf{n}}$ and $\hat{\tau}$. Subscripts denote partial differentiation. Specifying the pressure, its normal derivative, or a combination of the two at outflow is also a possibility:

$$\alpha p + \beta \hat{n} \cdot \nabla p = g. \tag{5}$$

For a discussion of allowable boundary conditions see, e.g. [9].

---

[*] This work was performed under the auspices of the U.S. Department of Energy by University of California Lawrence Livermore National Laboratory under contract No. W-7405-ENG-48

## 2 Projection methods

Projection methods, or "fractional step" methods, as they are sometimes called, advance the momentum equation (1) and enforce the continuity condition (2) in separate steps [1,4,10,13]. These methods make use of the Hodge decomposition theorem, which states that any vector function $\mathbf{v}(\mathbf{x})$ can be decomposed into a divergence-free part $\mathbf{u}$ plus the gradient of a scalar potential $\phi$, i.e.

$$\mathbf{v}(\mathbf{x}) = \mathbf{u}(\mathbf{x}) + \nabla\phi(\mathbf{x}) \tag{6}$$

with $\nabla \cdot \mathbf{u} = 0$, where furthermore, using a suitable inner product, $(\mathbf{u}, \nabla\phi) = 0$, i.e. the two parts are orthogonal. In order for the decomposition to be unique, boundary conditions must be specified as well. For the purposes of this paper, we choose to specify the normal component of the velocity, i.e.

$$\hat{\mathbf{n}} \cdot \mathbf{u} = u_b. \tag{7}$$

Thus, the divergence-free part of an arbitrary vector $\mathbf{v}$ can be obtained by a projection onto the orthogonal subspace of divergence-free vectors by removing the gradient of an appropriately chosen scalar potential $\phi$. The notation

$$\mathbf{u} = \mathbf{P}(\mathbf{v}) \tag{8}$$

is sometimes used to express this projection.

Using this information, one is naturally led to an approach whereby an approximation to the momentum equation

$$\mathbf{u}^*_t + (\mathbf{u} \cdot \nabla)\mathbf{u} + \nabla q = \nu\nabla^2\mathbf{u}^*, \tag{9}$$

is advanced for some interval of time $t_o <= t <= t_1$, and the divergence-free velocity is computed when needed using the projection

$$\mathbf{u}(t) = \mathbf{P}(\mathbf{u}^*(t)). \tag{10}$$

Here $\nabla q(\mathbf{x}, t)$ is some approximation to the pressure gradient, which may even be zero (see e.g. [10]). As a practical matter, the projection is effected by deriving an elliptic equation as follows. Applying the Hodge decomposition theorem, we can write

$$\mathbf{u}^* = \mathbf{u} + \nabla\phi. \tag{11}$$

Taking the divergence of (11) and applying (2), results in the elliptic constraint equation for $\phi$

$$\nabla^2\phi = \nabla \cdot \mathbf{u}^*. \tag{12}$$

The pressure can be recovered at any time using the formula

$$\nabla p = \nabla(q + \phi_t) - \nu\nabla^2\nabla\phi, \tag{13}$$

which is derived by substituting (11) into (9) and comparing with the original momentum equation (1). Common practice for projection methods is to advance (9) for a single timestep, compute $\mathbf{u}$ at the end of the timestep using the projection, and then replace $\mathbf{u}^*$ with this new value of $\mathbf{u}$ at the beginning of the next timestep. When $\mathbf{u}^*$ is reset to $\mathbf{u}$ at the end of each timestep, it always stays relatively close to $\mathbf{u}$ during the computation. Various projection methods discussed in the literature differ in their approximation of the advective terms $(\mathbf{u} \cdot \nabla)\mathbf{u}$, the approximation used for $q$, whether (9) is advanced explicitly or implicitly, and how the pressure gradient update formula (13) is approximated.

An alternative class of methods does not reset $\mathbf{u}^*$ at the end of each timestep, but allows it to evolve over the period of the computation. Such methods, known variously as "magnetization", "impulse" or "gauge" methods have been developed by Buttke [3], Cortez [5,6], E and Liu [7,8], Recchioni and Russo [11] and Summers and Chorin [12]. One advantage of such methods is that the approximation $q = 0$ can be used, and the pressure never need be determined unless pressure-dependent boundary conditions are required. Note also that the demonstrated success of these methods indicates that the length of time that the approximate momentum equation (9) can be integrated need not be restricted to a single timestep. These methods will not be discussed further in the present paper.

## 3   Boundary conditions when pressure term is included in predictor

Boundary conditions are required on $\phi$ in order to solve (12). In addition, boundary conditions on $\mathbf{u}^*$ are required if (9) is to be advanced implicitly and also to compute the right-hand side of (12). Since both $\mathbf{u}^*$ and $\phi$ are auxiliary variables, the original problem formulation does not tell how to set their boundary values. However it is clear that any boundary conditions that are specified must satisfy (11) as a constraint, i.e.

$$\nabla\phi|_{\partial\Omega} = (\mathbf{u}^* - \mathbf{u})|_{\partial\Omega}. \tag{14}$$

Since both $\mathbf{u}^*$ and $\mathbf{u}$ are known at the point that the elliptic equation (12) is solved, and given our choice for the boundary condition form (7), the appropriate boundary condition for this equation is therefore

$$\hat{\mathbf{n}} \cdot \nabla\phi|_{\partial\Omega} = \hat{\mathbf{n}} \cdot (\mathbf{u}^* - \mathbf{u})|_{\partial\Omega}. \tag{15}$$

The boundary conditions on $\mathbf{u}^*$ are more problematic since they are needed before $\phi$ is computed. As reported in [2], the correct choice depends on the approximation for $\nabla q$ used in (9). Bell, Colella and Glaz [1] advance (9) using Crank-Nicolson for time integration, approximate the advective terms with a Godunov procedure, and use the time-centered pressure gradient from the previous timestep to approximate $\nabla q$:

$$\frac{(\mathbf{u}^* - \mathbf{u}^n)}{\Delta t} + ((\mathbf{u} \cdot \nabla)\mathbf{u})^{n+\frac{1}{2}} + \nabla p^{n-\frac{1}{2}} = \frac{\nu}{2}(\nabla^2\mathbf{u}^* + \nabla^2\mathbf{u}^n) \tag{16}$$

Here, superscripts involving $n$ denote the time level. Since $\nabla p^{n-\frac{1}{2}} = \nabla p^{n+\frac{1}{2}} + \mathcal{O}(\Delta t)$, (16) implies that $\mathbf{u}^* = \mathbf{u}^{n+1} + \mathcal{O}(\Delta t^2)$, and so it appears plausible that a second-order method can be obtained simply by using the original specified boundary conditions for $\mathbf{u}$ as boundary conditions for $\mathbf{u}^*$, i.e.

$$\mathbf{u}^*|_{\partial\Omega} = \mathbf{u}^{n+1}|_{\partial\Omega}. \tag{17}$$

Substituting this condition into (15), it is apparent that the boundary condition for $\phi$ becomes particularly simple in this case as well:

$$\hat{\mathbf{n}} \cdot \nabla\phi|_{\partial\Omega} = 0. \tag{18}$$

In fact in [2], normal mode analysis is used to show that if an appropriate approximation to the pressure gradient update formula (13) is used, namely

$$\nabla p^{n+\frac{1}{2}} = \nabla p^{n-\frac{1}{2}} + \frac{1}{\Delta t}(\nabla\phi - \frac{\nu\Delta t}{2}\nabla^2\nabla\phi) \tag{19}$$

both the velocity $\mathbf{u}$ and the pressure $p$ converge to second order in $\Delta t$.

## 4    The pressure gradient update formula

Implementing (19) is problematical because of the term involving $\nabla^2\nabla\phi$. Near boundaries, for example, formulation of an appropriately accurate discretization of this term can be difficult. Removing this term altogether apparently involves only an $\mathcal{O}(\Delta t)$ perturbation to the formula for $\nabla p^{n+\frac{1}{2}}$. Again, by inspection of (9), It might therefore be reasonable to expect that using the approximation

$$\nabla p^{n+\frac{1}{2}} = \nabla p^{n-\frac{1}{2}} + \frac{1}{\Delta t}\nabla\phi, \tag{20}$$

as is done in [1] and related papers, would lead to a second-order result for the velocity. It remains the case, however, that this is a first order perturbation to the formula for the pressure, so second-order convergence of the pressure would not be expected. Since the pressure and velocity are coupled by the Navier-Stokes equations, it would also seem doubtful that the velocity would be computed to second-order by this method. Surprisingly, in [2], it has been both proven using normal mode analysis and demonstrated numerically that using (20) in conjunction with (16), while it leads to only first order convergence of the pressure, does result in second-order convergence of the velocity. The first order convergence of the pressure results from a spurious mode in the pressure that converges like $\mathcal{O}(\nu\Delta t)$. This mode can only be annihilated by using the update formula for the pressure given in (19).

## 5    Boundary conditions when pressure term is set to zero in predictor

Provided that boundary conditions can be chosen appropriately, any convenient approximation can be used for the $\nabla q$ term in (9) since the subsequent projection step will always pick out the correct potential gradient to give a divergence-free velocity. The method proposed by Kim and Moin [10] uses $q = 0$, for example. Finding appropriate boundary conditions for $\mathbf{u}^*$ and $\phi$ become more difficult in this case, since it is no longer true that $\mathbf{u}^*$ is an $\mathcal{O}(\Delta t^2)$ perturbation of $\mathbf{u}$. The normal mode analysis presented in [2] suggests again that as long as the compatibility condition (14) is satisfied, second-order convergence should be possible. Since the compatibility condition is only one constraint on the boundary conditions, a convenient choice such as $\hat{\mathbf{n}} \cdot \nabla\phi = 0$ seems desirable, leading to the proposed boundary conditions

$$\hat{\mathbf{n}} \cdot \mathbf{u}^*|_{\partial\Omega} = \hat{\mathbf{n}} \cdot \mathbf{u}^{n+1}|_{\partial\Omega} \tag{21a}$$

$$\hat{\tau} \cdot \mathbf{u}^*|_{\partial\Omega} = \tau \cdot (\mathbf{u}^{n+1} + \nabla\tilde{\phi}^{n+1})|_{\partial\Omega} \tag{21b}$$

$$\hat{\mathbf{n}} \cdot \nabla\phi^{n+1}|_{\partial\Omega} = 0 \tag{21c}$$

Here, $\nabla\tilde{\phi}^{n+1}$ is an approximation to $\nabla\phi^{n+1}$. According to the normal mode analysis [2], a first order approximation to $\nabla\phi^{n+1}$ is required in order to obtain second-order convergence, and so

$$\nabla\tilde{\phi}^{n+1} = \nabla\phi^n \tag{22}$$

is used. Numerical studies using these boundary conditions, however, demonstrate only first-order convergence for the pressure in many cases even when using the improved formula

$$p^{n+\frac{1}{2}} = \phi^{n+1} - \frac{\nu\Delta t}{2}\nabla^2\phi \tag{23}$$

for the pressure. The problem with using these boundary conditions lies in the fact that the function $\mathbf{u}^*$ is not smooth near boundaries in all cases. Since $\nabla^2\phi = \nabla\cdot\mathbf{u}^*$, (23) implies that the pressure will not be smooth either near boundaries.

Rather than choosing homogeneous Neumann boundary conditions for $\phi$ in this case, a better idea is to choose the boundary condition for $\mathbf{u}^*$ in such a way as to guarantee smoothness of that function up to the boundary. Instead of (21a), we extrapolate $\hat{\mathbf{n}} \cdot \mathbf{u}^*$ to the boundary using at least a second-order extrapolation formula. The boundary conditions become

$$E\hat{\mathbf{n}} \cdot \mathbf{u}^*|_{\partial\Omega} = 0 \tag{24a}$$

$$\hat{\tau} \cdot \mathbf{u}^*|_{\partial\Omega} = \tau \cdot (\mathbf{u}^{n+1} + \nabla\tilde{\phi}^{n+1})|_{\partial\Omega} \tag{24b}$$

$$\hat{\mathbf{n}} \cdot \nabla\tilde{\phi}|_{\partial\Omega} = \hat{\mathbf{n}}(\mathbf{u}^* - \mathbf{u})|_{\partial\Omega} \tag{24c}$$

While an inhomogenous Neumann condition is now required for the elliptic problem for $\phi$, the corresponding numerical experiments demonstrate fully second-order convergence for both the velocity and pressure, justifying the additional effort.

## 6 Summary of results

Table 1 summarizes the convergence results reported here and in [2]. The columns labeled "boundary conditions" indicate what the inhomogeneous terms are in the boundary conditions for $\mathbf{u}^*$ and $\hat{\mathbf{n}} \cdot \nabla \phi$. The columns labeled "difference approx." indicate the choice for $q$ and the formula used for the pressure update or evaluation. The number shown in the columns labeled "conv. rate" are the exponent in the observed convergence rate in time for the resulting combination of boundary conditions and difference approximation.

**Table 1.** The convergence rate for the pressure depends upon the choice of boundary conditions and pressure updates in the projection scheme.

| Boundary Conditions | | | | Difference Approx. | Conv. Rate | |
|---|---|---|---|---|---|---|
| $\hat{\mathbf{n}} \cdot \mathbf{u}^*\|_{\partial\Omega}$ | $\hat{\tau} \cdot \mathbf{u}^*\|_{\partial\Omega}$ | $\hat{\mathbf{n}} \cdot \nabla\phi^{n+1}\|_{\partial\Omega}$ | $q$ | $p$ update | $u$ | $p$ |
| $\hat{\mathbf{n}} \cdot \mathbf{u}^{n+1}$ | $\hat{\tau} \cdot \mathbf{u}^{n+1}$ | $0$ | $p^{n-\frac{1}{2}}$ | (20) | 2 | 1 |
| $\hat{\mathbf{n}} \cdot \mathbf{u}^{n+1}$ | $\hat{\tau} \cdot \mathbf{u}^{n+1}$ | $0$ | $p^{n-\frac{1}{2}}$ | (19) | 2 | 2 |
| $\hat{\mathbf{n}} \cdot \mathbf{u}^{n+1}$ | $\hat{\tau} \cdot (\mathbf{u}^{n+1} + \nabla\phi^n)$ | $0$ | $0$ | (23) | 2 | 1 |
| Extrapolate | $\hat{\tau} \cdot (\mathbf{u}^{n+1} + \nabla\phi^n)$ | $\hat{\mathbf{n}} \cdot (\mathbf{u}^* - \mathbf{u}^{n+1})$ | $0$ | (23) | 2 | 2 |

# 7   Acknowledgements

The author acknowledges extensive discussions with Michael Minion, Ricardo Cortez and Bill Henshaw during the preparation of [2] which form the basis for the present summary paper.

# References


1. J. B. BELL, P. COLELLA, AND H. M. GLAZ, *A second order projection method for the incompressible Navier-Stokes equations*, J. Comp. Phys., 85 (1989), pp. 257–283.

2. D. L. BROWN, R. CORTEZ, AND M. L. MINION, *Accurate projection methods for the incompressible Navier–Stokes equations*, J. Comp. Physics, 168 (2001), pp. 464–499.

3. T. F. BUTTKE, *Velicity methods: Lagrangian numerical methods which preserve the Hamiltonian structure of incompressible fluid flow*, in Vortex Flows and Related Numerical Methods, J. T. Beale, G.-H. Cottet, and S. Huberson, eds., Kluwer Academic Publishers, 1993, pp. 39–57. NATO ASI Series C, vol. 395.

4. A. J. CHORIN, *Numerical solution of the Navier-Stokes equations*, Math. Comp., 22 (1968), pp. 745–762.

5. R. CORTEZ, *Impulse-based Particle Methods for Fluid Flow*, PhD thesis, University of California, Berkeley, May 1995.

6. ———, *An impulse-based approximation of fluid motion due to boundary forces*, J. Comp. Phys., 123 (1996), pp. 341–353.

7. W. E AND J. GUO LIU, *Gauge method for viscous incompressible flows.* unpublished, 1996.

8. ———, *Finite difference schemes for incompressible flows in the velocity-impulse density formulation*, J. Comp. Phys., 130 (1997), pp. 67–76.

9. W. D. HENSHAW, H.-O. KREISS, AND L. REYNA, *A fourth-order accurate difference approximation for the incompressible Navier-Stokes equations*, Computers and Fluids, 23 (1994), pp. 575–593.

10. J. KIM AND P. MOIN, *Application of a fractional-step method to incompressible Navier-Stokes equations*, J. Comp. Phys., 59 (1985), pp. 308–323.

11. M. C. RECCHIONI AND G. RUSSO, *Hamilton-based numerical methods for a fluid-membrane interaction in two and three dimensions*, SIAM J. Sci. Comput., 19 (1998), pp. 861–892.

12. D. M. SUMMERS AND A. J. CHORIN, *Numerical vorticity creation based on impulse conservation*, Proc. Nat. Acad. Sci. USA, 93 (1996), pp. 1881–1885.

13. J. VAN KAN, *A second-order accurate pressure-correction scheme for viscous incompressible flow*, SIAM J. Sci. Comput., 7 (1986), pp. 870–891.

# A Split-Step Scheme for the Incompressible Navier-Stokes Equations

William D. Henshaw and N. Anders Petersson

Centre for Applied Scientific Computing,
Lawrence Livermore National Laboratory,
Livermore, CA, 94551,
henshaw@llnl.gov, andersp@llnl.gov

October 1, 2001

**Abstract.** We describe a split-step finite-difference scheme for solving the incompressible Navier-Stokes equations on composite overlapping grids. The split-step approach decouples the solution of the velocity variables from the solution of the pressure. The scheme is based on the velocity-pressure formulation and uses a method of lines approach so that a variety of implicit or explicit time stepping schemes can be used once the equations have been discretized in space. We have implemented both second-order and fourth-order accurate spatial approximations that can be used with implicit or explicit time stepping methods. We describe how to choose appropriate boundary conditions for the pressure to make the scheme accurate and stable. A divergence damping term is added to the pressure equation to keep the numerical dilatation small. Several numerical examples are presented.

## 1   Introduction

We consider solving the incompressible Navier-Stokes equations with finite difference methods on composite overlapping grids using an accurate and stable split-step approach that decouples the solution of the velocity from the solution of the pressure. Second-order and fourth-order accuracy has been achieved in space and second or higher order accuracy in time can be easily accomplished.

In primitive-variables the initial-boundary-value problem (IBVP) for the incompressible Navier-Stokes equations is

$$\begin{aligned}
\partial \mathbf{u}/\partial t + (\mathbf{u} \cdot \nabla)\mathbf{u} + \nabla p &= \nu \Delta \mathbf{u} + \mathbf{F}, & \text{for } \mathbf{x} \in \Omega, & \quad t > 0, \\
\nabla \cdot \mathbf{u} &= 0, & \text{for } \mathbf{x} \in \overline{\Omega}, & \quad t > 0, \\
B(\mathbf{u}, p) &= \mathbf{g}, & \text{for } \mathbf{x} \in \partial\Omega, & \quad t > 0, \\
\mathbf{u}(\mathbf{x}, 0) &= \mathbf{f}(\mathbf{x}), & \text{for } \mathbf{x} \in \Omega.
\end{aligned} \tag{1}$$

Here $\mathbf{u} = (u_1, u_2, u_3)$ is the velocity, $p$ is the kinematic pressure (pressure divided by the constant density), $\nu$ is the kinematic viscosity, $\mathbf{F}$ is the forcing per unit volume, $\Omega$ is a bounded open domain in $R^d$, $d = 2$ or $d = 3$, and $\partial\Omega$ is the boundary of $\Omega$. The initial conditions should satisfy $\nabla \cdot \mathbf{f} = 0$. There are $d$ boundary conditions denoted by $B(\mathbf{u}, p) = 0$. On a fixed wall, for example, the boundary conditions are $\mathbf{u} = 0$. We assume that all data are sufficiently smooth and compatible. Depending on the boundary conditions the pressure may only be determined up to a constant in which case we impose the average pressure to be zero. We refer to the formulation (1) as the "velocity-divergence" formulation.

An alternative form of this IBVP, which we call the "velocity-pressure" formulation, is

$$
\begin{aligned}
\partial \mathbf{u}/\partial t + (\mathbf{u} \cdot \nabla)\mathbf{u} + \nabla p &= \nu \Delta \mathbf{u} + \mathbf{F}, & &\text{for } \mathbf{x} \in \Omega, \quad t > 0, \\
\Delta p + J(\nabla \mathbf{u}) - \alpha(\mathbf{x})\nabla \cdot \mathbf{u} &= \nabla \cdot \mathbf{F}, & &\text{for } \mathbf{x} \in \Omega, \quad t \geq 0, \\
B(\mathbf{u}, p) &= \mathbf{g}, & &\text{for } \mathbf{x} \in \partial\Omega, \quad t > 0, \\
\nabla \cdot \mathbf{u} &= 0, & &\text{for } \mathbf{x} \in \partial\Omega, \\
\mathbf{u}(\mathbf{x}, 0) &= \mathbf{f}(\mathbf{x}), & &\text{for } \mathbf{x} \in \Omega,
\end{aligned}
\tag{2}
$$

where

$$
J(\nabla \mathbf{u}) = \sum_{m=1}^{d} \nabla u_m \cdot \frac{\partial \mathbf{u}}{\partial x_m}.
$$

Here, the equation $\nabla \cdot \mathbf{u} = 0$ in (1) has been replaced by an elliptic equation for the pressure, which is obtained by taking the divergence of the momentum equations, and using $\nabla \cdot \mathbf{u} = 0$. The Poisson equation for the pressure needs an additional boundary condition and the boundary condition on the dilatation in (2) fills this purpose. Although $\nabla \cdot \mathbf{u} = 0$ does not look like a boundary condition for $p$, once the equations have been discretized the connection will become apparent. We have also added the **divergence damping** term,

$$
\alpha(\mathbf{x})\nabla \cdot \mathbf{u} , \qquad \alpha \geq 0 ,
$$

to the pressure equation. In the continuous case this term has no effect, but in the discrete case this term will be important to keep the dilatation small. To see why this term might be important we can write down the equation satisfied by the dilatation, $\delta = \nabla \cdot \mathbf{u}$, formed by taking the divergence of the momentum equation,

$$
\partial \delta/\partial t + (\mathbf{u} \cdot \nabla)\delta = \nu \Delta \delta - \alpha \delta.
\tag{3}
$$

The term we have added to the pressure equation appears as a linear damping term in the evolution equation for the divergence. Note that $\delta(\mathbf{x}, t)$ will be identically zero for all times since the initial condition is $\delta(\mathbf{x}, 0) = 0$ and the boundary condition is $\delta(\mathbf{x}, t) = 0$. This observation motivates the extra boundary condition $\nabla \cdot \mathbf{u} = 0$, since it forces the dilatation to be zero everywhere for all times and thus guarantees that solutions of the velocity-pressure system (2) also satisfy the velocity-divergence equations (1).

The incompressible Navier-Stokes equations in primitive variables can be discretized in a variety of ways. Harlow and Welch [8] provided what was perhaps the first discretization with the MAC technique using staggered grids. Later the projection method was devised by Chorin[5] and independently by Temam[25]. The projection method was extended to an implicit fractional-step

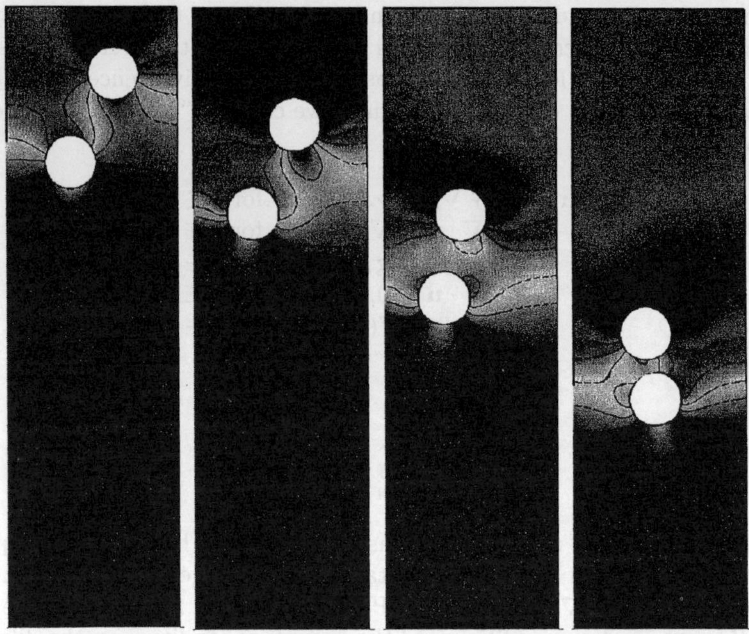

**Fig. 1.** Two falling bodies in an incompressible flow. The overlapping grids are recomputed at every time-step.

method by Kim and Moin[15]. The method of artificial compressibility, introduced by Chorin [4], is another popular approach. For example, this method was used by Kiris et al. [16] to compute the flow in an artifical heart. There have been numerous other approaches developed based on finite-difference, finite-volume, finite-element and spectral-element discretizations, such as [1][2][14][26][17][23][9], to name a few.

There are a number of fundamental issues that must be dealt with when designing a scheme for the incompressible Navier-Stokes equations:

- The pressure should be free of spurious oscillations. Straight-forward discretizations of (1) can lead to the checker-board instability (corresponding to a violation of the Babuška-Brezzi condition in finite-elements).
- Many approaches require extra boundary conditions, either for the pressure or for an intermediate velocity field, which can be non-trivial to choose.
- If the pressure is only determined up to a constant (for example when Neumann boundary conditions are enforced on $\partial\Omega$), there will be a compatibility condition on the data for the pressure equation.
- For efficiency it is useful to decouple the solution of the velocity from the solution of the pressure.
- The discrete divergence should be small, i.e., of the order of the truncation error in the numerical method.

There have been long discussions in the literature related to these issues, especially concerning boundary conditions for the pressure [7,18,14,23] and whether fractional-step projection methods are inherently first-order accurate in the pressure [19,24,22,6]. We refer to Brown et. al. [3] for a discussion of how to get second-order accuracy in the pressure with the fractional-step projection method. In the present paper, we summarize the results of our research, and describe how the above four issues are handled in our approach.

Here we describe a straight-forward approach that leads to an efficient second (or higher) order accurate scheme in both the velocity and the pressure. We use a method of lines approach to discretize the velocity-pressure formulation (2). We begin by discretizing in space. For ease of presentation we consider solving the equations in two space dimensions on a square grid $\mathbf{G}$, with grid spacing $h = 1/N$, for $N$ a positive integer:

$$\mathbf{G} = \{\mathbf{x_i} = (x_\mathbf{i}, y_\mathbf{i}) = (ih, jh) \qquad i, j = -1, 0, 1, \ldots, N+1\}$$

Here $\mathbf{i} = (i, j)$ is a multi-index. We include ghost points at the boundaries to aid in the discretization. To be specific, we consider a Dirichlet boundary condition for the velocity:

$$\mathbf{u}(\mathbf{x}, t) = \mathbf{g}(\mathbf{x}, t). \tag{4}$$

We first discretize in space. Let $(\mathbf{U_i}(t), P_\mathbf{i}(t))$ be the numerical approximation to $(\mathbf{u}(\mathbf{x}, t), p(\mathbf{x}, t))$ with $\mathbf{U_i}(t) = (U_\mathbf{i}(t), V_\mathbf{i}(t))$. The spatial approximation is

$$d\mathbf{U_i}/dt = -(\mathbf{U_i} \cdot \nabla_h)\mathbf{U_i} - \nabla_h P_\mathbf{i} + \nu \Delta_h \mathbf{U_i} + \mathbf{F_i}, \quad i, j = 1, 2, \ldots, N-1 \tag{5}$$

$$\Delta_h P_\mathbf{i} = \alpha_\mathbf{i} \nabla_h \cdot \mathbf{U_i} - J(\nabla_h \mathbf{U_i}) + \nabla_h \cdot \mathbf{F_i}, \qquad i, j = 0, 1, 2, \ldots, N \tag{6}$$

$$\mathbf{U_i} = \mathbf{g}(\mathbf{x_i}, t) \equiv (g^u(\mathbf{x_i}, t), g^v(\mathbf{x_i}, t)) \qquad i = 0, j = 0, 1, 2, \ldots, N \tag{7}$$

$$D_{0x} U_\mathbf{i} = -D_{0y} g^v_\mathbf{i} \qquad i = 0, j = 0, 1, 2, \ldots, N \tag{8}$$

The boundary conditions have only been specified at $x = 0$; similar expressions will hold at the other boundaries. The operators $\nabla_h$, $\Delta_h$ are the standard centered difference approximations to $\nabla$ and $\Delta$:

$$\nabla_h \cdot \mathbf{U_i} = D_{0x} \mathbf{U_i} + D_{0y} \mathbf{U_i}, \qquad \Delta_h \mathbf{U_i} = (D_{+x} D_{-x} + D_{+y} D_{-y}) \mathbf{U_i},$$

$$D_{0x} \mathbf{U_i} = \frac{\mathbf{U}_{i+1,j} - \mathbf{U}_{i-1,j}}{2h}, \qquad D_{0y} \mathbf{U_i} = \frac{\mathbf{U}_{i,j+1} - \mathbf{U}_{i,j-1}}{2h},$$

$$D_{+x} \mathbf{U_i} = \frac{\mathbf{U}_{i+1,j} - \mathbf{U}_{i,j}}{h}, \qquad D_{-x} \mathbf{U_i} = \frac{\mathbf{U}_{i,j} - \mathbf{U}_{i-1,j}}{h}.$$

We avoid the checker-board instability problem since we have directly discretized the pressure equation using a compact difference approximation. The discrete approximation will require extra numerical boundary conditions. Applying the general principle for deriving numerical boundary conditions described in [13], we use the equations themselves to tell how the solution should behave at the boundary. Note for example, that the pressure equation (6) is applied on the boundary, $i = 0$. As a numerical boundary condition, we could also apply the momentum equations on the boundary and thus determine the values for $\mathbf{U}_{-1,j}$. However, in order to keep the solution of the pressure decoupled from the velocity, we instead only apply the normal component of the momentum equation on the boundary,

$$\frac{\partial p}{\partial n} = \mathbf{n} \cdot (-\mathbf{g}_t - (\mathbf{g} \cdot \nabla)\mathbf{u} + \nu \Delta \mathbf{u}), \tag{9}$$

and extrapolate the tangential component of the velocity,

$$D_{+x}^3 V_{-1j} \equiv V_{-1j} - 3V_{0j} + 3V_{1j} - V_{2j} = 0. \tag{10}$$

We call (9) the div-grad pressure boundary condition. Note that by itself it adds no new information to the continuous PDE and cannot replace $\nabla \cdot \mathbf{u} = 0$ as the extra boundary condition required by the velocity-pressure formulation. After discretization, (9) becomes

$$D_{0x} P_{0j} = \nu D_{+x} D_{-x} U_{0j}, \quad j = 0, 1, \ldots, N, \tag{11}$$

where we assumed $\mathbf{g} = 0$ for simplicity. We can now see how $\nabla \cdot \mathbf{u} = 0$ provides a boundary condition for the pressure: the discrete divergence boundary condition (8) determines the ghost line value of the normal component of the velocity, $U_{-1,j}$, which is used in (11). By eliminating that ghost line value, we obtain a special stencil for the right hand side:

$$D_{0x} P_{0j} = \nu \frac{2}{h} D_{+x} U_{0j}, \quad j = 0, 1, \ldots, N.$$

We believe that the reason (9) is not more widely used is due to the fact that applying a discrete form of (9) can easily lead to a unstable method since the second-derivative of $\mathbf{u}$ appears on the right hand side. To achieve a stable scheme using (9), it is extremely important to also enforce the essential boundary condition $\nabla \cdot \mathbf{u} = 0$.

Equations (5-8, 11, 10) can be solved with a method of lines approach. If we wish to have a split-step scheme where the solution of the pressure equation is decoupled from the solution of the velocity components, we should choose a time stepping scheme for the velocity components that only involves the pressure from previous time steps. Let us introduce the operators $L = L_E + L_I$ representing various terms in the momentum equation:

$$L\mathbf{U_i} = -(\mathbf{U_i} \cdot \nabla_h)\mathbf{U_i} - \nabla_h P_\mathbf{i} + \nu \Delta_h \mathbf{U_i},$$
$$L_E \mathbf{U_i} \equiv -(\mathbf{U_i} \cdot \nabla_h)\mathbf{U_i} - \nabla_h P_\mathbf{i},$$
$$L_I \mathbf{U_i} \equiv \nu \Delta_h \mathbf{U_i}.$$

$L_I$ and $L_E$ will be the parts of the operator that we treat implicitly and explicitly, respectively.

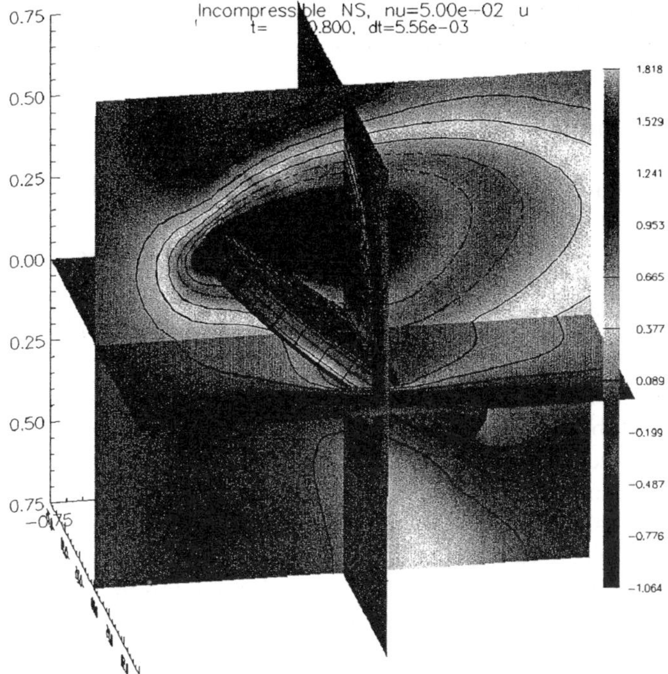



**Fig. 2.** Flow past a rotating disk.

## 1.1   An explicit scheme

As a first example we solve these equations with the explicit second-order Adams-Bashforth scheme. In this approach we first advance the velocity using

$$\frac{\mathbf{U}_\mathbf{i}^{n+1} - \mathbf{U}_\mathbf{i}^{n}}{\Delta t} = \frac{3}{2}(L\mathbf{U}_\mathbf{i}^{n} + \mathbf{F}_\mathbf{i}^{n}) - \frac{1}{2}(L\mathbf{U}_\mathbf{i}^{n-1} + \mathbf{F}_\mathbf{i}^{n-1}), \quad i,j = 1,\ldots,N-1, \quad (12)$$

$$\mathbf{U}_\mathbf{i}^{n+1} = \mathbf{g}(\mathbf{x}_\mathbf{i}, t^{n+1}), \qquad\qquad i = 0, j = 0,\ldots,N, \quad (13)$$

$$D_{0x}U_\mathbf{i}^{n+1} = -D_{0y}g^{v}(\mathbf{x}_\mathbf{i}, t^{n+1}), \qquad\qquad i = 0, j = 0,\ldots,N, \quad (14)$$

$$D_{+x}^{3}V_{-1j}^{n+1} = 0, \qquad\qquad j = 0,\ldots,N. \quad (15)$$

Here $\mathbf{U}_\mathbf{i}^{n} \approx \mathbf{u}(\mathbf{x}_\mathbf{i}, n\Delta t)$. These equations determine $\mathbf{U}_\mathbf{i}^{n+1}$ at all points including the ghost points. We then solve for the pressure from

$$\Delta_h P_\mathbf{i}^{n+1} - \alpha_\mathbf{i} \nabla_h \cdot \mathbf{U}_\mathbf{i}^{n+1} + J(\nabla_h \mathbf{U}_\mathbf{i}^{n+1}) = 0, \qquad i,j = 0,1,\ldots,N, \quad (16)$$

$$D_{0x}P_\mathbf{i}^{n+1} = \nu D_{+x}D_{-x}U_\mathbf{i}^{n+1} + B_p(\mathbf{U}_\mathbf{i}^{n+1}, \mathbf{g}_\mathbf{i}^{n+1}), \quad i = 0, j = 0,1,\ldots,N, \quad (17)$$

where the boundary forcing satisfies

$$B_p(\mathbf{U}, \mathbf{g}) = -\frac{\partial g^{u}}{\partial t} - g^{u}D_{0x}U - g^{v}D_{0y}g^{v} + \nu D_{+y}D_{-y}g^{u}.$$

To improve the stability properties of the time-integrator, we often use the above scheme as the predictor followed by a second-order Adams-Moulton corrector.

## 1.2   A semi-implicit scheme

As another example, we have implemented a semi-implicit method that combines a Crank-Nicholson treatment for the viscous terms and an Adams-Bashforth approach for the advection terms and pressure. Instead of equation (12) we use

$$\frac{\mathbf{U}_i^{n+1} - \mathbf{U}_i^n}{\Delta t} = \frac{3}{2}(L_E\mathbf{U}_i^n + \mathbf{F}_i^n) - \frac{1}{2}(L_E\mathbf{U}_i^{n-1} + \mathbf{F}_i^{n-1}) + \frac{1}{2}(L_I\mathbf{U}_i^{n+1} + L_I\mathbf{U}_i^n)$$

together with equations (13-15) to advance the velocity in time. Again, we can improve the stability properties by using the above scheme as the predictor followed by a second order Adams-Moulton corrector.

Since the boundary condition for the pressure (17) depends on $\nu/h^2$ and since the pressure is taken explicitly in the time stepping scheme, one might suspect that $\Delta t$ must depend on the ratio $h^2/\nu$. This is indeed the case when a discrete version of (9) is used as the boundary condition for the pressure. For implicit time stepping, a more stable pressure boundary condition is formed by explicitly removing the dilatation on the boundary in the highest order term, $\Delta\mathbf{u}$, by using the vector identity

$$\Delta\mathbf{u} = \nabla(\nabla \cdot \mathbf{u}) - \nabla \times \nabla \times \mathbf{u},$$

together with $\nabla(\nabla \cdot \mathbf{u}) = 0$, to give

$$\frac{\partial p}{\partial n} = \mathbf{n} \cdot (-\mathbf{g}_t - (\mathbf{g} \cdot \nabla\mathbf{u}) - \nu\nabla \times \nabla \times \mathbf{u}). \tag{18}$$

We call this the curl-curl boundary condition for the pressure. This boundary condition was apparently first advocated by Karniadakis, Israeli and Orszag [14] as the appropriate boundary condition for the pressure. Unlike equation (9) this new equation does add new information to the system and can be used as an alternative boundary condition to $\nabla \cdot \mathbf{u} = 0$. Note that the boundary condition (18) is specifying the normal derivative of the dilatation to be zero on the boundary since (18) was derived from (9) using

$$\frac{\partial}{\partial n}(\nabla \cdot \mathbf{u}) = 0. \tag{19}$$

Referring back to equation (3) we see that the above boundary condition, together with the initial condition $\delta(\mathbf{x}, 0) = 0$, will also force the dilatation to be zero for all time.

On the square grid $\mathbf{G}$, the discrete version of (18) becomes

$$D_{0x}P_i^{n+1} = -\nu D_{0x}D_{0y}V_i^{n+1} + B_p(\mathbf{U}_i^{n+1}, \mathbf{g}_i^{n+1}), \quad i = 0, j = 0, 1, \ldots, N, \tag{20}$$

Note that the stencil for the viscous term is significantly different in the discretized curl-curl (20) and div-grad (17) pressure boundary conditions; this is the source of the different stability properties of the two boundary conditions. A detailed stability analysis, Petersson [20], shows that the time step is only determined by the advection terms (i.e. is independent of $\nu$) if (17) is replaced by the curl-curl boundary condition (20). This is somewhat remarkable, since the pressure is treated explicitly.

We remark that in our implementation, we use the curl-curl pressure boundary condition both for the explicit and semi-implicit schemes.

It is straight-forward to build time stepping schemes that are accurate to any desired order in time. The reason for this is that in the approach we have taken we have effectively reduced the solution of the incompressible Navier-Stokes equations to solving a system of ODE's for $\mathbf{U_i}$:

$$\frac{d}{dt}\mathbf{U_i} = \mathbf{F}(\mathbf{U_i}, t),$$

since we can treat the pressure simply as a function of the velocity.

The difference approximations presented here can be easily extended to curvilinear grids and to composite overlapping grids. The approximations can be also be made fourth order accurate in space, see [10] for further details.

## 1.3 The compatibility condition for the pressure equation

If the pressure has a Neumann boundary condition on all boundaries the pressure equation will be singular and it is only solvable if the right hand side satisfies a compatibility condition on the velocity. In order to solve this singular system one could, for example, eliminate one equation and replace it with an equation that, for example, sets the value of $p$ at a point or sets the mean value for $p$. Rather than single out a particular equation to be removed we prefer to use a different approach which is better conditioned. If we denote the equation for the pressure as the matrix equation,

$$Ax = b$$

then we solve the augmented system

$$\begin{bmatrix} A & r \\ r^T & 0 \end{bmatrix} \begin{bmatrix} x \\ \beta \end{bmatrix} = \begin{bmatrix} b \\ 0 \end{bmatrix}.$$

Here $r$ is the right null vector of A; the vector with all components equal to one. It is well-known that this augmented system is non-singular and has an unique solution. The last equation will set the mean value of $p$ to zero.

## 1.4 The discrete divergence

In practice it is important to include the divergence damping term, $\alpha_i \nabla_h \cdot \mathbf{U_i}$, in the pressure equation (6). Alternatively, one could apply an extra explicit projection to the solution after every step, but this would require a significant amount of extra work. The damping coefficient $\alpha(\mathbf{x}, t)$ can be chosen to be quite large. When using an explicit time stepping approach, we choose

$$\alpha = C_d \, \frac{\nu}{h^2},$$

where the coefficient $C_d$ is usually taken to be about one. On a rectangular grid this makes $\alpha$ proportional to $1/\Delta t$. Note that on a curvilinear grid the coefficient will vary in space. One might wonder whether this divergence damping term, which is a potentially order one addition to the pressure equation, will destroy the accuracy of the method. In [12] we analyse this damping term using normal-mode stability analysis and show that the method retains it's accuracy even with this term. In practice we find that increasing $C_d$ will result in a decrease of the maximum dilatation (up to a point) but it can also increase the error in the pressure.

A von Neumann stability analysis shows that the allowable time step, $\Delta t$, will depend on the size of $\alpha$. For implicit methods we do not want the damping coefficient to significantly reduce the time step so it is necessary to limit the size of $\alpha$ by $C/\Delta t$ for a constant $C = O(1)$.

## 2  Numerical results

In this section we present some results from a numerical implementation of the scheme described in this paper. The method was implemented in the **OverBlown** flow solver. **OverBlown** is developed using the **Overture** object-oriented framework and can solve the incompressible Navier-Stokes equations on composite overlapping grids in two and three space dimensions. Both **OverBlown** and **Overture** are freely available from the internet at `http://www.llnl.gov/-casc/Overture`. Currently **OverBlown** only has a second-order accurate spatial approximation implemented, although a fourth-order accurate method has previously been developed in Fortran [10]. Details on the discretization and solution procedure, as well as more extensive convergence studies can be found in [11].

As a first test we use the method of analytic solutions to define the following exact solution to the (forced) two-dimensional incompressible Navier-Stokes equations,

$$u_{\text{true}}(x, y, t) = (x^2 + 2xy + y^2) \, T(t),$$

$$v_{\text{true}}(x, y, t) = (x^2 - 2xy - y^2) \, T(t),$$

$$p_{\text{true}}(x, y, t) = (x^2 + \frac{1}{2}xy + y^2 - 1) \, T(t),$$

$$T(t) = 1 + \frac{1}{2}t + \frac{1}{3}t^2.$$

The exact solution has been chosen to be divergence free to simplify the implementation. This test example is extremely useful for our purposes since this

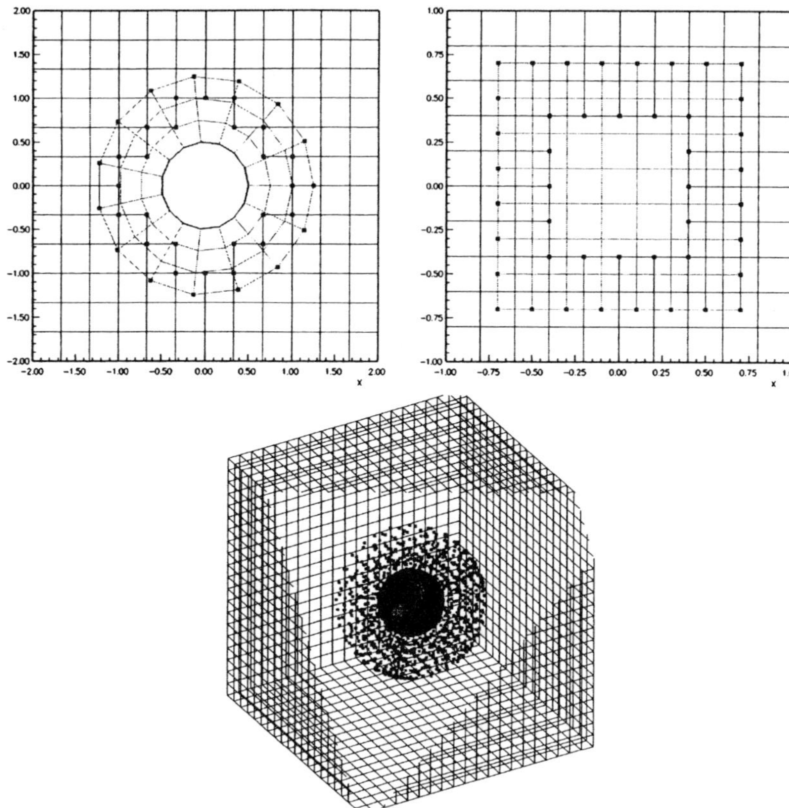

**Fig. 3.** Grids used for the convergence studies: circle-in-a-channel (top-left), square-in-a-square (top right), sphere-in-a-box (bottom left)

solution should also be an *exact solution to the discrete equations* on a rectangular grid. Indeed the errors obtained when solving this problem are of the order of the round-off error for grids consisting of a single square and also for a composite grid consisting of a rotated square in a square, as shown in Table 1. For debugging purposes this is an excellent test since, modulo roundoff errors, the numerical errors should be exactly zero at every time step. For these grids, it is therefore not necessary to run extensive convergence tests using grid refinement to check convergence rates.

**Table 1.** Maximum errors when the analytic solution is a quadratic polynomial, $\nu = .1$, $t = 1.0$. Since the grids are rectangular the method gives the exact answer to within roundoff. The grid "sis" is the square-in-a-square overlapping grid consisting of a square embedded in a larger square shown in figure (3). The 3D grid "rbib" is a rotated box within a box.

| grid | method | $\|p - p_e\|_\infty$ | $\|u - u_e\|_\infty$ | $\|v - v_e\|_\infty$ | $\|w - w_e\|_\infty$ | $\|\nabla \cdot \mathbf{u}\|_\infty$ |
|------|--------|------|------|------|------|------|
| square10 | explicit | $3.1\ 10^{-15}$ | $1.8\ 10^{-15}$ | $8.9\ 10^{-16}$ | | $6.2\ 10^{-15}$ |
| square10 | implicit | $1.2\ 10^{-14}$ | $9.8\ 10^{-15}$ | $3.8\ 10^{-15}$ | | $5.0\ 10^{-14}$ |
| sis | explicit | $1.3\ 10^{-14}$ | $3.6\ 10^{-15}$ | $3.2\ 10^{-15}$ | | $1.2\ 10^{-14}$ |
| box10 | explicit | $1.8\ 10^{-14}$ | $2.7\ 10^{-15}$ | $1.6\ 10^{-15}$ | $1.8\ 10^{-15}$ | $1.6\ 10^{-14}$ |
| rbib | explicit | $1.3\ 10^{-10}$ | $6.4\ 10^{-12}$ | $8.2\ 10^{-12}$ | $1.2\ 10^{-11}$ | $1.1\ 10^{-11}$ |

In three dimensions we use the exact solution

$$u_{\text{true}}(x, y, t) = (x^2 + 2xy + y^2 + xz)\, T(t),$$
$$v_{\text{true}}(x, y, t) = (x^2 - 2xy - y^2 + 3xz)\, T(t),$$
$$w_{\text{true}}(x, y, t) = (x^2 + y^2 - 2z^2)\, T(t),$$
$$p_{\text{true}}(x, y, t) = (x^2 + \frac{1}{2}xy + y^2 + z^2 - 1)\, T(t),$$

with results shown in Table 1 for a box (`box10`) and a rotated-box in a box (`rbib`).

On curvilinear grids, the above polynomial functions are not exact solutions to the discrete equations due to the variable coefficients in the metric terms. Here the order of accuracy can be checked by performing a grid refinement study, see Tables 2-3. In these runs, the convergence rates were estimated by least squares fits to the data.

The discrete pressure equation has an additional divergence damping term, $\alpha \nabla \cdot \mathbf{u}$. To illustrate the effect this term has on the solution we present some convergence studies. We force the equations so that the exact solution is known. In two space dimensions we use

$$\mathbf{u}_{\text{true}}(x, y, t) = \left(\ \sin^2(fx)\sin(2fy)\cos(2\pi t)\ ,\quad -\sin(2fx)\sin^2(fy)\cos(2\pi t)\ \right),$$
$$p_{\text{true}}(x, y, t) = \sin(fx)\sin(fy)\cos(2\pi t).$$

We solve the IBVP with and without the damping term turned on. The domain is taken to be the unit square with all boundaries being walls where the velocity is specified. The results for the fourth-order method are given in Tables 4 and 5, where the maximum errors in $\mathbf{u}$, $p$ and $\nabla \cdot \mathbf{u}$ are reported. The estimated convergence rate $\sigma$, error $\propto h^\sigma$, is also shown. Here, $\sigma$ is estimated by a least squares fit to the maximum errors given in the table.

The results show that although the methods are converging at the expected rates without the damping term, the errors are significantly reduced when damping term is used.

## 2.1 Remarks on the pressure boundary condition

To illustrate the benefits of using the curl-curl boundary condition (18) we compare the numerically determined largest stable time-step to that of the div-grad boundary condition (9), for the semi-implicit time-stepping scheme described in Section 1.2. The results are presented in Table 6. The grid is a circle-in-a-channel and the Reynolds number based on the diameter of the circular cylinder is $Re = 100$. Note that the allowable time step for the curl-curl boundary condition is independent of $\nu/h^2$ so that $\Delta t$ only depends on the advection terms (which are treated explicitly). In contrast, the div-grad boundary condition (9) requires a significantly smaller $\Delta t$ that depends on $\nu/h^2$.

In practice, it is not uncommon for application codes to use the simpler pressure boundary condition

$$\frac{\partial p}{\partial n} = 0. \tag{21}$$

Using this boundary condition, it is not difficult to attain a stable scheme; a likely reason for the popularity of the approach. On the other hand, a naïve implementation of the accurate condition $\partial p/\partial n = \nu \mathbf{n} \cdot \Delta \mathbf{u}$ can easily lead to an unstable method. For high-Reynolds number laminar boundary layer flow, equation (21) could be a good approximation since $\partial p/\partial n = O(Re^{-1/2})$ tends to zero as $Re \to \infty$. To evaluate this approximation at a finite Reynolds number, we computed the unsteady flow around two circular cylinders in a channel using both the curl-curl pressure boundary condition and (21). In this case, $Re = 10^3$ based on the diameter of one cylinder. As can be seen in Figure 4, the flow

**Table 2.** Maximum errors at $t = .5$ for a polynomial analytic solution with $\nu = .1$. The domain is discretized by the circle-in-a-channel grid shown in figure (3) where the coarsest grid ($g = 1$) has $13 \times 13 \bigcup 16 \times 3$ grid points. The time stepping was a second-order explicit predictor corrector method. The column entitled $h_1/h_g$ denotes the ratio of the grid spacing on grid 1 to the spacing on grid $g$.

| grid | $h_1/h_g$ | $\|p - p_e\|_\infty$ | $\|u - u_e\|_\infty$ | $\|v - v_e\|_\infty$ | $\|\nabla \cdot \mathbf{u}\|_\infty$ |
|---|---|---|---|---|---|
| $g = 1$ | 1 | $3.3 \ 10^0$ | $2.8 \ 10^{-1}$ | $4.0 \ 10^{-1}$ | $6.5 \ 10^{-1}$ |
| $g = 2$ | 2 | $3.0 \ 10^{-1}$ | $4.2 \ 10^{-2}$ | $4.4 \ 10^{-2}$ | $1.5 \ 10^{-1}$ |
| $g = 3$ | 4 | $6.0 \ 10^{-2}$ | $1.0 \ 10^{-2}$ | $8.2 \ 10^{-3}$ | $3.5 \ 10^{-2}$ |
| $g = 4$ | 8 | $7.0 \ 10^{-3}$ | $2.2 \ 10^{-3}$ | $1.4 \ 10^{-3}$ | $9.8 \ 10^{-3}$ |
| rate | | 2.9 | 2.3 | 2.7 | 2.0 |

**Table 3.** Maximum errors at $t = .1$ for a polynomial analytic solution with $\nu = .1$. The grid is a sphere-in-a-box. The coarse grid, g=1, consists of component grids with $17 \times 17 \times 17 \bigcup 12 \times 12 \times 4 \bigcup 12 \times 12 \times 4$ grid points. The column entitled $h_1/h_g$ denotes the ratio of the grid spacing on grid 1 to the spacing on grid $g$.

| grid | $h_1/h_g$ | $\|p - p_e\|_\infty$ | $\|u - u_e\|_\infty$ | $\|v - v_e\|_\infty$ | $\|w - w_e\|_\infty$ | $\|\nabla \cdot \mathbf{u}\|_\infty$ |
|---|---|---|---|---|---|---|
| $g = 1$ | 1 | $2.4 \times 10^0$ | $8.3 \times 10^{-2}$ | $7.1 \times 10^{-2}$ | $8.5 \times 10^{-2}$ | $6.0 \times 10^{-1}$ |
| $g = 2$ | 2 | $6.4 \times 10^{-1}$ | $2.6 \times 10^{-2}$ | $1.9 \times 10^{-2}$ | $2.0 \times 10^{-2}$ | $1.6 \times 10^{-1}$ |
| $g = 3$ | 3 | $2.6 \times 10^{-1}$ | $1.1 \times 10^{-2}$ | $7.7 \times 10^{-3}$ | $7.6 \times 10^{-3}$ | $7.8 \times 10^{-2}$ |
| rate | | 2.0 | 1.8 | 2.0 | 2.2 | 1.9 |

**Table 4.** Maximum errors at $t = 1.$ with the 4th order spatial approximation and $\nu = .05$. The divergence damping coefficient is $C_d = 1$. Compare these results to Table 5 where no divergence damping is used.

| grid | $\|\mathbf{u} - \mathbf{u}_e\|_\infty$ | $\|p - p_e\|_\infty$ | $\|\nabla \cdot \mathbf{u}\|_\infty$ |
|---|---|---|---|
| $20 \times 20$ | $9.3\ 10^{-4}$ | $8.0\ 10^{-3}$ | $2.2\ 10^{-2}$ |
| $30 \times 30$ | $1.2\ 10^{-4}$ | $1.4\ 10^{-3}$ | $2.4\ 10^{-3}$ |
| $40 \times 40$ | $2.8\ 10^{-5}$ | $4.3\ 10^{-4}$ | $5.1\ 10^{-4}$ |
| rate | 5.0 | 4.2 | 5.4 |

**Table 5.** Maximum errors at $t = 1.$ with the 4th order spatial approximation and $\nu = .05$. The divergence damping coefficient is $C_d = 0$. Compare these results to Table 4 where divergence damping is used.

| grid | $\|\mathbf{u} - \mathbf{u}_e\|_\infty$ | $\|p - p_e\|_\infty$ | $\|\nabla \cdot \mathbf{u}\|_\infty$ |
|---|---|---|---|
| $20 \times 20$ | $2.4\ 10^{-3}$ | $1.3\ 10^{-2}$ | $6.4\ 10^{-2}$ |
| $30 \times 30$ | $5.1\ 10^{-4}$ | $2.5\ 10^{-3}$ | $1.3\ 10^{-2}$ |
| $40 \times 40$ | $1.7\ 10^{-4}$ | $8.2\ 10^{-4}$ | $4.4\ 10^{-3}$ |
| rate | 3.8 | 4.0 | 3.9 |

is laminar on the leading sides and detached on the trailing sides of the cylinders. The pressure on the boundary of the upper cylinder using both pressure boundary conditions is presented in Figure 5. Clearly, the simplified pressure boundary condition (21) leads to an inaccurate boundary pressure. To validate the computational results, we use the laminar boundary layer results provided by Schlichting [21], pp.170–173, for 2–d flow around a circular cylinder. Schlichting

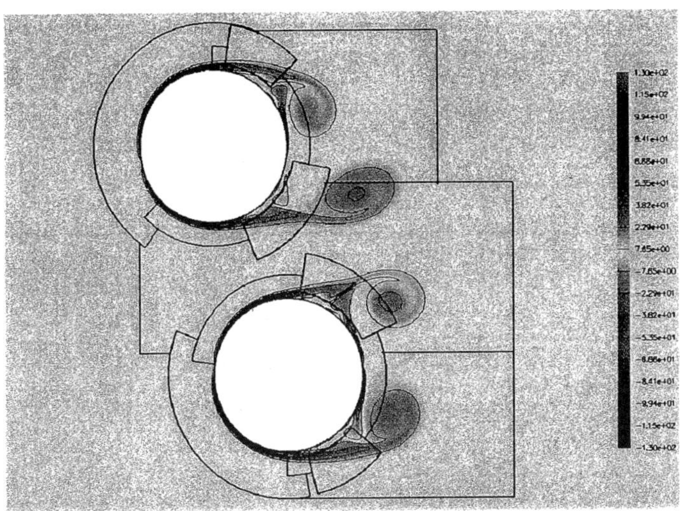

**Fig. 4.** The vorticity around two circular cylinders in a channel at time $t = 2$ plotted with 18 equally spaced contour lines between $\pm 130$. The Reynolds number based on the diameter of one cylinder is $Re = 10^3$.

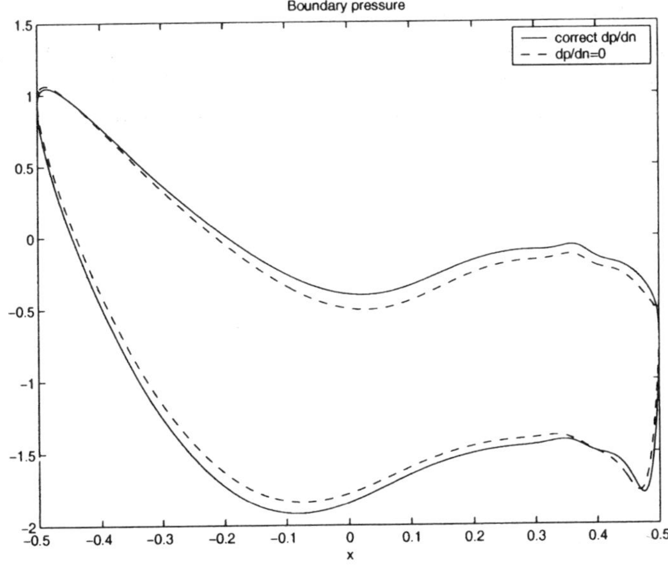

**Fig. 5.** The (kinematic) pressure on the boundary of the upper cylinder using the curl-curl pressure boundary condition (solid) and the simple boundary condition (21) (dashed) at time $t = 2$. The Reynolds number is $Re = 10^3$. Note that the inflow velocity $U_{in} = 1$ in this computation, so the pressure coefficient $C_p = 2p/U_{in}^2$, equals twice the kinematic pressure $p$.

**Table 6.** A comparison of the div-grad boundary condition (9) and the curl-curl boundary condition (18) for semi-implicit time-stepping. Shown are the largest stable time-steps and maximum errors for a forced computation. The time-step for the curl-curl boundary condition can be chosen much larger since it is does not depend on the ratio $\nu/h^2$. The coarse grid had $41 \times 21 \cup 21 \times 14$ points and the fine grid had $81 \times 41 \cup 41 \times 27$ points.

| BC | Grid | $\Delta t$ | $\|p - p_e\|_\infty$ | $\|u - u_e\|_\infty$ | $\|v - v_e\|_\infty$ | $\|\nabla \cdot \mathbf{u}\|_\infty$ |
|---|---|---|---|---|---|---|
| div-grad | coarse | $2.0 \cdot 10^{-3}$ | $5.8 \cdot 10^{-2}$ | $4.2 \cdot 10^{-2}$ | $8.6 \cdot 10^{-2}$ | $1.1 \cdot 10^{-1}$ |
| div-grad | fine | $5.0 \cdot 10^{-4}$ | $5.7 \cdot 10^{-3}$ | $8.8 \cdot 10^{-3}$ | $1.5 \cdot 10^{-2}$ | $2.4 \cdot 10^{-2}$ |
| curl-curl | coarse | $1.1 \cdot 10^{-2}$ | $5.8 \cdot 10^{-2}$ | $4.2 \cdot 10^{-2}$ | $8.6 \cdot 10^{-2}$ | $1.1 \cdot 10^{-1}$ |
| curl-curl | fine | $5.5 \cdot 10^{-3}$ | $5.8 \cdot 10^{-3}$ | $8.8 \cdot 10^{-3}$ | $1.5 \cdot 10^{-2}$ | $2.4 \cdot 10^{-2}$ |

provides a graph for the boundary shear stress,

$$\tau_0 = \mu \frac{\partial u^{(s)}}{\partial n},$$

along the leading side of the cylinder. (Here, $u^{(s)}$ denotes the tangential velocity component and $\mu$ is the dynamic viscosity.) By evaluating the slope of that graph, we obtain $\partial^2 u^{(s)}/\partial s \partial n$, where $s$ is the arclength along the boundary of the cylinder. This term is proportional to the normal derivative of the pressure, since

$$\frac{\partial p}{\partial n} = -\nu \mathbf{n} \cdot \nabla \times \nabla \times \mathbf{u} = -\nu \frac{\partial^2 u^{(s)}}{\partial s \partial n}. \tag{22}$$

In Figure 6, we plot (22) for $Re = 10^3$. At this Reynolds number, the normal derivative of the pressure is not small even in the laminar region on the leading side of the cylinder. On the trailing side of the cylinder, where the flow is detached, the normal derivative is of the order $O(1)$. We believe this will be the case also for higher Reynolds numbers, since the assumptions of laminar boundary layer theory do not apply for detached flows. We conclude that it is important to use one of the accurate boundary conditions (9) or (18) for laminar flows at low and medium Reynolds numbers as well as detached flows at any Reynolds number.

As an illustration of some more advanced applications of our approach, in Figure 1 we show the solution to an incompressible flow containing two rigid cylinders. The cylinders fall under the influence of gravity and their motion is determined by the forces exerted by the fluid. In this computation, the overlapping grids are recomputed at every time step. Figure 2 shows the flow past a rotating disk and Figure 7 shows the flow past a valve.

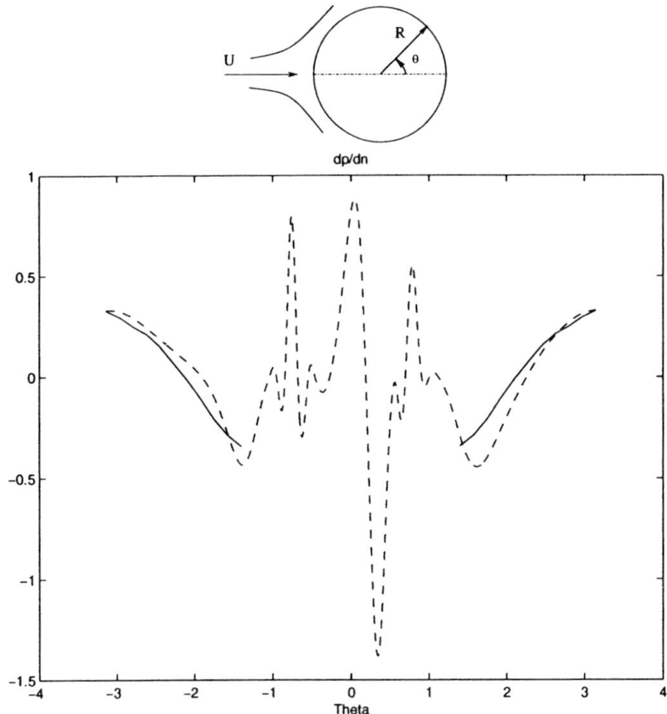



**Fig. 6.** The normal derivative of the pressure along the boundary of the circular cylinder for $Re \equiv 2UR/\nu = 10^3$. The solid line was evaluated from the slope of the boundary shear stress for a single cylinder, provided by Schlichting [21]. The dashed line was calculated along the upper cylinder in the flow shown in Figure 4.

## 3  Software availability

The **OverBlown** flow solver and **Overture** are freely available from the internet at http://www.llnl.gov/casc/Overture.

**Acknowledgments:** The authors thank Heinz-Otto Kreiss for many interesting discussions on the present subject.

This work was performed under the auspices of the U.S. Department of Energy by the University of California, Lawrence Livermore National Laboratory under contract No. W-7405-Eng-48.

## References

1. S. ABDALLAH, *Numerical solutions for the pressure poisson equation with Neumann boundary conditions using a non-staggered grid, I*, J. Comp. Phys., 70 (1987), pp. 182–192.

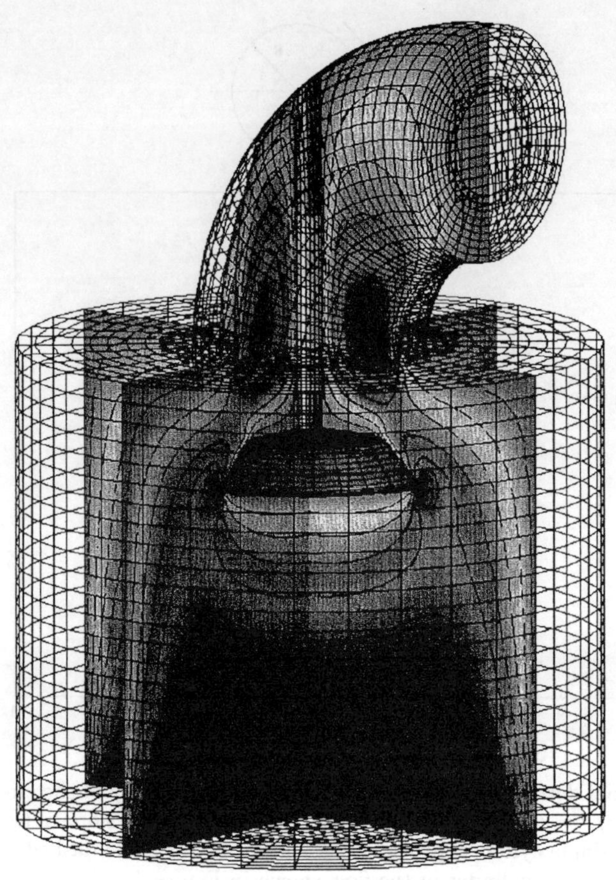

**Fig. 7.** Flow past a valve.

2. J. B. BELL, P. COLELLA, AND H. M. GLAZ, *A second-order projection method for the incompressible Navier-Stokes equations*, J. Comp. Phys., 85 (1989), pp. 257–283.

3. D. L. BROWN, R. CORTEZ, AND M. L. MINION, *Accurate projection methods for the incompressible Navier–Stokes equations*, J. Comp. Phys., 168 (2001), pp. 464–499.

4. A. J. CHORIN, *A numerical method for solving incompressible viscous flow problems*, J. Comp. Phys., 2 (1967), pp. 12–26.

5. ———, *Numerical solution of the Navier-Stokes equations*, Math. Comp., 22 (1968), pp. 745–762.

6. W. E AND J. GUO LIU, *Projection method II: Godunov-Ryabenki analysis*, SIAM J. of Numer. Anal., 33 (1996), pp. 1597–1621.

7. P. M. GRESHO AND R. L. SANI, *On the pressure boundary conditions for the incompressible Navier-Stokes equations*, International Journal for Numerical Methods in Fluids, 7 (1987), pp. 1111–1145.

8. F. HARLOW AND J. WELCH, *Numerical calculation of time-dependent viscous incompressible flow of fluid with free surface*, J. Comp. Phys., 8 (1965), pp. 2182–2189.

9. W. HEINRICHS, *Splitting techniques for the unsteady Stokes equations*, SIAM J. of Numer. Anal., 35 (1998), pp. 1646–1662.

10. W. HENSHAW, *A fourth-order accurate method for the incompressible Navier-Stokes equations on overlapping grids*, J. Comp. Phys., 113 (1994), pp. 13–25.

11. ———, *OverBlown: A fluid flow solver for overlapping grids, user guide*, Research Report UCRL-MA-134288, Lawrence Livermore National Laboratory, 1999.

12. W. HENSHAW AND H.-O. KREISS, *Analysis of a difference approximation for the incompressible Navier-Stokes equations*, Research Report LA-UR-95-3536, Los Alamos National Laboratory, 1995.

13. W. HENSHAW, H.-O. KREISS, AND L. REYNA, *A fourth-order accurate difference approximation for the incompressible Navier-Stokes equations*, Comput. Fluids, 23 (1994), pp. 575–593.

14. G. KARNIADAKIS, M. ISRAELI, AND S. A. ORSZAG, *High-order splitting methods for the incompressible Navier-Stokes equations*, J. Comp. Phys., 97 (1991), pp. 414–443.

15. J. KIM AND P. MOIN, *Application of a fractional-step method to incompressible Navier-Stokes equations*, J. Comp. Phys., 59 (1985), pp. 308–323.

16. C. KIRIS, D. KWAK, S. ROGERS, AND I. CHANG, *Computational approach for probing the flow through artificial heart devices*, ASME J. of Biofluidmechanical Engineering, 119 (1997), pp. 452–460.

17. K. KORCZAK AND A. T. PATERA, *An isoparametric spectral element method for solution of the Navier-Stokes equations in complex geometry*, J. Comp. Phys., 62 (1986), pp. 361–382.

18. S. A. ORSZAG, M. ISRAELI, AND M. O. DEVILLE, *Boundary conditions for incompressible flows*, J. Sci. Comp., 1 (1986), pp. 75–111.

19. J. B. PEROT, *An analysis of the fractional step method*, J. Comput. Phys., 108 (1993), pp. 51–58.

20. N. A. PETERSSON, *Stability of pressure boundary conditions for Stokes and Navier-Stokes equations*, J. Comp. Phys., 172 (2001), pp. 40–70.

21. H. SCHLICHTING, *Boundary-Layer Theory*, McGraw-Hill, New York, 1979. Seventh Edition.

22. J. SHEN, *On error estimates of the projection methods for the Navier-Stokes equations: second-order schemes*, Math. Comp., 65 (1996), pp. 1039–1065.

23. J. STRIKWERDA, *Finite difference methods for the Stokes and Navier-Stokes equations*, SIAM J. Sci. Stat. Comput., 5 (1984), pp. 56–68.

24. J. C. STRIKWERDA AND Y. S. LEE, *The accuracy of the fractional step method*, SIAM J. Numer. Anal., 37 (1999), pp. 37–47.

25. R. TEMAM, *Sur l'approximation de la solution des equation de Navier-Stokes par la méthode des fractionnarires II*, Arch. Rational Mech. Anal., 33 (1969), pp. 377–385.

26. J. A. WRIGHT AND W. SHYY, *A pressure-based composite grid method for the Navier-Stokes equations*, J. Comp. Phys., 107 (1993), pp. 225–238.

# Higher-Order Semi-Implicit Projection Methods

Michael L. Minion [1]

ABSTRACT  A semi-implicit form of the method of spectral deferred corrections is applied to the solution of the incompressible Navier-Stokes equations. A methodology for constructing semi-implicit projection methods with arbitrarily high order of temporal accuracy in both the velocity and pressure is presented. Three variations of projection methods are discussed which differ in the manner in which the auxiliary velocity and the pressure are calculated. The presentation will make clear that projection methods in general need not be viewed as fractional step methods as is often the practice. Two simple numerical examples are used to demonstrate fourth-order accuracy in time for an implementation of each variation of projection method.

## 1  Introduction

A great deal of effort has been dedicated to the design of numerical methods for the simulation of high Reynolds number flow, often including techniques for incorporating turbulence models to represent sub-grid scale effects. There are, however, many situations of interest in which the Reynolds number is "moderate", i.e. high enough so that the Stokes equations are not a reasonable model, but low enough so that the flow can be resolved on a computational grid and no subgrid scale model is needed. The modeling of biological systems such as the flow of blood in the heart or the swimming of organisms is just one such example. Often in these examples, it is desirable to compute the motion of the fluid with high accuracy since other components in the system which depend on the fluid are of interest (e.g. the motion of membranes or the passive transport of solutes). As the interest in numerical modeling in many fields in the applied sciences and the speed and memory capacity of computers continue to increase, the need for accurate, efficient, and adaptable numerical methods for moderately viscous flows should increase.

Numerical methods for viscous flows must treat viscous terms implicitly in order to avoid a severe time step restriction. This paper introduces semi-implicit projection methods with arbitrarily high temporal accuracy for

[1] This work was performed under the auspices of the U.S. Department of Energy by the University of California, Lawrence Livermore National Laboratory under contract No. W-7405-Eng-48. Support also provided by the NSF under Grant DMS-9973290.

the integration of the equations governing viscous incompressible flows. Furthermore, the methods require the solution of only two linear implicit equations for which efficient high-order solvers exist. Therefore the use of the current methods with complex geometry as well as spatial and temporal adaptivity is possible immediately.

The projection methods in this paper are based on a method of lines approach coupled with a semi-implicit method for integrating ordinary differential equations (ODEs) based on the method of spectral deferred corrections (SDC) [DGR00, Min01]. These so-called SISDC methods have several advantages over the more traditional semi-implicit multi-step and Runge-Kutta methods. Higher-order semi-implicit linear multi-step methods are not self starting, they typically require a severely restricted time step for stability, and they present difficulties when variable time stepping is required (see e.g. [ARW95, FHV97]). On the other hand, semi-implicit Runge-Kutta methods are self starting and generally have good stability properties, and efficient methods for orders up to five have been proposed. (See e.g. [SZ96, ARS97, KC01, CdFN01]). It has been well documented however, that when combined with a method of lines approach for PDEs, Runge-Kutta methods typically suffer from a loss of accuracy when time dependent boundary conditions are prescribed, unless special care is taken when imposing intermediate boundary conditions [SSVH86, CGAD95]. Suggestions for restoring full accuracy in certain cases have been proposed (e.g. [AGC96, Pat97, CP01]), but at present, no general strategy for semi-implicit Runge-Kutta methods has been developed.

The SDC method has the advantage that a relatively simple numerical method is used to compute a higher-order solution. This is accomplished by using the simple method to solve a series of correction equations during each time step, each of which increases the order of accuracy of the solution. This makes SISDC methods particularly attractive to problems possessing disparate time scales since a time-split approach can be used without being limited to lower-order accuracy. The method of spectral deferred corrections has been used already in this manner in the context of unsteady combustion [HRZ98, HRSZ99]. Furthermore, imposing correct boundary conditions for a semi-implicit method of lines approach for PDEs, although certainly not trivial, is more transparent for SISDC methods because of the simplicity of the underlying method. An overview of SISDC methods is presented in Sect. 3.

The strategy for the construction of SISDC projection methods relies on applying the method of lines approach to a non-standard form of the equations of motion. The connection between this form and projection methods is discussed in Sect. 2. Three variations of a generalized projection method will be presented in Sect. 4. Numerical examples set in an idealized geometry are presented in Sect. 5. Fourth-order convergence for both pressure and velocity is demonstrated in these examples.

# 2   The Equations for Incompressible Flow

In this Section, different formulations of the equations of motion of an incompressible fluid are discussed. In an $N$-dimensional bounded domain $\Omega$, the usual form of the equations of motion are given by the incompressible Navier-Stokes equations in terms of the velocity $\mathbf{u}$ and the pressure $p$. Denoting the kinematic viscosity by $\nu$, the Navier-Stokes equations are

$$\mathbf{u}_t + \nabla p = -(\mathbf{u} \cdot \nabla)\mathbf{u} + \nu\nabla^2\mathbf{u} \tag{1.1}$$
$$\nabla\cdot\mathbf{u} = 0 \tag{1.2}$$

with boundary conditions

$$\mathbf{u}|_{\partial\Omega} = \mathbf{u}_b. \tag{1.3}$$

Numerous equivalent formulations of these equations have been developed to facilitate their numerical approximation.

Projection methods first introduced by Chorin [Cho68, Cho69] are based on the observation that the left-hand side of Eq. (1.1) is a Hodge decomposition. Hence an equivalent projection formulation is given by

$$\mathbf{u}_t = \mathbf{P}\left[-(\mathbf{u} \cdot \nabla)\mathbf{u} - \nabla p + \nu\nabla^2\mathbf{u}\right] \tag{1.4}$$

where $\mathbf{P}$ is the operator which projects a vector field onto the space of divergence-free vector fields with appropriate boundary conditions. In general, the strategy employed for one time step in a generic projection method is to first approximate Eq. (1.1) without regard to the divergence constraint (1.2) to yield an "auxiliary velocity" $\mathbf{u}^*$, and then to project $\mathbf{u}^*$ onto the space of divergence-free fields to give the new velocity. For this reason, projection methods are often referred to as *fractional step* methods.

Second order, semi-implicit projection methods in which the advective terms in the equation for $\mathbf{u}^*$ are handled explicitly while the viscous terms are handled implicitly have been developed in [KM85, BCG89] and others. The most significant issue that must be resolved in these methods is the imposition of boundary conditions for $\mathbf{u}^*$ in the implicit step. Since the projection of $\mathbf{u}^*$ is usually implemented by the solution of a Poisson problem for which only one boundary condition can be prescribed, the boundary conditions for $\mathbf{u}^*$ must be chosen in such a way that the remaining velocity boundary conditions are satisfied. The boundary conditions on $\mathbf{u}^*$ also effect the accuracy of the pressure gradient. An overview as well as a further analysis of boundary conditions for second-order semi-implicit projection methods appears in [BCM01].

In order to develop higher-order numerical methods based on the projection method strategy, it is helpful to consider an alternative formulation of the incompressible Navier-Stokes equations based on an auxiliary variable equal to the fluid velocity plus the gradient of a scalar. Such formulations were first introduced by Oseledets [Ose89] and many variations have since been proposed for use in numerical methods for various problems [But93, Cor95, Cor96, EL96, EL97, RR98, SC96, CM00].

Consider two new variables, $\mathbf{m}$ and $\chi$ which are related to the fluid velocity by

$$\mathbf{m} = \mathbf{u} + \nabla\chi. \tag{1.5}$$

The vector field $\mathbf{m}$ and the potential $\chi$ can be chosen to satisfy evolution equations in such a way that the fluid velocity and pressure derived from them satisfy the Navier-Stokes equations. One possibility which facilitates the development of accurate projection methods is

$$\mathbf{m}_t + (\mathbf{u} \cdot \nabla)\mathbf{u} = -\nabla q + \nu \nabla^2 \mathbf{m} \tag{1.6}$$

$$\mathbf{u}|_{\partial\Omega} = \mathbf{u}_b. \tag{1.7}$$

where $\nabla q$ is an arbitrary gradient that will be discussed later and

$$\mathbf{u} = \mathbf{P}(\mathbf{m}). \tag{1.8}$$

In this formulation, the pressure has been eliminated from the equations; however, it can be recovered from the potential $\chi$ by enforcing the equivalence of equations (1.1) and (1.6) giving

$$p = q + \chi_t - \nu \nabla^2 \chi. \tag{1.9}$$

Initially, $\nabla\chi$ can be zero, (i.e. $\mathbf{m} = \mathbf{u}$), or $\nabla\chi$ can be non-zero so long as the choice is consistent with Eq. (1.5). Note that the boundary conditions are given in terms of $\mathbf{u}$, which by Eq. (1.5), implies that there is a coupling of the boundary conditions of $\mathbf{m}$ and $\nabla\chi$ (See [BCM01]).
Clearly if $\nabla q$ is defined as the pressure gradient and $\mathbf{m}$ is given initial and boundary values equal to $\mathbf{u}$, then Eq. (1.6) reduces to Eq. (1.1). If $\nabla q$ is identically zero, the resulting equations are equivalent to those used as the basis of the numerical method in [EL96]. As discussed in the next section, the key observation to make is that a single step of a projection method can be thought of as approximating Eq. (1.6) rather than Eq. (1.1).

# 3 Semi-Implicit Spectral Deferred Corrections

In [DGR00], the SDC method for the solution of ODEs is introduced. The strategy within each time step of the method is to use a simple numerical method (specifically forward or backward Euler) to calculate a provisional solution at specific points within the interval of the time step. Then, the same simple method is used to solve a series of correction equations each of which increases by one the order of accuracy of the provisional solution. The correction equation is cast in a form similar to the Picard integral iteration, hence the points at which the provisional solution and the corrections are calculated correspond to quadrature nodes. A semi-implicit version of the SDC method (SISDC) suitable for use with the Navier-Stokes equations will be briefly described below. For more complete details see [DGR00] and [Min01].
Consider the ODE

$$\phi'(t) = F(t, \phi(t)) = F_E(t, \phi(t)) + F_I(t, \phi(t)) \qquad t \in [a, b]$$
$$\phi(a) = \phi_a$$

where $F_E$ is a non-stiff term that will be handled explicitly, and $F_I$ is a stiff term that will be handled implicitly.

For a given time step on the interval $[t_n, t_{n+1}]$, the first step in the SISDC method is to compute a provisional solution $\tilde{\phi}(t_m)$ at the points $t_m$ for $m = 0 \ldots p$ with $t_n = t_0 < t_1 < \ldots < t_p = t_{n+1}$. Assuming that $\phi(t_n)$ is known and using the notation $\tilde{\phi}(t_m) \equiv \tilde{\phi}^m$, a first-order, semi-implicit numerical method for computing $\tilde{\phi}$ is

$$\tilde{\phi}^{m+1} = \tilde{\phi}^m + \Delta t_m [F_E(t_m, \tilde{\phi}^m) + F_I(t_{m+1}, \tilde{\phi}^{m+1})], \tag{1.10}$$

where $\Delta t_m = t_{m+1} - t_m$ and $\tilde{\phi}^0 = \tilde{\phi}(t_n)$. For clarity, the interval $[t_m, t_{m+1}]$ will be referred to as a *substep* as opposed to the *time step* $[t_n, t_{n+1}]$.

Following [DGR00], if one regards $\tilde{\phi}$ as a continuous function, a measure of the error in the provisional solution can be written

$$E(t, \tilde{\phi}) = \tilde{\phi}_0 + \int_{t_0}^{t} F(\tau, \tilde{\phi}(\tau)) d\tau - \tilde{\phi}(t).$$

Defining the correction to the provisional solution $\delta(t)$ by $\phi(t) = \tilde{\phi}(t) + \delta(t)$, some elementary algebra yields

$$\delta(t) = \int_{t_0}^{t} F(\tau, \tilde{\phi}(\tau) + \delta(\tau)) - F(\tau, \tilde{\phi}(\tau)) d\tau + E(t, \tilde{\phi}). \tag{1.11}$$

The SISDC method proceeds by computing a sequence of corrections $\delta^k(t)$, where each $\delta^k$ is computed by first approximating $E(t, \tilde{\phi})$ with an appropriate quadrature rule, and then using a semi-implicit method to approximate Eq. (1.11). After each $\delta^k$ is computed, it is added to $\tilde{\phi}$ to yield an updated provisional solution.

Specifically, (suppressing the $k$ index on $\delta$) a substep of the correction equation is

$$\begin{aligned} \delta^{m+1} &= \delta^m + \Delta t_m [F_E(t_m, \tilde{\phi}^m + \delta^m) - F_E(t_m, \tilde{\phi}^m) \\ &+ F_I(t_{m+1}, \tilde{\phi}^{m+1} + \delta^{m+1}) - F_I(t_{m+1}, \tilde{\phi}^{m+1})] \\ &+ I_m^{m+1}(\tilde{\phi}) - \tilde{\phi}^{m+1} + \tilde{\phi}^m \end{aligned} \tag{1.12}$$

where

$$I_m^{m+1}(\tilde{\phi}) \approx \int_{t_m}^{t_{m+1}} F(\tau, \tilde{\phi}(\tau)) d\tau.$$

As in [Min01], the points $t_m$ above are chosen to be the nodes of the standard Gauss-Labatto quadrature. Since the quadrature must be done for each subinterval $[t_m, t_{m+1}]$, there are actually $p$ quadrature rules of the form

$$I_m^{m+1}(\tilde{\phi}) = \sum_{l=0}^{p} q_l^m F(t_l, \tilde{\phi}^l) \tag{1.13}$$

for $m = 0 \ldots p - 1$.

To summarize, a complete SISDC time step consists of the computation of the first provisional solution $\tilde{\phi}$ using $p$ substeps of Eq. (1.10) followed by $K$ iterations of the correction Eq. (1.12), each of which also involves $p$ substeps. It can be shown that for $K$ iterations of the correction equation, the above procedure will produce an approximate solution with overall error $O(\Delta t^{K+1})$, provided the integrals in Eq. (1.13) have $O(\Delta t^{K+1})$ error as well. (See [Min01] for a further discussion of accuracy and the choice of integration nodes.)

### 3.1  Stability of SISDC Methods

To study the stability of the SISDC method for ODEs, it is most useful to consider the model problem

$$\phi'(t) = \lambda\phi = \alpha i\phi + \beta\phi.$$

For the SISDC method, $F_E(t, \phi) = \alpha i\phi$ and $F_I(t, \phi) = \beta\phi$. This is the relevant model problem when studying the stability of finite difference methods for advection-diffusion type PDEs with $\alpha$ corresponding to the advection term and $\beta$ corresponding to the stiff diffusion term. For a given $\lambda$, let $\phi^1(\lambda)$ denote the result of taking one time step with the SISDC method applied to the model problem with $\phi(0) = 1$ and $\Delta t = 1$. The stability region is then defined as the set of $\lambda$ for which $|\phi^1(\lambda)| < 1$.

Fig. 1 shows two stability regions for two different SISDC methods, both of which use three iterations of the correction equation (i.e. $K = 3$). The larger region, however, corresponds to using four substeps per time step while the smaller only three. Both versions are fourth-order accurate but as explained in Sect. 4, projection methods based on the SISDC method require that four substeps be completed in order to compute the pressure to fourth-order accuracy. Although the stability region for this version is larger, a total of 16 implicit solves are required as opposed to only 12 for the version with only 3 substeps. The stability diagrams clearly show that the time step is restricted solely by the explicit term in the equation. See [Min01] for more information on the relative size of stability diagrams for SISDC methods.

## 4   Projection Methods

To help set the stage for the introduction of the semi-implicit projection methods in the next section, a brief discussion of first-order semi-implicit projection methods is now presented. For the moment, assume that the velocity and pressure have been discretized in time and denote by $\mathbf{u}^n$ the approximation to the velocity at time $t_n$ (and likewise $p^n$). For further ease of presentation an unbounded domain is considered. Boundary conditions are briefly discussed in Sect. 4.2. A first-order semi-implicit temporal discretization of Eq. (1.4) is

$$\frac{\mathbf{u}^{n+1} - \mathbf{u}^n}{\Delta t} = \mathbf{P}[-(\mathbf{u}^n \cdot \nabla)\mathbf{u}^n - \nabla p^n + \nu\nabla^2\mathbf{u}^{n+1}]. \qquad (1.14)$$

Under many circumstances (depending on the boundary conditions that projected fields are required to satisfy), the pressure gradient term is annihilated by the projection and Eq. (1.14) is equivalent to

$$\frac{\mathbf{u}^{n+1} - \mathbf{u}^n}{\Delta t} = \mathbf{P}[-(\mathbf{u}^n \cdot \nabla)\mathbf{u}^n + \nu\nabla^2\mathbf{u}^{n+1}]. \qquad (1.15)$$

Because of the difficulty in solving either of the above equations directly, usually an approximate two-step procedure is used instead. Specifically for Eq. (1.14), let $\mathbf{u}^*$ and $\mathbf{u}^{n+1}$ be defined by

$$\frac{\mathbf{u}^* - \mathbf{u}^n}{\Delta t} = -(\mathbf{u}^n \cdot \nabla)\mathbf{u}^n - \nabla p^n + \nu\nabla^2\mathbf{u}^* \qquad (1.16)$$

$$\mathbf{u}^{n+1} = \mathbf{P}(\mathbf{u}^*). \qquad (1.17)$$

For Eq. (1.15), the first of these two equations becomes

$$\frac{\mathbf{u}^* - \mathbf{u}^n}{\Delta t} = -(\mathbf{u}^n \cdot \nabla)\mathbf{u}^n + \nu\nabla^2\mathbf{u}^*. \qquad (1.18)$$

The term *fractional step method* is often used interchangeably with the term *projection method* to describe such a two step procedure. Note however, that by equating $\mathbf{u}^*$ with $\mathbf{m}^{n+1}$ where $\mathbf{m}$ is defined by Eq. (1.6), Eqs. (1.16) and (1.17) are also a first-order semi-implicit temporal discretization of Eqs. (1.6) and (1.8). To approximate Eq. (1.16), $\nabla q$ should be set to $\nabla p^n$. while on the other hand, $\nabla q = 0$ for Eq. (1.18). In this sense, the projection methods above are not fractional step methods, and higher order semi-implicit projection methods can be constructed by considering the equations for $\mathbf{m}$ directly. Furthermore, the question of imposing boundary conditions for the intermediate quantity $\mathbf{u}^*$ is reduced to the question of imposing boundary conditions for $\mathbf{m}$. As mentioned above, there is a coupling between the boundary conditions for $\mathbf{m}$ and $\nabla\chi$ which must be respected in a numerical implementation in order to enforce the specified boundary conditions on $\mathbf{u}$.

## 4.1  SISDC Projection Methods

To construct fourth-order SISDC methods for the equations of incompressible flow, a method of lines approach is applied to Eq. (1.6). In the notation from Sect. 3, $\phi \equiv \mathbf{m}$ while

$$F_E = -(\mathbf{u} \cdot \nabla)\mathbf{u} - \nabla q$$

$$F_I = \nu\nabla^2\mathbf{m}.$$

To be more specific, the complete forward-backward Euler substep for computing the provisional solution in the SISDC method is

$$\tilde{\mathbf{m}}^{m+1} = \tilde{\mathbf{m}}^m + \Delta t(-[(\tilde{\mathbf{u}} \cdot \nabla)\tilde{\mathbf{u}}]^m - \nabla q + \nu\nabla^2\tilde{\mathbf{m}}^{m+1}) \qquad (1.19)$$

$$\tilde{\mathbf{u}}^{m+1} = \mathbf{P}(\tilde{\mathbf{m}}^{m+1}). \qquad (1.20)$$

All spatial derivatives are approximated with standard fourth-order centered-difference stencils. For each method below, $\mathbf{m}$ and $\mathbf{u}$ are evolved according to Eqs. (1.19-1.20) to yield the first approximate solution in the SISDC method. Three variants of this approach will be presented which differ in the choice of $\mathbf{m}$, $\nabla\chi$ and $\nabla q$. For all the methods, the term $\nabla q$ is held constant during the entire time step. One effect of this is that $\nabla q$ does not appear in the analog to Eq. (1.11). Letting $\hat{\mathbf{m}} = \tilde{\mathbf{m}} + \delta\mathbf{m}$ and $\hat{\mathbf{u}} = \tilde{\mathbf{u}} + \delta\mathbf{u}$, and rearranging Eq. (1.11) gives

$$
\begin{aligned}
\hat{\mathbf{m}}^{m+1} &= \hat{\mathbf{m}}^m + \Delta t[-(\hat{\mathbf{u}}^m \cdot \nabla)\hat{\mathbf{u}}^m + (\tilde{\mathbf{u}}^m \cdot \nabla)\tilde{\mathbf{u}}^m \\
&+ \nu\nabla^2\hat{\mathbf{m}}^{m+1} - \nu\nabla^2\tilde{\mathbf{m}}^{m+1}] + I_m^{m+1}(\tilde{\mathbf{m}}) \\
\hat{\mathbf{u}}^{m+1} &= \mathbf{P}(\hat{\mathbf{m}}^{m+1}),
\end{aligned}
$$

where $I_m^{m+1}$ is defined in Eq. (1.13). In each variation, the new pressure term is computed by a finite-difference approximation of Eq. (1.9) after the full time step is completed. Since the temporal derivative term in Eq. (1.9) is not centered in time, five values of $\chi_m$ are needed to achieve fourth-order accuracy (i.e. p $= 4$). However, since three iterations of the correction equation are performed, the overall accuracy of the method is still order four.

The first variation of the SISDC projection method will be referred to as the "BCM" method since it is similar in spirit to the projection method introduced in [BCM01]. At the beginning of each time step in this variation, $\mathbf{m}^0$ is reset to $\mathbf{u}^n$, and $\nabla q$ is set to $\nabla p^n$. As mentioned above, $\nabla q$ is held constant during the time step, so the pressure term does not appear in the correction equation.

The second method will be referred to as the "KM" method since it is similar in spirit to the projection method introduced in [KM85]. As before, $\mathbf{m}^0$ is reset to $\mathbf{u}^n$ at the beginning of each time step, however $\nabla q$ is identically zero throughout the time step. Hence the KM method is a so-called *pressure free* method. This also eliminates the $\nabla q$ term in the computation of the pressure by Eq. (1.9).

The last method will be referred to as the "EL" method since it is similar in spirit to the methods introduced in [EL96, EL97]. In this variation, $\mathbf{m}$ is never reset to $\mathbf{u}$, hence in time the difference between $\mathbf{m}$ and $\mathbf{u}$ (i.e. $\nabla\chi$) can grow to have magnitude $O(1)$. As in the KM method $\nabla q$ is identically zero throughout the time step.

If one adopts a slightly different definition of what constitutes a projection method, i.e. a numerical method in which the velocity is obtained by the projection of an auxiliary variable, then all three of the preceding numerical methods can be thought of as projection methods. The only difference then between the three is the procedure used to advance the auxiliary variable. In this remainder of this paper, the term projection method will be used in this somewhat looser sense.

## *4.2 Boundary conditions for semi-implicit projection methods*

The emphasis in the current work is on the construction of higher-order semi-implicit time marching schemes and the relationship between the form of the evolution equation for the velocity and the pressure update equation. For the most part, the subject of numerical boundary conditions has been deferred.

It is important to point out that accurately imposing prescribed slip or in-flow boundary conditions for these methods is by no means a trivial matter. In [Min01] it is shown that even for a simple PDE such as the heat equation, using the prescribed boundary conditions for the PDE as the boundary conditions for the implicit step in the SISDC method will cause a degradation of accuracy similar to that which occurs with Runge-Kutta methods [SSVH86, CGAD95].

The situation is more complicated for projection methods where boundary conditions for $\mathbf{m}$ cannot be fully prescribed *a priori*. As is the case with the second-order semi-implicit methods described in [BCM01], the form of the evolution equation for $\mathbf{m}$ has a direct implication on how accurately the coupling of the boundary conditions for $\mathbf{m}$ and $\mathbf{u}$ must be approximated. In the present context, the implication is that the closer $\nabla q$ approximates the pressure gradient, the closer the boundary conditions for $\mathbf{m}$ will be to the prescribed conditions on $\mathbf{u}$. A paper presenting specific details of imposing accurate boundary conditions for SISDC projection methods is in preparation.

# 5   Convergence Studies

To demonstrate the temporal accuracy of the SISDC projection methods, two numerical examples set in the simplified domain of the doubly-periodic unit square are considered. This removes the issue of imposing boundary conditions and allows the solution of the implicit equations by using the FFT. In the first example, the initial conditions are chosen so that the exact solution is known and convergence rates are calculated to corroborate the expected values. In the second example, the SISDC method is applied to the well studied problem of the roll-up of shear layers in a doubly-periodic geometry.

## 5.1   *Exact Solution*

For the first example, the projection methods are tested on the well-known travelling wave solution to the Navier-Stokes equations. Specifically,

$$
\begin{aligned}
u(x,y,t) &= 0.75 + 0.25\cos(2\pi(x-t))\sin(2\pi(y-t))e^{-8\pi^2\nu t} \\
v(x,y,t) &= 0.75 - 0.25\sin(2\pi(x-t))\cos(2\pi(y-t))e^{-8\pi^2\nu t} \\
p(x,y,t) &= -\frac{1}{64}(\cos(4\pi(x-t)) - \cos(4\pi(y-t)))e^{-16\pi^2\nu t}.
\end{aligned}
$$

The viscosity is set to $\nu = 0.1$, and the time step is chosen to be $\Delta t = 8.0\Delta x$. For each method, errors are computed at time 0.5 in the $L_\infty$ norm for the pressure and $u$-component of the velocity for runs on grids of size $N \times N$ for $N = 64, 128,$ and 256. Note that for the $256 \times 256$ runs,

$\nu \Delta t / \Delta x^2 = 204.8$. Convergence rates calculated from the finest two runs are listed in Table 1.1 and are indicative of fourth-order convergence.

To be certain that the convergence rates in the above numerical result reflect temporal accuracy as opposed to spatial accuracy, the same example is repeated using a slightly different SISDC method. In this case, only 3 sub-steps are taken within each iteration, but 3 iterations of the correction equation are performed. This means that only 4 values of $\chi$ are available at the end of each time-step with which to calculate the time derivative piece of the pressure in Eq. (1.9). Therefore, if temporal error is significant the accuracy of the pressure should drop to third order. The errors and convergence rates for this example are shown in Table 1.2 and demonstrate third-order accuracy. The errors for the velocity are not shown but remain $O(\Delta t^4)$ as expected.

## 5.2 Double Shear Layer

The second numerical example is performed on an unsteady flow within the same geometry and given by the perturbed shear layer initial conditions

$$
\begin{aligned}
u &= \begin{cases} \tanh(30(y - 0.25)), & \text{for } y \leq 0.5 \\ \tanh(30(0.75 - y)), & \text{for } y > 0.5 \end{cases} \\
v &= 0.05 \sin(2\pi(x + 0.25)).
\end{aligned}
$$

The viscosity is set to $\nu = 0.002$ for all runs. The exact solution for this problem is not known, therefore for each method, a reference solution on a $768 \times 768$ grid using $\Delta t = 2\Delta x$ is used as an exact solution. Three numerical runs using grids of size $64 \times 64$, $128 \times 128$, and $256 \times 256$ and with $\Delta t = 4\Delta x$ are used to estimate convergence rates. Errors are calculated in the $L_\infty$ norm at $t = 1.0$ by comparison with the reference solution. Convergence results for the velocity and the pressure shown in Table 1.3 again indicate fourth-order convergence.

The parameters for this example are the same as those used in a convergence study in [MB97] where the velocity computed by a spectral collocation method was used as a reference solution. (The spectral method does not compute the pressure.) The velocities from the reference solutions computed by the SISDC differ from this reference solution by an amount that is at least two orders of magnitude smaller than the errors used to compute the convergence rates in Table 1.3. This direct comparison leaves little doubt that the projection methods above are indeed converging to the exact solution at a fourth-order rate.

# 6 Conclusion

In this paper, a general strategy for constructing semi-implicit projection methods for the incompressible Navier-Stokes equations with higher-order temporal accuracy is presented. The strategy utilizes recently introduced semi-implicit methods for ODEs based on spectral deferred corrections. Three variations of a generalized projection method are presented and fourth-order accuracy is demonstrated on flows in a two-dimensional doubly-periodic domain.

In terms of implicit equations that need to be solved, the projection methods in this paper require only the solution of the Poisson equation associated with the projection and the implicit equation in each time-step which takes the form

$$(I - \nu\Delta t\nabla^2)u = f. \tag{1.21}$$

Both of these equations are linear, and higher-order solvers have been developed for each based on using the Fast Multipole Method to efficiently compute the convolution integral solution to the equation. For the two-dimensional Poisson equation, adaptive methods suitable for use with irregular domains have recently been implemented [EG00] that incorporate standard block rectangular mesh refinement (see e.g. [BO84]) and classical layer potentials for imposing boundary conditions. Versions with up to eighth-order spatial accuracy have been developed, and the computational cost per grid point of these methods is only a small factor times that for a uniform grid FFT-based solver. The analogous algorithms for Eq. (1.21) in a two-dimensional rectangular geometry have also been implemented [HGE01], and the complex geometry versions are currently being developed. A project to incorporate these solvers into projection methods for two-dimensional incompressible flow in complex geometry is also underway. Three dimensional versions of all the algorithms have been described, but the implementation is still in the future.

It is worth reiterating that despite the similarities of the three methods described in this paper and the comparable performance on the test problems presented, the issues concerning accurately imposing prescribed slip or inflow boundary conditions differ in a significant way between variations. A paper describing in detail these issues is in preparation.

# 7    References


[AGC96]   Saul Abarbanel, David Gottlieb, and Mark H. Carpenter. On the removal of boundary errors caused by Runge-Kutta integration of nonlinear partial differential equations. *SIAM J. Sci. Comput.*, 17:777–782, 1996.

[ARS97]   Uri M. Ascher, Steven J. Ruuth, and Raymond J. Spiteri. Implicit-explicit Runge-Kutta methods for time-dependent partial differential equations. *Appl. Num. Math.*, 25:151–167, 1997.

[ARW95]   Uri M. Ascher, Steven J. Ruuth, and Brian T. R. Wetton. Implicit-explicit methods for time-dependent PDE's. *SIAM J. of Numer. Anal.*, 32:797–823, 1995.

[BCG89]   J. B. Bell, P. Colella, and H. M. Glaz. A second order projection method for the incompressible Navier-Stokes equations. *J. Comput. Phys.*, 85(2):257–283, December 1989.

[BCM01]   D. L. Brown, R. Cortez, and M. Minion. Accurate projection methods for the incompressible Navier-Stokes equations. *J. Comput. Phys.*, 168, Apr. 2001.

[BO84]   M. J. Berger and J. Oliger. Adaptive mesh refinement for hyperbolic partial differential equations. *J. Comput. Phys.*, 53:484–512, 1984.

[But93]   T. Buttke. *Velicity methods: Lagrangian numerical methods which preserve the Hamiltonian structure of incompressible fluid flow*, pages 39–57. Kluwer Academic Publisher, 1993. NATO ASI Series C, vol. 395.

[CdFN01]   M. P. Calvo, J. de Frutos, and J. Novo. Linearly implicit Runge-Kutta methods for the advection-reaction-diffusion equations. *Appl. Num. Math.*, 37:535–549, 2001.

[CGAD95]   Mark H. Carpenter, David Gottlieb, Saul Abarbanel, and Wai-Sun Don. The theoretical accuracy of Runge-Kutta time discretization for the initial boundary value problem: A study of the boundary error. *SIAM J. Sci. Comput.*, 16:1241–1252, 1995.

[Cho68]   A. J. Chorin. Numerical solution of the Navier-Stokes equations. *Math. Comp.*, 22:742–762, 1968.

[Cho69]   A. J. Chorin. On the convergence of discrete approximations to the Navier-Stokes equations. *Math. Comp.*, 23:341–353, 1969.

[CM00]   Ricardo Cortez and Michael L. Minion. The blob projection method for immersed boundary problems. *J. Comput. Phys.*, 161:428–453, 2000.

[Cor95]   Ricardo Cortez. *Impulse-based Particle Methods for Fluid Flow*. PhD thesis, University of California, Berkeley, May 1995.

[Cor96]   Ricardo Cortez. An impulse-based approximation of fluid motion due to boundary forces. *J. Comput. Phys.*, 123:341–353, 1996.

[CP01]   M. P. Calvo and C. Palencia. Avoiding the order reduction of Runge-Kutta methods for linear initial boundary value problems. *Math. Comp.*, 2001. to appear.

[DGR00]   Alok Dutt, Leslie Greengard, and Vladimir Rokhlin. Spectral deferred correction methods for ordinary differential equations. *BIT*, 40(2):241–266, 2000.

[EG00]   F. Ethridge and L. Greengard. A new fast-multipole accelerated Poisson solver in two dimensions. 2000. submitted.

[EL96]   Wienan E and Jian-Guo Liu. Gauge method for viscous incompressible flows. Unpublished, 1996.

[EL97]   Wienan E and Jian-Guo Liu. Finite difference schemes for incompressible flows in the velocity-impulse density formulation. *J. Comput. Phys.*, 130:67–76, 1997.

[FHV97]   J. Frank, W. H. Hundsdorder, and J. G. Verwer. Stability of implicit-explicit linear multistep methods. *Appl. Num. Math.*, 25:193–205, 1997.

[HGE01]   J. Huang, L. Greengard, and F. Ethridge. A new fast-multipole accelerated Yukawa solver in two dimensions. 2001. In preparation.

[HRSZ99]   T. Hagstrom, K. Radhakrishnan, S. Steinberg, and R. Zhou. Simulation of unsteady combustion phenomena using complex models. In *35th AIAA/ASME/SAE/ASEE Joint propulsion conference and exhibit*, June 1999.

[HRZ98]   T. Hagstrom, K Radhakrishnan, and R. Zhou. Computation of steady and unsteady laminar flames: theory. In *34th AIAA/ASME/SAE/ASEE Joint propulsion conference and exhibit*, July 1998.

[KC01]   Christopher A. Kennedy and Mark H. Carpenter. Additive Runge-Kutta schemes for convection-diffusion-reaction equations. *Appl. Num. Math.*, 2001. Submitted.

[KM85]   J. Kim and P. Moin. Application of a fractional-step method to incompressible Navier-Stokes equations. *J. Comput. Phys.*, 59:308–323, 1985.

[MB97]   M. L. Minion and D. L. Brown. Performance of underresolved two-dimensional incompressible flow simulations II. *J. Comput. Phys.*, 138:734–765, 1997.

[Min01]   M. L. Minion. Higher-order semi-implicit methods for initial boundary value partial differential equations. In preparation.

[Ose89]   V. I. Oseledets. On a new way of writing the Navier-Stokes equation: The Hamiltonian formalism. *Commun. Moscow Math.Journal of Society*, 44(1):210–211, 1989.

[Pat97]   D. Pathria. The correct formulation of intermediate boundary condidtions for Runge-Kutta time integration of intitial boundary value problems. *SIAM J. Sci. Comput.*, 18(5):1255–1266, 1997.

[RR98]   M. C. Recchioni and G. Russo. Hamiltonian-based numerical methods for a fluid-membrane interaction in two and three dimensions. *SIAM J. Sci. Comput.*, 19:861–892, 1998.

[SC96]   D. M. Summers and A. J. Chorin. Numerical vorticity creation based on impulse conservation. *Proc. Nat. Acad. Sci. USA*, 93:1881–1885, 1996.

[SSVH86]   J. M. Sans-Serna, J. G. Verwer, and W. H. Hundsdorfer. Convergence and order reduction of Runge-Kutta schemes applied to evolutionary problems in partial differential equations. *Num. Math.*, 50:405–418, 1986.

[SZ96]    J. W. Shen and X. Zhong. Semi-implicit Runge-Kutta schemes for the non-autonomous differential equations in reactive flow computations. In *Proceedings of the 27th AIAA Fluid Dynamics Conference*. AIAA, June 1996.

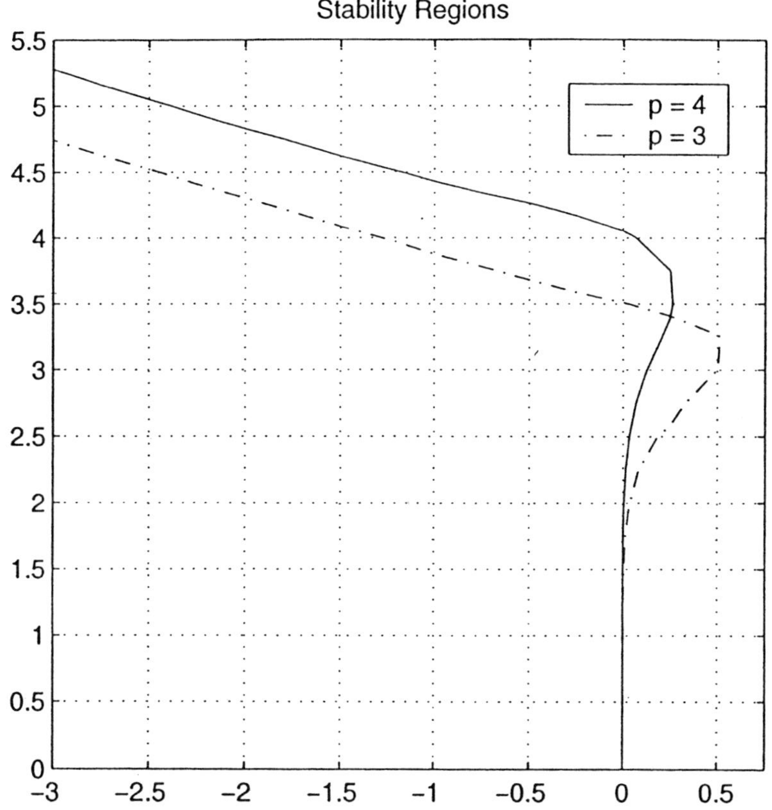



FIGURE 1. Stability regions for fourth-order SISDC methods. The two regions correspond to versions with $p = 3$ substeps per time step and $p = 4$ substeps.

| Errors for exact solution problem | | | | | |
|---|---|---|---|---|---|
| | | $64 \times 64$ | $128 \times 128$ | $256 \times 256$ | rate |
| BCM | u | 3.349e-5 | 1.966e-6 | 1.384e-7 | 3.83 |
| | $p$ | 6.508e-7 | 2.037e-8 | 9.266e-10 | 4.46 |
| KM | u | 3.349e-5 | 1.966e-6 | 1.384e-7 | 3.83 |
| | $p$ | 6.970e-7 | 2.092e-8 | 9.031e-10 | 4.53 |
| EL | u | 3.349e-5 | 1.966e-6 | 1.384e-7 | 3.83 |
| | $p$ | 4.097e-7 | 1.946e-8 | 1.084e-9 | 4.16 |

TABLE 1.1. $L_\infty$ errors for the $u$-component of velocity and the pressure for the exact solution test problem. The rates are computed from the errors in the $128 \times 128$ and $256 \times 256$ grids.

| Errors in the pressure | | | | |
|---|---|---|---|---|
| | $64 \times 64$ | $128 \times 128$ | $256 \times 256$ | rate |
| BCM | 1.963e-6 | 2.036e-7 | 2.246e-8 | 3.18 |
| KM | 2.816e-6 | 2.483e-7 | 2.624e-8 | 3.24 |
| EL | 2.704e-6 | 2.714e-7 | 2.848e-8 | 3.25 |

TABLE 1.2. $L_\infty$ errors for the pressure for the exact solution test problem with only 3 substeps per time step. The rates are computed from the errors in the $128 \times 128$ and $256 \times 256$ grids and reflect the fact that the temporal part of the pressure equation is being computed with only third-order accuracy.

| Errors for the shear layer problem | | | | | |
|---|---|---|---|---|---|
| | | $64 \times 64$ | $128 \times 128$ | $256 \times 256$ | rate |
| BCM | u | 8.311e-4 | 5.544e-5 | 3.477e-6 | 4.00 |
| | $p$ | 3.887e-4 | 2.576e-5 | 1.628e-6 | 3.98 |
| KM | u | 1.158e-3 | 8.000e-5 | 4.296e-6 | 4.22 |
| | $p$ | 1.185e-3 | 5.395e-5 | 2.662e-6 | 4.34 |
| EL | u | 3.816e-3 | 2.520e-4 | 1.574e-5 | 4.00 |
| | $p$ | 3.892e-3 | 2.543e-4 | 1.588e-5 | 4.00 |

TABLE 1.3. Errors for the pressure and $u$-component of the velocity for the shear layer test problem. The rates are computed from the errors in the $128 \times 128$ and $256 \times 256$ grids by comparison with the reference solutions.

# Numerical Solutions of the Incompressible Navier Stokes Equations Using Helmholz Velocity Decomposition


**K. Nikfetrat**

Mechanical and Manufacturing Technologies, British Columbia Institute of Technology, Burnaby, BC V5G 3H2, Canada

**M. Hafez**

Department of Mechanical and Aeronautical Engineering, University of California, Davis, CA 95616, U.S.A


## Abstract


A numerical method for solving the incompressible Navier Stokes equations based on splitting the velocity into a rotational and a potential part via a Helmholz decomposition is presented, using the Lamb form of the Navier Stokes equations. The scheme conserves both mass and momentum to machine accuracy and is not limited to two dimensions. No special handling is required for multipley connected domains, since a Poisson equation is solved for the pressure. Careful selection of farfield boundary conditions eliminate artificial sources of vorticity.

A finite difference discretization on a staggered grid is used to test the scheme. A segregated solution algorithm is adopted and shows that the scheme has good convergence properties. To demonstrate that the method is equally well suited for internal and external flow calculations, flow in a driven cavity, flow over a backward facing step, flow in a cascade and flow over a flat plate are computed.


## Introduction

Solutions of the incompressible Navier Stokes equations call for sophisticated algorithms to ensure a divergence free velocity field. The following observations set the incompressible Navier Stokes equations apart from their counter part for compressible flows:

1) The continuity equation looses it's dominant term, namely the time derivative of density, because the density remains constant.
2) In the incompressible limit, $Ma \rightarrow 0$, the thermodynamic pressure is replaced by a mechanical pressure which is not governed by an equation of state.
3) The energy equation decouples from the rest of the system and temperature becomes independent of the pressure.

To address these issues, many numerical methods for simulation of incompressible viscous flows have been developed since computers became available in the 1950's [1]. For two dimensional problems, the stream function vorticity formulation (or biharmonic equation) has been used based on finite difference or finite element approximations. However, the extension of this formulation to three dimensions in terms of a vector potential is not straightforward. Velocity/vorticity formulation is another attractive alternative[2].

In terms of primitive variables (without restriction to two dimensions), the artificial compressibility method of Chorin[3], Temam[4] and others was among the first attempts to solve steady Navier Stokes equations via adding artificial (time dependent) terms. Harlow and Welsh[5] introduced the MAC method, which uses a staggered grid and a Poisson equation for pressure, where the latter also serves to indirectly enforce mass conservation. Ghia, Hankey and Hodge[6] adopted this approach to a collocated grid and reported that mass could be conserved only to within the truncation error of the scheme, since the discrete form of the Poisson equation for pressure acts merely as a bound on the *temporal change* of the divergence of the velocity. To enforce mass conservation to machine accuracy, Chorin introduced his projection or splitting method[7], where mass conservation was enforced by solving a Poisson's equation for a pressure correction to the velocity field. The latter was obtained from the momentum equation by ignoring the pressure gradient based on the rational that the vortical components of the velocity are independent of the pressure distribution. A similar strategy was adopted by the Los Alamos group in their version of simple MAC (SMAC) method with homogeneous boundary conditions for a potential correction. These methods require special handling when applied to multipley connected domains, since the pressure is eliminated. Later, a family of methods were introduced by Patankar and Spalding[8], based on the pressure correction approach, they are known as SIMPLE, SIMPLER and SIMPLEC methods. In these methods, the pressure corrections are related to the velocity corrections by approximate forms of the momentum equation. Also a predictor/corrector strategy was proposed by Issa[9] in his Pressure Implicit with Splitting of Operators (PISO) method, where two or more Poisson's equations are solved at each time step. Gresho[10] proposed optimum projection methods using two or more Poisson's equations, one for a potential correction to the velocity field to conserve mass and the others for the pressure, and it's derivative with respect to time. Also, in Glowinski and Pironneau's[11] decomposition method, two Poisson's equations are solved, one for the potential and one for the pressure, with the pressure equation subject to Dirichlet boundary condition, which is evaluated by solution of an integral equation. Most of these methods use staggered grids or mixed finite elements. Modifications are necessary for using regular (collocated) grids or the same shape function for velocity and pressure, as for example in the least squares/Galerkin method of Tezduyar, Behr and Hughes[12].

In the present effort, the velocity vector is split into a potential function and a rotational part via a Helmholz decomposition. We use two Poisson's equations, one for the potential function and the other for pressure. Staggered grids are the natural choice for finite differences. In particular, some vector identities hold on the discrete level. For example the curl of a gradient vanishes, as it should. Mass and momentum are conserved to machine accuracy, and the correct pressure gradient is computed at the boundaries by using the normal momentum equation as a Neuman boundary condition. Hence, the tangential momentum equation is also satisfied at the boundaries and no special handling for multipley connected domains is necessary. In the present method, no relation is assumed between the potential and the pressure, unlike some methods, where the velocity correction is proportional to the pressure gradient, and the pressure itself is proportional to the potential function.

Other potential correction schemes treat the potential as a correction that is either *absorbed* into the velocity, or approaches zero on convergence. However in the present scheme the potential is neither absorbed nor is it a correction, it is a separate part of the velocity. The potential part of the velocity does not vanish on convergence. For flows where a physical velocity potential exists, the scheme can be interpreted as adding a rotational correction to a potential velocity field, in order to satisfy the no slip boundary condition, in contrast to traditional schemes where a potential correction is added to a rotational velocity field. The present method reduces to the potential formulation if viscosity and vorticity effects are dropped. For inviscid, irrotational flows, the velocity is the gradient of a potential function which is governed by a Laplace equation with Neumann boundary conditions at solid surfaces. The pressure can be obtained in terms of the velocity in closed form as an integral of the momentum equations. With viscous effects and a no slip condition at a solid surface, vorticity is generated and diffused. The vortical components of the velocity are obtained from the momentum equations (in Lamb's form) and to conserve mass, a Poisson's equation for the potential is solved, while pressure is obtained from another Poisson's equation which is the divergence of the momentum equations. Results are presented to demonstrate the feasibility of the formulation and the quality of the solution particularly near solid boundaries. Finally, a comparison with fully coupled methods is important. For example, Kwak etal[13] and Zienkiewicz etal[14] solved the primitive variables equations, where the continuity equation is augmented with artificial dissipation terms. In their work, all equations must be solved everywhere, and in regions where the flow is inviscid and irrotational, numerical vorticity is generated due to the truncation error, particularly if the grid is coarse. In the current approach, splitting of the velocity vector into a rotational and a potential part via a Helmholz decomposition and careful selection of farfield boundary conditions eliminate all numerical sources of vorticity.

## Formulation

The dimensionless equations that enforce conservation of mass and momentum for incompressible flows are:

continuity equation
$$\vec{\nabla} \cdot \vec{V} = 0 \tag{1}$$

Navier Stokes equations
$$\frac{\partial \vec{V}}{\partial \tau} - (\vec{V} \cdot \vec{\nabla})\vec{V} = -\vec{\nabla}p + \frac{1}{Re}\nabla^2 \vec{V} \tag{2}$$

The continuity equation is in fact a constraint. In order to conserve mass, in the present formulation the velocity vector is split into two parts:

$$\vec{V} = \vec{V}_r + \vec{\nabla}\phi \tag{3}$$

Since the curl of a gradient is zero, the two parts of the velocity in Eq. (3) can be interpreted as the rotational and irrotational parts in Helmholz decomposition. Substitution of Eq. (3) into the continuity equation results in a Poisson equation for $\phi$ :

$$\vec{\nabla} \cdot (\vec{V}_r + \vec{\nabla}\phi) = 0 \tag{4}$$

or

$$\nabla^2 \phi = -\vec{\nabla} \cdot \vec{V}_r$$

hence mass conservation can be enforced explicitly by solving the Poisson Eq. (4) for $\phi$.

Introduction of $\vec{\nabla}\phi$ into the system of equations affects every aspect of the scheme, including discretization and boundary conditions. Of immediate concern is the viscous term in the momentum equation that takes the form $\nabla^2(\vec{V}_r + \vec{\nabla}\phi)$. Evaluation of the discrete $\nabla^2(\vec{\nabla}\phi)$ term requires special handling at the boundaries. On the other hand the equation for vorticity is:

$$\vec{\omega} = \vec{\nabla} \times \vec{V} = \vec{\nabla} \times (\vec{V}_r + \vec{\nabla}\phi)$$
$$= \vec{\nabla} \times \vec{V}_r \tag{5}$$

According to Eq. (5) the contribution of the irrotational velocity can be eliminated if the vorticity $\vec{\omega}$ is introduced. The momentum equations can be written in the Lamb or rotational form as:

$$\frac{\partial \vec{V}}{\partial \tau} - \vec{V} \times \vec{\omega} = -\vec{\nabla}(p + \frac{1}{2}V^2) - \frac{1}{Re}\vec{\nabla} \times \vec{\omega} \tag{6a}$$

Eq. (6a) can be written in more compact form by noting that $p + 1/2\,V^2$ is the total pressure $P$:

$$\frac{\partial \vec{V}}{\partial \tau} - \vec{V} \times \vec{\omega} = -\vec{\nabla}P - \frac{1}{Re}\vec{\nabla} \times \vec{\omega} \tag{6b}$$

For steady flow two sets of Cauchy Riemann equations can be identified. Continuity equation and vorticity definition lead to $\vec{\nabla} \cdot \vec{V} = 0$ and $\vec{\nabla} \times \vec{\omega} = 0$. The steady state form of the vector equation (6b) in the directions tangent ($\hat{s}$) and normal ($\hat{n}$) to the velocity vector lead to:

$$\frac{\partial P}{\partial s} = -\frac{1}{Re}\frac{\partial \omega}{\partial n}$$

$$\frac{\partial P}{\partial n} = -\frac{1}{Re}\frac{\partial \omega}{\partial s} - V^2\omega$$

which are also a system of Cauchy Riemann equations.

Poisson equations for the static and total pressures can be obtained in the usual manner by taking the divergence of the momentum equations, in this case Eqs. (6a) or (6b). The result is:

$$\vec{\nabla} \cdot (\vec{\nabla}(p + \frac{1}{2}V^2) + \vec{V} \times \vec{\omega}) = 0 \tag{7a}$$

or

$$\vec{\nabla} \cdot (\vec{\nabla}P + \vec{V} \times \vec{\omega}) = 0 \tag{7b}$$

where the unsteady and viscous terms have been dropped because they are identically zero. (The unsteady term is eliminated because the velocity is divergence free and the viscous term because the gradient of a curl is zero.) Experience shows that if the velocity field is not divergence free, terms involving $\vec{\nabla} \cdot \vec{V}$ must be retained in the Poisson equation for pressure, this includes the viscous term[]. Since the current algorithm enforces a divergence free velocity field at each step, there is no need to keep any of the terms involving $\vec{\nabla} \cdot \vec{V}$ in the Poisson equation for pressure.

It must be emphasized that $\vec{\omega}$ in the above equations stands for $\vec{\nabla} \times \vec{V}_r$ and is not a new variable. The system of equations and unknowns is the following: the three components of Eq. (6a) or (6b) are solved for $u_r$, $v_r$ and $w_r$, Eq. (4) for $\phi$ and Eq. (7a) for $p$ or (7b) for $P$. In other words the momentum equations impose the correct vorticity in the velocity and the continuity equation makes it divergence free. In order to use the Lamb form of the Navier Stokes equations, the pressure is needed, since the pressure and vorticity constitute a Cauchy Riemann system as discussed before. Without the correct pressure, Eqs. (6a) and (6b) do not admit a divergence free velocity field.

The Helmholz decomposition of the velocity calls naturally for the Lamb form of the Navier Stokes equations. With the exception of spectral methods, use of the Lamb form is not popular in CFD. Burggraf[15] used a Poisson equation based on the divergence of the Lamb form to obtain the total pressure for the driven cavity problem, after he computed the velocity using a streamfunction vorticity formulation. A similar approach was used by Abdellah in part one of his two part paper [16], which focused on computation of the pressure from a known velocity field. However in part two of his paper, where he investigated a method for solving the Navier Stokes equations, he abandoned the Lamb form completely [17]. Peeters, Habashi and Nguyen [18] arrived at the Lamb form by decomposing the velocity into a rotational part and a gradient of the pressure, where the latter is multiplied by a small parameter. They reported difficulties in implementing the viscous term $\bar{\nabla} \times \bar{\omega}$, and instead used $\nabla^2 \bar{V}_r$. This amounts to neglecting the contribution of $\nabla^2 (\bar{\nabla}\phi)$ to the viscous term. The contribution of $\nabla^2 (\bar{\nabla}\phi)$ is not necessarily small. (If the contribution of $\nabla^2 (\bar{\nabla}\phi)$ had been kept, their solution would not have been smooth.) Such difficulties are not encountered in the present scheme and $\bar{\nabla} \times \bar{\omega}$ is used directly.

To prevent the production of artificial vorticity, it is recommended to use equations (6a) and (7a) in terms of static pressure rather than the pair (6b) and (7b) which are in terms of the total pressure. If the total pressure is computed from Eq. (7b), gradients of total pressure are propagated over the *entire* flow field, forcing equation (6b) to produce vorticity to balance the gradient of total pressure. Production of vorticity is undesirable in regions that should remain irrotational. On the other hand, in equation (6a), the gradient of the static pressure is balanced in part by the gradient of the square of the magnitude of the velocity, and it is not necessary to generate vorticity. This situation can be avoided if the total pressure is calculated from a convection/diffusion equation rather the a Poisson's equation. Flow fields that do not have vorticity free regions can be computed by using Eqs. (6a) and (7a), and several of the flow fields presented in this paper have been computed with this pair.

## Boundary Conditions

Decomposition of velocity into two parts allows more flexibility in applying boundary conditions, but at the same time leads to additional complexity. Careful consideration is required to ensure that the sum of the two parts of the velocity satisfies the boundary conditions and the physical meaning of each part is retained. The flux of vorticity at the boundaries is a crucial factor and can be modeled correctly only if the rotational velocities along the boundaries are correctly specified. The pressure along the boundaries is always obtained from the normal component of the momentum equation evaluated at the boundary.

In order to demonstrate the effect of the boundary conditions on the evolution of the two parts of the velocity, four flow configurations have been considered: driven cavity flow (with and without an obstacle), flow over a backward facing step, flow through a cascade and flow over a flat plate. The flow field in the driven cavity is predominantly rotational, whereas flow over a flat plate is predominantly irrotational outside the boundary layer.

## Driven Flow in a Square Cavity

Boundary conditions for this configuration are straightforward. Solid walls surround the domain on all sides and the no slip and no penetration conditions also apply to the central obstacle, if it is present. The potential part of the velocity must not penetrate a solid wall. Setting $\vec{\nabla}\phi \cdot \hat{n} = \partial\phi/\partial n = 0$ at a solid wall meets this condition and the normal component of the rotational velocity is also zero at the wall and both parts of the velocity satisfy the no penetration condition. This is not true for the no slip condition. Because the potential velocity can slip along the walls, the no slip condition is only satisfied by the sum of the potential and rotational parts. To summarize:

At a solid boundary:

$$\frac{\partial\phi}{\partial n} = 0$$

$$\vec{V}_r \cdot \hat{n} = 0 \qquad\qquad (9a)$$

$$\vec{V}_r \cdot \hat{s} + \frac{\partial\phi}{\partial s} = V_{wall}$$

The condition $\partial\phi/\partial n = 0$ results in a Neumann problem with homogeneous boundary conditions for $\phi$ and the integral constraint which takes the following form must be satisfied for an admissible solution:

$$-\iint(\nabla \cdot \vec{V}_r)\,dA = \oint\frac{\partial\phi}{\partial n}ds = 0$$

## Flow Over a Backward Facing Step

Along the lower and upper walls and along the solid surface of the upstream boundary no slip conditions (9a) are specified. Along the upper half of the upstream boundary a fully developed parabolic velocity profile is specified. The incoming velocity parts are chosen such that the potential flow is uniform and carries all the mass. To summarize:

Inflow condition

$$\frac{\partial\phi}{\partial x} = 1$$

$$u_r = 12(y - 2y^2) - 1 \qquad\qquad (9b)$$

$$v = 0$$

At the exit of the channel a fully developed flow exists, provided the channel is long enough. Several exit conditions can be specified for $\phi$, each leading to a different potential field. The simplest exit boundary condition is a uniform potential flow, obtained by imposing a uniform $\partial\phi/\partial x$ across the exit plane. To summarize:

Outflow condition

$$\frac{\partial \phi}{\partial x} = \int_{y_{min}}^{y_{max}} dy - (u_r)_m$$

$$\frac{\partial u}{\partial x} = 0$$

$$v = 0 \tag{9c}$$

where

$$(u_r)_m = \frac{\displaystyle\int_{y_{min}}^{y_{max}} u_r \, dy}{y_{max} - y_{min}}$$

$(u_r)_m$ is the mean value of the rotational part of the streamwise velocity evaluated over the exit plane.

## Flow Through a Cascade

This problem demonstrates uniform inflow and outflow conditions for channel type flows. Symmetry of the flow relative to the cascade centreline allows only half the flow field to be solved. To summarize:

Symmetry condition

$$\frac{\partial \phi}{\partial y} = 0$$

$$\frac{\partial u_r}{\partial y} = 0$$

$$v_r = 0 \tag{9d}$$

Along the channel walls, conditions (9a) are specified.

Far upstream of the cascade the flow is irrotational, while at the exit it is fully developed. A uniform, potential flow can be specified by requiring:

Inflow condition

$$\frac{\partial \phi}{\partial x} = 1$$

$$u_r = 0$$

$$\frac{\partial v_r}{\partial x} = 0 \tag{9e}$$

We noticed setting $v_r = 0$ does not guarantee irrotationality of the incoming flow. Failure to satisfy the irrotationality condition causes a shear layer to be formed at the inlet, which is convected downstream and adversely affects the entire flow field. Consequently, the vertical component of the incoming velocity can not be specified, it is part of the solution. If the inlet is sufficiently far from the channel, the computed value for $v$ will vanish. The outflow boundary condition is similar to condition (9c). Since $(u_r)_m$ is negative, the potential flow leaving the channel is faster than the potential flow that enters the channel. The potential part of the flow accelerates in the channel.

**Flow Over a Flat Plate**

This problem demonstrates inflow, outflow and farfield boundary conditions for external flows. Flow over a flat plate is essentially irrotational with a correction added to satisfy the no slip boundary condition along the plate. Development of the boundary layer over a flat plate is influenced strongly by the free stream condition, and if a Blasius type boundary layer is desired, the pressure drop must be small. Besides the boundary conditions, the size of the physical domain plays an important role in achieving satisfactory levels of convergence. The solution presented in this paper is obtained using a distance of 5 plate lengths before and above the plate. In terms of the maximum boundary layer thickness this is $100\ \delta s$.

Conditions for the rotational parts of the velocity are typical outflow conditions: $u_r$ is extrapolated form the interior and the derivative of $v_r$ is set to zero or extrapolated similar to $u_r$. To summarize:

Outflow condition

$$\frac{\partial^2 \phi}{\partial x^2} = 0$$

$$\frac{\partial^2 u_r}{\partial x^2} = 0 \tag{9f}$$

$$\frac{\partial v_r}{\partial x} = 0$$

Along far field

$$\phi = x$$

$$\frac{\partial u_r}{\partial y} = 0 \tag{9g}$$

$$v_r = 0$$

For pressure the normal component of the momentum equations is used:

$$\hat{n} \cdot \left( \frac{\partial \vec{V}}{\partial \tau} - \vec{V} \times \vec{\omega} + \vec{\nabla}P + \frac{1}{Re}\vec{\nabla} \times \vec{\omega} \right) = 0$$

which reduces to:

$$\frac{\partial P}{\partial n} = \hat{n} \cdot (\vec{V} \times \vec{\omega}) - \frac{1}{Re}\vec{\nabla} \times \vec{\omega} \tag{10a}$$

or if (6a) is used:

$$\frac{\partial p}{\partial n} = -\frac{1}{2}\frac{\partial V^2}{\partial n} + \hat{n}\cdot(\vec{V}\times\vec{\omega}) - \frac{1}{Re}\vec{\nabla}\times\vec{\omega} \qquad (10b)$$

Along solid walls the convective term $\hat{n}\cdot(\vec{V}\times\vec{\omega})$ is zero. The pressure equation is subject to pure Neumann boundary conditions and a great deal has been written about the implications of this [2][3][5] and the associated integral constraint that must be satisfied[1],[2],[4]. The appropriate form of the constraint for Eq. (8) is derived by examining the integral form the equation:

$$\iint(\vec{\nabla}\cdot\vec{\nabla}P)dA = \iint[\vec{\nabla}\cdot(\vec{V}\times\vec{\omega})]dA$$

Hence:

$$\oint(\vec{\nabla}P\cdot\hat{n})ds = \iint[\vec{\nabla}\cdot(\vec{V}\times\vec{\omega})]dA$$

Abdellah has demonstrated that the discrete sum of $(\vec{\nabla}\times\vec{\omega})\cdot\hat{n}$ along the boundaries is identically zero and the integral constraint is exactly satisfied [2]. There is no need to subtract a correction from the source term of the Poisson equation during each iteration, as was done in [5].

## Numerical Schemes

In two dimensions, four scalar equations have to be solved. Depending on whether the total or static pressure is used these equations have slightly different forms. The continuity equation is always:

$$\frac{\partial}{\partial x}(u_r + \frac{\partial\phi}{\partial x}) + \frac{\partial}{\partial y}(v_r + \frac{\partial\phi}{\partial y}) = 0 \qquad (11)$$

If static pressure is used the momentum and pressure equations are

$$\frac{\partial u}{\partial\tau} - v\omega + \frac{\partial p}{\partial x} + \frac{1}{2}\frac{\partial V^2}{\partial x} + \frac{1}{Re}\frac{\partial\omega}{\partial y} = 0 \qquad (12a)$$

$$\frac{\partial v}{\partial\tau} + u\omega + \frac{\partial p}{\partial y} + \frac{1}{2}\frac{\partial V^2}{\partial y} - \frac{1}{Re}\frac{\partial\omega}{\partial x} = 0 \qquad (13a)$$

$$\frac{\partial}{\partial x}(\frac{\partial p}{\partial x} + \frac{1}{2}\frac{\partial V^2}{\partial x} - v\omega) + \frac{\partial}{\partial y}(\frac{\partial p}{\partial y} + \frac{1}{2}\frac{\partial V^2}{\partial y} + u\omega) = 0 \qquad (14a)$$

or, if the total pressure is used

$$\frac{\partial u}{\partial\tau} - v\omega + \frac{\partial P}{\partial x} + \frac{1}{Re}\frac{\partial\omega}{\partial y} = 0 \qquad (12b)$$

$$\frac{\partial v}{\partial \tau} + u\omega + \frac{\partial P}{\partial y} - \frac{1}{Re}\frac{\partial \omega}{\partial x} = 0 \tag{13b}$$

$$\frac{\partial}{\partial x}\left(\frac{\partial P}{\partial x} - v\omega\right) + \frac{\partial}{\partial y}\left(\frac{\partial P}{\partial y} + u\omega\right) = 0 \tag{14b}$$

In these equations $\omega = \partial v_r/\partial x - \partial u_r/\partial y$, $u = u_r + \partial\phi/\partial x$ and $v = v_r + \partial\phi/\partial y$. Equations "b" are used for the driven cavity and channel flow problems, whereas equations "a" are used for flow over the flat plate.

At this point it is necessary to examine various options for arrangement of the variables. For incompressible flows the divergence free condition dictates how the variables should be arranged and how the equations should be discretized. The objective is to ensure that when the discrete divergence operator is applied to the velocities, the right hand side is zero.

$$\frac{u_{i+\frac{1}{2},j} - u_{i-\frac{1}{2},j}}{\Delta\xi}\xi_x + \frac{v_{i,j+\frac{1}{2}} - v_{i,j-\frac{1}{2}}}{\Delta\eta}\eta_y = 0 \tag{15}$$

Because part of the velocity is the gradient of a potential, components of the rotational velocity must either be stored or evaluated at nodes between the nodes where the potential is stored, so that the sum of the two velocity parts *exactly* satisfies the divergence free condition as expressed in Eq. (15). To meet this requirement, three options are available: 1) Use a staggered grid. 2) Use a non staggered nine point wide stencil. 3) Use a five point collocated arrangement, but use *only* averaged rotational velocities evaluated between the nodes. Option 3 is more attractive than option 1 when the equations are transformed to a non orthogonal generalized coordinate system, because all the variables are available at all the nodes. Option 3 will be investigated later when the current scheme is extended to non orthogonal generalized coordinates. Option 2 was abandoned because it encountered a number of difficulties. One difficulty was the odd even decoupling, in the solution of the Poisson equations. An interesting side effect of the decoupling was the dependence of the solutions for $\phi$ on the over relaxation parameter! The decoupling was even more acute in the case of the Poisson pressure equation. Moreover, application of a nine point stencil near the boundaries requires ghost points  Staggered grids not only avoid these problems, but have the advantage that the curl of a gradient is divergence free at the discrete level. For all these reasons, a staggered mesh was selected for this study. By aligning the physical boundaries with the vertical velocities on a staggered grid, the governing equations can be used to compute the flow variables at the ghost points using central differencing, thereby eliminating the need for ad hoc assumptions and one sided differencing. The staggered mesh is shown in Figures 1a and b. Figure 1a shows $u_r$ is stored at $i+1/2$, $v_r$ at $j-1/2$, while $\phi$ and

$P$ or $p$ are stored at $i,j$. The shaded region in Fig. 1b is the extent of the physical domain. The thick solid line is the extent of the grid points, it does not line up with the physical boundaries of the domain. The dashed lines are extensions at the end of which ghost points must be implemented.

The ideal algorithm for solving the four equations is a fully implicit direct solver, however the amount of memory required is prohibitive if small computers are used. For this reason, a fully implicit segregated solution procedure is used. First Eqs. (12a) and (13a) or (12b) and (13b) are solved coupled together for $u_r$ and $v_r$. The resulting velocity field has the correct vorticity, but is not divergence free. Next $u_r$ and $v_r$ are used in Eq. (11) to compute $\phi$. At this point the velocity field is divergence free and has the correct vorticity, but neither satisfies the momentum equation nor the no slip boundary condition. Finally either $p$ is computed from Eq. (14a) or $P$ from Eq. (14b), in order to find a pressure field that is compatible with the divergence free velocity field. At this stage the four unknowns are all updated, but the momentum equation is not satisfied, because it has an unsteady term, and because the boundary conditions are not satisfied. The process has to be repeated until the momentum equation ceases to modify $u_r$ and $v_r$. As the solution converges, fewer and fewer iterations are needed to solve the Poisson equations.

Every new value computed for $\phi$ drives the momentum equations to produce new values for $u_r$ and $v_r$ in an effort to satisfy the no slip boundary condition. Identical discrete divergence operators must be used for the two Poisson equations. Furthermore, the boundary conditions for $u_r, v_r$ and $\phi$ must be compatible. It is also crucial that the pressure be updated every time $\phi$ is updated and before the updated $\phi$ is used in the momentum equations. It is the pressure that forces the solutions for $u_r$ and $v_r$ and $\phi$ to converge. Other segregated schemes require the pressure solution to be under relaxed before it is used in the momentum equations [3][5]. The current scheme does not require under relaxation. The solution is considered converged when the right hand sides of the *steady state* momentum equations and the two Poisson's equations drop below $10^{-6}$.

The governing equations are discretized on an orthogonal stretched coordinate system. The Jacobian and metric terms of the transformation are evaluated using second order central differences. A fully implicit backward Euler time discretization scheme is used for the momentum equations. Since only steady state solutions are of interest, first order time accuracy is adequate. The two Poisson equations have no time dependence and are always central differenced in space. For high Reynolds numbers the momentum equations loose diagonal dominance if central differencing is used exclusively; some form of upwinding must be used. However it is not immediately clear how upwinding should be applied to $\bar{V} \times \bar{\omega} + \bar{\nabla}(V^2/2)$ in the Lamb form of the Navier Stokes equations, which replaces the flux term $(\bar{V} \cdot \bar{\nabla})\bar{V}$ in the standard form. In the present work, upwinding is limited to the $\bar{V} \times \bar{\omega}$ term, $\bar{\nabla}(V^2/2)$ and $\bar{\nabla}p$ are central differenced. A number of schemes can be considered to upwind difference $\bar{V} \times \bar{\omega}$. One sided differences can be used to evaluate $\omega$ at the centre of the momentum cells to high orders of accuracy. This option was dropped in favour of evaluating $\omega$ at cell boundaries rather than cell centres. In this manner $\omega$ is evaluated from a central difference formula, and a block pentadiagonal solver is not needed. The sign of the velocity determines which cell boundary $\omega$ must be computed at. In Fig 1c, the y-momentum cell, centered at $i, j - 1/2$ is highlighted with a gray background. For this cell, without upwinding $\omega = 1/2(\omega_{i+1/2,j-1/2} + \omega_{i-1/2,j-1/2})$, with upwinding and positive $u$, $\omega$ is evaluated at the left wall (point $i - 1/2, j - 1/2$) and with negative $u$, $\omega$ is evaluated at the right wall (point $i + 1/2, j - 1/2$). This upwind scheme is first order accurate, higher order schemes will be discussed elsewhere.

The details of the solution procedure for each of the discretized equations will now be discussed separately, starting with the momentum equations. The x and y momentum equations are centred at $i + 1/2, j$ and $i, j - 1/2$ respectively and solved coupled together for $u_r$ and $v_r$. The discrete form of the x momentum in terms of the total pressure (Eq. 12b) is shown below:

$$\frac{(u_r)_{i,j} - (u_r^n)_{i,j}}{\Delta t} + \frac{P_{i+1,j} - P_{i,j}}{\Delta\xi}\xi_x$$

$$- v_{i+\frac{1}{2}}\left\{\frac{((v_r)_{i+1,j+1} + (v_r)_{i+1,j}) - ((v_r)_{i,j+1} + (v_r)_{i,j})}{2\Delta\xi}\xi_x + \frac{(u_r)_{i,j+1} - (u_r)_{i,j-1}}{2\Delta\eta}\eta_y\right\}$$

$$+ \frac{1}{Re}\left\{\frac{\left(\dfrac{(v_r)_{i+1,j+1} - (v_r)_{i,j+1}}{\Delta\xi}\xi_x - \dfrac{(u_r)_{i,j+1} - (u_r)_{i,j}}{\Delta\eta}\eta_y\right) - \left(\dfrac{(v_r)_{i+1,j} - (v_r)_{i,j}}{\Delta\xi}\xi_x - \dfrac{(u_r)_{i,j} - (u_r)_{i,j-1}}{\Delta\eta}\eta_y\right)}{\Delta\eta}\eta_y\right\} = 0$$

The superscript $n+1$ for the new time level has been left out for clarity. It is noteworthy that $(u_r)_{i+1,j}$ and $(u_r)_{i-1,j}$ are not present in this equation. The same is true for $(v_r)_{i,j+1}$ and $(v_r)_{i,j-1}$ in the y momentum equation. This can result in loss of diffusion and prevent the system of equations from converging, if the momentum equations are not solved coupled together.

Along the southern boundary the boundary condition for $u_r$ is embedded into the x-momentum equation and the boundary condition for $v_r$ is enforced explicitly, for all three test problems. Along the eastern boundary, the condition for $v_r$ is embedded in the y-momentum equation and the condition for $u_r$ is imposed explicitly along a solid wall. For flow-through boundaries, both conditions are imposed explicitly. Northern and western boundary conditions are always imposed explicitly. In all cases second order accuracy is maintained at the boundaries, without requiring second order one sided differencing. By avoiding second order accurate one sided differences it is possible to use a block tridiagonal solver in a line relaxation algorithm to solve for $u_r$ and $v_r$ along lines of constant $j$, by sweeping in the $i$ direction, or vice versa. Several sweeps are needed at each step. For the sample problems considered, 10 sweeps per step proved sufficient. The block tridiagonal solver used in this study is based on the Thomas algorithm and does not perform pivoting. As a result, every single block row of the discrete coefficient matrix must have a sizable diagonal term. This restricts the size of the time step at high Reynolds numbers..

The discrete Poisson equation for $\phi$ is centred at $i,j$ and is obtained by expressing $u$ and $v$ in Eq. (15) in terms of their potential and rotational parts :

$$\frac{\left[(u_r)_{i+\frac{1}{2},j} + \dfrac{\phi_{i+1,j} - \phi_{i,j}}{\Delta\xi}(\xi_x)_{i+\frac{1}{2},j}\right] - \left[(u_r)_{i-\frac{1}{2},j} + \dfrac{\phi_{i,j} - \phi_{i-1,j}}{\Delta\xi}(\xi_x)_{i-\frac{1}{2},j}\right]}{\Delta\xi}(\xi_x)_{i,j} +$$

$$\frac{\left[(v_r)_{i,j+\frac{1}{2}} + \dfrac{\phi_{i,j+1} - \phi_{i,j}}{\Delta\eta}(\eta_y)_{i,j+\frac{1}{2}}\right] - \left[(v_r)_{i,j-\frac{1}{2}} + \dfrac{\phi_{i,j} - \phi_{i,j-1}}{\Delta\eta}(\eta_y)_{i,j-\frac{1}{2}}\right]}{\Delta\eta}(\eta_y)_{i,j} = 0 \qquad (16)$$

Eq. (16) is solved at all interior points and also along southern and eastern boundaries, where the boundary condition for $\phi$ is embedded. Along the northern and western boundaries the boundary condition for $\phi$ is enforced explicitly, since these points are outside the physical domain. In this manner, mass is conserved in the entire gray region in Fig. 1a, enforcement of boundary conditions does not compromise mass conservation for boundary cells. Eq. (16) is solved for $\phi$ using a point or preferably line SOR method. The optimum over relaxation factor was found by trial and error to be 1.8. As already mentioned, Eq. (16) must be solved to full convergence at each time step.

Eq. (14a) for p and Eq. (14b) for P are both centered at i, j and solved using the exact same discrete divergence operator used in Eq. (16). A point or line SOR method similar to the one used for Eq. (16) is used. The relaxation factor of 1.8 works best here also. Eq. (14b) is considerably more complex than Eq. (11) and takes about three times longer to evaluate per iteration. Eq. (14a) has even more terms than Eq. (14b), a strong argument for using (14b) whenever possible. Fortunately, Eqs. (14a) or (14b) do not need to be solved to full convergence at each step. The boundary conditions for Eqs. (14a) and (14b), obtained by discretizing Eqs. (10a) or (10b), are evaluated explicitly along the western and northern boundaries; they are embedded along the southern and eastern boundaries. The southeast corner requires the embedding of both the southern and the eastern boundary conditions. The *momentum* equations will not converge to machine accuracy, if pure Neumann conditions are not used for Eqs. (14a) or (14b).

No attempt is made to start the solution by supplying a reasonable initial guess for the unknowns. All unknowns are assigned a constant value everywhere as their initial condition.

## Discussion of Results

### Driven Flow in a Square Cavity

To demonstrate the ability of the present method to deal with multipley connected domains, results for a cavity with a central obstacle are presented in addition to results for the standard driven cavity problem. When a central obstacle is present, the driven cavity is a doubley connected domain. Three solutions are presented:

- Cavity without obstacle:
    1.  Reynolds number = 100, using a uniform $21 \times 21$ grid
    2.  Reynolds number = 400, using a stretched $31 \times 31$ grid

- Cavity with obstacle
    3.  Reynolds number = 100, using a uniform $31 \times 31$ grid

Since flow in a cavity is highly rotational and diffusive for the range of Reynolds numbers investigated, the total pressure is computed from Eq. (14) and used directly in the momentum equations. Central differencing is used exclusively. Figs. 2 through 6 show the velocities, total pressure, vorticity and flow directions for the two cases without an obstacle. Our calculations are in close agreement with published results [6][16][20]. Subtle details of the flow are captured correctly, down to the recirculation zones shown in Figs. 5a and 5b. Figs. 7 and 8 show the static and total pressure contours when a square obstacle of length $1/3$ is placed at the center of the cavity. Unlike methods that do not compute the pressure at each step, it is not necessary to impose additional constraints to exclude non physical solutions. The flow field and vorticity contours for this case are shown in Figs 9 and 10.

As a result of imposing the no penetration condition on $\phi$ at all boundaries, flow in a driven cavity does not have a potential part if $\bar{\nabla} \cdot \bar{V}_r = 0$ everywhere. In other words, the potential part of the velocity is produced by the source term $\bar{\nabla} \cdot \bar{V}_r$ exclusively. In these circumstances no physical significance can be attached to the potential part of the velocity and it's role can be interpreted as a correction to the rotational velocity, which helps conserve mass. The potential part plays an important role in some other flows discussed next.

## Flow Over a Backward Facing Step

The length of the channel is 30h, where h is the step height, 51 points are used across the channel and 220 points along its length. The solutions shown in Figures 11 and 12 where computed at a Reynolds number of 400 based on the channel height and average velocity of the incoming flow. Central differencing and total pressure are used. Our results are in good agreement with other published results, in particular the sizes of the recirculation zones are in close agreement with results published by Rogers and Kwak[19] and Kim and Moin[20]. The streamlines in Figure 11 are the result of post processing. The vertical scale has been expanded in the figures for ease of viewing.

## Flow Through a Cascade

The calculation is carried out for $Re = 10$ based on the channel diameter, which is quite small. 89 points are used in the streamwise direction and 15 points across the channel. The grid is stretched in the streamwise direction, uniform in the cross flow direction and extends 10 channel diameters upstream and 10 channel diameters downstream of the cascade inlet. As for the driven cavity flow, the total pressure is used. At this low Reynolds number the leading edge singularity is easily dealt with by the natural viscosity in the flow and central differencing is used exclusively. The leading edge can be placed at either the velocity or the pressure center. Figures 13 to 15 show the potential and velocities for the channel flow.

Similarly to the flow over the backward facing step, the potential part of the velocity is not just produced by the source term $\bar{\nabla} \cdot \bar{V}_r$. The potential part is larger everywhere except at the boundaries where the two parts are of equal magnitude but have the opposite sign. Figure 14 demonstrates the distortion in the uniform incoming potential flow caused by the source term.

## Flow Over a Flat Plate

Flow over a flat plate is presented for a Reynolds number of 10000, using upwind differencing for the $\bar{V} \times \bar{\omega}$ term of the momentum equation. The static rather than the total pressure is used. The leading edge of the plate is placed at the pressure centre. While this arrangement avoids using the vorticity at the leading edge, the pressure boundary condition requires the value of $\partial p/\partial y$ at the leading edge, which calls for evaluation of $\partial \omega/\partial x$ across the singularity, an operation that is easily accomplished numerically. $\partial \omega/\partial x$ is scaled by $1/Re$, which reduces it's contribution to $\partial p/\partial y$. As a consequence of the large Reynolds number, vorticity is confined to a narrow region close to the plate, leaving the vast majority of the domain *completely* vorticity free, hence it can be described by a potential.

Boundary conditions for the boundary layer equations are ineffective for this formulation of the full Navier Stokes equations. Traditional farfield boundary conditions, namely $u_\infty = 1$ and $p_\infty = \text{constant}$, are clearly inappropriate. If $u_\infty$ is kept constant, $\partial u_\infty/\partial x$ is zero and the divergence free condition requires that $\partial v_\infty/\partial y$ also be zero, implying a constant vertical velocity is introduced into the uniform farfield, resulting in an *increase* in the total pressure. This inconsistency does not appear in boundary layer theory, because the theory is not concerned with computing the inviscid flow outside the boundary layer, and the theory applies strictly only in the limit $Re \to \infty$. However a Navier Stokes solution of the *entire* flow field at *finite* Reynolds numbers can not accommodate an *increase* in the total pressure when in fact frictional *losses* near the plate should result in a *decrease* in the total pressure. If the farfield pressure and/or farfield velocity are fixed, the current scheme does not converge to machine accuracy. It is possible to achieve better convergence with fixed farfield pressure and velocity with larger domains, however for a reasonably sized domain, both the farfield pressure and velocity must be allowed to develop with the solution.

The grid consists of 61 points distributed uniformly in the x direction and 91 points stretched in the y direction. Figure 16 shows that the potential part of the velocity is hardly affected by the source term of the Poisson equation. This is in contrast to the potential for the cascade flow shown in Fig. 14, which is visibly modified by the rotational part. The flow is essentially a potential field with a rotational correction to satisfy the no slip boundary condition. Figs. 17 and 18 depict the velocity profile in the vicinity of the plate. The slight deviation from the Blasius solution near the edge of the boundary layer in Fig. 20 may be attributed to first order upwinding, approximate exit conditions and the non uniform pressure field. Another departure from the Blasius solution, visible upon close inspection of Figs. 17 and 20, is acceleration of the flow immediately outside the boundary layer. This phenomenon is the result of the divergence free condition. The maximum overshoot above the reference velocity $u_\infty$ is less than 1% in magnitude and occurs close to the edge of the boundary layer. The overshoot drops slowly as the distance from the edge of the boundary layer increases. The contours of the static pressure are shown in Figs. 21 and 22. The values have been scaled by the Reynolds number for better plotting resolution. The pressure gradient normal to the plate, depicted in Fig. 21 is not zero. The large pressure rise at the leading edge stagnation point is detectable in Fig. 22. Once again, this phenomenon is completely absent in the Blasius solution.

## Conclusions

Helmholz decomposition of the velocity along with the Lamb form of the Navier Stokes equations lead to an effective method for the solution of the incompressible Navier Stokes equations. The method is particularly attractive because no physical principles are compromised. This method differs from traditional projection type schemes, as the potential part of the velocity is not explicitly linked to the pressure and is capable of representing a true potential when such a field exists.

Mass and momentum are conserved to machine accuracy. Careful selection of boundary conditions eliminate artificial sources of vorticity. No special treatment for multipley connected domains is needed. The method is equally effective for internal and external flows at low and high Reynolds numbers. Furthermore, all boundary conditions are satisfied to machine accuracy. The price is the solution of one more Poisson equation.

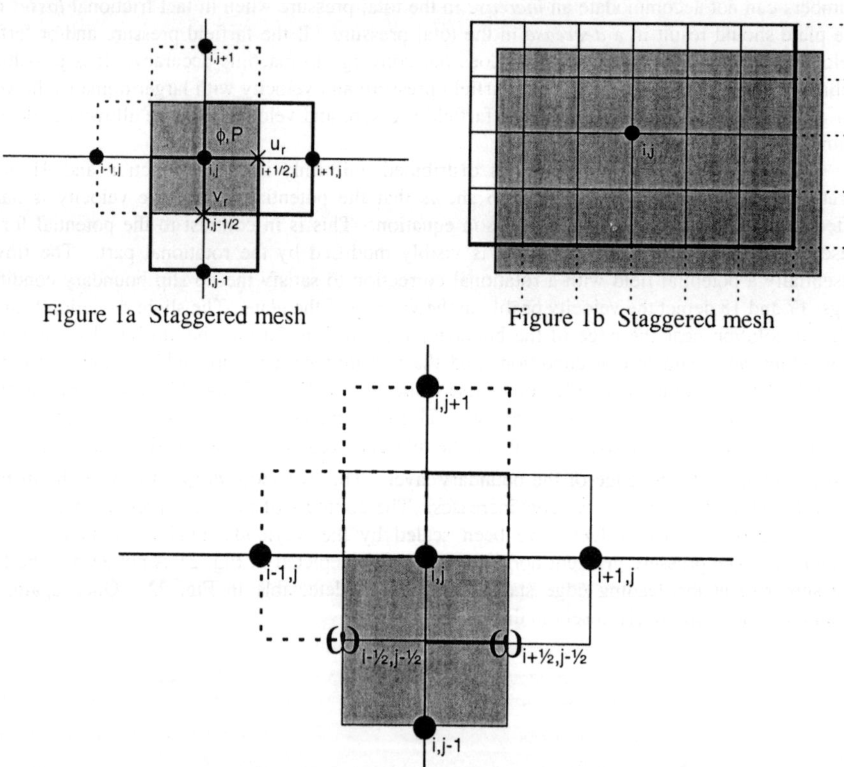


Figure 1a  Staggered mesh

Figure 1b  Staggered mesh



Figure 1c  Evaluation of Vorticity at the Walls of the y-momentum Cell

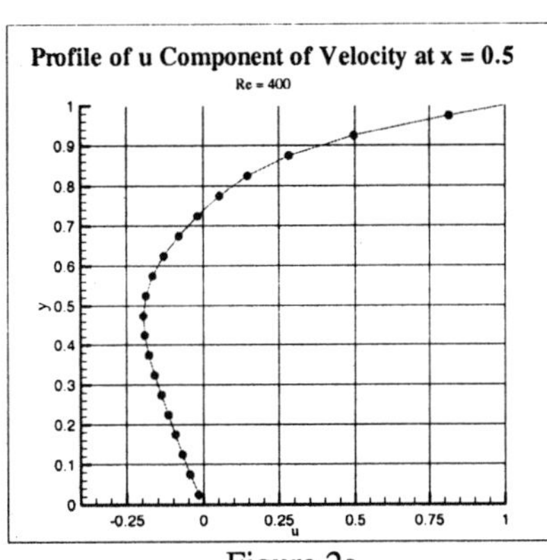


Figure 2a

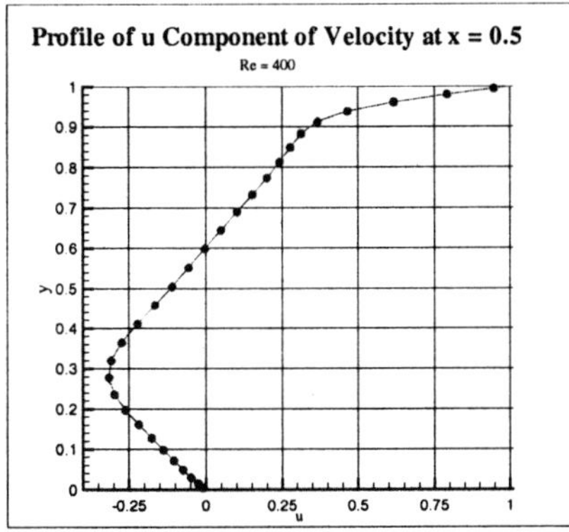


Figure 2b

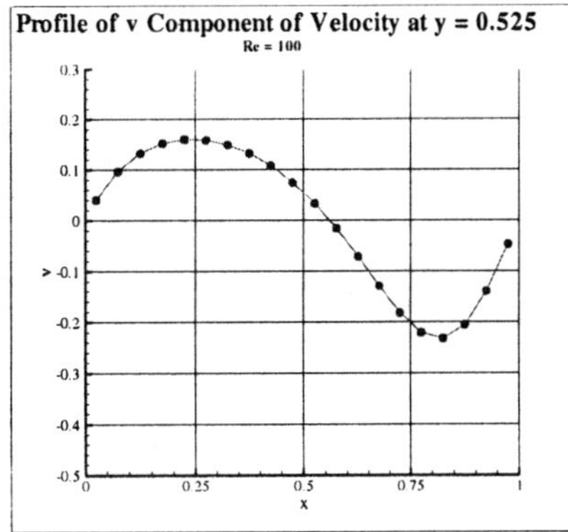


Figure 3a

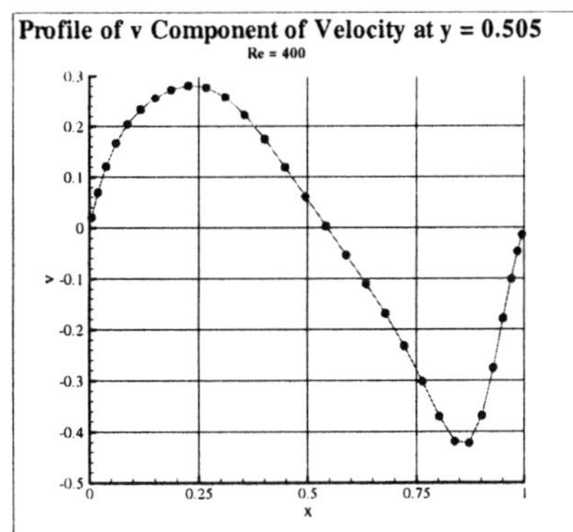


Figure 3b

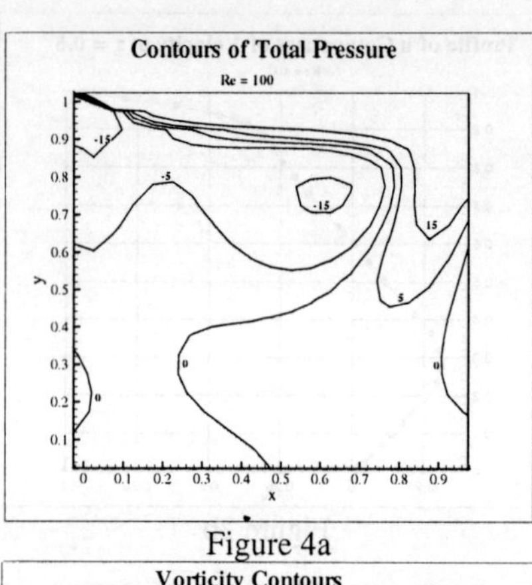

Figure 4a

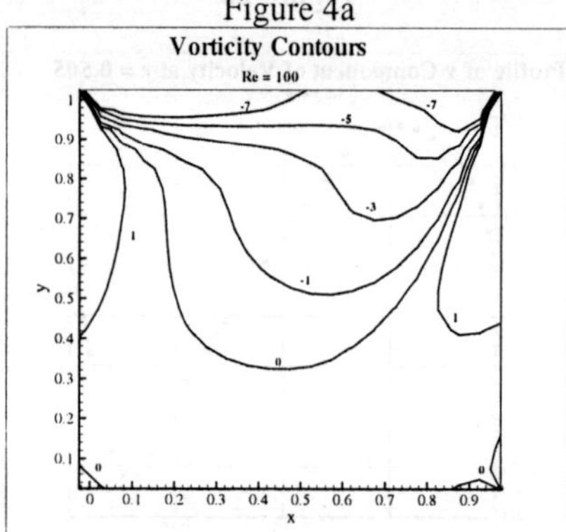

Figure 5a

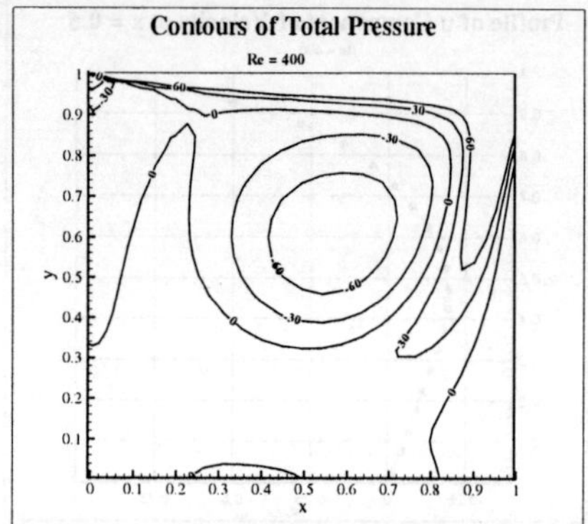

Figure 4b

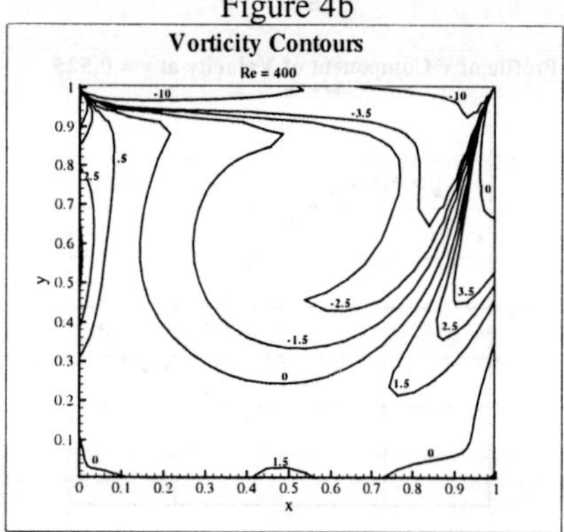

Figure 5b

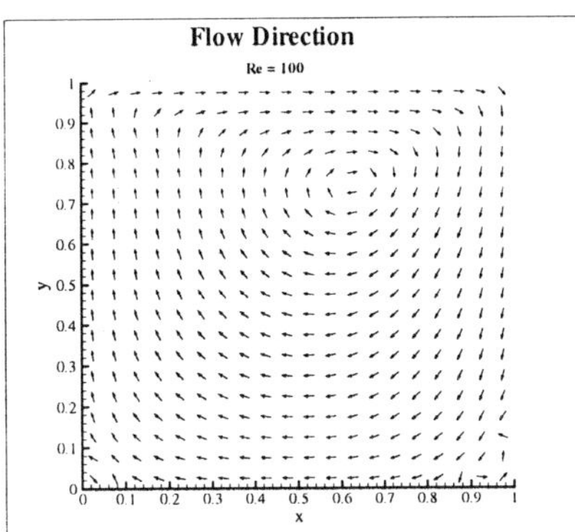

Figure 6a

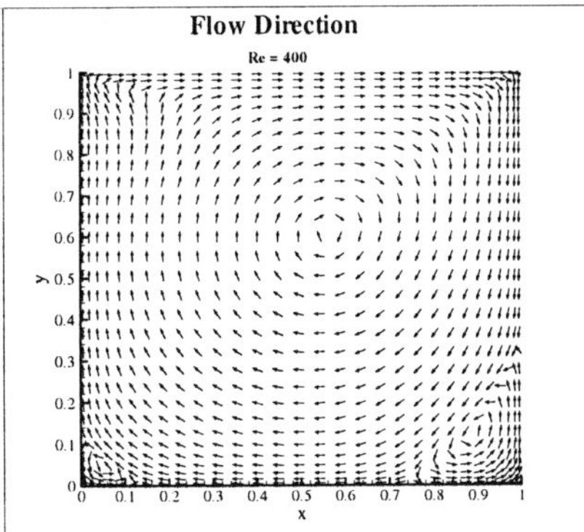

Figure 6b

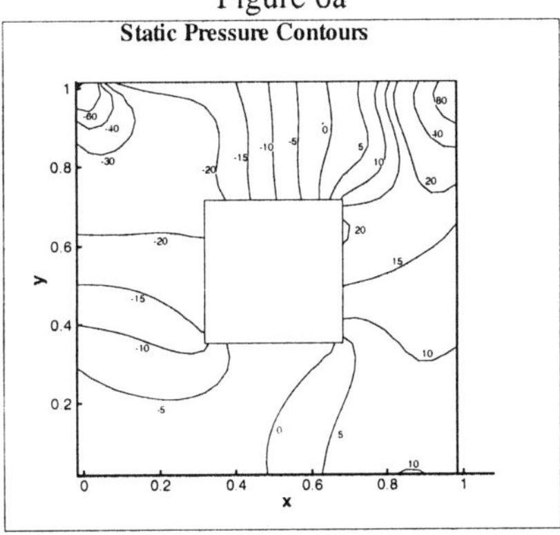

Figure 7

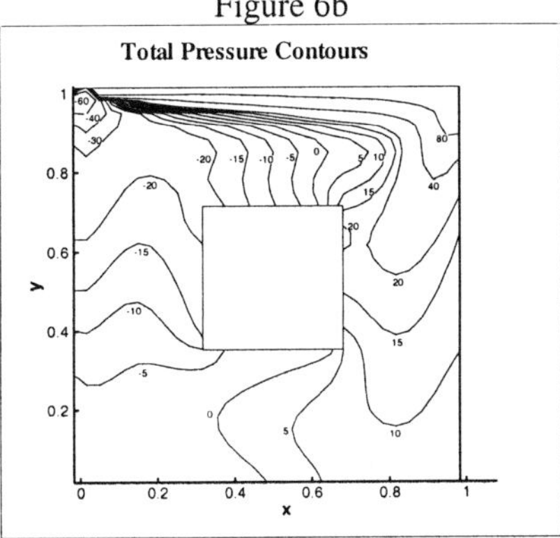

Figure 8

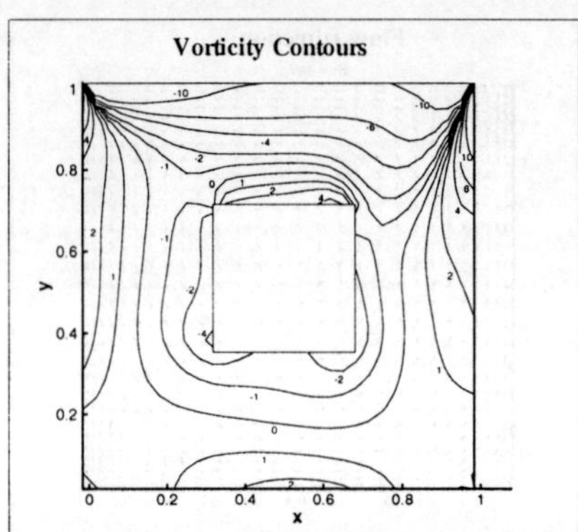

Figure 9

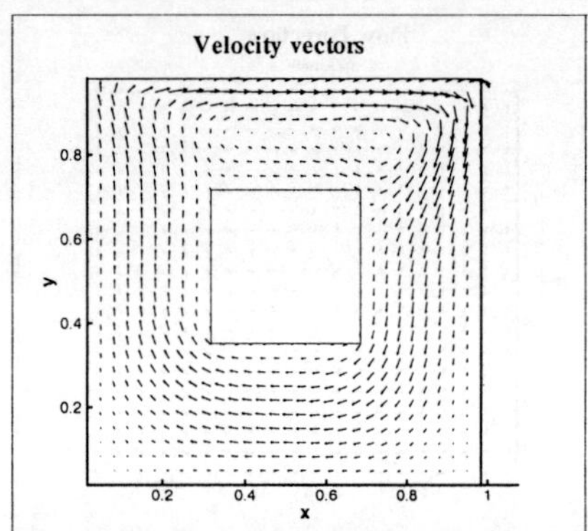

Figure 10

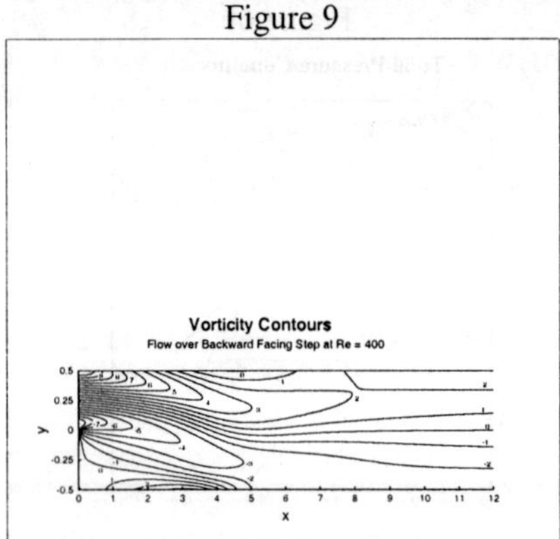

Figure 11

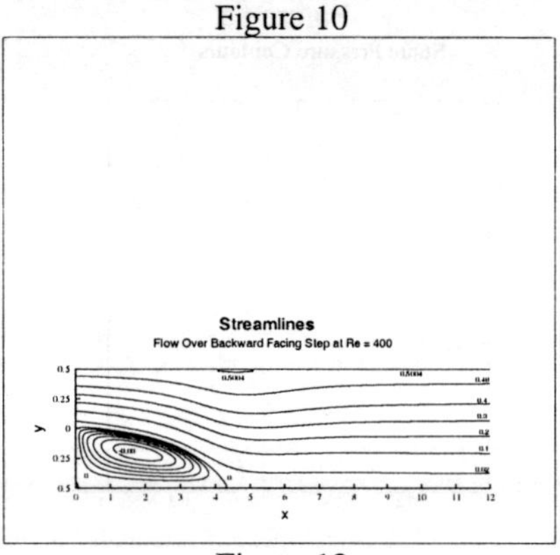

Figure 12

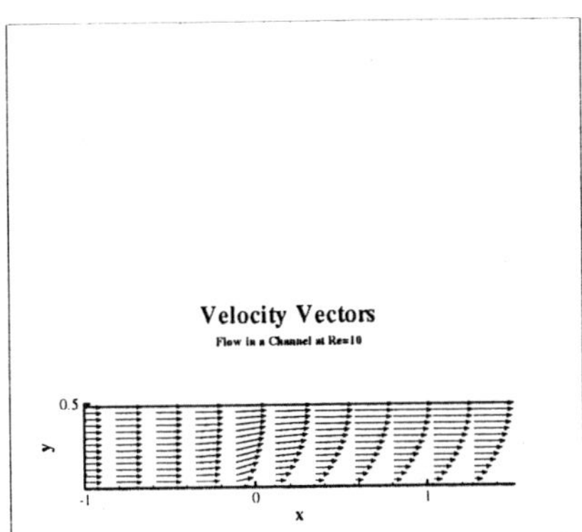

Figure 13

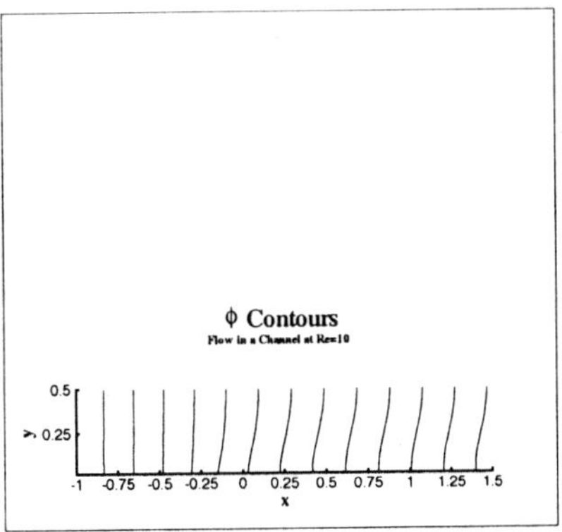

Figure 14

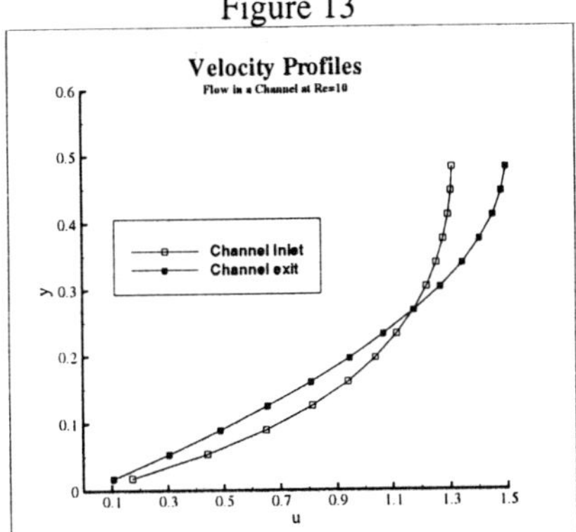

Figure 15

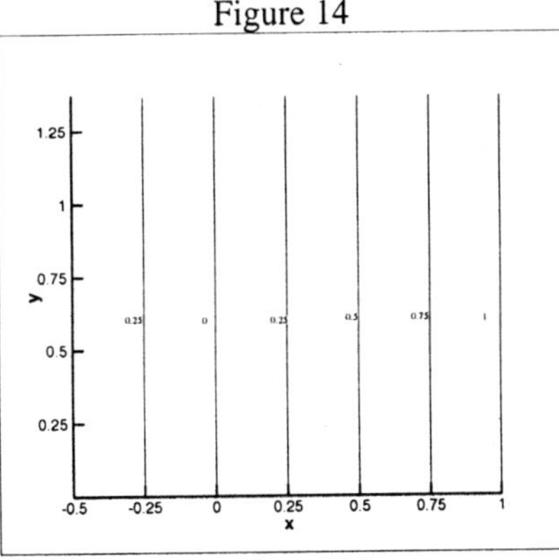

Figure 16

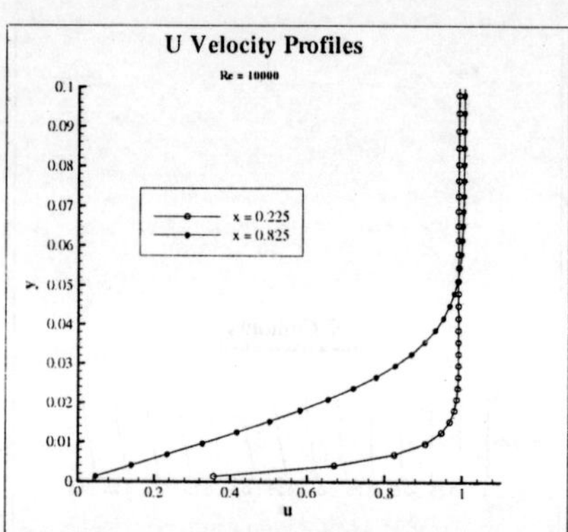

Figure 17

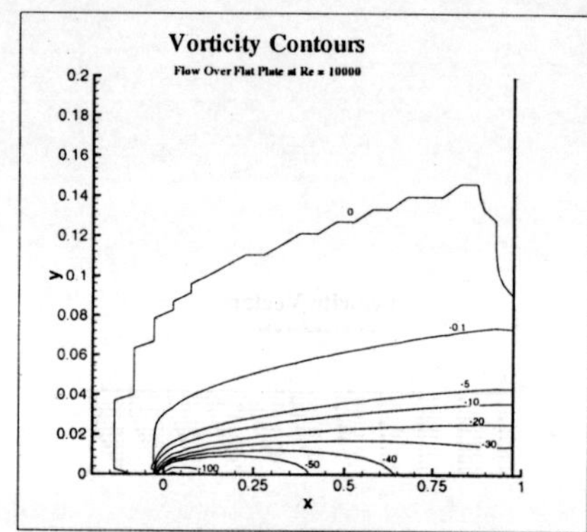

Figure 18

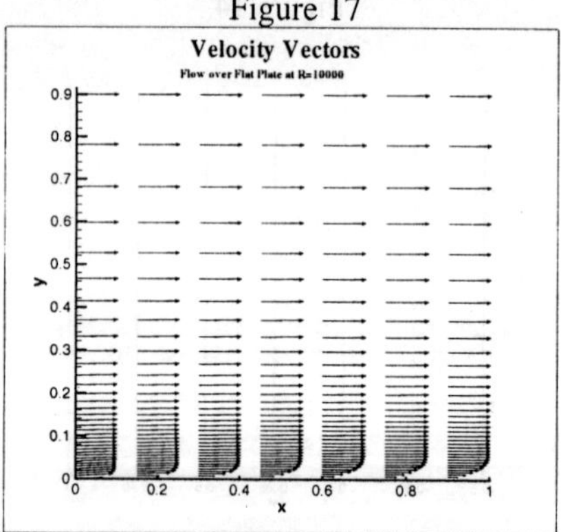
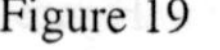

Figure 19

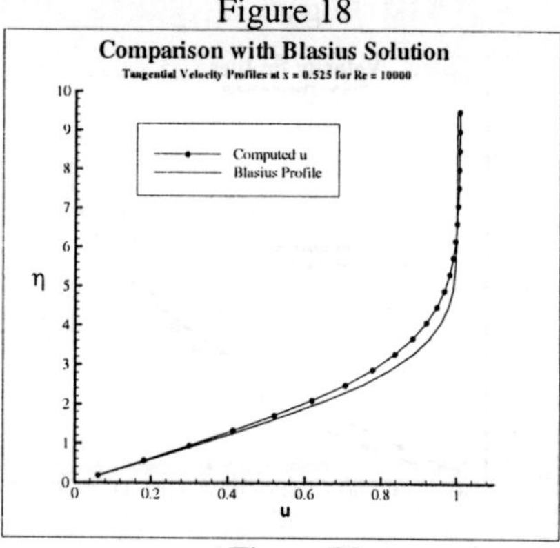

Figure 20

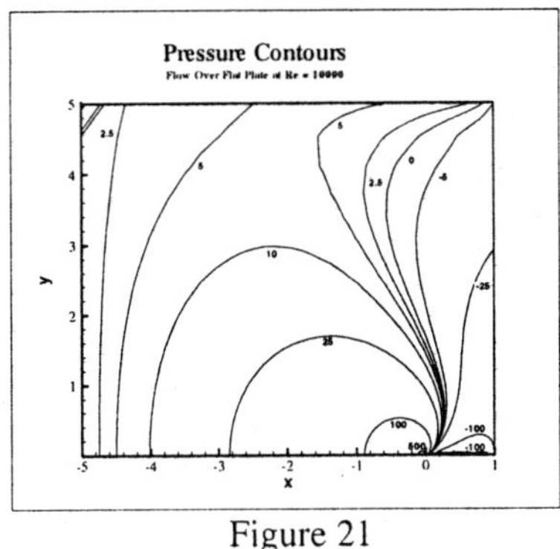

Figure 21

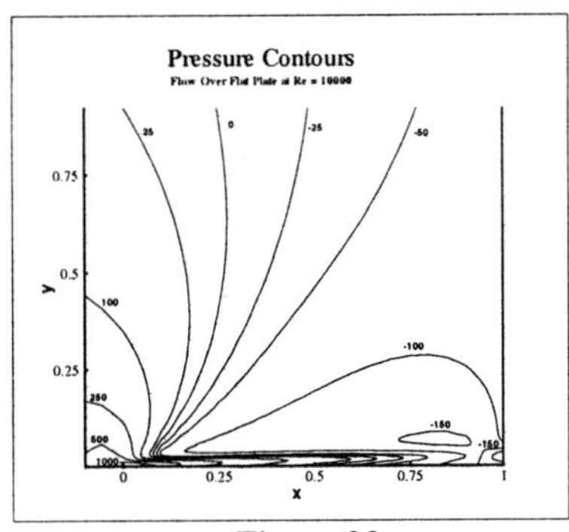

Figure 22

## References


[1]  R. Peyret and T. Taylor, Computational Methods for Fluid Flow, Springer Verlag, 1983.

[2]  F. Betagnolio and O. Darbe "Velocity-Vorticity Formulation of the Incompressible Navier Stokes Equations on Non Orthogonal Grids", Proceedings of ECCOMAS 96, Wiley and Sons

[3]  Chorin, A. J. "A Numerical Method for Solving Incompressible Viscous Flow Problems" J. Comp. Phys., Vol.2, No. 1, Aug. 1967, pp. 12-26.

[4]  Temam, R. Navier Stokes Equations, North Holland, Amsterdam, 1977

[5]  Harlow, F. H. and Welsh, J. E. "Numerical Calculation of Time Dependent Viscous Incompressible Flow of Fluids with Free Surface", Phys. Fluids. Vol. 8. 1965, pp 2182-2189

[6]  Ghia, K. N., Hankey, W. L. and Hodge, J. K. "Study of Incompressible Navier Stokes equations in Primitive Variables Using Implicit Numerical Technique", AIAA paper 77-648, Albuquerque, New Mexico.

[7]  Chorin, A. J. "Numerical Solution of Navier Stokes Equations", Mathematics of Computations, Vol 23, pp 34-45, 1968

[8]  Patankar, S. V. and Splading, D. B. "A Calculation Procedure for Heat, Mass and Momentum Transfer in Three Dimensional Parabolic Flows", Int. J. Heat Mass Transfer, vol. 15, 1972, pp. 1787-1806.

[9]    Issa, R. I. "Solution of the Implicitly Discretized Fluid Flow Equations by Operator Splitting", Journal  of Computational Physics, Vol 62, pp 40-65, 1985

[10]   Gresho, P. M. "On the Theory of Semi Implicit Projection Methods for Viscous Incompressible Flow and Its Implementation via a Finite Element Method that also Introduces a Nearly Consistent Mass Matrix – Part I: Theory", Int. Journal for Numerical Methods in Fluids, Vol. 11, pp. 587-620, 1990

[11]   Glowinski, R. and Pironneau, O. "On the mixed Finite Element Approximation of the Stokes Problem", Numerische Mathematik, Vol. 33, pp 397-424, 1979

[12]   Tezduyar, T. E., Behr, M. and Hughes, T., "Computational Fluid Dynamics Review", 1995, Wiley and Sons, 1995

[13]   Kwak, D., Kiris, C., Dacles-Mariani, J., Rogers, S. and Yoon, S. "Computational Fluid Dynamics Review 1998", Vol II, World Scientific, 1998.

[14]   Zienkiewicz, O., Nithiarasu, P., Cordina, R., Vazquez, P. and Ortiz, P. "The Characteristics Based Split Procedure: An Efficient and Accurate Algorithm for Fluid Problems", Int. Journal for Numerical Methods in Fluids, Vol 31 #1, Sep. 1999

[15]   Burggraf, O. R., J. Fluid mech. Vol. 24, 1966, pp. 113

[16]   Abdellah, S. "Numerical Solutions for pressure Poisson Equation with Neuman Boundary Conditions Using a Non Staggered Grid, I", J. Comp. Phys. Vol. 70, 1987, pp. 182-192

[17]   Abdellah, S. "Numerical Solutions for pressure Poisson Equation with Neuman Boundary Conditions Using a Non Staggered Grid, II", J. Comp. Phys. Vol. 70, 1987, pp. 183-202

[18]   Peeters, M. F., Habashi, W. G. and Nguyen, B. Q., "Finite Element Solution of the Incompressible Navier Stokes equations by a Helmholz Velocity Decomposition", Int. Journal for Numerical Methods in Fluids, vol. 13, 1991, pp. 135-144

[19]   Rogers, S. and Kwak, D., "An Upwind Differencing Scheme for the Incompressible Navier Stokes Equations", Applied Numerical Mathematics, Vol. 8, pp. 43-64, 1991

[20]   J. Kim and P. Moin, "Application of Fractional Step Methods to Incompressible Navier Stokes Equations", Journal of Computational Physics, 59, 308-323, 1985

# Three Dimensional Viscous Incompressible Flow Simulations Using Helmholz Velocity Decomposition

**Koorosh Nikfetrat**

Mechanical Design and Manufacturing Technologies, British Columbia Institute of Technology, Burnaby, BC, V5G 3H2, Canada

**Mohamed Hafez**

Department of Mechanical and Aeronautical Engineering, University of California, Davis, CA 95616, U.S.A

## Abstract

A recently proposed Helmholz Decomposition method for solving the incompressible Navier Stokes equations is applied to three dimensional flow in a driven cavity. An alternative equation in the form of a convection diffusion equation for the total pressure is implemented to overcome undesirable properties of the Poisson's equation for total pressure. Second order upwinding is used for the rotational term of the Lamb form of the momentum equations and the convection term of the pressure equation. A finite difference discretization on a staggered grid is used and a segregated solution procedure is employed.

Key Words:   Incompressible flow, Navier-Stokes equations, Helmholz decomposition, Three dimensional driven cavity flow.

## Introduction

This paper describes several developments of the Helmholz Decomposition method, described recently in [1] and [2]. The Helmholz Decomposition method solves the incompressible Navier Stokes equations numerically for the primitive variables. Primitive variables based numerical methods for solving incompressible flows can be placed in two categories: artificial compressibility methods with an artificial time derivative term of pressure in the continuity equation, proposed by Chorin [3], and projection methods with a potential function to achieve a divergence free velocity field. Projection methods can be traced back to the MAC method of Harlow and Welsh[4] and also to Chorin[5]. The current method is of the latter kind, since it relies on decomposition of velocity into rotational and irrotational (potential) parts. It requires solution of the continuity and momentum equations and an equation for either the static or total pressure. In this method the irrotational and rotational parts of velocity are stored separately and unlike other projection methods, the irrotational part is relevant to the inviscid potential for external and core internal flows. No explicit relationship between the velocity potential and pressure is assumed, eliminating the need for artificial terms and ad hoc parameters. The independence of the velocity potential and the pressure makes it possible to satisfy simultaneously both the no penetration and no slip conditions for velocity components. Consequently the conservation laws and boundary conditions are enforced to machine accuracy.

The Lamb form of the momentum equations is natural and convenient to use in the framework of the Helmholz Decomposition method [1]. With the momentum equations expressed in Lamb form, the kinetic energy and static pressure can be combined and replaced with the total pressure.

Indeed several of the solutions presented in [1] were obtained with the total pressure formulation. It was however pointed out that the Poisson equation that is traditionally used to obtain the total pressure, spreads gradients of the total pressure over the entire domain. According to Bernoulli's law, gradients of total pressure should however vanish in irrotational (inviscid) regions. In this paper an alternative to the Poisson equation for the total pressure is presented. The alternative formulation is a convection diffusion equation that has the property that the gradients of total pressure tend to zero as the Reynolds number tends to infinity, namely in irrotational regions. An additional benefit of the convection diffusion equation is a reduction in computational effort.

Other developments described in this paper include extension of the Helmholz Decomposition method to three dimensions. Modified treatment of boundary conditions have resulted in significantly improved convergence rates. Solutions are presented for driven flow in a cubic cavity at Reynolds numbers in the range 100 to 1000.

## Governing Equations

As in reference [1] all equations are dimensionless, the subscript $r$ donates rotational part, $\tau$ is dimensionless time, $\omega$ is the vorticity, $\phi$ a scalar potential, $P = p + \frac{1}{2}V^2$ the total pressure and Re the Reynolds number. Splitting the velocity vector into rotational and potential parts, and recognizing that the potential part does not contribute to vorticity, results in the following system of equations in three dimensions:

continuity equation:

$$\vec{\nabla} \cdot \left( \vec{\nabla}\phi + \vec{V}_r \right) = 0 \tag{1}$$

momentum equations in Lamb form:

$$\vec{M} \equiv \frac{\partial \vec{V}}{\partial \tau} - \vec{V} \times \vec{\omega} + \vec{\nabla}P + \frac{1}{\text{Re}}\vec{\nabla} \times \vec{\omega} = 0 \tag{2}$$

The three scalar components of the velocity vector are computed from:

$$u = u_r + \frac{\partial \phi}{\partial x} \quad v = v_r + \frac{\partial \phi}{\partial y} \quad w = w_r + \frac{\partial \phi}{\partial z} \tag{3}$$

and the three scalar components of the vorticity vector from:

$$\omega_1 = \frac{\partial w_r}{\partial y} - \frac{\partial v_r}{\partial z} \quad \omega_2 = \frac{\partial u_r}{\partial z} - \frac{\partial w_r}{\partial x} \quad \omega_3 = \frac{\partial v_r}{\partial x} - \frac{\partial u_r}{\partial y} \tag{4}$$

Traditionally, the divergence of the momentum equations serves as an equation for the total pressure. The resulting equation is of Poisson type:

$$\vec{\nabla} \cdot \vec{M} = \vec{\nabla} \cdot \left[ \vec{\nabla}P - \vec{V} \times \vec{\omega} \right] = 0 \tag{5}$$

Since it is difficult to obtain a solution with zero gradients from eq. (5), an alternative to eq. (5) is proposed.

An alternative equation for total pressure can be constructed by combining the mechanical energy equation with Poisson's equation. The mechanical energy equation is obtained by taking the dot product of the velocity vector with the momentum equations. Thus, the equation for total pressure reads:

$$\vec{V} \cdot \left[ -\vec{V} \times \vec{\omega} + \vec{\nabla} P + \frac{1}{\mathrm{Re}} \vec{\nabla} \times \vec{\omega} \right] - \frac{1}{\mathrm{Re}} \left\{ \vec{\nabla} \cdot \left[ \vec{\nabla} P - \vec{V} \times \vec{\omega} \right] \right\} = 0 \qquad (6)$$

$$\vec{V} \cdot \vec{\nabla} P + \frac{1}{\mathrm{Re}} \left\{ \vec{V} \cdot \left( \vec{\nabla} \times \vec{\omega} \right) - \left[ \vec{\nabla} \cdot \left( \vec{\nabla} P - \vec{V} \times \vec{\omega} \right) \right] \right\} = 0$$

Combining terms and denoting the magnitude of vorticity as $\omega$ results in the following form:

$$\vec{V} \cdot \vec{\nabla} P + \frac{1}{\mathrm{Re}} \left( \omega^2 - \vec{\nabla} \cdot \vec{\nabla} P \right) = 0 \qquad (7)$$

Equation (7) reduces to Bernoulli's equation in the limit $\mathrm{Re} \to \infty$.

## Discretization

Since only steady state solutions are of interest here, first order accurate time discretization is used. To ensure mass is conserved to machine accuracy at the discrete level, a staggered arrangement as shown in Fig. 1 is adopted. This type of arrangement eliminates the well documented decoupling between velocity and pressure. Several additional benefits are derived from a staggered arrangement: Mass and momentum can be conserved at the boundaries and as a result the integral constraint for Poisson's equations with Neumann boundary conditions is automatically satisfied. The curl of a divergence is identically zero on a staggered mesh, and finally extrapolation of variables to the nodes located outside the physical domain is not necessary.

Equation (1) is elliptic and thus it is central differenced, ensuring that the discrete divergence free velocity condition is satisfied to machine accuracy. To prevent numerical oscillations, upwinding is needed for the convection term in the momentum equations and also in the equation for total pressure. Upwinding is achieved by extrapolation from cell boundaries to cell centres, implemented via a Taylor series expansion at the upstream or downstream cell boundaries. Location and direction of extrapolation depend on the sign of the velocities as discussed in [2]. Accuracy depends on the number of terms in the Taylor series. A quantity $\Phi$ at the centre of a cell is evaluated from the following equation:

$$\Phi_{centre} = \Phi_{wall} \pm \lambda \frac{\Delta s}{2} \left[ \frac{1 \pm \kappa}{2} \left. \frac{\partial \Phi}{\partial s} \right|_{forward} + \frac{1 \mp \kappa}{2} \left. \frac{\partial \Phi}{\partial s} \right|_{backward} \right] + \cdots \qquad (8)$$

In eq. (8) $\lambda$ controls the number of terms and $\kappa$ the downwind influence or degree of upwinding. Discretization of the momentum equations will be discussed first, followed by a detailed description of the discretization of the convection diffusion equation for total pressure.

**Momentum equations**

Referring to Figure 1, the x momentum cell is centred at $i + \frac{1}{2}, j, k$, the y momentum cell at $i, j - \frac{1}{2}, k$ and the z momentum cell at $i, j, k + \frac{1}{2}$, $\omega_3$ is evaluated at $i + \frac{1}{2}, j - \frac{1}{2}, k$, $\omega_2$ at $i + \frac{1}{2}, j, k + \frac{1}{2}$ and $\omega_1$ at $i, j - \frac{1}{2}, k + \frac{1}{2}$.

Viscous terms are evaluated by central differences, facilitated by the convenient location of vorticity nodes on the faces of the momentum cells. The convection term of the momentum equations is $-\vec{V} \times \vec{\omega} + \vec{\nabla} P$ and must be evaluated at the centre of the momentum cells.

Evaluation of $\vec{\nabla} P$ at the momentum centres does not require upwinding on a staggered grid, thus $\vec{V} \times \vec{\omega}$ is the only term in need of upwinding. A second order upwind extrapolation of vorticity from the walls to the centres of momentum cells is employed to obtain the required vorticity values at the cell centres. The appropriate Taylor series are generated by substituting $\omega_1$ or $\omega_2$ or $\omega_3$ for $\Phi$ in eq. (8). For details see [2].

**Total pressure equation**

During transient stage of the computations, the convection term in eq. (7) will have a contribution from the unsteady term of the momentum equations. This unsteady term is replaced with an artificial time derivative term of the total pressure, to enable solution via a time marching scheme. Every term that appears in the total pressure equation must come from the momentum equations in exactly the same form, and the discrete operators used in the pressure equation must be identical to those used in the momentum equations. If the discretization of the pressure equation is not compatible with the discretization of the continuity and momentum equation, the total pressure obtained from the pressure equation will not drive the momentum equations towards convergence. To ensure compatibility, the equation for the total pressure should be reconstructed at the discrete level rather than from eq. (7). Unlike its continuous counterpart, the term $\vec{V} \cdot (\vec{V} \times \vec{\omega})$ is not zero at the discrete level, because the components of vorticity are evaluated at different locations on a staggered grid. Since the momentum equation is satisfied at each node, the following linear combination is guaranteed to be consistent with the momentum equations:

$$\frac{\partial P_{i,j,k}}{\partial \tau} + \alpha \left( \vec{V} \cdot \vec{M} \right)_{i,j,k} + \beta \left( \vec{\nabla} \cdot \vec{M} \right)_{i,j,k} = 0 \qquad (10)$$

Expanding the terms in (10) and using $\vec{M} = \vec{\nabla} P - \vec{V} \times \vec{\omega}$ for short hand notation yields the following discrete equation:

$$\frac{\partial P}{\partial \tau} + u_{i,j,k} (M_x)_{i,j,k} + v_{i,j,k} (M_y)_{i,j,k} + w_{i,j,k} (M_z)_{i,j,k}$$
$$+ \left( \omega + \frac{1}{\text{Re}} \right) \left\{ \left[ (\tilde{M}_x)_{i+\frac{1}{2},j,k} - (\tilde{M}_x)_{i-\frac{1}{2},j,k} \right] \xi_x \right.$$
$$+ \left[ (\tilde{M}_y)_{i,j+\frac{1}{2},k} - (\tilde{M}_y)_{i,j-\frac{1}{2},k} \right] \eta_y$$
$$\left. + \left[ (\tilde{M}_z)_{i,j,k+\frac{1}{2}} - (\tilde{M}_z)_{i,j,k-\frac{1}{2}} \right] \zeta_z \right\} = 0 \qquad (11)$$

The Poisson equation can be recovered from eq. (11) with $\alpha = 0$ and $\beta = 1$. To obtain a convection diffusion equation, $\alpha = 1$. The choice of $\beta$ is dictated by numerical considerations and the desire to diminish the influence of the Poisson equation in irrotational regions. In this study $\beta = \omega + 1/\mathrm{Re}$, where $\omega$ is the magnitude of the vorticity vector. The resulting equation has a strong contribution from the Poisson term in rotational regions and a contribution only from the convection term in irrotational regions. Upwinding is applied to the convection terms of eq. (11), as discussed before.

## Boundary Conditions

The sum of the rotational and potential parts of velocity must satisfy the physical boundary conditions. Consequently, at solid surfaces the no slip and no penetration conditions can be satisfied to machine accuracy and at the farfield velocity and vorticity can both be specified. In previous implementations of the Helmholz Decomposition method, [1] and [2], the boundary vorticities were computed in the same manner as the interior points, i.e. from the rotational velocities and eq. (4). This required knowledge of the tangential component of the rotational part of velocity on nodes located outside physical boundaries of the domain. While on a Cartesian mesh the tangential component of the rotational velocity is easily computed outside the domain, this is not the case for curved solid walls. To eliminate the need for splitting the velocity parts outside solid boundaries, the vorticity is computed from the full velocities at solid boundaries. The full velocity outside solid walls is easily specified by imposing the no slip condition. It must however be emphasized that at far field boundaries the advantage gained by specifying separate boundary conditions can be maintained, since the difficulty with nodes located outside solid walls is not encountered at the farfield.

In the staggered arrangement, physical boundaries line up with the normal components of the rotational velocity parts. Velocity boundary conditions for flow in a cubic cavity take the following form for the continuity and momentum equations:

$$\underbrace{\vec{V}_r \cdot \hat{n} = 0}_{on\ the\ wall} \qquad \underbrace{\vec{V} \cdot \hat{s} \;=\; V_{wall}}_{accross\ the\ wall} \tag{12}$$

The boundary condition for pressure is of Neumann type and derived from the normal component of the momentum equations. This choice of boundary condition ensures momentum is conserved at the discrete level at the boundaries. Conservation of momentum is particularly important at boundaries where mass crosses into or out of the domain e.g. inflow, outflow, suction or blowing and also at or near sharp edges. The normal component of the momentum equation is used for the pressure boundary condition and thus conserves momentum at the boundaries. The normal momentum equation reads:

$$\frac{\partial P}{\partial n} - \hat{n} \cdot \left( \vec{V} \times \vec{\omega} \right) \;=\; -\frac{1}{\mathrm{Re}} \hat{n} \cdot \left( \vec{\nabla} \times \vec{\omega} \right) \tag{13}$$

Equation (13) is used in this and previous work on the Helmholz Decomposition method [1] and [2]. It is exact and valid at all values of the Reynolds numbers and can be applied without modification along all boundaries, even when mass crosses the boundaries. In the limit $\mathrm{Re} \to \infty$, eq. (13) reduces to $\partial P / \partial n \;-\; \hat{n} \cdot (\vec{V} \times \vec{\omega}) \;=\; 0$, and if in addition vorticity is zero at the boundary, $\partial P/\partial n = 0$. This last approximation is appropriate for inviscid (Euler) flows.

## Solution Procedure

A segregated solution procedure is used to solve the system of algebraic equations resulting from the discretization of equations (1), (2) and (11). A fully implicit backward Euler time marching scheme is used to obtain the steady state solutions.

The solution cycle begins by solving the three momentum equations (eqs. (2)) coupled together for the rotational velocities, via a line relaxation algorithm. Application of line relaxation to the upwind differenced equations results in a block pentadiagonal structure for the coefficient matrices. Line relaxation as well as nonlinearity of the algebraic equations, make it necessary to perform several iterations on the solution of the momentum equations per timestep. To prevent directional bias, the momentum equations are solved by successively linking them along each of the three spatial dimensions, with a corresponding change in the directions of sweep. Six sweeps have been found to be more than adequate. This stage of the calculation imposes the correct vorticity on the velocity field. The velocity field at this stage is however not divergence free.

To make the velocity field divergence free, the continuity equation (eq. (1)) is solved for the potential. An SLOR scheme requiring a scalar tridiagonal solver is employed. The continuity equation is solved by linking the nodal variables along the direction that has the maximum number of nodes and sweeping repeatedly in the other directions. While for time accurate solutions the continuity equation must be solved to full convergence at every timestep, this requirement is relaxed for steady state calculations. This stage of the computation does not affect the vorticity. However the velocity field does not satisfy the momentum equations and velocity boundary conditions.

The pressure is updated to reflect the most recent updates to the vorticity and velocity fields. Equation (11) is solved via a relaxation scheme which is similar to the scheme used for the continuity equation, except that the bandwidth is five due to upwinding and there is no over relaxing. At this stage the pressure equation is not solved to full convergence, and for the results presented in the next section, 10 sweeps per time step over the domain proved sufficient. After this stage solution cycle is repeated.

The solution is considered converged when the right hand sides of all the governing equations drops below $10^{-6}$.

## Numerical Results

Driven flow in a cubic cavity has a complex, three dimensional structure, and due to its geometric simplicity is ideal for evaluation of numerical methods. To ensure existence of steady state laminar solutions, Reynolds numbers above 1000 are excluded here. The mesh sizes are given in Figs. 2a-d. The mesh for the first three test cases, Re=100, 400 and 600 is uniform, while for Re=1000 it is stretched slightly to cluster points near the walls. To maintain second order accuracy, stretching is 10% near the walls and drops rapidly as the distance to the walls increases.

Contours of the three components of vorticity are presented in Figs. 2a-d. They are in agreement with results presented in [6-10] and reveal the details of the flowfield, confirming adequate grid resolution. The two alternative pressure formulations produced identical results, as demonstrated by the static pressure contours for Re=1000 in Fig. 3. For all test cases, the solutions obtained with the two total pressure equations are identical.

The particle traces in Figs. 4 show the secondary circulation zone near the intersection of the lower and upstream planes. The intensity of secondary circulation currents increases with the Reynolds number, and the flow becomes more three dimensional.

## Concluding Remarks

Standard three dimensional viscous flow in a driven cavity is simulated using a Helmholz velocity decomposition. Unlike many of the projection methods, the potential part of the velocity is not related explicitly to the pressure. Moreover, a convection diffusion equation is introduced for the total pressure. The merits of the present formulation are discussed and the numerical results confirm the validity of the adopted approach.

## References

[1] Nikfetrat, K. and Hafez, M. Numerical solution of the incompressible Navier Stokes equations using Helmholz velocity decomposition. Computational Fluid Dynamics Journal, Vol. 9, No. 3, Special issue 2000 pp 266-280

[2] Nikfetrat, K. Upwind discretization schemes for the incompressible Navier Stokes equations. Proceedings of the First MIT Conference on Computational Fluid and Solid Mechanics, June 2001, Vol. 2, pp 934-939

[3] Chorin, A. J. A numerical method for solving incompressible viscous flow problems. Journal of Computational Physics, Vol. 2, No. 1 Aug. 1967, pp 12-26

[4] Harlow, F. H. and Welsh, J. E. Numerical calculation of time dependent viscous incompressible flow of fluids with free surfaces. Physics of Fluids Vol. 8 1965 pp 2182-2189

[5] Chorin, A. J. Numerical solution of Navier Stokes equations. Mathematics of Computations, Vol. 23, 1968 pp 34-45.

[6] Dennis, S. Ingham, D. and Cook, R. Finite difference methods for calculating steady incompressible flows in three dimensions. J. of Computational Physics, Vol. 33, 1979

[7] Koseff, J. et al. A three dimensional lid-driven cavity flow: experiment and simulation. Proceedings of the Third International Conference on Numerical Methods in Laminar and Turbulent Flows, Seattle, WA 1983

[8] Oswald, G., Ghia, K. and Ghia, U. A direct algorithm for solution of three dimensional unsteady Navier Stokes equations. AIAA paper 87-1139

[9] Oswald, G., Ghia, K. and Ghia, U. Direct method for solution of three dimensional unsteady incompressible Navier Stokes equations. Proceedings of the eleventh International Conference on Numerical Methods in Fluid Dynamics, Williamsburg, VA June 1988

[10] Rosenfeld, M. Kwak, D. and Vinokur M. Solution method for unsteady incompressible Navier Stokes equations in generalized coordinate systems. AIAA paper 88-0718

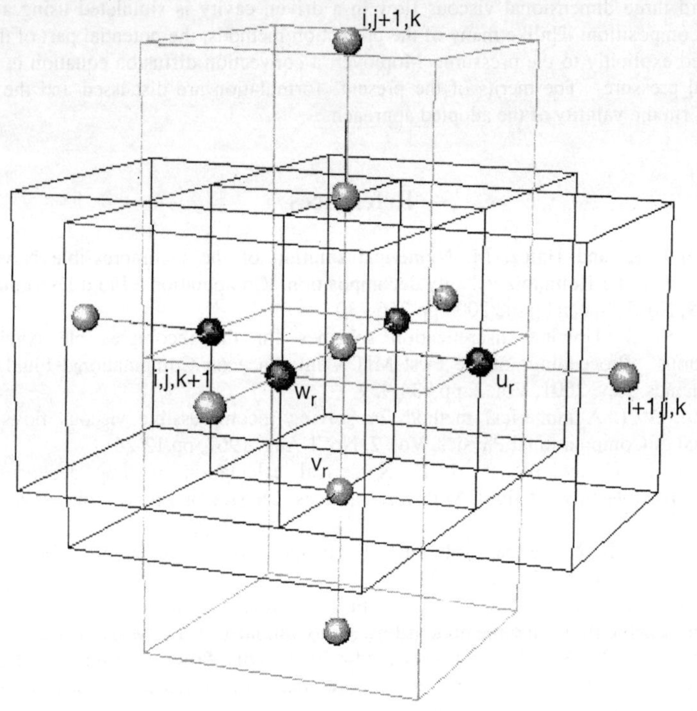



Fig. 1  Staggered arrangement

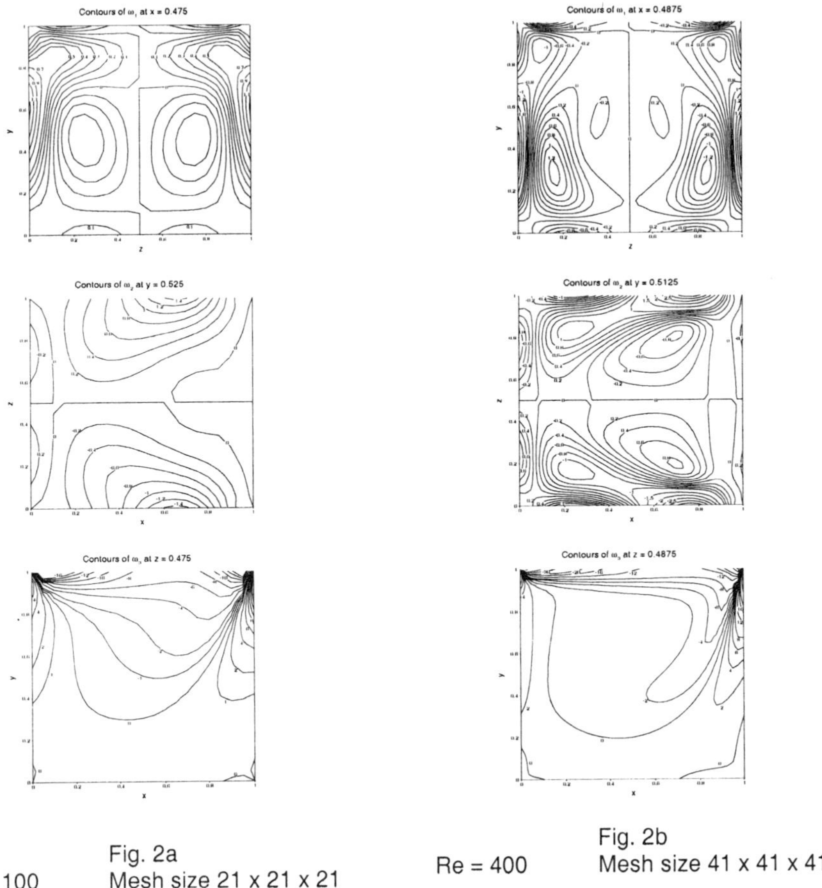

Fig. 2a
Re = 100    Mesh size 21 x 21 x 21

Fig. 2b
Re = 400    Mesh size 41 x 41 x 41

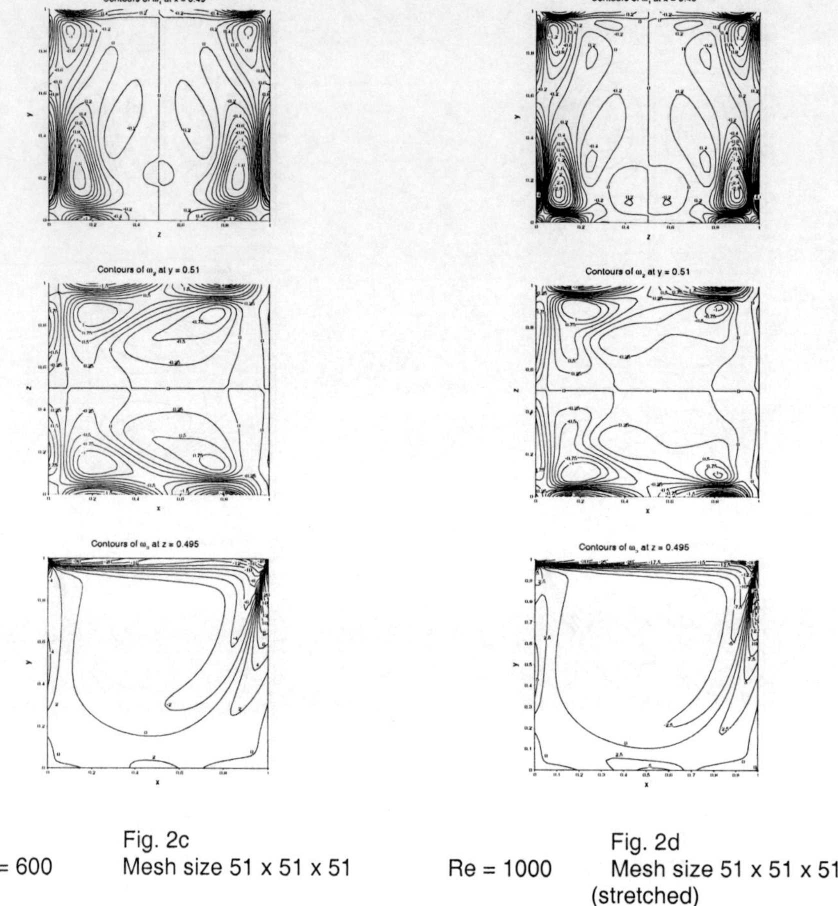

Fig. 2c
Re = 600    Mesh size 51 x 51 x 51

Fig. 2d
Re = 1000    Mesh size 51 x 51 x 51
(stretched)

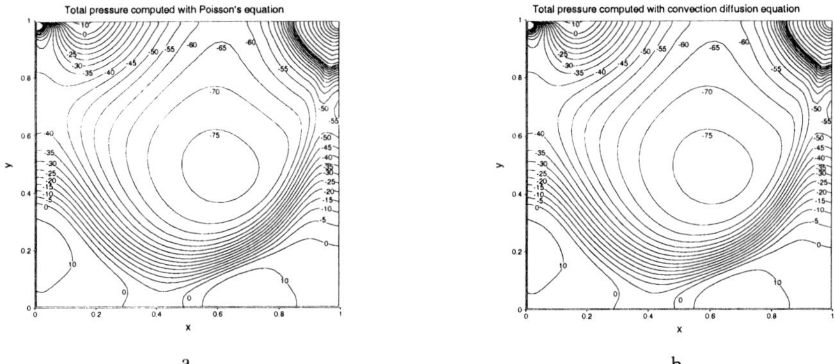

a

b

Fig. 3    Comparison of static pressure contours in the plane z=0.487005

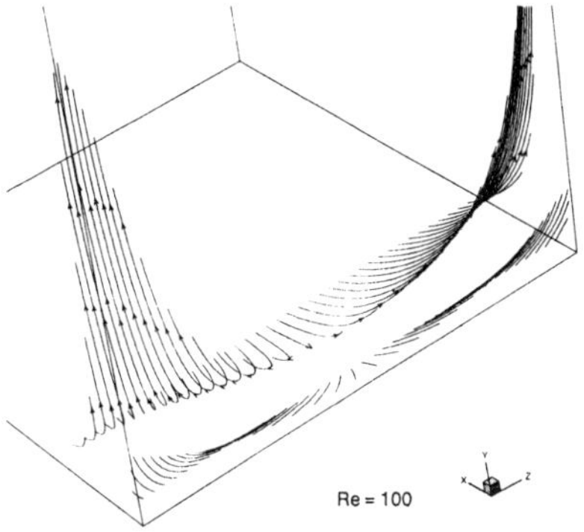

a

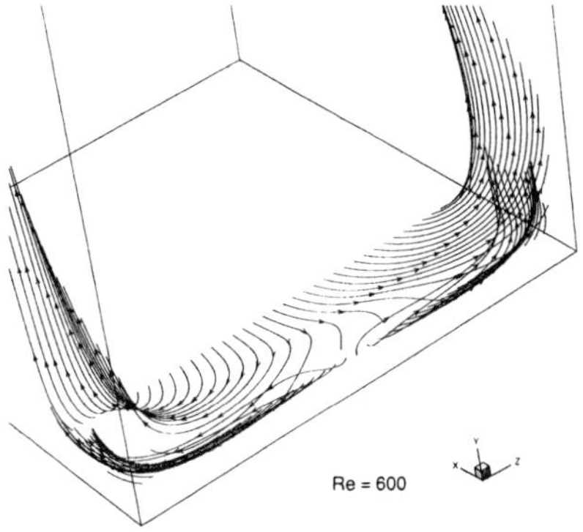

c

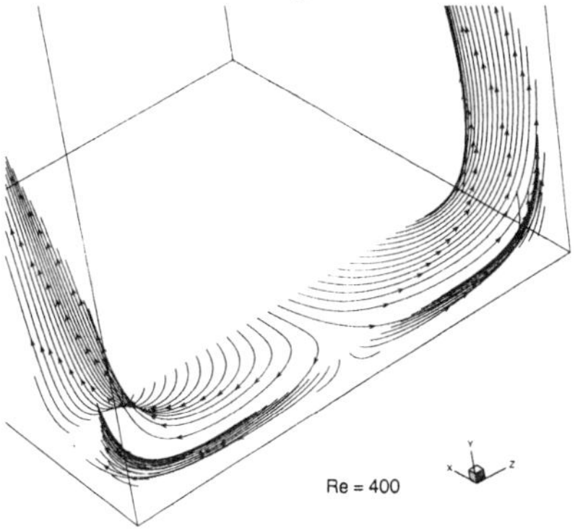

b

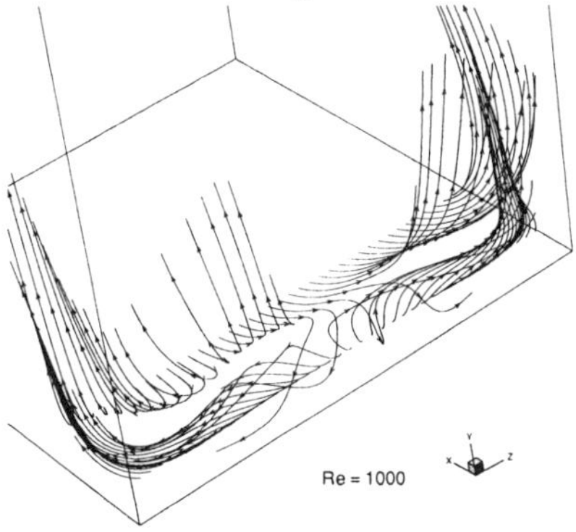

d

Fig. 4    Close-up of circulation region at the bottom of upstream wall

# Finite Element Methods

# On the Numerical Simulation of Incompressible Viscous Fluid Flow around Moving Rigid Bodies of Elliptical Shape

Roland Glowinski[1], L. Hector Juárez[2,1], and Tsorng-Whay Pan[1]

[1] Department of Mathematics, University of Houston, Houston, TX 77204, USA
[2] Departamento de Matemáticas, U.A.M.-I, México D.F. AP-532

**Abstract.** The main goal of this article is to discuss the numerical simulation of incompressible viscous fluid flow around moving rigid bodies of elliptical shapes, such as ellipses in two–dimensions, and ellipsoids in three–dimensions. Related simulations for circular and spherical bodies, by methods closely related to those described in this article, have been investigated by the authors in preceding publications, however, the non–sphericity introduces additional difficulties to be discussed in the article. Numerical results for two and three–dimensional test problems will be presented; they include the interaction of two ellipsoids settling in a narrow channel.

## 1 Introduction and Synopsis

In [1] to [9] the authors of this article, and other collaborators, have investigated the direct numerical simulation of the motion of rigid bodies interacting with a surrounding incompressible viscous fluid. In two–dimensions, quite general shapes have been considered; concerning however three–dimensional phenomena, only spherical bodies were considered by those authors. The main reason for the above situation is that non–sphericity introduces additional difficulties, both from the modeling and computational points of view. The main goal of this article is to address precisely these difficulties and to show that by an appropriate treatment of the angular momentum the fictitious domain based methodology discussed in the above references can be extended to handle the more complicated situations considered here. The content of this article is as follows:

In Section 2, we describe a global variational formulation (of the virtual power principle type) of the coupled flow/rigid body motion phenomena.

In Section 3, we discuss a fictitious domain variant of the above formulation; this formulation involves Lagrange multipliers vector–valued functions to force the rigid body motion of the moving bodies. The treatment of body/body and body/wall collisions (or near–collisions) is briefly addressed in this section.

In Section 4, we discuss first the finite element implementation of the computational methods discussed in Section 3, a particular attention being given to the fictitious domain part of the computational procedure, and then the time discretization by operator–splitting of the system of differential equations and algebraic relations modeling the coupled flow/rigid body motion phenomena is discussed.

In Section 5, we briefly address the solution of the various sub–problems encountered at each time step of the operator splitting scheme.

Finally, in Section 6 we present the results of numerical experiments concerning the solution of test problems in two and three–dimensions. We shall consider first the simulation of the flow of an incompressible viscous fluid in a two–dimensional channel, containing an elliptic rigid body rotating freely around its fixed center of mass; next, we shall consider the settling of a sphere, and then of two ellipsoids in a narrow channel filled with an incompressible viscous fluid.

## 2 The Prototypical Coupled Flow/Rigid Motion Problem and its Global Variational Formulation

In order to perform the *direct numerical simulation* of the interaction between rigid bodies and the surrounding fluid, we have developed a methodology combining *time discretization by operator splitting* with a *fictitious domain* (also called *domain embedding* by some) approach, and with other "ingredients" to be detailed latter (descriptions of this methodology can be found in, e.g., [1] to [9]). To begin with we shall consider our basic model problem, namely the settling of a *rigid body B* in a cavity $\Omega$ filled with a *Newtonian incompressible viscous fluid* (the geometrical aspects of the above problem are shown in Fig. 1). The coupled flow/rigid body motion is modeled by the *Navier–Stokes equations* (for the flow) coupled with the *Newton–Euler equations* (for the rigid body motion); the resulting model is as follows:

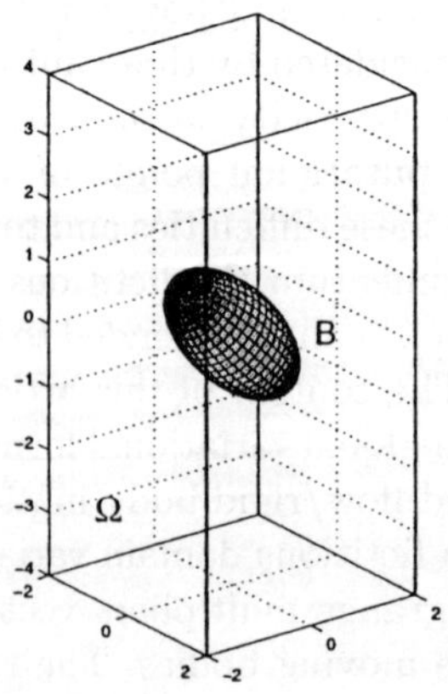

Fig. 1. The flow region with one rigid body

$$\rho_{\mathrm f}\left[\frac{\partial \boldsymbol u}{\partial t} + (\boldsymbol u \cdot \nabla)\boldsymbol u\right] - \mu_{\mathrm f}\Delta \boldsymbol u + \nabla p = \rho_{\mathrm f}\,\boldsymbol g \;\; in \;\; \Omega \setminus \overline{B}(t),\; t \in (0,T), \tag{1}$$

$$\nabla \cdot \boldsymbol u = 0 \;\; in \;\; \Omega \setminus \overline{B}(t),\; t \in (0,T), \tag{2}$$

$$\boldsymbol u(0) = \boldsymbol u_0(\boldsymbol x), \; (with \;\; \nabla \cdot \boldsymbol u_0 = 0), \tag{3}$$

$$\boldsymbol u = \boldsymbol g_0 \;\; on \;\; \Gamma \times (0,T), \;\; with \;\; \int_\Gamma \boldsymbol g_0 \cdot \boldsymbol n \, d\Gamma = 0, \tag{4}$$

$$\frac{\mathrm d\boldsymbol G}{\mathrm dt} = \boldsymbol V, \tag{5}$$

$$M_{\mathrm B}\frac{\mathrm d\boldsymbol V}{\mathrm dt} = M_{\mathrm B}\,\boldsymbol g + \boldsymbol F_{\mathrm H} + \boldsymbol F^{\mathrm r}, \tag{6}$$

$$\frac{\mathrm d(\boldsymbol I_{\mathrm B}\,\omega)}{\mathrm dt} = \boldsymbol T_{\mathrm H} + \overrightarrow{\boldsymbol G \boldsymbol x_{\mathrm r}} \times \boldsymbol F^{\mathrm r}, \tag{7}$$

$$\boldsymbol u(\boldsymbol x,t) = \boldsymbol V(t) + \omega(t) \times \overrightarrow{\boldsymbol G(t)\boldsymbol x}, \; \forall \, \boldsymbol x \in \partial B(t),\; t \in (0,T). \tag{8}$$

In (1)–(8) we have:

- $(0,T)$ is a time interval $(0 < T \le \infty)$.
- $\rho_{\mathrm f}$ and $\mu_{\mathrm f}$ are the fluid *density* and *viscosity*, respectively.
- $\boldsymbol u$ and $p$ are the flow *velocity* and *pressure*, respectively.
- $\boldsymbol g$ denotes *gravity*.
- $\Gamma$ is the external boundary of the flow region.
- $\boldsymbol n$ is the unit normal vector outward to the flow region.
- $\boldsymbol G$, $\boldsymbol V$ and $\omega$ are the *center of mass*, the *velocity of the center of mass*, and the *angular velocity*, respectively.
- $M_{\mathrm B}$ (resp., $\boldsymbol I_{\mathrm B}$) is the *mass* (resp., the *inertia tensor*) of the rigid body.
- $\boldsymbol F_{\mathrm H}$ and $\boldsymbol T_{\mathrm H}$ are, respectively, the *resultant* and the *torque* of the hydrodynamical forces acting on the rigid body, i.e.,

$$\boldsymbol F_{\mathrm H} = -\int_\gamma \sigma \boldsymbol n \, d\gamma, \;\; \boldsymbol T_{\mathrm H} = -\int_\gamma \overrightarrow{\boldsymbol G \boldsymbol x} \times \sigma \boldsymbol n \, d\gamma, \tag{9}$$

with $\gamma = \partial B$, and the stress tensor $\sigma$ defined by

$$\sigma = \mu_{\mathrm f}(\nabla \boldsymbol u + (\nabla \boldsymbol u)^{\mathrm t}) - p\boldsymbol I, \tag{10}$$

where $\boldsymbol I$ is the $3 \times 3$ identity matrix.
- $\boldsymbol F^{\mathrm r}$ is a short range force whose role is to prevent those particle/particle and particle/wall penetrations which may happen during the numerical simulation. There is then a resulting torque in (7) acting on the point $\boldsymbol x_{\mathrm r}$ where $\boldsymbol F^{\mathrm r}$ applies on $B$.
- The (no–slip) condition (8) expresses that the flow and the rigid body motion velocities coincide on the boundary $\gamma$ of the rigid body $B$.

182

Above, in (7) we found preferable to deal with the *kinematic angular momentum* $I_\mathrm{B}\,\omega$ making the formulation more conservative. Equations (1)–(8) are completed by the following initial conditions,

$$G(0) = G_0, \quad V(0) = V_0, \quad \omega(0) = \omega_0. \tag{11}$$

To solve system (1)–(8), (11) we can use, for example, *Arbitrary Lagrange–Euler (ALE)* methods as in [10], [11], and [12], or *fictitious domain methods*, which allow the flow calculation on a fixed grid, as in [1] to [9]. The fictitious domain methods that we advocate have some common features with the *immersed boundary method* of Ch. Peskin (see, e.g., [13] and [14]) but also some significant differences in the sense that we take systematically advantage of *distributed vector–valued Lagrange multipliers* to force the rigid body motion inside the particle, which seems still to be a relatively novel approach in this context, and whose possibilities have not been fully explored yet (an interesting application concerning the solution of blood flow around heart valves can be found in [15]). As with the methods in [13] and [14], our approach takes advantage of the fact that the flow can be computed on a grid which does not have to vary in time, a substantial simplification indeed. A preliminary – and most important – step in that direction is the following *global variational formulation* of problem (1)–(8), (11) (easily obtained by applying the *virtual power principle*):

$$\rho_\mathrm{f} \int_{\Omega \setminus \overline{B}(t)} \left[ \frac{\partial u}{\partial t} + (u \cdot \nabla)u \right] \cdot v \, dx + 2\mu_\mathrm{f} \int_{\Omega \setminus \overline{B}(t)} D(u) : D(v) \, dx$$

$$-\int_{\Omega \setminus \overline{B}(t)} p\nabla \cdot v \, dx + M_\mathrm{B} \frac{dV}{dt} \cdot Y + \frac{d}{dt}(I_\mathrm{B}\omega) \cdot \theta = \rho_\mathrm{f} \int_{\Omega \setminus \overline{B}(t)} g \cdot v \, dx \tag{12}$$

$$+ M_\mathrm{B} g \cdot Y + F^\mathrm{r} \cdot Y + \overrightarrow{Gx_\mathrm{r}} \times F^\mathrm{r} \cdot \theta, \quad \forall \{v, Y, \theta\} \in W_0(t), \ t \in (0, T),$$

$$\int_{\Omega \setminus \overline{B}(t)} q\nabla \cdot u(t) \, dx = 0, \quad \forall q \in L^2(\Omega \setminus \overline{B}(t)), \ t \in (0, T), \tag{13}$$

$$u = g_0 \ on \ \Gamma \times (0, T), \tag{14}$$

$$u(x, t) = V(t) + \omega(t) \times \overrightarrow{G(t)x}, \quad \forall x \in \partial B(t), \tag{15}$$

$$\frac{dG}{dt} = V, \tag{16}$$

to be completed by the following *initial conditions*

$$u(x, 0) = u_0(x), \quad \forall x \in \Omega \setminus \overline{B}(0), \tag{17}$$

$$B(0) = B_0, \quad G(0) = G_0, \quad V(0) = V_0, \omega(0) = \omega_0. \tag{18}$$

In relations (12) to (18):

- 

$$W_0(t) = \{ \{v, Y, \theta\} | v \in (H^1(\Omega \setminus \overline{B}(t)))^3, \ v = 0 \ on \ \Gamma,$$
$$Y \in \mathbb{R}^3, \ \theta \in \mathbb{R}^3, v(x, t) = Y + \theta \times \overrightarrow{G(t)x} \ on \ \partial B(t)\}.$$

- We have denoted functions such as $x \to \phi(x,t)$ by $\phi(t)$.
- We have used the following notation

$$a \cdot b = \sum_{k=1}^{d} a_k b_k, \ \forall a = \{a_k\}_{k=1}^{d}, \ b = \{b_k\}_{k=1}^{d},$$
$$A : B = \sum_{k=1}^{d} \sum_{l=1}^{d} a_{kl} b_{kl}, \ \forall A = (a_{kl})_{1 \le k, l \le d}, \ B = (b_{kl})_{1 \le k, l \le d}.$$

- It is reasonable to assume that $u(t) \in (H^1(\Omega \setminus \overline{B}(t)))^3$ and $p(t) \in L^2(\Omega \setminus \overline{B}(t))$.

## 3   A distributed Lagrange multiplier based fictitious domain formulation.

In broad terms our goal is to find a methodology such that

(a) A fixed mesh can be used for flow computations.
(b) The rigid body position is obtained from the solution of the Newton–Euler equations of motion, and
(c) The time discretization is done by operator splitting methods in order to treat individually – and in principle optimally – the various operators occurring in the mathematical model.

To achieve such a goal we proceed as follows:

(i) We fill the rigid body with the surrounding fluid.
(ii) We assume that the fluid inside the rigid body has a rigid body motion.
(iii) We use (i) and (ii) to modify the variational formulation (12)–(18).
(iv) We force the rigid body motion inside the moving body via a vector–valued Lagrange multiplier defined (distributed) over the body.
(v) We combine (iii) and (iv) to derive a variational formulation involving Lagrange multipliers to force the rigid body motion inside the moving body.

We suppose (for simplicity) that $B$ is made of an *homogeneous material* of density $\rho_B$; then, taking into account the fact that any rigid body motion velocity field $v$ verifies $\nabla \cdot v = 0$ and $D(v) = 0$, steps (i) to (iii) yield the following variant of formulation (12)–(18):

For a.e. $t > 0$, find $u(t)$, $p(t)$, $\{V(t), \ G(t), \ \omega(t)\}$ such that

$$\rho_f \int_{\Omega} \left[ \frac{\partial u}{\partial t} + (u \cdot \nabla)u \right] \cdot v \, dx - \int_{\Omega} p \nabla \cdot v \, dx + 2\mu_f \int_{\Omega} D(u) : D(v) \, dx$$

$$+ (1 - \rho_f/\rho_B) \left[ M_B \frac{dV}{dt} \cdot Y + \frac{d}{dt}(I_B \omega) \cdot \theta \right] \tag{19}$$

$$= \rho_f \int_{\Omega} g \cdot v \, dx + (1 - \rho_f/\rho_B) M_B g \cdot Y + F^r \cdot Y + \overrightarrow{Gx_r} \times F^r \cdot \theta,$$

$$\forall \{v, Y, \theta\} \in \tilde{W}_0(t), \ t \in (0, T),$$

$$\int_{\Omega} q \nabla \cdot u \, dx = 0, \ \forall q \in L^2(\Omega), \tag{20}$$

$$u = g_0 \text{ on } \Gamma \times (0, T), \tag{21}$$

$$u(x, t) = V(t) + \omega(t) \times \overrightarrow{G(t)x}, \ \forall x \in B(t), \tag{22}$$

$$\frac{dG}{dt} = V, \tag{23}$$

$$B(0) = B_0, \ V(0) = V_0, \ \omega(0) = \omega_0, G(0) = G_0, \tag{24}$$

$$u(x, 0) = u_0(x), \ \forall x \in \Omega \setminus \overline{B}_0 \text{ and } u(x, 0) = V_0 + \omega_0 \times \overrightarrow{G_0 x}, \ \forall x \in \overline{B}_0 \tag{25}$$

with, in relation (19), space $\tilde{W}_0(t)$ defined by

$$\tilde{W}_0(t) = \{\{v, Y, \theta\} | v \in (H_0^1(\Omega))^3, \ Y \in \mathbb{R}^3, \ \theta \in \mathbb{R}^3,$$
$$v(x, t) = Y + \theta \times \overrightarrow{G(t)x} \text{ in } B(t)\}.$$

Concerning $u$ and $p$ it makes sense to assume that $u(t) \in (H^1(\Omega))^3$ and $p(t) \in L_0^2(\Omega) = \{q | q \in L^2(\Omega), \int_\Omega q \, dx = 0\}$.

In order to relax the *rigid body motion constraint* (22) we are going to employ a *Lagrange multiplier* $\lambda$ so that $\lambda(t) \in \Lambda(t)$ with

$$\Lambda(t) = (H^1(B(t)))^3. \tag{26}$$

We obtain, thus, the following *fictitious domain formulation with Lagrange multipliers*:

For a.e. $t > 0$, find $u(t)$, $p(t)$, $\{V(t), G(t), \omega(t), \lambda(t)\}$, such that

$$u(t) \in (H^1(\Omega))^3, \ u(t) = g_0(t) \text{ on } \Gamma, \ p(t) \in L_0^2(\Omega),$$
$$V(t) \in \mathbb{R}^3, \ G(t) \in \mathbb{R}^3, \ \omega(t) \in \mathbb{R}^3, \ \lambda(t) \in \Lambda(t), \tag{27}$$

*and*

$$\rho_f \int_\Omega \left[\frac{\partial u}{\partial t} + (u \cdot \nabla)u\right] \cdot v \, dx - \int_\Omega p\nabla \cdot v \, dx + 2\mu_f \int_\Omega D(u) : D(v) \, dx$$

$$+ (1 - \rho_f/\rho_B)\left[M_B \frac{dV}{dt} \cdot Y + \frac{d}{dt}(I_B \omega) \cdot \theta\right] - <\lambda, \ v - Y - \theta \times \overrightarrow{Gx}> \tag{28}$$

$$= \rho_f \int_\Omega g \cdot v \, dx + (1 - \rho_f/\rho_B)M_B g \cdot Y + F^r \cdot Y + \overrightarrow{Gx_r} \times F^r \cdot \theta,$$

$$\forall v \in (H_0^1(\Omega))^3, \ \forall Y \in \mathbb{R}^3, \ \forall \theta \in \mathbb{R}^3,$$

$$\int_\Omega q\nabla \cdot u \, dx = 0, \ \forall q \in L^2(\Omega), \tag{29}$$

$$<\mu, \ u(t) - V(t) - \omega(t) \times \overrightarrow{G(t)x}> = 0, \ \forall \mu \in \Lambda(t), \tag{30}$$

$$\frac{dG}{dt} = V, \tag{31}$$

$$V(0) = V_0, \ G(0) = G_0, \ \omega(0) = \omega_0, \ B(0) = B_0, \tag{32}$$

$$u(x,0) = u_0(x), \ \forall x \in \Omega \setminus \overline{B}_0 \ \text{and} \ u(x,0) = V_0 + \omega_0 \times \overrightarrow{G_0 x}, \ \forall x \in \overline{B}_0. \quad (33)$$

The two most natural choices for $< \cdot, \cdot >$ are defined by

$$< \mu, v >= \int_{B(t)} (\mu \cdot v + \delta^2 \nabla \mu : \nabla v) \, dx, \ \forall \mu \ \text{and} \ v \in \Lambda(t), \quad (34)$$

$$< \mu, v >= \int_{B(t)} (\mu \cdot v + \delta^2 D(\mu) : D(v)) \, dx, \ \forall \mu \ \text{and} \ v \in \Lambda(t), \quad (35)$$

with $\delta$ a *characteristic length* (the diameter of $B$, for example).

*Remark 1.* In (27)–(33), only the center of mass, the translation velocity of the center of mass and the angular velocity of the particle are considered. Knowing these two velocities and the center of mass of the particle, one is able to translate and rotate the particle in space by tracking two extra points $x_1$ and $x_2$ in each particle; they both verify the rigid body motion equation

$$\frac{dx_i}{dt} = V(t) + \omega(t) \times \overrightarrow{G(t)x_i}, \ \ x_i(0) = x_{i,0}, \ i = 1, 2. \quad (36)$$

In practice we shall track two orthogonal normalized vectors rigidly attached to the body $B$ and originating from the center of mass $G$.

*Remark 2.* An approach with some similarities to ours has been developed by S. Schwarzer et al (see ref. [16]) in a finite difference framework. In the above reference (dedicated to the simulation of particulate flow), the interaction between the rigid body and the fluid is forced via a *penalty method*, instead of the multiplier technique used in the present article; also minor particle–particle penetration is allowed and no enforcement of the rigid body motion inside the region occupied by the particle is done.

*Remark 3.* Since, in (28), $u$ is *divergence free* and satisfies Dirichlet boundary conditions on $\Gamma$, we have

$$2 \int_{\Omega} D(u) : D(v) \, dx = \int_{\Omega} \nabla u : \nabla v \, dx, \ \forall v \in (H_0^1(\Omega))^3,$$

a substantial simplification indeed, from a *computational point of view*, which is another plus for the fictitious domain approach used here.

*Remark 4.* Using High Energy Physics terminology, the multiplier $\lambda$ can be viewed as a *gluon* whose role is to force the rigidity inside $B$ by matching the velocity fields of two continua. More precisely, the multipliers $\lambda$ are mathematical objects of the *mortar* type, very close to those used in *domain decomposition methods* to match local solutions at interfaces or on overlapping regions (see ref. [17]). Indeed, the $\lambda$'s in the present article have genuine mortar properties since their role is to force a fluid to behave like a rigid solid inside the space region occupied by the moving bodies.

*Remark 5.* The *neutrally buoyant case* (where $\rho_B = \rho_f$) is more complicated to handle, however, as shown in [18], the methods discussed in this article where $\rho_B \neq \rho_f$ can be modified to handle this more delicate situation (see [18] for details).

## 3.1 On the treatment of collisions.

In the above sections, we have considered the flow motion of fluid/rigid body mixtures and given various mathematical models of this phenomenon. Actually, with the mathematical model that we have considered it is not known if collisions can take place in finite time (in fact several scientists strongly believe that lubrication forces prevent these collisions in the case of viscous fluids). However, collisions take place in Nature and also in actual numerical simulations if special precautions are not taken. In the particular case of rigid bodies moving in a viscous fluid, under the effect of gravity and hydrodynamical forces, we shall assume that the collisions taking place are *smooth* ones in the sense that if two rigid bodies collide (resp., if a rigid body hits the boundary) the rigid body velocities (resp., the rigid body and boundary velocities) coincide at the points of contact. From the smooth nature of these collisions the only precaution to be taken will be to avoid the overlapping of the regions occupied by the rigid bodies. To achieve this goal, we include in the right–hand sides of the *Newton–Euler equations* (6) and (7) modeling the rigid body motion a *short range repulsive force*.

If $B_1$ and $B_2$ are two rigid bodies of general shape with the points $y_1$ and $y_2$ from $B_1$ and $B_2$, respectively, realizing the shortest distance $d$ between $B_1$ and $B_2$ ($d = |y_1 - y_2|$), we shall require the repulsion force $F^r$ between $B_1$ and $B_2$ to satisfy the following properties:

(i) To be parallel to $\overrightarrow{y_1 y_2}$.
(ii) To verify

$$|F^r| = 0 \text{ if } d \geq \rho,$$
$$|F^r| = c/\epsilon \text{ if } d = 0,$$

(iii) $|F^r|$ has to behave as in Fig. 2 for $0 \leq d \leq \rho$,

with $c$ a scaling factor, $\epsilon$ a "small" positive number, and $\rho$ the *range* of the repulsion force. For the simulations discussed in the following sections, we have taken $\rho \simeq h_\Omega$ ($h_\Omega$ is the *space discretization step* used for approximating the *velocity*). Body/wall collisions can be treated in a similar way.

## 4 Space and time discretization of problem (27)–(33)

### 4.1 Space discretization

For simplicity, we assume that $\Omega \subset \mathbb{R}^3$ is a rectangular parallelepiped. Concerning the *space approximation* of problem (27)–(33) by a *finite element method*, we have

$$W_h = \{v_h | \ v_h \in (C^0(\overline{\Omega}))^3, \ v_h|_T \in (P_1)^3, \ \forall T \in \mathcal{T}_h\}, \tag{37}$$

$$W_{0h} = \{v_h | \ v_h \in W_h, \ v_h = 0 \ on \ \Gamma\}, \tag{38}$$

$$L_h^2 = \{q_h | \ q_h \in C^0(\overline{\Omega}), \ q_h|_T \in P_1, \ \forall T \in \mathcal{T}_{2h}\}, \tag{39}$$

$$L_{0h}^2 = \{q_h | \ q_h \in L_h^2, \ \int_\Omega q_h \, dx = 0\} \tag{40}$$

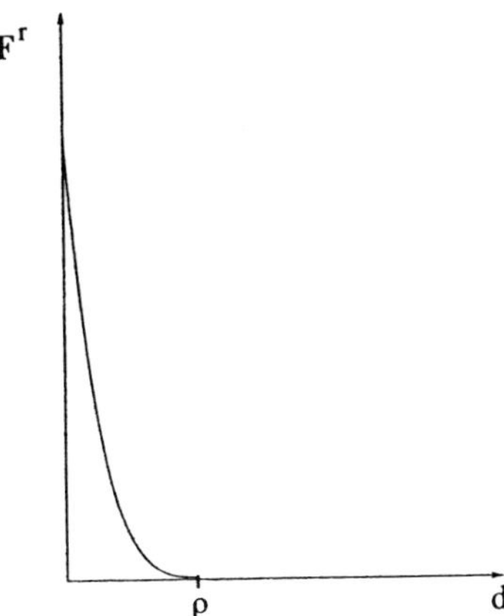

**Fig. 2.** Repulsion force behavior

where $\mathcal{T}_h$ is a tetrahedrization of $\Omega$, $\mathcal{T}_{2h}$ is twice coarser than $\mathcal{T}_h$, and $P_1$ is the space of the polynomials in three variables of degree $\leq 1$. A finite dimensional space approximating $\Lambda(t)$ is as follows: let $\{\xi_i\}_{i=1}^N$ be a set of points from $\overline{B}(t)$ which cover $\overline{B}(t)$ (uniformly, for example); we define then

$$\Lambda_h(t) = \{\mu_h |\ \mu_h = \sum_{i=1}^N \mu_i\ \delta(x - \xi_i),\ \mu_i \in \mathbb{R}^3,\ \text{for}\,i = 1, ..., N\}, \qquad (41)$$

where $\delta(\cdot)$ is the Dirac measure at $x = 0$. Then we shall use $< \cdot, \cdot >_h$ defined by

$$< \mu_h, v_h >_h = \sum_{i=1}^N \mu_i \cdot v_h(\xi_i),\ \forall \mu_h \in \Lambda_h(t),\ v_h \in W_h. \qquad (42)$$

A typical choice of points for defining (41) is a collection of grid points for velocity field covered by the interior of the particle $B(t)$, whose distance to the boundary of $B(t)$ is great than, e.g., $h_\Omega$ and selected points from the surface of $B(t)$. An example of choice of surface points is shown in Fig. 3.

Using the above finite dimensional spaces leads to the following approximation of problem (27)–(33):

For a.e. $t > 0$, find $u_h(t)$, $p_h(t)$, $\{V(t),\ G(t),\ \omega(t),\ \lambda_h(t)\}$, such that

$$u_h(t) \in W_h,\ u_h(t) = g_{0h}(t)\ on\ \Gamma,\ p_h(t) \in L_{0h}^2,$$
$$V(t) \in \mathbb{R}^3,\ G(t) \in \mathbb{R}^3,\ \omega(t) \in \mathbb{R}^3,\ \lambda_h(t) \in \Lambda_h(t), \qquad (43)$$

*and*

$$\rho_f \int_\Omega \left[ \frac{\partial u_h}{\partial t} + (u_h \cdot \nabla)u_h \right] \cdot v\, dx - \int_\Omega p_h \nabla \cdot v\, dx + \mu_f \int_\Omega \nabla u_h : \nabla v\, dx$$

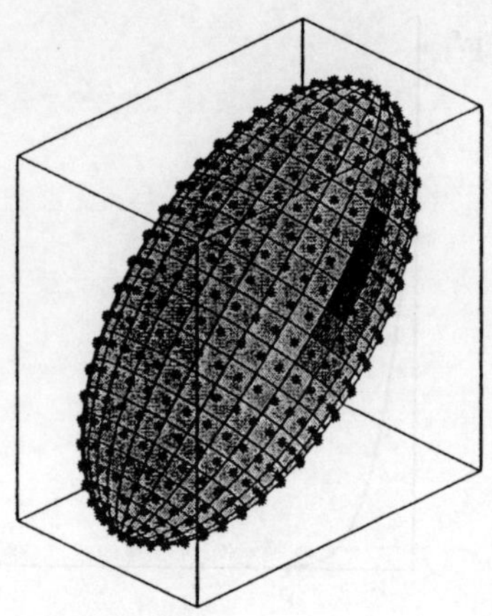

**Fig. 3.** An example of grids covering the surface of $B(t)$

$$+(1 - \rho_{\rm f}/\rho_{\rm B}) \left[ M_{\rm B}\frac{{\rm d}V}{{\rm d}t} \cdot Y + \frac{{\rm d}}{{\rm d}t}(I_{\rm B}\omega) \cdot \theta \right] - < \lambda_{\rm h}, \; v - Y - \theta \times \overrightarrow{Gx} >_{\rm h} \quad (44)$$

$$= \rho_{\rm f} \int_{\Omega} g \cdot v \, dx + (1 - \rho_{\rm f}/\rho_{\rm B}) M_{\rm B} g \cdot Y + F^{\rm r} \cdot Y + \overrightarrow{Gx_{\rm r}} \times F^{\rm r} \cdot \theta,$$

$$\forall v \in W_{0\rm h}, \; \forall Y \in \mathbb{R}^3, \; \forall \theta \in \mathbb{R}^3,$$

$$\int_{\Omega} q \nabla \cdot u_{\rm h} \, dx = 0, \; \forall q \in L_{\rm h}^2(\Omega), \tag{45}$$

$$< \mu, \; u_{\rm h}(t) - V(t) - \omega(t) \times \overrightarrow{G(t)x} >_{\rm h} = 0, \; \forall \mu \in \Lambda_{\rm h}(t), \tag{46}$$

$$\frac{{\rm d}G}{{\rm d}t} = V, \tag{47}$$

$$V(0) = V_0, \; G(0) = G_0, \; \omega(0) = \omega_0, \; B(0) = B_0, \tag{48}$$

$$u_{\rm h}(x,0) = u_{0\rm h}(x), \; \forall x \in \Omega \setminus \overline{B}_0 \; and \; u_{\rm h}(x,0) = V_0 + \omega_0 \times \overrightarrow{G_0x}, \; \forall x \in \overline{B}_0. \tag{49}$$

In (43), $g_{0\rm h}$ is an approximation of $g_0$ belonging to

$$\gamma W_h = \{z_{\rm h} | \; z_{\rm h} \in (C^0(\Gamma))^3, \; z_{\rm h} = \tilde{z_{\rm h}}|_\Gamma \; with \; \tilde{z_{\rm h}} \in W_{\rm h}\}$$

and verifying $\int_\Gamma g_{0\rm h} \cdot n \, d\Gamma = 0$.

## 4.2 Time discretization by operator–splitting of problem (43)–(49)

Following A. Chorin (refs. [19]–[21]), most "modern" Navier–Stokes solvers are based on *operator splitting* schemes (see, e.g., refs. [22], [23]) in order to force the incompressibility condition via a Stokes solver or a $L^2$–projection method. This approach still applies to the initial value problem (43)–(49) which contains five numerical difficulties to each of which can be associated a specific operator, namely

(a) The incompressibility condition and the related unknown pressure.
(b) An advection term.
(c) A diffusion term.
(d) The rigid body motion of $B(t)$ and the related multiplier $\lambda(t)$.
(e) The collision terms.

The operators in (a) and (d) are essentially *projection operators*. ¿From an abstract point of view, problem (43)–(49) is a particular case of the following class of initial value problems

$$\frac{d\phi}{dt} + \sum_{i=1}^{5} A_i(\phi, t) = f, \quad \phi(0) = \phi_0, \tag{50}$$

where the operators $A_i$ can be *multivalued*. Among the many operator–splitting methods which can be employed to solve problem (50) we advocate (following, e.g., [24]) the very simple one below; it is only *first order accurate* but its low order accuracy is compensated by good stability and robustness properties. Actually, this scheme can be made *second order accurate by symmetrization* (see, e.g., [25], [26] for the application of *symmetrized splitting schemes* to the solution of the Navier–Stokes equations).

**A fractional step scheme à la Marchuk–Yanenko:** With $\Delta t (> 0)$ a *time discretization step*, applying the *Marchuk–Yanenko scheme* to the initial value problem (50) leads to

$$\phi^0 = \phi_0, \tag{51}$$

*and for $n \geq 0$, compute $\phi^{n+1}$ from $\phi^n$ via*

$$\frac{\phi^{n+i/5} - \phi^{n+(i-1)/5}}{\Delta t} + A_i(\phi^{n+i/5}, t^{n+1}) = f_i^{n+1}, \tag{52}$$

*for $i = 1, 2, 3, 4, 5$ with $t^n = n\Delta t$ and $\sum_{i=1}^{5} f_i^{n+1} = f^{n+1}$.*

### 4.3 Application of the Marchuk–Yanenko scheme to the solution of problem (36) and (43)–(49)

Applying scheme (51), (52) to problem (36) and (43)–(49), we have the following scheme after dropping some of the subscripts h:

$$u^0 = u_0, \; V^0 = V_0, \; \omega^0 = \omega_0, \; G^0 = G_0, \; x_1^0 = x_{1,0}, \text{ and } x_2^0 = x_{2,0} \text{ given;} \tag{53}$$

for $n \geq 0$, knowing $u^n$, $V^n$, $\omega^n$ $G^n$, $x_1^n$ and $x_1^n$, compute $u^{n+1/5}$, $p^{n+1/5}$ via the solution of

$$\rho_f \int_\Omega \frac{u^{n+1/5} - u^n}{\Delta t} \cdot v \, dx - \int_\Omega p^{n+1/5} \nabla \cdot v \, dx = 0, \; \forall v \in W_{0h}, \tag{54}$$

$$\int_\Omega q \nabla \cdot u^{n+1/5} \, dx = 0, \; \forall q \in L_h^2, \tag{55}$$

$$u^{n+1/5} \in W_h, \; u^{n+1/5} = g_{0h}^{n+1} \text{ on } \Gamma, \; p^{n+1/5} \in L_{0h}^2.$$

Then compute $u^{n+2/5}$ via the solution of

$$\int_\Omega \frac{\partial u}{\partial t} \cdot v \, dx + \int_\Omega (u^{n+1/5} \cdot \nabla) u \cdot v \, dx = 0, \tag{56}$$

$$\forall v \in W_{0h}^{n+1,-}, \text{ a.e. on } (t^n, t^{n+1}),$$

$$u(t^n) = u^{n+1/5}, \tag{57}$$

$$u(t) \in W_h, \ u(t) = g_{0h}^{n+1} \text{ on } \Gamma_-^{n+1} \times (t^n, t^{n+1}); \tag{58}$$

$$\text{and set } u^{n+2/5} = u(t^{n+1}), \tag{59}$$

where

(a) $\Gamma_-^{n+1} = \{x | x \in \Gamma, \ g_{0h}^{n+1}(x) \cdot n(x) < 0\}$,
(b) $W_{0h}^{n+1,-} = \{v | v \in W_h, \ v = 0 \text{ on } \Gamma_-^{n+1}\}$.

Next, compute $u^{n+3/5}$ via the solution of

$$\rho_f \int_\Omega \frac{u^{n+3/5} - u^{n+2/5}}{\Delta t} \cdot v \, dx + \alpha \mu_f \int_\Omega \nabla u^{n+3/5} : \nabla v \, dx = \rho_f \int_\Omega g \cdot v \, dx,$$

$$\forall v \in W_{0h}; \ u^{n+3/5} \in W_h, \ u^{n+3/5} = g_{0h}^{n+1} \text{ on } \Gamma, \tag{60}$$

Now predict the motion of the center of mass and the angular velocity of the rigid body via

$$\frac{dG}{dt} = V(t), \tag{61}$$

$$\frac{dx_i}{dt} = V(t) + \omega(t) \times \overrightarrow{G(t)x_i}, \text{ for } i = 1, 2, \tag{62}$$

$$(1 - \rho_f/\rho_B) M_B \frac{dV}{dt} = (1 - \rho_f/\rho_B) M_B g + F^r, \tag{63}$$

$$(1 - \rho_f/\rho_B) \frac{d(I_B \omega)}{dt} = \overrightarrow{Gx_r} \times F^r, \tag{64}$$

$$G(t^n) = G^n, \ V(t^n) = V^n, \ (I_B \omega)^n = (I_B \omega)(t^n), \tag{65}$$

$$x_1(t^n) = x_1^n, \ x_2(t^n) = x_2^n,$$

for $t^n < t < t^{n+1}$.

Then set $G^{n+4/5} = G(t^{n+1})$, $V^{n+4/5} = V(t^{n+1})$, $(I_B \omega)^{n+4/5} = (I_B \omega)(t^{n+1})$, $x_1^{n+4/5} = x_1(t^{n+1})$, $x_2^{n+4/5} = x_2(t^{n+1})$, and $u^{n+4/5} = u^{n+3/5}$.

With the center $G^{n+4/5}$, $x_1^{n+4/5}$ and $x_2^{n+4/5}$ obtained at the above step, we enforce the rigid body motion in the region $B(t^{n+4/5})$ occupied by the rigid body

$$\rho_f \int_\Omega \frac{u^{n+1} - u^{n+4/5}}{\Delta t} \cdot v \, dx + \beta \mu_f \int_\Omega \nabla u^{n+1} : \nabla v \, dx$$

$$+ (1 - \frac{\rho_f}{\rho_B}) M_B \frac{V^{n+1} - V^{n+4/5}}{\Delta t} \cdot Y + (1 - \frac{\rho_f}{\rho_B}) \frac{(I_B \omega)^{n+1} - (I_B \omega)^{n+4/5}}{\Delta t} \cdot \theta \tag{66}$$

$$=<\lambda^{n+4/5},\ v-Y-\theta\times\overrightarrow{G^{n+4/5}x}>_{\mathrm{h}},\ \forall v\in W_{0\mathrm{h}}, Y\in\mathbb{R}^3, \theta\in\mathbb{R}^3;$$

$$u^{n+1}\in W_{\mathrm{h}}, u^{n+1}=g_{0\mathrm{h}}^{n+1}\text{ on }\Gamma,\ \lambda^{n+4/5}\in\Lambda_{\mathrm{h}}^{n+4/5},$$

$$V^{n+1}\in\mathbb{R}^3,\ \omega^{n+1}\in\mathbb{R}^3,$$

$$<\mu,\ u^{n+1}-V^{n+1}-\omega^{n+1}\times\overrightarrow{G^{n+4/5}x}>_{\mathrm{h}}=0,\ \forall\mu\in\Lambda_{\mathrm{h}}^{n+4/5}. \tag{67}$$

In above (53)–(67), $g_{0\mathrm{h}}^{n+1}=g_{0\mathrm{h}}(t^{n+1})$, $\Lambda_{\mathrm{h}}^{n+s}=\Lambda_{\mathrm{h}}(t^{n+s})$ and $\alpha+\beta=1$. In the numerical simulation, we usually choose $\alpha=1$ and $\beta=0$.

## 5 On the solution of the subproblems (54)–(55), (56)–(58), (60), (61)–(65), and (66)–(67)

The discrete degenerated quasi–Stokes problem (54)–(55) is solved by an Uzawa/ preconditioned conjugate gradient algorithm as in [27], where the discrete elliptic problems from the preconditioning are solved by a matrix–free fast solver from FISHPAK due to Adams et al. in [28]. The advection problem (56)–(58) for the velocity field is solved by a wave–like equation method as in [26]. Problem (60) is a classical discrete elliptic problem which can be solved by the same matrix–free fast solver.

System (61)–(65) is a system of ordinary differential equations thanks to operator splitting. For its solution one can choose a time step smaller than $\Delta t$, (i.e., we can divide $\Delta t$ into smaller steps) to predict the translation velocity of the center of mass, the angular velocity of the particle, the position of the center of mass and the regions occupied by each particle so that the repulsion forces can be effective to prevent particle–particle and particle–wall overlapping. At each subcycling time step, keeping the distance constant between any pair of points $x_1$ and $x_2$ in each particle is important since we are dealing with rigid bodies. We have applied the following approach to satisfy the above constraint:

- Translate $x_1$ and $x_2$ according to the new position of the mass center at each subcycling time step.
- Rotate $Gx_1$ and $Gx_2$, the relative positions of $x_1$ and $x_2$ to the center of mass $G$, by the following Crank–Nicolson scheme (a Runge–Kutta scheme of order 2, in fact):

$$\frac{Gx_i^{\text{new}}-Gx_i^{\text{old}}}{\tau}=\omega\times\frac{Gx_i^{\text{new}}+Gx_i^{\text{old}}}{2} \tag{68}$$

for $i=1,2$ with $\tau$ as a subcycling time step. By (68), we have $|Gx_i^{\text{new}}|^2=|Gx_i^{\text{old}}|^2$ for $i=1,2$ and $|Gx_2^{\text{new}}-Gx_1^{\text{new}}|^2=|Gx_2^{\text{old}}-Gx_1^{\text{old}}|^2$ (i.e., scheme (68) is distance and in fact shape preserving).

*Remark 6.* In order to activate the short range repulsion force, we have to find the shortest distance between two ellipsoids. Unlike the cases for spheres, it is not trivial to locate the points from the ellipsoid surfaces realizing the shortest between two ellipsoids. There is no explicit formula for such distance. In practice, we first choose a set of points from the surface of each ellipsoid. Then we find the point among the chosen points from each surface at which the distance is the shortest We repeat this (kind of relaxation) process in the neighborhood of the newly located point on each surface of ellipsoid until convergence, usually obtained in very few iterations.

For the shortest distance between the wall and ellipsoid, there exists an explicit formula. To check whether two ellipsoids overlap each other, there exists an algorithm used by people working on computer graphics and in robotics (e.g., see, [29]).  □

After solving (61)–(65), the rigid body motion is enforced in $B(t^{n+4/5})$, via (66)–(67). At the same time those hydrodynamical forces acting on the particles are also taken into account in order to update the translation and angular velocities of the particles. To solve (66)–(67), we use a conjugate gradient algorithm as discussed in [1]. Since we have taken $\beta = 0$ in (66) for the simulation, we actually do not need to solve any non–trivial linear systems for the velocity field; this saves a lot of computing time. To get the angular velocity $\omega^{n+1}$, computed via

$$\omega^{n+1} = (I_{\mathrm{B}}^{n+4/5})^{-1}(I_{\mathrm{B}}\,\omega)^{n+1}, \tag{69}$$

we need to have $I_{\mathrm{B}}^{n+4/5}$, the inertia of the particle $B(t^{n+4/5})$. We first compute the inertia $I_0$ in the coordinate system attached to the particle. Then via the center of mass $G^{n+4/5}$ and points $x_1^{n+4/5}$ and $x_2^{n+4/5}$, we have the rotation transformation $Q$ ($QQ^{\mathrm{T}} = Q^{\mathrm{T}}Q = I_{\mathrm{B}}$, $\mathrm{det}Q=1$) which transforms vectors expressed in the particle frame to vectors in the flow domain coordinate system and $I_{\mathrm{B}}^{n+s} = QI_0Q^{\mathrm{T}}$. Actually in order to update matrix $Q$ we can also use *quaternion* techniques, as shown, in the review paper [30].

# 6    Numerical experiments

## 6.1    Freely rotating elliptic rigid body in a two–dimensional channel

We consider the flow past a freely rotating two–dimensional elliptic body in a channel $\Omega = (-15, 40) \times (-4, 4)$ filled with a viscous incompressible liquid of density $\rho_{\mathrm{f}} = 1$. The flow is assumed to move from left to right, and acts on a rigid elliptic body of density $\rho_{\mathrm{B}} = 1.1$, with center fixed at $(0,0)$. The lengths of the major and minor axes of the elliptic body are $a = 1$ and $b = 0.5$, respectively. The motion of the body is described by its angular position $\theta$, defined as the the angle of the major axes with the horizontal, and its angular velocity $\omega = \mathrm{d}\theta/\mathrm{d}t$. This problem is solved for Reynolds numbers $Re = 20$, 100 and 200. The Reynolds number is defined as $Re = \rho_{\mathrm{f}}Ua/\mu_{\mathrm{f}}$, where $U$ is the maximum inlet velocity at the upstream boundary. As initial conditions we choose $\theta^0 = 0$, $\omega^0 = 0$, and $u^0 = 0$. These problems are solved on a non–regular mesh, with refinement in the region where the ellipse rotates, as shown in Fig. 4(a). The collocation points inside the elliptic body as well as the collocation points chosen on its boundary are shown in Fig. 4(b). The time step is $\Delta t = 0.001$ for all cases. For these

**Fig. 4.** (a) Non–regular mesh for the velocity. (b) Collocation points: mesh nodes inside the elliptic rigid body and points chosen on its boundary.

meshes we are unable to use the fast elliptic solvers. Instead we use a sparse matrix algorithm based on Markowitz' method [33].

For $Re = 20$ the elliptic body rotates counterclockwise so that its broad side tends to be perpendicular to the flow direction, as qualitatively expected [7]. The flow tends to a steady state and the elliptic body remains in its stable vertical

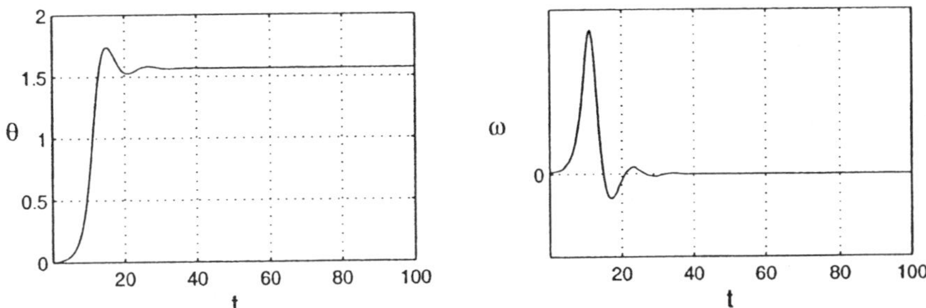

**Fig. 5.** Time history of the angle and of angular velocity of the ellipse. $Re = 20$.

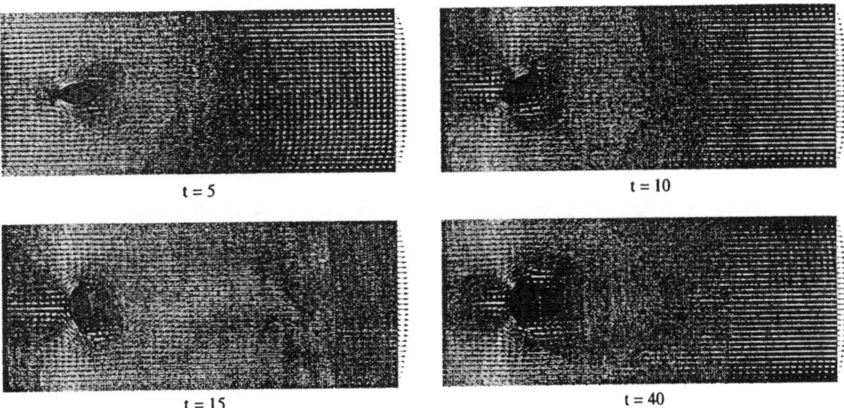



**Fig. 6.** Visualization of the pressure and velocity vector field at different times. $Re = 20$.

position $\theta = \pi/2$, as shown in Figs. 5 and 6.

For $Re = 100$ and 200 unsteady oscillatory solutions are obtained. The main feature of these solutions is periodic oscillation of the elliptic rigid body around its vertical position, as shown in Figs. 7 and 9. These oscillations are associated with vortex shedding behind the elliptic body (see Figs. 8 and 10). Both, the amplitude of the oscillations and the angular velocity of the elliptic body, are larger at higher Reynolds number. Concerning the frequency, it is a common practice to find the Strouhal number defined as $St = 2fa/U$, where $f$ is the

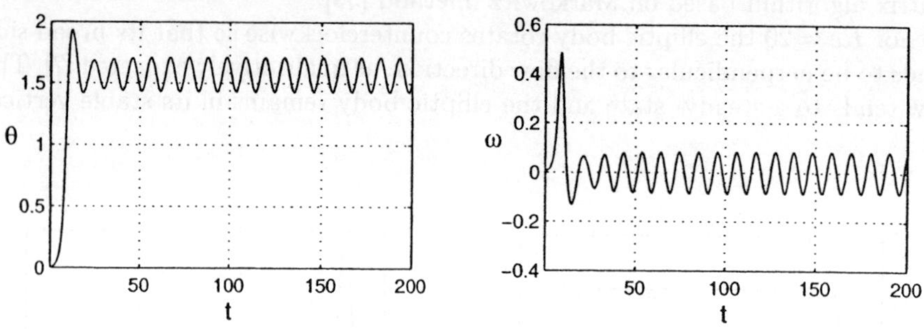

Fig. 7. Time history of the angle and of angular velocity of the ellipse. $Re = 100$.

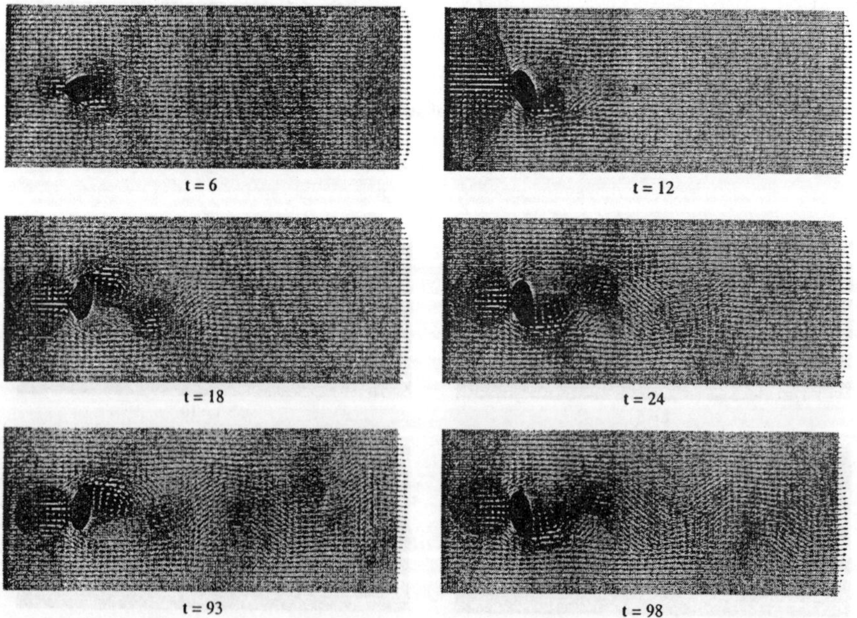



Fig. 8. Visualization of the pressure and velocity vector field at different times. $Re = 100$.

frequency of the oscillations, and $U$ and $a$ are defined as before. For the cases considered here we obtain $St = 0.19$ and $St = 0.22$ for $Re = 100$ and $Re = 200$, respectively. These results are consistent with previous results about a freely rotating cylinder in a channel [32].

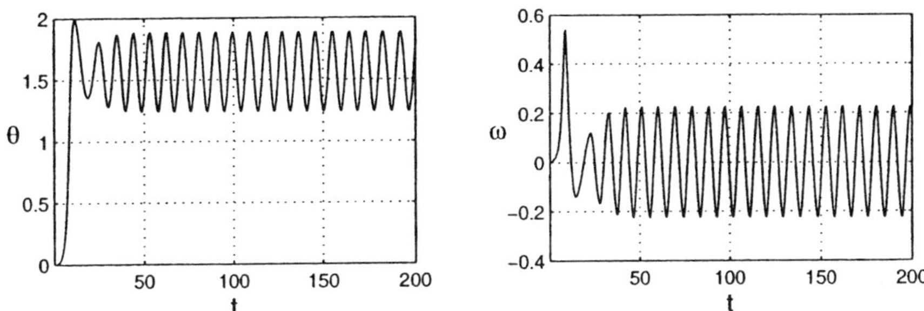

**Fig. 9.** Time history of the angle and of angular velocity of the ellipse. $Re = 200$.

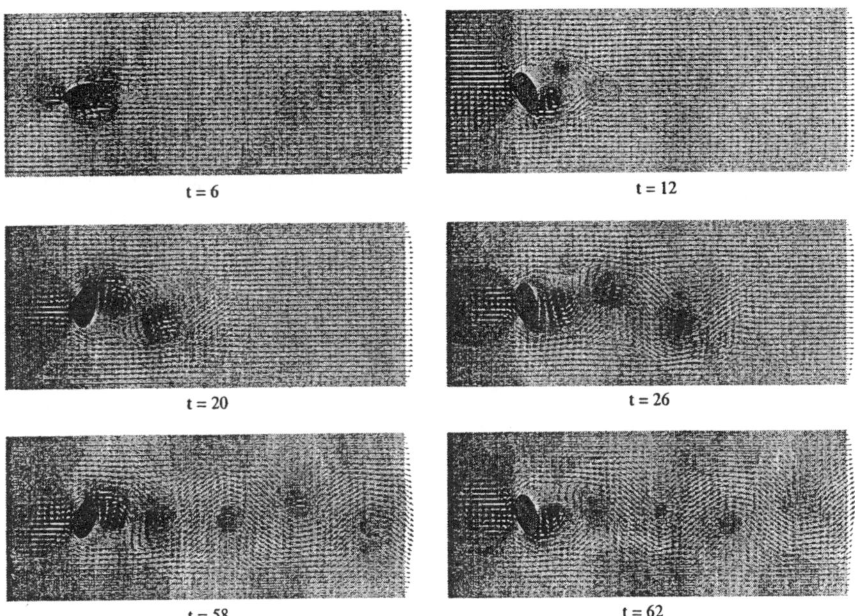


**Fig. 10.** Visualization of the pressure and velocity vector field at different times. $Re = 200$.

As the last two–dimensional example we present the results of the free motion of a elliptic pendulum in an incompressible viscous fluid. The hydrodynamic elliptic pendulum is defined in the following way: The center $G$ of the elliptic

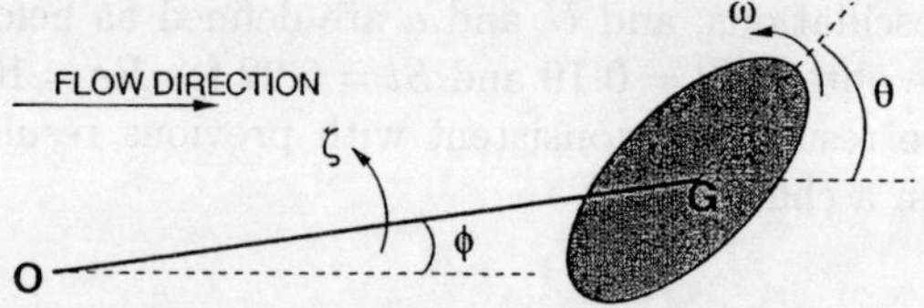



**Fig. 11.** A pendulum with a free to rotate elliptic body in a viscous fluid.

rigid body is constrained to move on a circular trajectory around the fixed axis of rotation $O$, as shown in Fig. 11; simultaneously the elliptic body rotates freely around its center $G$. This pendulum is placed in a channel where the flow direction is from left to right. The position and orientation of the elliptic body are known at any time by the angles $\theta$ and $\phi$, respectively. The velocity of the rigid body is known through the angular velocities $\omega = d\theta/dt$ and $\zeta = d\phi/dt$. The elliptic body has the same dimensions as in the previous examples. The axis of rotation of the pendulum is located at $O = (-8, 0)$, and the length of the pendulum is $l_{\rm p} = |\overrightarrow{OG}| = 8$. The initial conditions are $\theta^0 = 0$, $\phi^0 = 0$, $\omega^0 = 0$, $\zeta^0 = 0$, and $u^0 = 0$. We solve the problem for $Re = 200$, where $Re$ is defined as before. The main feature of the solution is periodic oscillation of the pendulum around its axes $O$ coupled with periodic oscillation of the elliptic body around its center $G$, as shown in Fig. 12. The amplitude of rotation of the

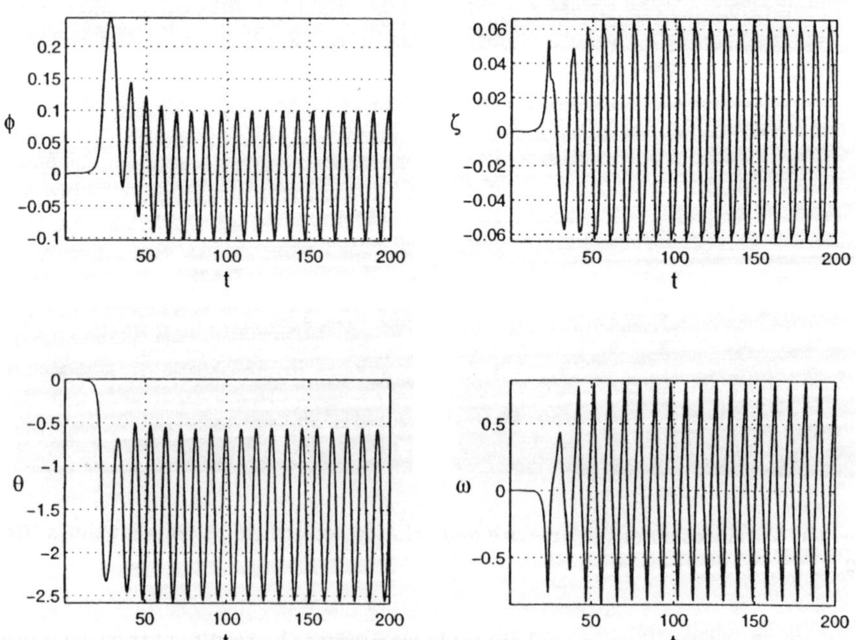

**Fig. 12.** Time history of the angle and angular velocity of the pendulum (top), and of the angle and angular velocity of the elliptic rigid body (bottom) at $Re = 200$.

rigid body around its center is about fifteen times larger than the amplitude of rotation of the pendulum for this particular case. In fact, we expect that the longer $l_p$ the shorter the amplitude of oscillation of the pendulum [34]. These oscillations have the same frequency as the oscillations of the free to rotate elliptic body for $Re = 200$ (see Figure 9). This corroborates that the oscillations are strongly associated with vortex shedding behind the elliptic body. The result is a synchronized periodic motion of both the pendulum and body. Fig. 13 shows – at different times – some snapshots of the pressure and velocity vectors in a region near the body. However, a distinctive characteristic of the solution here is that that the amplitude of rotation and angular velocity of the elliptic body around its center $G$ are significantly higher than those obtained for a freely rotating elliptic body placed at the center of the channel. This phenomenon is due to the unsymmetric shear, produced by the presence of the channel walls, on the two sides of the elliptic body when it is off the center line of the channel. When the elliptic body moves up the higher shear on the upper part of the body introduce a net rotation (in addition to the induced rotation by vortex shedding) in the clockwise direction, while when it moves down the body gets an additional net rotation in the counterclockwise direction. Again this is consistent with previous results found in [31] and [32].

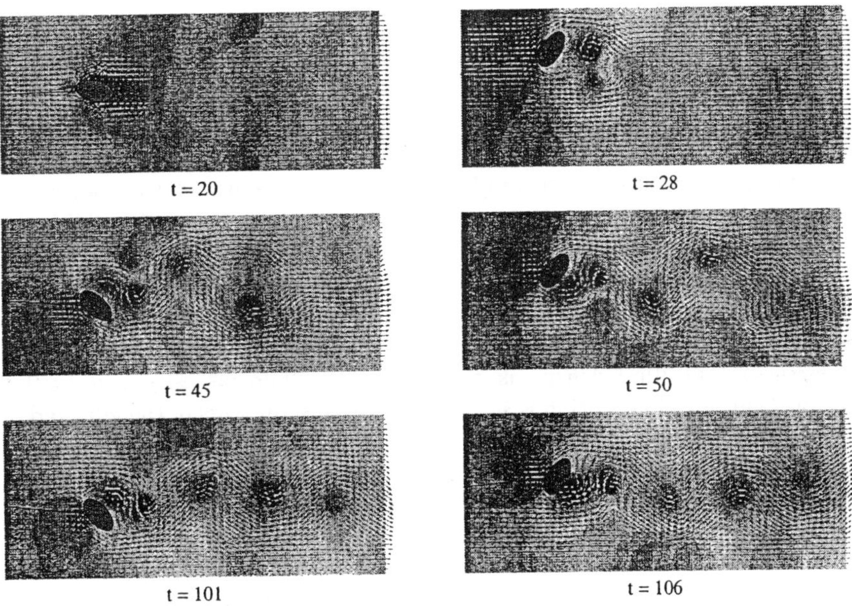



**Fig. 13.** Visualization of the pressure and velocity vector field at different times. Hydrodynamic pendulum, $Re = 200$.

## 6.2 Settling of a sphere in a narrow channel

In order to validate the simulation results, we have considered the test case of the settling of a sphere in a narrow channel filled with a Newtonian fluid. The computational domain is $\Omega = (0, 16D) \times (0, D) \times (0, 46D)$ with $D = 0.349\text{cm}$. The diameter of the ball is 0.317 cm. The fluid density is $\rho_f = 1\text{g/cm}^3$ and the fluid viscosity is $\mu_f = 0.009\text{cm}^2/\text{sec}$. The initial condition of flow field is $u = 0$. The initial velocity and initial angular velocity of the sphere are $0$. The density of the ball is $\rho_B = 1.194\text{g/cm}^3$. The initial positions of the ball center is $(8D, D/2, 46D - 0.317)$. The mesh sizes for the velocity field are $h_v = D/16$ and $D/20$, respectively. The mesh size for the pressure is $h_p = 2h_v$. The time step is $\Delta t = 0.001$.

We have chosen collocation points from the surface of the ball by dividing the latitudinal angle of the ball uniformly and, for each angle, choosing points from its associate circle on the surface (plus those grid points of the velocity field in the region occupied by the ball, whose distance is $0.5h_v$ away from the surface of the ball). We have shown the projection of the flow field on the $xz$-plane passing through the center of the ball at $t = 1.8$ in Fig. 14 for two different mesh sizes. The averaged particle Reynolds numbers (based on the length of the diameter) for $h_v = D/16$ and $D/20$ are about 252 and 261, respectively. The wave lengths of the vortex shedding behind the ball and the Reynolds numbers are close to the experimental results obtained by E. Ramos at the University of Texas at Austin.

## 6.3 Two ellipsoids sedimenting side–by–side in a narrow channel

It had been observed experimentally that when two ellipsoid–like long bodies sediment side–by–side in a narrow channel filled with a Newtonian fluid, they interact periodically with each other. To reproduce this phenomenon, we consider the following test case. The computational domain is $\Omega = (0, 1) \times (0, 0.25) \times (0, 4)$ initially, then it moves down with the lower center of two ellipsoids (see, e.g., [31] for adjusting the computational domain according to the position of the particle). The fluid density is $\rho_f = 1$ and the fluid viscosity is $\mu_f = 0.01$. The flow field initial condition is $u = 0$. The three semi–axes of the ellipsoid are 0.2, 0.1 and 0.1. The initial velocity and angular velocity of the ellipsoid are $0$. The density of the ellipsoid is $\rho_B = 1.25$. The initial positions of the centers are $(0.22, 0.125, 0.75)$ and $(0.78, 0.125, 0.75)$, respectively. The frames rigidly initially attached to the ellipsoids are

$$\{(\cos \pi/3, 0, \sin \pi/3), (0, 1, 0), (\cos 5\pi/6, 0, \sin 5\pi/6)\}$$

and

$$\{(\cos(-\pi/3), 0, \sin(-\pi/3)), (0, 1, 0), (\cos \pi/6, 0, \sin \pi/6)\},$$

respectively (see Fig. 15). The mesh size for the velocity field (resp., pressure) is $h_v = 1/80$ (resp., $h_p = 2h_v$). The time step is $\Delta t = 0.001$. In the simulation, we obtained results of two ellipsoids interacting with each other as shown in Fig. 15

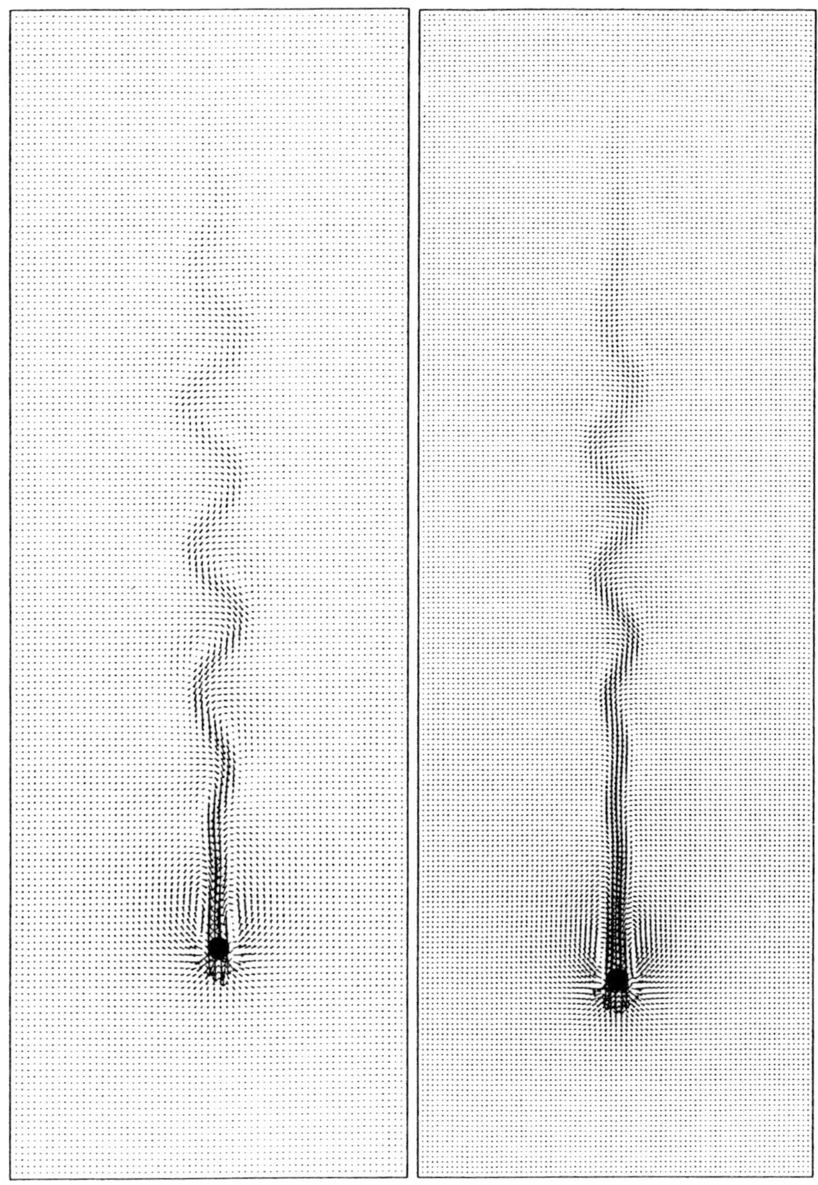

**Fig. 14.** Projection of the flow field on the $xz$-plane passing through the center of the ball at $t = 1.8$: $h_v = D/16$ (left) and $h_v = D/20$ (right).

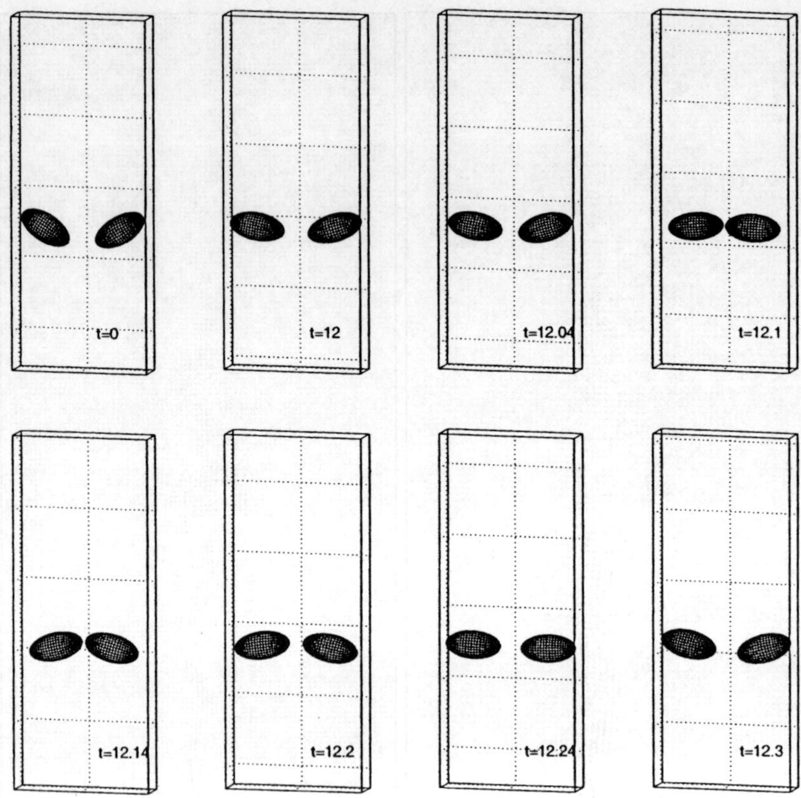

**Fig. 15.** Position of ellipsoids at $t = 0$, 12, 12.04, 12.1, 12.14 12.2, 12.24, and 12.3 (from left to right and from top to bottom).

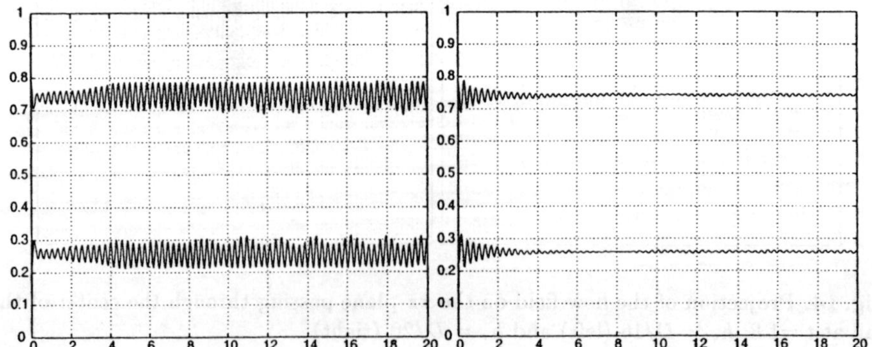

**Fig. 16.** Histories of the $x$–coordinates of the centers of two ellipsoids of long axes 0.4 (left) and the $x$–coordinates of the centers of two ellipsoids of long axes 0.36 (right).

(the computation was performed in a moving frame of reference, so the ellipsoids appear not moving downward), which is in good agreement with experimental results, qualitatively. The averaged terminal velocity is about 2.497, so the averaged particle Reynolds number is 99.88 based on the length of the long axis (which is 0.4). In Fig. 16, we can see the very strong interaction between two ellipsoids of long axes 0.4. We also have tested the case with two ellipsoids of long axes 0.36 and found that they settle in the channel with very weak interaction between each other (see Fig. 16).

## Acknowledgments

We acknowledge the helpful comments and suggestions of R. Bai, G.P. Galdi, J. He, D.D. Joseph, and V.I. Paulsen. We acknowledge also the support of NSF (grants DMS–9973318, and CCR–9902035), Texas Board of Higher Education (ARP grant 003652–0383–1999), and DOE/LASCI (grant R71700K–292–000–99).

## References


1. R. Glowinski, T.-W. Pan, T. Hesla, D.D. Joseph: Int. J. Multiphase Flow **25**, 755 (1999)
2. R. Glowinski, T.-W. Pan, T. Hesla, D.D. Joseph, J. Periaux: Int. J. Num. Meth. in Fluids **30**, 1043 (1999)
3. T.-W. Pan: C. R. Acad. Sci. Paris, Série IIb **327**, 1035 (1999)
4. R. Glowinski, T.-W. Pan, T.I. Hesla, D.D. Joseph, J. Periaux: Comput. Methods Appl. Mech. Engrg. **184**, 241 (2000)
5. R. Glowinski, T.-W. Pan, T.I. Hesla, D.D. Joseph, J. Periaux: J. Comput. Phys. **169**, 363 (2001)
6. T.-W. Pan, D.D. Joseph, R. Glowinski: J. Fluid Mech. **434**, 23 (2001)
7. L.H. Juarez: C. R. Acad. Sci. Paris, Série IIb **329**, 221 (2001)
8. T.-W. Pan, D.D. Joseph, R. Bai, R. Glowinski and V. Sarin: J. Fluid Mech. **451**, 169 (2002)
9. R. Glowinski: 'Finite Element Methods for Incompressible Viscous Flow'. In : *Handbook of Numerical Analysis*. ed. by P.G. Ciarlet and J.L. Lions (North-Holland, Amsterdam) (to appear)
10. H.H. Hu: Int. J. Multiphase Flow **22**, 335 (1996)
11. A. Johnson, T. Tezduyar: Comp. Meth. Appl. Mech. Eng. **145**, 301 (1997)
12. B. Maury, R. Glowinski: C.R. Acad. Sc. Paris, Serie I, **324**, 1079 (1997)
13. C.S. Peskin: J. Comp. Phys. **25**, 220 (1977)
14. C.S. Peskin, D.M. McQueen: J. Comp. Phys. **37**, 113 (1980)
15. J. de Hart: Fluid–Structure Interaction in the Aortic Heart Valve: a Three-Dimensional Computational Analysis. Ph.D Thesis, Technische Universiteit Eindhoven, Eindhoven, the Netherlands (2002)
16. K. Hofler, M. Muller, S. Schwarzer, B. Wachman: 'Interacting particle–liquid systems'. In *High Performance Computing in Science and Engineering*, ed. by E. Krause, W. Jager (Springer-Verlag, Berlin, 1998) pp. 54–64

17. Y. Achdou, Y. Kuznetsov: 'Algorithms for the mortar element method'. In *Domain Decomposition Methods in Sciences and Engineering*, ed. by R. Glowinski, J. Periaux, Z.-C. Shi, O. Widlund ( Wiley, Chichester, 1997) pp.33–42

18. T.-W. Pan, R. Glowinski: 'Direct simulation of the motion of neutrally buoyant circular cylinders in plane Poiseuille flow'. J. Comput. Phys. (to appear)

19. A. J. Chorin: J. Comp. Phys. **2**, 12 (1967)

20. A. J. Chorin: Math. Comp. **23**, 341 (1968)

21. A. J. Chorin: J. Fluid Mech. **57**, 785 (1973)

22. R. Glowinski, O. Pironneau: Annual Rev. Fluid Mech. **24**, 167 (1992)

23. S. Turek: *Efficient Solvers for Incompressible Flow Problems: An Algorithmic and Computational Approach.* (Springer–Verlag, Berlin, 1999)

24. G. I. Marchuk: 'Splitting and Alternating Direction Methods'. In *Handbook of Numerical Analysis, Vol. I,*, ed. by P. G. Ciarlet, J. L. Lions (North–Holland, Amsterdam, 1990) pp. 197–462

25. J. T. Beale, A. Majda: Math. Comp. **37**, 243 (1981)

26. E. J. Dean, R. Glowinski: C.R. Acad. Sci. Paris Série 1 **325**, 789 (1997)

27. R. Glowinski, T.-W. Pan, J. Periaux: Comput. Methods Appl. Mech. Engrg. **151**, 181 (1998)

28. J. Adams, P. Swarztrauber, R. Sweet: *FISHPAK: A package of Fortran subprograms for the solution of separable elliptic partial differential equations.* (The National Center for Atmospheric Research, Boulder, CO, 1980)

29. E. Rimon, S. Boyd: 'Efficient distance computation using best ellipsoid fit' In *The IEEE International symposium on intelligent control Glasgow, UK* (IEEE, 1992) pp. 360–365

30. J.C. K. Chou: IEEE transaction on robotics and automation **8**, 53 (1992)

31. H.H. Hu, D.D. Joseph, and M. Crochet: Theoret. Comput. Fluid Dynamics **3**, 285 (1992)

32. H. Juarez, R. Scott, R. Metcalfe, B. Bagheri: Comput. & Fluids **29**, 547 (2000)

33. S. Pissanetsky: *Sparse Matrix Technology.* (Academic Press, 1984)

34. Y. Roux, E. Rivoalen, B. Marichal: C. R. Acad. Sci. Paris, Série IIb **328**, 829 (2000)

# Steady Incompressible Inviscid and Viscous Flow Simulation Using Unstructured Tetrahedral Meshes

Kenneth Morgan, Dhemi Harlan, Oubay Hassan, Kaare Sørensen, and Nigel Weatherill

## 1 Introduction

Incompressible flow modelling is an area of significant industrial interest and has received considerable attention from researchers for many years. In this area, the computational techniques that have been developed allow for the modelling of complex geometrical configurations by, generally, employing multiblock or unstructured grids of hexahedral cells [1,2]. However, a comparatively small amount of effort has been devoted to the development of equal order velocity–pressure formulations for implementation on general tetrahedral grids [3–8], presumably because of the difficulties associated with central difference type schemes for this class of problems [9,10].

In this context, we are currently interested in extending the range of application of the unstructured mesh based cell vertex finite volume solution techniques that we have recently been developing for compressible high speed aerodynamic flows [11] into the incompressible regime. The method by which this may be achieved for steady incompressible flows was first successfully demonstrated in the case of quadrilateral or hexahedral meshes [12,13] and employs the artificial compressibility concept of Chorin [14]. This approach has, more recently, also been demonstrated for unstructured triangular [15] and tetrahedral meshes [8].

In this chapter, we present an implementation of the method for the solution of the 3D steady incompressible laminar Navier–Stokes equations on general tetrahedral grids. A cell vertex finite volume method is employed and steady solutions are computed by the use of explicit multi–stage time stepping. The computational performance of the approach is improved by the use of an agglomerated multigrid procedure, to accelerate convergence to steady state, and by parallelisation. The numerical performance of the method is demonstrated for several different examples.

## 2 Governing Equations

The flows of interest in this chapter are governed by the 3D steady incompressible laminar Navier Stokes equations. Relative to a cartesian coordinate

204

system $Ox_1x_2x_3$, these equations are expressed, for an arbitrary volume $\mathcal{V}$ with closed boundary surface $\mathcal{S}$, in the non–dimensional integral form

$$\int_{\mathcal{S}} F^j n_j \mathrm{d}\mathcal{S} = \frac{1}{\mathrm{Re}} \int_{\mathcal{S}} \frac{\partial V}{\partial x_j} n_j \mathrm{d}\mathcal{S} \qquad j = 1, 2, 3 \qquad (1)$$

where the summation convention is employed and

$$V = \begin{bmatrix} 0 \\ u_1 \\ u_2 \\ u_3 \end{bmatrix} \qquad F^j = \begin{bmatrix} u_j \\ u_1 u_j + p\delta_{1j} \\ u_2 u_j + p\delta_{2j} \\ u_3 u_j + p\delta_{3j} \end{bmatrix} \qquad (2)$$

Here, $u_j$ and $p$ denote the component of the fluid velocity in the direction $x_j$ and the pressure in the fluid respectively. In addition, $\delta_{ij}$ is the Dirac delta function, $\mathrm{Re}$ is the flow Reynolds number and $n_j$ is the component, in direction $x_j$, of the unit outward normal vector to $\mathcal{S}$. This equation system is modified, using the artificial compressibility approach [14], to enable the modelling of steady incompressible flows. This requires the addition of an artificial pseudo–time dependent term to the governing equations (1) and we now seek steady solutions of the equation set

$$\int_{\mathcal{V}} P^{-1} \frac{\partial U}{\partial \tau} \mathrm{d}\mathcal{V} + \int_{\mathcal{S}} F^j n_j \mathrm{d}\mathcal{S} = \frac{1}{\mathrm{Re}} \int_{\mathcal{S}} \frac{\partial V}{\partial x_j} n_j \mathrm{d}\mathcal{S} \qquad j = 1, 2, 3 \qquad (3)$$

where $\tau$ denotes the (pseudo) time and

$$U = \begin{bmatrix} p \\ u_1 \\ u_2 \\ u_3 \end{bmatrix} \qquad P = \begin{bmatrix} \beta^2 & 0 & 0 & 0 \\ 0 & 1 & 0 & 0 \\ 0 & 0 & 1 & 0 \\ 0 & 0 & 0 & 1 \end{bmatrix} \qquad (4)$$

The matrix $P$ acts as a pre–conditioner for the equation system, with the value of the parameter $\beta^2$ selected so as to optimise the convergence to steady state [12,16]. Based upon the results of numerical experiments, the value

$$\beta^2 = \max\left(0.3, 2.5u_j^2\right) \qquad (5)$$

is adopted [12], where it is understood that the non–dimensionalisation that has been used is such that the magnitude of the free stream velocity is unity.

## 3  Mesh Generation

The boundaries of the solution domain are triangulated using an advancing front procedure [17]. For inviscid flow simulations, a Delaunay mesh generator, using automatic point creation, is then employed to discretise the solution domain into an unstructured assembly of tetrahedral elements [18].

The size of the elements generated, on the boundary and in the domain, is controlled by a user defined function, constructed using a background grid in combination with point, line and planar sources [19]. For viscous flows, the discretisation of the boundaries is followed by the generation of layers of stretched tetrahedral elements on solid surfaces, using the advancing layers method [20]. Following the generation of these layers of elements, the remainder of the mesh is again generated using the Delaunay procedure.

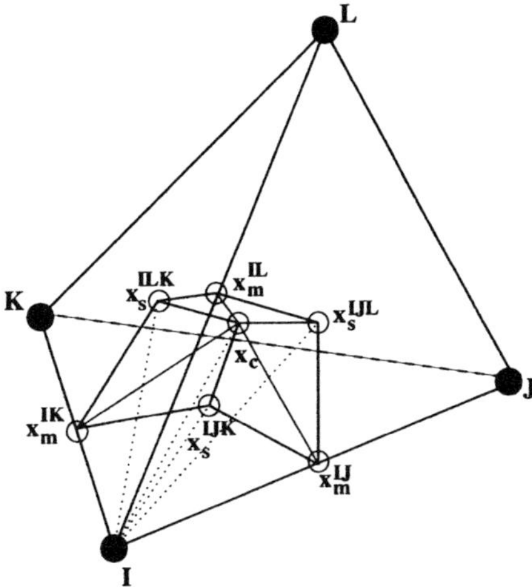

**Fig. 1.** Illustration of that portion of the dual mesh volume $\Omega_I$ surrounding node $I$ that is contained within one tetrahedral element

## 4 Finite Volume Solution Algorithm

### 4.1 Construction of the Dual Mesh

The numerical solution of the governing equations on an unstructured tetrahedral mesh is achieved by employing an edge based cell vertex finite volume approach. With this approach, each vertex, or node, $I$ of the tetrahedral mesh is associated with a volume $\Omega_I$, with surface $\Gamma_I$, of a dual mesh. Equation (3) is then applied, over each dual mesh volume in turn, in the form

$$\Omega_I P_{II}^{-1} \frac{\mathrm{d}U_I}{\mathrm{d}t} + \int_{\Gamma_I} F^j n_j \mathrm{d}\Gamma = \frac{1}{\mathrm{Re}} \int_{\Gamma_I} \frac{\partial V}{\partial x_j} n_j \mathrm{d}\Gamma \qquad j = 1, 2, 3 \quad (6)$$

The dual mesh is constructed from the unstructured tetrahedral mesh by connecting edge midpoints, element centroids and face centroids. Each edge of the tetrahedral mesh is associated with a segment of the dual mesh interface between the nodes joined by the edge. This segment may be regarded as a surface made up of an assembly of triangular facets, where each facet is connected to the midpoint of the edge, a neighbouring element centroid and the centroid of an element face connected to the edge. The portion of the dual mesh volume $\Omega_I$ surrounding node $I$ that is contained within a single tetrahedron is illustrated in Figure 1. In this Figure, the mid point of the edge between nodes $I$ and $J$ is at $x_m^{IJ}$, the centroid of the face with vertices $I$, $J$ and $K$ is at $x_s^{IJK}$ and the element centroid is at $x_c$. With this dual mesh definition, the control volume can be viewed as an assembly of tetrahedra and Figure 2 illustrates the form of the complete volume $\Omega_I$ of the dual mesh

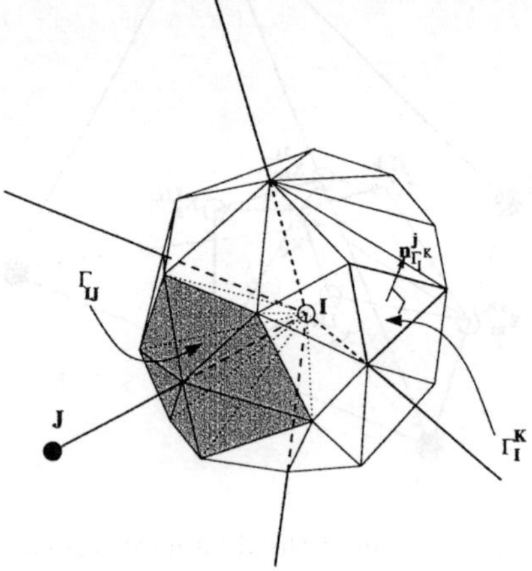


**Fig. 2.** The volume $\Omega_I$ of the dual mesh that surrounds node $I$

that surrounds the internal node $I$. The dual is constructed by a closed set of planar triangular facets $\{\Gamma_I^K\}$, where each facet only touches a single edge. The set of facets touching the edge between nodes $I$ and $J$ is denoted by $\Gamma_{IJ}$. The intersection of a volume of the dual mesh with the computational domain boundary is illustrated in Figure 3. The dual mesh associated with the boundary edge between nodes $I$ and $J$ is made up of the two triangular facets with vertices at node $I$, the surface element centroids and the midpoint of the edge.

## 4.2  Integration of the Fluxes

The integrals over $\Gamma_I$ in equation (6) are evaluated by summing contributions from each segment of the dual mesh interface associated with each edge in the mesh that is connected to node $I$. The integration of the inviscid flux over the segment of the dual mesh associated with an edge is performed by assuming the flux to be constant over the segment and equal to its value at

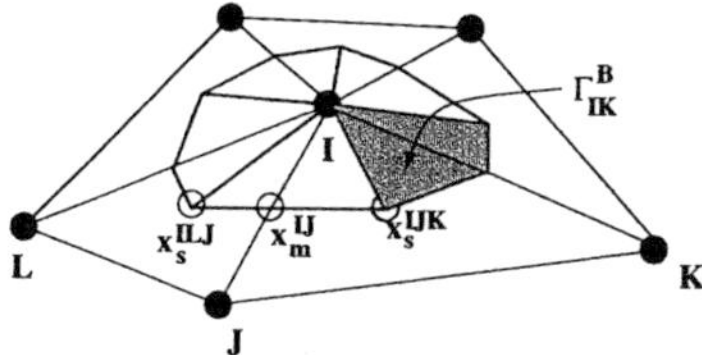


**Fig. 3.** Illustration of the construction of the boundary faces of the dual mesh surrounding a boundary node $I$

the midpoint of the edge. Then, in equation (6), the integral of the inviscid flux over the control volume surface associated with node $I$ is evaluated as

$$\int_{\Gamma_I} F^j n_j \mathrm{d}\Gamma = \sum_{J\in\Lambda_I} \frac{C^j_{IJ}}{2}\left(F^j_I + F^j_J\right) + \sum_{J\in\Lambda^B_I} D^j_{IJ} F^j_I \tag{7}$$

where the second term on the right hand side only appears if $I$ is a boundary node. Here, $\Lambda_I$ denotes the set of nodes connected to node $I$ by an edge and $\Lambda^B_I$ denotes the set of nodes connected to node $I$ by an edge on the computational boundary. The coefficients in equation (7) are calculated by

$$C^j_{IJ} = \sum_{K\in\Gamma_{IJ}} \Gamma^K_I n^j_{\Gamma^K_I} \qquad\qquad D^j_{IJ} = \sum_{K\in\Gamma^B_{IJ}} \Gamma^K_I n^j_{\Gamma^K_I}, \tag{8}$$

where $n^j_{\Gamma^K_I}$ is the outward normal to the facet $\Gamma^K_I$ from the viewpoint of node $I$ and $\Gamma^B_{IJ}$ is the set of dual mesh faces on the boundary of the computational domain touching the edge between nodes $I$ and $J$. When the facet $\Gamma^K_I$ forms part of the computational boundary, $n^j_{\Gamma^K_I}$ is to be interpreted as a component of the normal to the facet in the direction out of the computational domain. From equation (8), it is apparent that the internal edge coefficients are antisymmetric, which means that they only need to be stored once for every edge. It also follows that the scheme is consistent, in the sense that the surface integral of a constant flux is zero. These expressions guarantee that the numerical scheme is conservative. It can also be demonstrated that $D_{IJ} = D_{JI}$, provided the edge midpoints are used in the dual mesh construction. This means that the boundary coefficients also need to be stored only once for each edge.

The viscous fluxes involve integrals of the solution gradients and these can be treated in several ways within the finite volume framework. Here, the most straightforward approach is adopted, in which the derivatives of $V$ are computed first as

$$W_I^j = \left.\frac{\partial V}{\partial x_j}\right|_I = \frac{1}{\Omega_I}\left[\sum_{J\in\Lambda_I}\frac{C_{IJ}^j}{2}(V_I+V_J) + \sum_{J\in\Lambda_I^B}D_{IJ}^j V_I\right] \tag{9}$$

In equation (6), the integral of the viscous fluxes is then evaluated as

$$\int_{\Gamma_I}\frac{\partial V}{\partial x_j}n_j\mathrm{d}\Gamma = \sum_{J\in\Lambda_I}\frac{C_{IJ}^j}{2}\left(W_I^j + W_J^j\right) + \sum_{J\in\Lambda_I^B}D_{IJ}^j W_I^j \tag{10}$$

This results in a computational stencil that stretches over five points in each direction.

## 4.3 Artificial Dissipation

To produce a practical scheme for the simulation of incompressible flows, the physical inviscid fluxes in equation (7) are modified, in a consistent fashion, by the addition of artificial dissipative fluxes. This is achieved by an approach of JST type [21] in which the added artificial dissipation flux is constructed as

$$H_I \equiv \sum_{J\in\Lambda_I}H_{IJ} = \sum_{J\in\Lambda_I}\mathcal{D}_{IJ}(T_J - T_I) \tag{11}$$

where

$$T_I = \sum_{K\in\Lambda_I}(U_K - U_I) \tag{12}$$

and $\mathcal{D}_{IJ}$ is the dissipation matrix. A scalar version of this scheme is used here, in which the dissipation matrix is defined in the the diagonal form

$$\mathcal{D}_{IJ} = \epsilon_4|\lambda_{IJ}|I \tag{13}$$

where $I$ is the unit matrix. The value $\epsilon_4 = 0.2$ is normally adopted with [8]

$$|\lambda_{IJ}| = \frac{\lambda_1}{2}\left[1 + \left(\frac{\lambda_2}{\lambda_1}\right)^\nu\right] \qquad 0 \le \nu \le 1 \tag{14}$$

where

$$\lambda_i = \|u_i\| + \sqrt{\|u_i\|^2 + \beta^2\|\ell\|^2} \qquad i = 1,2$$
$$\ell = (\ell_1,\ell_2,\ell_3) \qquad \ell_j = C_{IJ}^j/(C_{IJ}^k C_{IJ}^k) \tag{15}$$

Here, for each edge $IJ$, $u_1$ and $u_2$ are evaluated by splitting the velocity according to

$$u_1 = (u\cdot\ell)\ell \qquad u_2 = u - u_1 \tag{16}$$

The values of $u$ and $\beta$ used in these expressions are obtained as the average of the values of these quantities at nodes $I$ and $J$. In practice, the value $\nu = 0$ is employed for inviscid flow computations while the value $\nu = 0.5$ is used for viscous computations on stretched grids.

## 4.4 Temporal Discretisation

At each node $I$, the spatially discretised equations may be expressed in the form

$$\frac{\mathrm{d}U_I}{\mathrm{d}\tau} = R_I \tag{17}$$

and these equations are advanced in (pseudo) time using an $m$ stage explicit scheme [21]. This means that the solution is advanced through one time step, from time $\tau_I^n$ to $\tau_I^{n+1} = \tau_I^n + \Delta\tau_I$, as

$$
\begin{aligned}
U_I^{(0)} &= U_I^n \\
U_I^{(k)} &= U_I^n + \alpha_k \mathcal{C} \Delta\tau_I R_I^{(k-1)} \qquad k = 1, \cdots, m \\
U_I^{n+1} &= U_I^{(m)}
\end{aligned}
\tag{18}
$$

where $\alpha_1, \alpha_2, \cdots, \alpha_m$ are appropriately defined coefficients and $\mathcal{C}$ denotes the CFL number. As is it the computation of steady state solutions that is of interest here, computational performance can be improved by selecting the value of the (pseudo) time step $\Delta\tau_I$ separately for each node $I$, according to the local satisfaction of a stability criterion obtained by using the approach of Giles [22].

## 5  Multigrid Acceleration

Explicit multi–stage time stepping coupled with multigrid acceleration has been shown to be an effective method for the solution of the equations of incompressible flow by the artificial compressibility method [8,13]. In the approach that is followed here, the coarse meshes needed for the multigrid procedure are automatically generated from the fine grid by agglomeration [11,23]. Agglomeration does not generate meshes in the classical sense, but rather generates a series of edge based data structures that can be used by the multigrid solver. The agglomeration works with the dual mesh, by merging the control volumes to achieve the required local coarseness ratio. The approach is purely edge based and can be applied after the edge data structure has been constructed. In addition, since no mesh generation issues are raised, the approach is completely stable. The process is also fast, requiring typically the cpu time equivalent of 1–3 multigrid cycles and involves little memory overhead. In addition, the nested meshes that are produced simplify the inter–grid mappings that are necessary within the multigrid procedure. When highly anisotropic meshes are used, the convergence rate can be improved by employing a directional agglomeration procedure for generation of the coarser meshes [11].

The solution is advanced using a three stage explicit time stepping scheme, with the addition of a V–cycle multigrid. A volume weighted restriction operator is employed and prolongation is by injection. The value 1.8 is assigned to the CFL number.

# 6  Parallelisation

Parallelisation of the solution procedure is necessary to make best use of multi–processor computer platforms. In the parallelisation approach used here, the mesh is created sequentially and then split into the desired number of computational domains. The domain splitting is performed using the METIS library [24], which employs a multilevel graph partitioning strategy. This procedure essentially colours the nodes in the mesh according to the domain to which they belong. A preprocessor renumbers the nodes in the domains and generates the communication data required. The edges of the global mesh are placed in the domain of the nodes attached to the edge, or if the nodes belong to different domains, as occurs with an inter–domain edge, the edge is placed in the domain with the smallest domain number. The edges are renumbered in each domain, placing the inter–domain edges first in the list.

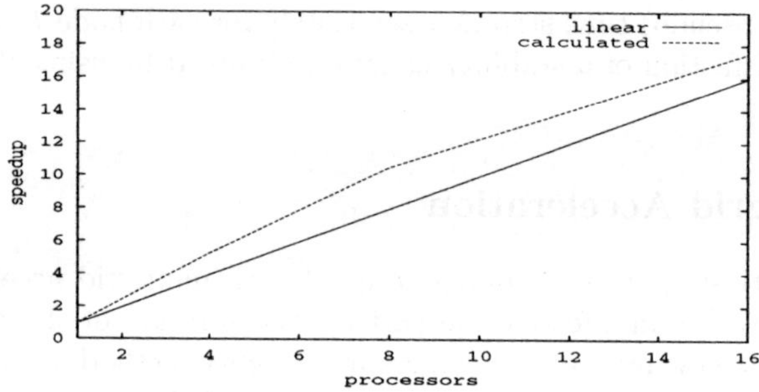



**Fig. 4.** Effectiveness of the parallelisation strategy, showing actual speed–up obtained and linear speed–up, as the number of processors is increased

On the inter–domain boundaries, the nodes are duplicated so that each domain stores a copy of the nodes appearing in the edge list. These ghost nodes are allocated a place in memory to store the nodal contributions from adjacent domains. If the node belongs to the domain, it is termed a real node. The communication data structure consists of two registers for each domain combination on each grid level. These registers are termed the ghost node register and real node register and simply store the local node indexes of the ghost and real nodes respectively. The data structure of the boundary edges is created in an analogous fashion.

The solution procedure consists, essentially, of loops over edges and loops over nodes. The parallelisation process can, therefore, be described by just considering these two loops. Edge loops appear wherever boundary integrals are performed and the parallel version can be summarized as:

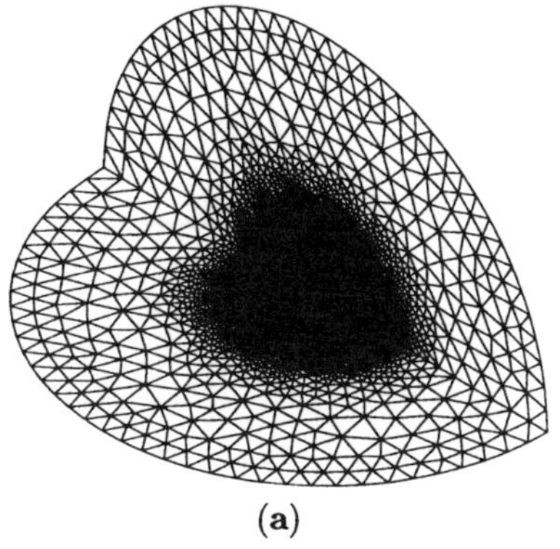
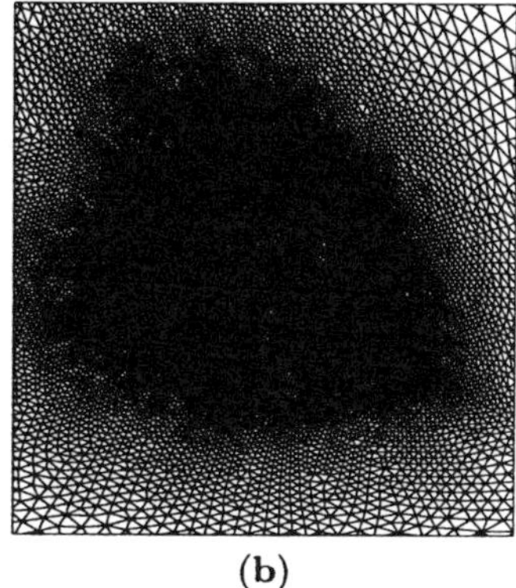

**Fig. 5.** Inviscid flow over a sphere: (**a**) the mesh on the sphere and on the symmetry planes and (**b**) a detail of the view of the mesh on the sphere and on the symmetry planes

1. the increment from the current domain on the communication nodes is calculated by looping over the internal and boundary communication edges;
2. the increments of the ghost nodes in the domain are sent to the neighboring domains using the ghost node register;
3. the internal and boundary edge increments internal to the domain are calculated;
4. the ghost increments from the neighboring domains are received using the real node register and added to the increments of the real nodes; the complete increment has now been found for the real nodes of each domain;
5. the real node register is used to send the node increments to the neighboring domains with corresponding ghost nodes;
6. the complete increments are received using the ghost node register and placed in the ghost nodes of the domain, overwriting the previous values of the increment in these nodes.

The approach ensures that the first send is performed as soon as possible, thus enabling the communication to take place while the edge increments internal to the domain are being calculated.

Node loops usually do not require any communication since the ghost node increments are updated within the edge loop procedure. There are, however, some exceptions where the information stored in the ghost nodes are incorrect and need to be updated from the real node values of the neighboring domain. A good example of such a case is the intergrid mappings used in the multigrid procedure. The ghost nodes do not receive all contributions from the current domain nodes of another mesh level and need to be updated from the domain

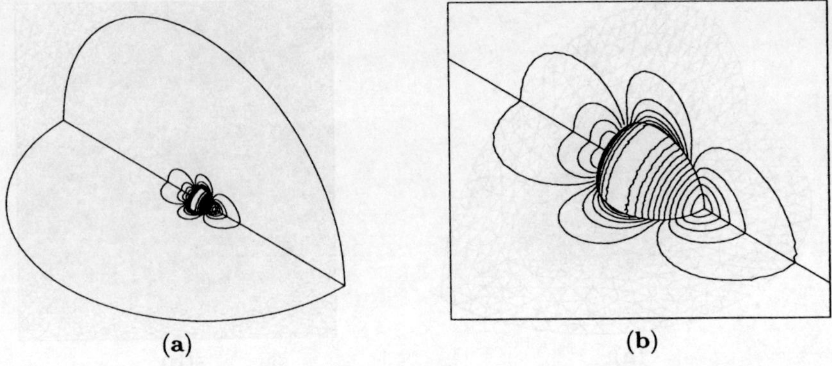


**Fig. 6.** Inviscid flow over a sphere: (**a**) a view of the pressure contour distribution on the sphere and on the symmetry planes and (**b**) a detail of the view of the pressure contour distribution on the sphere and on the symmetry planes

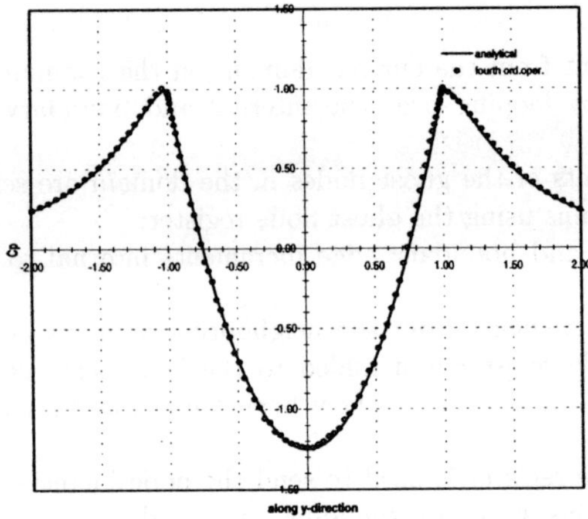


**Fig. 7.** Inviscid flow over a sphere: comparison of computed and exact distributions of pressure coefficient

where the node is real. This is done by first sending the true values using the real node register and then receiving and placing them in the connected domains using the ghost node register.

Two parallelisation strategies for the multigrid accelerated procedure were investigated. The first approach retains the colouring of the fine mesh for the coarse mesh levels. This is incorporated by only allowing merging of control volumes in the same domain in the agglomeration implementation and can thus be termed a local agglomeration approach. The other approach is a global approach, in which agglomeration is allowed across domain boundaries. In this case, the coarse mesh super node is associated with the domain to which the largest number of sub nodes belong. While requiring communication to be performed in the intergrid mappings, this approach yields an implementation equivalent to a sequential code. Neither of these approaches guarantees balanced coarse meshes, but this is of secondary importance as the coarse meshes are significantly smaller than the finest mesh. Comparisons between the local and global approaches have demonstrated the latter to be superior for all test cases considered. This is believed to be due to the fact that the constraint of agglomerating within the domain may severely degrade the coarse mesh quality. This results in reduced multigrid efficiency, requiring more cycles for convergence. In addition, the extra communication load in the intergrid mappings was found to be insignificant. The global agglomeration scheme is therefore used in the computations that are presented.

The effectiveness of the parallelisation strategy is demonstrated for a multigrid simulation involving five grids, with the fine grid containing approximately 7 million cells. Using R14000 processors, the speed–up obtained as the number of processors is increased is shown in Figure 4. It can be observed that super linear speed–up is achieved on a machine with 16 processors. This is due to the reduction in the the number of cache misses when multiple processors are employed.

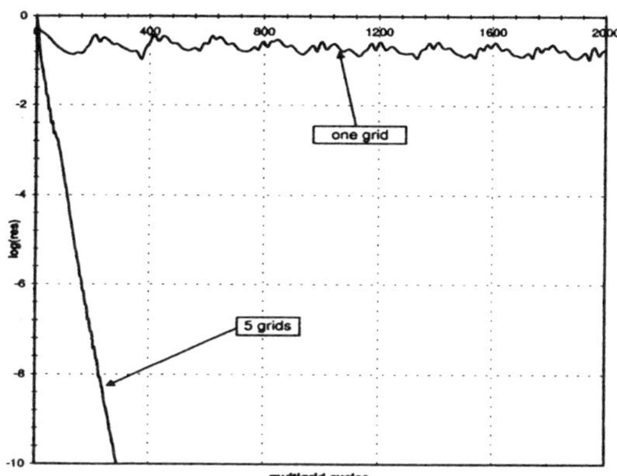


**Fig. 8.** Inviscid flow over a sphere: the variation in the magnitude of the residual, on the fine mesh alone and using 5 grids, against the iteration number

## 7  Numerical Examples

A number of examples are included which will illustrate the numerical performance of this procedure. The inviscid implementation has total memory requirements of about 0.7kbytes per vertex for the single grid version and

0.85kbytes per vertex for the multigrid version. The corresponding memory requirements for the viscous program are 0.8kbytes per vertex and 0.95kbytes per vertex respectively. A speed–up of around 25 is generally obtained when the multigrid acceleration is employed.

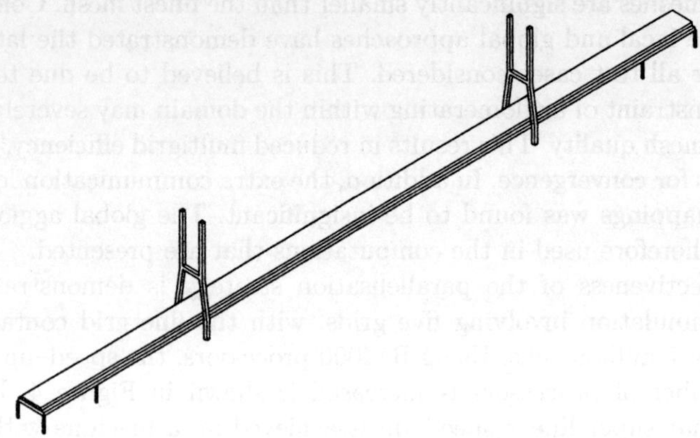

**Fig. 9.** Inviscid flow over a bridge: the CAD definition of the bridge surface

## 7.1 Inviscid Flow Past a Sphere

The first example, used to verify the performance and accuracy of the approach, involves the simulation of inviscid flow over a sphere. By making use of the symmetry of the problem, only the flow over one quarter of a sphere is considered. The radius of the sphere is unity and the far field boundary is located at a distance of 12 from the centre of the sphere. The multigrid procedure is used with five grids, containing 94 673, 15 737, 2 598, 464 and 104 nodes. A view of the finest mesh on the sphere and on the symmetry planes is given in Figure 5(a) and a detail, showing the mesh in the vicinity of the sphere, appears in Figure 5(b). The minimum spacing of 0.04 on the fine mesh was employed in the vicinity of the sphere and this increased to 0.08 near the outer boundary. Corresponding plots showing the computed pressure contour distribution are given in Figure 6(a) and (b). The accuracy of the method is demonstrated in Figure 7 which presents a comparison between the computed and the exact distribution of the pressure coefficient along the symmetry axis and the intersection of the sphere surface with the meridian plane. The effectiveness of the multigrid implementation can be observed in Figure 8, which shows the convergence rate with respect to the number of iterations.

 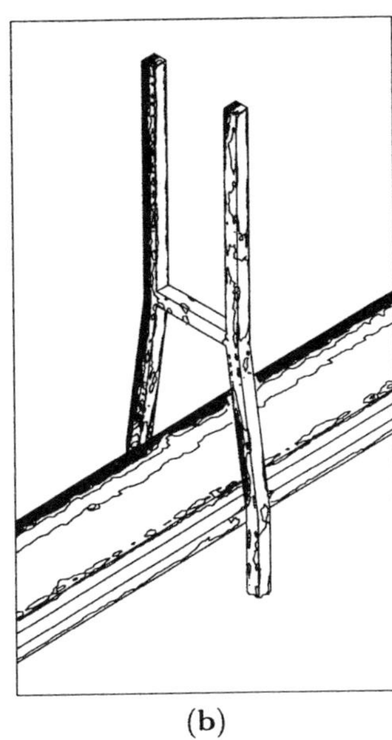

(a)          (b)

**Fig. 10.** Inviscid flow over a bridge: (a) a detail of the surface mesh and (b) a detail of the computed pressure contour distribution

## 7.2 Inviscid Flow Over a Bridge

To demonstrate the predictive capability of the method, the next example involves the simulation of the inviscid flow over a road bridge configuration. The surface definition of the bridge is shown in Figure 9 and this is produced with AutoCAD, using the published definition of the geometry of the central section of Second Severn Crossing [25]. A sequence of five meshes is produced containing 958 181, 154 059, 24 229, 4 319 and 1 009 nodes respectively. For the flow simulation, the wind direction is taken to be at right angles to the bridge deck. A detail of the surface mesh on the bridge is shown in Figure 10(a) and the computed distribution of the pressure contours on the same section is shown in Figure 10(b).

## 7.3 Viscous Flow Past a Cylinder

This example simulates the laminar viscous flow over a circular cylinder. Although this is a two dimensional problem, the simulation is performed in three dimensions, using the computational domain illustrated in Figure 11(a). The radius of the cylinder is taken to be 0.5 and the computational boundary is located at a distance of 5 ahead of, below and above the cylinder. The downstream boundary is located at a distance of 15 from the centre of the cylinder. A sequence of five meshes is employed with the multigrid procedure, with the fine mesh consisting of 286 415 nodes and the coarser meshes

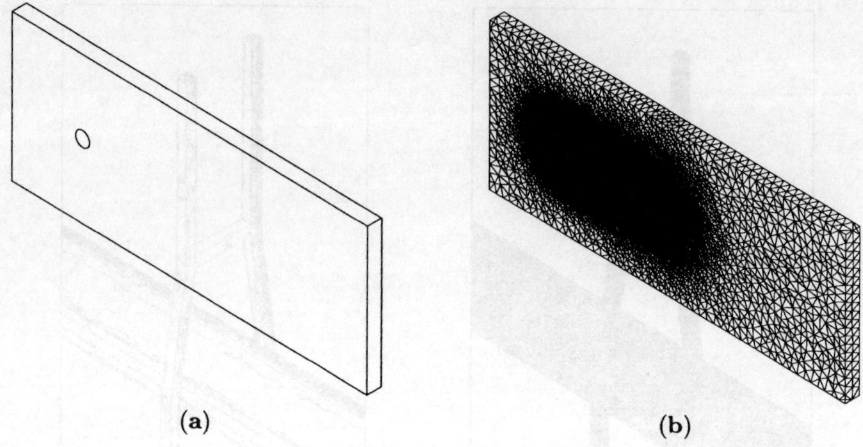

**Fig. 11.** Viscous flow over a circular cylinder: (a) the definition of the surfaces of the computational domain and (b) the discretisation of these surfaces

having 35 617 nodes, 4 772 nodes, 995 nodes and 247 nodes on the coarsest mesh. A view of the fine mesh discretisation of the computational boundaries is given in Figure 11(b). A detail of the surface mesh on the symmetry plane and on the cylinder is given in Figure 12(a). On the finest mesh, the minimum mesh spacing is 0.03 in the vicinity of the cylinder and this is increased to 0.06 near the outer boundaries. The computed distribution of the pressure contours on the symmetry plane and over the cylinder surface, for the case Re=40, is shown in Figure 12(b). The computed distribution of the pressure coefficient over the cylinder is shown to be in good agreement with experimental measurements [26–28] in Figure 13.

## 7.4 Viscous Flow Past a Sphere

The final example consists of the simulation of steady laminar viscous flow over a sphere. As in the corresponding inviscid simulation, only the flow over one quarter of the sphere is modelled. The computational domain and the surface mesh is the same as that employed for the corresponding inviscid simulation. For the multigrid process, a sequence of five meshes is employed with the fine mesh consisting of 94 673 nodes and the coarser meshes having 15 737 nodes, 2 598 nodes, 464 nodes and 104 nodes on the coarsest mesh. The computed distribution of the pressure contours on the symmetry planes and over the sphere surface, for the case Re=40, is shown in Figure 14(a). The corresponding computed distribution of the pressure coefficient along the line of intersection of the sphere surface with the meridian plane is shown in Figure 14(b). The effectiveness of the multigrid implementation for this

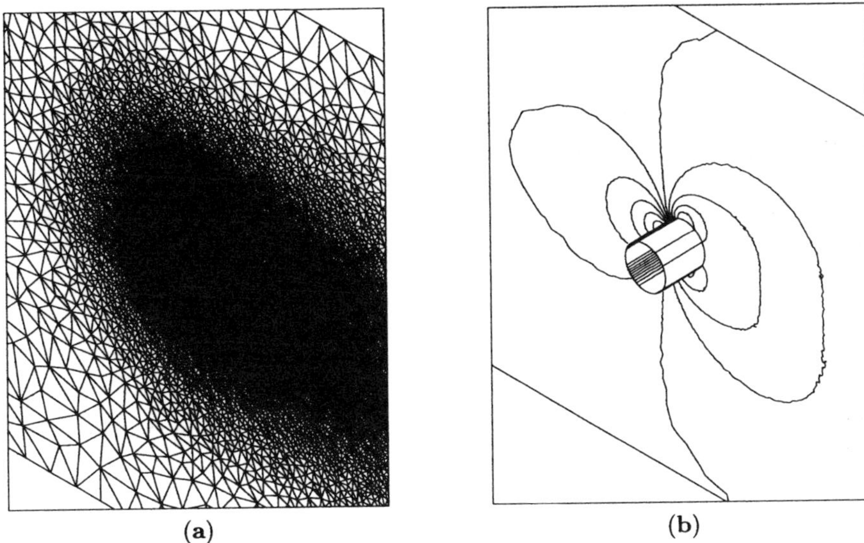



**Fig. 12.** Viscous flow over a circular cylinder: (a) a view of the mesh on the cylinder surface and a symmetry plane and (b) the computed pressure contour distribution for the case Re=40

problem can be observed in Figures 15, which show the convergence rate with respect to the number of iterations for the case Re=40.

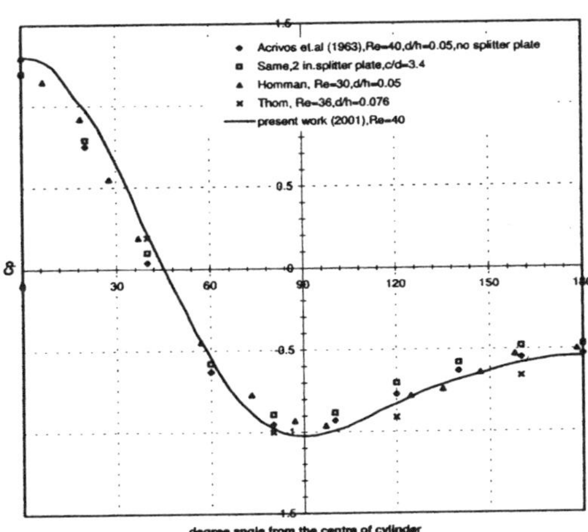



**Fig. 13.** Viscous flow over a circular cylinder: comparison between the computed and experimentally observed distributions of the pressure coefficient on the cylinder for the case Re=40

# 8    Conclusions

The artificial compressibility approach, coupled with explicit multi stage time stepping, agglomerated multigrid acceleration and parallelisation, has been shown to produce a computationally efficient procedure for the simulation of 3D steady incompressible laminar viscous flows on general unstructured tetrahedral grids. Current work is directed towards extending the range of applicability of the procedure by the addition of a turbulence modelling capability.

## Acknowledgements

Kaare Sørensen acknowledges the financial support provided by The Research Council of Norway, project number 125676/410.

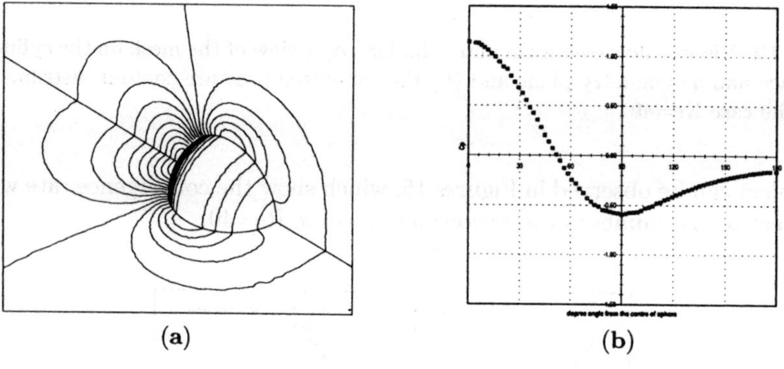

(a)            (b)

**Fig. 14.** Viscous flow over a sphere: (a) a view of the computed pressure contours and (b) the computed pressure coefficient on the sphere for the case Re=40

## References

1. S. E. Rogers, D. Kwak, C. Kiris: AIAA J. **29**, 603 (1991)
2. C. Kiris, D. Kwak: *AIAA Paper 96-2089: Numerical Solution of Incompressible Navier–Stokes Equations Using a Fractional–Step Approach.* (AIAA, Washinton 1996)
3. L.P. Franca, T.J.R. Hughes, R. Stenberg: 'Stabilized Finite Element Methods'. In: *Incompressible Computational Fluid Dynamics.* ed. by M. Gunzburger, R. A. Nicolaides (Cambridge University Press 1993) pp. 87–107

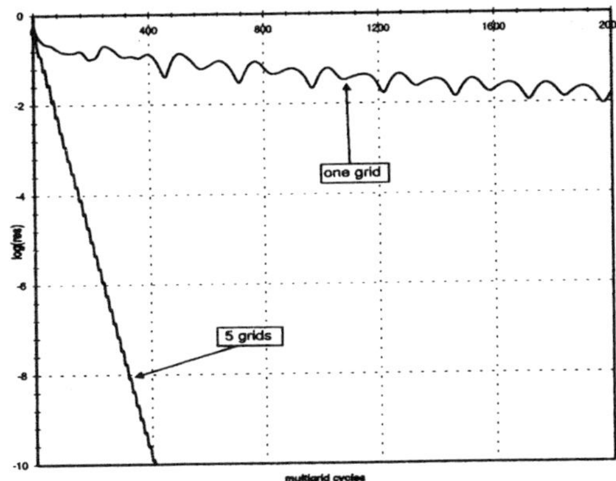


**Fig. 15.** Viscous flow over a sphere at Re=40: the variation in the magnitude of the residual, on the fine mesh alone and using 5 grids, against iteration number


4. T. Tezduyar: Arch. Comp. Meth. Engng. **8**, 83 (2001)
5. R. Löhner: *Applied CFD Techniques*, (John Wiley & Sons, Chichester 2001)
6. O. C. Zienkiewicz, R. Codina: Int. J. Num. Meth. Fluids **20**, 869 (1996)
7. M. Hafez: Int. J. Num. Meth. Fluids **20**, 713 (1996)
8. J. Peraire, K. Morgan, J. Peiró: 'The Simulation of 3D Incompressible Flows Using Unstructured Grids'. In: *Frontiers of Computational Fluid Dynamics— 1994.* ed. by D. A. Caughey, M. M. Hafez (John Wiley & Sons, Chichester 1994) pp. 281–301
9. O. C. Zienkiewicz, R. L. Taylor: *The Finite Element Method*, 5th edn. (Butterworth Heinemann, Oxford 2000)
10. O. Pironneau: *Finite Element Methods for Fluids*, (John Wiley and Sons, Chichester 1989)
11. K. A. Sørensen: *A Multigrid Accelerated Procedure for the Solution of Compressible Fluid Flow on Unstructured Hybrid Meshes.* PhD Thesis, University of Wales, Swansea (2002)
12. A. Rizzi, L. Eriksson: J. Fluid Mech. **153**, 275 (1985)
13. J. Farmer, L. Martinelli, A. Jameson: *AIAA Paper 93–0767: A Fast Multigrid Method for Solving Incompressible Hydrodynamics Problems with Free Surfaces*, (AIAA, Washington 1993)
14. A. Chorin: J. Comp. Phys. **2**, 12 (1967)
15. J. Dreyer: *Finite Volume Solutions to the Unsteady Incompressible Euler Equations on Unstructured Triangular Meshes.* M.S. Thesis, MAE Department, Princeton University (1990)
16. E. Turkel: J. Comp. Phys. **72**, 277 (1987)
17. J. Peraire, K. Morgan, J. Peiró: 'Unstructured Finite Element Mesh Generation and Adaptive Procedures for CFD'. In: *AGARD Conference Proceedings No:464—Application of Mesh Generation to Complex 3-D Configurations*, (AGARD, Paris 1989) pp. 18.1–18.12
18. N. P. Weatherill, O. Hassan: Int. J. Num. Meth. Engng **37**, 2005 (1994)
19. K. Morgan, J. Peraire, J. Peiró, O. Hassan: 'Unstructured Grid Methods for High Speed Compressible Flows'. In: *The Mathematics of Finite Elements and*

*Applications: Highlights 1993*. ed. by J. R. Whiteman (John Wiley & Sons, Chichester 1994) pp. 215–241
20. O. Hassan, K. Morgan, E. J. Probert, J. Peraire: Int. J. Num. Meth. Engng **39**, 549 (1996)
21. A. Jameson, W. Schmidt, E. Turkel: *AIAA Paper 81–1259: Numerical Solutions of the Euler Equations by Finite Volume Methods*, (AIAA, Washington 1981)
22. M. Giles: *CFD Laboratory Report CFDL-TR-87-1: Energy Stability Analysis of Multi-Step Methods on Unstructured Meshes* (MIT, Cambridge, Massachusetts, 1987)
23. M. Lallemand, H. Steve, A. Dervieux: Comp. Fluids **21**, 397 (1992)
24. G. Karypis, V. Kumar: J. Par. Distr. Comp. **48**, 96 (1998)
25. P. R. Head, P. Iley, E. T. Bradley, Civ. Engng. **120**, Special Issue 2, 3 (1997)
26. A. S. Grove, F. H. Shair, E. E. Petersen, A. Acrivos: J. Fluid Mech. **19**, 60 (1964)
27. F. Homann: Z. angew. Math. Phys. **6**, 153 (1936)
28. A. Thom: Proc. Roy. Soc. A **141**, 651 (1933)

# Stabilized Finite Element Formulations and Interface-Tracking and Interface-Capturing Techniques for Incompressible Flows

Tayfun E. Tezduyar
Mechanical Engineering, Rice University – MS 321
6100 Main Street, Houston, TX 77005, USA

## Summary

We provide an overview of the interface-tracking and interface-capturing techniques we have developed in recent years for computation of flow problems with moving boundaries and interfaces. Both classes of techniques are based on stabilized formulations, and determination of the stabilization parameters used in these formulations is also highlighted here. The interface-tracking techniques are based on the Deforming-Spatial-Domain/Stabilized Space-Time formulation, where the mesh moves to track the interface. The interface-capturing techniques, which were developed for two-fluid flows, are based on the stabilized formulation, over non-moving meshes, of both the flow equations and the advection equation governing the time-evolution of an interface function marking the interface location. For interface-capturing techniques, to increase the accuracy in representing the interface, the Enhanced-Discretization Interface-Capturing Technique can be used. We also provide and overview of some of the additional ideas developed to increase the scope and accuracy of these two classes of techniques.

## 1. Introduction

In computation of flow problems with moving boundaries and interfaces, depending on the complexity of the interface and other aspects of the problem, we can use an interface-tracking or interface-capturing technique. An interface-tracking technique requires meshes that "track" the interfaces. The mesh needs to be updated as the flow evolves. In an interface-capturing technique for two-fluid flows, the computations are based on fixed spatial domains, where an interface function, marking the location of the interface, needs to be computed to "capture" the interface. The interface is captured within the resolution of the finite element mesh covering the area where the interface is. This approach can be seen as a special case of interface representation techniques, where the interface is somehow represented over a non-moving fluid mesh, the main point being that the fluid mesh does not move to "track" the interfaces. A consequence of the mesh not moving to "track" the interface is that for fluid-solid interfaces, independent of how well the interface geometry is represented, the resolution of the boundary layer will be limited by the resolution of the fluid mesh where the interface is.

The interface-tracking and interface-capturing techniques we describe here are based on the streamline-upwind/Petrov-Galerkin (SUPG) [1], Galerkin/least-squares (GLS) [2], and pressure-stabilizing/Petrov-Galerkin (PSPG) [3] formulations. These stabilization techniques prevent numerical oscillations and other instabilities in solving problems with high Reynolds and/or Mach numbers and shocks and strong boundary layers, as well as when using equal-order interpolation functions for velocity and pressure and other unknowns. Furthermore, this class of stabilized formulations substantially improve the convergence rate in iterative solution of the large, coupled nonlinear equation system that needs to be solved at every time step of a flow computation. Such nonlinear systems are typically solved with the Newton-Raphson method, which involves, at its every iteration step, solution of a large, coupled linear equation system. It is in iterative solution of such linear equation systems that using a good stabilized method makes substantial difference in convergence, and this was pointed out in [4].

The Deforming-Spatial-Domain/Stabilized Space-Time (DSD/SST) formulation [3], developed for moving boundaries and interfaces, is an interface-tracking technique, where the finite element formulation of the problem is written over its space-time domain. At each time step the locations of the interfaces are calculated as part of the overall solution. As the spatial domain occupied by the fluid changes its shape in time, mesh needs to be updated. In general, this is accomplished by moving the mesh with the motion of the nodes governed by the equations of elasticity, and full or partial remeshing (i.e., generating a new set of elements, and sometimes also a new set of nodes) as needed.

In computation of two fluid-flows (we mean this category to include free-surface flows) with interface-tracking techniques, sometimes the interface might be too complex or unsteady to track while keeping the frequency of remeshing at an acceptable level. Not being able to reduce the frequency of remeshing in 3D might introduce overwhelming mesh generation and projection costs, making the computations with the interface-tracking technique no longer feasible. In such cases, interface-capturing techniques, which do not normally require costly mesh update steps, could be used with the understanding that the interface will not be represented as accurately as we would have with an interface-tracking technique. Because they do not require mesh update, the interface-capturing techniques are more flexible than the interface-tracking techniques. However, for comparable levels of spatial discretization, interface-capturing methods yield less accurate representation of the interface. These methods can be used as practical alternatives in carrying out the simulations when compromising the accurate representation of the interfaces becomes less of a concern than facing major difficulties in updating the mesh to track such interfaces. The desire to increase the accuracy of our interface-capturing techniques without adding a major computational cost lead us to seeking techniques with a different kind of "tracking". The Enhanced-Discretization Interface-Capturing Technique (EDICT) was first introduced in [5] to increase accuracy in representing an interface. We will describe the EDICT more in a later section. In later sections, we will also describe some of the additional ideas and methods developed to increase the scope and accuracy of the interface-tracking and interface-capturing techniques.

## 2.   Governing Equations

Let $\Omega_t \subset \mathbb{R}^{n_{sd}}$ be the spatial fluid mechanics domain with boundary $\Gamma_t$ at time $t \in (0, T)$, where the subscript $t$ indicates the time-dependence of the spatial domain. The Navier-Stokes equations of incompressible flows can be written on $\Omega_t$ and $\forall t \in (0, T)$ as

$$\rho(\frac{\partial \mathbf{u}}{\partial t} + \mathbf{u} \cdot \nabla \mathbf{u} - \mathbf{f}) - \nabla \cdot \boldsymbol{\sigma} = 0, \tag{1}$$

$$\nabla \cdot \mathbf{u} = 0, \tag{2}$$

where $\rho$, $\mathbf{u}$ and $\mathbf{f}$ are the density, velocity and the external force, respectively. The stress tensor $\boldsymbol{\sigma}$ is defined as

$$\boldsymbol{\sigma}(p, \mathbf{u}) = -p\mathbf{I} + 2\mu\boldsymbol{\varepsilon}(\mathbf{u}). \tag{3}$$

Here $p$ is the pressure, $\mathbf{I}$ is the identity tensor, $\mu = \rho\nu$ is the viscosity, $\nu$ is the kinematic viscosity, and $\boldsymbol{\varepsilon}(\mathbf{u})$ is the strain-rate tensor:

$$\boldsymbol{\varepsilon}(\mathbf{u}) = \frac{1}{2}((\boldsymbol{\nabla}\mathbf{u}) + (\boldsymbol{\nabla}\mathbf{u})^T). \tag{4}$$

The essential and natural boundary conditions for Eq. (1) are represented as

$$\mathbf{u} \;=\; \mathbf{g} \text{ on } (\Gamma_t)_{\mathrm{g}}, \quad \mathbf{n}\cdot\boldsymbol{\sigma} = \mathbf{h} \text{ on } (\Gamma_t)_{\mathrm{h}}, \tag{5}$$

where $(\Gamma_t)_g$ and $(\Gamma_t)_h$ are complementary subsets of the boundary $\Gamma_t$, $\mathbf{n}$ is the unit normal vector, and $\mathbf{g}$ and $\mathbf{h}$ are given functions. A divergence-free velocity field $\mathbf{u}_0(\mathbf{x})$ is specified as the initial condition.

If the problem does not involve any moving boundaries or interfaces, the spatial domain does not need to change with respect to time, and the subscript $t$ can be dropped from $\Omega_t$ and $\Gamma_t$. This might be the case even for flows with moving boundaries and interfaces, if in the formulation used the spatial domain is not defined to be the part of the space occupied by the fluid(s). For example, we can have a fixed spatial domain, and model the fluid-fluid interfaces by assuming that the domain is occupied by two immiscible fluids, A and B, with densities $\rho_A$ and $\rho_B$ and viscosities $\mu_A$ and $\mu_B$. In modeling a free-surface problem where Fluid B is irrelevant, we assign a sufficiently low density to Fluid B. An interface function $\phi$ serves as a marker identifying Fluid A and B with the definition $\phi = \{1$ for Fluid A and 0 for Fluid B$\}$. The interface between the two fluids is approximated to be at $\phi = 0.5$. In this context, $\rho$ and $\mu$ are defined as

$$\rho = \phi\rho_A + (1 - \phi)\rho_B, \quad \mu = \phi\mu_A + (1 - \phi)\mu_B. \tag{6}$$

The evolution of the interface function $\phi$, and therefore the motion of the interface, is governed by a time-dependent advection equation, written on $\Omega$ and $\forall t \in (0, T)$ as

$$\frac{\partial\phi}{\partial t} + \mathbf{u}\cdot\boldsymbol{\nabla}\phi = 0. \tag{7}$$

## 3. Stabilized Formulations and Stabilization Parameters

### 3.1 Background

The SUPG, GLS and PSPG formulations stabilize the method without introducing excessive numerical dissipation. In these formulations, selection of the stabilization parameter, which is almost universally known as $"\tau"$, has attracted a significant amount of attention and research. This stabilization parameter involves a measure of the local length scale (also known as "element length") and other parameters such as the local Reynolds and Courant numbers. Various "element length"s and $"\tau"$s were proposed starting with those in [1] and [6], followed by the one introduced in [7], and those proposed in the subsequently reported SUPG, GLS and PSPG methods. A number of $"\tau"$s, dependent upon spatial and temporal discretizations, were introduced and tested in [8]. More recently, $"\tau"$s which are applicable to higher-order elements were proposed in [9].

In this section, we propose ways to determine the stabilization parameter $"\tau"$. The parameters we propose are calculated from the element-level matrices and vectors, and these automatically take into account the local length scales as well as the advection field and the element-level Reynolds number.

224

## 3.2 Advection-Diffusion Equation

Let us consider over a domain $\Omega$ with boundary $\Gamma$ the following time-dependent advection-diffusion equation, written on $\Omega$ and $\forall t \in (0, T)$ as

$$\frac{\partial \phi}{\partial t} + \mathbf{u} \cdot \nabla \phi - \nabla \cdot (\nu \nabla \phi) = 0, \tag{8}$$

where $\phi$ represents the quantity being transported (e.g., temperature, concentration, interface function), and $\nu$ is the diffusivity. The essential and natural boundary conditions associated with Eq. (8) are represented as

$$\phi = g \quad \text{on } \Gamma_g, \quad \mathbf{n} \cdot \nu \nabla \phi = h \quad \text{on } \Gamma_h. \tag{9}$$

A function $\phi_0(\mathbf{x})$ is specified as the initial condition.

Let us assume that we have constructed some suitably-defined finite-dimensional trial solution and test function spaces $\mathcal{S}_\phi^h$ and $\mathcal{V}_\phi^h$. The stabilized finite element formulation of Eq. (8) can then be written as follows: find $\phi^h \in \mathcal{S}_\phi^h$ such that $\forall w^h \in \mathcal{V}_\phi^h$:

$$\int_\Omega w^h \left( \frac{\partial \phi^h}{\partial t} + \mathbf{u}^h \cdot \nabla \phi^h \right) d\Omega + \int_\Omega \nabla w^h \cdot \nu \nabla \phi^h d\Omega - \int_{\Gamma_h} w^h h d\Gamma$$

$$+ \sum_{e=1}^{n_{el}} \int_{\Omega^e} \tau_{\text{SUPG}} \mathbf{u}^h \cdot \nabla w^h \left( \frac{\partial \phi^h}{\partial t} + \mathbf{u}^h \cdot \nabla \phi^h - \nabla \cdot (\nu \nabla \phi^h) \right) d\Omega = 0. \tag{10}$$

Here $n_{el}$ is the number of elements, $\Omega^e$ is the domain for element $e$, and $\tau_{\text{SUPG}}$ is the SUPG stabilization parameter.

Let us use the notation $\mathbf{b} : \int_{\Omega^e} (\ldots) d\Omega : \mathbf{b}_v$ to denote the element-level matrix $\mathbf{b}$ and element-level vector $\mathbf{b}_v$ corresponding to the element-level integration term $\int_{\Omega^e} (\ldots) d\Omega$. We now define the following element-level matrices and vectors:

$$\mathbf{m} : \qquad \int_{\Omega^e} w^h \frac{\partial \phi^h}{\partial t} d\Omega \qquad : \mathbf{m}_v, \tag{11}$$

$$\mathbf{c} : \qquad \int_{\Omega^e} w^h \mathbf{u}^h \cdot \nabla \phi^h d\Omega \qquad : \mathbf{c}_v, \tag{12}$$

$$\mathbf{k} : \qquad \int_{\Omega^e} \nabla w^h \cdot \nu \nabla \phi^h d\Omega \qquad : \mathbf{k}_v, \tag{13}$$

$$\tilde{\mathbf{k}} : \qquad \int_{\Omega^e} \mathbf{u}^h \cdot \nabla w^h \, \mathbf{u}^h \cdot \nabla \phi^h d\Omega \qquad : \tilde{\mathbf{k}}_v, \tag{14}$$

$$\tilde{\mathbf{c}} : \qquad \int_{\Omega^e} \mathbf{u}^h \cdot \nabla w^h \frac{\partial \phi^h}{\partial t} d\Omega \qquad : \tilde{\mathbf{c}}_v. \tag{15}$$

We define the element-level Reynolds and Courant numbers as follows:

$$Re = \frac{\|\mathbf{u}^h\|^2}{\nu} \frac{\|\mathbf{c}\|}{\|\tilde{\mathbf{k}}\|}, \tag{16}$$

$$Cr_u = \frac{\Delta t}{2} \frac{\|\mathbf{c}\|}{\|\mathbf{m}\|}, \tag{17}$$

$$Cr_\nu = \frac{\Delta t}{2} \frac{\|\mathbf{k}\|}{\|\mathbf{m}\|}, \tag{18}$$

$$Cr_{\tilde{\nu}} = \frac{\Delta t}{2} \tau_{\text{SUPG}} \frac{\|\tilde{\mathbf{k}}\|}{\|\mathbf{m}\|}, \tag{19}$$

where $\|\mathbf{b}\|$ is the norm of matrix $\mathbf{b}$. The components of element-matrix-based $\tau_{\text{SUPG}}$ are defined as follows:

$$\tau_{\text{S1}} = \frac{\|\mathbf{c}\|}{\|\tilde{\mathbf{k}}\|}, \tag{20}$$

$$\tau_{\text{S2}} = \frac{\Delta t}{2}\frac{\|\mathbf{c}\|}{\|\tilde{\mathbf{c}}\|}, \tag{21}$$

$$\tau_{\text{S3}} = \tau_{\text{S1}}Re = \left(\frac{\|\mathbf{c}\|}{\|\tilde{\mathbf{k}}\|}\right)Re. \tag{22}$$

To construct $\tau_{\text{SUPG}}$ from its components we propose the form

$$\tau_{\text{SUPG}} = \left(\frac{1}{\tau_{\text{S1}}^r} + \frac{1}{\tau_{\text{S2}}^r} + \frac{1}{\tau_{\text{S3}}^r}\right)^{-\frac{1}{r}}, \tag{23}$$

which is based on the inverse of $\tau_{\text{SUPG}}$ being defined as the $r$-norm of the vector with components $\frac{1}{\tau_{\text{S1}}}$, $\frac{1}{\tau_{\text{S2}}}$ and $\frac{1}{\tau_{\text{S3}}}$. We note that the higher the integer $r$ is, the sharper the switching between $\tau_{\text{S1}}$, $\tau_{\text{S2}}$ and $\tau_{\text{S3}}$ becomes.

The components of the element-vector-based $\tau_{\text{SUPG}}$ are defined as follows:

$$\tau_{\text{SV1}} = \frac{\|\mathbf{c}_\mathbf{v}\|}{\|\tilde{\mathbf{k}}_\mathbf{v}\|}, \tag{24}$$

$$\tau_{\text{SV2}} = \frac{\|\mathbf{c}_\mathbf{v}\|}{\|\tilde{\mathbf{c}}_\mathbf{v}\|}, \tag{25}$$

$$\tau_{\text{SV3}} = \tau_{\text{SV1}}Re = \left(\frac{\|\mathbf{c}_\mathbf{v}\|}{\|\tilde{\mathbf{k}}_\mathbf{v}\|}\right)Re. \tag{26}$$

With these three components,

$$(\tau_{\text{SUPG}})_{\text{v}} = \left(\frac{1}{\tau_{\text{SV1}}^r} + \frac{1}{\tau_{\text{SV2}}^r} + \frac{1}{\tau_{\text{SV3}}^r}\right)^{-\frac{1}{r}}. \tag{27}$$

## 3.3 Navier-Stokes Equations of Incompressible Flows

Given Eqs. (1)-(2), let us assume that we have some suitably-defined finite-dimensional trial solution and test function spaces for velocity and pressure: $\mathcal{S}_\mathbf{u}^h$, $\mathcal{V}_\mathbf{u}^h$, $\mathcal{S}_p^h$ and $\mathcal{V}_p^h = \mathcal{S}_p^h$. The stabilized finite element formulation of Eqs. (1)-(2) can then be written as follows: find $\mathbf{u}^h \in \mathcal{S}_\mathbf{u}^h$ and $p^h \in \mathcal{S}_p^h$ such that $\forall \mathbf{w}^h \in \mathcal{V}_\mathbf{u}^h$ and $q^h \in \mathcal{V}_p^h$:

$$\int_\Omega \mathbf{w}^h \cdot \rho \left(\frac{\partial \mathbf{u}^h}{\partial t} + \mathbf{u}^h \cdot \nabla \mathbf{u}^h - \mathbf{f}\right) d\Omega + \int_\Omega \boldsymbol{\varepsilon}(\mathbf{w}^h) : \boldsymbol{\sigma}(p^h, \mathbf{u}^h) d\Omega - \int_{\Gamma_h} \mathbf{w}^h \cdot \mathbf{h}^h d\Gamma$$

$$+ \int_\Omega q^h \nabla \cdot \mathbf{u}^h d\Omega + \sum_{e=1}^{n_{el}} \int_{\Omega^e} \frac{1}{\rho}[\tau_{\text{SUPG}}\rho \mathbf{u}^h \cdot \nabla \mathbf{w}^h + \tau_{\text{PSPG}}\nabla q^h] \cdot$$

$$\left[\rho\left(\frac{\partial \mathbf{u}^h}{\partial t} + \mathbf{u}^h \cdot \nabla \mathbf{u}^h\right) - \nabla \cdot \boldsymbol{\sigma}(p^h, \mathbf{u}^h) - \rho\mathbf{f}\right] d\Omega$$

$$+ \sum_{e=1}^{n_{el}} \int_{\Omega^e} \tau_{\text{LSIC}}\nabla \cdot \mathbf{w}^h \rho \nabla \cdot \mathbf{u}^h d\Omega = 0. \tag{28}$$

226

Here $\tau_{\text{PSPG}}$ and $\tau_{\text{LSIC}}$ are the PSPG and LSIC (least-squares on incompressibility constraint) stabilization parameters.

We now define the following element-level matrices and vectors:

$$\mathbf{m} : \qquad \int_{\Omega^e} \mathbf{w}^h \cdot \rho \frac{\partial \mathbf{u}^h}{\partial t} d\Omega \qquad : \mathbf{m}_{\text{v}}, \tag{29}$$

$$\mathbf{c} : \qquad \int_{\Omega^e} \mathbf{w}^h \cdot \rho(\mathbf{u}^h \cdot \nabla \mathbf{u}^h) d\Omega \qquad : \mathbf{c}_{\text{v}}, \tag{30}$$

$$\mathbf{k} : \qquad \int_{\Omega^e} \boldsymbol{\varepsilon}(\mathbf{w}^h) : 2\mu\boldsymbol{\varepsilon}(\mathbf{u}^h) d\Omega \qquad : \mathbf{k}_{\text{v}}, \tag{31}$$

$$\mathbf{g} : \qquad \int_{\Omega^e} (\nabla \cdot \mathbf{w}^h) p^h d\Omega \qquad : \mathbf{g}_{\text{v}}, \tag{32}$$

$$\mathbf{g}^{\text{T}} : \qquad \int_{\Omega^e} q^h (\nabla \cdot \mathbf{u}^h) d\Omega \qquad : \mathbf{g}^{\text{T}}_{\text{v}}, \tag{33}$$

$$\tilde{\mathbf{k}} : \qquad \int_{\Omega^e} (\mathbf{u}^h \cdot \nabla \mathbf{w}^h) \cdot \rho(\mathbf{u}^h \cdot \nabla \mathbf{u}^h) d\Omega \qquad : \tilde{\mathbf{k}}_{\text{v}}, \tag{34}$$

$$\tilde{\mathbf{c}} : \qquad \int_{\Omega^e} (\mathbf{u}^h \cdot \nabla \mathbf{w}^h) \cdot \rho \frac{\partial \mathbf{u}^h}{\partial t} d\Omega \qquad : \tilde{\mathbf{c}}_{\text{v}}, \tag{35}$$

$$\tilde{\gamma} : \qquad \int_{\Omega^e} (\mathbf{u}^h \cdot \nabla \mathbf{w}^h) \cdot \nabla p^h d\Omega \qquad : \tilde{\gamma}_{\text{v}}, \tag{36}$$

$$\beta : \qquad \int_{\Omega^e} \nabla q^h \cdot \frac{\partial \mathbf{u}^h}{\partial t} d\Omega \qquad : \beta_{\text{v}}, \tag{37}$$

$$\gamma : \qquad \int_{\Omega^e} \nabla q^h \cdot (\mathbf{u}^h \cdot \nabla \mathbf{u}^h) d\Omega \qquad : \gamma_{\text{v}}, \tag{38}$$

$$\theta : \qquad \int_{\Omega^e} \nabla q^h \cdot \nabla p^h d\Omega \qquad : \theta_{\text{v}}, \tag{39}$$

$$\mathbf{e} : \qquad \int_{\Omega^e} (\nabla \cdot \mathbf{w}^h) \rho (\nabla \cdot \mathbf{u}^h) d\Omega \qquad : \mathbf{e}_{\text{v}}. \tag{40}$$

The element-level Reynolds and Courant numbers are defined the same way as they were defined before, as given by Eqs. (16)-(19). The components of the element-matrix-based $\tau_{\text{SUPG}}$ are defined the same way as they were defined before, as given by Eqs. (20)-(22). $\tau_{\text{SUPG}}$ is constructed from its components the same way as it was constructed before, as give by Eq. (23). The components of the element-vector-based $\tau_{\text{SUPG}}$ are defined the same way as they were defined before, as given by Eqs. (24)-(26). The construction of $(\tau_{\text{SUPG}})_{\text{V}}$ is also the same as it was before, given by Eq. (27).

The components of the element-matrix-based $\tau_{\text{PSPG}}$ are defined as follows:

$$\tau_{\text{P1}} = \frac{\|\mathbf{g}^{\text{T}}\|}{\|\gamma\|}, \tag{41}$$

$$\tau_{\text{P2}} = \frac{\Delta t}{2} \frac{\|\mathbf{g}^{\text{T}}\|}{\|\beta\|}, \tag{42}$$

$$\tau_{\text{P3}} = \tau_{\text{P1}} Re = \left( \frac{\|\mathbf{g}^{\text{T}}\|}{\|\gamma\|} \right) Re. \tag{43}$$

$\tau_{\mathrm{PSPG}}$ is constructed from its components as follows:

$$\tau_{\mathrm{PSPG}} = \left( \frac{1}{\tau_{\mathrm{P1}}^{r}} + \frac{1}{\tau_{\mathrm{P2}}^{r}} + \frac{1}{\tau_{\mathrm{P3}}^{r}} \right)^{-\frac{1}{r}}. \tag{44}$$

The components of the element-vector-based $\tau_{\mathrm{PSPG}}$ are defined as follows:

$$\tau_{\mathrm{PV1}} = \tau_{\mathrm{P1}}, \tag{45}$$

$$\tau_{\mathrm{PV2}} = \tau_{\mathrm{PV1}} \frac{\|\gamma_{\mathrm{v}}\|}{\|\beta_{\mathrm{v}}\|}, \tag{46}$$

$$\tau_{\mathrm{PV3}} = \tau_{\mathrm{PV1}} Re. \tag{47}$$

With these components,

$$(\tau_{\mathrm{PSPG}})_{\mathrm{v}} = \left( \frac{1}{\tau_{\mathrm{PV1}}^{r}} + \frac{1}{\tau_{\mathrm{PV2}}^{r}} + \frac{1}{\tau_{\mathrm{PV3}}^{r}} \right)^{-\frac{1}{r}}. \tag{48}$$

The element-matrix-based $\tau_{\mathrm{LSIC}}$ is defined as follows:

$$\tau_{\mathrm{LSIC}} = \frac{\|\mathbf{c}\|}{\|\mathbf{e}\|}. \tag{49}$$

We define the element-vector-based $\tau_{\mathrm{LSIC}}$ as:

$$(\tau_{\mathrm{LSIC}})_{\mathrm{v}} = \tau_{\mathrm{LSIC}}. \tag{50}$$

**Remark 1** *We can also calculate a separate $\tau$ for each element node, or degree of freedom, or element equation. In that case, each component of $\tau$ would be calculated separately for each element node, or degree of freedom, or element equation. For this, we first represent an element matrix $\mathbf{b}$ in terms of its row matrices: $\mathbf{b}_1, \mathbf{b}_2, \ldots, \mathbf{b}_{n_{ex}}$. If we want a separate $\tau$ for each element node, then $\mathbf{b}_1, \mathbf{b}_2, \ldots, \mathbf{b}_{n_{ex}}$, would be the row matrices corresponding to each element node, with $n_{ex} = n_{en}$, where $n_{en}$ is the number of element nodes. If we want a separate $\tau$ for each degree of freedom, then $\mathbf{b}_1, \mathbf{b}_2, \ldots, \mathbf{b}_{n_{ex}}$, would be the row matrices corresponding to each degree of freedom, with $n_{ex} = n_{dof}$, where $n_{dof}$ is the number of degrees of freedom. If we want a separate $\tau$ for each element equation, then $\mathbf{b}_1, \mathbf{b}_2, \ldots, \mathbf{b}_{n_{ex}}$ would be the row matrices corresponding to each element equation, with $n_{ex} = n_{ee}$, where $n_{ee}$ is the number of element equations. Based on this, the components of $\tau$ would be calculated using the norms of these row matrices, instead of the element matrices. For example, a separate $\tau_{\mathrm{S1}}$ for each element node would be calculated by using the expression $(\tau_{\mathrm{S1}})_a = \frac{\|\mathbf{c}_a\|}{\|\mathbf{k}_a\|}, \ a = 1, 2, \ldots, n_{en}$.*

**Remark 2** *The concept of calculating a separate $\tau$ for each element node or equation can be extended to calculating a separate $\tau$ for each global node or equation. This can be accomplished by first representing a global matrix in terms of its row matrices associated with the global nodes or equations, and then by calculating the components of $\tau$ using the norms of these global row matrices. With this approach, applying the class of stabilization techniques described in this paper to element-free methods would become more direct.*

For the purpose of comparison, we define here also the stabilization parameters that are based on an earlier definition of the length scale $h$ [7]:

$$h_{\mathrm{UGN}} = 2 \|\mathbf{u}^h\| \left( \sum_{a=1}^{n_{en}} |\mathbf{u}^h \cdot \nabla N_a| \right)^{-1}, \tag{51}$$

where $N_a$ is the interpolation function associated with node $a$. The stabilization parameters are defined as follows:

$$\tau_{\text{SUGN1}} = \frac{h_{\text{UGN}}}{2\|\mathbf{u}^h\|}, \tag{52}$$

$$\tau_{\text{SUGN2}} = \frac{\Delta t}{2}, \tag{53}$$

$$\tau_{\text{SUGN3}} = \frac{h_{\text{UGN}}^2}{4\nu}, \tag{54}$$

$$(\tau_{\text{SUPG}})_{\text{UGN}} = \left(\frac{1}{\tau_{\text{SUGN1}}^2} + \frac{1}{\tau_{\text{SUGN2}}^2} + \frac{1}{\tau_{\text{SUGN3}}^2}\right)^{-\frac{1}{2}}, \tag{55}$$

$$(\tau_{\text{PSPG}})_{\text{UGN}} = (\tau_{\text{SUPG}})_{\text{UGN}}, \tag{56}$$

$$(\tau_{\text{LSIC}})_{\text{UGN}} = \frac{h_{\text{UGN}}}{2}\|\mathbf{u}^h\|\,z. \tag{57}$$

Here $z$ is given as follows:

$$z = \begin{cases} \left(\frac{Re_{\text{UGN}}}{3}\right) & Re_{\text{UGN}} \le 3, \\ 1 & Re_{\text{UGN}} > 3, \end{cases} \tag{58}$$

where $Re_{\text{UGN}} = \frac{\|\mathbf{u}^h\| h_{\text{UGN}}}{2\nu}$.

Comparisons between the performances of these earlier stabilization parameters and the ones proposed here can be found in [10]. These comparisons show that, especially for special element geometries, the performances are similar.

As a potential alternative or complement to the LSIC stabilization, we propose the Discontinuity-Capturing Directional Dissipation (DCDD) stabilization. In introducing the DCDD stabilization, we first define the unit vectors $\mathbf{s}$ and $\mathbf{r}$:

$$\mathbf{s} = \frac{\mathbf{u}^h}{\|\mathbf{u}^h\|}, \quad \mathbf{r} = \frac{\nabla\|\mathbf{u}^h\|}{\|\,\nabla\|\mathbf{u}^h\|\,\|}, \tag{59}$$

and the element-level matrices and vectors $\mathbf{c}_r$, $\tilde{\mathbf{k}}_r$, $(\mathbf{c}_r)_{\text{v}}$, and $(\tilde{\mathbf{k}}_r)_{\text{v}}$:

$$\mathbf{c}_r : \quad \int_{\Omega^e} \mathbf{w}^h \cdot \rho(\mathbf{r} \cdot \nabla\mathbf{u}^h)d\Omega \quad : (\mathbf{c}_r)_{\text{v}}, \tag{60}$$

$$\tilde{\mathbf{k}}_r : \quad \int_{\Omega^e} (\mathbf{r} \cdot \nabla\mathbf{w}^h) \cdot \rho(\mathbf{r} \cdot \nabla\mathbf{u}^h)d\Omega \quad : (\tilde{\mathbf{k}}_r)_{\text{v}}. \tag{61}$$

Then the DCDD stabilization is defined as

$$S_{\text{DCDD}} = \sum_{e=1}^{n_{el}} \int_{\Omega^e} \rho\nu_{\text{DCDD}}\nabla\mathbf{w}^h : \left(\left[\mathbf{rr} - (\mathbf{r} \cdot \mathbf{s})^2\mathbf{ss}\right] \cdot \nabla\mathbf{u}^h\right) d\Omega, \tag{62}$$

where the element-matrix-based and element-vector-based DCDD viscosities are:

$$\nu_{\text{DCDD}} = |\mathbf{r} \cdot \mathbf{u}^h| \frac{\|\mathbf{c}_r\|}{\|\tilde{\mathbf{k}}_r\|}, \tag{63}$$

$$(\nu_{\text{DCDD}})_{\text{v}} = |\mathbf{r} \cdot \mathbf{u}^h| \frac{\|(\mathbf{c}_r)_{\text{v}}\|}{\|(\tilde{\mathbf{k}}_r)_{\text{v}}\|}. \tag{64}$$

An approximate version of the expression given by Eq. (63) can be written as

$$\nu_{\mathrm{DCDD}} = |\mathbf{r} \cdot \mathbf{u}^h| \frac{h_{\mathrm{RGN}}}{2},$$
(65)

where

$$h_{\mathrm{RGN}} = 2 \left( \sum_{a=1}^{n_{en}} |\mathbf{r} \cdot \nabla N_a| \right)^{-1}.$$
(66)

A different way of determining $\nu_{\mathrm{DCDD}}$ can be expressed as

$$\nu_{\mathrm{DCDD}} = \tau_{\mathrm{DCDD}} \|\mathbf{u}^h\|^2,$$
(67)

where

$$\tau_{\mathrm{DCDD}} = \frac{h_{\mathrm{DCDD}}}{2\|\mathbf{U}\|} \frac{\| \nabla\|\mathbf{u}^h\| \| h_{\mathrm{DCDD}}}{\|\mathbf{U}\|}.$$
(68)

Here $\mathbf{U}$ represents a global velocity scale, and $h_{\mathrm{DCDD}}$ can be calculated by using the expression

$$h_{\mathrm{DCDD}} = 2 \frac{\|\mathbf{c}_r\|}{\|\tilde{\mathbf{k}}_r\|},$$
(69)

or the approximation

$$h_{\mathrm{DCDD}} = h_{\mathrm{RGN}}.$$
(70)

Combining Eqs. (67) and (68), we obtain

$$\nu_{\mathrm{DCDD}} = \frac{1}{2} \left( \frac{\|\mathbf{u}^h\|}{\|\mathbf{U}\|} \right)^2 (h_{\mathrm{DCDD}})^2 \| \nabla\|\mathbf{u}^h\| \|.$$
(71)

It was shown by Mittal [11] that flow computations with the SUPG and PSPG formulations (based on stabilization parameters very much like those given by Eqs. (51)-(58)) and very high aspect-ratio elements in the boundary layers might in some cases exhibit convergence problems and inaccuracies. As a remedy, Mittal [11] proposed to use an element length definition that represents the "minimum dimension" of an element, and gives the minimum edge length in the special case of a rectangular element.

We propose to re-define $\tau_{\mathrm{PSPG}}$ by modifying the definitions of $\tau_{\mathrm{P3}}$ and $\tau_{\mathrm{PV3}}$ given by Eqs. (43) and (47). We propose to accomplish that by using the expressions

$$\tau_{\mathrm{P3}} = \tau_{\mathrm{P1}} \frac{\|\mathbf{c}\|}{\nu \|\tilde{\mathbf{k}}_r\|}, \qquad \tau_{\mathrm{PV3}} = \tau_{\mathrm{PV1}} \frac{\|\mathbf{c}\|}{\nu \|\tilde{\mathbf{k}}_r\|},$$
(72)

or the approximations

$$\tau_{\mathrm{P3}} = \tau_{\mathrm{P1}} \, Re \left( \frac{h_{\mathrm{RGN}}}{h_{\mathrm{UGN}}} \right)^2, \qquad \tau_{\mathrm{PV3}} = \tau_{\mathrm{PV1}} \, Re \left( \frac{h_{\mathrm{RGN}}}{h_{\mathrm{UGN}}} \right)^2.$$
(73)

These modifications can also be applied to $\tau_{\mathrm{S3}}$ and $\tau_{\mathrm{SV3}}$ given by Eqs. (22) and (26).

## 4. DSD/SST Finite Element Formulation

In the DSD/SST method, the finite element formulation of the governing equations is written over a sequence of $N$ space-time slabs $Q_n$, where $Q_n$ is the slice of the space-time domain between the time levels $t_n$ and $t_{n+1}$. At each time step, the integrations involved in the finite element formulation are performed over $Q_n$. The space-time finite element interpolation functions are continuous within a space-time slab, but discontinuous from one space-time slab to another. Typically we use first-order polynomials as interpolation functions. The notation $(\cdot)_n^-$ and $(\cdot)_n^+$ denotes the function values at $t_n$ as approached from below and above respectively. Each $Q_n$ is decomposed into space-time elements $Q_n^e$, where $e = 1, 2, \ldots, (n_{el})_n$. The subscript $n$ used with $n_{el}$ is to account for the general case in which the number of space-time elements may change from one space-time slab to another. The Dirichlet- and Neumann-type boundary conditions are enforced over $(P_n)_g$ and $(P_n)_h$, the complementary subsets of the lateral boundary of the space-time slab. The finite element trial function spaces $(\mathcal{S}_{\mathbf{u}}^h)_n$ for velocity and $(\mathcal{S}_p^h)_n$ for pressure, and the test function spaces $(\mathcal{V}_{\mathbf{u}}^h)_n$ and $(\mathcal{V}_p^h)_n = (\mathcal{S}_p^h)_n$ are defined by using, over $Q_n$, first-order polynomials in both space and time.

The DSD/SST formulation is written as follows: given $(\mathbf{u}^h)_n^-$, find $\mathbf{u}^h \in (\mathcal{S}_{\mathbf{u}}^h)_n$ and $p^h \in (\mathcal{S}_p^h)_n$ such that $\forall \mathbf{w}^h \in (\mathcal{V}_{\mathbf{u}}^h)_n$ and $q^h \in (\mathcal{V}_p^h)_n$:

$$\int_{Q_n} \mathbf{w}^h \cdot \rho \left( \frac{\partial \mathbf{u}^h}{\partial t} + \mathbf{u}^h \cdot \nabla \mathbf{u}^h - \mathbf{f}^h \right) dQ + \int_{Q_n} \boldsymbol{\varepsilon}(\mathbf{w}^h) : \boldsymbol{\sigma}(p^h, \mathbf{u}^h) dQ$$

$$- \int_{(P_n)_h} \mathbf{w}^h \cdot \mathbf{h}^h dP + \int_{Q_n} q^h \nabla \cdot \mathbf{u}^h dQ + \int_{\Omega_n} (\mathbf{w}^h)_n^+ \cdot \rho \left( (\mathbf{u}^h)_n^+ - (\mathbf{u}^h)_n^- \right) d\Omega$$

$$+ \sum_{e=1}^{(n_{el})_n} \int_{Q_n^e} \frac{\tau_{\mathrm{LSME}}}{\rho} \mathbf{L}(q^h, \mathbf{w}^h) \cdot \left[ \mathbf{L}(p^h, \mathbf{u}^h) - \rho \mathbf{f}^h \right] dQ$$

$$+ \sum_{e=1}^{n_{el}} \int_{Q_n^e} \tau_{\mathrm{LSIC}} \nabla \cdot \mathbf{w}^h \rho \nabla \cdot \mathbf{u}^h dQ = 0, \tag{74}$$

where

$$\mathbf{L}(q^h, \mathbf{w}^h) = \rho \left( \frac{\partial \mathbf{w}^h}{\partial t} + \mathbf{u}^h \cdot \nabla \mathbf{w}^h \right) - \nabla \cdot \boldsymbol{\sigma}(q^h, \mathbf{w}^h), \tag{75}$$

and $\tau_{\mathrm{LSME}}$ and $\tau_{\mathrm{LSIC}}$ are the stabilization parameters (see [12, 13]). This formulation is applied to all space-time slabs $Q_0, Q_1, Q_2, \ldots, Q_{N-1}$, starting with $(\mathbf{u}^h)_0^- = \mathbf{u}_0$. For an earlier, detailed reference on this stabilized formulation see [3].

## 5. Mesh Update Methods

How the mesh should be updated depends on several factors, such as the complexity of the interface and overall geometry, how unsteady the interface is, and how the starting mesh was generated. In general, the mesh update could have two components: moving the mesh for as long as it is possible, and full or partial remeshing (i.e., generating a new set of elements, and sometimes also a new set of nodes) when the element distortion becomes too high.

In mesh moving strategies, the only rule the mesh motion needs to follow is that at the interface the normal velocity of the mesh has to match the normal velocity of the fluid.

Beyond that, the mesh can be moved in any way desired, with the main objective being to reduce the frequency of remeshing. In 3D simulations, if the remeshing requires calling an automatic mesh generator, the cost of automatic mesh generation becomes a major reason for trying to reduce the frequency of remeshing. Furthermore, when we remesh, we need to project the solution from the old mesh to the new one. This introduces projection errors. Also, in 3D, the computing time consumed by this projection step is not a trivial one. All these factors constitute a strong motivation for designing mesh update strategies which minimize the frequency of remeshing.

In some cases where the changes in the shape of the computational domain allow it, a special-purpose mesh moving method can be used in conjunction with a special-purpose mesh generator. In such cases, simulations can be carried out without calling an automatic mesh generator and without solving any additional equations to determine the motion of the mesh. One of the earliest examples of that, 2D computation of sloshing in a laterally vibrating container, can be found in [3]. Extension of that concept to 3D parallel computation of sloshing in a vertically vibrating container can be found in [4].

In general, however, we use an automatic mesh moving scheme [14] to move the nodal points, as governed by the equations of linear elasticity. The motion of the internal nodes is determined by solving these additional equations, with the boundary conditions for these mesh motion equations specified in such a way that they match the normal velocity of the fluid at the interface. Similar mesh moving techniques were used earlier by other researchers (see for example [15]).

In our mesh moving method based on linear elasticity, the structured layers of elements generated around solid objects (to fully control the mesh resolution near solid objects and have more accurate representation of the boundary layers) move "glued" to these solid objects, undergoing a rigid-body motion. No equations are solved for the motion of the nodes in these layers, because these nodal motions are not governed by the equations of elasticity. This results in some cost reduction. But more importantly, the user has full control of the mesh resolution in these layers. For early examples of automatic mesh moving combined with structured layers of elements undergoing rigid-body motion with solid objects, see [4]. Earlier examples of element layers undergoing rigid-body motion, in combination with deforming structured meshes, can be found in [3].

In computation of flow problems with fluid-solid interfaces where the solid is deforming, the motion of the fluid mesh near the interface cannot be represented by a simple rigid-body motion. Depending on the deformation mode of the solid, we may have to use the automatic mesh moving technique described earlier in this section. In such cases, presence of very thin fluid elements near the solid surface creates a challenge for the automatic mesh moving technique. In Solid-Extension Mesh Moving Technique (SEMMT), we propose to treat those very thin fluid elements almost like an extension of the solid elements. In the SEMMT, in solving the equations of elasticity governing the motion of the fluid nodes, we assign a much higher rigidity to these thin elements, compared to the other fluid elements. This could be implemented in two ways; we can solve the elasticity equations for the nodes connected to the thin elements separate from the elasticity equations for the other nodes, or together. If we solve them separately, for the thin elements, as boundary conditions at the interface with the other elements, we would use traction-free boundary conditions.

## 6. Enhanced-Discretization Interface-Capturing Technique

In EDICT, we start with the basic approach of an interface-capturing technique such as the volume of fluid (VOF) method [16]. The Navier-Stokes equations are solved over a non-moving mesh together with the time-dependent advection equation governing the evolution of the interface function $\phi$. In writing the stabilized finite element formulation for the EDICT (see [17]), the notation we use here for representing the finite-dimensional function spaces is very similar to the notation we used in the section where we described the DSD/SST formulation. The trial function spaces corresponding to velocity, pressure and interface function are denoted, respectively, by $(\mathcal{S}_{\mathbf{u}}^h)_n$, $(\mathcal{S}_p^h)_n$, and $(\mathcal{S}_\phi^h)_n$. The weighting function spaces corresponding to the momentum equation, incompressibility constraint and time-dependent advection equation are denoted by $(\mathcal{V}_{\mathbf{u}}^h)_n$, $(\mathcal{V}_p^h)_n$ $(= (\mathcal{S}_p^h)_n)$, and $(\mathcal{V}_\phi^h)_n$. The subscript $n$ in this case allows us to use different spatial discretizations corresponding to different time levels.

The stabilized formulations of the flow and advection equations can be written as follows: given $\mathbf{u}_n^h$ and $\phi_n^h$, find $\mathbf{u}_{n+1}^h \in (\mathcal{S}_{\mathbf{u}}^h)_{n+1}$, $p_{n+1}^h \in (\mathcal{S}_p^h)_{n+1}$, and $\phi_{n+1}^h \in (\mathcal{S}_\phi^h)_{n+1}$, such that, $\forall \mathbf{w}_{n+1}^h \in (\mathcal{V}_{\mathbf{u}}^h)_{n+1}$, $\forall q_{n+1}^h \in (\mathcal{V}_p^h)_{n+1}$, and $\forall \psi_{n+1}^h \in (\mathcal{V}_\phi^h)_{n+1}$:

$$\int_\Omega \mathbf{w}_{n+1}^h \cdot \rho \left( \frac{\partial \mathbf{u}^h}{\partial t} + \mathbf{u}^h \cdot \boldsymbol{\nabla} \mathbf{u}^h - \mathbf{f}^h \right) d\Omega + \int_\Omega \boldsymbol{\varepsilon}(\mathbf{w}_{n+1}^h) : \boldsymbol{\sigma}(p^h, \mathbf{u}^h) d\Omega$$

$$- \int_{\Gamma_h} \mathbf{w}_{n+1}^h \cdot \mathbf{h}^h d\Gamma + \int_\Omega q_{n+1}^h \boldsymbol{\nabla} \cdot \mathbf{u}^h d\Omega$$

$$+ \sum_{e=1}^{n_{el}} \int_{\Omega^e} \left[ \tau_{\mathrm{SUPG}} \mathbf{u}^h \cdot \boldsymbol{\nabla} \mathbf{w}_{n+1}^h + \frac{\tau_{\mathrm{PSPG}}}{\rho} \boldsymbol{\nabla} q_{n+1}^h \right] \cdot \left[ \mathbf{L}(p^h, \mathbf{u}^h) - \rho \mathbf{f}^h \right] d\Omega$$

$$+ \sum_{e=1}^{n_{el}} \int_{\Omega^e} \tau_{\mathrm{LSIC}} \boldsymbol{\nabla} \cdot \mathbf{w}_{n+1}^h \rho \boldsymbol{\nabla} \cdot \mathbf{u}^h d\Omega = 0, \tag{76}$$

$$\int_\Omega \psi_{n+1}^h \left( \frac{\partial \phi^h}{\partial t} + \mathbf{u}^h \cdot \boldsymbol{\nabla} \phi^h \right) d\Omega$$

$$+ \sum_{e=1}^{n_{el}} \int_{\Omega^e} \tau_\phi \mathbf{u}^h \cdot \boldsymbol{\nabla} \psi_{n+1}^h \left( \frac{\partial \phi^h}{\partial t} + \mathbf{u}^h \cdot \boldsymbol{\nabla} \phi^h \right) d\Omega = 0. \tag{77}$$

In this formulation, $\tau_{\mathrm{SUPG}}$, $\tau_{\mathrm{PSPG}}$ and $\tau_\phi$ are the stabilization parameters [3,12].

To increase the accuracy, we use function spaces corresponding to enhanced discretization at and near the interface. A subset of the elements in the base mesh, Mesh-1, are identified as those at and near the interface. A more refined mesh, Mesh-2, is constructed by patching together second-level meshes generated over each element in this subset. The interpolation functions for velocity and pressure will all have two components each: one coming from Mesh-1 and the second one coming from Mesh-2. To further increase the accuracy, we construct a third-level mesh, Mesh-3, for the interface function only. The construction of Mesh-3 from Mesh-2 is very similar to the construction of Mesh-2 from Mesh-1. The interpolation functions for the interface function will have three components, each coming from one of these three meshes. We re-define the subsets over which we build Mesh-2 and Mesh-3 not every time step but with sufficient frequency to keep the interface enveloped in. We need to avoid this envelope being too wide or too narrow.

## 7. Extensions of EDICT to Other Classes of Applications

Extension of EDICT to other classes of problems was first reported in [18] for computation of compressible flows with shocks. This extension is based on re-defining the "interface" to mean the shock front. In this approach, at and near the shock fronts, we use enhanced discretization to increase the accuracy in representing those shocks. Later, the EDICT was extended to computation of vortex flows. The results were first reported in [19, 20]. In this case, the definition of the interface is extended to mean regions where the vorticity magnitude is larger than a specified value.

Here we propose to extend EDICT to computation of flow problems with boundary layers. In this extension, the "interface" means solid surfaces with boundary layers. In 3D problems with complex geometries and boundary layers, mesh generation poses a serious challenge. This is because accurate resolution of the boundary layer requires elements that are very thin in the direction normal to the solid surface. This needs to be accomplished without having a major increase in mesh refinement also in the tangential directions or creating very distorted elements. Otherwise, we might be increasing the computational cost excessively or decreasing the numerical accuracy unacceptably. In the Enhanced-Discretization Mesh Refinement Technique (EDMRT), we propose two different ways of using the EDICT concept to increase the mesh refinement in the boundary layers in a desirable fashion.

In the EDICT-Clustered-Mesh-2 approach, Mesh-2 is constructed by patching together clusters of second-level meshes generated over each element of Mesh-1 designated to be one of the "boundary layer elements". Depending on the type of these boundary layer elements in Mesh-1, Mesh-2 could be structured or unstructured, with hexahedral, tetrahedral or triangle-based prismatic elements. In the EDICT-Layered-Mesh-2 approach, a thin but multi-layered and more refined Mesh-2 is "laid over" the solid surfaces. Depending on the geometric complexity of the solid surfaces and depending on whether we prefer the same type elements as those we used in Mesh-1, the elements in mesh-2 could be hexahedral, tetrahedral or triangle-based prismatic elements.

The EDMRT, as an EDICT-based boundary layer mesh refinement strategy, would allow us accomplish our objective without facing the implementational difficulties associated with elements having variable number of nodes.

In the Enhanced-Discretization Space-Time Technique (EDSTT), we propose to use enhanced time-discretization in the context of a space-time formulation. The motivation behind this is to have a flexible way of carrying out time-accurate computations of fluid-structure interactions where we find it necessary to use smaller time steps for the structural dynamics part of the problem. There would be two ways of formulating EDSTT. In the EDSTT-Single-Mesh (EDSTT-SM) approach, a single space-time mesh, unstructured both in space and time, would be used to enhance the time-discretization in regions of the fluid domain near the structure. This, in general, might require a fully unstructured 4D mesh generation. In the EDSTT-Multi-Mesh (EDSTT-MM) approach, multiple space-time meshes, all structured in time, would be used to enhance the time-discretization in regions of the fluid domain near the structure. In a way, this would be the space-time version of the EDMRT. This approach would not require a fully unstructured 4D mesh generation, and therefore would not pose a mesh generation difficulty.

In general, EDSTT can be used in time-accurate computations where we require smaller time steps in some parts of the fluid domain (for example, where the spatial element sizes are small or where there is a fluid-fluid interface).

Whether we are using a space-time formulation or a semi-discrete formulation, at every time step of the computation, we need to solve a coupled, nonlinear equation system, and we use the Newton-Raphson method for this purpose. Sometimes, some parts of the computational domain may offer more of a challenge for the Newton-Raphson method than the others. This might happen, for example, at the fluid-solid interface in a fluid-structure interaction problem, and in such cases the nonlinear convergence might become even a bigger challenge if the structure is going through some sort of buckling or wrinkling. It might also happen at a fluid-fluid interface, for example, if the interface is very unsteady. In the Enhanced-Iteration Nonlinear Solution Technique (EINST), as a variation of the Newton-Raphson method, we propose to use sub-iterations in the parts of the domain where we are facing a nonlinear convergence challenge. This could be implemented, for example, by identifying the nodes of the zones where we need enhanced iterations, and performing multiple iterations for those nodes for each iteration we perform for all other nodes.

A coupled, linear equation system needs to be solved at every step of the Newton-Raphson sequence. In the class of computations we typically carry out, this equation system would be too large to solve with a direct method. Therefore we solve it iteratively. In these iterations, we use a preconditioning matrix, which is essentially an approximation to the original matrix of the coupled, linear equation system. Because of its simplicity and parallel computation efficiency, in most cases we use a diagonal matrix as the approximation. In some challenging cases, this simple approach might not lead to a satisfactory level of convergence at some locations, in the parts of the domain posing the challenge. This might happen, for example, in a fluid-structure interaction problem, where the structure or the fluid zones near the structure might be suffering from convergence problems, the situation might become worse if the structure is going through buckling or wrinkling. It might also happen at a fluid-fluid interface. We might also face this difficulty in the SEMMT described in the section on mesh update methods, if the elasticity equations for the nodes connected to the thin elements are solved together with the elasticity equations for the other nodes. In the Enhanced-Approximation Linear Solution Technique (EALST), we propose to use stronger approximations for the parts of the domain where we are facing convergence challenges. This could be implemented, for example, by identifying the elements covering the zones where we need enhanced approximation, and reflecting this in defining the element-level constituents of the approximation matrix. For example, for the elements that need stronger approximations, we can use as the element-level approximation matrix the full element-level matrix, while for all other elements we use a diagonal element-level matrix.

## 8.   Mixed Interface-Tracking/Interface-Capturing Technique

In computation of flow problems with fluid-solid interfaces, an interface-tracking technique, where the fluid mesh moves to track the interface, would allow us to have full control of the resolution of the fluid mesh in the boundary layers. With an interface-capturing technique (or an interface representation technique in the more general case), on the other hand, independent of how well the interface geometry is represented, the resolution of the fluid mesh in the boundary layer will be limited by the resolution of the fluid mesh where the interface is. In computation of flow problems with fluid-fluid interfaces where the interface is too complex or unsteady to track while keeping the remeshing frequency under control, interface-capturing techniques, with enhanced-discretization as needed, could be used as more flexible alternatives. Sometimes we may need to solve flow problems with both fluid-solid interfaces and complex or unsteady fluid-fluid interfaces.

MITICT was introduced in [21], primarily for fluid-object interactions with multiple fluids. The class of applications we were targeting were fluid-particle-gas interactions and free-surface flow of fluid-particle mixtures. However, the MITICT can be applied to a larger class of problems, where it is more effective to use an interface-tracking technique to track the solid-fluid interfaces and an interface-capturing technique to capture the fluid-fluid interfaces. The interface-tracking technique is the DSD/SST formulation (but could as well be the Arbitrary Lagrangian-Eulerian method or other moving mesh methods). The interface-capturing technique rides on this, and is based on solving over a moving mesh, in addition to the Navier-Stokes equations, the advection equation governing the time-evolution of the interface function. The additional DSD/SST formulation is for this advection equation:

$$\int_{Q_n} \psi^h \left( \frac{\partial \phi^h}{\partial t} + \mathbf{u}^h \cdot \nabla \phi^h \right) dQ + \int_{\Omega_n} (\psi^h)_n^+ \left( (\phi^h)_n^+ - (\phi^h)_n^- \right) d\Omega$$

$$+ \sum_{e=1}^{(n_{el})_n} \int_{Q_n^e} \tau_\phi \left( \frac{\partial \psi^h}{\partial t} + \mathbf{u}^h \cdot \nabla \psi^h \right) \left( \frac{\partial \phi^h}{\partial t} + \mathbf{u}^h \cdot \nabla \phi^h \right) dQ = 0. \tag{78}$$

This equation, together with Equation (74), constitute a mixed interface-tracking/interface-capturing technique that would track the solid-fluid interfaces and capture the fluid-fluid interfaces that would be too complex or unsteady to track with a moving mesh. The interface-capturing part of MITICT can be upgraded to the EDICT formulation for more accurate representation of the interfaces captured.

The MITICT can also be used for computation of fluid-structure interactions with multiple fluids or for flows with mechanical components moving in a mixture of two fluids. In more general cases, the MITICT can be used for classes of problems that involve both interfaces that can be accurately tracked with a moving mesh method and interfaces that are too complex or unsteady to be tracked and therefore require an interface-capturing technique.

## 9.  Edge-Tracked Interface Locator Technique

The Edge-Tracked Interface Locator Technique (ETILT) was introduced in [21], to have an interface-capturing technique with better volume conservation properties and sharper representation of the interfaces. To this end, we first define a second finite-dimensional representation of the interface function, namely $\phi^{he}$. With $\phi^{he}$, interfaces are represented as collection of positions along element edges crossed by the interfaces. Nodes belong to "chunks" of Fluid A or Fluid B. An edge either belongs to a chunk of Fluid A or Fluid B or is an interface edge. Each element is either filled fully by a chunk of Fluid A or Fluid B, or is shared by a chunk of Fluid A and a chunk of Fluid B. If an element is shared like that, the shares are determined by the position of the interface along the edges of that element. The base finite element formulation is essentially the one described by Equations (76) and (77). Although the ETILT can be used in combination with the

EDICT, we assume that we are working here with the plain, non-EDICT versions of Equations (76) and (77).

At each time step, given $\mathbf{u}_n^h$ and $\phi_n^{he}$, we determine $\mathbf{u}_{n+1}^h$, $p_{n+1}^h$, and $\phi_{n+1}^{he}$. The definitions of $\rho$ and $\mu$ are modified to use the edge-based representation of the interface function: $\rho^h = \phi^{he}\rho_A + (1 - \phi^{he})\rho_B$, $\mu^h = \phi^{he}\mu_A + (1 - \phi^{he})\mu_B$. In marching from time level $n$ to $n + 1$, we first calculate $\phi_n^h$ from $\phi_n^{he}$ by a least-squares projection:

$$\int_\Omega \psi^h \left(\phi_n^h - \phi_n^{he}\right) d\Omega = 0. \tag{79}$$

To calculate $\phi_{n+1}^h$, we use Equation (77). From $\phi_{n+1}^h$, we calculate $\phi_{n+1}^{he}$ by a combination of a least-squares projection:

$$\int_\Omega (\psi_{n+1}^{he})_P \left((\phi_{n+1}^{he})_P - \phi_{n+1}^h\right) d\Omega = 0, \tag{80}$$

and corrections to enforce volume conservation for all chunks of Fluid A and Fluid B, taking into account the mergers between the chunks and the split of chunks. This volume conservation condition can symbolically be written as $VOL\left(\phi_{n+1}^{he}\right) = VOL\left(\phi_n^{he}\right)$. Here the subscript $P$ is used for representing the intermediate values following the projection, but prior to the corrections for volume conservation. These projections and volume corrections are embedded in the iterative solution technique, and are carried out at each iteration (see [21]).

## 10. Numerical Examples

### 10.1 Free-Surface Flow Past a Bridge Support

The free-stream Reynolds and Froude numbers are 10 million and 0.564. The mesh has 230,480 prism-based space-time elements. The DSD/SST formulation is used with algebraic mesh update. Figure 1 shows, at an instant, the cylinder together with the free-surface color-coded with the velocity magnitude. For more on this simulation see [22].

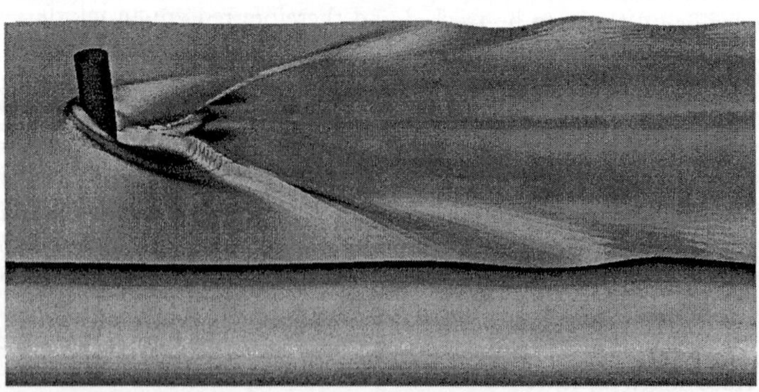

Figure 1. Free-surface flow past a bridge support. The bridge support and the free-surface color-coded with the velocity magnitude.

## 10.2   2D Sloshing in a Container

A container partially filled with water is subjected to a horizontal acceleration of 0.2g. We first compute with the DSD/SST formulation, with 6,000 quadrilateral elements. We refer to this as Solution-IT. Solution-1 is obtained by using Mesh-1, with 30,000 triangular elements. Solution-2 is obtained with the EDICT, where all functions come from Mesh-1 $\oplus$ Mesh-2. Solution-3 is obtained with the functions for velocity and pressure coming from Mesh-1 $\oplus$ Mesh-2, and for the interface function from Mesh-1 $\oplus$ Mesh-2 $\oplus$ Mesh-3. Figure 2 shows shows the results. For more details see [23].

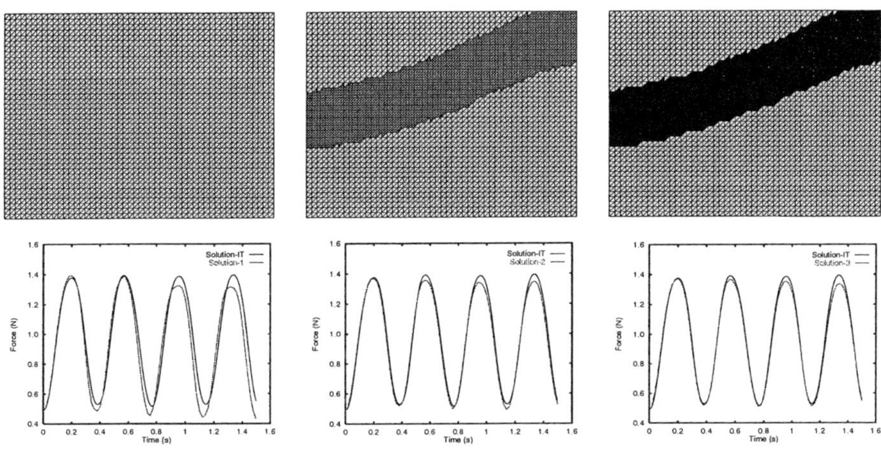

Figure 2. 2D sloshing in a container. Pictures shows, at the top, at t = 0.2 s, Mesh-1 together with Mesh-2 and Mesh-3 (both shown on top of Mesh-1); and at the bottom, the time histories of the horizontal forces exerted on the container.

## 11.   Concluding Remarks

In this paper, we provided an overview of the stabilized finite element interface-tracking and interface-capturing techniques we have developed in recent years for computation of flow problems with moving boundaries and interfaces. We also provided an overview of the stabilized formulations underlying these techniques and determination of the stabilization parameters used in these formulations. The interface-tracking techniques are based on the DSD/SST formulation, where the mesh moves to track the interface. The interface-capturing techniques, which were developed for two-fluid flows, are based on the stabilized formulation, over non-moving meshes, of both the flow equations and the advection equation governing the time-evolution of an interface function marking the location of the interface. For interface-capturing techniques, the EDICT increases the accuracy in representing the interface. The MITICT was developed for the classes of problems that involve both interfaces that can be accurately tracked with a moving mesh method and interfaces that are too complex or unsteady to be tracked and therefore require an interface-capturing technique. The ETILT was developed to improve the interface-capturing techniques with better volume conservation properties and sharper representation of the interfaces.

## Acknowledgment

This work was supported by the US Army Natick Soldier Center (contract no. DAAD16-00-C-9222) and by NASA Johnson Space Center (grant no. NAG9-1059).

# References

[1] A.N. Brooks and T.J.R. Hughes, "Streamline upwind/Petrov-Galerkin formulations for convection dominated flows with particular emphasis on the incompressible Navier-Stokes equations", *Computer Methods in Applied Mechanics and Engineering*, **32** (1982) 199–259.

[2] T.J.R. Hughes, L.P. Franca, and G.M. Hulbert, "A new finite element formulation for computational fluid dynamics: VIII. the Galerkin/least-squares method for advective-diffusive equations", *Computer Methods in Applied Mechanics and Engineering*, **73** (1989) 173–189.

[3] T.E. Tezduyar, "Stabilized finite element formulations for incompressible flow computations", *Advances in Applied Mechanics*, **28** (1991) 1–44.

[4] T. Tezduyar, S. Aliabadi, M. Behr, A. Johnson, and S. Mittal, "Parallel finite-element computation of 3D flows", *IEEE Computer*, **26** (1993) 27–36.

[5] T.E. Tezduyar, S. Aliabadi, and M. Behr, "Enhanced-Discretization Interface-Capturing Technique", in Y. Matsumoto and A. Prosperetti, editors, *Proceedings of the ISAC '97 High Performance Computing on Multiphase Flows*, 1–6, Japan Society of Mechanical Engineers, 1997.

[6] T.E. Tezduyar and T.J.R. Hughes, "Finite element formulations for convection dominated flows with particular emphasis on the compressible Euler equations", in *Proceedings of AIAA 21st Aerospace Sciences Meeting*, AIAA Paper 83-0125, Reno, Nevada, (1983).

[7] T.E. Tezduyar and Y.J. Park, "Discontinuity capturing finite element formulations for nonlinear convection-diffusion-reaction problems", *Computer Methods in Applied Mechanics and Engineering*, **59** (1986) 307–325.

[8] T.E. Tezduyar and D.K. Ganjoo, "Petrov-Galerkin formulations with weighting functions dependent upon spatial and temporal discretization: Applications to transient convection-diffusuion problmes", *Computer Methods in Applied Mechanics and Engineering*, **59** (1986) 49–71.

[9] L.P. Franca, S.L. Frey, and T.J.R. Hughes, "Stabilized finite element methods: I. Application to the advective-diffusive model", *Computer Methods in Applied Mechanics and Engineering*, **95** (1992) 253–276.

[10] T.E. Tezduyar and Y. Osawa, "Finite element stabilization parameters computed from element matrices and vectors", *Computer Methods in Applied Mechanics and Engineering*, **190** (2000) 411–430.

[11] S. Mittal, "On the performance of high aspect-ratio elements for incompressible flows", *Computer Methods in Applied Mechanics and Engineering*, **188** (2000) 269–287.

[12] T.E. Tezduyar and Y. Osawa, "Methods for parallel computation of complex flow problems", *Parallel Computing*, **25** (1999) 2039–2066.

[13] M. Behr and T.E. Tezduyar, "Finite element solution strategies for large-scale flow simulations", *Computer Methods in Applied Mechanics and Engineering*, **112** (1994) 3–24.

[14] T.E. Tezduyar, M. Behr, S. Mittal, and A.A. Johnson, "Computation of unsteady incompressible flows with the finite element methods – space-time formulations, iterative strategies and massively parallel implementations", in P. Smolinski, W.K. Liu, G. Hulbert, and K. Tamma, editors, *New Methods in Transient Analysis*, AMD-Vol.143, ASME, New York, (1992) 7–24.

[15] D.R. Lynch, "Wakes in liquid-liquid systems", *Journal of Computational Physics*, **47** (1982) 387–411.

[16] C. W. Hirt and B. D. Nichols, "Volume of fluid (VOF) method for the dynamics of free boundaries", *Journal of Computational Physics*, **39** (1981) 201–225.

[17] T.E. Tezduyar, S. Aliabadi, and M. Behr, "Enhanced-Discretization Interface-Capturing Technique (EDICT) for computation of unsteady flows with interfaces", *Computer Methods in Applied Mechanics and Engineering*, **155** (1998) 235–248.

[18] S. Mittal, S. Aliabadi, and T.E. Tezduyar, "Parallel computation of unsteady compressible flows with the EDICT", *Computational Mechanics*, **23** (1999) 151–157.

[19] T.E. Tezduyar, Y. Osawa, K. Stein, R. Benney, V. Kumar, and J. McCune, "Numerical methods for computer assisted analysis of parachute mechanics", in *Proceedings of 8th Conference on Numerical Methods in Continuum Mechanics (CD-ROM)*, Liptovsky Jan, Slovakia, (2000).

[20] T.E. Tezduyar, Y. Osawa, K. Stein, R. Benney, V. Kumar, and J. McCune, "Computational methods for parachute aerodynamics", to appear in *Proceedings of Computational Fluid Dynamics for the 21st Century*, Kyoto, Japan, 2000.

[21] T.E. Tezduyar, "Finite element methods for flow problems with moving boundaries and interfaces", to appear in *Archives of Computational Methods in Engineering*, 2001.

[22] I. Güler, M. Behr, and T.E. Tezduyar, "Parallel finite element computation of free-surface flows", *Computational Mechanics*, **23** (1999) 117–123.

[23] T.E. Tezduyar and S. Aliabadi, "EDICT for computation of unsteady flows with interfaces", in *Modeling and Simulation Based Engineering (eds. S. Atluri and P. O'Donoghue)*, *Proceedings of International Conference on Computational Engineering Science*, Atlanta, Georgia, (1998).

# Computation of Moving Boundaries and Interfaces and Stabilization Parameters

Tayfun E. Tezduyar
Mechanical Engineering, Rice University – MS 321
6100 Main Street, Houston, TX 77005, USA

## Summary

The interface-tracking and interface-capturing techniques we developed in recent years for computation of flow problems with moving boundaries and interfaces rely on stabilized formulations such as the streamline-upwind/Petrov-Galerkin (SUPG) and pressure-stabilizing/Petrov-Galerkin (PSPG) methods. The interface-tracking techniques are based on the Deforming-Spatial-Domain/Stabilized Space-Time formulation, where the mesh moves to track the interface. The interface-capturing techniques, typically used with non-moving meshes, are based on a stabilized semi-discrete formulation of the Navier-Stokes equations, combined with a stabilized formulation of the advection equation governing the time-evolution of an interface function marking the interface location. We provide an overview of the interface-tracking and interface-capturing techniques, and highlight how we determine the stabilization parameters used in the stabilized formulations.

## 1. Introduction

The finite element techniques we have developed in recent years for computation of flow problems with moving boundaries and interfaces (such as free-surface and two-fluid interface flows, fluid-particle and fluid-structure interactions, and flows with moving mechanical components) can be categorized into interface-tracking and interface-capturing techniques (see [1–3]). Depending on the complexity of the interface and other aspects of the problem, we can use one class of techniques or the other, or both in some cases, as it was pointed out in [1–3]. An interface-tracking technique requires meshes that "track" the interfaces. The mesh needs to be updated as the flow evolves. In interface-capturing techniques, such as one designed for two-fluid flows, the computations are based on spatial domains that are typically not moving or deforming. An interface function, marking the location of the interface, needs to be computed to "capture" the interface over the non-moving mesh. The interface is captured within the resolution of the finite element mesh covering the area where the interface is. This approach can be seen as a special case of interface representation techniques, where the interface is somehow represented over a non-moving fluid mesh, the main point being that the fluid mesh does not move to "track" the interfaces. A consequence of the mesh not moving to "track" the interface is that for fluid-solid interfaces, independent of how well the interface geometry is represented, the resolution of the boundary layer will be limited by the resolution of the fluid mesh where the interface is.

The Deforming-Spatial-Domain/Stabilized Space-Time (DSD/SST) formulation [4], developed for moving boundaries and interfaces, is an interface-tracking technique, where the finite element formulation of the problem is written over its space-time domain. At each time step the locations of the interfaces are calculated as part of the overall solution. As the spatial domain occupied by the fluid changes its shape in time, mesh needs to be updated. In general, this is accomplished by moving the mesh with the motion of the nodes governed by the equations of elasticity, and full or partial remeshing (i.e., generating a new set of elements, and sometimes also a new set of nodes) as needed.

In computation of two fluid-flows with interface-tracking techniques, sometimes the interface might be too complex or unsteady to track while keeping the frequency of remeshing at an acceptable level. Not being able to reduce the frequency of remeshing in 3D might introduce overwhelming mesh generation and projection costs, making the computations with the interface-tracking technique no longer feasible. In such cases, interface-capturing techniques, which do not normally require costly mesh update steps, could be used with the understanding that the interface will not be represented as accurately as we would have with an interface-tracking technique. In other words, for comparable levels of spatial discretization, interface-capturing techniques yield less accurate representation of the interface. However, these techniques can be used as practical alternatives in carrying out the simulations when compromising the accurate representation of the interfaces becomes less of a concern than facing major difficulties in updating the mesh to track such interfaces. To increase the accuracy of an interface-capturing technique without adding a major computational cost, we developed the Enhanced-Discretization Interface-Capturing Technique (EDICT), first introduced in [5]. How EDICT can be used in several other contexts, such as shock-capturing in compressible flows or sub-time-stepping in fluid-structure interactions, is highlighted in [1–3]. Methods developed to increase the scope and accuracy of the interface-tracking and interface-capturing techniques are also highlighted in [1–3].

Our interface-tracking and interface-capturing techniques are based on the streamline-upwind/Petrov-Galerkin (SUPG) [6, 7], Galerkin/least-squares (GLS) [8], and pressure-stabilizing/Petrov-Galerkin (PSPG) [4] formulations. In the interface-capturing techniques, stabilized semi-discrete formulations are used for both the Navier-Stokes equations of incompressible flows and the advection equation governing the time-evolution of an interface function marking the interface location. These stabilization techniques prevent numerical oscillations and other instabilities in solving problems with advection-dominated flows and when using equal-order interpolation functions for velocity and pressure. Furthermore, this class of stabilized formulations substantially improve the convergence rate in iterative solution of the large, coupled nonlinear equation system that needs to be solved at every time step of a flow computation. Such nonlinear systems are typically solved with the Newton-Raphson method, which involves, at its every iteration step, solution of a large, coupled linear equation system. In iterative solution of such linear equation systems using a good stabilized method makes substantial difference in convergence, as it was pointed out in [9].

The SUPG, GLS and PSPG formulations stabilize the method without introducing excessive numerical dissipation. In these formulations, judicious selection of the stabilization parameter, which is almost always known as "$\tau$", plays an important role in determining the accuracy of the formulation. This stabilization parameter involves a measure of the local length scale (also known as "element length") and other parameters such as the local Reynolds and Courant numbers. Various "element length"s and "$\tau$"s were proposed starting with those in [7] and [10], followed by the one introduced in [11], and those proposed in the subsequently reported SUPG, GLS and PSPG methods. A number of "$\tau$"s, dependent upon spatial and temporal discretizations, were introduced and tested in [12]. More recently, "$\tau$"s which are applicable to higher-order elements were proposed in [13].

Ways to calculate "$\tau$"s from the element-level matrices and vectors were first introduced in [14]. These new definitions are expressed in terms of the ratios of the norms of the relevant matrices or vectors. They automatically take into account the local length scales, advection field and the element-level Reynolds number. Based on these definitions, a "$\tau$" can be calculated for each element, or even for each element node or degree of freedom or element equation. Certain variations and complements of these new "$\tau$"s were introduced in [15, 3]. In this paper, we describe the element-matrix-based and element-vector-based "$\tau$"s designed for the semi-discrete and space-time formulations of the advection-diffusion and Navier-Stokes equations. We also describe approximate versions of these "$\tau$"s, which are based on the local length scales for the advection- and diffusion-dominated limits.

## 2.  Governing Equations

Let $\Omega_t \subset I\!\!R^{n_{sd}}$ be the spatial fluid mechanics domain with boundary $\Gamma_t$ at time $t \in (0, T)$, where the subscript $t$ indicates the time-dependence of the spatial domain. The Navier-Stokes equations of incompressible flows can be written on $\Omega_t$ and $\forall t \in (0, T)$ as

$$\rho(\frac{\partial \mathbf{u}}{\partial t} + \mathbf{u} \cdot \nabla \mathbf{u} - \mathbf{f}) - \nabla \cdot \boldsymbol{\sigma} = 0, \tag{1}$$

$$\nabla \cdot \mathbf{u} = 0, \tag{2}$$

where $\rho$, $\mathbf{u}$ and $\mathbf{f}$ are the density, velocity and the external force, respectively. The stress tensor $\boldsymbol{\sigma}$ is defined as

$$\boldsymbol{\sigma}(p, \mathbf{u}) = -p\mathbf{I} + 2\mu\boldsymbol{\varepsilon}(\mathbf{u}). \tag{3}$$

Here $p$ is the pressure, $\mathbf{I}$ is the identity tensor, $\mu = \rho\nu$ is the viscosity, $\nu$ is the kinematic viscosity, and $\boldsymbol{\varepsilon}(\mathbf{u})$ is the strain-rate tensor:

$$\boldsymbol{\varepsilon}(\mathbf{u}) = \frac{1}{2}((\nabla\mathbf{u}) + (\nabla\mathbf{u})^T). \tag{4}$$

The essential and natural boundary conditions for Eq. (1) are represented as

$$\mathbf{u} = \mathbf{g} \text{ on } (\Gamma_t)_g, \qquad \mathbf{n} \cdot \boldsymbol{\sigma} = \mathbf{h} \text{ on } (\Gamma_t)_h, \tag{5}$$

where $(\Gamma_t)_g$ and $(\Gamma_t)_h$ are complementary subsets of the boundary $\Gamma_t$, $\mathbf{n}$ is the unit normal vector, and $\mathbf{g}$ and $\mathbf{h}$ are given functions. A divergence-free velocity field $\mathbf{u}_0(\mathbf{x})$ is specified as the initial condition.

If the problem does not involve any moving boundaries or interfaces, the spatial domain does not need to change with respect to time, and the subscript $t$ can be dropped from $\Omega_t$ and $\Gamma_t$. This might be the case even for flows with moving boundaries and interfaces, if in the formulation used the spatial domain is not defined to be the part of the space occupied by the fluid(s). For example, we can have a fixed spatial domain, and model the fluid-fluid interfaces by assuming that the domain is occupied by two immiscible fluids, A and B, with densities $\rho_A$ and $\rho_B$ and viscosities $\mu_A$ and $\mu_B$. In modeling a free-surface problem where Fluid B is irrelevant, we assign a sufficiently low density to Fluid B. An interface function $\phi$ serves as a marker identifying Fluid A and B with the definition $\phi = \{1$ for Fluid A and $0$ for Fluid B$\}$. The interface between the two fluids is approximated to be at $\phi = 0.5$. In this context, $\rho$ and $\mu$ are defined as

$$\rho = \phi\rho_A + (1 - \phi)\rho_B, \qquad \mu = \phi\mu_A + (1 - \phi)\mu_B. \tag{6}$$

The evolution of the interface function $\phi$, and therefore the motion of the interface, is governed by a time-dependent advection equation, written on $\Omega$ and $\forall t \in (0, T)$ as

$$\frac{\partial \phi}{\partial t} + \mathbf{u} \cdot \nabla \phi = 0. \tag{7}$$

## 3. Stabilized Formulations and Stabilization Parameters

### 3.1 Advection-Diffusion Equation

Let us consider over a domain $\Omega$ with boundary $\Gamma$ the following time-dependent advection-diffusion equation, written on $\Omega$ and $\forall t \in (0, T)$ as

$$\frac{\partial \phi}{\partial t} + \mathbf{u} \cdot \nabla \phi - \nabla \cdot (\nu \nabla \phi) = 0, \tag{8}$$

where $\phi$ represents the quantity being transported (e.g., temperature, concentration, interface function), and $\nu$ is the diffusivity. The essential and natural boundary conditions associated with Eq. (8) are represented as

$$\phi = \mathrm{g} \text{ on } \Gamma_{\mathrm{g}}, \qquad \mathbf{n} \cdot \nu \nabla \phi = \mathrm{h} \text{ on } \Gamma_{\mathrm{h}}, \tag{9}$$

A function $\phi_0(\mathbf{x})$ is specified as the initial condition.

Let us assume that we have constructed some suitably-defined finite-dimensional trial solution and test function spaces $\mathcal{S}_\phi^h$ and $\mathcal{V}_\phi^h$. The stabilized finite element formulation of Eq. (8) can then be written as follows: find $\phi^h \in \mathcal{S}_\phi^h$ such that $\forall w^h \in \mathcal{V}_\phi^h$:

$$\int_\Omega w^h \left( \frac{\partial \phi^h}{\partial t} + \mathbf{u}^h \cdot \nabla \phi^h \right) d\Omega + \int_\Omega \nabla w^h \cdot \nu \nabla \phi^h d\Omega - \int_{\Gamma_{\mathrm{h}}} w^h \mathrm{h} d\Gamma$$

$$+ \sum_{e=1}^{n_{el}} \int_{\Omega^e} \tau_{\mathrm{SUPG}} \mathbf{u}^h \cdot \nabla w^h \left( \frac{\partial \phi^h}{\partial t} + \mathbf{u}^h \cdot \nabla \phi^h - \nabla \cdot (\nu \nabla \phi^h) \right) d\Omega = 0. \tag{10}$$

Here $n_{el}$ is the number of elements, $\Omega^e$ is the domain for element $e$, and $\tau_{\mathrm{SUPG}}$ is the SUPG stabilization parameter.

Let us use the notation $\mathbf{b} : \int_{\Omega^e} (\dots) d\Omega : \mathbf{b}_\mathrm{v}$ to denote the element-level matrix $\mathbf{b}$ and element-level vector $\mathbf{b}_\mathrm{v}$ corresponding to the element-level integration term $\int_{\Omega^e} (\dots) d\Omega$. We now define the following element-level matrices and vectors:

$$\mathbf{m} : \qquad \int_{\Omega^e} w^h \frac{\partial \phi^h}{\partial t} d\Omega \qquad : \mathbf{m}_\mathrm{v}, \tag{11}$$

$$\mathbf{c} : \qquad \int_{\Omega^e} w^h \mathbf{u}^h \cdot \nabla \phi^h d\Omega \qquad : \mathbf{c}_\mathrm{v}, \tag{12}$$

$$\mathbf{k} : \qquad \int_{\Omega^e} \nabla w^h \cdot \nu \nabla \phi^h d\Omega \qquad : \mathbf{k}_\mathrm{v}, \tag{13}$$

$$\tilde{\mathbf{k}} : \qquad \int_{\Omega^e} \mathbf{u}^h \cdot \nabla w^h \, \mathbf{u}^h \cdot \nabla \phi^h d\Omega \qquad : \tilde{\mathbf{k}}_\mathrm{v}, \tag{14}$$

$$\tilde{\mathbf{c}} : \qquad \int_{\Omega^e} \mathbf{u}^h \cdot \nabla w^h \frac{\partial \phi^h}{\partial t} d\Omega \qquad : \tilde{\mathbf{c}}_\mathrm{v}. \tag{15}$$

We define the element-level Reynolds and Courant numbers as follows:

$$Re \;=\; \frac{\|\mathbf{u}^h\|^2}{\nu}\frac{\|\mathbf{c}\|}{\|\tilde{\mathbf{k}}\|}, \tag{16}$$

$$Cr_u \;=\; \frac{\Delta t}{2}\frac{\|\mathbf{c}\|}{\|\mathbf{m}\|}, \tag{17}$$

$$Cr_\nu \;=\; \frac{\Delta t}{2}\frac{\|\mathbf{k}\|}{\|\mathbf{m}\|}, \tag{18}$$

$$Cr_{\tilde{\nu}} \;=\; \frac{\Delta t}{2}\tau_{\mathrm{SUPG}}\frac{\|\tilde{\mathbf{k}}\|}{\|\mathbf{m}\|}, \tag{19}$$

where $\|\mathbf{b}\|$ is the norm of matrix $\mathbf{b}$.

The components of element-matrix-based $\tau_{\mathrm{SUPG}}$ are defined as follows:

$$\tau_{\mathrm{S1}} \;=\; \frac{\|\mathbf{c}\|}{\|\tilde{\mathbf{k}}\|}, \tag{20}$$

$$\tau_{\mathrm{S2}} \;=\; \frac{\Delta t}{2}\frac{\|\mathbf{c}\|}{\|\tilde{\mathbf{c}}\|}, \tag{21}$$

$$\tau_{\mathrm{S3}} \;=\; \tau_{\mathrm{S1}}Re = \left(\frac{\|\mathbf{c}\|}{\|\tilde{\mathbf{k}}\|}\right)Re. \tag{22}$$

To construct $\tau_{\mathrm{SUPG}}$ from its components we propose the form

$$\tau_{\mathrm{SUPG}} = \left(\frac{1}{\tau_{\mathrm{S1}}^r} + \frac{1}{\tau_{\mathrm{S2}}^r} + \frac{1}{\tau_{\mathrm{S3}}^r}\right)^{-\frac{1}{r}}, \tag{23}$$

which is based on the inverse of $\tau_{\mathrm{SUPG}}$ being defined as the $r$-norm of the vector with components $\frac{1}{\tau_{\mathrm{S1}}}$, $\frac{1}{\tau_{\mathrm{S2}}}$ and $\frac{1}{\tau_{\mathrm{S3}}}$. We note that the higher the integer $r$ is, the sharper the switching between $\tau_{\mathrm{S1}}$, $\tau_{\mathrm{S2}}$ and $\tau_{\mathrm{S3}}$ becomes.

The components of the element-vector-based $\tau_{\mathrm{SUPG}}$ are defined as follows:

$$\tau_{\mathrm{SV1}} \;=\; \frac{\|\mathbf{c}_{\mathrm{V}}\|}{\|\tilde{\mathbf{k}}_{\mathrm{V}}\|}, \tag{24}$$

$$\tau_{\mathrm{SV2}} \;=\; \frac{\|\mathbf{c}_{\mathrm{V}}\|}{\|\tilde{\mathbf{c}}_{\mathrm{V}}\|}, \tag{25}$$

$$\tau_{\mathrm{SV3}} \;=\; \tau_{\mathrm{SV1}}Re = \left(\frac{\|\mathbf{c}_{\mathrm{V}}\|}{\|\tilde{\mathbf{k}}_{\mathrm{V}}\|}\right)Re. \tag{26}$$

With these three components,

$$(\tau_{\mathrm{SUPG}})_{\mathrm{V}} = \left(\frac{1}{\tau_{\mathrm{SV1}}^r} + \frac{1}{\tau_{\mathrm{SV2}}^r} + \frac{1}{\tau_{\mathrm{SV3}}^r}\right)^{-\frac{1}{r}}. \tag{27}$$

**Remark 1** *The definition of $\tau_{\mathrm{SUPG}}$ given by Eq. (27) can be seen as a nonlinear definition because it depends on the solution. However, in marching from time level $n$ to $n+1$ the element vectors can be evaluated at level $n$. This might be preferable in some cases, as it spares us from ending up with a nonlinear semi-discrete equation system.*

## 3.2 Navier-Stokes Equations of Incompressible Flows

Given Eqs. (1)-(2), let us assume that we have some suitably-defined finite-dimensional trial solution and test function spaces for velocity and pressure: $\mathcal{S}_{\mathbf{u}}^h$, $\mathcal{V}_{\mathbf{u}}^h$, $\mathcal{S}_p^h$ and $\mathcal{V}_p^h = \mathcal{S}_p^h$. The stabilized finite element formulation of Eqs. (1)-(2) can then be written as follows: find $\mathbf{u}^h \in \mathcal{S}_{\mathbf{u}}^h$ and $p^h \in \mathcal{S}_p^h$ such that $\forall \mathbf{w}^h \in \mathcal{V}_{\mathbf{u}}^h$ and $q^h \in \mathcal{V}_p^h$:

$$
\int_\Omega \mathbf{w}^h \cdot \rho \left( \frac{\partial \mathbf{u}^h}{\partial t} + \mathbf{u}^h \cdot \nabla \mathbf{u}^h - \mathbf{f} \right) d\Omega + \int_\Omega \boldsymbol{\varepsilon}(\mathbf{w}^h) : \boldsymbol{\sigma}(p^h, \mathbf{u}^h) d\Omega - \int_{\Gamma_h} \mathbf{w}^h \cdot \mathbf{h}^h d\Gamma
$$

$$
+ \int_\Omega q^h \nabla \cdot \mathbf{u}^h d\Omega + \sum_{e=1}^{n_{el}} \int_{\Omega^e} \frac{1}{\rho} [\tau_{\mathrm{SUPG}} \rho \mathbf{u}^h \cdot \nabla \mathbf{w}^h + \tau_{\mathrm{PSPG}} \nabla q^h] \cdot
$$

$$
\left[ \rho \left( \frac{\partial \mathbf{u}^h}{\partial t} + \mathbf{u}^h \cdot \nabla \mathbf{u}^h \right) - \nabla \cdot \boldsymbol{\sigma}(p^h, \mathbf{u}^h) - \rho \mathbf{f} \right] d\Omega
$$

$$
+ \sum_{e=1}^{n_{el}} \int_{\Omega^e} \tau_{\mathrm{LSIC}} \nabla \cdot \mathbf{w}^h \rho \nabla \cdot \mathbf{u}^h d\Omega = 0. \tag{28}
$$

Here $\tau_{\mathrm{PSPG}}$ and $\tau_{\mathrm{LSIC}}$ are the PSPG and LSIC (least-squares on incompressibility constraint) stabilization parameters.

We now define the following element-level matrices and vectors:

$$
\mathbf{m} : \qquad \int_{\Omega^e} \mathbf{w}^h \cdot \rho \frac{\partial \mathbf{u}^h}{\partial t} d\Omega \qquad : \mathbf{m}_{\mathrm{v}}, \tag{29}
$$

$$
\mathbf{c} : \qquad \int_{\Omega^e} \mathbf{w}^h \cdot \rho(\mathbf{u}^h \cdot \nabla \mathbf{u}^h) d\Omega \qquad : \mathbf{c}_{\mathrm{v}}, \tag{30}
$$

$$
\mathbf{k} : \qquad \int_{\Omega^e} \boldsymbol{\varepsilon}(\mathbf{w}^h) : 2\mu\boldsymbol{\varepsilon}(\mathbf{u}^h) d\Omega \qquad : \mathbf{k}_{\mathrm{v}}, \tag{31}
$$

$$
\mathbf{g} : \qquad \int_{\Omega^e} (\nabla \cdot \mathbf{w}^h) p^h d\Omega \qquad : \mathbf{g}_{\mathrm{v}}, \tag{32}
$$

$$
\mathbf{g}^{\mathrm{T}} : \qquad \int_{\Omega^e} q^h (\nabla \cdot \mathbf{u}^h) d\Omega \qquad : \mathbf{g}_{\mathrm{v}}^{\mathrm{T}}, \tag{33}
$$

$$
\tilde{\mathbf{k}} : \quad \int_{\Omega^e} (\mathbf{u}^h \cdot \nabla \mathbf{w}^h) \cdot \rho(\mathbf{u}^h \cdot \nabla \mathbf{u}^h) d\Omega \quad : \tilde{\mathbf{k}}_{\mathrm{v}}, \tag{34}
$$

$$
\tilde{\mathbf{c}} : \qquad \int_{\Omega^e} (\mathbf{u}^h \cdot \nabla \mathbf{w}^h) \cdot \rho \frac{\partial \mathbf{u}^h}{\partial t} d\Omega \qquad : \tilde{\mathbf{c}}_{\mathrm{v}}, \tag{35}
$$

$$
\tilde{\gamma} : \qquad \int_{\Omega^e} (\mathbf{u}^h \cdot \nabla \mathbf{w}^h) \cdot \nabla p^h d\Omega \qquad : \tilde{\gamma}_{\mathrm{v}}, \tag{36}
$$

$$
\beta : \qquad \int_{\Omega^e} \nabla q^h \cdot \frac{\partial \mathbf{u}^h}{\partial t} d\Omega \qquad : \beta_{\mathrm{v}}, \tag{37}
$$

$$
\gamma : \qquad \int_{\Omega^e} \nabla q^h \cdot (\mathbf{u}^h \cdot \nabla \mathbf{u}^h) d\Omega \qquad : \gamma_{\mathrm{v}}, \tag{38}
$$

$$
\theta : \qquad \int_{\Omega^e} \nabla q^h \cdot \nabla p^h d\Omega \qquad : \theta_{\mathrm{v}}, \tag{39}
$$

$$
\mathbf{e} : \qquad \int_{\Omega^e} (\nabla \cdot \mathbf{w}^h) \rho (\nabla \cdot \mathbf{u}^h) d\Omega \qquad : \mathbf{e}_{\mathrm{v}}. \tag{40}
$$

**Remark 2** *In the definition of the element-level matrices listed above, we assume that* $\mathbf{u}^h$ *appearing in the advective operator (i.e. in* $\mathbf{u}^h \cdot \nabla \mathbf{u}^h$ *and* $\mathbf{u}^h \cdot \nabla \mathbf{w}^h$*) is evaluated at time level n rather than* $n + 1$*. The definition would essentially be the same if we, alternatively, assumed that it is evaluated at time level* $n + 1$ *but nonlinear iteration level i rather than* $i + 1$*. Except, in the first option, in the advective operator we use* $(\mathbf{u}^h)_n$*, whereas in the second option we use* $(\mathbf{u}^h)^i_{n+1}$*. The second option can be seen as a nonlinear definition. The first option might be preferable in some cases, as it spares us from another level of nonlinearity coming from the way* $\tau$ *is defined. In the definition of the element-level-vectors, we face the same choices in terms of the evaluation of* $\mathbf{u}^h$ *in the advective operator.*

The element-level Reynolds and Courant numbers are defined the same way as they were defined before, as given by Eqs. (16)-(19). The components of the element-matrix-based $\tau_{\text{SUPG}}$ are defined the same way as they were defined before, as given by Eqs. (20)-(22). $\tau_{\text{SUPG}}$ is constructed from its components the same way as it was constructed before, as give by Eq. (23). The components of the element-vector-based $\tau_{\text{SUPG}}$ are defined the same way as they were defined before, as given by Eqs. (24)-(26). The construction of $(\tau_{\text{SUPG}})_{\text{V}}$ is also the same as it was before, given by Eq. (27).

The components of the element-matrix-based $\tau_{\text{PSPG}}$ are defined as follows:

$$\tau_{\text{P1}} = \frac{\|\mathbf{g}^{\text{T}}\|}{\|\gamma\|}, \tag{41}$$

$$\tau_{\text{P2}} = \frac{\Delta t}{2} \frac{\|\mathbf{g}^{\text{T}}\|}{\|\beta\|}, \tag{42}$$

$$\tau_{\text{P3}} = \tau_{\text{P1}} Re = \left( \frac{\|\mathbf{g}^{\text{T}}\|}{\|\gamma\|} \right) Re. \tag{43}$$

$\tau_{\text{PSPG}}$ is constructed from its components as follows:

$$\tau_{\text{PSPG}} = \left( \frac{1}{\tau_{\text{P1}}^r} + \frac{1}{\tau_{\text{P2}}^r} + \frac{1}{\tau_{\text{P3}}^r} \right)^{-\frac{1}{r}}. \tag{44}$$

The components of the element-vector-based $\tau_{\text{PSPG}}$ are defined as follows:

$$\tau_{\text{PV1}} = \tau_{\text{P1}}, \tag{45}$$

$$\tau_{\text{PV2}} = \tau_{\text{PV1}} \frac{\|\gamma_{\text{V}}\|}{\|\beta_{\text{V}}\|}, \tag{46}$$

$$\tau_{\text{PV3}} = \tau_{\text{PV1}} Re. \tag{47}$$

With these components,

$$(\tau_{\text{PSPG}})_{\text{V}} = \left( \frac{1}{\tau_{\text{PV1}}^r} + \frac{1}{\tau_{\text{PV2}}^r} + \frac{1}{\tau_{\text{PV3}}^r} \right)^{-\frac{1}{r}}. \tag{48}$$

The element-matrix-based $\tau_{\text{LSIC}}$ is defined as follows:

$$\tau_{\text{LSIC}} = \frac{\|\mathbf{c}\|}{\|\mathbf{e}\|}. \tag{49}$$

We define the element-vector-based $\tau_{\text{LSIC}}$ as:

$$(\tau_{\text{LSIC}})_{\text{V}} = \tau_{\text{LSIC}}. \tag{50}$$

**Remark 3** *We can also calculate a separate $\tau$ for each element node, or degree of freedom, or element equation. In that case, each component of $\tau$ would be calculated separately for each element node, or degree of freedom, or element equation. For this, we first represent an element matrix $\mathbf{b}$ in terms of its row matrices: $\mathbf{b}_1, \mathbf{b}_2, \ldots, \mathbf{b}_{n_{ex}}$ and an element vector $\mathbf{b}_{\mathrm{v}}$ in terms of it subvectors: $(\mathbf{b}_{\mathrm{v}})_1, (\mathbf{b}_{\mathrm{v}})_2, \ldots, (\mathbf{b}_{\mathrm{v}})_{n_{ex}}$. If we want a separate $\tau$ for each element node, then $\mathbf{b}_1, \mathbf{b}_2, \ldots, \mathbf{b}_{n_{ex}}$ and $(\mathbf{b}_{\mathrm{v}})_1, (\mathbf{b}_{\mathrm{v}})_2, \ldots, (\mathbf{b}_{\mathrm{v}})_{n_{ex}}$ would be the row matrices and subvectors corresponding to each element node, with $n_{ex} = n_{en}$, where $n_{en}$ is the number of element nodes. If we want a separate $\tau$ for each degree of freedom, then $\mathbf{b}_1, \mathbf{b}_2, \ldots, \mathbf{b}_{n_{ex}}$ and $(\mathbf{b}_{\mathrm{v}})_1, (\mathbf{b}_{\mathrm{v}})_2, \ldots, (\mathbf{b}_{\mathrm{v}})_{n_{ex}}$ would be the row matrices and subvectors corresponding to each degree of freedom, with $n_{ex} = n_{dof}$, where $n_{dof}$ is the number of degrees of freedom. If we want a separate $\tau$ for each element equation, then $\mathbf{b}_1, \mathbf{b}_2, \ldots, \mathbf{b}_{n_{ex}}$ and $(\mathbf{b}_{\mathrm{v}})_1, (\mathbf{b}_{\mathrm{v}})_2, \ldots, (\mathbf{b}_{\mathrm{v}})_{n_{ex}}$ would be the row matrices and subvectors corresponding to each element equation, with $n_{ex} = n_{ee}$, where $n_{ee}$ is the number of element equations. Based on this, the components of $\tau$ would be calculated using the norms of these row matrices or subvectors, instead of the element matrices or vectors. For example, a separate $\tau_{\mathrm{S1}}$ or $\tau_{\mathrm{SV1}}$ for each element node would be calculated by using the expression*

$$\left(\tau_{\mathrm{S1}}\right)_a = \frac{\|\mathbf{c}_a\|}{\|\tilde{\mathbf{k}}_a\|}, \quad a = 1, 2, \ldots, n_{en} \tag{51}$$

*or*

$$\left(\tau_{\mathrm{SV1}}\right)_a = \frac{\|(\mathbf{c}_{\mathrm{v}})_a\|}{\|(\tilde{\mathbf{k}}_{\mathrm{v}})_a\|}, \quad a = 1, 2, \ldots, n_{en}. \tag{52}$$

**Remark 4** *The concept of calculating a separate $\tau$ for each element node or equation can be extended to calculating a separate $\tau$ for each global node or equation. This can be accomplished by first representing a global matrix or vector in terms of its row matrices or subvectors associated with the global nodes or equations, and then by calculating the components of $\tau$ using the norms of these global row matrices or subvectors. With this approach, applying the class of stabilization techniques described in this paper to element-free methods would become more direct.*

For the purpose of comparison, we define here also the stabilization parameters that are based on an earlier definition of the length scale $h$ [11]:

$$h_{\mathrm{UGN}} = 2\,\|\mathbf{u}^h\| \left(\sum_{a=1}^{n_{en}} |\mathbf{u}^h \cdot \nabla N_a|\right)^{-1}, \tag{53}$$

where $N_a$ is the interpolation function associated with node $a$. The stabilization parameters are defined as follows:

$$\tau_{\mathrm{SUGN1}} = \frac{h_{\mathrm{UGN}}}{2\|\mathbf{u}^h\|}, \tag{54}$$

$$\tau_{\mathrm{SUGN2}} = \frac{\Delta t}{2}, \tag{55}$$

$$\tau_{\mathrm{SUGN3}} = \frac{h_{\mathrm{UGN}}^2}{4\nu}, \tag{56}$$

$$(\tau_{\mathrm{SUPG}})_{\mathrm{UGN}} = \left(\frac{1}{\tau_{\mathrm{SUGN1}}^2} + \frac{1}{\tau_{\mathrm{SUGN2}}^2} + \frac{1}{\tau_{\mathrm{SUGN3}}^2}\right)^{-\frac{1}{2}}, \tag{57}$$

$$(\tau_{\mathrm{PSPG}})_{\mathrm{UGN}} \;=\; (\tau_{\mathrm{SUPG}})_{\mathrm{UGN}}, \tag{58}$$

$$(\tau_{\mathrm{LSIC}})_{\mathrm{UGN}} \;=\; \frac{h_{\mathrm{UGN}}}{2}\|\mathbf{u}^h\|\, z. \tag{59}$$

Here $z$ is given as follows:

$$z = \begin{cases} \left(\dfrac{Re_{\mathrm{UGN}}}{3}\right) & Re_{\mathrm{UGN}} \le 3, \\ 1 & Re_{\mathrm{UGN}} > 3, \end{cases} \tag{60}$$

where $Re_{\mathrm{UGN}} = \frac{\|\mathbf{u}^h\| h_{\mathrm{UGN}}}{2\nu}$.

Comparisons between the performances of these earlier stabilization parameters and the ones proposed here can be found in [14]. These comparisons show that, especially for special element geometries, the performances are similar.

**Remark 5** *The expression for $\tau_{\mathrm{SUGN1}}$ can be written more directly as*

$$\tau_{\mathrm{SUGN1}} = \left( \sum_{a=1}^{n_{en}} |\mathbf{u}^h \cdot \boldsymbol{\nabla} N_a| \right)^{-1}, \tag{61}$$

*and based on that, the expression for $h_{\mathrm{UGN}}$ can be written as*

$$h_{\mathrm{UGN}} = \frac{\tau_{\mathrm{SUGN1}}}{2\|\mathbf{u}^h\|}. \tag{62}$$

*A rationale for $\tau_{\mathrm{SUGN1}}$ given by Eq. (61) can be provided based on Remark 3 and Eq. (52). For that, we apply Eq. (52) to the advection-diffusion equation:*

$$(\tau_{\mathrm{SV1}})_a \;=\; \left| \int_{\Omega^e} N_a \left(\mathbf{u}^h \cdot \boldsymbol{\nabla}\phi^h\right) d\Omega \right| \left/ \left| \int_{\Omega^e} \left(\mathbf{u}^h \cdot \boldsymbol{\nabla} N_a\right)\left(\mathbf{u}^h \cdot \boldsymbol{\nabla}\phi^h\right) d\Omega \right| \right. . \tag{63}$$

*Assuming one-point integration, we can write:*

$$(\tau_{\mathrm{SV1}})_a \;=\; \frac{|N_a|}{|\mathbf{u}^h \cdot \boldsymbol{\nabla} N_a|}. \tag{64}$$

*Let us define $\tau_{\mathrm{SUGN1}}$ to be the weighted average (weighted with $|\mathbf{u}^h \cdot \boldsymbol{\nabla} N_a|$) of these nodal $\tau_{\mathrm{SV1}}$ values:*

$$\tau_{\mathrm{SUGN1}} \;=\; \left( \sum_{a=1}^{n_{en}} |\mathbf{u}^h \cdot \boldsymbol{\nabla} N_a| \right)^{-1} \sum_{a=1}^{n_{en}} |N_a| . \tag{65}$$

*For linear, bilinear and trilinear elements $|N_a| = N_a$ and therefore $\sum_{a=1}^{n_{en}} |N_a| = 1$. Consequently:*

$$\tau_{\mathrm{SUGN1}} = \left( \sum_{a=1}^{n_{en}} |\mathbf{u}^h \cdot \boldsymbol{\nabla} N_a| \right)^{-1}. \tag{66}$$

As a potential alternative or complement to the LSIC stabilization, we propose the Discontinuity-Capturing Directional Dissipation (DCDD) stabilization. In describing the DCDD stabilization, we first define the unit vectors $\mathbf{s}$ and $\mathbf{r}$:

$$\mathbf{s} = \frac{\mathbf{u}^h}{\|\mathbf{u}^h\|}, \qquad \mathbf{r} = \frac{\boldsymbol{\nabla}\|\mathbf{u}^h\|}{\|\,\boldsymbol{\nabla}\|\mathbf{u}^h\|\,\|}, \tag{67}$$

and the element-level matrices and vectors $\mathbf{c}_r$, $\tilde{\mathbf{k}}_r$, $(\mathbf{c}_r)_v$, and $(\tilde{\mathbf{k}}_r)_v$:

$$\mathbf{c}_r : \quad \int_{\Omega^e} \mathbf{w}^h \cdot \rho(\mathbf{r} \cdot \nabla \mathbf{u}^h) d\Omega \qquad : (\mathbf{c}_r)_v \ , \tag{68}$$

$$\tilde{\mathbf{k}}_r : \quad \int_{\Omega^e} (\mathbf{r} \cdot \nabla \mathbf{w}^h) \cdot \rho(\mathbf{r} \cdot \nabla \mathbf{u}^h) d\Omega \quad : (\tilde{\mathbf{k}}_r)_v \ . \tag{69}$$

Then the DCDD stabilization is defined as

$$S_{\mathrm{DCDD}} = \sum_{e=1}^{n_{el}} \int_{\Omega^e} \rho \nu_{\mathrm{DCDD}} \nabla \mathbf{w}^h : \left( \left[ \mathbf{rr} - (\mathbf{r} \cdot \mathbf{s})^2 \mathbf{ss} \right] \cdot \nabla \mathbf{u}^h \right) d\Omega, \tag{70}$$

where the element-matrix-based and element-vector-based DCDD viscosities are:

$$\nu_{\mathrm{DCDD}} = |\mathbf{r} \cdot \mathbf{u}^h| \frac{\|\mathbf{c}_r\|}{\|\tilde{\mathbf{k}}_r\|} , \tag{71}$$

$$(\nu_{\mathrm{DCDD}})_v = |\mathbf{r} \cdot \mathbf{u}^h| \frac{\|(\mathbf{c}_r)_v\|}{\|(\tilde{\mathbf{k}}_r)_v\|} . \tag{72}$$

An approximate version of the expression given by Eq. (71) can be written as

$$\nu_{\mathrm{DCDD}} = |\mathbf{r} \cdot \mathbf{u}^h| \frac{h_{\mathrm{RGN}}}{2}, \tag{73}$$

where

$$h_{\mathrm{RGN}} = 2 \left( \sum_{a=1}^{n_{en}} |\mathbf{r} \cdot \nabla N_a| \right)^{-1} . \tag{74}$$

A different way of determining $\nu_{\mathrm{DCDD}}$ can be expressed as

$$\nu_{\mathrm{DCDD}} = \tau_{\mathrm{DCDD}} \|\mathbf{u}^h\|^2, \tag{75}$$

where

$$\tau_{\mathrm{DCDD}} = \frac{h_{\mathrm{DCDD}}}{2\|\mathbf{U}\|} \frac{\| \nabla \|\mathbf{u}^h\| \| h_{\mathrm{DCDD}}}{\|\mathbf{U}\|} . \tag{76}$$

Here $\mathbf{U}$ represents a global velocity scale, and $h_{\mathrm{DCDD}}$ can be calculated by using the expression

$$h_{\mathrm{DCDD}} = 2 \frac{\|\mathbf{c}_r\|}{\|\tilde{\mathbf{k}}_r\|} , \tag{77}$$

or the approximation

$$h_{\mathrm{DCDD}} = h_{\mathrm{RGN}} \ . \tag{78}$$

Combining Eqs. (75) and (76), we obtain

$$\nu_{\mathrm{DCDD}} = \frac{1}{2} \left( \frac{\|\mathbf{u}^h\|}{\|\mathbf{U}\|} \right)^2 (h_{\mathrm{DCDD}})^2 \| \nabla \|\mathbf{u}^h\| \| \ . \tag{79}$$

It was shown by Mittal [16] that flow computations with the SUPG and PSPG formulations (based on stabilization parameters very much like those given by Eqs. (53)-(60)) and very high aspect-ratio elements in the boundary layers might in some cases exhibit convergence problems and inaccuracies. As a remedy, Mittal [16] proposed to use an element length definition that represents the "minimum dimension" of an element and gives the minimum edge length in the special case of a rectangular element.

In [3], we proposed to re-define $\tau_{\mathrm{PSPG}}$ by modifying the definitions of $\tau_{\mathrm{P3}}$ and $\tau_{\mathrm{PV3}}$ given by Eqs. (43) and (47). We proposed to accomplish that by using the expressions

$$\tau_{\mathrm{P3}} = \tau_{\mathrm{P1}} \, \frac{\|\mathbf{c}\|}{\nu \, \|\tilde{\mathbf{k}}_{\mathrm{r}}\|} \;, \qquad \tau_{\mathrm{PV3}} = \tau_{\mathrm{PV1}} \, \frac{\|\mathbf{c}\|}{\nu \, \|\tilde{\mathbf{k}}_{\mathrm{r}}\|} \;, \tag{80}$$

or the approximations

$$\tau_{\mathrm{P3}} = \tau_{\mathrm{P1}} \, Re \left(\frac{h_{\mathrm{RGN}}}{h_{\mathrm{UGN}}}\right)^2 , \qquad \tau_{\mathrm{PV3}} = \tau_{\mathrm{PV1}} \, Re \left(\frac{h_{\mathrm{RGN}}}{h_{\mathrm{UGN}}}\right)^2 . \tag{81}$$

In [3], we further stated that these modifications can also be applied to $\tau_{\mathrm{S3}}$ and $\tau_{\mathrm{SV3}}$ given by Eqs. (22) and (26). Here we write that explicitly:

$$\tau_{\mathrm{S3}} = \tau_{\mathrm{S1}} \, \frac{\|\mathbf{c}\|}{\nu \, \|\tilde{\mathbf{k}}_{\mathrm{r}}\|} \;, \qquad \tau_{\mathrm{SV3}} = \tau_{\mathrm{SV1}} \, \frac{\|\mathbf{c}\|}{\nu \, \|\tilde{\mathbf{k}}_{\mathrm{r}}\|} \;, \tag{82}$$

$$\tau_{\mathrm{S3}} = \tau_{\mathrm{S1}} \, Re \left(\frac{h_{\mathrm{RGN}}}{h_{\mathrm{UGN}}}\right)^2 , \qquad \tau_{\mathrm{SV3}} = \tau_{\mathrm{SV1}} \, Re \left(\frac{h_{\mathrm{RGN}}}{h_{\mathrm{UGN}}}\right)^2 . \tag{83}$$

However, if we are dealing with just an advection-diffusion equation, rather than the Navier-Stokes equations of incompressible flows, then the definition of the unit vector $\mathbf{r}$ changes as follows:

$$\mathbf{r} = \frac{\nabla|\phi^h|}{\|\,\nabla|\phi^h|\,\|} \;. \tag{84}$$

We also propose to re-define $\tau_{\mathrm{SUGN3}}$ given by Eq. (56) as follows:

$$\tau_{\mathrm{SUGN3}} = \frac{h_{\mathrm{RGN}}^2}{4\nu} \;. \tag{85}$$

Furthermore, we propose to replace $(\tau_{\mathrm{LSIC}})_{\mathrm{UGN}}$ given by Eq. (59) as follows:

$$(\tau_{\mathrm{LSIC}})_{\mathrm{UGN}} = (\tau_{\mathrm{SUPG}})_{\mathrm{UGN}} \, \|\mathbf{u}^h\|^2 \;. \tag{86}$$

**Remark 6** *The "element length"s $h_{\mathrm{UGN}}$ (given by Eq. (53)) and $h_{\mathrm{RGN}}$ (Eq. (74)) can be viewed as the local length scales corresponding to the advection- and diffusion-dominated limits, respectively.*

# 4.  DSD/SST Finite Element Formulation

In the DSD/SST method, the finite element formulation of the governing equations is written over a sequence of $N$ space-time slabs $Q_n$, where $Q_n$ is the slice of the space-time domain between the time levels $t_n$ and $t_{n+1}$. At each time step, the integrations involved in the finite element formulation are performed over $Q_n$. The space-time finite element interpolation functions are continuous within a space-time slab, but discontinuous from one space-time slab to another. Typically we use first-order polynomials as interpolation functions. The notation $(\cdot)_n^-$ and $(\cdot)_n^+$ denotes the function values at $t_n$ as approached from below and above respectively. Each $Q_n$ is decomposed into space-time elements $Q_n^e$, where $e = 1, 2, \ldots, (n_{el})_n$. The subscript $n$ used with $n_{el}$ is to account for the general case in which the number of space-time elements may change from one space-time slab to another. The Dirichlet- and Neumann-type boundary conditions are enforced over $(P_n)_{\mathrm{g}}$ and $(P_n)_{\mathrm{h}}$, the complementary subsets of the lateral boundary of the space-time slab. The finite element trial function spaces $(\mathcal{S}_{\mathbf{u}}^h)_n$ for velocity and $(\mathcal{S}_p^h)_n$ for pressure, and the test function spaces $(\mathcal{V}_{\mathbf{u}}^h)_n$ and $(\mathcal{V}_p^h)_n = (\mathcal{S}_p^h)_n$ are defined by using, over $Q_n$, first-order polynomials in both space and time.

The DSD/SST formulation is written as follows: given $(\mathbf{u}^h)_n^-$, find $\mathbf{u}^h \in (\mathcal{S}_{\mathbf{u}}^h)_n$ and $p^h \in (\mathcal{S}_p^h)_n$ such that $\forall \mathbf{w}^h \in (\mathcal{V}_{\mathbf{u}}^h)_n$ and $q^h \in (\mathcal{V}_p^h)_n$:

$$
\int_{Q_n} \mathbf{w}^h \cdot \rho \left( \frac{\partial \mathbf{u}^h}{\partial t} + \mathbf{u}^h \cdot \nabla \mathbf{u}^h - \mathbf{f}^h \right) dQ + \int_{Q_n} \boldsymbol{\varepsilon}(\mathbf{w}^h) : \boldsymbol{\sigma}(p^h, \mathbf{u}^h) dQ
$$

$$
- \int_{(P_n)_{\mathrm{h}}} \mathbf{w}^h \cdot \mathbf{h}^h dP + \int_{Q_n} q^h \nabla \cdot \mathbf{u}^h dQ + \int_{\Omega_n} (\mathbf{w}^h)_n^+ \cdot \rho \left( (\mathbf{u}^h)_n^+ - (\mathbf{u}^h)_n^- \right) d\Omega
$$

$$
+ \sum_{e=1}^{(n_{el})_n} \int_{Q_n^e} \frac{\tau_{\mathrm{LSME}}}{\rho} \mathbf{L}(q^h, \mathbf{w}^h) \cdot \left[ \mathbf{L}(p^h, \mathbf{u}^h) - \rho \mathbf{f}^h \right] dQ
$$

$$
+ \sum_{e=1}^{n_{el}} \int_{Q_n^e} \tau_{\mathrm{LSIC}} \nabla \cdot \mathbf{w}^h \rho \nabla \cdot \mathbf{u}^h dQ = 0, \tag{87}
$$

where

$$
\mathbf{L}(q^h, \mathbf{w}^h) = \rho \left( \frac{\partial \mathbf{w}^h}{\partial t} + \mathbf{u}^h \cdot \nabla \mathbf{w}^h \right) - \nabla \cdot \boldsymbol{\sigma}(q^h, \mathbf{w}^h), \tag{88}
$$

and $\tau_{\mathrm{LSME}}$ and $\tau_{\mathrm{LSIC}}$ are the stabilization parameters (see [17]). This formulation is applied to all space-time slabs $Q_0, Q_1, Q_2, \ldots, Q_{N-1}$, starting with $(\mathbf{u}^h)_0^- = \mathbf{u}_0$. For an earlier, detailed reference on this stabilized formulation see [4].

**Remark 7**

**Space-Time Extension of the $\tau$ Calculations Described in Section 3.**

Let us write a DSD/SST formulation that is slightly different than the one given by Eq. (87). We do that by neglecting the $(\tau_{\mathrm{LSME}}/\rho) \, \nabla \cdot (2\mu \boldsymbol{\varepsilon}(\mathbf{w}^h))$ term and replacing $\tau_{\mathrm{LSME}}$ with $\tau_{\mathrm{SUPG}}$ and $\tau_{\mathrm{PSPG}}$:

$$
\int_{Q_n} \mathbf{w}^h \cdot \rho \left( \frac{\partial \mathbf{u}^h}{\partial t} + \mathbf{u}^h \cdot \nabla \mathbf{u}^h - \mathbf{f}^h \right) dQ + \int_{Q_n} \boldsymbol{\varepsilon}(\mathbf{w}^h) : \boldsymbol{\sigma}(p^h, \mathbf{u}^h) dQ
$$

$$- \int_{(P_n)_h} \mathbf{w}^h \cdot \mathbf{h}^h \, dP + \int_{Q_n} q^h \nabla \cdot \mathbf{u}^h \, dQ + \int_{\Omega_n} (\mathbf{w}^h)_n^+ \cdot \rho \left( (\mathbf{u}^h)_n^+ - (\mathbf{u}^h)_n^- \right) d\Omega$$

$$+ \sum_{e=1}^{(n_{el})_n} \int_{Q_n^e} \frac{1}{\rho} \left[ \tau_{\mathrm{SUPG}} \rho \left( \frac{\partial \mathbf{w}^h}{\partial t} + \mathbf{u}^h \cdot \nabla \mathbf{w}^h \right) + \tau_{\mathrm{PSPG}} \nabla q^h \right] \cdot \left[ \mathbf{L}(p^h, \mathbf{u}^h) - \rho \mathbf{f}^h \right] dQ$$

$$+ \sum_{e=1}^{n_{el}} \int_{Q_n^e} \tau_{\mathrm{LSIC}} \nabla \cdot \mathbf{w}^h \rho \nabla \cdot \mathbf{u}^h \, dQ = 0. \tag{89}$$

For extensions of the $\tau$ calculations based on matrix norms, we define the space-time augmented versions of the element-level matrices and vectors given by Eqs. (30), (34), and (38):

$$\mathbf{c}_A : \qquad \int_{Q_n^e} \mathbf{w}^h \cdot \rho \left( \frac{\partial \mathbf{u}^h}{\partial t} + \mathbf{u}^h \cdot \nabla \mathbf{u}^h \right) dQ \qquad : (\mathbf{c}_A)_v, \tag{90}$$

$$\tilde{\mathbf{k}}_A : \qquad \int_{Q_n^e} \left( \frac{\partial \mathbf{w}^h}{\partial t} + \mathbf{u}^h \cdot \nabla \mathbf{w}^h \right) \cdot \rho \left( \frac{\partial \mathbf{u}^h}{\partial t} + \mathbf{u}^h \cdot \nabla \mathbf{u}^h \right) dQ \quad : (\tilde{\mathbf{k}}_A)_v, \tag{91}$$

$$\gamma_A : \qquad \int_{Q_n^e} \nabla q^h \cdot \left( \frac{\partial \mathbf{u}^h}{\partial t} + \mathbf{u}^h \cdot \nabla \mathbf{u}^h \right) dQ \qquad : (\gamma_A)_v. \tag{92}$$

The components of element-matrix-based $\tau_{\mathrm{SUPG}}$ are defined as follows:

$$\tau_{\mathrm{S12}} = \frac{\|\mathbf{c}_A\|}{\|\tilde{\mathbf{k}}_A\|}, \tag{93}$$

$$\tau_{\mathrm{S3}} = \tau_{\mathrm{S12}} \frac{\|\mathbf{c}_A\|}{\nu \, \|\tilde{\mathbf{k}}_r\|}, \tag{94}$$

where $\tilde{\mathbf{k}}_r$ is the space-time version (i.e. integrated over the space-time element domain $Q_n^e$) of the element-level matrix given by Eq. (69). To construct $\tau_{\mathrm{SUPG}}$ from its components we propose the form

$$\tau_{\mathrm{SUPG}} = \left( \frac{1}{\tau_{\mathrm{S12}}^r} + \frac{1}{\tau_{\mathrm{S3}}^r} \right)^{-\frac{1}{r}}. \tag{95}$$

The components of the element-vector-based $\tau_{\mathrm{SUPG}}$ are defined as follows:

$$\tau_{\mathrm{SV12}} = \frac{\|(\mathbf{c}_A)_v\|}{\|(\tilde{\mathbf{k}}_A)_v\|}, \tag{96}$$

$$\tau_{\mathrm{SV3}} = \tau_{\mathrm{SV12}} \frac{\|\mathbf{c}_A\|}{\nu \, \|\tilde{\mathbf{k}}_r\|}. \tag{97}$$

From these two components,

$$(\tau_{\mathrm{SUPG}})_v = \left( \frac{1}{\tau_{\mathrm{SV12}}^r} + \frac{1}{\tau_{\mathrm{SV3}}^r} \right)^{-\frac{1}{r}}. \tag{98}$$

The components of element-matrix-based $\tau_{\mathrm{PSPG}}$ are defined as follows:

$$\tau_{\mathrm{P12}} = \frac{\|\mathbf{g}^{\mathrm{T}}\|}{\|\gamma_A\|}, \tag{99}$$

$$\tau_{\mathrm{P3}} = \tau_{\mathrm{P12}} \frac{\|\mathbf{c}_A\|}{\nu \, \|\tilde{\mathbf{k}}_r\|}, \tag{100}$$

where $\mathbf{g}^{\mathrm{T}}$ is the space-time version of the element-level matrix given by Eq. (33). To construct $\tau_{\mathrm{PSPG}}$ from its components we propose the form

$$\tau_{\mathrm{PSPG}} = \left(\frac{1}{\tau_{\mathrm{P12}}^{r}} + \frac{1}{\tau_{\mathrm{P3}}^{r}}\right)^{-\frac{1}{r}} . \tag{101}$$

The components of the element-vector-based $\tau_{\mathrm{PSPG}}$ are defined as follows:

$$\tau_{\mathrm{PV12}} = \frac{\|\mathbf{g}_{\mathrm{v}}^{\mathrm{T}}\|}{\|(\gamma_{\mathrm{A}})\mathbf{v}\|} , \tag{102}$$

$$\tau_{\mathrm{PV3}} = \tau_{\mathrm{PV12}}\frac{\|\mathbf{c}_{\mathrm{A}}\|}{\nu \, \|\tilde{\mathbf{k}}_{\mathrm{r}}\|} . \tag{103}$$

From these components,

$$(\tau_{\mathrm{PSPG}})_{\mathrm{v}} = \left(\frac{1}{\tau_{\mathrm{PV12}}^{r}} + \frac{1}{\tau_{\mathrm{PV3}}^{r}}\right)^{-\frac{1}{r}} . \tag{104}$$

The element-matrix-based $\tau_{\mathrm{LSIC}}$ is defined as

$$\tau_{\mathrm{LSIC}} = \frac{\|\mathbf{c}_{\mathrm{A}}\|}{\|\mathbf{e}\|} , \tag{105}$$

where $\mathbf{e}$ is the space-time version of the element-level matrix given by Eq. (40).

The element-vector-based $\tau_{\mathrm{LSIC}}$ is defined as

$$(\tau_{\mathrm{LSIC}})_{\mathrm{v}} = \tau_{\mathrm{LSIC}} . \tag{106}$$

The space-time versions of $\tau_{\mathrm{SUGN1}}$, $\tau_{\mathrm{SUGN2}}$, $\tau_{\mathrm{SUGN3}}$, $(\tau_{\mathrm{SUPG}})_{\mathrm{UGN}}$, $(\tau_{\mathrm{PSPG}})_{\mathrm{UGN}}$, and $(\tau_{\mathrm{LSIC}})_{\mathrm{UGN}}$, given respectively by Eqs. (54), (55), (85), (57), (58), and (86), are defined as follows:

$$\tau_{\mathrm{SUGN12}} = \left(\sum_{a=1}^{n_{en}}\left|\frac{\partial N_{a}}{\partial t} + \mathbf{u}^{h} \cdot \nabla N_{a}\right|\right)^{-1} , \tag{107}$$

$$\tau_{\mathrm{SUGN3}} = \frac{h_{\mathrm{RGN}}^{2}}{4\nu} , \tag{108}$$

$$(\tau_{\mathrm{SUPG}})_{\mathrm{UGN}} = \left(\frac{1}{\tau_{\mathrm{SUGN13}}^{2}} + \frac{1}{\tau_{\mathrm{SUGN3}}^{2}}\right)^{-\frac{1}{2}} , \tag{109}$$

$$(\tau_{\mathrm{PSPG}})_{\mathrm{UGN}} = (\tau_{\mathrm{SUPG}})_{\mathrm{UGN}} , \tag{110}$$

$$(\tau_{\mathrm{LSIC}})_{\mathrm{UGN}} = (\tau_{\mathrm{SUPG}})_{\mathrm{UGN}} \|\mathbf{u}^{h}\|^{2} . \tag{111}$$

Here, $n_{en}$ is the number of nodes for the space-time element, and $N_{a}$ is the space-time interpolation function associated with node $a$.

## 5.  Test Computations

### 5.1  2D Advection Skew to Mesh

Here we compare the performance of some of the stabilization parameters for a 2D advection-diffusion problem with negligible diffusivity and with advection skew to the mesh. Figure 1 shows the problem set up. The advection direction is 30 degrees from the $x$-axis. The domain is square, the mesh is uniform with $20\times20$ square elements, and $\|\mathbf{u}^h\|\Delta t/\Delta x = 2.0$. The stabilization parameters tested are: $(\tau_{\text{SUPG}})_{\text{UGN}}$ (given by Eq. (57)), $\tau_{\text{SUPG}}$ (Eq. (23)), and $(\tau_{\text{SUPG}})_{\text{V}}$ (Eq. (27)). Figure 2 shows the solution along

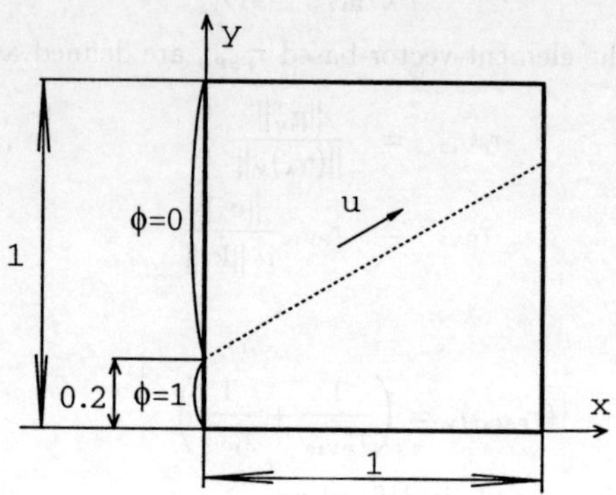


Figure 1. 2D Advection Skew to Mesh. Problem set up.

$x = 0.5$. The exact solution is for the pure advection case. We observe that the solutions

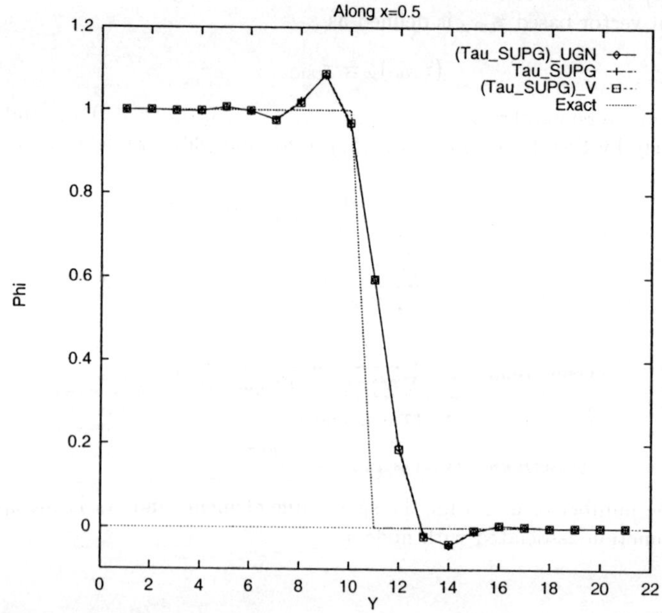


Figure 2. 2D Advection Skew to Mesh. Solution along $x = 0.5$, obtained with three different stabilization parameters, compared to the exact solution.

obtained with the stabilization parameters tested are almost identical. For more details on this test computation, see [14].

255

## 5.2  2D Incompressible Flow Past a Cylinder

In this case we compare the stabilization parameters in computation of 2D incompressible flow past a cylinder at $Re = 100$. The mesh near the cylinder is shown in Figure 3. In this computation $U_\infty \Delta t / R = 0.1$, where $U_\infty$ is the free-stream velocity and $R$ is the cylinder radius. Figure 4 shows, at later stages of the computation, time history of the

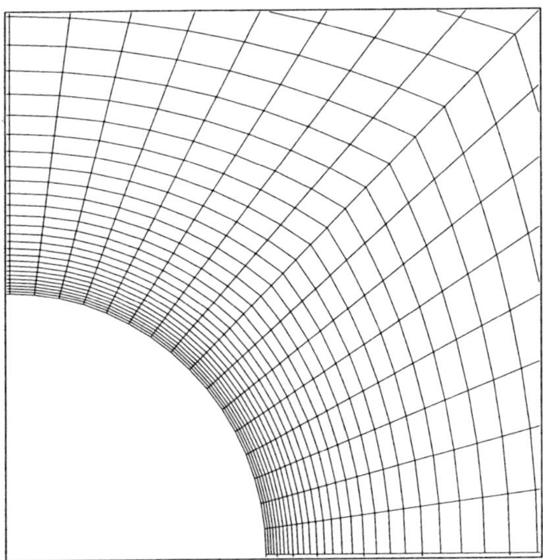

Figure 3. 2D Incompressible Flow Past a Cylinder. Mesh near cylinder.

drag and lift coefficients. For $(\tau)_{\text{UGN}}$, $\tau$, and $(\tau)_{\text{V}}$, respectively, the average drag coefficient is 1.422, 1.421, and 1.417, and the average Strouhal number is 0.167, 0.169, and 0.168. Here $(\tau)_{\text{UGN}}$ represents the group of stabilization parameters $(\tau_{\text{SUPG}})_{\text{UGN}}$ (given by Eq. (57)), $(\tau_{\text{PSPG}})_{\text{UGN}}$ (Eq. (58)), and $(\tau_{\text{LSIC}})_{\text{UGN}}$ (Eq. (59)); $\tau$ represents $\tau_{\text{SUPG}}$ (Eq. (23)), $\tau_{\text{PSPG}}$ (Eq. (44)), and $\tau_{\text{LSIC}}$ (Eq. (49)); and $(\tau)_{\text{V}}$ represents $(\tau_{\text{SUPG}})_{\text{V}}$ (Eq. (27)), $(\tau_{\text{PSPG}})_{\text{V}}$ (Eq. (48)), and $(\tau_{\text{LSIC}})_{\text{V}}$ (Eq. (50)). Figure 5 shows the values of the stabilization parameters along the vertical line passing through the cylinder center, starting from the upper cylinder surface. For more details on this test computation, see [14].

# 6.  Concluding Remarks

We provided an overview of the interface-tracking and interface-capturing techniques we developed for computation of flow problems with moving boundaries and interfaces. These techniques rely on stabilized formulations such as the streamline-upwind/Petrov-Galerkin and pressure-stabilizing/Petrov-Galerkin methods. The interface-tracking techniques are based on the Deforming-Spatial-Domain/Stabilized Space-Time formulation, where the mesh moves to track the interface. The interface-capturing techniques, typically used with non-moving meshes, are based on a stabilized semi-discrete formulation of the Navier-Stokes equations, combined with a stabilized formulation of an advection equation. The advection equation governs the time-evolution of an interface function

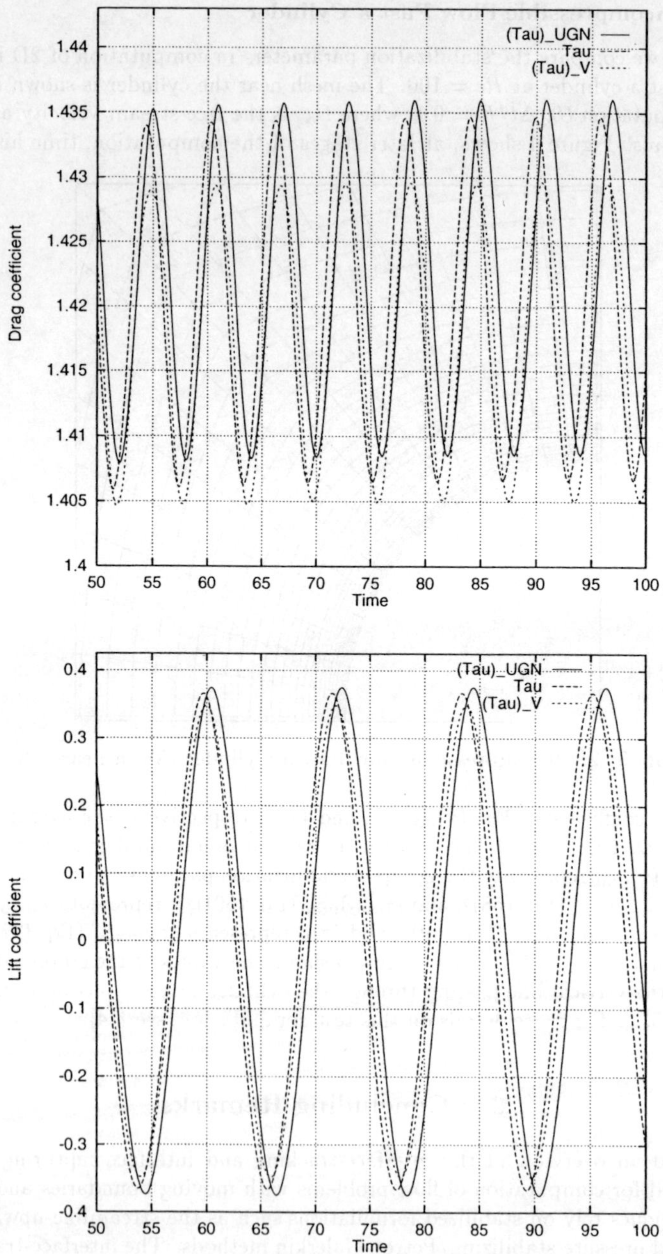

Figure 4. 2D Incompressible Flow Past a Cylinder. Time history of the drag (upper) and lift (lower) coefficients.

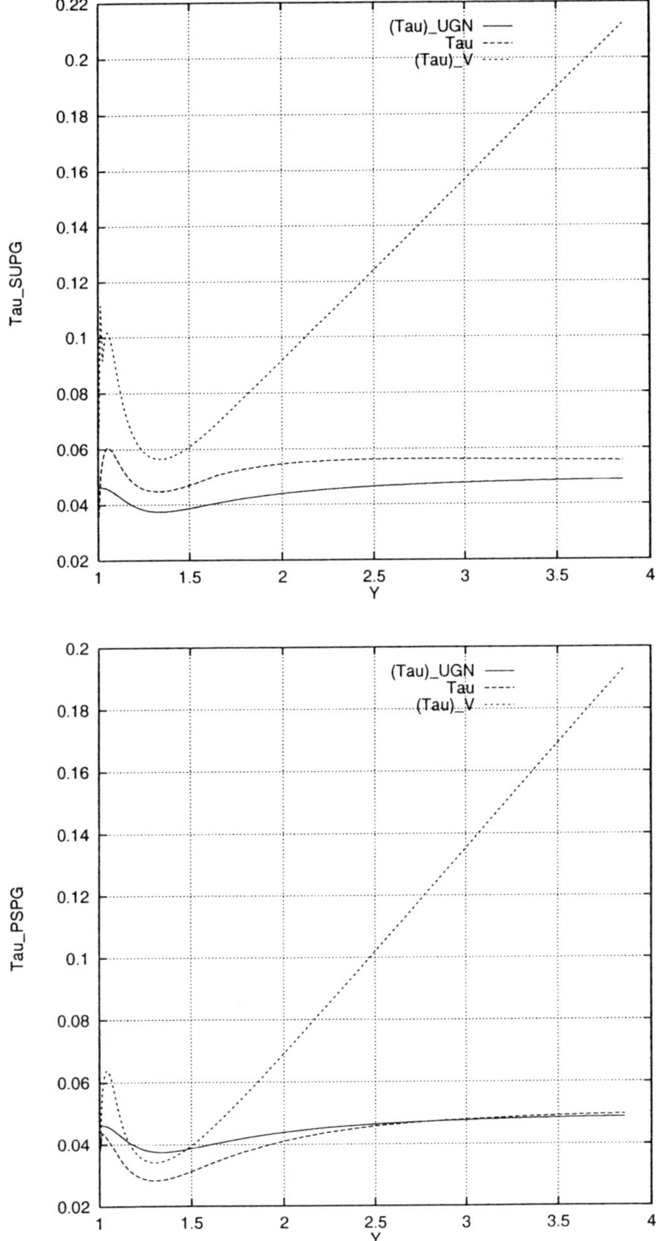

Figure 5. 2D Incompressible Flow Past a Cylinder. Stabilization parameters: $\tau_{\text{SUPG}}$ (upper) and $\tau_{\text{PSPG}}$ (lower).

marking the interface location. We highlighted how we determine the stabilization parameters ("$\tau$"s) used in the stabilized formulations. For the Navier-Stokes equations and the advection equation, we described the element-matrix-based and element-vector-based "$\tau$"s designed for semi-discrete and space-time formulations. These "$\tau$" definitions are expressed in terms of the ratios of the norms of the relevant matrices or vectors. They automatically take into account the local length scales, advection field and the element-level Reynolds number. Based on these definitions, a "$\tau$" can be calculated for each element, or even for each element node or degree of freedom or element equation. We also described certain variations and complements of these new "$\tau$"s, including the approximate versions that are based on the local length scales for the advection- and diffusion-dominated limits.

### Acknowledgment

This work was supported by the US Army Natick Soldier Center (contract no. DAAD16-00-C-9222) and by NASA Johnson Space Center (grant no. NAG9-1059).

## References

[1] T.E. Tezduyar, "Finite element methods for flow problems with moving boundaries and interfaces", *Archives of Computational Methods in Engineering*, **8** (2001) 83–130.

[2] T.E. Tezduyar, "Finite element interface-tracking and interface-capturing techniques for flows with moving boundaries and interfaces", in *Proceedings of the ASME Symposium on Fluid-Physics and Heat Transfer for Macro- and Micro-Scale Gas-Liquid and Phase-Change Flows (CD-ROM)*, ASME Paper IMECE2001/HTD-24206, ASME, New York, New York, (2001).

[3] T.E. Tezduyar, "Stabilized finite element formulations and interface-tracking and interface-capturing techniques for incompressible flows", to appear in *Proceedings of the Workshop on Numerical Simulations of Incompressible Flows*, Half Moon Bay, California, 2001.

[4] T.E. Tezduyar, "Stabilized finite element formulations for incompressible flow computations", *Advances in Applied Mechanics*, **28** (1991) 1–44.

[5] T.E. Tezduyar, S. Aliabadi, and M. Behr, "Enhanced-Discretization Interface-Capturing Technique", in Y. Matsumoto and A. Prosperetti, editors, *Proceedings of the ISAC '97 High Performance Computing on Multiphase Flows*, 1–6, Japan Society of Mechanical Engineers, 1997.

[6] T.J.R. Hughes and A.N. Brooks, "A multi-dimensional upwind scheme with no cross-wind diffusion", in T.J.R. Hughes, editor, *Finite Element Methods for Convection Dominated Flows*, AMD-Vol.34, 19–35, ASME, New York, 1979.

[7] A.N. Brooks and T.J.R. Hughes, "Streamline upwind/Petrov-Galerkin formulations for convection dominated flows with particular emphasis on the incompressible Navier-Stokes equations", *Computer Methods in Applied Mechanics and Engineering*, **32** (1982) 199–259.

[8] T.J.R. Hughes, L.P. Franca, and G.M. Hulbert, "A new finite element formulation for computational fluid dynamics: VIII. the Galerkin/least-squares method for advective-diffusive equations", *Computer Methods in Applied Mechanics and Engineering*, **73** (1989) 173–189.

[9] T. Tezduyar, S. Aliabadi, M. Behr, A. Johnson, and S. Mittal, "Parallel finite-element computation of 3D flows", *IEEE Computer*, **26** (1993) 27–36.

[10] T.E. Tezduyar and T.J.R. Hughes, "Finite element formulations for convection dominated flows with particular emphasis on the compressible Euler equations", in *Proceedings of AIAA 21st Aerospace Sciences Meeting*, AIAA Paper 83-0125, Reno, Nevada, (1983).

[11] T.E. Tezduyar and Y.J. Park, "Discontinuity capturing finite element formulations for nonlinear convection-diffusion-reaction problems", *Computer Methods in Applied Mechanics and Engineering*, **59** (1986) 307–325.

[12] T.E. Tezduyar and D.K. Ganjoo, "Petrov-Galerkin formulations with weighting functions dependent upon spatial and temporal discretization: Applications to transient convection-diffusuion probllmes", *Computer Methods in Applied Mechanics and Engineering*, **59** (1986) 49–71.

[13] L.P. Franca, S.L. Frey, and T.J.R. Hughes, "Stabilized finite element methods: I. Application to the advective-diffusive model", *Computer Methods in Applied Mechanics and Engineering*, **95** (1992) 253–276.

[14] T.E. Tezduyar and Y. Osawa, "Finite element stabilization parameters computed from element matrices and vectors", *Computer Methods in Applied Mechanics and Engineering*, **190** (2000) 411–430.

[15] T.E. Tezduyar, "Adaptive determination of the finite element stabilization parameters", in *Proceedings of the ECCOMAS Computational Fluid Dynamics Conference 2001 (CD-ROM)*, Swansea, Wales, United Kingdom, (2001).

[16] S. Mittal, "On the performance of high aspect-ratio elements for incompressible flows", *Computer Methods in Applied Mechanics and Engineering*, **188** (2000) 269–287.

[17] T.E. Tezduyar and Y. Osawa, "Methods for parallel computation of complex flow problems", *Parallel Computing*, **25** (1999) 2039–2066.

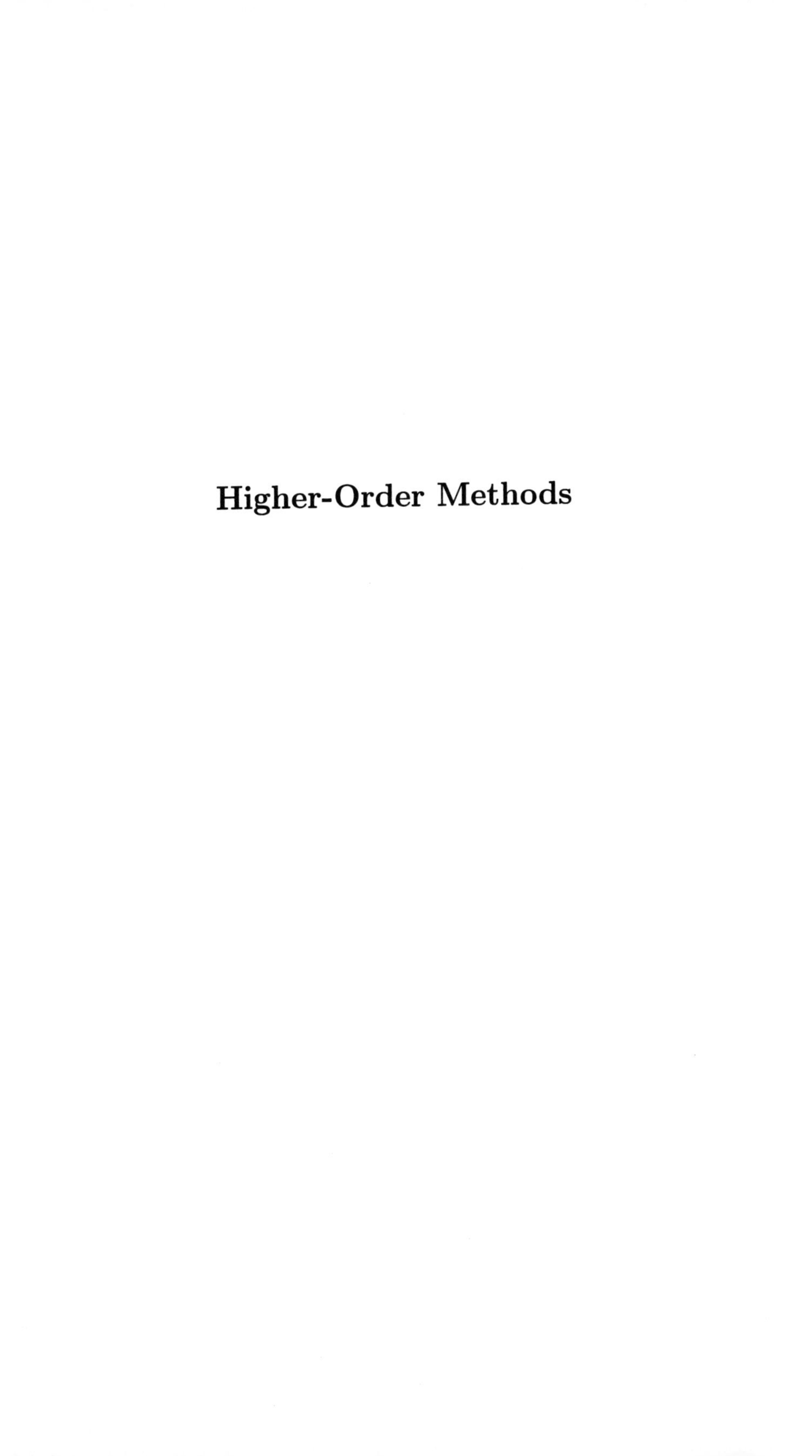

# Higher-Order Methods

# A fourth order difference method for the incompressible Navier-Stokes equations

Bertil Gustafsson, Per Lötstedt and Anders Göran
Department of Scientific Computing, Uppsala University
Uppsala, Sweden

## Abstract

We develop a difference method for the solution of the incompressible Navier-Stokes equations on a staggered grid. It is based on a Padé type fourth order approximation in space and a semi-implicit second order method in time. The method does not allow for any parasitic odd-even oscillatory solutions. We carry out a complete stability analysis for the periodic case, and construct an iterative method for efficient solution of the systems of equations to be solved for each timestep. A few numerical experiments are included for illustration of the efficiency of the iterative method.

## 1  Introduction

The work presented in this paper is part of a larger effort to develop a code for direct simulation of turbulence in 3D in curvilinear coordinates. In this development process, we intend to use a spectral method in one space direction, which leaves us with a number of 2D-problems to be solved by a difference method, which is fourth order accurate in space on a staggered grid, and (initially) second order accurate in time.

Hence, we will first consider the 2D Navier-Stokes equations in Cartesian coordinates

$$
\begin{aligned}
u_t + uu_x + vu_y + p_x &= \nu(u_{xx} + u_{yy}), \\
v_t + uv_x + vv_y + p_y &= \nu(v_{xx} + v_{yy}), \\
u_x + v_y &= 0.
\end{aligned}
$$

Here $u$ and $v$ are the velocity components in the $x$- and $y$-direction, respectively, and $p$ is the pressure. In [2] we have given a preliminary presentation of the method, mainly based on an analysis of the Stokes equations, which turn out to be of fundamental significance for the Navier-Stokes equations. By defining $\mathbf{v} = (u\ v)^T$ and the vectors

$$
Navier(\mathbf{v}) = \left[ \begin{array}{c} uu_x + vu_y \\ uv_x + vv_y \end{array} \right], \quad Stokes(\mathbf{v}, p) = \left[ \begin{array}{c} p_x - \nu(u_{xx} + u_{yy}) \\ p_y - \nu(v_{xx} + v_{yy}) \end{array} \right],
$$

we can write the Navier-Stokes equations in the form

$$\mathbf{v}_t + Navier(\mathbf{v}) + Stokes(\mathbf{v},p) = 0,$$
$$div\,\mathbf{v} = 0.$$

For the time-discretization we use a semi-implicit method based on the backwards differentiation method combined with an extrapolation method for the advection terms obtaining second order accuracy in time:

$$\frac{3}{2}\mathbf{v}^{n+1} + \Delta t\, Stokes(\mathbf{v}^{n+1},p^{n+1}) = 2\mathbf{v}^n - \frac{1}{2}\mathbf{v}^{n-1}$$
$$-\Delta t(2\,Navier(\mathbf{v}^n) - Navier(\mathbf{v}^{n-1})),$$
$$div\,\mathbf{v}^{n+1} = 0. \tag{1.1}$$

We need to solve for $(\mathbf{v}^{n+1}, p^{n+1})$ in each timestep $\Delta t = t_{n+1} - t_n$. Clearly, a fast iterative method for the Stokes equations will provide a fast method also for the Navier-Stokes equations.

In this paper we present our method, analyze its stability and construct a solver of the system of linear equations. The discretization is given here for a Cartesian grid but the generalization to curvilinear grids and numerical examples are found in [1]. Other high order finite difference approximations are found in e.g. [6], [7], [11]. Compared to these methods, our method needs no damping terms or additional artificial viscosity for stability. The solution of the resulting linear system for second order accuracy is discussed in e.g. [3], [10]. Our iterative solver is designed to be efficient for the combination of second order accuracy in time and fourth order in space.

## 2 Stability analysis for the linear periodic case

In this section we present a stability analysis for the complete system and staggered grids. There seem to be no published complete stability analysis even for the second order case, so we begin with that one.

### 2.1 Second order accuracy

In [8] there is a stability analysis of two other semi-implicit schemes for a scalar equation and second order approximation in time. Here we give an analysis of the scheme in (1.1) for the complete linearized Navier-Stokes system for the staggered grid method presented in [2]. The variables are represented at different points as shown in Figure 2.1.

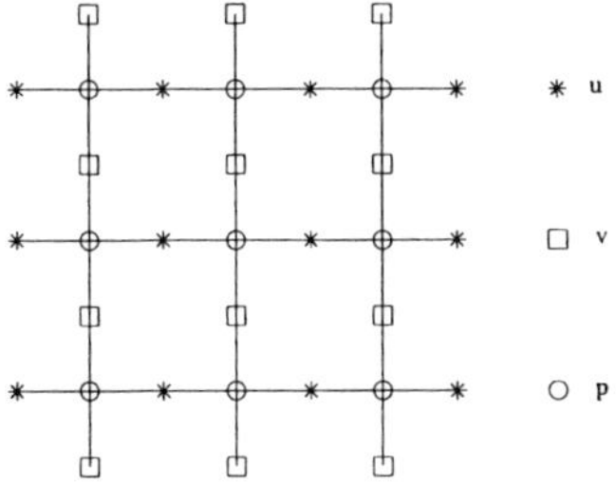

Figure 2.1: The staggered grid.

The Stokes part of the equations are approximated in the most compact way using the staggered structure of the grid. However, for the advection terms, we must use approximations over two intervals, e.g.:

$$uu_x \approx u(x,y,t)\frac{u(x+\Delta x,y,t)-u(x-\Delta x,y,t)}{2\Delta x}\,,$$

$$vv_y \approx v(x,y,t)\frac{v(x,y+\Delta y,t)-v(x,y-\Delta y,t)}{2\Delta y}\,.$$

The linearized system is obtained by letting $u=a$, $v=b$ in the advection terms, where $a$ and $b$ are constants. With the notation

$$\xi_1 = \omega_1\Delta x\,,\ \ \xi_2 = \omega_2\Delta y\,,\ \ 0 \le |\xi_1|, |\xi_2| \le \pi\,,$$

we get the Fourier transformed system

$$\frac{3}{2}\hat{u}^{n+1} - 2\hat{u}^n + \frac{1}{2}\hat{u}^{n-1} + \frac{a\Delta t}{\Delta x}i\sin\xi_1(2\hat{u}^n - \hat{u}^{n-1}) + \frac{b\Delta t}{\Delta y}i\sin\xi_2(2\hat{u}^n - \hat{u}^{n-1})$$

$$+\frac{2\Delta t}{\Delta x}i\sin\frac{\xi_1}{2}\hat{p}^{n+1} = -4\nu\left(\frac{\Delta t}{\Delta x^2}\sin^2\frac{\xi_1}{2} + \frac{\Delta t}{\Delta y^2}\sin^2\frac{\xi_2}{2}\right)\hat{u}^{n+1}\,,$$

$$\frac{3}{2}\hat{v}^{n+1} - 2\hat{v}^n + \frac{1}{2}\hat{v}^{n-1} + \frac{a\Delta t}{\Delta x}i\sin\xi_1(2\hat{v}^n - \hat{v}^{n-1}) + \frac{b\Delta t}{\Delta y}i\sin\xi_2(2\hat{v}^n - \hat{v}^{n-1})$$

$$+\frac{2\Delta t}{\Delta y}i\sin\frac{\xi_2}{2}\hat{p}^{n+1} = -4\nu\left(\frac{\Delta t}{\Delta x^2}\sin^2\frac{\xi_1}{2} + \frac{\Delta t}{\Delta y^2}\sin^2\frac{\xi_2}{2}\right)\hat{v}^{n+1}\,,$$

$$\frac{1}{\Delta x}i\sin\frac{\xi_1}{2}\hat{u}^{n+1} + \frac{1}{\Delta y}i\sin\frac{\xi_2}{2}\hat{v}^{n+1} = 0.$$

By solving the last equation for $\hat{v}$ (assuming $\xi_2 \ne 0$), using the trigonometric identity $\sin\xi = 2\sin(\xi/2)\cos(\xi/2)$ and the notation

$$\hat{\mathbf{U}} = (\hat{u},\ \hat{p})^T,\ \ \lambda_1 = \frac{\Delta t}{\Delta x}\,,\ \ \lambda_2 = \frac{\Delta t}{\Delta y}\,,\ \ \{s_j = \sin\frac{\xi_j}{2}\,,\ c_j = \cos\frac{\xi_j}{2}\}\,,\ j = 1,2\,,$$

we have

$$\begin{bmatrix} \gamma & 2i\lambda_1 s_1 \\ \gamma\frac{s_1}{s_2} & -\frac{2i\lambda_2^2 s_2}{\lambda_1} \end{bmatrix}\hat{\mathbf{U}}^{n+1} + \begin{bmatrix} \alpha & 0 \\ \alpha\frac{s_1}{s_2} & 0 \end{bmatrix}\hat{\mathbf{U}}^n + \begin{bmatrix} \beta & 0 \\ \beta\frac{s_1}{s_2} & 0 \end{bmatrix}\hat{\mathbf{U}}^{n-1} = 0\,, \quad (2.1)$$

where

$$\alpha = -2 + 4a\lambda_1 i s_1 c_1 + 4b\lambda_2 i s_2 c_2\,, \quad \beta = \frac{1}{2} - 2a\lambda_1 i s_1 c_1 - 2b\lambda_2 i s_2 c_2\,, \quad (2.2)$$

$$\gamma = \frac{3}{2} + 4\nu\left(\frac{\Delta t}{\Delta x^2}s_1^2 + \frac{\Delta t}{\Delta y^2}s_2^2\right). \qquad (2.3)$$

The amplification factor $z$ for this two-step scheme is given by

$$Det\left[\begin{array}{cc} \gamma z^2 + \alpha z + \beta & 2i\lambda_1 s_1 z^2 \\ \frac{s_1}{s_2}(\gamma z^2 + \alpha z + \beta) & -\frac{2i\lambda_2^2 s_2}{\lambda_1} z^2 \end{array}\right] = -2iz^2 q(z)\left(\frac{\lambda_2^2 s_2}{\lambda_1} + \frac{\lambda_1 s_1^2}{s_2}\right) = 0\,,$$

$$q(z) = \gamma z^2 + \alpha z + \beta. \qquad (2.4)$$

By assumption $s_2 \neq 0$. (If $s_2 = 0$ we solve for $\hat{u}$ and get the same equation but with $s_1 \to s_2$, $s_2 \to s_1$. The trivial case $s_1 = s_2 = 0$ implies $z_1 = 1$, $z_2 = 1/3$.) Hence, we can assume that the last factor is nonzero, and furthermore we can disregard the case $z = 0$. Accordingly we are left with the scalar quadratic equation $q(z) = 0$ in (2.4) and $\alpha$, $\beta$, $\gamma$ given by (2.3).

We have computed the roots $z_1, z_2$ of (2.4) numerically for $\Delta x = \Delta y = h$. Then $\max(|z_1|, |z_2|)$ is determined at $20 \times 20$ points in the $a\lambda$-$b\lambda$ plane for all $\xi_j$ on a discrete $12 \times 12$ lattice in $(-\pi, \pi) \times (-\pi, \pi)$. In Fig. 2.2, the stability regions of the scheme with $\max|z_j| \leq 1$ are plotted for different $\theta = \nu\lambda/h$. For equal size of the discretization errors in space and time it is natural to take $\Delta t^2 \sim h^2$, and we obtain $\lambda = O(1)$. Then $\theta \sim \nu/h$ which is proportional to the inverse of the cell Reynolds number. Here we expect $h$ to be of the same order as $\nu$.

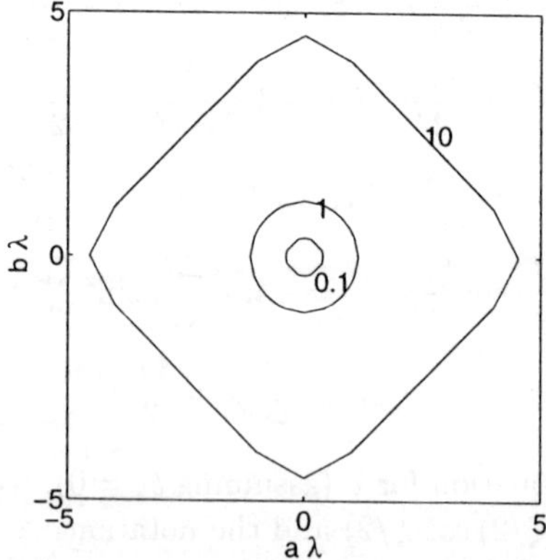


Figure 2.2: The stability regions of the second order method for $\theta = 0.1, 1, 10$.

The second order method is stable inside the contours for each $\theta$. The conclusion is that for given $a$, $b$ and $\nu$ we can obtain stability by selecting sufficiently small $h$ and $\Delta t$. Then $\theta$ will be large enough so that $(a\lambda, b\lambda)$ is inside the stability region.

We shall discuss the well known phenomenon of parasitic odd-even oscillatory solutions, and we will show, that our scheme does not allow for such solutions. This is true despite the fact that we are forced to use non-staggered approximations for the advection terms. For convenience we consider the steady state case, and in Fourier space we look for nontrivial solutions of the system

$$\begin{bmatrix} \eta & 0 & 2\lambda_1 i \sin\frac{\xi_1}{2} \\ 0 & \eta & 2\lambda_2 i \sin\frac{\xi_2}{2} \\ \lambda_1 i \sin\frac{\xi_1}{2} & \lambda_2 i \sin\frac{\xi_2}{2} & 0 \end{bmatrix} \begin{bmatrix} \hat{u} \\ \hat{v} \\ \hat{p} \end{bmatrix} = 0,$$

$$\eta = a\lambda_1 i \sin\xi_1 + b\lambda_2 i \sin\xi_2 + 4\nu\left(\frac{\lambda_1 \sin^2\frac{\xi_1}{2}}{\Delta x} + \frac{\lambda_2 \sin^2\frac{\xi_2}{2}}{\Delta y}\right).$$

The odd-even oscillations correspond to $\xi_1 = \xi_2 = \pi$, and we have

$$Det \begin{bmatrix} 4\nu(\frac{\lambda_1}{\Delta x} + \frac{\lambda_2}{\Delta y}) & 0 & 2\lambda_1 i \\ 0 & 4\nu(\frac{\lambda_1}{\Delta x} + \frac{\lambda_2}{\Delta y}) & 2\lambda_2 i \\ \lambda_1 i & \lambda_2 i & 0 \end{bmatrix} = 8\nu\left(\frac{\lambda_1}{\Delta x} + \frac{\lambda_2}{\Delta y}\right)(\lambda_1^2 + \lambda_2^2) \neq 0,$$

showing that no such solution exists.

In the regular non-staggered grid approximation, the elements $\sin(\xi_1/2)$ and $\sin(\xi_2/2)$ in the last row and column are substituted by $\sin\xi_1$ and $\sin\xi_2$, respectively. The correponding problem for $\xi_1 = \xi_2 = \pi$ is in this case

$$\begin{bmatrix} 4\nu(\frac{\lambda_1}{\Delta x} + \frac{\lambda_2}{\Delta y}) & 0 & 0 \\ 0 & 4\nu(\frac{\lambda_1}{\Delta x} + \frac{\lambda_2}{\Delta y}) & 0 \\ 0 & 0 & 0 \end{bmatrix} \begin{bmatrix} \hat{u} \\ \hat{v} \\ \hat{p} \end{bmatrix} = 0,$$

which obviously has the nontrivial solution $\hat{u} = \hat{v} = 0$, $\hat{p} = 1$.

## 2.2 Fourth order accuracy

The complete form of the approximation including boundary conditions is given in [2], but we limit ourselves here to the periodic case. The compact Padé type approximation have the form $Pf'_j = Qf_j$ for derivatives of first order, and $Rf''_j = Sf_j$ for derivatives of second order. Because of the staggered grid, the first order operators are different depending on what variable they are applied to. We use the following operators:

$$P_1 f'_{j+1/2} = \frac{1}{24}(f'_{j-1/2} + 22f'_{j+1/2} + f'_{j+3/2}),$$

$$Q_1 f_{j+1/2} = \frac{1}{\Delta x}(f_{j+1} - f_j),$$

$$P_2 f'_j = \frac{1}{24}(f'_{j-1} + 22f'_j + f'_{j+1}), \quad Q_2 f_{j+1/2} = \frac{1}{\Delta x}(f_{j+1/2} - f_{j-1/2}),$$

$$\tilde{P} f'_j = \frac{1}{6}(f'_{j-1} + 4f'_j + f'_{j+1}), \quad \tilde{Q} f_j = \frac{1}{2\Delta x}(f_{j+1} - f_{j-1}),$$

$$R f''_j = \frac{1}{12}(f''_{j-1} + 10f''_j + f''_{j-1}), \quad S f_j = \frac{1}{\Delta x^2}(f_{j-1} - 2f_j + f_{j+1}).$$

Note that the right hand side operators $Q$ and $S$ approximate the first and second derivatives respectively to second order accuracy, and they are identical to those used in the previous subsection.

The operators defined above are one-dimensional, we use the notation $P_{1x}$ etc to indicate the coordinate direction. By using a fourth order accurate averaging operator $E$ where it is necessary, the complete scheme is

$$\frac{3}{2}u^{n+1} + \Delta t\left(P_{1x}^{-1}Q_{1x}p^{n+1} - \nu(R_x^{-1}S_x + R_y^{-1}S_y)u^{n+1}\right) =$$
$$2u^n - \frac{1}{2}u^{n-1} - 2\Delta t(u^n\tilde{P}_x^{-1}\tilde{Q}_xu^n + (Ev^n)\tilde{P}_y^{-1}\tilde{Q}_yu^n) +$$
$$\Delta t(u^{n-1}\tilde{P}_x^{-1}\tilde{Q}_xu^{n-1} + (Ev^{n-1})\tilde{P}_y^{-1}\tilde{Q}_yu^{n-1}),$$

$$\frac{3}{2}v^{n+1} + \Delta t\left(P_{1y}^{-1}Q_{1y}p^{n+1} - \nu(R_x^{-1}S_x + R_y^{-1}S_y)v^{n+1}\right) = \qquad (2.5)$$
$$2v^n - \frac{1}{2}v^{n-1} - 2\Delta t((Eu^n)\tilde{P}_x^{-1}\tilde{Q}_xv^n + v^n\tilde{P}_y^{-1}\tilde{Q}_yv^n) +$$
$$\Delta t((Eu^{n-1})\tilde{P}_x^{-1}\tilde{Q}_xv^{n-1} + v^{n-1}\tilde{P}_y^{-1}\tilde{Q}_yv^{n-1}),$$

$$P_{2x}^{-1}Q_{2x}u^{n+1} + P_{2y}^{-1}Q_{2y}v^{n+1} = 0.$$

For the stability analysis, we Fourier transform the equations. The second order scheme treated in the previous subsection can be written in the form (2.5), but with all the $P$- and $R$-operators being substituted by the identity operator $I$. Hence, we use the system obtained for the second order case, and divide by the Fourier transform $\hat{P}$ and $\hat{R}$ of $P$ and $R$ respectively at the proper places. We obtain for $j = 1, 2$

$$\frac{3}{2}\hat{u}^{n+1} - 2\hat{u}^n + \frac{1}{2}\hat{u}^{n-1} + \frac{3a\lambda_1 i\sin\xi_1}{2+\cos\xi_1}(2\hat{u}^n - \hat{u}^{n-1})$$
$$+\frac{3b\lambda_2 i\sin\xi_2}{2+\cos\xi_2}(2\hat{u}^n - \hat{u}^{n-1}) + \frac{24\lambda_1 i\sin(\xi_1/2)}{11+\cos\xi_1}\hat{p}^{n+1}$$
$$= -4\nu\left(\frac{6\Delta t}{\Delta x^2}\frac{\sin^2(\xi_1/2)}{5+\cos\xi_1} + \frac{6\Delta t}{\Delta y^2}\frac{\sin^2(\xi_2/2)}{5+\cos\xi_2}\right)\hat{u}^{n+1},$$

$$\frac{3}{2}\hat{v}^{n+1} - 2\hat{v}^n + \frac{1}{2}\hat{v}^{n-1} + \frac{3a\lambda_1 i\sin\xi_1}{2+\cos\xi_1}(2\hat{v}^n - \hat{v}^{n-1})$$
$$+\frac{3b\lambda_2 i\sin\xi_2}{2+\cos\xi_2}(2\hat{v}^n - \hat{v}^{n-1}) + \frac{24\lambda_2 i\sin(\xi_2/2)}{11+\cos\xi_2}\hat{p}^{n+1}$$
$$= -4\nu\left(\frac{6\Delta t}{\Delta x^2}\frac{\sin^2(\xi_1/2)}{5+\cos\xi_1} + \frac{6\Delta t}{\Delta y^2}\frac{\sin^2(\xi_2/2)}{5+\cos\xi_2}\right)\hat{v}^{n+1},$$

$$\frac{12i\sin(\xi_1/2)}{\Delta x(11+\cos\xi_1)}\hat{u}^{n+1} + \frac{12i\sin(\xi_2/2)}{\Delta y(11+\cos\xi_2)}\hat{v}^{n+1} = 0.$$

By redefining the parameters $\alpha, \beta, \gamma$ in (2.3) as

$$\alpha = -2 + \frac{12a\lambda_1 is_1 c_1}{1+2c_1^2} + \frac{12b\lambda_2 is_2 c_2}{1+2c_2^2}, \quad \beta = \frac{1}{2} - \frac{6a\lambda_1 is_1 c_1}{1+2c_1^2} - \frac{6b\lambda_2 is_2 c_2}{1+2c_2^2},$$
$$\gamma = \frac{3}{2} + 4\nu\left(\frac{3\Delta t s_1^2}{\Delta x^2(2+c_1^2)} + \frac{3\Delta t s_2^2}{\Delta y^2(2+c_2^2)}\right), \qquad (2.6)$$

we get

$$\begin{bmatrix} \gamma & \frac{12i\lambda_1 s_1}{5+c_1^2} \\ \gamma\frac{s_1}{s_2} & -\frac{2i\lambda_2^2 s_2(5+c_2^2)}{\lambda_1(5+c_1^2)} \end{bmatrix} \hat{\mathbf{U}}^{n+1} + \begin{bmatrix} \alpha & 0 \\ \alpha\frac{s_1}{s_2} & 0 \end{bmatrix} \hat{\mathbf{U}}^{n} + \begin{bmatrix} \beta & 0 \\ \beta\frac{s_1}{s_2} & 0 \end{bmatrix} \hat{\mathbf{U}}^{n-1} = 0 \,.$$

$$(2.7)$$

Looking for the amplification factor $z$, we find that it is given by the same equation (2.4) as for the second order case

$$\gamma z^2 + \alpha z + \beta = 0 \,, \qquad (2.8)$$

but now with the parameters given by (2.6). Assuming that $\Delta x \sim h$, $\Delta y \sim h$, the error due to the time discretization is of $O(\Delta t^2)$ and the spatial error is of $O(h^4)$. For a balance between these two we take $\Delta t \sim h^2$. Hence, $\theta = \nu \Delta t / h^2 \sim \nu$, which is small for high Reynolds numbers.

We have again calculated the stability regions numerically from (2.8) for different $\theta$ as in Fig. 2.2, and the result is shown in Fig. 2.3.

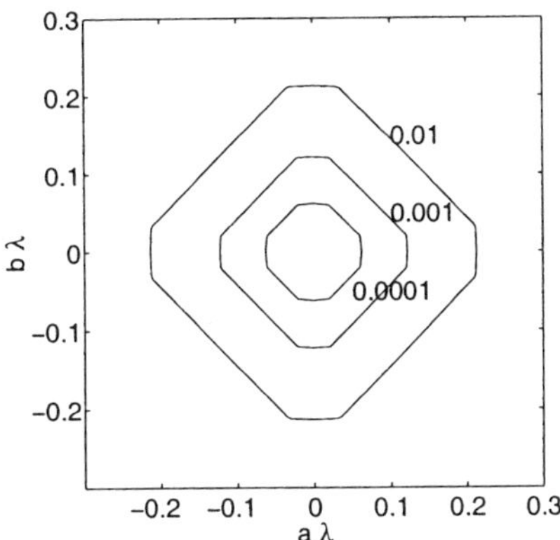


Figure 2.3: The stability regions of the higher order method for $\theta = 10^{-4}, 10^{-3}, 10^{-2}$.

The method is stable for $a\lambda$ and $b\lambda$ inside the contours. We find that the stability region shrinks as $\theta$ decreases, i.e. for smaller $\nu$. Better resolution and smaller $h$ are required for smaller $\nu$. If $h \sim \nu$, then $a\lambda$ and $b\lambda$ will stay inside the stability region when $\nu$ decreases, since $\lambda = \Delta t / h \sim h$.

Similar to the second order case, we can show that there are no parasitic odd-even oscillatory solutions for the fourth order approximation either. For convenience, we again consider the steady state case, and in Fourier space we look for nontrivial solutions of the system

$$\begin{bmatrix} \eta & 0 & \frac{24}{11+\cos\xi_1}\lambda_1 i \sin\frac{\xi_1}{2} \\ 0 & \eta & \frac{24}{11+\cos\xi_2}\lambda_2 i \sin\frac{\xi_2}{2} \\ \frac{12}{11+\cos\xi_1}\lambda_1 i \sin\frac{\xi_1}{2} & \frac{12}{11+\cos\xi_2}\lambda_2 i \sin\frac{\xi_2}{2} & 0 \end{bmatrix} \begin{bmatrix} \hat{u} \\ \hat{v} \\ \hat{p} \end{bmatrix} = 0,$$

$$\eta = \frac{3a\lambda_1 i \sin\xi_1}{2+\cos\xi_1} + \frac{3b\lambda_2 i \sin\xi_2}{2+\cos\xi_2} + 4\nu\left(\frac{6\lambda_1 \sin^2\frac{\xi_1}{2}}{(5+\cos\xi_1)\Delta x} + \frac{6\lambda_2 \sin^2\frac{\xi_2}{2}}{(5+\cos\xi_2)\Delta y}\right).$$

With $\xi_1 = \xi_2 = \pi$ corresponding to the odd-even parasitic solution and

$$\eta_0 = 6\nu\left(\frac{\lambda_1}{\Delta x} + \frac{\lambda_2}{\Delta y}\right),$$

we get

$$Det \begin{bmatrix} \eta_0 & 0 & \frac{12}{5}\lambda_1 i \\ 0 & \eta_0 & \frac{12}{5}\lambda_2 i \\ \frac{6}{5}\lambda_1 i & \frac{6}{5}\lambda_2 i & 0 \end{bmatrix} = \frac{72}{25}\eta_0(\lambda_1^2 + \lambda_2^2) \neq 0,$$

showing that no non-trivial solution exists.

## 3 The iterative solver

At each timestep, we have to solve a system of equations of the type

$$MU = g, \tag{3.1}$$

cf. (1.1), where

$$M = \begin{bmatrix} A & G \\ D & 0 \end{bmatrix}, \quad U = U^{n+1} = \begin{bmatrix} v^{n+1} \\ p^{n+1} \end{bmatrix}, \quad g = g(U^n, U^{n-1}) = \begin{bmatrix} g_1 \\ g_2 \end{bmatrix}.$$

Here we have eliminated the boundary values, such that $U$ represents the velocity and the pressure at inner points, and $g$ contains not only the values of $U$ at the previous time levels, but also the boundary data att the new time level. The submatrix $A$ has the form

$$A = \frac{3}{2}I - \nu\Delta tL, \tag{3.2}$$

where $L$ represents the discrete Laplacian. It follows from the previous section that $\nu\Delta tL$ is of $O(\nu)$.

For solving the system (3.1), we use an approximate factorization constructed in [10]

$$\tilde{M} = \begin{bmatrix} A & 0 \\ D & \frac{2}{3}DG \end{bmatrix} \begin{bmatrix} I & \frac{2}{3}G \\ 0 & -I \end{bmatrix} = \begin{bmatrix} A & \frac{2}{3}AG \\ D & 0 \end{bmatrix}. \tag{3.3}$$

The difference between the original matrix and its approximation is

$$\tilde{M} - M = \begin{pmatrix} 0 & -\frac{2}{3}\nu\Delta t LG \\ 0 & 0 \end{pmatrix},$$ (3.4)

which is small for small $\nu$.

In the outer iteration we solve for the correction $\delta\mathbf{U}$:

$$\begin{aligned} &1.\ \mathbf{r}^0 = \mathbf{g} - M\mathbf{U}^0,\ \ k = 0, \\ &2.\ \tilde{M}\delta\mathbf{U}^k = \mathbf{r}^k, \\ &3.\ \mathbf{U}^{k+1} = \mathbf{U}^k + \delta\mathbf{U}^k,\ \ \mathbf{r}^{k+1} = \mathbf{g} - M\mathbf{U}^{k+1}, \\ &4.\ k+1 \to k,\ \ \text{goto } 2. \end{aligned}$$ (3.5)

One or two outer iterations usually suffice. In step 2, the algorithm computes the corrections using the factorization (3.3) and forward and back substitution:

$$\begin{aligned} &1.\ A\mathbf{y}_1 = \mathbf{g}_1, \\ &2.\ \tfrac{2}{3}DG y_2 = g_2 - D\mathbf{y}_1, \\ &3.\ \delta p = -y_2, \\ &4.\ \delta\mathbf{u} = \mathbf{y}_1 - \tfrac{2}{3}G y_2. \end{aligned}$$ (3.6)

Two systems of equations have to be solved in (3.6):

$$A\mathbf{y}_1 = \mathbf{g}_1, \tag{3.7}$$

$$DG y_2 = \frac{3}{2}(g_2 - D\mathbf{y}_1). \tag{3.8}$$

The first equation (3.7) is solved by fixed point iteration and a preconditioner $B$:

$$\begin{aligned} &1.\ \mathbf{r}^0 = \mathbf{g}_1 - A\mathbf{y}_1^0,\ \ k = 0, \\ &2.\ \delta\mathbf{y}_1^k = B\mathbf{r}^k, \\ &3.\ \mathbf{y}_1^{k+1} = \mathbf{y}_1^k + \delta\mathbf{y}_1^k,\ \ \mathbf{r}^{k+1} = \mathbf{g}_1 - A\mathbf{y}_1^{k+1}, \\ &4.\ k+1 \to k,\ \ \text{goto } 2. \end{aligned}$$ (3.9)

Recalling that $A$ defined in (3.2) has the property $A = \frac{3}{2}I + \mathcal{O}(\nu)$, we construct $B$ as an approximate inverse of $A$ as either

$$B_1 = \frac{2}{3}I \text{ or } B_2 = \frac{2}{3}I + \frac{4}{9}\nu\Delta t L. \tag{3.10}$$

The iteration matrix is $I - BA$, and the convergence rate is determined by the spectral radius $\rho(I - BA)$.

The second equation (3.8) resembles the Poisson equation. However, the matrix $DG$ is not symmetric due to the boundary conditions, and furthermore, the elements should not be computed explicitly, since we then would end up with dense matrices. Therefore, we solve the equation using the Bi-CGSTAB algorithm [5]. This method is suitable for nonsymmetric matrices with eigenvalues away from the imaginary axis, it does not need the transpose of DG, and the equivalent of only about five solution vectors is required as workspace. The Bi-CGSTAB algorithm only needs the vector $DGy$ for an arbitrary $y$ in its inner iterations and not $DG$ explicitly.

However, the convergence of this method when applied to (3.8) is slow without preconditioning. Incomplete LU factorization (ILU) [5] is often efficient in improving the convergence rate for elliptic equations, but we cannot apply it directly to $DG$ since its elements are not explicitly available. The second order accurate divergence and gradient discretizations $D_2G_2$ mentioned in subsection 2.1 can easily be computed, and the ILU factors $L$ and $U$ are based on this second order stencil, satisfying

$$LU \approx D_2G_2 \approx DG. \tag{3.11}$$

Now (3.8) is solved using Bi-CGSTAB with the ILU preconditioning (3.11). This is the most time consuming part of the iterative solver.

## 4  Numerical results

| | $N_x = N_y = 20$ | | $N_x = N_y = 40$ | |
|---|---|---|---|---|
| **Re** | $\rho(I - B_1A)$ | $\rho(I - B_2A)$ | $\rho(I - B_1A)$ | $\rho(I - B_2A)$ |
| 1 | 0.44 | 0.19 | 1.9 | 3.5 |
| 10 | 0.044 | 0.0019 | 0.19 | $4.3 \cdot 10^{-7}$ |
| 100 | 0.0044 | $1.9 \cdot 10^{-5}$ | 0.019 | $4.9 \cdot 10^{-9}$ |
| 1000 | 0.00044 | $1.9 \cdot 10^{-7}$ | 0.0019 | $3.1 \cdot 10^{-9}$ |
| 2500 | 0.00018 | $3.1 \cdot 10^{-8}$ | 0.00074 | $5.9 \cdot 10^{-10}$ |

Table 4.1: Spectral radius of the iteration matrix

In the numerical experiments the iteration (3.5) is applied to the solution of the Navier-Stokes equations in a straight channel with Orr-Sommerfeld inflow and outflow boundary conditions. (The full flow computation in a curvilinear grid is presented in [1]). The right hand side $\mathbf{g}$ in (3.1) is determined by two Orr-Sommerfeld solutions at $t_n$ and $t_{n-1}$ and the boundary conditions. The viscosity is $\nu = 1/250$ and the time step is $\Delta t = 10^{-3}$. Two grids are used having $20 \times 20$ and $40 \times 40$ inner grid points. The algorithm is implemented in Matlab using their Bi-CGSTAB routine and ILU factorization in these experiments. The ILU algorithm has a threshold parameter $\varepsilon_{ILU}$ which determines the size of the elements to include in the factorization (3.11). The iterations are interrupted when the relative residual is sufficiently small. The iteration (3.9) converges very quickly in at most 3 iterations so that $\|\mathbf{r}^k\|/\|\mathbf{g}_1\| < 10^{-6}$. This is explained by the size of $\rho(I - BA)$, which is shown for $B_j$, $j = 1, 2$ in (3.10) and different Reynolds numbers $\nu^{-1}$ and grid sizes in Table 4.1.

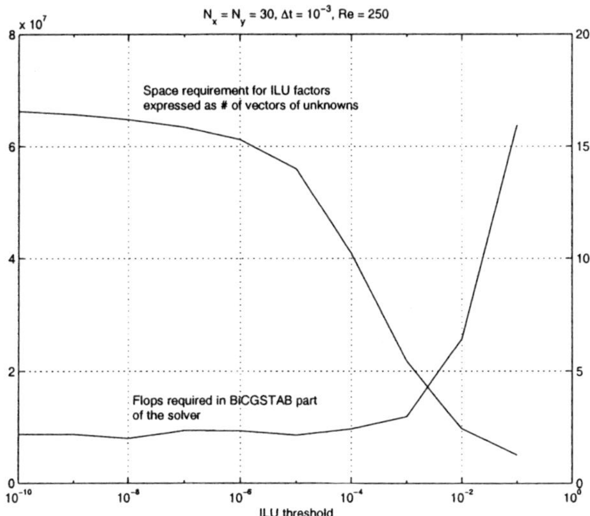

Figure 4.1: Flops and storage requirements for solution of (3.8) by Bi-CGSTAB and 2nd order ILU preconditioner.

The solution of (3.8) dominates the cost of each outer iteration in (3.5). The smaller $\varepsilon_{ILU}$ is in the preconditioner, the more accurate the factorization of the matrix is, and the faster the convergence is measured in number of iterations. On the other hand, a smaller $\varepsilon_{ILU}$ requires larger storage. This is illustrated in Fig. 4.1. The number of flops for convergence is on the left axis and the storage measured in number of solution vectors is on the right axis. The abscissa is the $\varepsilon_{ILU}$ threshold. We find in Fig. 4.1 that $\varepsilon_{ILU} = 10^{-3}$ is a good compromise.

The number of iterations and the number of flops for convergence so that the relative residual is below $10^{-6}$, and the memory requirements are shown in Table 4.2 for the problem of size $40 \times 40$ for different $\varepsilon_{ILU}$, and the factorization based on the true $DG$ (for comparison computed explicitly) and based on $D_2 G_2$ (which is easily determined). The ILU factorization using $D_2 G_2$ not only requires less memory but it is also more efficient.

In Fig. 4.2 the sensitivity of the iterations to the time step $\Delta t$ is depicted. The work in step 2 of (3.6) is about 10 times larger than in step 1 for all time steps. We also see a growth of the number of flops as $\Delta t$ grows as we can expect from an analysis of the iterations.

Finally, the algorithm (3.5) is compared to the restarted GMRES algorithm with 25 as restart length [5] applied directly to (3.1) without preconditioning

| ILU threshold | Iterations | | Flops ($\cdot 10^6$) | | Storage | |
|---|---|---|---|---|---|---|
| | 4th | 2nd | 4th | 2nd | 4th | 2nd |
| $10^{-1}$ | - | - | - | - | 1.4 | 1.3 |
| $10^{-3}$ | 6.5 | 10.0 | 3.2 | 3.1 | 9.3 | 5.7 |
| $10^{-5}$ | 2.5 | 6.5 | 4.0 | 3.4 | 39 | 18 |
| $10^{-7}$ | 1.0 | 6.0 | 5.1 | 3.5 | 73 | 21 |
| $10^{-9}$ | 1.0 | 6.0 | 5.7 | 3.5 | 104 | 22 |

Table 4.2: Comparison of 4th and 2nd order ILU factorizations. Failure to converge is indicated with - in the iterations column. Flops denotes total number of flops in the Bi-CGSTAB solver, and storage is measured with a solution vector $\mathbf{U}$ as the unit.

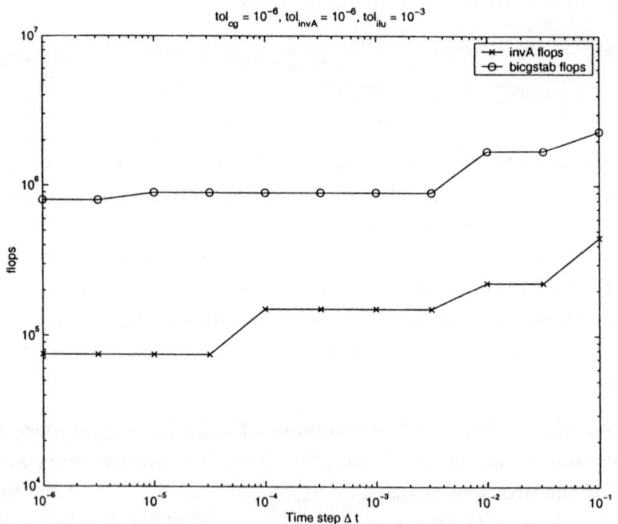


Figure 4.2: Flops requirement for (3.7) and (3.8) as $\Delta t$ is varied.

and to the solution of the Poisson-like pressure equation (see [3])

$$DA^{-1}Gp = g_2 - DA^{-1}\mathbf{g}_1. \tag{4.1}$$

The pressure $p$ in (4.1) is calculated by Bi-CGSTAB preconditioned by ILU factorization. Then $\mathbf{v}$ follows from the first equation in (3.1). The difference between solving (4.1) directly by Bi-CGSTAB and (3.8) by Bi-CGSTAB is that in each inner iteration on (4.1) we have to invert $A$ by using (3.9). In (3.5) we invoke (3.9) only once in each outer iteration. In Table 4.3 we show the number of iterations, number of flops, and storage requirements. Our algorithm is the winner in this test.

|  | (3.5) | (4.1) | GMRES(25) |
|---|---|---|---|
| Iterations | 1/1/4 | 10 | 25 |
| Residual | $2.10 \cdot 10^{-8}$ | $1.10 \cdot 10^{-7}$ | $1.15 \cdot 10^{-6}$ |
| Flops | $2.45 \cdot 10^{7}$ | $1.02 \cdot 10^{8}$ | $1.53 \cdot 10^{9}$ |
| Storage | 6.15 | 6.15 | 25 |
| $A^{-1}$ calls | 1 | 43 | - |
| $DG$ calls | 21 | - | - |
| $DA^{-1}G$ calls | - | 41 | - |

Table 4.3: Comparison of (3.5), (4.1), and GMRES applied to a $40 \times 40$ problem. Storage is measured in number of vectors of unknowns. Iterations for (3.5) are presented as {outer/A/DG}.

More examples and evaluations of variations of parameters are found in [4].

## Acknowledgment

The work presented here is part of a larger project for direct simulation of turbulence on curvilinear grids. Other participants in this project are Arnim Brueger, Dan Henningson, Arne Johansson, Wendy Kress and Jonas Nilsson. The project is supported by Swedish Research Council for Engineering Science, Proj. No 281-98-522.

## References

[1] A. BRUEGER, J. NILSSON, A compact higher order finite difference method for the incompressible Navier-Stokes equations, manuscript (2001).

[2] B. GUSTAFSSON, J. NILSSON, Fourth order methods for the Stokes and Navier-Stokes equations on staggered grids, in *Frontiers of Computational Fluid Dynamics-2000*, eds. D. A. Caughey, M. M. Hafez, World Scientific Publishing (2001).

[3] J. CAHOUET, J.-P. CHABARD, *Some fast 3D finite element solvers for the generalized Stokes problem*, Int. J. Numer. Meth. Fluids, 8 (1988), p. 869–895.

[4] A. GÖRAN, *Preconditioned iterative solution of the incompressible Navier-Stokes equations*, MSc thesis, Dept of Scientific Computing, Uppsala University, Uppsala, Sweden, 2001.

[5] A. GREENBAUM, *Iterative methods for solving linear systems*, SIAM, Philadelphia, 1997.

[6] W. D. HENSHAW, *A fourth-order accurate method for the incompressible Navier-Stokes equations on overlapping grids*, J. Comput. Phys., 113 (1994), p. 13–25.

[7] M. LI, T. TANG, B. FORNBERG, *A compact fourth-order finite difference scheme for the steady incompressible Navier-Stokes equations*, Int. J. Numer. Meth. Fluids, 20 (1995), p. 1137–1151.

[8] A. LUNDBLADH, D. S. HENNINGSON, A. V. JOHANSSON, *An efficient spectral integration method for the solution of the time-dependent Navier-Stokes equations*, Report FFA-TN 1992-28, Aeronautical Research Institute of Sweden, Bromma, Sweden, 1992.

[9] J. NILSSON, *Initial-boundary-value problems for the Stokes and Navier-Stokes equations on staggered grids*, PhD thesis, Dept of Scientific Computing, Uppsala University, Uppsala, Sweden, 2000.

[10] J. B. PEROT, *An analysis of the fractional step method*, J. Comput. Phys., 108 (1993), p. 51–58.

[11] J. C. STRIKWERDA, *High-order-accurate schemes for incompressible viscous flow*, Int. J. Numer. Meth. Fluids, 24 (1997), p. 715–734.

# Solving the Navier-Stokes Equations using B-spline Numerical Methods

Olivier Botella[1,2] and Karim Shariff[3]

[1] Center for Turbulence Research, Stanford University, Stanford CA 94305, USA
[2] LEMTA, 2 Avenue de la forêt de Haye, B.P. 160, 54504 Vandœuvre, France
[3] NASA-Ames Research Center, Moffett Field CA 94035, USA

**Abstract.** B-splines are basis functions for piecewise polynomials having a high level of derivative continuity. They possess attractive properties for complex flow simulations: they have compact support, provide a straightforward handling of boundary conditions and grid nonuniformities, yield numerical schemes with high resolving power, and the order of accuracy is a mere input parameter. This paper discusses some basic approximation properties of B-spline schemes, and their relationship with more conventional numerical schemes. Then, we present a spline collocation method for solving the incompressible Navier-Stokes equations in velocity-pressure formulation by means of the fractional step method.

## 1   Introduction

In simple geometries, direct numerical simulation (DNS) and large eddy simulation (LES) has reached a sufficient level of maturity for conducting genuine "numerical experiments" of flows in the transitional and turbulent regime. They have proved to be invaluable tools for the study of these physical phenomena. The success of such investigations depends greatly on highly accurate numerical methods that display the following features : high-order asymptotic accuracy, kinetic energy preservation in the inviscid limit, and "spectral-like" resolving power for representing a broader range of scales for a given grid size. Other key factors in the success of such simulations has been the simplicity of the computational domain, and the cartesian structure of the grid which allow the use of low-cost algorithms suitable for high performance computing.

As flow geometries of interest become more complex, numerical methods used in DNS and LES have to evolve. As observed in [12], turbulence simulations in complex geometries have not yet reached a high level of fidelity. These simulations demand more algorithmic development, since they can seldom be conducted with numerical methods on tensor product cartesian grids.

Thus there is a need to formulate accurate numerical schemes that are flexible with respect to the computational domain while preserving, as much as possible, the resolution properties that were instrumental in the success of simple geometry computations.

Basic splines, or B-splines for short, form a basis for one-dimensional piecewise polynomial functions, also termed spline functions. They were first introduced in [6] for smoothing of statistical data. Swartz & Wendroff [22] were among the first investigators to reveal their favorable properties for solving partial differential equations. B-spline approximation of differential problems is usually performed by means of a weighted residual method (e.g. [17]), of which the Galerkin method and collocation method are particular cases.

A B-spline method for incompressible flow problems was first developed in the mid 70's [19]. In the last decade, B-splines have been used in the nonhomogeneous direction for the simulation of turbulent shear flows (e.g. [13]). In this context, their attractiveness over the more traditional Chebyshev method is grid flexiblity, ease of applying boundary conditions and compactness, while their high resolving power allows "spectral-like" accuracy. A survey of resolution properties of B-spline schemes can be found in [15]. Other fluid dynamics applications are reviewed in [4] : most of these are restricted to structured cartesian grids.

Recently, a B-spline method has been formulated for the simulation of more complex turbulent flows. Kravchenko *et al.* [14] developed a Galerkin method for LES simulations of flow past a circular cylinder. A key factor in the success of these simulations was the use of spline bases on block-structured grids developed by Shariff & Moser [20]. Unlike standard domain decomposition methods, this technique maintains the high degree of continuity of B-splines at mesh block boundaries. This feature preserves the resolution properties of conventional B-spline schemes, and allows a substantial reduction of the cost of a simulation when a refined grid is used in physically significant flow regions only.

The purpose of this paper is twofold : highlighting the properties of B-spline methods for fluid dynamics problems, and presenting the work of one of the authors on the subject [2,3]. We also refer to [4] for a survey on these topics.

## 2   Background on B-spline Numerical Schemes

Let us list some salient properties of B-spline schemes and then go to discuss their relationship to more conventional numerical methods.

(i) A function $u(x)$ can be represented in a B-spline basis as

$$u(x) = \sum_{i=1}^{N} u_i B_i^k(x), \tag{1}$$

where $B_i^k(x)$ is a B-spline function and $u_i$ are coefficients to be determined. B-splines functions are essentially characterized by their polynomial order $k$ ($k$ = degree + 1) and the level of derivative continuity enforced across cells, or "knots", of the partitioned computational domain. The polynomial order dictates the asymptotic order of accuracy while the level of continuity affects resolving power.

(ii) B-splines have minimal compact support. More precisely, a B-spline function of order $k$ is nonzero over $k$ consecutive grid intervals. This property leads to sparse banded matrix representation of differential operators.

(iii) The availability of efficient algorithms for evaluating B-splines and their derivatives has been a landmark for calculating with B-splines [7]. These algorithms allow the systematic formulation of B-spline schemes of $\mathcal{O}(N^{-k})$ accuracy on nonuniform grids, with $k$ being a mere input parameter.

(iv) B-splines of maximal continuity, also termed smoothest B-splines, yield numerical schemes with "spectral-like" resolving power, allowing accurate approximation of problems with a broad range of scales [22,15]. This property is highly attractive for fluid flows in the transitional and turbulent regimes.

(v) Boundary conditions of various types are easily incorporated into a B-spline basis. Approximation of boundary value problems does not require special treatment at boundaries, as is found necessary for high-order finite difference methods (e.g. [5]).

(vi) A multidimensional approximation on cartesian (or mapped cartesian) grids is obtained by a tensor product of one-dimensional B-splines. Tensor product properties allow the implementation of fast (line by line) algorithms for the following operations: matrix-vector products, transforms between coefficient and data space, and solution of separable elliptic equations (e.g. [4]). These properties are usually associated with finite-difference and spectral methods.

We now turn our attention to the relationship between B-spline and other commonly used schemes.

On cartesian grids, there is a strong relationship between B-spline collocation schemes and compact, or Hermitian, finite difference methods (e.g. [17]). Compact schemes are often termed implicit schemes, since a linear combination of nodal derivatives equates a linear combination of nodal values. Compact schemes are known to have not only higher order accuracy than explicit schemes of the same stencil length, but also to have better resolving power for the same order of accuracy. These boons come at the expense of solving a "mass" matrix problem for evaluating nodal derivatives. A B-spline approximation with the collocation or Galerkin method shares this pecularity since it yields a non-diagonal mass matrix. A non-diagonal mass matrix has a profound impact on the efficiency of time-stepping methods for evolution equations, in particular for the incompressible Navier-Stokes equations in pressure-velocity form. This issue is addressed in the next section for the smoothest B-spline collocation method, where an accurate sparse approximate inverse of the mass matrix developed in [2] allows the formulation of explicit, or semi-consistent, B-spline schemes.

For periodic uniform grids, Kwok *et al.* [15] observed that compact schemes have similar, but somewhat better resolution power than smoothest B-spline collocation methods of same stencil width. However, this particular configuration does not address issues related to the grid nonuniformity and the treatment of boundaries. As stated by properties (iii) and (v), spline methods have a straightforward formulation, whereas a finite difference method requires, for example, special differencing schemes at the boundary, whose stable and accurate formulation is a difficult task [5].

The classical finite element method (e.g. [10]) uses linear or quadratic polynomials mostly, and therefore a low $C^0$ continuity is enforced. Higher continuity as in the Hermite finite element method can be obtained through the introduction of nodal derivatives as additional degrees of freedom. The high interest in the finite element method is due to its ability to handle complex geometries through isoparametric mapping of its elements. Its weakness for turbulence simulation lies in its low approximation order and resolving power.

Higher-order approximations by means of $h/p$ finite elements [23] and spectral elements [12] are typically obtained by increasing the polynomial order within the elements, while maintaining the $C^0$ continuity at element interfaces. An analogous concept for the spline method is to be found in de Boor's collocation method [7] which uses B-splines of arbitrary order $k$, with low degree of continuity at interfaces (e.g. $C^1$ for Poisson's equation), irrespective of the value of $k$. This method has seen considerable development over the last decades under the rubric 'orthogonal spline collocation method' (see [1] for a review).

A totally different concept governs higher-order approximations with smoothest B-splines. The maximal $C^{k-2}$ continuity that is enforced is now dependent on the B-spline order $k$, and increases with the order of the basis. As observed in [3] and [15], the high resolving power of smoothest B-spline schemes is a direct result of this higher continuity. However, these methods have not yet reached the level of flexibility of the isoparametric formulation of the finite element method for arbitrary geometries. Most multidimensional spline applications employ a tensor product representation, on separable coordinates (e.g. polar, cylindrical, and spherical) which admit a global mapping into a rectangle. Spline methods continue to be developed however, and one step towards the alleviation of the tensor product limitation is represented by the Shariff & Moser algorithm [20] mentioned above. Efforts are also underway in the sphere of approximation theory towards constructing spline bases of higher continuity on triangulations (see e.g. [8]) which would be attractive for fluid flow problems in arbitrary geometries.


## 3 Collocation Method for the Navier-Stokes Equations

In this part, we present a spline collocation method for solving the incompressible Navier-Stokes equations in velocity-pressure formulation by means of the fractional step method (e.g. [17]). The first sub-section is devoted to the spline collocation treatment of spurious pressure oscillations. Then, we present a semi-consistent fractional step scheme which, by taking advantage of local spline interpolation, allows cost-effective unsteady computations. Finally, some numerical examples are presented. We refer to [2,3] for a complete description of these techniques.

## 3.1 The Staggered Spline Collocation Method

For ease of discussion, we consider the following first-order scheme, with the nonlinear term discarded:

$$\frac{\overline{v} - v^n}{\Delta t} - \frac{1}{\mathrm{Re}} \nabla^2 \overline{v} + \nabla p^n = 0, \tag{2a}$$

$$\overline{v}_{|\partial\Omega} = 0, \tag{2b}$$

where $v$ and $p$ are respectively the velocity and the pressure, Re is the Reynolds number and,

$$\frac{v^{n+1} - \overline{v}}{\Delta t} + \nabla \left(p^{n+1} - p^n\right) = 0, \tag{3a}$$

$$\nabla \cdot v^{n+1} = 0, \tag{3b}$$

$$v^{n+1}_{|\partial\Omega} = 0. \tag{3c}$$

The prediction step (2a)-(2b) amounts to solving a Helmholtz equation for the provisional velocity $\overline{v}$. To obviate the need of artificial boundary conditions on the pressure, the projection step (3a)-(3c) is left written as a Div-Grad problem instead of casting it as a Poisson equation (e.g. [11]).

The computational domain is $\Omega = ]a, b[^2$ and the spatial discretization is based on tensor product smoothest B-splines. A stable discretization of the velocity and pressure must be achieved in order to verify the *Inf-Sup* condition of the Div-Grad problem. If this condition is not verified, we may observe that the pressure operator associated to this problem is rank deficient, resulting in spurious pressure oscillations (e.g. [10,11]). Indeed, it has been observed in [3] that the identical B-spline representation of velocity and pressure yields four zero eigenvalues, one corresponding to the constant pressure mode, the three others being spurious.

To overcome this difficulty, one has to represent the velocity and pressure with distinct spline bases as

$$v = \sum_{i,l=1}^{N} v_{i,l} B_i^k(x) B_l^k(y), \qquad p = \sum_{i,l=1}^{N-2} p_{i,l} \widetilde{B}_i^{k-1}(x) \widetilde{B}_l^{k-1}(y), \tag{4}$$

The compatible pressure basis $\{\widetilde{B}_i^{k-1}(x), i = 1, \ldots, N - 2\}$ is of order one less than the velocity basis $\{B_i^k(x), i = 1, \ldots, N\}$, with pressure knots, or cells, staggered with respect to those of the velocity. The collocation approximation of (2a)-(3c) is then obtained by discretizing the equations on a single collocation grid $\{(x_j, y_m)\}$, based on the velocity basis. A representation of the staggered bases is shown in Fig.1.

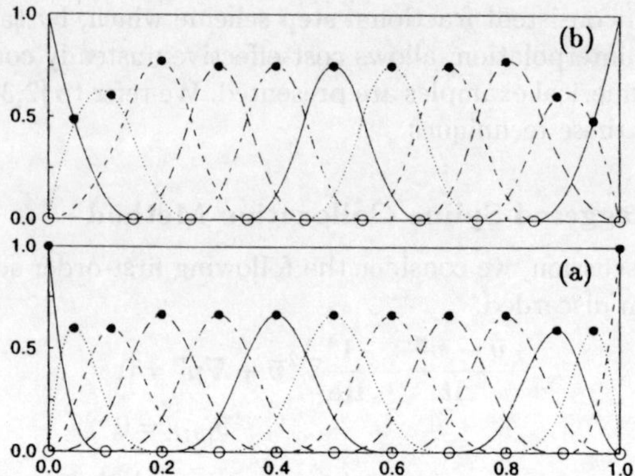

**Fig. 1.** Staggered B-spline bases for (a) the velocity and (b) the pressure on a uniform distribution of knots ($\circ$), with $k = 4$. The collocation points ($\bullet$) are defined here as the maximum of the velocity B-splines

### 3.2 Semi-Consistent Fractional Step Scheme

The second major difficulty is related to the form of the linear systems when using a time-advancement based on the fractional step technique. To obtain a sparse pressure equation, which is mandatory for cost-effective computations, a fractional step scheme with a semi-consistent approximation of the mass matrix has to be developed.

Following the methodology of Gresho & Chan [9] introduced for the finite element method, we consider the spline collocation discretization of Eqs. (2a)-(3c) defined by

$$\mathcal{M}\frac{\overline{U} - U^n}{\Delta t} - \mathcal{K}\overline{U} + \mathcal{M}\Phi^{-1}\widetilde{\mathcal{D}}P^n = 0, \tag{5}$$

and

$$\mathcal{M}\frac{U^{n+1} - \overline{U}}{\Delta t} + \mathcal{M}\Phi^{-1}\widetilde{\mathcal{D}}(P^{n+1} - P^n) = 0, \tag{6a}$$

$$\mathcal{D}U^{n+1} = 0, \tag{6b}$$

where $\Delta t$ is the time step, $U$ and $P$ are vectors representing the unknown spline coefficients of the velocity and the pressure respectively, $\mathcal{M}$ is the collocation mass matrix, $\mathcal{D}$ and $\widetilde{\mathcal{D}}$ are discrete divergence and gradient operators and $\mathcal{K}$ is the discrete viscous operator. It is important to note that, as in the finite element method, the matrix $\mathcal{M}$ is non-diagonal. In the following, $\mathcal{M}$ is referred to as the "consistent" mass matrix.

In addition, we introduces the square matrix $\Phi^{-1}$ such that, when $\Phi^{-1} = \mathcal{M}^{-1}$, Scheme (5)-(6b) corresponds to the discretization of Eqs. (2a)-(3c) by the staggered spline collocation discretization of the next sub-section. This method is subsequently referred to as the consistent method (CM).

When $\Phi^{-1} \neq \mathcal{M}^{-1}$ is sparse, Scheme (5)-(6b) leads to the semi-consistent (SCM) method, which is an alteration of the CM scheme which leads to a full decoupling of the velocity and pressure. Note however that the divergence equation (6b) is identical for both methods, and expresses that the continuity condition be satisfied at the inner collocation points.

The high interest of the SCM scheme is that the projection step (6a),(6b) yields the pressure equation

$$\mathcal{A}(P^{n+1} - P^n) = \mathcal{D}\overline{U}/\Delta t, \tag{7}$$

where the pressure operator $\mathcal{A} = \mathcal{D}\Phi^{-1}\widetilde{\mathcal{D}}$ is sparse since $\Phi^{-1}$ is sparse, resulting in a pressure equation that can be efficiently solved by standard iterative methods for elliptic problems.

In contrast, the pressure operator is dense for the CM method, and Eqs. (6a),(6b) may only be solved by Uzawa iterations (e.g. [18]). As it will be seen in Sect. 3.4, this solution method is prohibitively costly.

The principle of the fractional step scheme (5)-(6b) was introduced in [9] for the finite element method with $\Phi^{-1} = \mathcal{M}_L^{-1}$ which is, in general, a spatially first-order approximation. Higher-order accuracy can be reached with $\Phi^{-1} = \mathcal{M}_A^{-1}$, where $\mathcal{M}_A^{-1}$ are the sparse approximate inverses of the mass matrix described in the next sub-section. It is then possible to recover the level of accuracy of the CM method when the order of $\mathcal{M}_A^{-1}$ is consistent with the order of the B-spline basis [2].

## 3.3 Sparse Approximate Inverse of the Consistent Mass Matrix

For ease of discussion, we describe the construction of the approximate inverse in one space dimension. The two-dimensional approximate inverse is then constructed by using tensor product properties, i.e.

$$\mathcal{M}_A^{-1} = \bar{\mathcal{M}}_A^{-1} \otimes \ddot{\mathcal{M}}_A^{-1},$$

where one-dimensional operators are overlined.

Consider the consistent spline interpolation of the function $f(x)$, which involves finding the coefficients of the spline function (1) that takes on the values of $f(x)$ on the collocation grid $\{x_j\}$, i.e.

$$\sum_{i=1}^{N} u_i B_i^k(x_j) = f(x_j), \quad j = 1, \ldots, N. \tag{8}$$

Thus, the determination of the spline coefficients requires solution of the linear system

$$\bar{\mathcal{M}} U = F, \tag{9}$$

where $\bar{\mathcal{M}}$ is the one-dimensional consistent mass matrix. An *ad hoc* diagonalization of this system, commonly used in the finite element method, consists in summing the rows of the mass matrix and putting the result on the diagonal (the row-sum lumped mass approximation, see e.g. [10]). For the spline collocation method, the row-sum lumped mass matrix is $\bar{\mathcal{M}}_L = \bar{\mathcal{I}}$, i.e. the identity matrix, since the row elements sum up to one thanks to the partition of unity property.

A generalization of this technique, yielding both computational efficiency and high-order accuracy, is obtained by application of local spline interpolation (e.g. [7,16]). In this technique, spline coefficients are not determined by inverting a matrix but rather as the linear combination of the values of the function and/or its derivatives at a small number of data points. Of special interest is the local scheme of Lyche & Schumaker [16] which is based on point evaluations. The $i^{\text{th}}$ spline coefficient is determined as

$$u_i = \sum_{j=1}^{k_l} \beta_{ij}\, f(\tau_{ij}), \tag{10}$$

where $k_l \leq k$ is the order of this scheme, and $\{\tau_{ij},\, j = 1, \ldots, k_l\}$ are distinct data points that we chose among the set of collocation points. The coefficients $\{\beta_{ij}\,;\, i = 1, \ldots, N,\, j = 1, \ldots, k_l\}$ are then determined such that scheme (10) reproduces polynomials of order $k_l \leq k$. This local scheme is $\mathcal{O}(N^{-k_l})$ accurate in the maximum norm [2,16].

Scheme (10) can be written in matrix form as

$$U = \bar{\mathcal{M}}_A^{-1} F, \tag{11}$$

where the coefficients $\{\beta_{ij}\}$ define the entries of $\bar{\mathcal{M}}_A^{-1}$, the approximate inverse of the consistent mass matrix. The matrix $\bar{\mathcal{M}}_A^{-1}$ is sparse, with at most $k_l$ nonzero entries per row. Thus, the linear system solution required by the consistent approximation is now replaced by a sparse matrix-vector multiplication involving the same right-hand-side.

Local interpolation can be viewed as a high-order generalization of the mass-lumping technique: the lowest order approximate inverse is $\bar{\mathcal{M}}_A^{-1} = \bar{\mathcal{M}}_L^{-1} = \bar{\mathcal{I}}$, and the increase of number $k_l$ of data points raises its accuracy until, when $k_l = k$, the order of accuracy of the consistent interpolation is reached.

Note that the local interpolant defined by (10) leaves freedom in the choice of data points. The positioning of these points influences greatly its pointwise accuracy, even though a $O(N^{-k_l})$ rate of convergence is asymptotically reached. The Fourier analysis of [2] shows that better accuracy is found when the data points $\{\tau_{ij},\, j = 1, \ldots, k_l\}$ are positioned as close as possible to $x_i$.

## 3.4  Numerical Examples

As an illustration, we solve the projection step (6a),(6b) with splines of order $k = 4$ on a uniform grid. To preserve the order of accuracy of the CM method,

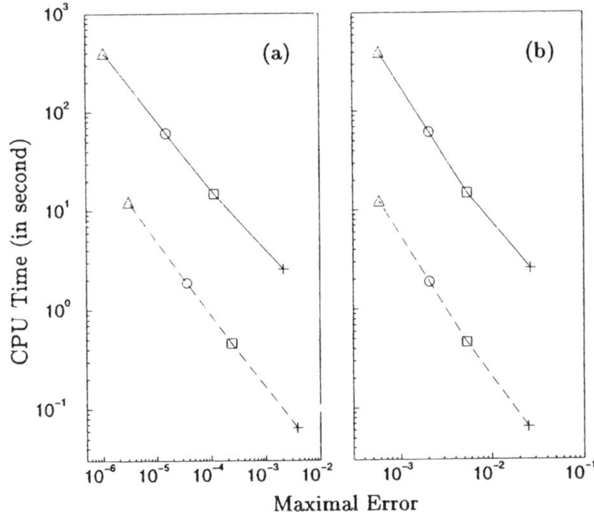

**Fig. 2.** CPU time for the solution of the projection step with respect to the maximal error on (*a*) the velocity and (*b*) the pressure for the CM (———) and SCM (– – – –) approximation. The spatial discretization is defined by $k = 4$, $k_1 = 4$ (for SCM method) and $N = 28$ (+), 53 ($\square$), 83 ($\circ$), 153 ($\triangle$)

namely $O(N^{-k})$ for the velocity and $O(N^{2-k})$ for the pressure [3], the SCM method uses the approximate inverse of order $k_1 = 4$. To give a fair evaluation of the computational efficiency of the methods, the CM and SCM equations are solved by similar iterative techniques. For various values of $N$, Fig. 2 displays the CPU cost of the solution to these systems against the maximal error reached. We observe that the SCM method requires only a fraction of the CPU time of the CM method to reach the same level of accuracy : for this example, the SCM method cuts down the CPU time of the CM method by a factor of 25 for both velocity and pressure.

The benchmark Navier-Stokes tests of Ref. [2] have assessed the high accuracy of the SCM method and its robustness for unsteady computations. The accuracy and the robustness of the SCM solver is best illustrated on the computation of the time-periodic flow in the regularized driven cavity [17], which happens after the first Hopf bifurcation. We considered as reference solution the spectral computation of Ref. [21] at $Re = 12000$, on a $65 \times 65$ Chebyshev grid with $\Delta t = 5 \times 10^{-3}$. For comparison purposes, the SCM computation uses the same discretization parameters. The spatial discretization employs velocity splines of order $k = 6$ in each direction, with an approximate inverse of order $k_1 = 4$, yielding a fourth-order spatially accurate method. Fig. 3 displays the time evolution of the kinetic energy on nearly half a million time-steps. The periodic state is asymptotically reached with the same period $T = 3.085 \pm \Delta t$ measured in [21]. This result shows the ability of the SCM method to conserve kinetic energy on a long time integration, and to reproduce spectral results with a similar coarse spatial resolution.

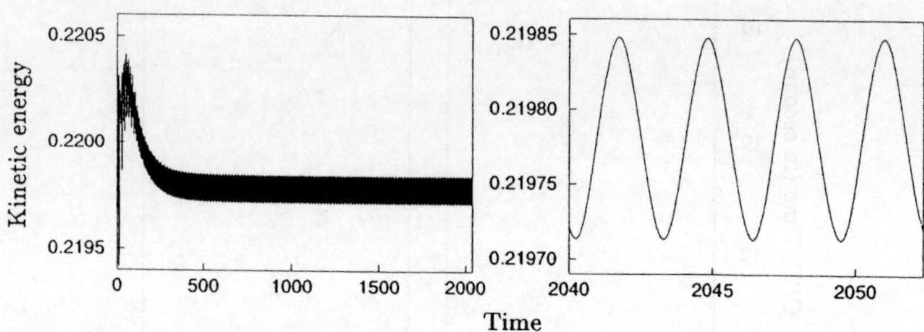

**Fig. 3.** Time evolution of the kinetic energy in the driven cavity at $Re = 12000$ computed by the SCM fractional-step scheme on a $65 \times 65$ grid with $\Delta t = 5 \times 10^{-3}$

## 4  Concluding remarks

The main interest in using B-splines is that they provide high resolution power approaching that of spectral methods and similar to that of compact schemes.

B-spline and compact schemes are implicit methods, i.e. nodal derivatives are obtained as a solution to a linear "mass" system involving nodal values. On cartesian meshes, the use of tensor product properties renders the cost of B-spline schemes similar to that of compact schemes. The presence of a non-diagonal mass matrix may hamper the efficiency of implicit methods when applied to evolution equations or, as presented above, in calculating the pressure in incompressible flow simulations. The use of semi-consistent B-spline schemes allows a substantial reduction of the cost of this approach, and is a decisive step towards accurate and cost-effective spline computations of incompressible turbulent flows.

The success of computations in simple to moderately complex geometries encourages the development of spline methods on arbitrary meshes, that would provide wider flexibily than tensor product B-splines, and for which the SCM Navier-Stokes solver would be the building block. In these respects, the development of multivariate splines of higher continuity and accurate sparse approximate inverses represents key elements into this endeavor.

## References

1. B. Bialecki and G. Fairweather (2001) *J. Comp. Appl. Math.*, **128**, 55-82.

2. O. Botella (2000) *CTR Annual Research Briefs 2000*, Center for Turbulence Research, NASA Ames/Stanford Univ., 149-167.

3. O. Botella (2002) *Computers & Fluids*, **31**, 397-420.

4. O. Botella and K. Shariff. B-spline methods in fluid dynamics. To be published in *Int. J. Comput. Fluid Dynamics*.

5. M. H. Carpenter, D. Gottlieb, and S. Abarbanel (1993) *J. Comp. Phys.*, **108**, 272-295.

6. H. B. Curry and I. J. Schoenberg (1947) *Abstr. Bull. Amer. Math. Soc.*, **53**, 1114.

7. C. de Boor (1978) *A Practical Guide to Splines*. Springer, New York.

8. C. de Boor (1993) *Acta Numerica*, **2**, 65-109.

9. P. M. Gresho and S. T. Chan (1990) *Int. J. Numer. Methods Fluids*, **11**, 621-659.

10. P. M. Gresho and R. L. Sani (1998) *Incompressible Flow and the Finite Element Method*. Wiley, Chichester.

11. J. Guermond (1996) *Modél. Math. Anal. Numér.*, **30**, 637-667.

12. G. E. Karniadakis (1999) *Fluid Dyn. Res.*, **24**, 343-362.

13. A. G. Kravchenko, P. Moin, and R. D. Moser (1996) *J. Comput. Phys.*, **127**, 412-423.

14. A. G. Kravchenko, P. Moin, and K. R. Shariff (1999) *J. Comput. Phys.*, **151**, 757-789.

15. W. Y. Kwok, R. D. Moser, and J. Jiménez (2001) *J. Comp. Phys.*, **174**, 510-551.

16. T. Lyche and L. L. Schumaker (1975) *J. Approx. Theory*, **15**, 294-325.

17. R. Peyret and T. D. Taylor (1983) *Computational Methods for Fluid Flow*. Springer, New York.

18. A. Quarteroni and A. Valli (1994) *Numerical Approximation of Partial Differential Equations*. Springer, Berlin.

19. S. G. Rubin and R. A. Graves (1975) *Comput. Fluids*, **3**, 1-36.

20. K. Shariff and R. D. Moser (1998) *J. Comput. Phys.*, **145**, 471-488.

21. J. Shen (1991) *J. Comput. Phys.*, **95**, 228-245.

22. B. Swartz and B. Wendroff (1974) In *Conference on the Numerical Solution of Differential Equations (1973 : Dundee, Scotland)*, G. A. Watson ed., pp. 153-163, Springer, Berlin.

23. B. Szabó and I. Babuška (1991) *Finite Element Analysis*. Wiley, New York.

# Viscous Flow Simulation Using Compact-Differencing and Filtering Schemes

Miguel R. Visbal

Air Force Research Laboratory
Wright-Patterson Air Force Base OH 45387

**Abstract.** This paper describes the development and application of a high-order finite-difference approach for the simulation of viscous flows on stretched, curvilinear and dynamic grids. The solver utilizes $4^{th}$- and $6^{th}$-order compact-differencing schemes for the spatial discretization, coupled with both explicit and implicit time-marching methods. Up to $10^{th}$-order, Pade-type low-pass spatial filter operators are incorporated to eliminate the spurious high-frequency modes which otherwise arise due to the lack of inherent dissipation in the spatial scheme. Special attention is given to proper metric evaluation procedures for three-dimensional moving and curvilinear meshes so that the advantages of the high-order approach are retained in practical calculations. The improvements derived from the new high-order methodology are highlighted by considering several applications, including: vortex advection, shear-layer instability, Large-Eddy-Simulations of isotropic turbulence and channel flow, as well as the fluid-structure interaction of boundary layer instabilities with an elastic panel.

## 1 Introduction

Despite significant progress in computational sciences, challenges persist in the numerical simulation of turbulent flows where a broad range of spatio-temporal scales must be accurately resolved. In order to reduce the severe computational requirements of standard low-order schemes, higher-order formulations, as well as massively parallel approaches are being actively pursued. Due to their spectral-like resolution and ease of extension to multiple disciplines, high-order compact schemes[10] represent an attractive choice for reducing dispersion, anisotropy and dissipation errors associated with spatial discretizations. Until recently, these schemes have mostly been used, in conjunction with explicit time-integration methods, to address viscous flows on simple Cartesian-like geometries.

Recent work [4,5,26–28] has extended the use of compact algorithms to more practical applications. This has been achieved through the incorporation of enhanced high-order (up to 10th-order) low-pass filtering techniques, accurate and robust near-boundary formulations, consistent metric evaluation procedures, multi-domain implementation strategies, and sub-iterative implicit time-advancement methods. The purpose of this paper is to summarize some of these advancements, and to illustrate – through the application to increasingly complex flow problems – the improved accuracy, robustness and versatility of the compact/filtering approach for viscous flow simulation.

## 2  Governing Equations

In order to develop a procedure suitable for nonlinear fluid dynamic and aeroelastic applications over complex geometries, the full Navier-Stokes equations are selected and are cast in strong conservative form after introducing a time-dependent curvilinear coordinate transformation $(x, y, z, t) \rightarrow (\xi, \eta, \zeta, \tau)$[25]. In vector notation, and in terms of non-dimensional variables, these equations are:

$$\frac{\partial}{\partial \tau}\left(\frac{U}{J}\right) + \frac{\partial \hat{F}}{\partial \xi} + \frac{\partial \hat{G}}{\partial \eta} + \frac{\partial \hat{H}}{\partial \zeta} = \frac{1}{Re}[\frac{\partial \hat{F}_v}{\partial \xi} + \frac{\partial \hat{G}_v}{\partial \eta} + \frac{\partial \hat{H}_v}{\partial \zeta}] \tag{1}$$

Here $U = \{\rho, \rho u, \rho v, \rho w, \rho E\}$ denotes the solution vector and $J = \partial\left(\xi, \eta, \zeta, \tau\right)/\partial\left(x, y, z, t\right)$ is the transformation Jacobian. The inviscid fluxes $\hat{F}, \hat{G}$ and $\hat{H}$ are:

$$\hat{F} = \begin{bmatrix} \rho\hat{U} \\ \rho u\hat{U} + \hat{\xi}_x p \\ \rho v\hat{U} + \hat{\xi}_y p \\ \rho w\hat{U} + \hat{\xi}_z p \\ (\rho E + p)\hat{U} - \hat{\xi}_t p \end{bmatrix} \tag{2}$$

$$\hat{G} = \begin{bmatrix} \rho\hat{V} \\ \rho u\hat{V} + \hat{\eta}_x p \\ \rho v\hat{V} + \hat{\eta}_y p \\ \rho w\hat{V} + \hat{\eta}_z p \\ (\rho E + p)\hat{V} - \hat{\eta}_t p \end{bmatrix} \tag{3}$$

$$\hat{H} = \begin{bmatrix} \rho\hat{W} \\ \rho u\hat{W} + \hat{\zeta}_x p \\ \rho v\hat{W} + \hat{\zeta}_y p \\ \rho w\hat{W} + \hat{\zeta}_z p \\ (\rho E + p)\hat{W} - \hat{\zeta}_t p \end{bmatrix} \tag{4}$$

where

$$\hat{U} = \hat{\xi}_t + \hat{\xi}_x u + \hat{\xi}_y v + \hat{\xi}_z w \tag{5}$$

$$\hat{V} = \hat{\eta}_t + \hat{\eta}_x u + \hat{\eta}_y v + \hat{\eta}_z w \tag{6}$$

$$\hat{W} = \hat{\zeta}_t + \hat{\zeta}_x u + \hat{\zeta}_y v + \hat{\zeta}_z w \tag{7}$$

$$E = \frac{T}{(\gamma - 1)M_r^2} + \frac{1}{2}(u^2 + v^2 + w^2). \tag{8}$$

Here, $\hat{\xi}_x = J^{-1}\partial\xi/\partial x$ with similar definitions for the other metric quantities. The viscous fluxes, $\hat{F}_v, \hat{G}_v$ and $\hat{H}_v$ can be found, for instance, in Ref. [1]. In the expressions above, $u, v, w$ are the Cartesian velocity components, $\rho$ the density, $p$ the pressure, and $T$ the temperature. The perfect gas relationship $p = \rho T/\gamma M_r^2$ is also assumed. All flow variables have been normalized by their respective reference values except for pressure which has been nondimensionalized by $\rho_r u_r^2$.

# 3 Numerical Method

## 3.1 Spatial Discretization

A finite-difference approach is employed to discretize the governing equations, and all discrete quantities are therefore assumed to be pointwise in nature. This choice is motivated by the relative ease of formal extension to higher-order accuracy.

For any scalar quantity, $\phi$, such as a metric, flux component or flow variable, the spatial derivative $\phi'$ is obtained along a coordinate line in the transformed plane by solving the tridiagonal system:

$$\alpha\phi'_{i-1} + \phi'_i + \alpha\phi'_{i+1} = b\frac{\phi_{i+2} - \phi_{i-2}}{4} + a\frac{\phi_{i+1} - \phi_{i-1}}{2} \tag{9}$$

where $\alpha$, $a$ and $b$ determine the spatial properties of the algorithm. This formula yields the compact five-point, sixth-order $C6$, and three-point fourth-order $C4$ schemes with $\alpha = \frac{1}{3}$, $a = \frac{14}{9}$, $b = \frac{1}{9}$ and $\alpha = \frac{1}{4}$, $a = \frac{3}{2}$, $b = 0$ respectively. Equation (9) also incorporates the standard explicit fourth-order $E4$ ($\alpha = 0$, $a = 4/3$ and $b = -1/3$) and second-order $E2$ ($\alpha = 0$, $a = 1$, $b = 0$) schemes. At boundary points $1, 2, IL-1$ and $IL$, higher-order one-sided formulas are utilized which retain the tridiagonal form of the equation set. These are described in more detail in Refs. [4,26].

The derivatives of the inviscid fluxes are obtained by forming the fluxes at the nodes and differentiating each component with the above formulas. Viscous terms are obtained by first computing derivatives of the primitive variables. Subsequently, the components of the viscous flux are constructed at each node and differentiated by a second application of the same scheme. Although this approach may not be as accurate and robust as that in which a Pade-type scheme is employed directly for the second-derivative, it is significantly cheaper to implement in curvilinear coordinates. As demonstrated in Ref. [26], successive differentiation yields an accurate and stable method in conjunction with the added low-pass filter procedure described in Section 3.2.

In Fig. 1, the dispersion-error characteristics for several schemes are shown in the context of the 1-D linear advection equation. Here $w(= 2\pi k\Delta x/L)$ and $w'$ denote the scaled and modified scaled wavenumbers respectively. The $C6$ method exhibits the least dispersion error whereas the $E2$ scheme has significant phase-speed errors over a wide range of wavenumbers. Note that the three-point $C4$ scheme is superior to the five-point $E4$ and even to the seven-point explicit sixth-order scheme ($E6$). Although not shown, the third-order upwind-biased MUSCL scheme [23] exhibits similar dispersion properties as $E4$, but has a dominant dissipation error. Based on the resolving power of the various schemes displayed in Fig. 1, it is expected that the chosen numerical procedure will have a significant impact on LES/DNS applications. For instance, in performing LES employing a dynamic subgrid-scale (SGS) model, a test filter operation is typically performed with a filter width corresponding to $w = \pi/2$ for which significant errors are clearly present in low-order approaches.

## 3.2  Low-Pass Spatial Filter

Compact-difference discretizations, like other centered schemes, are non-dissipative and therefore susceptible to numerical instabilities due to the unrestricted growth of high-frequency modes. These difficulties originate from several sources including mesh non-uniformity, approximate boundary conditions and nonlinear flow features. In LES (where the physical viscous dissipation at the Kolmogorov scale is not represented), the use of a non-dissipative spatial scheme typically leads to the pile-up of energy at the high-wave numbers of the mesh, and ultimately, to numerical instability.

In order to extend the compact discretization approach to practical applications, a high-order low-pass spatial filtering technique [3,26] is incorporated. This low-pass filter is designed to provide dissipation *only* at the high scaled wavenumbers where the spatial discretization already exhibits significant dispersion errors. If a typical component of the solution vector is denoted by $\phi$, filtered values at interior points $\hat{\phi}$ in transformed space satisfy,

$$\alpha_f \hat{\phi}_{i-1} + \hat{\phi}_i + \alpha_f \hat{\phi}_{i+1} = \Sigma_{n=0}^{N} \frac{a_n}{2} \left( \phi_{i+n} + \phi_{i-n} \right) \tag{10}$$

Equation (10) is based on templates proposed in Ref. [10] and with proper choice of coefficients, provides a $2N$th-order formula on a $2N + 1$ point stencil. The $N + 1$ coefficients, $a_0, a_1, \ldots a_N$, are derived in terms of $\alpha_f$ with Taylor- and Fourier-series analyses and are given in Refs. [4,5].

The dissipation characteristcs of the filter operator as function of scaled wavenumber are given by the corresponding spectral function (denoted as $SF$). It can be shown [5] that:

$$SF(w) = \frac{\Sigma_{n=0}^{N} a_n \cos(nw)}{1 + 2\alpha_f \cos(w)} \tag{11}$$

For proper behavior of $SF$, the adjustable parameter $\alpha_f$ must be in the range $-0.5 < \alpha_f < 0.5$, with higher values of $\alpha_f$ corresponding to a less dissipative filter. Also, for $\alpha_f = 0.0$ (spatially) explicit filter formulas [24] are obtained. On uniform meshes, these symmetric filters are non-dispersive (*i.e.* $SF$ is real), do not amplify any waves ($SF \leq 1$), preserve constant functions ($SF(0) = 1.0$), and completely eliminate the odd-even mode ($SF(\pi) = 0.0$).

The spectral response of the second-, sixth- and tenth-order implicit ($\alpha_f = 0.49$) and explicit ($\alpha_f = 0$) filters are shown in Fig. 2. The optimized filters exhibit fairly sharp cutoff characteristics for various orders of accuracy. By contrast, the explicit filters display significant degradation of the spectral response as the order of accuracy is reduced. The dissipation error inherent in a third-order upwind-biased [23] approximation of the first derivative is also shown in Fig. 2 for the purpose of comparison. Unlike the high-order low-pass filter, the dissipation of the upwind-biased scheme does not exhibit a sharp cutoff but instead applies over a wide range of wavenumbers. This is also the case for standard numerical damping approaches (*e.g.* Ref. [15]), unless the damping term is derived based on spectral considerations as described in Ref. [21].

In this work, the filter operator is applied to the conserved variables along each transformed coordinate direction and after each time step. For the near-boundary points, the filtering strategies described in Refs. [5,26] are employed. The impact of filtering on the accuracy and stability of the high-order approach has been investigated in Refs. [5,26,27] for several applications including nonuniform grids, approximate boundary treatments and nonlinear governing equations.

## 3.3 Time Integration

The compact differencing scheme has been coupled with both explicit and implicit time-integration methods. The classical fourth-order four-stage Runge-Kutta scheme ($RK4$), implemented in low-storage form, is utilized primarily for wave propagation problems. For the highly stretched meshes employed in LES of wall-bounded flows, the stability constraint of explicit time-marching methods is too restrictive and the use of an implicit approach becomes necessary. For this purpose, an implicit approximately-factored scheme [2] is incorporated and augmented through the use of Newton-like sub-iterations. In delta form, and for second order temporal accuracy (denoted as $BW2$), the scheme may be written as

$$
\begin{aligned}
&\left[J^{-1^{p+1}} + \phi^i \Delta\tau \delta_\xi^{(2)} \left(\frac{\partial \hat{F}^p}{\partial U} - \frac{1}{Re}\frac{\partial \hat{F}_v^p}{\partial U}\right)\right] J^{p+1} \times \\
&\left[J^{-1^{p+1}} + \phi^i \Delta\tau \delta_\eta^{(2)} \left(\frac{\partial \hat{G}^p}{\partial U} - \frac{1}{Re}\frac{\partial \hat{G}_v^p}{\partial U}\right)\right] J^{p+1} \times \\
&\left[J^{-1^{p+1}} + \phi^i \Delta\tau \delta_\zeta^{(2)} \left(\frac{\partial \hat{H}^p}{\partial U} - \frac{1}{Re}\frac{\partial \hat{H}_v^p}{\partial U}\right)\right] \Delta U \\
&= -\phi^i \Delta\tau \left[J^{-1^{p+1}} \frac{(1+\phi)U^p - (1+2\phi)U^n + \phi U^{n-1}}{\Delta\tau}\right. \\
&\left. + U^p(1/J)_\tau{}^p + \delta_\xi\left(\hat{F}^p - \frac{1}{Re}\hat{F}_v^p\right) + \right. \\
&\left. \delta_\eta\left(\hat{G}^p - \frac{1}{Re}\hat{G}_v^p\right) + \delta_\zeta\left(\hat{H}^p - \frac{1}{Re}\hat{H}_v^p\right)\right]
\end{aligned}
\tag{12}
$$

where $\partial \hat{F}/\partial U$ etc are flux Jacobians, $\delta$ represents the spatial difference operator and $\Delta U = U^{p+1} - U^p$. The method combines the approximate factorization procedure of Ref. [2] with the diagonalized simplification of Ref. [16]. Note that while the derivatives of the flux Jacobians have been obtained to second-order accuracy (denoted with the superscript (2)), those on the right hand side, $i.e.$, in the residual, are evaluated with the compact-difference higher-order method. Nonlinear artificial dissipation terms [15], not explicitly shown in Eq (12), are also appended to the implicit operator to enhance stability. In order to reduce errors associated with these simplifications, a sub-iteration strategy is employed. Thus, for the first sub-iteration, $p = 1$, $U^p = U^n$ and as $p \to \infty$, $U^p \to U^{n+1}$. Typically, three sub-iterations are applied per time step. By changing the number of time levels employed to evaluate the time-derivative term appearing in the RHS of Eq (12), first- ($BW1$) and third-order ($BW3$) accurate forms of the implicit algorithm can be constructed. As demonstrated in Ref. [28], as well as in the examples below, second-order temporal accuracy provides adequate accuracy.

## 3.4 Treatment of Coordinate Transformation Metrics

The extension of high-order schemes to curvilinear and deforming meshes requires a careful evaluation of the spatial and temporal metric expressions arising from the coordinate transformation. Failure to enforce metric cancellation and freestream preservation in the finite-difference discretization of the strong-conservation form of the governing equations can catastrophically degrade the fidelity of higher-order approaches [5,26,27].

In deriving the strong-conservation flow equations, the following metric identities have been implicitly invoked,

$$I_1 = (\hat{\xi}_x)_\xi + (\hat{\eta}_x)_\eta + (\hat{\zeta}_x)_\zeta = 0 \tag{13}$$

$$I_2 = (\hat{\xi}_y)_\xi + (\hat{\eta}_y)_\eta + (\hat{\zeta}_y)_\zeta = 0 \tag{14}$$

$$I_3 = (\hat{\xi}_z)_\xi + (\hat{\eta}_z)_\eta + (\hat{\zeta}_z)_\zeta = 0 \tag{15}$$

$$I_4 = (1/J)_\tau + (\hat{\xi}_t)_\xi + (\hat{\eta}_t)_\eta + (\hat{\zeta}_t)_\zeta = 0 \tag{16}$$

where subscripts denote partial derivatives. The first three identities constitute a differential statement of surface conservation for a closed cell. The last metric identity $(I_4)$ is referred to in the literature as the Geometric Conservation Law $(GCL)$ [22] and becomes important in the case of dynamic meshes. In a finite-difference discretization, these identities must be satisfied numerically in order to ensure freestream preservation.

To numerically enforce identities $I_1$, $I_2$ and $I_3$ (Eqs. (13)-(15)), the transformation metrics are evaluated in the manner described in Refs. [5,27]. This approach adopts the formulation developed in the context of lower-order methods [22], in which the metric relation, for example,

$$\hat{\xi}_x = y_\eta z_\zeta - y_\zeta z_\eta \tag{17}$$

is evaluated by considering its analytically equivalent "conservative" form:

$$\hat{\xi}_x = (y_\eta z)_\zeta - (y_\zeta z)_\eta \tag{18}$$

Similar expressions are employed for the remaining metric terms. As demonstrated in Refs. [5,27], freestream preservation and improved accuracy is achieved on general 3-D meshes when the transformation metrics are cast in the form of Eq (18), and the derivatives are evaluated with the same high-order formulas employed for the fluxes.

In order to satisfy the $GCL$ identity of Eq (16), the time-derivative term in Eq (1) is split using chain-rule differentiation as follows:

$$(U/J)_\tau = (1/J)U_\tau + U(1/J)_\tau \tag{19}$$

The first term involves the inverse Jacobian $J^{-1}$, which is evaluated in a standard fashion using the instantaneous values of the grid coordinates. The second term, which includes the time-derivative of the Jacobian, requires special treatment. Rather than attempting to compute the time-derivative of $J^{-1}$ directly from the

294

grid coordinates at various time levels (either analytically or numerically), we simply invoke Eq (16) to evaluate $(1/J)_\tau$, i.e.

$$(1/J)_\tau = -[(\xi_t/J)_\xi + (\eta_t/J)_\eta + (\zeta_t/J)_\zeta] \tag{20}$$

where

$$\begin{aligned}
\xi_t/J &= -[x_\tau(\xi_x/J) + y_\tau(\xi_y/J) + z_\tau(\xi_z/J)] \\
\eta_t/J &= -[x_\tau(\eta_x/J) + y_\tau(\eta_y/J) + z_\tau(\eta_z/J)] \\
\zeta_t/J &= -[x_\tau(\zeta_x/J) + y_\tau(\zeta_y/J) + z_\tau(\zeta_z/J)]
\end{aligned} \tag{21}$$

As shown in Ref. [28], this strategy ensures freestream preservation and high-fidelity on rapidly deforming meshes for either analytic or numerically-evaluated grid speed terms.

## 4   Results

In order to highlight the capabilities of the previous numerical approach, several examples are now presented addressing the simulation of physical phenomena pertinent to unsteady flows, turbulence and fluid-structure interactions. These computational results have been obtained employing several numerical approaches. The centered spatial schemes (*E2, E4, C4, C6*) have been defined in Section 3.1. The low-pass filters employed (Section 3.2) are designated by appending the filter order to the scheme. For instance, *C4F8* denotes the fourth-order compact scheme combined with an $8^{th}$-order filter. The second-order algorithm (*E2*) incorporates, instead of a low-pass filter, the standard fourth-order scalar dissipation [8,15]. For the purpose of comparison, a third-order *MUSCL*-based upwind-biased scheme [23,18] is also considered.

### 4.1   Vortex Advection on a Curvilinear Mesh

The first case examined consists of the inviscid unsteady flow due to a convecting vortex in an otherwise uniform flow with freestream Mach number $M_\infty = 0.1$. This problem compares the capabilities of the various schemes to accurately advect vortical structures on a smooth but highly non-orthogonal curvilinear mesh (Fig. 3). The initial flow condition is imposed by prescribing a vortex, centered about the location $(x_c, y_c)$, and satisfying the following relations [14]:

$$\begin{aligned}
u &= \frac{-C(y - y_c)}{R^2}\exp(-r^2/2) \\[2mm]
v &= \frac{C(x - x_c)}{R^2}\exp(-r^2/2) \\[2mm]
p_\infty - p &= \frac{\rho C^2}{2R^2}\exp(-r^2) \\[2mm]
r^2 &= \frac{(x - x_c)^2 + (y - y_c)^2}{R^2}
\end{aligned}$$

where $u, v, p$ and $R$ denote the Cartesian velocity components, static pressure and vortex core radius respectively. The nondimensional vortex strength parameter $C/(U_\infty R)$ was chosen to be 0.02. The density was assumed constant, and pressure was obtained by integration of $\partial p/\partial r = \rho u_\theta^2/r$ about the vortex center.

The mesh was constructed analytically according to the expression:

$$x_{i,j} = x_{min} + \Delta x(i-1)$$
$$y_{i,j} = y_{min} + \Delta y_o[(j-1) + Asin(n\pi x/L + j\phi/JL)]$$
$$\Delta x = L/(IL-1), \Delta y_o = H/(JL-1)$$
$$1 \leq i \leq IL, 1 \leq j \leq JL$$

where the amplitude, frequency and phase shift parameters were specified as $A = 1.0$, $n = 8$ and $\phi = 3\pi/2$. Three different levels of spatial resolution were employed (namely, $\Delta x/R = \Delta y_o/R = 0.4, 0.2$ and $0.1$). Some of the grid quality parameters for the medium mesh are: maximum stretching ratio of 1.12, ratio of maximum to minimum cell size of 2.3, and maximum and mean deviation from orthogonality (skewness) of $56.4°$ and $40.1°$ respectively. Therefore it is apparent that the vortex is convected through regions of the mesh where stretching and pronounced skewness are present (see Figure 3).

Grid convergence plots of the maximum error in $v$-velocity component are shown in Figure 4 at a non-dimensional time $tU_\infty/R = 8.0$. The superior behavior of the high-order compact algorithm in terms of absolute error and order of accuracy is clearly apparent. The improved solution obtained with the compact algorithm ($C6F10$) relative to the standard second-order scheme ($E2$) is also shown in Fig. 5 in terms of the $y$-component of velocity along a $j$-constant grid line through the center of the vortex.

Since the present curvilinear grid is generated analytically, the convecting vortex case was also computed with analytically-determined metric values. As shown in Fig. 5 for the compact sixth-order scheme, the use of the exact metric expressions results in significant errors which are of course exaggerated on this highly distorted mesh. The accuracy improvement realized by discrete evaluation of the metrics with the same difference formula as for the other terms in the governing equations, has been discussed in Ref. [3].

## 4.2 Evolution of Small-Amplitude Disturbance in a Shear Layer

This case evaluates the capability of the higher-order scheme to capture the temporal growth of a small-amplitude, normal-mode disturbance in an inviscid shear layer. The parallel base flow consists of a hyperbolic tangent velocity profile, $u = 0.5(1 + tanhy)$ and $v = 0$ (here $u$, $v$ and $y$ are non-dimensional). To permit comparison with incompressible theory, the freestream Mach number is set to a small value ($M_\infty = 0.05$). A detailed inviscid linear stability analysis of the hyperbolic-tangent profile has been given by Michalke [11]. Following

Ref. [11], a disturbance is introduced of the form

$$\hat{u} = \frac{\epsilon \partial \Psi}{\partial y} \qquad \hat{v} = -\frac{\epsilon \partial \Psi}{\partial x}$$
$$\Psi = \text{Re}[\phi(y)e^{i\alpha(x-ct)}]$$
$$\phi(y) = \phi_r + i\phi_i, c = c_r + ic_i$$

where $\Psi$ is the perturbation streamfunction, $\phi$ the eigenfunction, $\alpha$ the wave number, and $\epsilon = 0.001$ the perturbation amplitude. The phase velocity $c_r$ is independent of wavenumber and is equal to 0.5. For the present test case, the most amplified disturbance is considered which corresponds to $\alpha = 0.4446$ and $c_i = 0.2133$ (see Ref. [11]). The eigenfunctions $\phi_r(y)$ and $\phi_i(y)$, required to describe the initial disturbance, are computed by solving Rayleigh stability equation with a simple relaxation procedure. The perturbation flow kinetic energy $E(t)$ is defined as

$$E(t) = \int_0^L \int_{-\infty}^{\infty} (\hat{u}^2 + \hat{v}^2) dy dx$$

where the wavelength $L = 2\pi/\alpha$ corresponds to the streamwise extent of the computational domain, on which periodic conditions are applied. For small perturbation amplitudes, the temporal energy growth predicted by the linear theory is given in normalized form as $E(t)/E(0) = e^{2\alpha c_i t}$.

Calculations for the shear-layer instability were performed on three separate grids but only results corresponding to the coarsest mesh ($33 \times 45$) are included. All cases were computed with the $RK4$ time-marching scheme and a non-dimensional time-step $\Delta t = 0.0005$. Results obtained with several schemes are compared with linear stability theory in Fig. 6 in terms of the normalized amplification rate. The second-order scheme with damping is characterized by unacceptably large energy growth rates. The third-order upwind method results in significant improvement with predicted growth rates that are somewhat lower than the theoretical value, a fact attributable to the dissipative nature of the upwind algorithm. The superior accuracy of the compact fourth- and sixth-order schemes on this crude mesh is evident from the predicted growths rates which are in excellent agreement with theory. For this 2-D flow stability problem, the use of the 4th-order compact algorithm permits a reduction of grid points by more than a factor of four when compared to the upwind method.

## 4.3 Decay of Compressible Isotropic Turbulence

The case of decaying isotropic turbulence has been used previously to investigate compressible formulations of SGS models for large-eddy simulations by Moin et al. [12] and Spyropoulos and Blaisdell [20] among others. Although this test represents a very simple example of turbulent flow, it permits an evaluation of the dissipation characteristics of the numerical schemes in the absence of mean flow inhomogeneity.

The isotropic turbulence simulations correspond to the compressible low-Reynolds number conditions denoted as "Case 6" in Ref. [20]. The initial three-dimensional turbulence spectrum is defined as $E_{3D} \propto k^4 \exp\left[-2(k/k_p)^2\right]$, where $k$ is the magnitude of the wavenumber vector, and $k_p = 4$ is the wavenumber at the peak in the spectrum. Root-mean-square(RMS) levels of the velocity, density, and temperature are established by adjusting the proportionality constant in the spectra. The velocity fluctuations are specified so that the initial turbulent Mach number $M_t = 0.4$, defined as the ratio of the RMS magnitude of the fluctuating velocity to the mean speed of sound. The turbulent Reynolds number is equal to 2157.

The computational domain of size $(2\pi \times 2\pi \times 2\pi)$ in physical space is discretized with a $32^3$ mesh of uniformly distributed grid points. Periodic conditions are enforced at all domain boundaries, and are implemented numerically employing an overlap of five extra points in each coordinate direction in order to facilitate the application of the implicit time-marching procedure. Unless noted otherwise, the solutions are obtained employing the explicit ($RK4$) time-marching algorithm and the sixth-order compact scheme ($C6$) combined with a $10^{th}$-order low-pass filter ($F10, \alpha_f = 0.49$). The time step is specified as $\Delta t = 0.05$ which yields an initial maximum $CFL$ number of approximately 0.8. This $\Delta t$ corresponds to approximately 250 time steps per eddy turnover time $\tau_o$. Unless indicated, results are obtained by solving the unfiltered Navier-Stokes equations without the inclusion of an SGS model. In this situation, dissipation is provided by the high-order low-pass spatial filter which prevents the pile-up of energy at the high wavenumbers represented by the mesh.

**Effect of Spatial Scheme**

The time history of the resolved (volume-averaged) turbulent kinetic energy $K = \left\langle \rho(u'^2 + v'^2 + w'^2) \right\rangle$ is shown in Fig. 7 for several spatial discretizations. Spectral DNS results from Spyropoulos and Blaisdell [20] are also shown for the purpose of comparison. The Pade-type schemes ($C6F10, C4F8$) are observed to be in good agreement with the DNS data. Reducing the fidelity of the spatial discretization just to a fourth-order explicit method ($E4$) combined with and explicit (*i.e.*, $\alpha_f = 0.0$) eighth-order filter produces excessive dissipation of turbulent kinetic energy. This error becomes more pronounced for the second-order and third-order, upwind-biased algorithms.

The origin of the excessive dissipation encountered with the lower-order approximations can be clarified by examining the instantaneous three-dimensional energy spectra, shown in Fig. 8, at $t/\tau_o = 0.2985$. The highest wavenumber represented on this $32^3$ mesh corresponds to $k_c = 16$. It is observed that the lower-fidelity schemes damp the high-wavenumber content of the energy. For the upwind-biased and $E2$ schemes, this occurs for $k > 4.5$ ($w > 0.28\pi$) which corresponds approximately to $PPW < 7$. Since significant energy is still present above this wavenumber, a more rapid decay of turbulent kinetic energy ensues (Fig. 7). The spectra for the compact/filtering schemes display good agreement with the DNS results up to $k \approx 12$ ($w \approx 0.75\pi$). Also, the spectra for the DNS and compact schemes display a limited inertial subrange ($E(k) \propto k^{-5/3}$), not apparent in the low-order spatial discretizations.

An investigation of the effect of the low-pass filter order was also performed in order to provide some guidance in the selection of the filter operator needed to maintain acceptable accuracy. In principle, for proper resolution of the low wavenumbers, the filter order should be equal or greater than the corresponding order of accuracy of the spatial discretization. Computations were performed using the $6^{th}$-order compact scheme in combination with various filter operators ranging from second- to tenth-order. The corresponding decay of turbulent kinetic energy is shown in Fig. 9. The results for *F6*, *F8* and *F10* are in good agreement with each other and with the DNS solution. Excessive dissipation starts to become apparent for *F4*, and is quite significant for *F2* at which point the compact/filtering solution provides no improved accuracy relative to the second-order approach (Fig. 7).

## Effect of Time-Marching Scheme

Although the implicit time-integration method is not needed for the efficient computation of isotropic turbulence decay on a uniform mesh, this test case was selected for the examination of accuracy issues of the implicit sub-iterative approach. For this purpose, computations were performed using both the first- and second-order versions of the implicit algorithm (denoted as *BW1* and *BW2*, respectively). The number of sub-iterations was set equal to two for both implicit schemes, and the time step previously employed with the explicit (*RK4*) solver was retained. The time histories of the turbulent kinetic energy are shown in Fig. 10. Results obtained with the first-order implicit method (*BW1*) are found to be extremely dissipative which renders the scheme unsuitable for LES. On the other hand, the second-order method (*BW2*) produces results which are in good agreement with the previous explicit calculation. Reasonable agreement was also found between the *BW2* and *RK4* results in terms of the energy spectra (not shown) for wavenumbers up to $k \approx 10$. This value of $k$ is very close to the wavenumber beyond which the spatial discretization itself ceases to be sufficiently accurate. Although the isotropic turbulence decay was not computed employing the third-order implicit method (*BW3*), preliminary calculations of benchmark acoustic problems indicate only a marginal gain in accuracy relative to the second-order scheme (accompanied also by a loss of robustness). The second-order sub-iterative approach therefore appears to be a good compromise in terms of accuracy and stability.

## Turbulence Decay on a Deforming Curvilinear Mesh

Since one of the objectives of this work is to develop a high-order computational approach suitable for DNS/LES on curvilinear and dynamic meshes, the decay of isotropic turbulence was also computed on a deforming curvilinear grid. A dynamically deforming mesh, shown in Fig. 11, was generated analytically by varying the $(x, y)$ coordinates of the original Cartesian grid according to the expressions

$$x_{i,j,k}(\tau) = x_{min} + \Delta x_o \left[(i-1) + A_x f \sin(2\pi\omega\tau) \right.$$
$$\left. \sin\frac{n_{xy}\pi(j-1)\Delta y_o}{L_y}\sin\frac{n_{xz}\pi(k-1)\Delta z_o}{L_z}\right]$$
$$y_{i,j,k}(\tau) = y_{min} + \Delta y_o \left[(j-1) + A_y f \sin(2\pi\omega\tau) \right.$$
$$\left. \sin\frac{n_{yx}\pi(i-1)\Delta x_o}{L_x}\sin\frac{n_{yz}\pi(k-1)\Delta z_o}{L_z}\right] \tag{22}$$

$$i = 1\ldots IL;\, j = 1\ldots JL;\, k = 1\ldots KL$$
$$\Delta x_o = \frac{L_x}{IL-1};\, \Delta y_o = \frac{L_y}{JL-1};\, \Delta z_o = \frac{L_z}{KL-1}$$

with the specified parameters $IL = JL = KL = 33$, $A_x = A_y = 1.5$, $L_x = L_y = L_z = 2\pi$ and $n_{xy} = n_{yz} = \ldots = 6$. A blending factor $f$ (defined in terms of the product of three squared-sine functions) is added to limit the deformation to the interior of the domain. The frequency of oscillation was set to $\omega = 0.266$ which gave approximately two cycles of the grid oscillation during the time of the computation. At $\tau = 0$, the grid is undeformed, simplifying the specification of the initial conditions. At the phase of maximum distortion (shown in Fig. 11), significant skewing of the grid is apparent. The grid speeds $(x_\tau, y_\tau)$ were obtained analytically by direct differentiation of Eq (22). The maximum value of the grid speed was approximately $0.2a_\infty$. The spatial and temporal transformation metric expressions were evaluated according to the procedure described in Section 3.4. The time history of the turbulent kinetic energy computed with the $C6F10$ scheme on the dynamic mesh is displayed in Fig. 12. The high-order compact/filtering procedure is shown to retain its fidelity despite the significant imposed dynamic mesh distortions.

## Comparison of Compact/Filtering Approach with Standard SGS models

Finally, we compare in this section the accuracy of the present high-order low-pass filtering approach for the unfiltered Navier-Stokes equations with the standard LES method employing the Smagorinsky (with $C_s = 0.092$) and the dynamic Smagorinsky subgrid-scale models. The time-histories of turbulent kinetic energy obtained with the various SGS approaches are shown in Fig. 13. First, it should be noted that all models were found to be numerically unstable without the inclusion of the low-pass filter operator. Therefore, in all computations, the $10^{th}$-order Pade-type filter was used in conjunction with the baseline $C6$ scheme. The decay of $K$ is seen to be quite similar for the Smagorinsky and dynamic models, and both display excessive dissipation relative to the DNS. It is also apparent that for this case better results are obtained with the compact/filtering approach for the unfiltered equations (*i.e.*, without the inclusion of an SGS model). A partial explanation of this behavior can be offered from the examination of the energy spectra (not shown). Unlike the high-order filter (which acts only at high wavenumber), the standard eddy-viscosity SGS models were found to dissipate energy over a wide range of wavenumbers including the resolved scales. This problem emanates from the inability of the model to effectively discriminate between resolved and unresolved scales and cannot be corrected by simply adjusting the constant in the model.

## 4.4 Turbulent Channel Flow

The next case considered is turbulent channel flow which is commonly employed in LES studies of wall-bounded flows (*e.g.*, Refs. [19]-[13]). The Reynolds number based on channel height, $Re_h = 6600$ ($Re_\tau \approx 180$), corresponds to the incompressible DNS of Kim et al. [9]. Since the present code solves the compressible form of the Navier-Stokes equations, a low Mach number, $M_\infty = 0.1$, is specified.

The streamwise and spanwise dimensions of the computational domain are set to $L_x/h = 2\pi$ and $L_z/h = \pi$ respectively. The mesh of size $61 \times 61 \times 61$ has constant spacing in the $x$ and $z$ directions, and is geometrically stretched in $y$ (with a stretching factor equal to 1.15). In terms of wall units, $\Delta x^+ = 47.1$ , $\Delta y^+_{wall} = 0.45$, $\Delta y^+_{max} = 25.8$ and $\Delta z^+ = 23.4$.

Periodic boundary conditions are applied for all variables in the streamwise and spanwise directions. At the channel walls, the no slip condition is satisfied, along with a constant surface temperature, and a vanishing normal pressure gradient. Due to the periodic streamwise boundary conditions, an artificial source term is introduced [17] in order to provide a driving mechanism which mimics an imposed constant pressure gradient.

Channel flow computations were performed using the following spatial discretizations: *C6F10* scheme with $\alpha_f = 0.49$, the standard second-order *E2* scheme with damping, as well as a third-order upwind-biased approach. No SGS model was incorporated. The iterative implicit second-order solver ($BW2$) was employed with a baseline nondimensional time step $\Delta t = 0.001$ and three sub-iterations. In terms of wall units, this time step corresponds to $\Delta t^+ = \Delta t u_\tau^2/\nu = 0.025$ which should be sufficiently accurate.

A comparison of the channel flow solutions for several spatial schemes is given in Figs. 14 and 15 in terms of the mean velocity and spanwise velofity fluctuations. On this level of resolution, the prediction of the mean velocity profile deteriorates significantly when switching to the lower-order second and upwind-biased schemes. Unlike the *C6F10*, which is in good agreement with the log-law, the *E2* and upwind methods display a profile which resembles more a laminar flow. Indeed, examination of the spanwise velocity fluctuations shows the suppression of turbulence caused by the lower-fidelity algorithms. Significant improvements are therefore achieved when employing the high-order compact/filtering approach.

A representative instantaneous flow structure computed with the compact algorithm is shown in Fig. 16 in terms of contours of streamwise velocity on both a transverse and a longitudinal plane close to the surface ($y^+ = 8.3$). The low-speed streaks typical of wall-bounded turbulent flows are clearly observed. This fine structure was not captured with the low-order approaches.

The present test case permits a comparison of the relative efficiency of explicit and implicit time-marching procedures for the simulation of turbulent wall-bounded flows. The results given above were computed with a time step $\Delta t = 0.001$, $\Delta t^+ \approx 0.025$ which corresponds to a maximum $CFL$ number of approximately 8.3. Numerical instability was encountered with the $C6F10\text{-}BW2$ method for $\Delta t = 0.004$, $\Delta t^+ \approx 0.05$. The $RK4$ scheme was found to be very inefficient on this highly stretched mesh. The explicit method was unstable for a time step as low as 0.0001 which corresponds to a $CFL \approx 0.8$. Assuming that a $\Delta t = 5 \times 10^{-5}$, $CFL \approx 0.4$ could be used, this value would still be a factor of 20 smaller that the corresponding time step for the implicit solver. The effect of $\Delta t$ on the solution computed with the implicit scheme was also investigated. The number of sub-iterations was held constant and equal to three. Over the range $0.0005 < \Delta t < 0.002$, the effect of time step on the mean velocity profile (Fig. 17) is observed to be negligible. The turbulence fluctuations obtained with the two smallest time steps (not shown) were also found to be in close agreement with each other [30]. Based on these channel flow results, the second-order iterative implicit procedure appears to be a sufficiently accurate and efficient approach for the simulation of compressible wall-bounded flows.

## 4.5 Boundary-Layer Transition Over a Flexible Panel

As an example of a simulation of multi-disciplinary physics with the present methodology, we consider the fluid-structure interaction between a transitional boundary-layer and an elastic panel. The complex flow-acousto-structural interactions which may arise when a boundary layer flows over an elastic or compliant surface are not only intriguing, but also of importance in a variety of practical situations. These include panel flutter, transition delay, drag and noise reduction, turbulent flow sensing and modification, and control of unsteady separation by means of deforming flexible surfaces.

The flow configuration considered is depicted in Fig. 18(a). The elastic panel is embedded in a flat surface along which a laminar boundary layer develops. The panel of length $a$ and thickness $h$ extends over the region $0.5 \leq x/a \leq 1.5$. The leading-edge region of the plate ($-0.5 \leq x/a \leq 0.0$) is formed by an ellipse of half-thickness $0.05h$ (*i.e.* aspect ratio 10). The computational mesh of size $315 \times 187$ contained 101 points in the streamwise direction over the flexible panel. The grid is stretched rapidly beyond the region of interest in order to provide, in combination with the high-order low-pass filter, an effective treatment of the farfield radiation conditions. Further details on the computational mesh, boundary conditions, and effect of spatial and temporal resolution are given in Ref. [29].

In the present aeroelastic application, the shape of the deforming panel is a function of time, and must be obtained as part of the solution based on the panel structural response to the fluid loading. This response is computed by solving the (non-linear) von Karman plate equations which are required when the magnitude of the deflections are of the order of the panel thickness. A detailed description of the structural equations, their numerical solution procedure, and the fluid-structural coupling is provided in Ref. [6]. At each subiteration of the implicit time-marching method, the shape of the panel is updated by the structural solver. Based on the new boundary coordinates, a new fluid dynamic mesh is constructed by propagating the panel deformations into the entire field using a blending procedure [29]. The sub-iterative approach effectively eliminates lag effects between the fluid and structural modules.

For brevity, only select results obtained at $M_\infty = 0.8$ and $Re_a = 10^5$ are presented here to highlight the ability of the method to capture the complicated unsteady phenomena under the influence of flow-induced surface deformation. At low values of the dynamic pressure, the panel exhibits (static) divergence, and a steady flow is obtained despite the adverse pressure gradient induced by the downward deflection of the panel. At higher dynamic pressures, however, a travelling-wave-flutter phenomenon is observed as summarized in Fig. 18b, c. The instantaneous panel shapes (not shown) display a seventh-mode oscillation with a dominant nondimensional frequency $St = fa/U_\infty = 1.52$ which is substantially higher than the fundamental frequency of the elastic plate. The high-mode flexural deflections are observed to travel along the panel and to reflect at the panel edges. These high-frequency fluctuations result in a pronounced acoustic radiation pattern above the vibrating plate, shown in Fig. 18b in terms of the instantaneous pressure field. Downstream of the flexible surface, a regular train of vortical disturbances is observed (Fig. 18c) with characteristic wavelength and frequency compatible with those of Tollmien-Schlichting (T-S) instability. The travelling wave flutter appears to originate from the coupling of the T-S waves with the panel high-mode transverse fluctuations, and this convective instability ceases below a critical value of Reynolds number[29].

## 5   Conclusions

Significant progress has been achieved in the extension of high-order compact schemes for the solution of the full Navier-Stokes equations on non-uniform, curvilinear and dynamic meshes. Fourth- and sixth-order compact spatial discretizations have been coupled with an implicit iterative time-marching approach as required for efficient solution of wall-bounded flows. The difficulties associated with high-frequency spurious oscillations – inherent to centered non-dissipative discretizations – have been overcome through the application of high-order (up to 10th-order) low-pass spatial filter operators. The proper numerical treatment of metric expressions arising from a general time-varying curvilinear coordinate transformation has been described. This aspect represents a critical issue in order to eliminate metric-cancellation errors associated with the solution of the strong-conservation form of the governing equations on practical 3-D geometries. Application to several unsteady inviscid and viscous flows has demonstrated the superior resolving power of the compact/filtering approach relative to standard second- and fourth-order centered, as well as third-order upwind-biased approximations. For LES of isotropic turbulence and turbulent channel flow, the compact/filtering approach (without the inclusion of an SGS model) provides similar or improved results as compared to standard SGS models. Finally, the potential of the method for applications requiring dynamically-deforming meshes (*e.g.* aeroelasticity, flight dynamics) has been demonstrated by considering the simulation of self-excited oscillations generated by transitional boundary-layer flow over a flexible panel.

*Acknowledgments* The author is grateful for AFOSR sponsorship under task monitored by Maj. W. Hilbun and Dr. T. Beutner. This work was also supported in part by a grant of HPC time from the DoD HPC Shared Resource Centers at ASC, ERDC and NAVO. Several helpful conversations with Drs. D. Gaitonde, D. Rizzetta and G. Blaisdell are gratefully acknowledged.

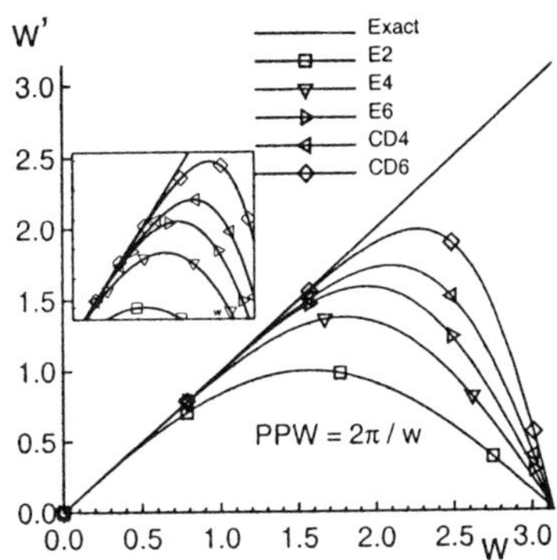


**Fig. 1.** Dispersion-error characteristics of various spatial discretizations for 1-D advection equation.

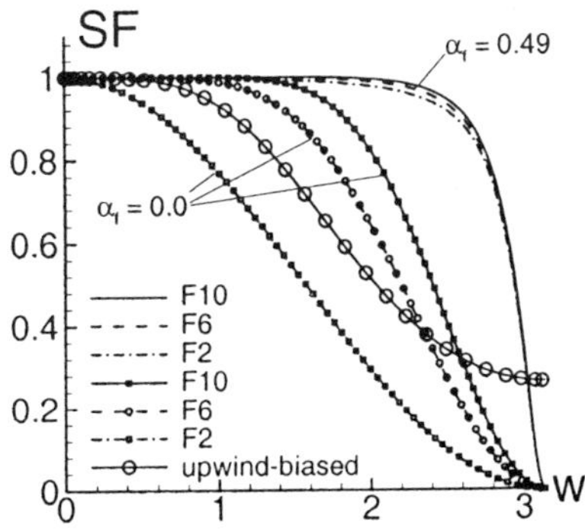


**Fig. 2.** Spectral response of interior low-pass spatial filters.

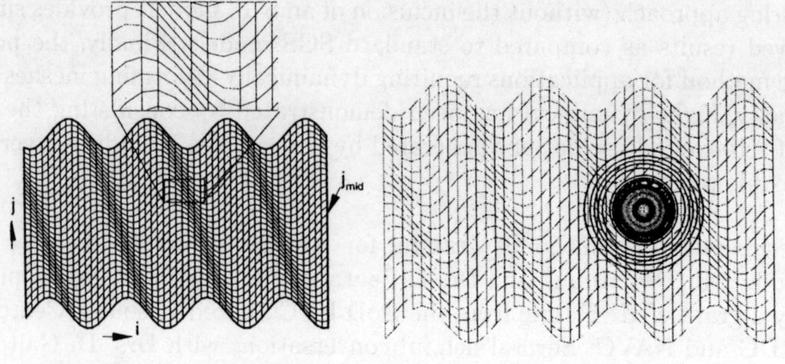

**Fig. 3.** Curvilinear mesh for vortical advection problem

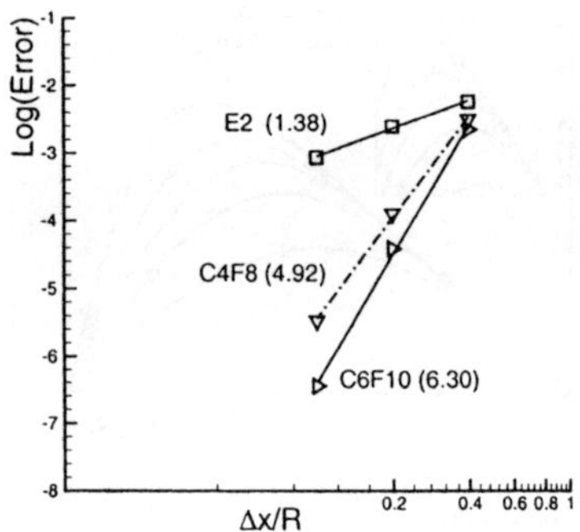

**Fig. 4.** Order of accuracy on curvilinear grid for vortical advection problem

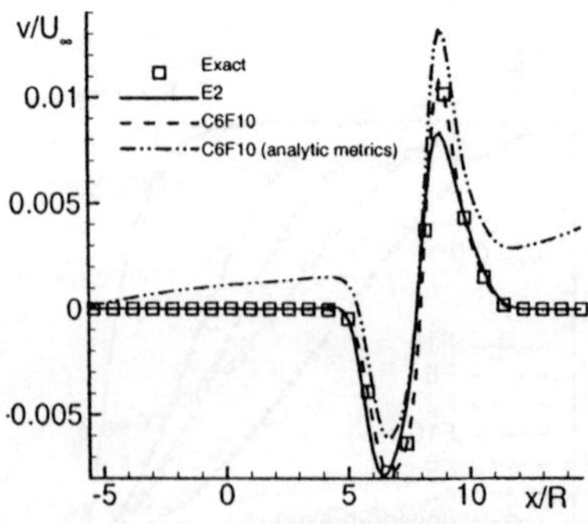

**Fig. 5.** $v$-velocity for vortical advection problem on curvilinear grid along $j = j_{mid}$, $(tU_\infty/R = 8, \Delta x/R = 0.2)$

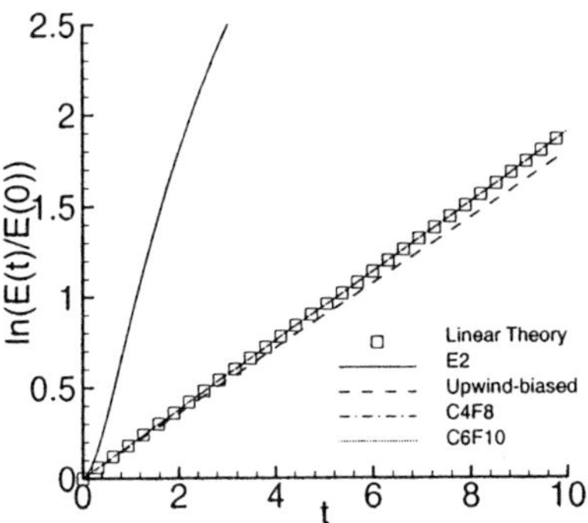

**Fig. 6.** Perturbation energy growth for inviscid shear layer instability

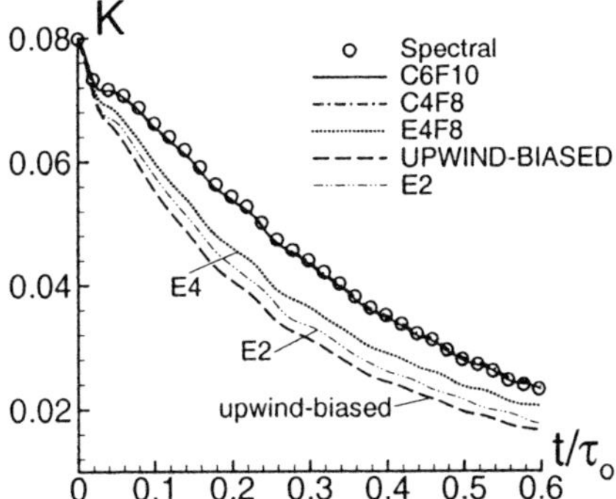

**Fig. 7.** Effect of spatial discretization on time history of turbulent kinetic energy for decaying isotropic turbulence.

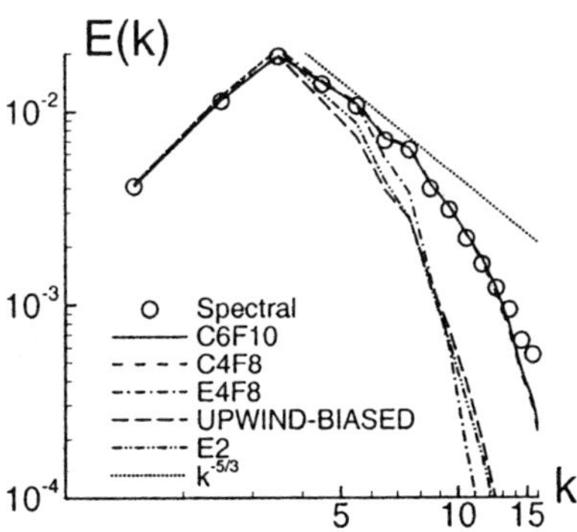

**Fig. 8.** Effect of spatial discretization on instantaneous 3-D energy spectra at $t/\tau_o = 0.2985$.

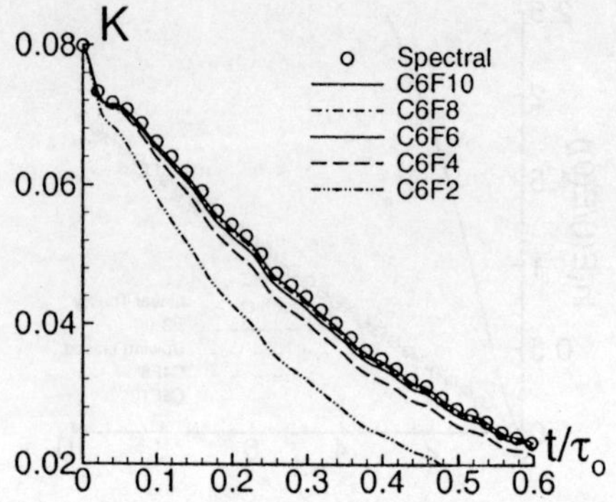


**Fig. 9.** Effect of low-pass filter order on time history of turbulent kinetic energy for decaying isotropic turbulence.

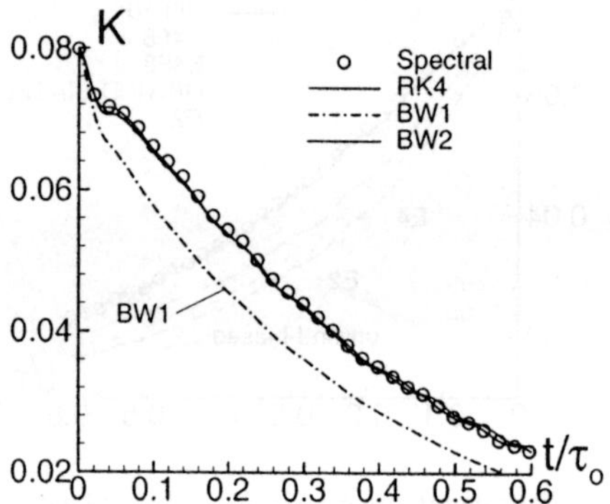


**Fig. 10.** Effect of time-integration scheme on turbulent kinetic energy for decaying isotropic turbulence, ($C6F10$).

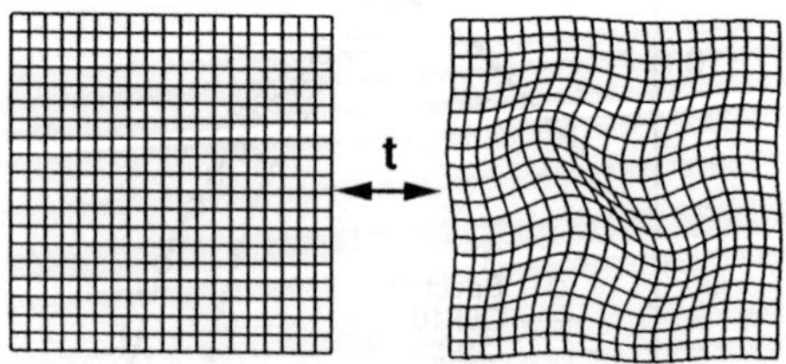


**Fig. 11.** Dynamically deforming curvilinear mesh for isotropic turbulence simulation.

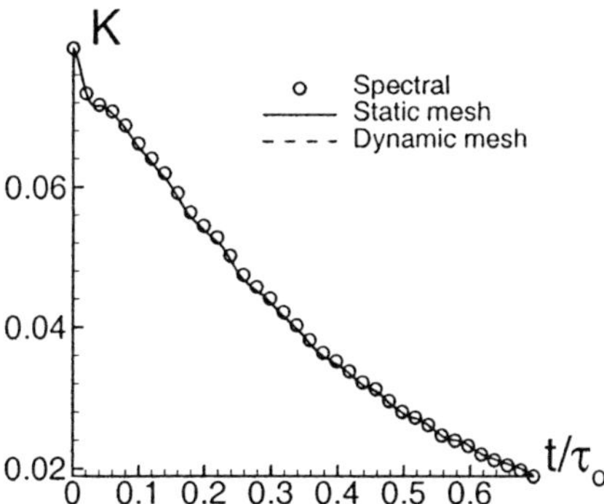


**Fig. 12.** Time history of turbulent kinetic energy for decaying isotropic turbulence on deforming mesh, ($C6F10 - RK4$ scheme).

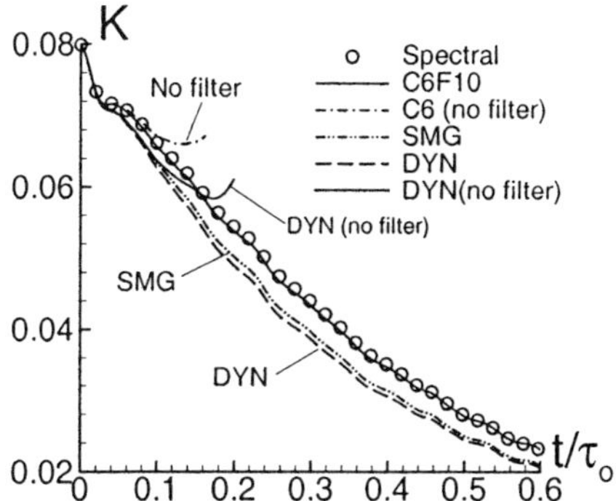


**Fig. 13.** Time history of turbulent kinetic energy for decaying isotropic turbulence using several SGS models, ($C6F10 - RK4$ scheme).

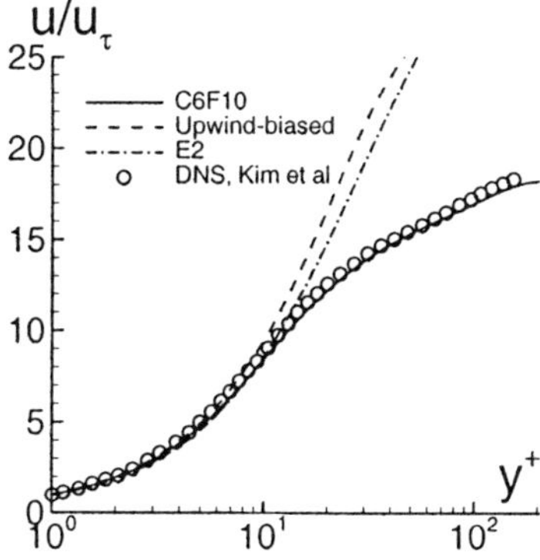


**Fig. 14.** Effect of spatial discretization on computed mean streamwise velocity profile for turbulent channel flow.

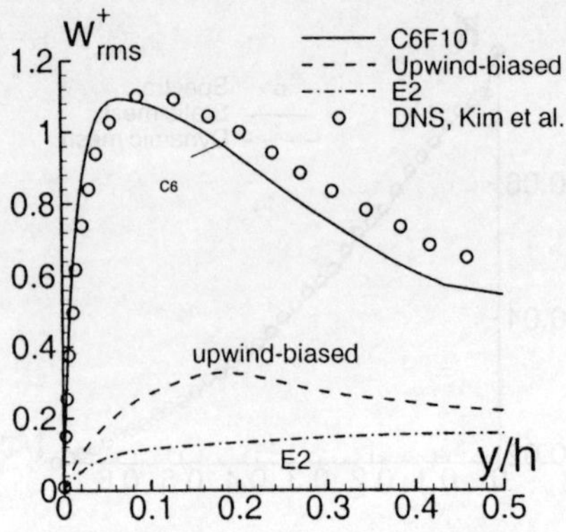



**Fig. 15.** Effect of spatial discretization on computed spanwise velocity fluctuations for turbulent channel flow.

**Fig. 16.** Instantaneous turbulent channel flow structure computed with $C6F10-BW2$ scheme.

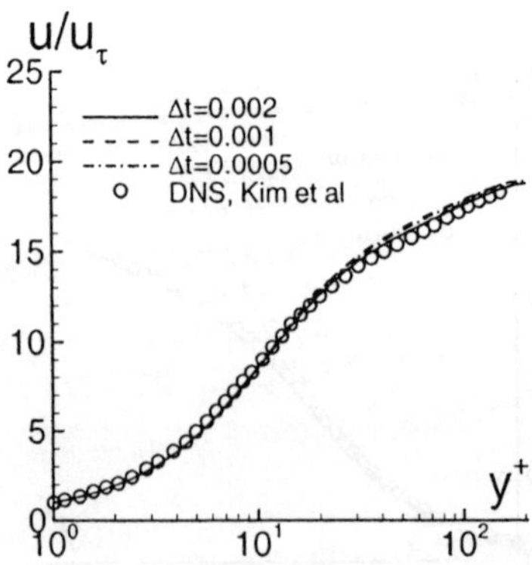



**Fig. 17.** Effect of computational time step on mean streamwise velocity profile for turbulent channel flow, $(C6F10 - BW2$ scheme).

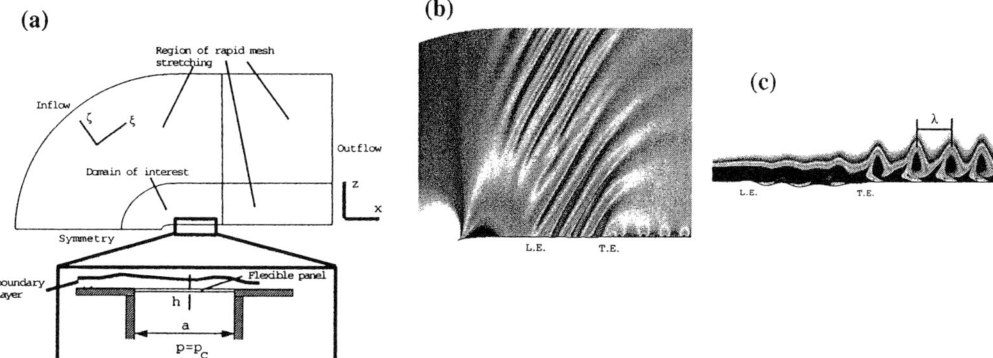

**Fig. 18.** Transitional Boundary-Layer Flow Over a Fluttering Elastic Panel, ($C6F10-BW2$ scheme).

# References


1. Anderson, D., Tannehill, J. and Pletcher, R.: *Computational Fluid Mechanics and Heat Transfer* (McGraw-Hill Book Company 1984)
2. Beam, R. and Warming, R.: 'An Implicit Factored Scheme for the Compressible Navier-Stokes Equations', *AIAA J.*, **16/4** (1978)
3. Gaitonde, D., Shang, J.S. and Young, J.L.: 'Practical Aspects of High-Order Accurate Finite-Volume Schemes for Electromagnetics', AIAA Paper 97-0363 (1997)
4. Gaitonde, D. and Visbal, M.: 'High-Order Schemes for Navier-Stokes Equations: Algorithm and Implementation into FDL3DI, Technical Report AFRL-VA-WP-TR-1998-3060, Air Force Research Laboratory, Wright-Patterson AFB (1998)
5. Gaitonde, D. and Visbal, M.: 'Further Development of a Navier-Stokes Solution Procedure Based on Higher-Order Formulas', AIAA Paper 99-0557 (1999)
6. Gordnier, R. and Visbal, M.: 'Development of a Three-Dimensional Viscous Aeroelastic Solver for Nonlinear Panel Flutter', AIAA Paper 2000-2337 (2000)
7. Gordnier, R. and Visbal, M.: 'Computation of Three-Dimensional Nonlinear Panel Flutter', AIAA Paper 2001-0571 (2000)
8. Jameson, A. Schmidt, W. and Turkel, E.: 'Numerical Solutions of the Euler Equations by a Finite Volume Method Using Runge-Kutta Time Stepping Schemes', AIAA Paper 81-1259 (1981)
9. Kim, J., Moin, P. and Moser, R.: 'Turbulent Statistics in Fully Developed Channel Flow at Low Reynolds Number', *J. Fluid Mechanics*, **117** (1992)
10. Lele, S.: 'Compact Finite Difference Schemes with Spectral-like Resolution', *J. Comp. Physics*, **103** (1992)
11. Michalke, A.: 'On the inviscid instability of the hyperbolic-tangent velocity profile', *J. Fluid Mechanics.*, **19** (1964)
12. Moin, P., Squires W., Cabot, W. and Lee, S.: A Dynamic Subgrid-Scale Model for Compressible Turbulence and Scalar Transport', *Physics of Fluids A*, **3/11** (1991)
13. Piomelli, U.: 'High Reynolds Number Calculations Using the Dynamic Subgrid-Scale Stress Model', *Physics of Fluids*, **5/6** (1993)
14. Poinsot, T. and Lele, S.: 'Boundary Conditions for Direct Simulations of Compressible Viscous Flows', *J. Comp. Physics*, **101** (1992)

15. Pulliam, T.: 'Artificial Dissipation models for the Euler Equations', *AIAA J.*, **24/12** (1986)

16. Pulliam, T. and Chaussee, D.: 'A Diagonal Form of an Implicit Approximate-Factorization Algorithm', *J. Comp. Physics*, **39/2** (1981)

17. Rizzetta, D., Visbal, M. and Blaisdell, G.: 'Application of a High-Order Compact Difference Scheme to Large-Eddy and Direct Numerical Simulation', AIAA Paper 99-3714 (1999)

18. Roe, P.: 'Approximate Riemann Solvers, Parameter Vectors and Difference Schemes', *J. Comp. Physics*, **43/2** (1981)

19. Schumann, U.: 'Subgrid-Scale Model for Finite Difference Simulations of Turbulent Flows in Plane Channels and Annuli', *J. Comp. Physics*, **18/4** (1975)

20. Spyropoulos, E. and Blaisdell, G.: 'Evaluation of the Dynamic Model for Simulations of Compressible Decaying Isotropic Turbulence', *AIAA J.*, **34/5** (1996)

21. Tam, C.: 'Computational Aeroacoustics: Issues and Methods', *AIAA J.*, **33/10** (1995)

22. Thomas, P. and Lombard, C.: 'Geometric Conservation Law and its Application to Flow Computations on Moving Grids', *AIAA J.*, **17/10** (1979)

23. van Leer, B.: 'Towards the Ultimate Conservative Difference Scheme. V. A Second-Order Sequel to Godunov's Method', *J. Comp. Physics*, **32** (1979)

24. Vichnevetsky, R. and Bowles, J.: *Fourier Analysis of Numerical Approximations of Hyperbolic Equations*, SIAM Studies in Applied Mathematics (1982)

25. Vinokur, M.: 'Conservation Equations of Gasdynamics in Curvilinear Coordinate Systems', *J. Comp. Physics*, **14** (1974)

26. Visbal, M. and Gaitonde, D.: 'High-Order Accurate Methods for Complex Unsteady Subsonic Flows', *AIAA J.*, **37/10** (1999)

27. Visbal, M. and Gaitonde, D.: 'Very High-Order Spatially Implicit Schemes for Computational Acoustics on Curvilinear Meshes', *J. Comp. Acoustics*, **9/4** (2001)

28. Visbal, M. and Gordnier, R.: 'A High-Order Flow Solver for Deforming and Moving Meshes', AIAA Paper 2000-2619 (2000)

29. Visbal, M. and Gordnier, R.: 'Direct Numerical Simulation of the Interaction of a Boundary Layer with a Flexible Panel', AIAA Paper 2001-2721 (2001)

30. Visbal, M. and Rizzetta, D.: 'Large-Eddy Simulation on general Geometries Using Compact Differencing and Filtering Schemes', AIAA Paper 2002-0288 (2002)

# Incompressible Flow Simulation by Using Multidirectional Finite Difference Scheme

Kunio Kuwahara[1], Satoko Komurasaki[2], and Angel Bethancourt[3]

[1] Institute of Space and Astronautical Science, Yoshinodai, Sagamihara, Kanagawa 229-8510, Japan
[2] Nihon University, Kanda-Surugadai, Chiyoda-ku, Tokyo 101-8308, Japan
[3] Institute of Computational Fluid Dynamics, Haramachi, Meguro-ku, Tokyo 152-0011, Japan

**Abstract.** High-Reynolds-number flows are simulated by solving the incompressible Navier-Stokes equations. A finite-difference method with third-order upwinding is employed without incorporating a turbulence model. In addition, a multi-directional formulation is used to improve the accuracy of the solution. The validity of this method is thoroughly discussed, and examples are presented to show the applicability of the present approach to a variety of problems using Cartesian and body-fitted coordinates systems. Some of the examples include: a) flows around bluff-bodies using Cartesian coordinates system, and b) flows past subsonic two- and three-dimensional airfoils using O-type coordinates system. The results show that typical flow mechanisms are clearly captured by the present direct numerical simulations. For the airfoil simulations, computed lift coefficients are found to agree quite well with the experimental data even for attack angles above the stall angle.

## 1 Introduction

Most of the high-Reynolds-number flow simulations have been based either on the Reynolds-averaged Navier-Stokes equations using a turbulence model or a large-eddy simulation incorporating a Smagorinsky-type model (Deardorff, 1970). However, neither of the above approaches are suitable for high-Reynolds-number-flow computation because the effect of turbulence mixing is usually replaced by second-order diffusion. This diffusion model is similar in form to viscous diffusion; therefore, in the end, we are simply computing a locally low-Reynolds-number flow.

There are some real direct numerical simulation in which most of the small-scale structure are resolved, but they are limited to relatively small Reynolds numbers. For high Reynolds number flows, which include most flows of practical interest, the required grid size to resolve the small structures is so enourmous that solutions attempts are deemed impractical. Because we are left with no option but to use coarser grid systems, development of schemes able to capture the large-scale structures are of primary importance. Luckily, in many applications, these large structures are the most important in defining the features of the flow.

Quite a few simulations (see Kuwahara, 1992) show that the large structures of high-Reynolds-number flows can be captured using a relatively coarse grid if the numerical instability associated with high-Reynolds-number-flow simulations is suppressed. Most successful simulations in these approaches are based on the third-order upwind formulation (Kawamura and Kuwahara, 1984). An approach similar in philosophy but different in method is adopted by Boris et. al. (1992).

In the present paper, we summarize the third-order upwind scheme for high-Reynolds-number-flow computations, as well as the a multi-directional finite-difference method, which was developed to increase the accuracy of the solution.

## 2  Computational method

The governing equations are the incompressible Navier-Stokes equations. In three-dimensional Cartesian coordinates system, they are as follows:

$$\frac{\partial u_j}{\partial x_j} = 0, \tag{1}$$

$$\frac{\partial u_i}{\partial t} + u_j \frac{\partial u_i}{\partial x_j} = \frac{\partial p}{\partial x_i} + \frac{\partial}{\partial x_j}\left\{ \frac{1}{Re}\left( \frac{\partial u_i}{\partial x_j} + \frac{\partial u_j}{\partial x_i}\right)\right\}. \tag{2}$$

For high-Reynolds-number flows, time-dependent computations are required owing to the strong unsteadiness. A finite-difference method is employed to discretize the basic equations, and the resulting algebraic equations are solved using the projection method (Chorin, 1968, and Takami and Kuwahara, 1974). This method can be written in the following form when applied to the above equations:

$$w + \mathrm{grad}\,p = F, \tag{3}$$

$$\mathrm{div}\,w = 0, \tag{4}$$

where

$$w = w_i, \quad F = F_i \quad (i = 1, 2, 3),$$

$$w_i = \frac{\partial u_i}{\partial t}, \tag{5}$$

$$F_i = u_j \frac{\partial u_i}{\partial x_j} + \frac{\partial}{\partial x_j}\left\{ \frac{1}{Re}\left( \frac{\partial u_i}{\partial x_j} + \frac{\partial u_j}{\partial x_i}\right)\right\}, \tag{6}$$

and, after applying the Helmholtz's decomposition,

$$w^\nu = F \quad \mathrm{Grad}\,p^\nu, \tag{7}$$

$$p^{\nu+1} = p^\nu \quad \varepsilon \left( \mathrm{Div}\,w^\nu + \frac{1}{\Delta t}\mathrm{Div}\,u\right), \tag{8}$$

where the index $\nu$ denotes the $\nu$th iterations and $\varepsilon$ is a relaxation constant. Symbols Grad and Div are suitable difference approximations to the operators grad and div. The last term with $1/\Delta t$ in eq.(8) is the correction term for preventing the accumulation of the error in $w$. If $p$ and $\varepsilon$ are regarded as the temperature and time increment, eq.(8) is considered to be the heat equation discretized by the Euler explicit method. The value of $\varepsilon$ is determined based on stability requirements. Equation (8) becomes the Poisson equation for the pressure

$$\mathrm{Div}(\mathrm{Grad}p) = \mathrm{Div}\boldsymbol{F} + \frac{1}{\Delta t}\mathrm{Div}\boldsymbol{u} \tag{9}$$

after convergence, when $p^{\nu+1}$ is equal to $p^{\nu}$. Although this method is essentially the same as the MAC method, mathematical basis is more secured and extension to compressibility is easier (Kuwahara, 1984).

All the spatial derivative terms are discretized using central difference approximation except for the convection terms, for which a third-order upwind scheme is used to stabilize the computation. This is the most important point for high-Reynolds-number computations, and the details are given below.

If grid resolution is not sufficient to resolve the small-scale structure, which is usually the case for high Reynolds number flows, a type of error (called aliasing error) due to the nonlinearity of the problem, introduces a strong numerical instability into the solution. Therefore, it is very important to filter out the high-frequency part of the original function when digitizing a continuous function into a finite number of the values. If this is not done, aliasing error can render the approximation meaningless. In practice, a turbulence model or a large eddy simulation is usually used to get rid of this instability. The increase of the diffusion coefficient due to the added turbulent viscosity reduce the aliasing error and suppress the numerical instability. However, in most of the models, this added diffusion has the same form as the viscous diffusion and it is usually much larger than the viscous diffusion. Therefore, the effect of physical diffusion is concealed, and the dependency of the flow on the Reynolds number is not properly captured during computations.

Another alternative to stabilize the computation is to use an upwind scheme. The first-order upwind scheme is widely used because of its very good stability properties, but the leading numerical error caused by this upwinding is second order and similar to the physical diffusion. For the reasons just mentioned above, this type of discretization should be avoided. On the other hand, The second-order upwind scheme has a dispersion type leading error, which generally makes the computation unstable. In order to obtain a stable computation, the order of accuracy (odd or even) of the nonlinear convective terms is of extreme importance. In case of even order of accuracy, the leading numerical-error term is the odd-order derivative which is dispersive. Then, once an error is introduced into the solution, the error never diffuses but moves around in the computational domain until the computation blows up; therefore, no stable solution can be obtained in this case. On the other hand, in case of odd order of accuracy, the leading error term is the even-order derivative which is diffusive. This makes the computation very stable by providing a dissipative mechanism for the aliasing error.

A third-order upwind scheme has been found to be most suitable for high-Reynolds-number-flow computation. The leading numerical error terms are the fourth-order derivative terms, where the effects of the second-order numerical diffusions are carefully removed. The numerical diffusion of fourth-order derivatives is of short range and does not conceal the effect of molecular diffusion, and have the added advantage to stabilize the computation.

One simple explanation of why the fourth-order diffusion does not conceal the effect of second-order diffusion can be given starting with the finite-difference representation of the fourth-order diffusion term as follows:

$$(u_{k+2} - 4u_{k+1} + 6u_k - 4u_{k-1} + u_{k-2})/(\Delta x)^4 \tag{10}$$

This can be written as,

$$\frac{4}{(\Delta x)^2} \left( \frac{u_{k+2} - 2u_k + u_{k-2}}{(2\Delta x)^2} - \frac{u_{k+1} - 2u_k + u_{k-1}}{(\Delta x)^2} \right). \tag{11}$$

The two terms in eq.(11) represent second-order diffusion and their effects are mutually canceled except near the point $k$. This interpretation indicates that fourth-order diffusion is very independent from second-order diffusion. In general, the effects of lower-order diffusion are not concealed by higher-order diffusion.

Similarly, fifth-order upwinding is possible and some computations have been done but it requires seven points in each direction to approximate the local derivative, implying a wider range of analyticity to the solution of the equations. High-Reynolds-number flows are not so analytical, therefore it is concluded that third-order upwinding is sufficient for most practical applications. There are several implementations of third-order upwind schemes. The one employed in this paper is described as follows. Initially, the one-sided second-order finite difference approximation is employed for the convection terms.

$$u\frac{\partial u}{\partial x} \approx \begin{cases} u_k \cdot \left( \dfrac{3u_k - 4u_{k-1} + u_{k-2}}{2\Delta x} \right) & (u_k > 0) \\[2em] u_k \cdot \left( \dfrac{-u_{k+2} + 4u_{k+1} - 3u_k}{2\Delta x} \right) & (u_k < 0) \end{cases} \tag{12}$$

We can rewrite the above equations to a symmetrical form by using the formula:

$$\frac{u_k + |u_k|}{2} = \begin{cases} u_k & (u_k > 0) \\ 0 & (u_k < 0) \end{cases}, \quad \frac{u_k - |u_k|}{2} = \begin{cases} 0 & (u_k > 0) \\ u_k & (u_k < 0) \end{cases}.$$

Then, eq.(12) can be rewritten as,

$$u\frac{\partial u}{\partial x} \approx \frac{u_k + |u_k|}{2} \left( \frac{3u_k - 4u_{k-1} + u_{k-2}}{2\Delta x} \right)$$

$$+ \frac{u_k - |u_k|}{2} \left( \frac{-u_{k+2} + 4u_{k+1} - 3u_k}{2\Delta x} \right).$$

Finally, we obtain

$$u\frac{\partial u}{\partial x} \approx u_k \cdot \frac{-u_{k+2} + 4(u_{k+1} - u_{k-1}) + u_{k-2}}{4\Delta x}$$

$$+ |u_k| \cdot \frac{u_{k+2} - 4u_{k+1} + 6u_k - 4u_{k-1} + u_{k-2}}{4\Delta x}. \tag{13}$$

If the right hand side terms are expanded into Taylor series, they become

$$u\frac{\partial u}{\partial x} - \frac{1}{3}(\Delta x)^2 u\frac{\partial^3 u}{\partial x^3} + O((\Delta x)^4) \quad \text{and} \quad \frac{1}{4}(\Delta x)^3 u\frac{\partial^4 u}{\partial x^4} + O((\Delta x)^5). \tag{14}$$

respectively. Therefore, the leading error of eq.(13) is of order $(\Delta x)^2$ and its coefficient includes third-order derivative. As mentioned previously, odd order-derivatives are not desirable, however, this error term can be eliminated if the first term of eq.(13) is replaced by

$$u_k \cdot (-u_{k+2} + 8(u_{k+1} - u_{k-1}) + u_{k-2})/12\Delta x. \tag{15}$$

As a result, the present third-order upwind scheme is represented by a five point stencil as follows:

$$u\frac{\partial u}{\partial x} \approx u_k \cdot \frac{-u_{k+2} + 8(u_{k+1} - u_{k-1}) + u_{k-2}}{12\Delta x}$$
$$+|u_k| \cdot \frac{u_{k+2} - 4u_{k+1} + 6u_k - 4u_{k-1} + u_{k-2}}{12\Delta x}. \tag{16}$$

This scheme is called the Kawamura-Kuwahara scheme (1984). There are several versions of third order upwind schemes, and they can be grouped using the following equation,

$$u\frac{\partial u}{\partial x} \approx u_k \cdot \frac{-u_{k+2} + 8(u_{k+1} - u_{k-1}) + u_{k-2}}{12\Delta x}$$
$$+\alpha|u_k| \cdot \frac{u_{k+2} - 4u_{k+1} + 6u_k - 4u_{k-1} + u_{k-2}}{12\Delta x}, \tag{17}$$

where $\alpha$ determines the corresponding version. We compared the Kawamura-Kuwahara scheme with UTOPIA and QUICK schemes (Leonard, 1979) by solving the energy spectrum in one-dimensional Burgers turbulence and found that the present scheme give the best results (fig.1). For a very fine grid (4096 points), the three schemes are in complete agreement with each other and theoretical prediction. However, after reducing the number of grid points, the differences become clear, and only the present method preserves the good agreement when compared with the results of very fine computations.

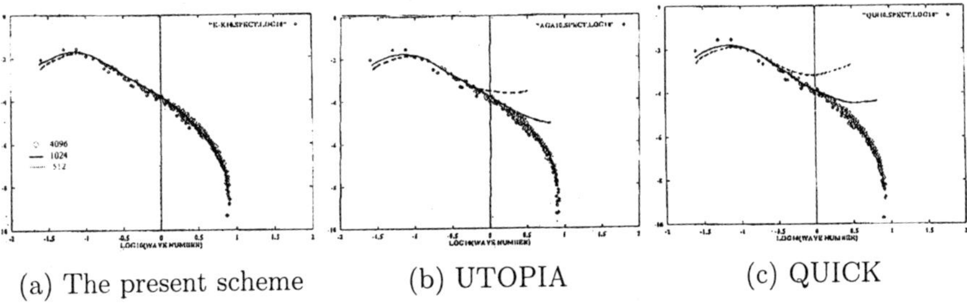

(a) The present scheme     (b) UTOPIA     (c) QUICK

Fig.1 Energy spectrum for Burgers turbulence.

There is another important problem in high-order upwind schemes. That is, the accuracy decreases when the flow direction is not well parallel to one of the coordinate lines. If a generalized coordinate system is used near the boundary, the flow direction and one of the coordinate lines are almost parallel, and the problem is not serious. However, in general, flow direction is not always parallel to a coordinate line and the problem becomes very important.

To overcome this problem we introduced the multi-directional upwind formulation. This method is summarized as follows:
For 2-dimensional computations, when structured grid points are given, the black points in fig.2(a) are usually used to approximate the derivatives at the central point (system A). If an additional 45 degrees-rotated local grid system is introduced, the white points in fig.2(b) can be used to approximate the derivatives at the central point (system B). In order to improve the derivative value at the central point, we mix the derivative values calculated from both systems (A and B) at a proper ratio. We adopt the ratio A : B = 2/3 : 1/3. Using this ratio, the resulting finite-difference scheme for the Laplacian coincides with the well-known 9-points formula with fourth-order accuracy. This method improves the rotational invariance of the coordinate system, then, those flows where flow direction is not parallel to the grid direction are better simulated.

For 3-dimensional computations, three different grid systems are introduced. Each grid system is obtained by rotating a 2-D plane (system A) around one of the Cartesian axis. One of those systems is shown in fig.2(c), and the others grid systems can be obtained similarly. In the same way as it was done for 2-dimensional computations, white circle points are used instead of black ones on the $x'$-$y'$ plane, and ordinary ones in the $z$ direction. Using the present formulation, three different values are obtained at each grid point. A simple mean averaged is used to obtain the final value considering that the physical phenomena are equivalent in each grid system. The rotational invariance of the coordinate system can be improved by means of this method.

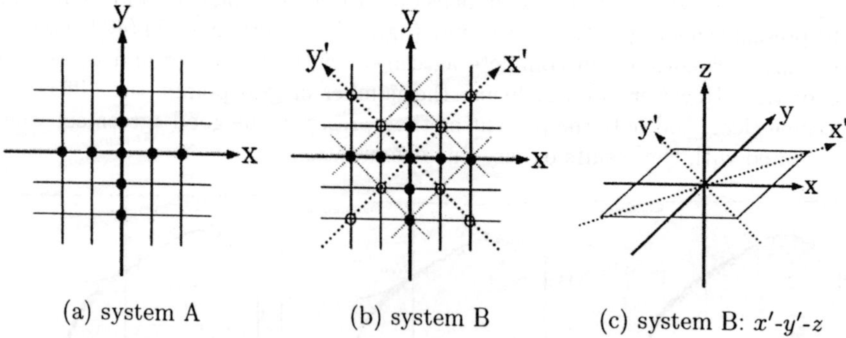

(a) system A      (b) system B      (c) system B: $x'$-$y'$-$z$

**Fig.2** Grid for multi-directional scheme.

This method has another advantage. In the MAC method, an staggered mesh is used to remove the unphysical oscillation of the pressure. This oscillation is caused by the decoupling of the computed values within the nearest two points.

This decoupling become less significant if a third-order upwind scheme is used because of the five-point differencing, but some effects still remains. However, if the multi-directional finite-difference method is employed, every point becomes tightly coupled and the oscillation disappears. Therefore, a non-staggered mesh system is employed where the velocity and pressure are defined at the same grid location.

For the temporal integration of the Navier-Stokes equations, the Crank-Nicolson implicit scheme is utilized. This scheme has second-order accuracy in time. These equations and the Poisson equation are iteratively solved at each time step by the successive over relaxation (SOR) method coupled with a multi-grid method.

## 3  Computational results

### I. Circular cylinder

The dependence of the drag coefficient on the Reynolds number is shown in fig.3. Numerical simulations are carried out employing a grid size of 32*16, 64*32 and 128*64. For Reynolds number less than 100, computational and experimental results agree very well. At high Reynolds number, even the 64*32 computation can capture the drag crisis qualitatively. As expected, a higher grid resolution (128*64) produces a closer agreement when compared with experiments.

The grid crisis is a phenomenom that occurs for flows around bluff bodies (i.e. spheres, circular cylinders). It manifests by a sudden decrease of the drag coefficient at about Reynolds number 400000. It is related to the transition from laminar to turbulent flow (Schlichting, 1979), causing the point of separation to move further downstream. This occurrence causes a reduction in the wake of the cylinder and the pressure drag. Figure 4 captures the phenomena just described.

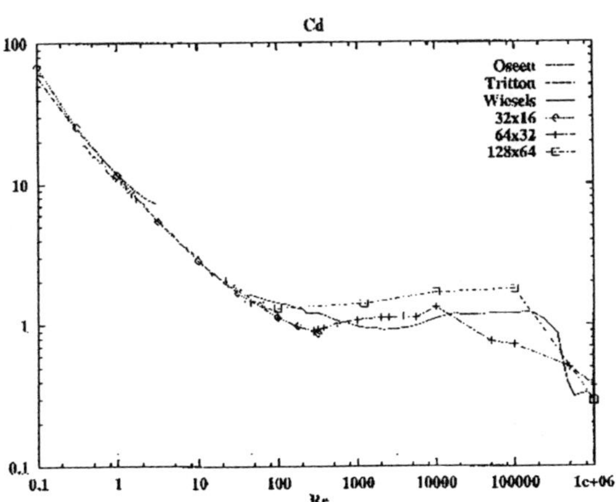


**Fig.3**  Drag coefficients of circular cylinder
(Kuwahara, 1999).

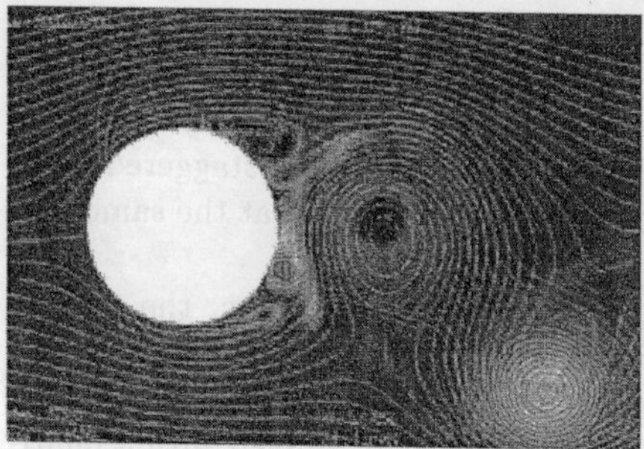 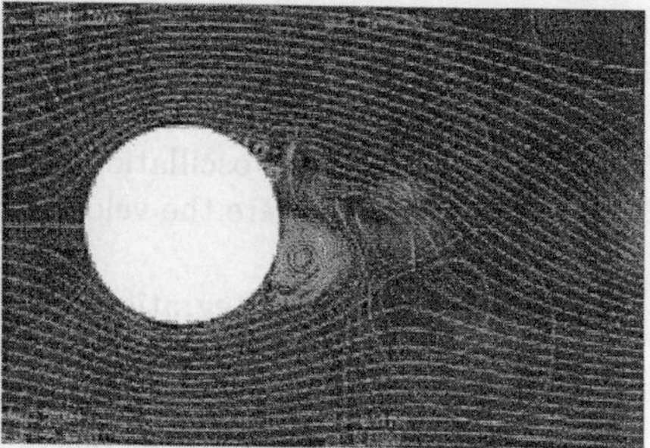

(a) Re=50000, before drag crisis    (b) Re=1000000, after drag crisis

**Fig.4** Flow past a circular cylinder, streamlines and pressure contours, background color shading shows, the vorticity distribution. Instantaneous flow field, 128*256 grid system.

## II. Rearrangement of Karman vortex street

Two-dimensional flow past a bluff body usually produces a Karman Vortex street. This vortex street is dissipated but eventually rearranges itself into another Karman Vortex street of different wavelength. However, this phenomenon occurs far beyond the body. Therefore, to successfully simulate its occurrence, a long wake region (of about 1000 times the length of the body size) must be included in the computational domain. In order to limit the number of grid point in the direction of the wake to a couple of thousands, the body has to be represented by a few grid points. This becomes possible by using the present multi-directional finite-difference method. Figure 5 shows the Karman Vortex street and its rearrangement. The body is represented by only three grid points, the total grid size is 2059*257. The upper part of figure shows the velocity component perpendicular to the flow direction.

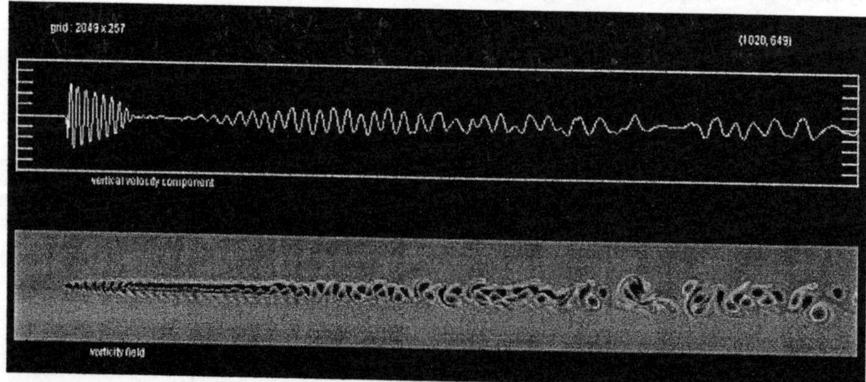


**Fig.5** Rearrangement of Karman vortex street. 2058*256 Cartesian grid system.

## III. Flow around an airfoil

Flow around an airfoil is a standard problem, but one for which unsteady computations have rarely been done. This becomes a necessity in order to fully understand the flow features, specially at high angles of attack, because of the large regions of flow separation and vortex shedding appearing in the flow field.

When performing airfoil simulations, a grid of C-topology is usually used to avoid the singularity encountered at the trailing edge of the airfoil. However, the generation of the C-grid is not an easy task, specially at very high angle of attack, this being another reason for the difficulty to simulate the flows at very high angle of attack. Additionally, a C-grid needs an unnecessary concentration of grid points in the near wake region beginning at the trailing edge. This make the computation more unstable. On the other hand, a grid of O-topology will provide a better representation if convergence of the computation can be assured. The introduction of the multi-directional finite-difference method makes this possible by providing a formulation that achieve a very stable solution even near the singular points. Stability of the O-grid computation around an airfoil allows to compute a flow around a circular arc-airfoil which has a singularity at the leading edge (fig.6), or the computation of a flow around an arc-airfoil of negative thickness (fig.6(b)).

Two- and three-dimensional numerical simulations of flows around an NACA0012 airfoil at high angles of attack are also carried out. The fully developed solution for 2-d flow is used as an initial condition for the 3-d computation to save computation time. The 2-d computations exhibit an excellent agreement with the experimental results (Abbott and Von Doenhoff, 1959) before the stall is reached. However, beyond the stall angle, there exits a definite discrepancy between the experimental and computational results (Kuwahara and Komurasaki, 2000). For 3-d computations, computed lift coefficients are found to be in good agreement when compared with experiments (Kuwahara and Komurasaki, 2001). Typical calculations were performed in a 128*64 grid for 2-d cases and 128*64*16 grid for 3-d cases.

Figure 7(a) shows the drag, lift and moment coefficients cd, cl, cm for 2-d and 3-d computations. For the 2-d case, the value of the lift coefficient cl is overpredicted for angles of attack beyond the stall angle. On the other hand, for the 3-d case, the lift coefficient does not change much when compared to 2-D simulations provided that the stall angle is not reached. After the stall angle is reached and surpased, a three-dimensional flow quickly develops, and the lift decreases accordingly. As it can be seen, the final results are in good agreement with the experimental ones. Figure 7(b) shows the history of drag, lift and moment coefficients for the 2-d to 3-d computations for an angle of attack 18 degrees (stall angle). The left side of graph shows the time evolution of the coefficients for the 2-D simulation, and the righnt side the results of the 3-D simulations. The drop on the lift coefficient is obvious.

Figures 8(a) and (b) describe the flow fields for the 2- and 3-d computations, respectively. In these figures, flow fields at angle of attack ranging from 14.0 to 20.0 degrees are visualized.

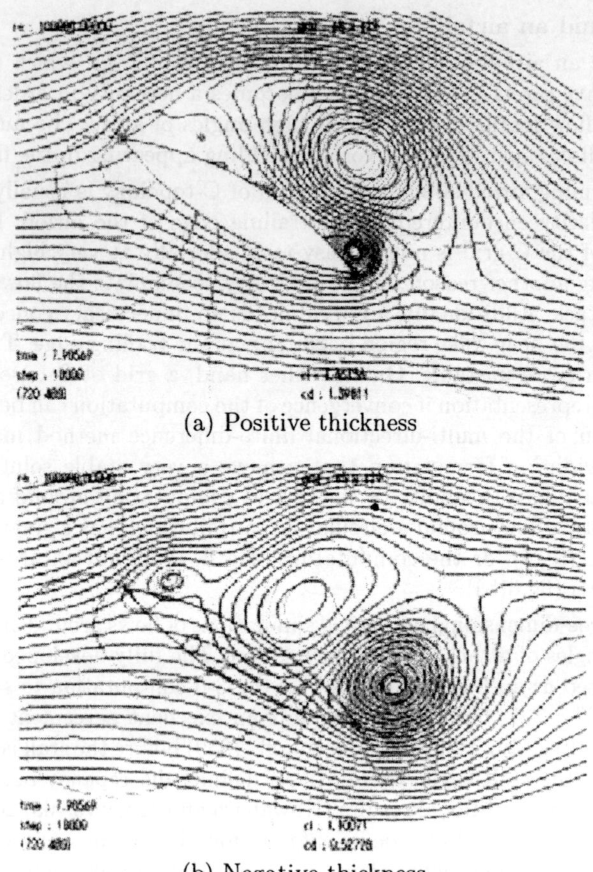

(a) Positive thickness

(b) Negative thickness

**Fig.6** Flow around a circular arc-airfoil at Re $= 10^5$, 128*64 O-grid system, angle of attack $= 30$ degrees, instantaneous streamlines and pressure contours.

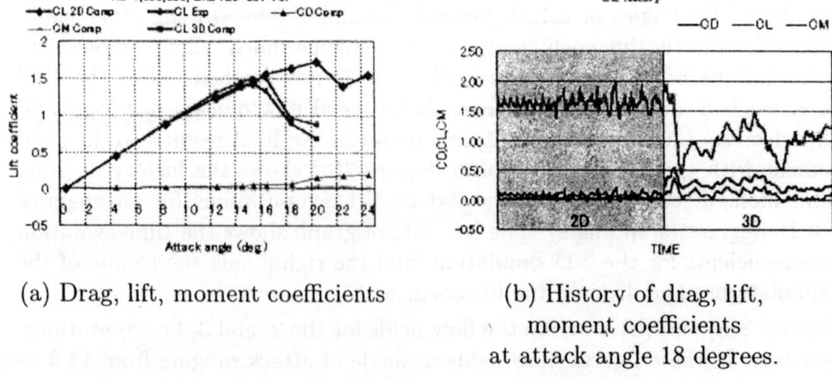

(a) Drag, lift, moment coefficients

(b) History of drag, lift, moment coefficients at attack angle 18 degrees.

**Fig.7** Computed force coefficients around NACA0012 airfoil.

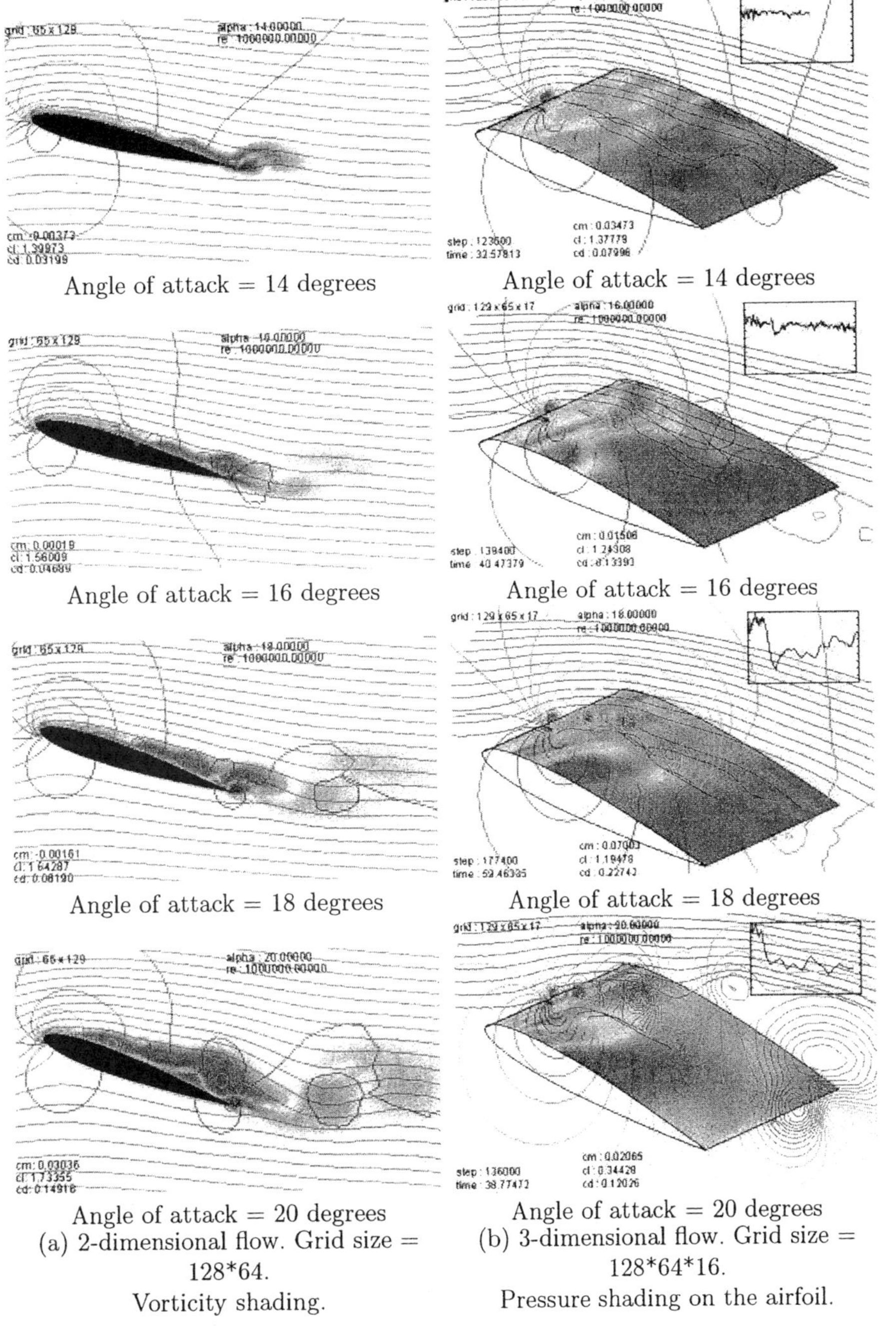

Angle of attack = 14 degrees

Angle of attack = 14 degrees

Angle of attack = 16 degrees

Angle of attack = 16 degrees

Angle of attack = 18 degrees

Angle of attack = 18 degrees

Angle of attack = 20 degrees
(a) 2-dimensional flow. Grid size = 128*64.
Vorticity shading.

Angle of attack = 20 degrees
(b) 3-dimensional flow. Grid size = 128*64*16.
Pressure shading on the airfoil.

**Fig.8** Flow around NACA0012 airfoil, at Re=$10^6$, instantaneous streamlines and pressure contours.

## IV. Transition in bluff body wake

It is very difficult to compute transitional flow by using a turbulence model. Even large-eddy simulation can not handle this type of problem because it assumes that the flow is turbulent from the beginning. However, transition phenomena is very important at high-Reynolds number flow. The way that the present formulation is constructed, allows for the computation of any transitional flow without having to explicitly consider it during the discretization.

Figures 9 and 10 show flows around a ski jumper and a protoceratopus, respectively. These can be considered bluff bodies, therefore, their wake turbulence is very similar to the sphere. Figure 11 exhibits a flow around a car. This is a more streamlined body, and accordingly its turbulent area is more reduced. In figs.12 and 13, development of turbulence behind a sphere is shown at Re=2.0*$10^5$ and 2.0*$10^3$, with the spheres impulsively started from rest. The visualization was done by showing the surface pressure of the body and the volume rendering of absolute value of the vorticity.

The calculation for the sphere and the protoceratopus were performed on a $288 * 144 * 144(= 5,971,968)$ and $256 * 128 * 128(= 4,194,304)$ grid system, respectively. These numbers are not big enough to resolve the smallest structures within the flow field but good enough to obtain the large structures in the transitional stage.

In these computations, a Cartesian system with equal grid spacing is used. Body fitting type grid system is not only difficult to create but uniformity of the grid is not possible. To see the behavior of the transition to turbulence, this uniformity is crucial.

Fig.9 Flow around a ski jumper at Re=2.0*$10^5$, 288*144*144 grid system.

Fig.10 Flow around a protoceratopus at Re=$10^5$, 256*128*128 grid system.

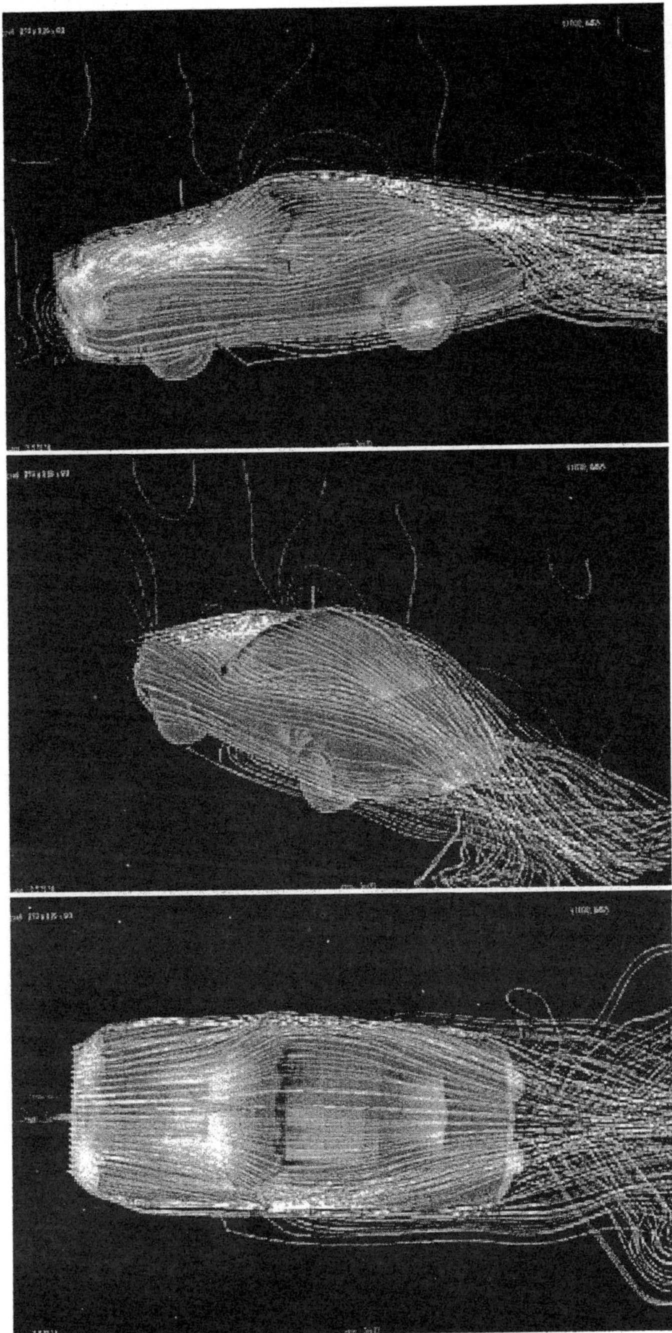

Fig.11   Flow past a car, 256*128*96 grid system.
Stream lines and pressure contour lines.

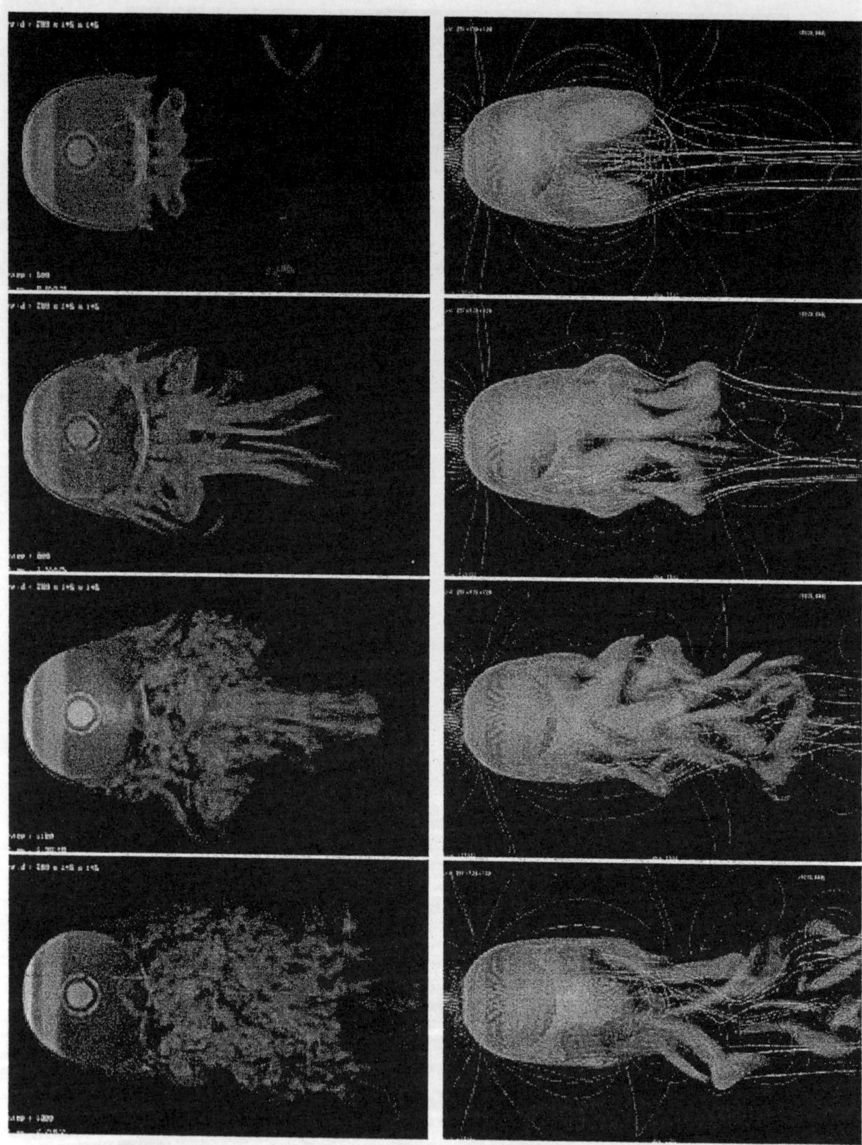

Fig.12 Development of turbulence
behind a sphere at Re=200000,
288*144*144 grid system.

Fig.13 Development of turbulence
behind a sphere at Re=2000,
256*128*128 grid system.

# 4  Visualization

It is undeniable that advances in computer power have had great impact in many aspects of our lives. Computational Fluid Dynamics (CFD) is another field that has taken advantage of the continuous development of computer hardware. Using the CFD approach, the set of partial differential equations (Navier-Stokes equations) governing the fluid flow are solved numerically, and knowledge of the flow structures can be obtained. However, when trying to improve the resolution of the problem at hand, the size of the problem becomes an issue. If this is coupled with a time-dependent solution and/or a 3D problem (as it is usually the case), the volume of data generated is considerable. Then, it is clear that without a visualization system, understanding of the underlying flow mechanisms is very difficult. By visualizing the flow fields, and overall picture of the flow field is attained and critical regions (or structures) are quickly identified.

Recognizing all of the above, our interest is focuse in obtaining access to a real-time visualization and animation tool. The gains are obvious if it is recognized that large computations involve a large amount of CPU time. Therefore, having immediate access to the generated data saves substantial amount of time while debugging because errors can easily be spotted. Additionally, for unsteady flow (as occurs at high Reynolds number), an animation tool is deemed necessary to properly understand the transient features of the flow.

All graphics for this article were generated using the visualization package (Clef2D and Clef3D) developed at the Institute of Computational Fluid Dynamics.

# 5  Conclusions

- For most practical applications, the large structures define the most important features of the flow, therefore, it is termed that resolution of the small scale structures are not needed for high Reynolds number flows.
- Standard turbulent models introduce additional viscous diffusion into the solution, masking the effects of physical diffusion. In order to avoid the above, the present formulation does not include an explicit turbulence model.
- To avoid numerical instabilities associated with aliasing errors, a third-order upwind difference scheme is incorporated into the formulation.
- A multi-directional finite-difference method is enforced to improve the accuracy of the computation, by making the solution less dependent on the direction of the flow.

# References


1. Abbott, I. H. and Von Doenhoff, A. E., 1959, "Theory of Wing Sections," Dover Pub., pp462
2. Boris, J.P., Grinstein, F.F., Oran, E.S. and Kolbe, R.L., 1992, "New insights into large eddy simulation," Fluid Dynamics Research 10, pp. 199-228

3. Chorin, A., J., 1968, Math. Comp. 22 pp. 745

4. Deardorff, J.W., 1970, "A Numerical Study of Three-Dimensional Turbulent Channel Flow at Large Reynolds Numbers," J. of Fluid Mechanics, Vol. 41, Part 2, pp. 453-480

5. Kawamura, T. and Kuwahara, K., 1984, "Computation of high Reynolds number flow around a circular cylinder with surface roughness," AIAA Paper 84-0340.

6. Kuwahara, K., 1984, "Computation of Thermal Convection with a Large Temperature Difference," Proc. International Conf. on Applied Numerical Modeling, Taiwan, R.O.C.

7. Kuwahara, K., 1992, "Flow Simulation on Supercomputers and Its Visualization," International Journal of High Speed Computing, Vol.4, No.1, pp. 49-70.

8. Kuwahara, K., 1999, "Unsteady Flow Simulation and Its Visualization," AIAA Paper 99-3405

9. Kuwahara, K. and Komurasaki, S., 2000, "Semi-Direct Simulation of a Flow around a Subsonic Airfoil," AIAA Paper 2000-2656

10. Kuwahara, K. and Komurasaki, S., 2001, "Direct Simulation of a Flow around a Subsonic Airfoil," AIAA Paper 2001-2545

11. Leonard, B.P., 1979, "A Stable and Accurate Convective Modeling Procedure Based on Quadratic Upstream Interpolation," Computer Methods in Applied Mechanics and Engineering, Vol. 19, pp. 59-89

12. Schlichting, H., 1979, "Boundary-Layer Theory", McGraw-Hill, $7^{th}$ edition

13. Takami, H. and Kuwahara, K., 1974, "Numerical Study of Three-Dimensional Flow within a Cubic Cavity," J. Phys. Soc. Japan, Vol. 37, No. 6

# Innovative Methods

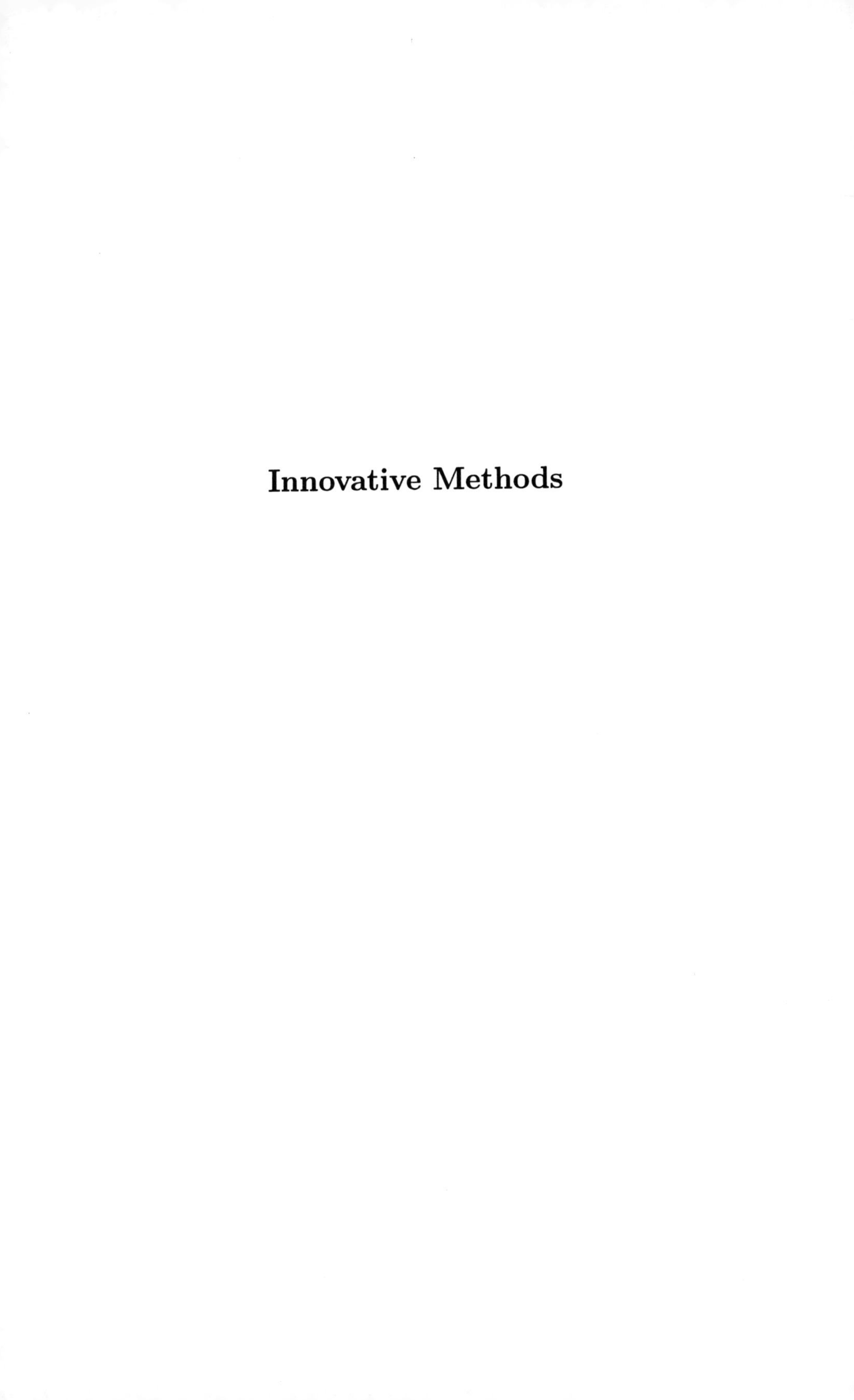

# Use of Lattice Boltzmann Method for Computing Incompressible Flows

Nobuyuki Satofuka[1], Mitsuru Ishikura[1], and Yutaka Ishikawa[1]

Department of Mechanical and System Engineering,
Kyoto Institute of Technology,
Matsugasaki, Sakyo-Ku, Kyoto 606-8585, Japan

**Abstract.** This paper deals with the lattice Boltzmann approach to CFD, especially to computation of incompressible flow problems. The approach starts from the Boltzmann equation instead of the Navier-Stokes equations on which the conventional CFD is based. Introducing BGK type simplified collision term and choosing appropriate discrete molecular velocity sets, the lattice Boltzmann BGK (LBGK) equation can be derived from the discrete velocity Boltzmann equation using suitable finite difference approximation. Calculations are carried out for a wide variety of both 2-D and 3-D, steady and unsteady incompressible flow problems. Results are compared with those of the Navier-Stokes equations. The lattice Boltzmann method is able to reproduce the dynamics of incompressible flows and could be an alternative to solving the Navier-Stokes equations.

## 1  Introduction

The Navier-Stokes equations can be derived as the macroscopic behavior of hard sphere particles with a Maxwellian velocity distribution function whose evolution is governed by the Boltzmann equation. However, much simpler kinetic models can result in the same macroscopic behavior. This idea led Frisch, Hasslacher and Pomeau to the invention of a novel technique called lattice gas automata (LGA) [1] for solving the Navier-Stokes equations. Although the LGA method has provided a fast and efficient way for solving partial differential equations, there exist some fundamental problems in this method in simulating realistic fluid flows obeying the Navier-Stokes equations. Beside its intrinsic noisy character which makes the computational accuracy difficult to achieve, it contains certain properties even in the fluid limit. The lattice gas fluid momentum equations cannot be reduced to the Navier-Stokes equations because of two fundamental problems. The first is the non-Galilean invariance property due to the density dependence of the convection coefficient. This limits the validity of the LGA method only a strict incompressible region. Second the pressure has an explicit and unphysical velocity dependence. To avoid some of these problems, several lattice Boltzmann (LB) models have been proposed [2]   [4]. The main feature of the LB method is to replace the particle occupation variables $n_i$ (Boolean variables) by the single-particle distribution function (real variables) $f_i = \langle n_i \rangle$, where

$\langle\ \rangle$ denotes a local ensemble average, in the evolution equation, i.e., the lattice Boltzmann equation. The LB model proposed by Chen et al [5] and Qian et al [6] applies the single relaxation time approximation first introduced by Bhatnager, Gross, and Krook in 1954 [7], to greatly simplify the collision operator. This model is called the lattice BGK (LBGK) model.

In the present paper, the LBGK method is used to simulate two- and three-dimensional, steady and unsteady incompressible flow problems, i.e., square driven cavity flow, decay of homogeneous isotropic turbulence and three-dimensional duct flow with/without sudden expansion. Detailed comparison of the accuracy, physical fidelity, and efficiency between the LBGK method and traditional Finite Difference Method (FDM) for solution of the incompressible Navier-Stokes equations are presented. Comparisons between the square lattice and the triangular (FHP) lattice are also discussed. The incompressible flows past a circular cylinder and a sphere are also simulated.

This paper is organized as follows. In Section 2 the lattice Boltzmann methods simulating the Navier-Stokes equations are discussed. The lattice Boltzmann simulation of two-dimensional square driven cavity flow is presented in Section 3. The two-dimensional decaying homogeneous isotropic turbulence is simulated by the LGBK method in section 4. Accuracy and efficiency of the lattice Boltzmann method in comparison with the conventional higher-order FDM approach are also discussed. Three-dimensional duct flow with/without sudden expansion is presented in section 5. The flows past a circular cylinder and a sphere are also presented in section 6. The final section contains concluding remarks.

## 2   Lattice Boltzmann Method

### 2.1  Square lattice model

In this section an outline is given of the LB methods with BGK model for the collision operator. A square lattice with unit spacing is used on which each node has eight nearest neighbors connected by eight links as shown in Fig. 1. Particles can only reside on the nodes and move to their nearest neighbors along these links in the unit time. There are two types of moving particles. Particles of type 1 move along the axes with speed $|\mathbf{e}_\alpha|$ of $c$ ($\alpha = 1, \cdots, 4$) and particles of type 2 move along the diagonal directions with speed $|\mathbf{e}_\alpha|$ of $\sqrt{2}c$ ($\alpha = 5, \cdots, 8$), where $c$ is particle speed $\delta x/\delta t$, $\delta x$ and $\delta t$ are the lattice separation and the time step, respectively. Rest particles with speed zero ($\alpha{=}0$) are also allowed at each node. The occupation of the three types of particles is represented by the distribution function $f_\alpha(\mathbf{x}, t)$, which is the probability of finding a particle at node $\mathbf{x}$ and time $t$ with velocity $\mathbf{e}_\alpha$. According to Bhatnagar, Gross, and Krook (BGK) [8], the collision operator is simplified using the single time relaxation approximation. Hence, the lattice Boltzmann BGK (LBGK) equation (in lattice unit) is

$$f_\alpha(\mathbf{x} + \mathbf{e}_\alpha \delta t, t + \delta t) \quad f_\alpha(\mathbf{x}, t) = \quad \frac{1}{\tau}[f_\alpha(\mathbf{x}, t) \quad f_\alpha^{(eq)}(\mathbf{x}, t)] \tag{1}$$

where $f_\alpha^{(eq)}(\mathbf{x}, t)$ is the equilibrium distribution and $\tau$ is the single relaxation time. The density per node $\rho$ and the macroscopic velocity $\mathbf{u}$ are defined in terms of the particle distribution function by

$$\rho = \sum f_\alpha, \quad \rho\mathbf{u} = \sum f_\alpha \mathbf{e}_\alpha \tag{2}$$

A suitable equilibrium distribution can be chosen in the following form for particles of each type

$$f_\alpha^{(eq)} = w_\alpha \rho\{1 + \frac{3(\mathbf{e}_\alpha \cdot \mathbf{u})}{c^2} + \frac{9(\mathbf{e}_\alpha \cdot \mathbf{u})^2}{2c^4} \quad \frac{3\mathbf{u}^2}{2c^2}\} \tag{3}$$

$$w_\alpha = \begin{cases} \frac{4}{9} & \alpha = 0 \\ \frac{1}{9} & \alpha = 1, \cdots, 4 \\ \frac{1}{36} & \alpha = 5, \cdots, 8 \end{cases} \tag{4}$$

where $w_\alpha$ is weighting parameter. The relaxation time is related to the viscosity by

$$\nu = \frac{2\tau \quad 1}{6} \tag{5}$$

where $\nu$ is the kinematic viscosity measured in lattice units.

The equilibrium populations are determined by assuming that they can be expressed as a power series in velocity and density of the form:

$$f_\alpha^{(eq)} = A_\alpha(\rho) + B_\alpha(\rho)\mathbf{e}_\alpha \cdot \mathbf{u} + C_\alpha(\rho)(\mathbf{e}_\alpha \cdot \mathbf{u})^2 + D(\rho)_\alpha \mathbf{u}^2 \tag{6}$$

A Chapman-Enskog procedure is then applied to determine the macroscopic behavior of this model. The values of $A_\alpha$, $B_\alpha$, $C_\alpha$ and $D_\alpha$ are chosen so that the macroscopic behavior matches the Navier-Stokes equations to as high an order as possible. The resulting continuity and momentum equations follow.

$$\frac{\partial \rho}{\partial t} + \frac{\partial \rho u_\beta}{\partial x_\beta} + O(\varepsilon^2) = 0 \tag{7}$$

$$\rho\frac{\partial u_\alpha}{\partial t} + \rho u_\beta \frac{\partial u_\alpha}{\partial x_\beta} = \quad \frac{\partial p}{\partial x_\alpha} + \frac{\partial}{\partial x_\beta}\left(\mu\left(\frac{\partial u_\beta}{\partial x_\alpha} + \frac{\partial u_\alpha}{\partial x_\beta}\right)\right) + O(\varepsilon^2) + O(M_a^3) \tag{8}$$

Characteristic dimensionless parameters are the Mach number of $\sqrt{3}U/c$ where $U$ is a characteristic macroscopic flow speed, the Knudsen number which is proportional to $c\tau/L$ where $L$ is a macroscopic flow length, and the Reynolds number of $\rho U L/\mu$.

To impose pressure boundary conditions explicitly, it is better to transform the density distribution function $f_\alpha$ to the pressure distribution function $p_\alpha$ using $\rho c_s^2 f_\alpha$, then Eq.(1) becomes

$$p_\alpha(\mathbf{x} + \mathbf{e}_\alpha \delta t, t + \delta t) \quad p_\alpha(\mathbf{x}, t) = \quad \frac{1}{\tau}[p_\alpha(\mathbf{x}, t) \quad p_\alpha^{(eq)}(\mathbf{x}, t)] \tag{9}$$

The pressure per node $p$ and the macroscopic velocity $\mathbf{u}$ are defined in terms of the pressure distribution function by

$$p = \sum p_\alpha, \quad \rho_0 c_s^2 \mathbf{u} = \sum p_\alpha \mathbf{e}_\alpha \tag{10}$$

where $\rho_0$ is the initial density. A suitable equilibrium distribution can be chosen in the following form for particles of each type:

$$p_\alpha^{(eq)} = w_\alpha[p + \rho_0\{\frac{(\mathbf{e}_\alpha \cdot \mathbf{u})}{c^2} + \frac{3}{2}\frac{(\mathbf{e}_\alpha \cdot \mathbf{u})^2}{c^4} \quad \frac{1}{2}\frac{\mathbf{u}^2}{c^2}\}] \tag{11}$$

where $w_\alpha$ is the same as Eq.(4).

## 2.2 Triangular (FHP) lattice model

Another lattice model commonly used in two-dimensional LB simulation is a triangular lattice (FHP) model. There are two types of particles on each node of the FHP model: rest particles and moving particles with unit velocity $\mathbf{e}_i$ along six directions as shown in Fig.2. The equilibrium distributions for the FHP model are given as,

$$f_0^{(0)} = d_0 \quad \rho \mathbf{u}^2 = \rho\alpha \quad \rho \mathbf{u}^2 \tag{12}$$

$$f_i^{(0)} = d + \frac{1}{3}\rho\left[(\mathbf{e}_i \cdot \mathbf{u}) + 2(\mathbf{e}_i \cdot \mathbf{u})^2 \quad \frac{1}{2}\mathbf{u}^2\right]$$

$$= \frac{\rho \quad \rho\alpha}{6} + \frac{1}{3}\rho\left[(\mathbf{e}_i \cdot \mathbf{u}) + 2(\mathbf{e}_i \cdot \mathbf{u})^2 \quad \frac{1}{2}\mathbf{u}^2\right] \tag{13}$$

where $\alpha$ is an adjustable parameter. If the ratio of rest and moving particles is defined as $\lambda = d_0/d$, the pressure is determined by the isothermal equation of state,

$$p = 3d = \frac{(1 \quad \alpha)\rho}{2} = \frac{3}{\lambda + 6}\rho \tag{14}$$

and the speed of sound is

$$c_s^2 = \frac{1 \quad \alpha}{2} = \frac{3}{\lambda + 6} \tag{15}$$

The viscosity is related to the relaxation time through an equation of the form

$$\nu = \frac{2\tau \quad 1}{8} \tag{16}$$

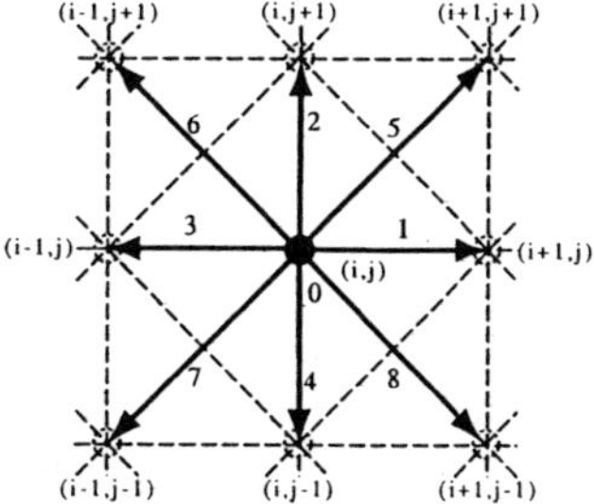

**Fig. 1.** Square lattice model.

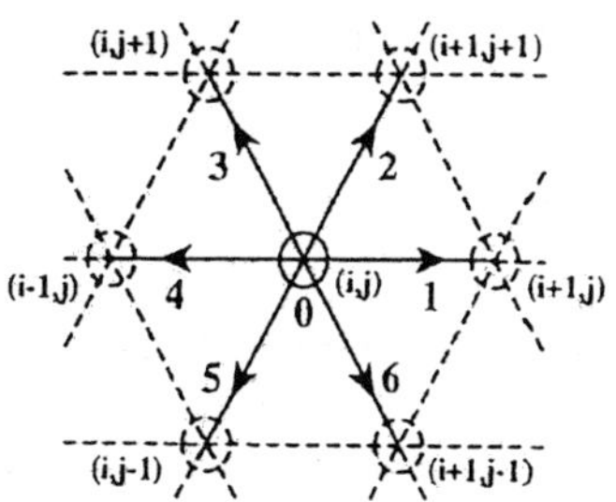

**Fig. 2.** Triangular lattice model.

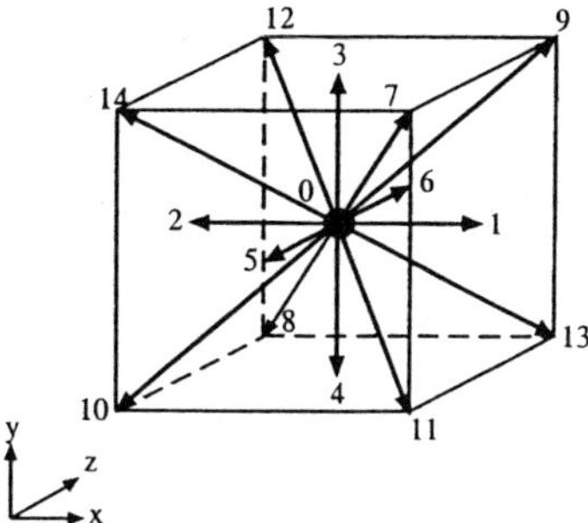

**Fig. 3.** Square lattice model.

## 2.3 Cubic lattice model

For three dimensional flow problems, a cubic lattice [9] with unit spacing is used on which each node has fourteen nearest neighbors connected by fourteen links as shown in Fig. 3. There are two types of moving particles. Particles of type 1 move along the axes with speed $|\mathbf{e}_\alpha|$ of $c$ ($\alpha = 1, \cdots, 6$) and particle of type 2 move along the links to the corners with speed $|\mathbf{e}_\alpha|$ of $\sqrt{3}c$ ($\alpha = 7, \cdots, 14$). Rest particles with speed zero ($\alpha=0$) are also allowed at each node.

A suitable equilibrium distribution can be the same as the square lattice model of Eq.(11). Then $w_\alpha$ becomes

$$w_\alpha = \begin{cases} \frac{2}{9} & \alpha = 0 \\[2mm] \frac{1}{9} & \alpha = 1, \cdots, 6 \\[2mm] \frac{1}{72} & \alpha = 7, \cdots, 14 \end{cases} \tag{17}$$

The relaxation time is related to the viscosity by

$$\tau = \frac{6\nu + 1}{2} \tag{18}$$

# 3  Square Driven Cavity Flow

## 3.1  Description of the test problem

The problem considered first is two-dimensional viscous flow in a cavity. The present simulation uses Cartesian coordinates with the origin located at lower left corner. The top boundry moves from left to right with velocity $U$ as shown in Fig. 4. The fluid motion generated in this cavity is an example of closed streamline problems that are of theoretical importance because they are part of a broader field of steady, separated flows.

Initially the velocities at all nodes, except the top nodes, are set to zero. The $x$-velocity of the top is $U$ and $y$-velocity is zero. Uniform fluid density $\rho = 2.1$ is imposed initially. Then the equilibrium particle distribution function, $f_\alpha^{(eq)}$, is calculated using Eq.(3), and $f_\alpha$ is set equal to $f_\alpha^{(eq)}$ for all nodes at $t = 0$. Bounce back boundary conditions are used on the three stationary wall. At the top boundary some of the distribution functions are unknown. These unknowns are calculated in the similar way to Noble et al [11].

## 3.2  Simulation of cavity flow

Most numerical simulation of two-dimensional cavity flow uses a vorticity-stream function formulation. One of the most comprehensive study of cavity flows to date is that by Ghia, Ghia and Shin [12].

Numerical simulations were carried out using the LBGK method for $Re = 100$ and $400$ on a $65 \times 65$ square lattice (65 lattice nodes and 64 lattice units in one side). The Reynolds number used in the present simulation is defined as $Re = UL_N/\nu$, where $U$ is the uniform velocity of the top plate, $L_N$ is the number of lattice units along one side of the cavity, and $\nu$ is the kinetic viscosity as given in Eq.(5) and Eq.(16).

Fig. 5 shows vorticity contour plots at a converged steady state. The LBGK solution (solid line) is compared with that of the second order FD solution (dashed line) of the Navier-Stokes equations in the vorticity-stream function formulation. Agreement is generally good. Velocity components along a vertical and horizontal center lines are shown Fig. 6 for $Re = 100$ (a) and $Re = 400$ (b), in which solid line shows a reference second order FD solution obtained on a $129 \times 129$ uniform grid.

Simulations were also carrried out on a $65 \times 75$ triangular (FHP) lattice. The velocity profiles change from curved at $Re = 100$ to linear for $Re = 400$ in the central core of the cavity. Agreement between the square lattice BGK method and the FD method is very close while significant difference is noticed in the FHP lattice simulation at $Re = 400$.

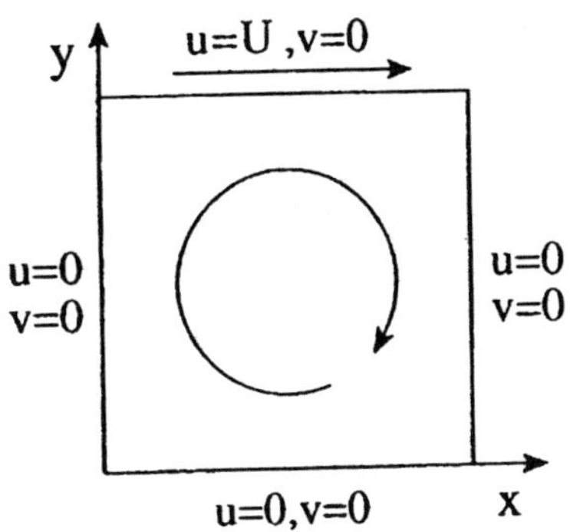



**Fig. 4.** Two-dimensional driven cavity flow.

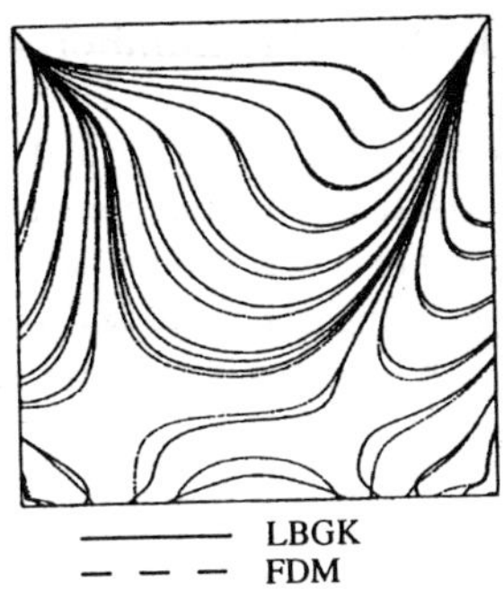



**Fig. 5.** Vorticity contours($Re = 100$, $129 \times 129$ grid points).

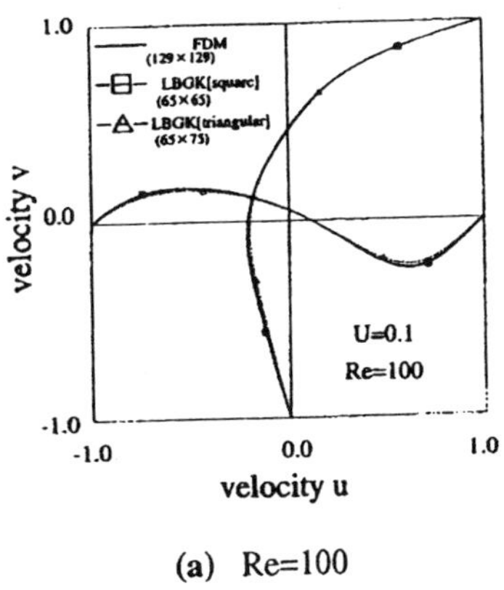

**(a)** Re=100

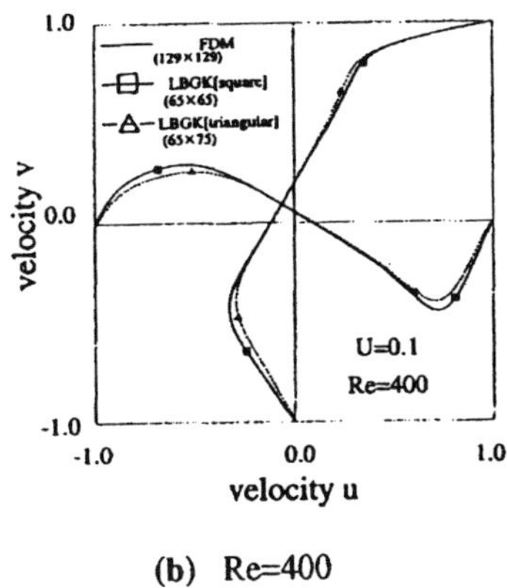

**(b)** Re=400

**Fig. 6.** Velocity components along vertical and horizontal center lines.

# 4  Two-dimensional Homogeneous Isotropic Turbulence

## 4.1  Initial and boundary conditions

The initial condition of the vorticity is randomly determined by satisfying the relation,

$$E(k) = \frac{1}{2} \sum_{|k'\ k|\leq 1/2} |\,\tilde{\omega}(k_1, k_2)\,|^2 / k'^2 = \frac{2}{3} k \exp\left(\frac{2}{3}k\right) \tag{19}$$

where $\tilde{\omega}$ denotes the vorticity in the Fourier space, $k'^2 = k_1^2 + k_2^2$, and $k_1$ and $k_2$ are the wave numbers. The periodic boundary conditions are imposed in the $x$ and $y$ directions. The computational domain is square, $(0,0) \leq (x,y) < (2\pi, 2\pi)$.

## 4.2  High Reynolds number simulation

As a large-scale direct numerical simulation of high Reynolds number homogeneous isotropic turbulence, simulation for the case with $\nu = 0.0001$ is carried out. This corresponds to the initial integral scale Reynolds number $R_L = 25500$, which is expressed as $R_L = \Omega/\nu\eta^{\frac{1}{3}}$. $\Omega$ and $\eta$ denote the total energy and the enstrophy dissipation rate, which are defined as

$$\Omega = \int_0^\infty E(k)dk, \quad \eta = 2\nu \int_0^\infty k^4 E(k)dk \tag{20}$$

The number of lattice nodes is $1025 \times 1025$. Fig. 7 shows comparison of vorticity contour plots at $t = 3.0$ between the square lattice BGK method and the 10th-order FDM. Although slight difference in vorticity contours is noticeable at late time, strikingly similar features can be found for the LBGK simulation, as compared with the solutions by the 10th-order FDM.

Time history of (a) the total energy $\Omega$ and (b) the enstrophy dissipation rate $\eta$ is shown in Fig. 8. In contrast to $\Omega$ which is monotonically decreasing in this case, $\eta$ can be amplified, as much as dissipated, and not monotonic as shown in Fig. 8. Wave number spectra of $k^3 E(k)$ at $t = 3.0$ are compared in Fig. 9. From this figure it is seen that two methods yield quite a similar answer in terms of the statistical behavior of the flow. With the present lattice of $1025 \times 1025$ nodes, the inertial range of two-dimensional turbulence can be resolved. The spectrum shown in Fig. 9 indicates that there is a range of wave number $k \leq 50$ for which $k^3 E(k)$ is roughly constant so that $E(k)$ is proportional to $k^{-3}$.

Computational cost of the LBGK method for this high Reynolds number simulation on SGI POWER ONYX 10000 is compared in table 1 with that of the 10th-order FDM. As far as efficiency is concerned, the LBGK method requires less than half CPU time per characteristic time of that of the FDM.

## 5  Three-dimensional Duct Flow

### 5.1  Description of the problem and boundary condition

Numerical simulations of the laminar flow development are carried out in a duct as shown in Fig. 10 and also in a square duct which undergo a sudden expansion with uniform step height equal to 0.5 times the width of the inlet duct as shown in Fig. 11. The noslip wall boundary conditions, $u = v = w = 0$, are used as in the three-dimensional simulations. At the inlet we assume uniform axial velocity $u = 0.1$, and the other velocity components are set to zero. The pressure at the outflow boundary is kept constant value of $p = 1.0$.

**Fig. 7.** Comparison of vorticity contours between LBGK(square) and FDM(10th-order) at $t = 3.0$, $\nu = 0.0001$.

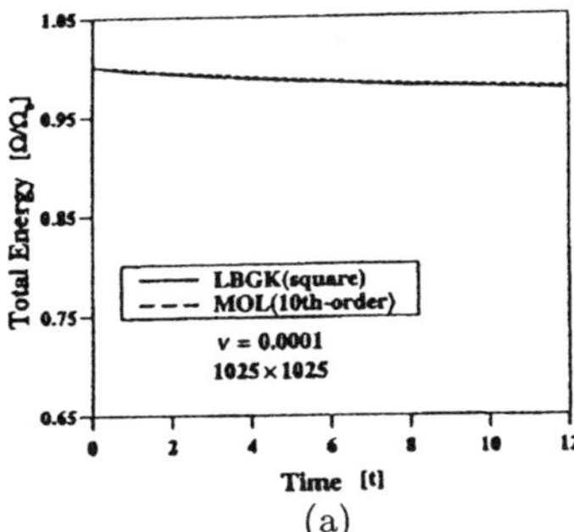



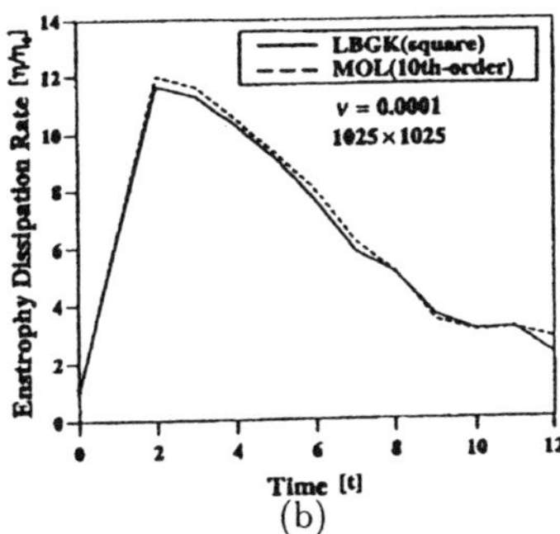



**Fig. 8.** Time history of (a) total energy $\Omega$ and (b) energy dissipation rate $\eta$ computed by LBGK(square) and FDM(10th-order) for $\nu = 0.001$.

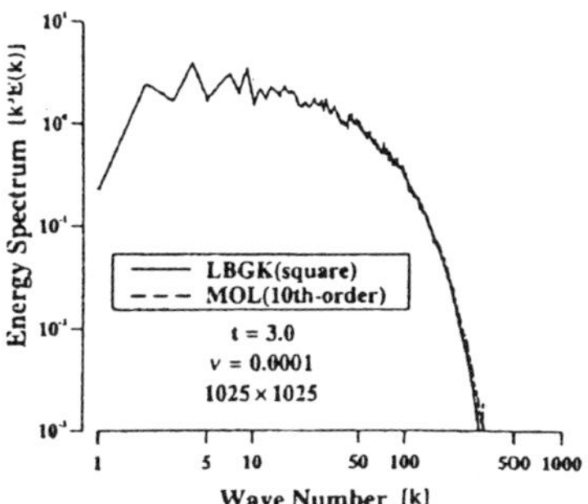



**Fig. 9.** Wave number dependence of energy spectrum $k^3 E(k)$ computed by LBGK(square) and FDM(10th-order) at $t = 3.0$, $\nu = 0.0001$.

**Table 1.** Comparison of computational cost

|  | LBGK | FDM(10th-order) |
|---|---|---|
| $\sqrt{\langle u^2 \rangle}$ | 0.04 | 1.0 |
| Time steps | 4047 | 1000 ($\Delta t = 0.001$) |
| $T_{NS}$ | 1.0 | 1.0 |
| CPU time | 63006 | 136639 |
| $\langle$ ratio $\rangle$ | $\langle 1 \rangle$ | $\langle 2.17 \rangle$ |
| CPU time/Time step | 15.47 | 136.64 |
| $\langle$ ratio $\rangle$ | $\langle 1 \rangle$ | $\langle 8.83 \rangle$ |

Computer: Silicon Graphics ONYX 10000
Number of lattice nodes: 1025×1025
Kinetic viscosity: 0.0001
Reynolds number: 25500

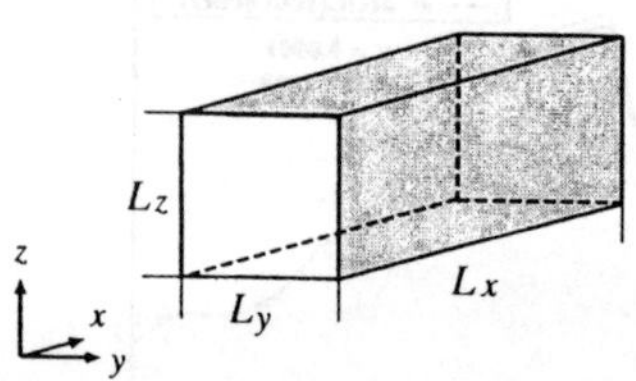

**Fig. 10.** Computational domain of duct without sudden expansion.

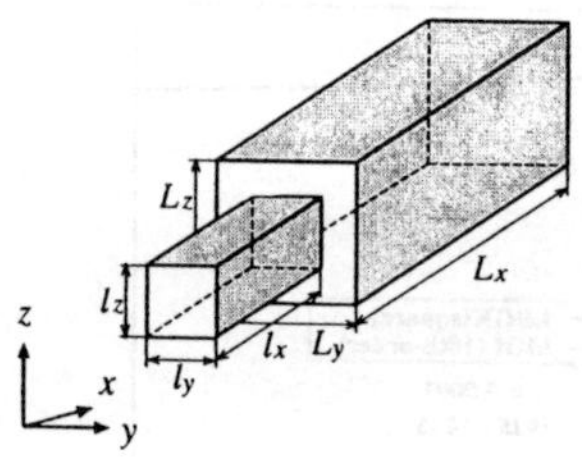

**Fig. 11.** Computational domain of duct with sudden expansion.

## 5.2 Results of simulations

In the case of a square duct without sudden expansion simulations were carried out by using a uniform cubic lattice tabulated in Table 2. Fig. 12 and Fig. 13 show velocity distributions at $x$-$z$ and $y$-$z$ cross sections of the square duct flow, and center-line velocity profiles at several longitudinal positions for $Re = 200$ computed on a $641 \times 65 \times 65$ lattice. The transitional process from uniform to fully developed velocity profile is well captured. Fig. 14 and Fig. 15 show velocity and pressure distributions at $x$-$z$ and $y$-$z$ cross sections for $Re = 400$. Fig. 16 shows velocity distributions at $x$-$z$ and $y$-$z$ cross sections of sudden expansion flow for the case with $Re = 100$ computed on a $160 \times 33 \times 33$ lattice for upstream duct and $161 \times 65 \times 65$ lattice for downstream. Fig. 17 shows center-line velocity profiles at $x = 5$ and $x = 10$. In this case the velocity profile is not yet fully developed due to a shortage of duct length.

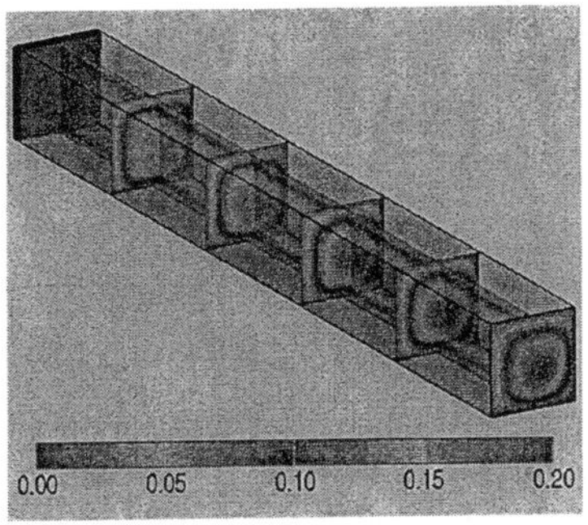

**Fig. 12.** Velocity profiles for the mainstream direction.

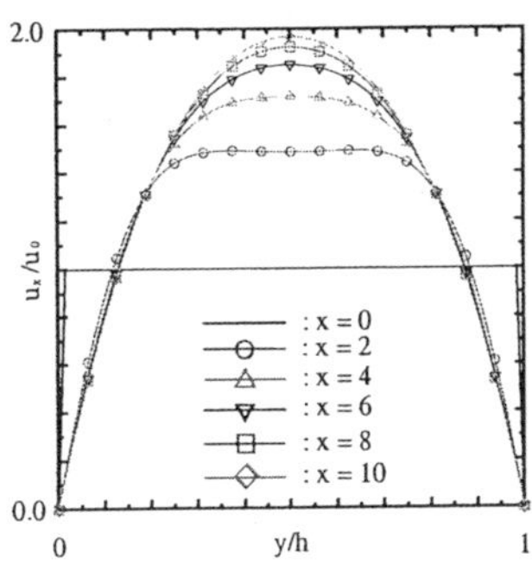



**Fig. 13.** Development of velocity profiles with $x$.

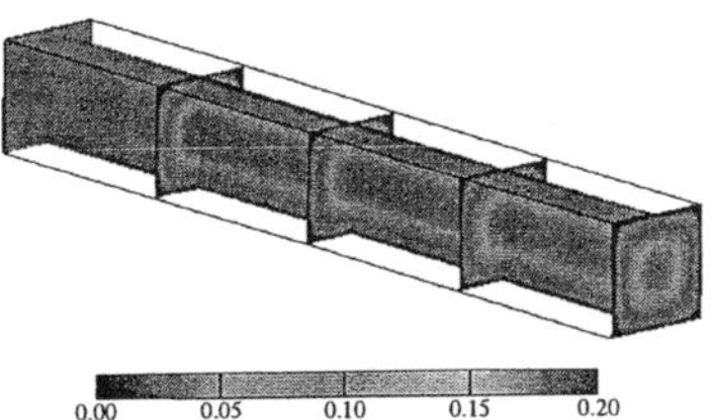

**Fig. 14.** Velocity profiles for the mainstream direction.

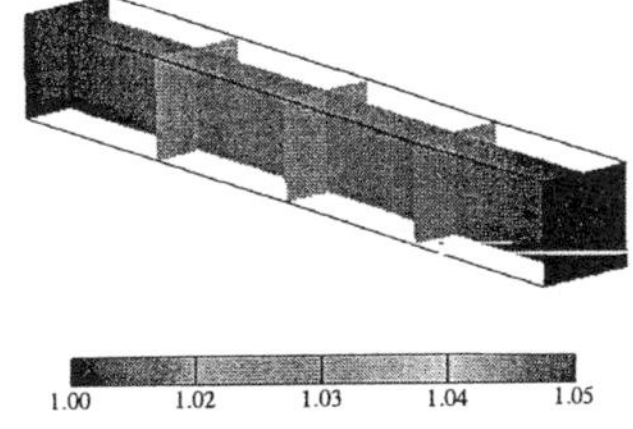



**Fig. 15.** Pressure profiles for the mainstream direction.

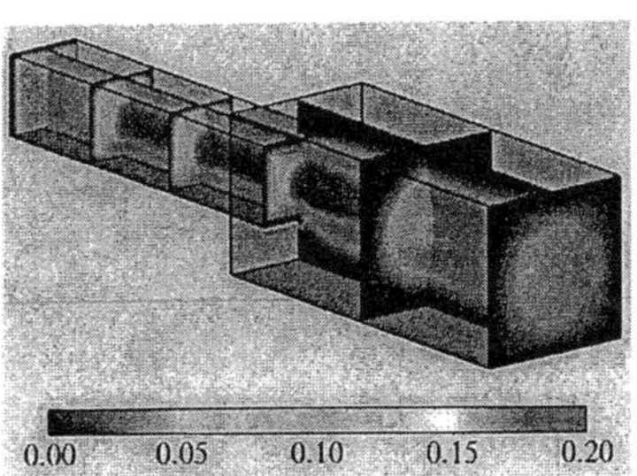

**Fig. 16.** Velocity profiles for the mainstream direction.

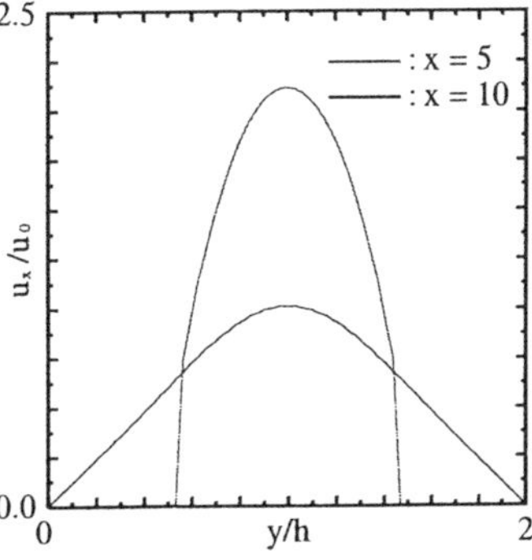



**Fig. 17.** Development of velocity profiles with $x$.

# Simulation of Structure-Fluid Interaction by Universal Solver CIP for Solid, Liquid and Gas in Cartesian Grid

Takashi Yabe, Youichi Ogata and Takao Kawai

Tokyo Institute of Technology,Tokyo 152-8552, Japan

**Abstract.** We present a review of the CIP method, which is a kind of semi-Lagrangian scheme and has been extended to treat incompressible flow in the framework of compressible fluid. Since it uses primitive Euler representation, it is suitable for multi-phase analysis. The recent version of this method guarantees the exact mass conservation even in the framework of semi-Lagrangian scheme. Comprehensive review is given for the strategy of the CIP method that has a compact support and subcell resolution including front capturing algorithm with functional transformation.

## 1   INTRODUCTION

Recent development of simulation techonology made possible the combined analysis of solid, liquid and gas. However, simultaneous treatment of materials undergoing phase state transition is still not well established.This is because the conventional combined analysis relies on the separate treatment of each phase. For example, overset grid is the easiest way to treat the structure-fluid interaction but it fails for a severely distorted structure or for melting of structure. A universal treatment of all phases by one simple algorithm is essential. In order to attack the problems mentioned above, we must first find a method to treat a sharp interface and to solve the interaction of compressible gas with incompressible liquid or solid.

Toward this goal, we take Eulerian-approach based on the CIP(cubic interpolated propagation) method[1–3] which does not need adaptive grid system and therefore removes the problems of grid distortion caused by structural break up and topology change. The material surface can be captured almost by one grid throughout the computation. Furthermore, the code can treat all the phases of matter from solid state through liquid and two phase state to gas without restriction on the time step from high sound speed.

Although the CIP can give accurate results keeping conservation error in quite low level, the semi-Lagrangian form can not guaranttee the exact mass conservation by itself. Recent version of the CIP-CSL[4,5] can overcome this difficulty and povide exactly conservative semi-Lagrangian scheme. Since these scheme do not use the cubic polynomial but use different orders of polynomial, we redefine the name of these CIP families as "Constrained Interpolation Profile" and still keep the abbreviation, CIP. This means that various constraints such as the time evolution of spatial gradient, that is used in the original CIP method, or spatially integrated conservative quantities can be used to construct the profile. In this paper, we shall give a review of the CIP method and related schemes to attack these important subjects.

## 2 CIP Method

### 2.1 Advection Processes

Although the nature is in a continuous world, digitization process is unavoidable in order to be implemented in numerical simulations. Primary goal of numerical algorithm will be to retrieve the lost information inside the grid cell between these digitized points. Most of numerical schemes proposed before, however, did not take care of real solution inside the grid cell and resolution has been limited to the grid size. The CIP method proposed by one of the authors tries to construct a solution inside the grid cell close enough to this real solution of the given equation with some constraints[3]. A simple advection equation

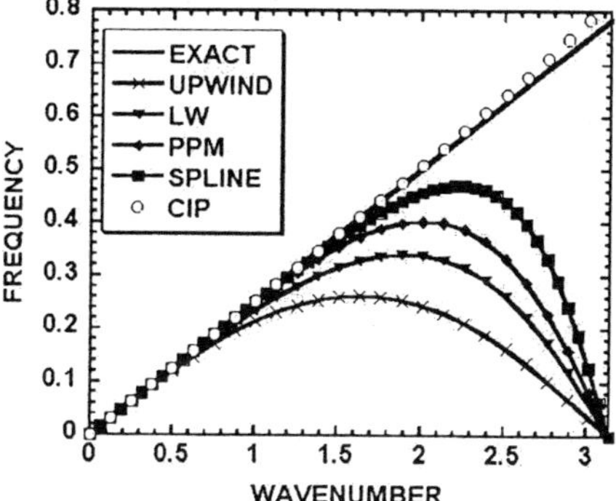


**Fig. 1.** Phase error of various schemes such as 1st order upwind, Lax-Wendroff, PPM, Spline, and CIP.

$$\frac{\partial f}{\partial t} + u\frac{\partial f}{\partial x} = 0. \tag{1}$$

is calculated by $f^{n+1} = F(x - u\Delta t), g^{n+1} = dF(x - u\Delta t)/dx$ in the CIP where $g$ stands for $\partial f/\partial x$. If two values of $f$ and $g$ are given at two grid points, the profile between these points can be interpolated by cubic polynomial $F(x) = ax^3 + bx^2 + cx + d$. Thus, the profile at n+1 step can be obtained as

$$f_i^{n+1} = a_i\xi^3 + b_i\xi^2 + g_i^n\xi + f_i^n,$$
$$g_i^{n+1} = 3a_i\xi^2 + 2b_i\xi + g_i^n, \tag{2}$$

where we define $\xi = u\Delta t$. The coefficients of polynomial $a, b$ are given by continuity requirement[3].

It would be interesting to examine phase error of various schemes using the method proposed by Purnell[6] and Utsumi $et.al.$[7]. Figure 1 summarizes those results. As is well known, phase speed of conventional schemes depart from the exact one, that is shown by the solid line, around $k\Delta x = \pi/2$. Surprisingly, however, the CIP can reproduce the correct phase speed even up to $k\Delta x = \pi$. This is remarkable because $k\Delta x = \pi$ means that one wavelength is described by 2 grid size. Let us consider the case where values of the 3 points are zero. Even in this case, one wave can exist. The CIP gives correct spatial gradients, which are non-zero at these points, and therefore it recognize the existence of the wave inside the grid cell. Any code that uses only the information of the value, that is zero now, can not correctly recognize the wave even if higher order polynomial is employed. The superiority of the CIP also appears in numerical damping and dissipation [7].

The importance of propagated gradients can be clearly demonstrated in comparison with the cubic spline[6]. Although the cubic spline uses the same cubic polynomial and the information of gradient as the CIP, it can not reproduce the result of the CIP, because the gradient of the spline is determined merely from smoothness requirement. As is easily recognized, such a constraint that is independent of the original equation will not help to retrieve the profile inside the grid cell. It is important to note that the computation time does not seriously increase by the addition of spatial gradient because coefficients of cubic polynomial can be used also for gradient calculation. Actually the CPU time required for a simple advection problem in one-dimension is Cubic-Lagrange/CIP=1.0, Spline/CIP=1.68 (Thomas method is used to solve matrix), RCIP/CIP=1.77, PPM/CIP=2.31. RCIP is the rational function CIP[3].

## 2.2 Burgers Turbulence

The CIP method is applied to Burgers equation :

$$\frac{\partial u}{\partial t} + u\frac{\partial u}{\partial x} = \nu\frac{\partial^2 u}{\partial x^2}. \tag{3}$$

After imposed random-noise initial perturbation, we observe the power spectrum calculated from various schemes as shown in Fig.2 where $\nu = 1/250$ and velocity is $u = 20 + \delta u$ and $\delta u$ is random number of $\pm 5$ size. The CIP can reproduce the spectrum regardless of mesh size. It should be noticed that the end point of the highest $k$ corresponds to $k\Delta x = \pi$ which corresponds to the wavelength of of the interfaces. In our method, rather than the original variable $f$ itself, its transformation, say $F(f)$, is calculated by the CIP method. We specify $F(f)$ to be a function of $f$ only, which means that the new function $F(f)$ is also governed by the same equation as (1). Hence, we have

$$\frac{\partial F(f)}{\partial t} + \mathbf{u} \cdot \nabla F(f) = 0, \tag{4}$$

and all the algorithms proposed for $f$ (schemes for advection equation) can be used to $F(f)$. Hopefully, by the considerable simplicity, this kind of techniques would be very attractive for practical implementation. We here use a transformation of a tangent function for $F(f)$.

Although $f$ experiences a rapid change from 0 to 1 at the interface, $F(f)$ shows a quite regular behavior. Because most of the values of $F(f)$ are concentrated near $f = 0$ and 1, the function transformation improves locally the spatial resolution near the large gradients. Thus, the sharp discontinuity can be described quite easily. This method does not involve any interface construction procedure and is quite economical in computational complexity. It should be notified that the present method is more attractive in 3-D computation since the extension of the scheme to 3-D is straightforward.

Figure 3 shows an example of this treatment. The initial sharpness of a bullet is well preserved and the discontinuities are advected with a correct speed.

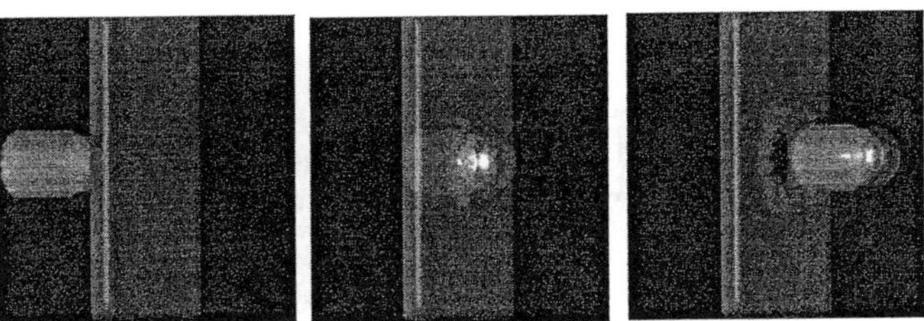

**Fig. 3.** A solid bullet penetrates through water curtain. The fixed uniform Cartesian grid of $80 \times 60 \times 40$ is used.

## 2.4 Conservative Semi-Lagrangian Scheme

It is quite useful to look for conservative semi-Lagrangian scheme because semi-Lagrangian method can be quite effective on parallel computers, suits for multiphase flow, and enables the advection calculation with large time step free from CFL condition. Although the semi-Lagrangian scheme has been successfully used in short-term atmospheric problems, the loss of exact conservation makes the scheme inappropriate for long-term problems and oceanic problems.

Furthermore, there exist subjects that require exact conservation of mass. One of the typical example is the black-hole formation and emission of gravity wave.In this case, small fraction of mass is converted into gravity wave and strict mass conservation is essential.Another example is plasma simulation in which Vlasov equation in six-dimensional phase space must be solved and total density of particle must be conserved, otherwise large electric field appears. The CIP method can be constructed to exactly conserve the total mass in Vlasov system of equally spaced grid in phase space[14]. However, the use of non-equally spaced grid or other general grids can save the computational cost and is worthy for investigation.

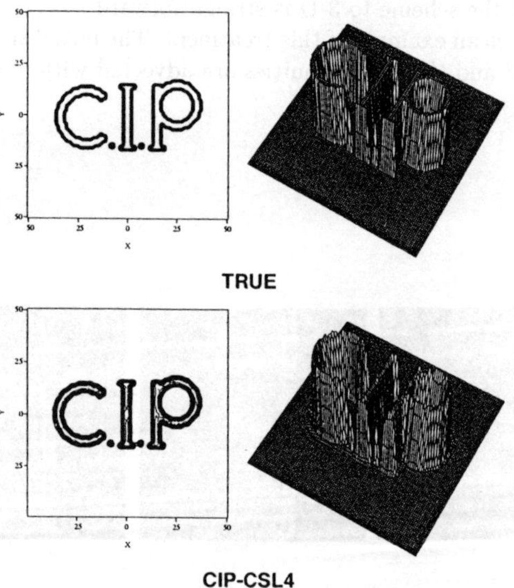

**TRUE**

**CIP-CSL4**

**Fig. 4.** Contour plots and three-dimensional views after one complete revolution of a solid-body which consists of three characters of "C.I.P" and all the lines composing the characers are thiner than 3 grid points. The dot of only one-grid size is preserved after one revolution.

In this sense, the effort to establish exact conservation in semi-Lagrangian form would be a challenging task. Toward this goal, we proposed two schemes CIP-CSL4[4] and CIP-CSL2[5]. In CIP-CSL4, the fourth-order polynomial is chosen as an interpolation function and the time development of $f$ and $g$ is calculated simply by shifting the interpolation function $F_i(x)$ by $u\Delta t$ in the same way as Eq.(2) of the CIP method. Since the order of polynomial is increased, we have now freedom to choose another constraint to determine the polynomial. We use the constraint to the integrated value to guarantee the mass conservation. Thus this scheme is able to describe a structure of one-grid size as shown in Fig.4 even without tangent transformation.

The CIP-CSL2 uses a set of equations :

$$\frac{\partial D}{\partial t} + u\frac{\partial D}{\partial x} = 0, \quad \frac{\partial f}{\partial t} + \frac{\partial(uf)}{\partial x} = 0. \tag{5}$$

These two equations are similar to the equations for $f$ and $g$ in Eq.(1) by replacing $f$ by $D = \int f dx$ and $g$ by $f = \partial D/\partial x$. Therefore, we can use all the procedure of the CIP method for the integral value of $f$. We can impose the conservation of integrated value easily by this change. The CIP-CSL2 is easily extended to multi-dimensions and can be put into a compact form[15].

## 3  Semi-Lagrangian Approach to Hydrodynamic Equations

### 3.1  Pressure-Projection Algorithm in Primitive Euler Scheme

Before presenting a method to solve all the phases of materials, we must at first construct a unified equation to descrive all the phases. For this purpose we use the following set of hydrodynamic-type equations :

$$\frac{\partial \mathbf{f}}{\partial t} + (\mathbf{u} \cdot \nabla)\mathbf{f} = \mathbf{S} \tag{6}$$

Here, $\mathbf{f} = (\rho, \mathbf{u}, T)$, $\mathbf{S} = (-\rho\nabla \cdot \mathbf{u} + Q_m, -\nabla p/\rho + Q_u, -P_{TH}\nabla \cdot \mathbf{u}/\rho C_v + Q_E)$, and $\rho$ is the density, $\mathbf{u}$ the velocity, $p$ the pressure, $T$ the temperature, $Q_m$ represents the mass source term, $Q_u$ represents viscosity, elastic stress tensor, surface tension etc., and $Q_E$ represents viscous heating, thermal conduction and heat source. Here, $C_v$ is the specific heat for constant volume and we define $P_{TH} = T(\partial p/\partial T)_\rho$ which is derived from the first principle of thermodynamics.

The underlying physics included in the above equations for continuum dynamics is complex and may include processes which possess quite different time scales of variation. It should be expedient to separate the solution procedure into several fractional steps.

The CIP method uses the primitive Euler method to solve Eq.(6) with a separate treatment of advection term like conventional semi-Lagrangian formulation, thus the formulation into a simultaneous solution of incompressible and compressible fluid is readily obtained.

Original CCUP (CIP-Combined Unified Procedure) method was proposed only for a special equation of state, but here we rebuild it with more general EOS [3]. That is , for small change of density and temperature, the pressure change can be linearly proportional to them as

$$\Delta p = \left(\frac{\partial p}{\partial \rho}\right)_T \Delta \rho + \left(\frac{\partial p}{\partial T}\right)_\rho \Delta T \tag{7}$$

where $\Delta p$ means the pressure change $p^{n+1} - p^*$ during one time step and $*$ is the profile after advection. This applies also to $\rho, T$. From this relation, once $\Delta \rho, \Delta T$ are predicted, $\Delta p$ will be predicted based on Eq.(7). Needless to say, $\partial p/\partial \rho, \partial p/\partial T$ are given by EOS. Since the CIP separates the non-advection terms from the advection, we can concentrate on the non-advection terms related to sound waves which are the primary cause of the difficulty in liquid having large sound speed. Thus we obtain

$$\nabla \left(\frac{1}{\rho^*} \nabla p^{n+1}\right) = \frac{p^{n+1} - p^*}{\Delta t^2 (\rho C_s^2 + \frac{P_{TH}^2}{\rho C_v T})} + \frac{\nabla \cdot \mathbf{u}^*}{\Delta t} \tag{8}$$

This equation shows that, at the sharp discontinuity, $\mathbf{n} \cdot (\nabla p/\rho)$ is continuous. Since $\nabla p/\rho$ is the acceleration, it is essential that this term is continuous since the density changes by several orders of magnitude at the boundary between liquid and gas. In this case, the denominator of $\nabla p/\rho$ changes by several orders and pressure gradient must be caulcuated accurately enough to ensure the continuous change of acceleration. The equations derived by the ICE[16] in the framework of conservative Euler seems to be quite similar to Eq.(7) but the continuity of $\nabla p/\rho$ in the formers is not guaranteed.

## 3.2  Application to Skimmer

Figures 5 shows an example of fluid-structure interaction calculated by the procedure given in the previous section. A thin aluminum cylinder is thrown over water surface. As our experience tells, the cylinder slides over the surface and then jumps up. Figure also shows the experimental result at the same condition. Two figures show several similarity. The swelling up of water at the circumference has a triangular shape and the water was drawn by the bottom of the aluminum disk after the disk detached itself from the water surface.

During the process, the attack angloe of the disk changed in time. Since the disk can move being influenced by the sum of ambient pressure acting on the disk surface, we can observe the time evolution in the simulation. We found the similarity of this time evolution of attack angle between experiment and simulation.

Since the mesh used here is $190 \times 60 \times 67$, the resolution is poor in the splashing of water in the front side.

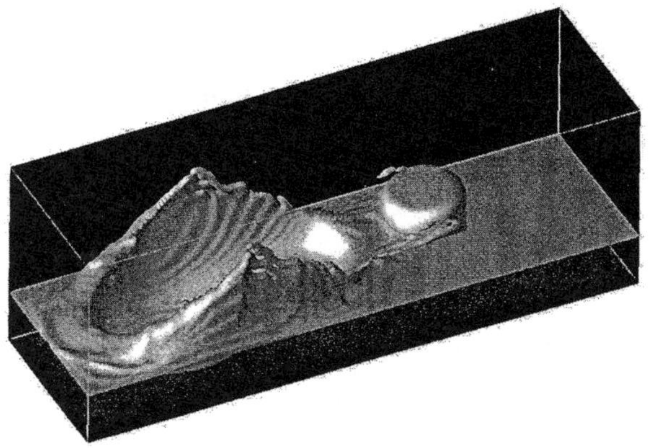

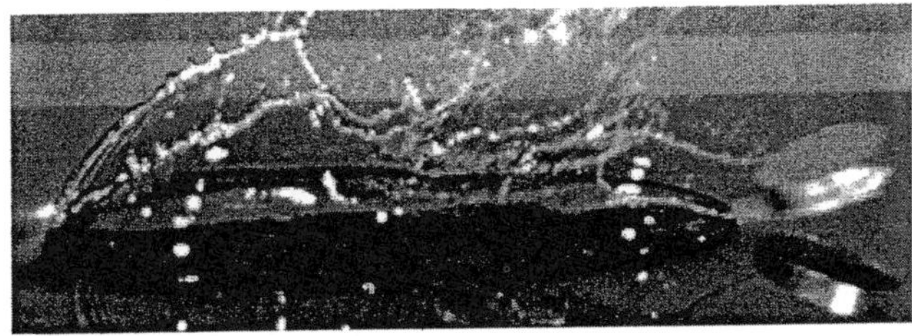

**Fig. 5.** Flat bottom (Top :Simulation, Bottom:Experiment).

## 4 Matrix-Free Incompressible-Flow Calculation

Since the CIP is superior in the calculation of advection equation, we can apply the procedure to conventional characteristic equations. Then, if this equation can be solved with large CFL, compressible and incompressible flow can be solved without matrix solution of pressure. As a simple application, we deal with shallow water equation. Let $h$ be the height of water surface, then shallow water equations are simply written in one dimension as

$$\frac{\partial}{\partial t}\begin{pmatrix} h \\ u \end{pmatrix} + \begin{pmatrix} u & h \\ g & u \end{pmatrix} \frac{\partial}{\partial x}\begin{pmatrix} h \\ u \end{pmatrix} \equiv \frac{\partial \mathbf{W}}{\partial t} + \mathbf{A}\left(\mathbf{W}\right)\frac{\partial \mathbf{W}}{\partial x} = 0. \tag{9}$$

Equation (9) leads to two equations for the Riemann invariants,

$$\frac{D^{\pm}}{Dt}\left(\Gamma \pm \frac{u}{2}\right) = 0, \quad \frac{D^{\pm}}{Dt} \equiv \frac{\partial}{\partial t} + \lambda^{\pm}\frac{\partial}{\partial x}. \tag{10}$$

where $\Gamma \equiv \sqrt{gh}$, $\lambda^{\pm} = u \pm \Gamma$ and $g$ represents the gravity. The form of Eq.(10) is the same as Eq.(1), so the CIP method can be applied to Eq.(10). Since the phase speed of the gravity waves $\Gamma$, which is equivalent to sound waves in hydro-dynamics, is so fast, time step is restricted by the CFL($= \lambda^{\pm}\Delta t/\Delta x$) condition of gravity wave when conventional explicit Eulerian schemes are employed. There-fore, the semi-Lagrangian approach[17][18] to the characteristic equations has been used for solving gravity wave to make time step be much longer than CFL condition.

This algorithm is tested with initial perturbation of $h$ in Gaussian form.

$$h(x, t = 0) = 1.0 + 0.01 \cdot exp\{-\left(\frac{x - X_{max}/2}{50.0}\right)\},\tag{11}$$

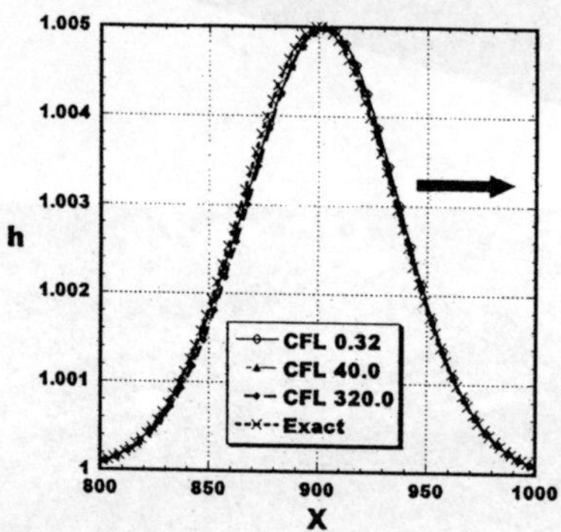

**Fig. 6.** The height of water wave at $t = 400.0$ calculated by the CIP. The results with various CFL numbers are compared. The scheme works even for CFL=320.

The mesh size of $\Delta x = 0.25$ and the number of meshes is 4000. The numerical solution is shown in Fig.6 and agrees well with analytical solution even for very large CFL=320.0.

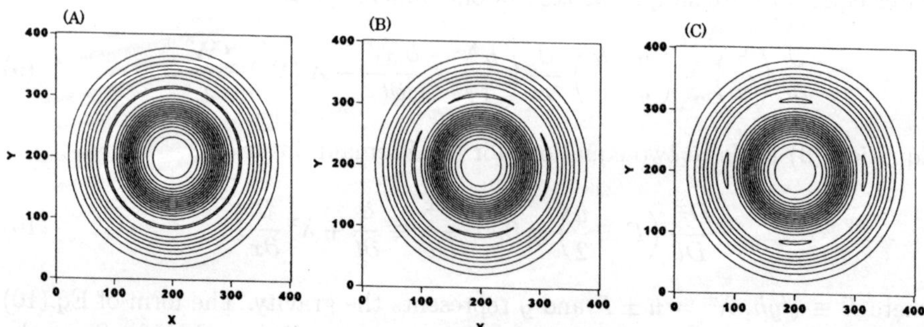

**Fig. 7.** The height contour of water wave at t=100.0.(A) CFL=1.4, (B)5.0, (C)14.0.

Next, we shall extend the scheme to two dimensions. We use directional splitting here and repeatedly apply the procedure of one dimension to each direction. The initial condition is set as follows.

$$h(x, y, t = 0) = 1.0 + 0.001 \cdot \exp\{-\left(\frac{x - X_{max}/2}{50.0}\right) - \left(\frac{y - Y_{max}/2}{50.0}\right)\}. \quad (12)$$

Mesh size is $\Delta x = \Delta y = 1.0$, the number of meshes is $(NX, NY) = (400, 400)$. We take CFL=1.4, 5.0 and 14.0, and calculate up to $t = 100.0$. The contours for each CFL are shown in Fig.8.

Even in two dimensions, the method can give a symmetrical wave propagation under large CFL condition. Therefore we conclude that the present method is applicable to the cases in which the change of characteristic speed can be neglected for one time-step, that is, perturbations are small. This does not restrict the applicability of the present method. If the speed of gravity wave is much larger than fluid velocity, variation of characteristic speed becomes smaller and large CFL can be used. The same can be said for incompressible flow because sound speed is much faster than fluid velocity.

Considering the conventional solution to incompressible flow, the use of the Poisson equation for pressure and its implicit solution do not trace the exact movement of sound wave profile, and the wave form in the solution above Lipschitz condition($\Delta u \Delta t/\Delta x = 1$) is smeared out by strong damping. If the phenomena in the time scale of $\Delta u \Delta t/\Delta x < 1$ are important for the system, these should be correctly treated in the time scale of $\Delta u \Delta t/\Delta x < 1$. If not, we can still use larger time step $\Delta u \Delta t/\Delta x >> 1$ suffering the exact shape of solutions. Since the present scheme is applicable to multi-dimensional shallow water equations, it is possible to apply it to hydrodynamics directly in future.

# References


1. H.Takewaki,A.Nishiguchi and T.Yabe : J. Comput. Phys. **61**, 261 (1985)
2. T.Yabe and T.Aoki : Comput.Phys.Commun. **66**, 219 (1991)
3. T.Yabe, F.Xiao and T.Utsumi: J. Comput. Phys. **169**, 556 (2001)
4. R.Tanaka,T.Nakamura and T.Yabe : Comput. Phys. Commun. **126**, 232 (2000)
5. T.Yabe, R.Tanaka, T.Nakamura and F.Xiao : Mon. Wea.Rev. **129**, 332(2001)
6. Purnell,D.K. : Mon. Wea. Rev. **104**, 42 (1975)
7. T.Utsumi, T.Kunugi, and T.Aoki : Comput. Phys. Commun. **101**, 9 (1996)
8. T.Kawamura and K.Kuwahara : AIAA Paper **84-0340** (1984)
9. K.Kuwahara : AIAA Paper **99-3405** (1999)
10. C.W.Hirt and B.D.Nichols : J. Comp. Phys. **39**, 201(1981)
11. S.O.Unverdi and G.A. Tryggvasson : J. Comp. Phys. **100**, 25(1992)
12. S. Osher and J.A.Sethian : J. Comp. Phys. **79**, 12(1988)
13. T.Yabe and F.Xiao : Comput. Math. Applic. **29**, 15(1995)
14. T.Nakamura and T.Yabe : Comput.Phys.Commun. **120**, 122 (1999)
15. K.Takizawa, T.Yabe and T.Nakamura: Comput.Phys.Commun. (2002) in press
16. F.H.Harlow and A.A.Amsden : J.Comput.Phys. **3**, 80 (1968)
17. J.R.Bates and A.Mcdonald : Mon.Wea.Rev. **110**, 1831 (1982)
18. G.Erbes : Mon.Wea.Rev. **121**, 3443 (1993)

# Vorticity Confinement - Recent Results: Turbulent Wake Simulations and A New, Conservative Formulation

J. Steinhoff [*], M. Fan [†], and L. Wang [‡]

## Abstract

Vorticity Confinement has been shown, over the last several years, to be a way to quickly and cheaply approximate incompressible flows over complex configurations at high Reynolds number. For these flows, there are many important salient features that are reproduced by the method in a simple way, on relatively coarse uniform Eulerian computational grids with fast low-order computational methods. No complex, high order schemes and no extensive refinement or body conforming grid are required. The basic features of the flow that are very effectively simulated all involve vortical regions: thin boundary layers on solid surfaces that are attached or separate, thin vortex sheets and thin filaments that can convect over long distances with no significant physical diffusion, and, finally, turbulent wakes, including small scale effects. These features are very difficult to treat with conventional schemes. The main point is that almost all of the vortical regions in these flows are either thin, so that their internal structure is not important (analogous to shocks or contact discontinuities in compressible flow), or contain small scale structures, such as turbulent wakes. All of these vortical structures are, of course, embedded in an incompressible irrotational flow field, which is also efficiently computed.

First, the salient features of Vorticity Confinement solutions will be discussed, to give the reader an understanding of the characteristics of the method. Then, a short presentation of the original formulation of the method will be given and a number of recent papers giving more details referenced. This will be followed by some recent results where the method is used to treat the small scales in turbulent wakes in LES-type simulations. Many other recent results concern free vortices and attached flow over complex configurations, as well as validation studies. Papers describing these will be referred to. Finally, a new,

[*]Professor, The University of Tennessee Space Institute, Tullahoma, TN

[†]Research Scientist, The University of Tennessee Space Institute, Tullahoma, TN

[‡]Graduate Research Assistant, The University of Tennessee Space Institute, Tullahoma, TN

simpler formulation will be described that is effective for both Vorticity Confinement as well as convection of thin streams of passive scalars. Preliminary results of this new formulation for convecting vortices and scalars in 2-D will then be presented.

# 1  Introduction

Vorticity Confinement has been shown, over the last several years, to be a way to quickly and cheaply approximate incompressible flows over complex configurations at high Reynolds number[1-10]. For these flows, there are many important salient features that are reproduced by the method in a simple way, on relatively coarse uniform Eulerian computational grids with fast low-order computational methods. No complex, high order schemes and no extensive refinement or body conforming grid are required. The basic features of the flow that are very effectively simulated all involve vortical regions: thin boundary layers on solid surfaces that are attached or separate, thin vortex sheets and thin filaments that can convect over long distances with no significant physical diffusion, and, finally, turbulent wakes, including small scale effects. These features are very difficult to treat with conventional schemes. The main point is that almost all of the vortical regions in these flows are either thin, so that their internal structure is not important (analogous to shocks or contact discontinuities[11] in compressible flow), or contain small scale structures, such as turbulent wakes. All of these vortical structures are, of course, embedded in an incompressible irrotational flow field, which is also efficiently computed.

The main idea behind Vorticity Confinement is that, at high Reynolds number, the above small vortical scales are modeled on the grid in an inherently discrete, simple way that is much more efficient than if model partial differential equations (pde's) were first formulated and then resolved with finite difference approximations. This modeling involves nonlinear terms in the discrete equations that result in a stable negative total diffusion for the vorticity in a band of length scales. This negative diffusion acts to confine the small scale vorticity to a thickness of about 2 grid cells: Below this scale, the structure is dominated by a positive diffusion. On scales significantly larger, the negative diffusion causes the structure to contract. The equilibrium solution results from a balance of these two effects. An important feature is that the irrotational flow regions, as well as the very large vortical scales that are present in the turbulent wakes are not significantly affected by these Vorticity Confinement terms and are accurately resolved and convected by the method, which reduces to a conventional CFD method at these large scales.

First, the salient features of Vorticity Confinement solutions will be discussed, to give the reader an understanding of the characteristics of the method. Then, a short presentation of the original formulation of the method will be given and a number of recent papers giving more details referenced. This will be followed by some recent results where the method is used to treat the small scales in turbulent wakes in LES-type simulations[2]. Many other recent results concern free vortices and attached flow over complex configurations, as well as validation studies. Papers describing these will be referred to. Finally, a new, simpler formulation will be described that is effective for both Vorticity Confinement as well as convection of thin streams of passive scalars. Preliminary results of this new formulation for convecting vortices and scalars in 2-D will then be presented[1].

## 2 Vorticity Confinement: Salient Features

There are several different ways to characterize Vorticity Confinement - they are all important in understanding how it works:

1. It is a way of treating flow with small vortical scales as "weak solutions" of the Euler equations, where the small scales are treated as regularized singular regions. In this way it is analogous to shock capturing-quantities are conserved, not at each point but integrated through the vortical feature, which is spread over a few grid cells. The confinement terms serve as a simple implicit model of the internal structure and have little effect on the flow external to the feature, which is accurately treated. Well known examples of this type of treatment include, besides shock capturing, vortex filament and vortex lattice schemes for convecting vortex sheets. Also, the use of panel methods for external flows, where thin vortical boundary layers are simply treated involves similar ideas[12-16]. An important example of this independence of the flow outside the features involves vortex filaments with near-axisymmetric cores with radius of curvature much larger than core radius, where the flow is approximately 2-D in planes normal to the filaments.

2. The current Vorticity Confinement method can be treated as a "zeroth order" approximation for the thin vortical features. As in well-known interacting boundary layer schemes, the basic solution can be treated as a first term in an asymptotic expansion. Higher order terms would involve modeling some of the effects of the structure, such as displacement thickness and skin friction. In addition to boundary layers, such secondary effects could include long-term spreading of aircraft trailing vortices[17] and modeling of any effects of axial flow on vortex breakdown. For most of the problems that we have treated until now with Vorticity Confinement, these effects have not been very important and the "zeroth order" approach was sufficient. Cases where such higher order effects have already been treated include long-term trailing vortex simulation[17], and a study involving flow separating from a curved surface[3] where the boundary layer was turbulent and affected the separation location.

It should be mentioned that, of course, a very large amount of work has been done, over the last few decades, on turbulent boundary layer modeling using conventional pde-based models, (mostly calibrating Reynolds averaged

"RANS" models). The same is true of the small scales in turbulent wakes, using large eddy simulation (LES). By comparison, very little turbulent modeling has been done based on Vorticity Confinement, i.e., by adding "higher order" additional discrete terms with adjustable parameters to the basic method. For these reasons we cannot claim that Vorticity Confinement turbulence models currently exist for these "first order" effects. We are only stating that Vorticity Confinement appears to be a promising approach and a framework for modeling, because of its simplicity and efficiency. Additional work should be done to develop confinement-based turbulence models.

3. For convecting vortex filaments, the internal dynamics is, effectively, treated as "fast" dynamics which approximately satisfies the confinement terms separately from the other, conventional CFD terms used in the method (these will be described below). This fast dynamics is then slaved to the basic fluid dynamics as the filaments are "slowly" convected through the flow.

4. The Vorticity Confinement method can be thought of as incorporating both conventional CFD ideas for solving for the "smooth", large scale features of the flow, and intrinsically discrete ideas for solving for the small scale features: The confinement terms result in intrinsically discrete thin, small scale solitary-wave structures that "live" on the computational lattice. In this way, it is similar to lattice gas approaches. For the large scales, as mentioned, the confinement terms have little effect and the method reverts to conventional CFD, which is well-known to be effective for these scales.

For small scales, the method corrects the well known difficulty of treating thin features (with, of course, large gradients) inherent in conventional CFD schemes. For example, very fine body conforming grids are required with conventional "Navier-Stokes" turbulence modeling methods for boundary layers, and even then, the result is a pde-based *model* solution. Thin boundary layers are very simply treated with confinement as intrinsically discrete structures.

On the other hand, while intrinsically discrete models such as "lattice gas" can, perhaps, be efficiently formulated for thin regions such as turbulent boundary layers, their treatment of large scales is very inefficient. For example, lattice gas schemes only converge statistically to smooth laminar Navier-stokes solutions; Samples of the solution at large numbers of grid points ($N_s$), over large numbers of time steps ($N_t$) must be used in any small region in time and space in order to estimate the solution there. For large scales, therefore, these methods converge like $\sqrt{N_s N_t}$ and can be thought of as "half-order" accurate, as opposed to conventional CFD, for which very high order methods have been developed[18].

# 3  Vorticity Confinement: Basic Formulation

A specific solver will be described which employs Vorticity Confinement. However, Vorticity Confinement can be implemented in a pre-existing flow solver, for both incompressible and compressible flow, by adding a term to the discretized momentum conservation equations[8].

For general unsteady incompressible flows, the governing equations with the Vorticity Confinement term are discretizations of the continuity equation and the momentum equations, with an added term:

$$\nabla \cdot \vec{q} = 0$$

$$\partial_t \vec{q} = -(\vec{q} \cdot \nabla)\vec{q} + \nabla(p/\rho) + [\mu \nabla^2 \vec{q} - \epsilon \vec{s}]h^2/\Delta t \tag{3.0.1}$$

where $\vec{q}$ is the velocity vector, $p$ is the pressure, $\rho$ is the density, and $\mu$ is a diffusion coefficient that includes numerical effects (we assume physical diffusion is much smaller). For the last term, $\epsilon \vec{s}$, $\epsilon$ is a numerical coefficient that, together with $\mu$, controls the size and time scales of the convecting vortical regions or vortical boundary layers. For this reason, we refer to the two terms in the brackets as "confinement terms". Also, $h$ is the grid cell size and $\Delta t$ the time step.

The first confinement term, or diffusion term is usually specified explicitly, but can be implicitly present in the solver, such as in a lower order upwind scheme. There are also many possible forms for the second confinement term. First, the original one used in this and earlier studies will be described. Then, a new more elegant, simpler form will be described, together with a simple demonstration.

# 4 Original Vorticity Confinement

## 4.1 Formulation

We define the second confinement term as

$$\vec{s} = \frac{1}{h}\hat{n} \times \vec{\omega}$$

where

$$\hat{n} = \frac{\nabla \eta}{|\nabla \eta|}$$

and the vorticity vector is given by

$$\vec{\omega} = \nabla \times \vec{q}$$

and

$$\eta = |\vec{\omega}|$$

In general, for boundary layers and convecting vortex filaments, since computed flow fields *external* to the vortical regions are not very sensitive to the details of the internal structure as long as they are smooth, they are not sensitive to the parameters $\epsilon$ and $\mu$ over a wide range of values. Hence, the issues involved in setting them are similar to those involved in setting numerical parameters in other analogous computational fluid dynamics schemes, such as artificial dissipation in conventional shock capturing schemes. For wake flows, $\epsilon$ and $\mu$ can be used to approximately simulate finite Reynolds number effects, since they control the intensity of the small vortical scales. This is an area of study that is just beginning - some results of which will be reported here.

An important feature of the Vorticity Confinement method is that the confinement terms are non-zero only in the vortical regions, since both the first (diffusion) term and the second (anti-diffusion) term vanish outside those regions (care has to be taken in the numerical implementation to preserve this feature).

Another important feature concerns the total change induced by the confinement correction in mass, vorticity and momentum, integrated over a cross section of a convecting vortex. It can be shown[5,9] that mass is conserved because of the pressure projection step in the (incompressible) solver and vorticity is explicitly conserved because of the vanishing of the correction outside the vortical regions. Momentum is almost exactly conserved. An extension of the original method[10] explicitly conserves the momentum at the expense of some additional complications. The new, simpler formulation, described below in Sec.5.2 explicitly conserves momentum with no required extensions.

## 4.2  Results of Wake Study

Two test cases are presented in this section involving the wake structure of a 3-D circular and square cylinder.

### 4.2.1  3-D Circular Cylinder

Flow over a 3-D circular cylinder was calculated to assess the ability of Vorticity Confinement to accurately model the wake flow behind a blunt body. A long cylinder was "immersed" in a uniform $141 \times 101 \times 61$ Cartesian grid in the streamwise, normal and spanwise directions, respectively. Periodic conditions were imposed at the lateral boundaries. The diameter of the cylinder was 15 grid cells. Comparisons between experiment and computed results are given in Figs. 1 and 2. In the figures, the origin of the coordinate system used is located in the center of the cylinder, and all distances are non-dimensionalized by the diameter. The experimental results of Lourenco and Shih[19] at a Reynolds number of about 3,900, were compared to.

The diffusion coefficient $\mu$ was held constant for this study. The confinement coefficient, $\epsilon$, was adjusted to impose different levels of confinement. This resulted in different levels of the intensity of the small vortical scales in the wake, and approximately simulated different Reynolds numbers.

Figure 1 depicts the mean streamwise velocities resulting from three values of the confinement coefficient. Figures a) and b) depict the result of two levels of confinement, where the flow is measured along lines traversing the wake (normal to the cylinder axis and the mean stream) at three different locations in the wake of the cylinder. For both values of $\epsilon$ the agreement can be seen to be very good, indicating that the effect of the confinement parameter is small over a range of values. Results from a case without confinement ($\epsilon = 0$) are depicted in Fig. 2c. Without confinement, the flow field is dominated by diffusive effects that are not counterbalanced by the anti-diffusive confinement term and approximates a steady, low Reynolds number flow.

The ability of Vorticity Confinement to model the turbulent wake was assessed by computing the rms streamwise velocity fluctuations in the wake region. Comparisons of these fluctuations with the experimental data were made along the same lines in the wake as for the mean velocity, for the same three values of $\epsilon$. Fig. 2a ($\epsilon = 0.25$) shows very good agreement. This is the same value that shows the best agreement for the mean velocity, as can be seen in Fig. 1. The effect of increasing confinement to 0.5 is depicted in Fig. 2b. In general, the effect of increased confinement is to thin the shear layer comprising the wake boundary, and to increase the fluctuation of the time-dependent flow from the mean flow. Fig. 3c depicts results without confinement. Without confinement, the flow can be seen to be steady and exhibits none of the fluctuations that occur when confinement is used.

Figure 3 depicts isosurfaces of vorticity magnitude for the same three levels of confinement. The use of confinement results in chaotic flow patterns, as would be expected in three-dimensional turbulent flows. This does not occur in two-dimensional simulations, indicating that this chaotic behavior is not due to numerical instability created by the confinement. Increasing confinement increases the chaotic nature of the flow in 3-D and reduces the characteristic size of the vortical structures, analogous to what would be expected with an increase in Reynolds number. Clearly, the use of confinement allows small-scale time-dependent wake structures to be generated on extremely coarse grids with the small-scale structure captured over only $1 \sim 2$ grid cells. Also, these small-scale structures serve as a viscous sink for turbulent energy, as in physical turbulence.

### 4.2.2   3-D Square Cylinder

Flow over a square cylinder was also calculated. As in the circular case, the cylinder was "immersed" in a uniform $141 \times 101 \times 61$ Cartesian grid and periodic conditions imposed at the lateral boundaries. The diameter (length of each side) of the cylinder was also 15 grid cells. The same coordinate system was used as for the circular cylinder. Results of the computations were compared to the experimental results of Lyn et al.[20] at a Reynolds number of about 21,400.

As in the circular cylinder case, the diffusion coefficient $\mu$ was set to 0.15.

The confinement coefficient, $\epsilon$, was adjusted to impose different levels of confinement so as to approximate the effects of different Reynolds numbers.

Figure 4 depicts the comparison with experimental data of the time-averaged streamwise velocity along a streamwise line extending downstream from the middle of the leeward face of the cylinder. Results of two values of the confinement coefficient are plotted. Figure 6 shows the time-averaged velocity along a line traversing the wake (normal to the cylinder axis and the mean stream), as in the circular case, at $x = 1$. Symbols represent the experimental data. Our numerical results agree well with the experimental data for $\epsilon = 0.35$ and $\epsilon = 0.25$.

Comparisons of the computed turbulence level with the experimental results also show reasonably good agreement in Fig. 6a. As in the circular case, the same value of $\epsilon$ that gave good agreement with experiment for RMS velocity also gave good agreement formean velocity. The effect of decreasing confinement is shown in Fig. 6b. Comparing this case with the previous 3-D circular cylinder case, it is easy to see that a higher value of $\epsilon$ is required for better results. This is apparently because the Reynolds number has increased from about 3,900 in the 3-D circular cylinder case to roughly 21,400 in the present 3-D square cylinder case.

### 4.2.3   Synopsis of Cylinder Study

In subsequent studies, the correlation between $\epsilon$ and Reynolds number will be studied by comparisons between computation and experiment for other cases and a useful calibration of $\epsilon$ determined. Also, it should be emphasized that our agreement with experiment is closer than many much finer, conforming grid studies with much more complex LES pde models[21], each requiring a number of empirical coefficients. As explained, just one coefficient is adjusted here, on a simple, coarse uniform Cartesian grid.

# 5   New Vorticity Confinement

## 5.1   Formulation

Because additional corrections must be added to make the original confinement explicitly conserve momentum, a new, simpler formulation that does not have these problems has been developed. This new, intrinsically discrete formulation is presented in this section.

A more detailed description is presented in [1]. First, a formulation for scalar confinement is given. Then, a velocity-based (primitive variable) Vorticity Confinement correction is given that reduces to the scalar confinement in terms of vorticity when the curl is taken.

### 5.1.1  Scalar Formulation

The scalar formulation presented here is related to that presented in [8] in 1-D.
We start with the form for a scalar:

$$\phi^{n+1} = \phi^n - \Delta t \nabla \cdot \vec{q}\phi^n + \delta\phi^n \tag{5.1.1}$$

where the confinement correction:

$$\delta\phi^n = h^2[\mu\nabla^2\phi^n - \epsilon\nabla^2\Phi^n]$$

where

$$\Phi^n = \left[\frac{\sum_l C_l(\phi^n + \delta)^{-1}}{\sum_l C_l}\right]^{-1} \tag{5.1.2}$$

where the sum is over a set of grid nodes near and including the node where $\Phi$
is computed, and $\delta$ is a small positive constant ($\sim 10^{-8}$) to prevent problems
due to finite precision. The coefficients, $C_l$, can be varied but good results are
obtained by simply setting them to 1. For example, in 2-D, one possibility is

$$\Phi_{ij}^n = \left[\frac{\sum_{\alpha=-1}^{+1}\sum_{\beta=-1}^{+1}(\phi_{i+\alpha,j+\beta}^n + \delta)^{-1}}{N}\right]^{-1} \tag{5.1.3}$$

where N neighboring terms are taken.

Here, we assume $\phi^n \geq 0$. Negative values can also be accommodated with a
small extension. Both $\mu$ and $\epsilon$ are positive.

An important feature is that all terms are homogeneous of degree 1 in Eq.
5.1.1. This is important because the confinement not depend on the scale of
the quantity being confined. Another important feature is the nonlinearity.
It is easy to show that a linear combination of terms of different order in the
derivatives cannot lead to a stable confinement for any finite range of coefficients.

For smooth $\phi$ fields (large scales), the last term represents a negative diffusion. If $\mu \leq \epsilon$, where $N$ is the number of terms in Eq. 5.1.2, the total diffusion
is negative. However, the iteration of Eq. 5.1.1 is still stable and converges for
values of $\epsilon$ up to about $5\mu$, resulting in an effective negative diffusion coefficient
in the long wavelength limit. Also, for $\vec{q} = 0$, the discrete, converged solution
can be given exactly in terms of (translationally invariant) sech functions.

For convection using a conservative discretization, the total amplitude

$$< \Phi > \equiv \sum_{ij} \phi_{ij}^n$$

is independent of $n$. If we define the centroid of an isolated region,

$$< \vec{X} >^n \equiv \sum_{ij} \vec{x}_{ij}^n \phi_{ij}^n / < \Phi >$$

and the weighted mean velocity

$$< \vec{Q} >^n \equiv \sum_{ij} \vec{q}_{ij}^n \phi_{ij}^n / < \Phi >$$

where $\vec{x}_{ij}$ is the (fixed) position vector of node $(i,j)$, and $\phi_{ij}$ and $\vec{q}_{ij}$ are the scalar value and the velocity at that node, then the centroid evolves according to:

$$< \vec{X} >^{n+1} = < \vec{X} >^n + \Delta t < \vec{Q} >^n$$

For vortices, the self-induced velocity which is included in the above sum exactly cancels and, as in the passive scalar case, the $\vec{q}_{ij}$ can be taken to be an externally applied (irrotational) velocity. The above result then still holds.

Since we are, at this point, only interested in the "expectation values" of scalars or vorticities for thin features and that the features remain compact, spread over only a few cells, this Ehrenfest-type relation is exactly what we need. Only the variables of importance are, effectively, solved for. This shows that the features, when isolated, evolve as particles with essentially no internal dynamics. However, we keep the very important Eulerian feature that the number of features is not fixed. We could, for example, create additional solitary waves by inserting a source: No additional computational markers need be created, as in Lagrangian schemes. For this study, we show that features can automatically merge and reduce in number. This will be seen in the results of Sec. 5.2. As for the earlier Vorticity Confinement method, this property is crucial for the general treatment of interacting vortical regions, especially in 3-D[4,7,9,10].

### 5.1.2 Vorticity Formulation

The next step involves letting $\phi$ be the magnitude of vorticity and deriving an equation for the corresponding velocity correction that leads to Eq. 5.1.1 when the curl is taken (in 2-D). This correspondence is still close for a 3-D vortex filament if the radius of curvature is large so that the flow is approximately 2-D in sections normal to the filament axis. This will result in a new formulation of Vorticity Confinement.

We simply define

$$\vec{q}^{n+1} = \vec{q}^n + \Delta t \nabla \cdot \vec{q}^n q^n + \delta \vec{q}^n \tag{5.1.4}$$

where the confinement correction:

$$\delta \vec{q}^n = h^2(\mu \nabla^2 \vec{q}^n + \epsilon \nabla \times \vec{w}^n)$$

360

or, for $\nabla \cdot \vec{q}^{\,n} = 0$ ,

$$\delta \vec{q}^{\,n} = -h^2 \nabla \times (\mu \vec{\omega}^n - \epsilon \vec{w}^n)$$

where

$$\vec{\omega}^n = \nabla \times \vec{q}^{\,n}$$

and

$$\vec{w}^n = \frac{\vec{\omega}^n}{|\vec{\omega}^n| + \delta} \left[ \frac{\sum_l (|\vec{\omega}^n_l| + \delta)^{-1}}{N} \right]^{-1} \tag{5.1.5}$$

where the sum is the same as in Eq. 5.1.3.

## 5.2   Results

All results presented are in 2-D. Also, all convection cases use a conservative second-order centered convection term together with a centered diffusion term and are first-order-explicit in time. In all the result plots, axes are labeled with grid node location. Plots of amplitude are made using dense contours with white grid lines superimposed.

### 5.2.1   Passive Scalar Convection

A good test involves convection of a small passive scalar pulse in solid-body rotating flow[22]. For this case a convection step was followed by a single confinement step (including diffusion). The full summation of Eq. 5.1.2 with $C_l = 1$ (Eq. 5.1.3) leads to a spreading of the pulses in the direction normal to their motion (This spreading does not occur for vortices, for reasons described in Sec. 5.2.2). This scalar spreading seemed to be due to the inclusion of "downwind" values, which is known to cause instability. By setting $C_l = 0$ for downwind points, the problem is almost completely cured, and only a very weak spreading remains.

Results are shown in Fig. 7a for a single pulse convecting through the 1st rotation ($2\pi$) and in Fig. 7b through the 10th rotation. Contours are plotted from initial maximum value to 1/3 of that value. The final image in Fig. 7b was after 12,560 time steps. This represented a travel of 1,256 cells or about 433 pulse diameters. The open circle labels the starting location. The time step was not set to exactly divide the circuit so we would not expect the final image to exactly coincide with the initial one. However, the final position is within plottable accuracy of the calculated angle and radius. The final radius can be seen to be very close to the initial one. This is somewhat surprising since the method was only first order in time and second order in space. We have not yet attempted to reduce the small, residual cross-flow spreading. This could easily be accomplished with a small amount of additional negative diffusion in the cross-stream direction. For the case computed, we used $\mu = 0.15$ and $\epsilon = 0.75$.

Finally, a run was made with no confinement and a lower, (minimal) diffusion set for stability. The maximum amplitude decreased to below the plotting threshold after only $\pi/8$ radians of travel (position of the first image in Fig. 2a).

## 5.2.2  Vorticity Convection

The incompressible fluid dynamic equations with Vorticity Confinement, Eq. 3.0.1, were discretized and solved. Each time step, the same convection scheme used for the above scalar was used. A standard pressure projection method was used to enforce incompressibility[23]. This used a staggered grid for pressure and velocity and a box scheme for the pressure gradient. The required Poisson equation for pressure was solved using a direct method "FishPack". Vorticity Confinement terms - Eqs 5.1.4 and 5.1.5, were used, with $\mu = 0.05$ and $\epsilon = 1.5\mu$. Results were very close to those for $\mu = 0.2$ and $\epsilon = 1.5\mu$, implying that the method is not very sensitive to the values of these parameters. (Further investigation of this point should be done). The time step, based on grid cell size and circumferential velocity near the vortex cores, was 0.4 and the grid had $128 \times 128$ cells. For vortices, the full summation (Eq. 5.1.3) with all $C_l = 1$ was used, without excluding downwind points. This exclusion, useful for scalar convection, was not important here because the large circulating velocities near the vortex cores rapidly convected any cross-stream spread vorticity around the cores, keeping them close to axisymmetric. The confinement then kept the cores compact.

First, interacting vortices of opposite sign were computed with no external free-stream-velocity, convecting only under their own induced velocities. To approximate far-field boundary conditions, each time step the total amplitudes and centroids of both positive and negative vorticity were computed. The velocities induced by corresponding point vortices at these locations were computed and imposed on the boundaries. Dirichlet conditions were used for pressure.

Results are presented in Fig. 8 for a sequence of time steps. The computed vortex centroids are convecting at the same velocity, to plottable accuracy, as if they were point vortices. The self-induced velocity in each vortex, although several times the convecting velocity of the centroid, automatically cancels and has no effect because the method conserves momentum, as explained in Sec. 5.1.1. The final images shown result from 1,800 time steps, yet the vortices are just as compact as initially. It should be mentioned that if the vortices are significantly closer, there will be an interaction and eventually vorticity will be exchanged. This is to be expected since there are "tails" of vorticity that (rapidly) decrease with radius beyond the region shown.

Vortices with the same sign were computed as they rotated around each other under their induced velocities. The vortices were initially (and finally) separated by 14 cells. First, the computation was done with no Vorticity Confinement, and with minimum diffusion required for convective stability. Contours are displayed in Fig. 9 corresponding to vorticity values between initial maximum and 1/4 of that value. It can be seen that the vorticity rapidly diffuses. The computation was repeated with Vorticity Confinement. Values of $\mu$ and $\epsilon$ were those used above for opposite sign vortices. Vorticity contours between initial maximum and 1/4 maximum are displayed for 3 different times in the 1st loop around each other in Fig. 9a, and in the 20th loop in Fig. 9b. The behavior is as expected: The vortices can be seen to be essentially the same at the end (after 7,200 time steps) as initially.

As stated, the main reason for using an Eulerian method as opposed to Lagrangian is to allow vortices to merge and otherwise change topology, so that complex, realistic vortical configurations can be treated. This has been amply demonstrated by the original Vorticity Confinement method both in 2-D and 3-D[4,5,7,9,10]. To demonstrate that the new confinement method does not inhibit this merging we repeat the above computation, but with the vortices initially 5 cells apart. This is a repeat of a calculation using the original Vorticity Confinement method[5]. It is known from vorticity contour of "waterbag" methods[24] that corotating vortices become unstable and merge when they are initially closer than a few diameters. It can be seen in the vorticity contour plots of Fig. 10 that this merging automatically occurs, as expected, with the new Vorticity Confinement method.

# 6   Conclusion

The ability of the original Vorticity Confinement method to efficiently compute blunt body turbulent wakes was demonstrated. This involved an LES-type representation but with the small scales treated as inherently discrete structures. The computations involved a circular and square cylinder in 3-D.

A new Eulerian Vorticity Confinement has been introduced. Its advantages over the original Vorticity Confinement method are that it is simpler and easier to analyse and explicitly conserves total momentum. The method accurately convects thin features, either vortices or scalars, over very long distances, even though they are confined to only 2-3 grid cells. As in shock and other discontinuity capturing schemes, the details of the internal structure of the feature are not solved for, but implicitly modeled (the features are treated as weak solutions). Integral quantities, such as total amplitude and centroid, however, are accurately solved for. As such, the method essentially represents the features as nonlinear discrete solitary waves that live on the computational lattice. It is argued that the new method is a rational generalization of 1-D discontinuity confinement schemes to multiple dimensions.

Examples are presented for thin, convecting passive scalars and vortices in 2-D incompressible flow.

## Acknowledgements

The first author acknowledges many helpful discussion with Barry Merriman and Stanley Osher. The work was supported by the Army Research Office and the University of Tennessee Space Institute.

# References


[1] Steinhoff, J., Fan, M., and Wang L., "Convection of Concentrated Vortices and Passive Scalars As Solitary Waves", submitted to *Journal of Scientific Computing*, May, 2002.

[2] Fan, M., Wenren, Y., Dietz, W., Xiao, M., Steinhoff, J., "Computing Blunt Body Flows On Coarse Grids Using Vorticity Confinement", to appear in *Journal of Fluids Engineering*, December, 2002.

[3] Wenren,Y., Fan, M., Dietz, W., Hu, G., Braun, C., Steinhoff, J., and Grossman, B., "Efficient Eulerian Computation of Realistic Rotorcraft Flows Using Vorticity Confinement", AIAA-01-0996 (2001).

[4] Steinhoff, J., Wenren, Y., Braun, C., Wang, L., and Fan, M. "Application of Vorticity Confinement to the Prediction of the Flow over Complex Bodies", pp. 197-226 in *Frontiers of Computational Fluid Dynamics 2002* (D.A. Caughey and M.M. Hafez eds.); World Scientific, (2002)

[5] Steinhoff, J., Mersch, T., and Wenren, Y., "Computational Vorticity Confinement: Two Dimensional Incompressible Flow", Proceedings of the Sixteenth Southeastern Conference on Theoretical and Applied Mechanics. 1992.

[6] Steinhoff, J., Wang, Clin, Underhill, D., Mersch, T., and Wenren, Y., "Computational Vorticity Confinement: A Non-Diffusive Eulerian Method for Vortex-Dominated Flows," UTSI preprint, 1992.

[7] Steinhoff, J., "Vorticity Confinement: A New Technique for Computing Vortex Dominated Flows," In D.A. Caughey and M.M. Hafez, editors,*Frontiers of Computational Fluid Dynamics*. John Wiley & Sons, 1994.

[8] Steinhoff, J., Puskas, E., Babu, S., Wenren, Y., and Underhill, D., "Computation of Thin Features Over Long Distances Using Solitary Waves," AIAA Proceedings, 13th Computational Fluid Dynamics Conference, July, 1997.

[9] Steinhoff, J. and Underhill, D., "Modification of the Euler Equations for Vorticity Confinement Application to the Computation of Interacting Vortex Rings", *Physics of Fluids*, 6, 1994.

[10] Wenren, Y., Steinhoff, J., Wang, L., Fan, M., and Xiao, M., "Application of Vorticity Confinement to the Prediction of the Flow over Complex Bodies," AIAA Paper No. 2000-2621, Fluids 2000, Denver, CO, June, 2000.

[11] Harten, A., "The Artificial Compression Method for Computation of Shocks and Contact Discontinuities III, Self-Adjusting Hybrid Schemes", *Mathematics of Computation*, Vol. 32, No. 142. April 1978.

[12] Hoeijmakers, W.M., and Vaatstra, W., "A Higher Order Panel Method Applied to Vortex Sheet Roll-Up", *AIAA Journal*, Vol.21, 1983.

[13] Rosenhead, *Proc. Roy. Soc., A, 134*, 1931, p.170.

[14] Westwater, *Reports and Memoranda, 1692*, 1936.

[15] R. Krasny, "Vortex Sheet Computations: Roll-Up, Wakes, Separation", *Lectures in Applied Mathematics, 28*, 1991.

[16] A. Leonard, "Vortex Methods for Flow Simulation", *J. Comput. Phys., 37*, 289 (1980).

[17] Steinhoff, J., and Wenren, Y., "Computation of Generation and Convection of Aircraft Vortices", NASA SBIR Contract NAS1-97068 Final Report, Flow Analysis Inc., 1997.

[18] Hirsch, C., *Numerical Computation of Internal and External Flows*, John Wiley & Sons, 1990.

[19] Ma, X., Karamanos, G., and Karniadakis, G., "Dynamics and Low-Dimensionality of a Turbulent Near Wake," *J. Fluid Mech.*, 410, pp. 29-65, 2000.

[20] Lyn, D.A., Einav, S., Rodi, W., and Park, J.H., "A Laser-Doppler Velocimetry Study of Ensemble-Averaged Characteristics of the Turbulent Near Wake of a Square Cylinder ", *J. Fluid Mech.*, 304, pp. 285-319, 1995.

[21] Rodi, W., "Large-Eddy Simulation and Statistical Turbulence Models: Complementary Approaches", *New Tools In Turbulence Modeling*, Eds. Metais, O. and Ferziger, J.. Springer-Verlag 1997.

[22] Orszag, S.A., "Numerical Simulation of Incompressible Flows Within simple Boundaries: Accuracy", *J. Fluid Mech.*, 49, 1971.

[23] Kim, J., and Moin, P., "Application of a Fractional Step Method to Incompressible Navier-Stokes Equations", *Journal of Computational Physics*, Vol. 59, 1995.

[24] Overman, E.A. and Zabusky, N.J., "Evolution and Merger of Isolated Vortex Structures", *Journal of Physical Fluids*, Vol. 25, 1982.

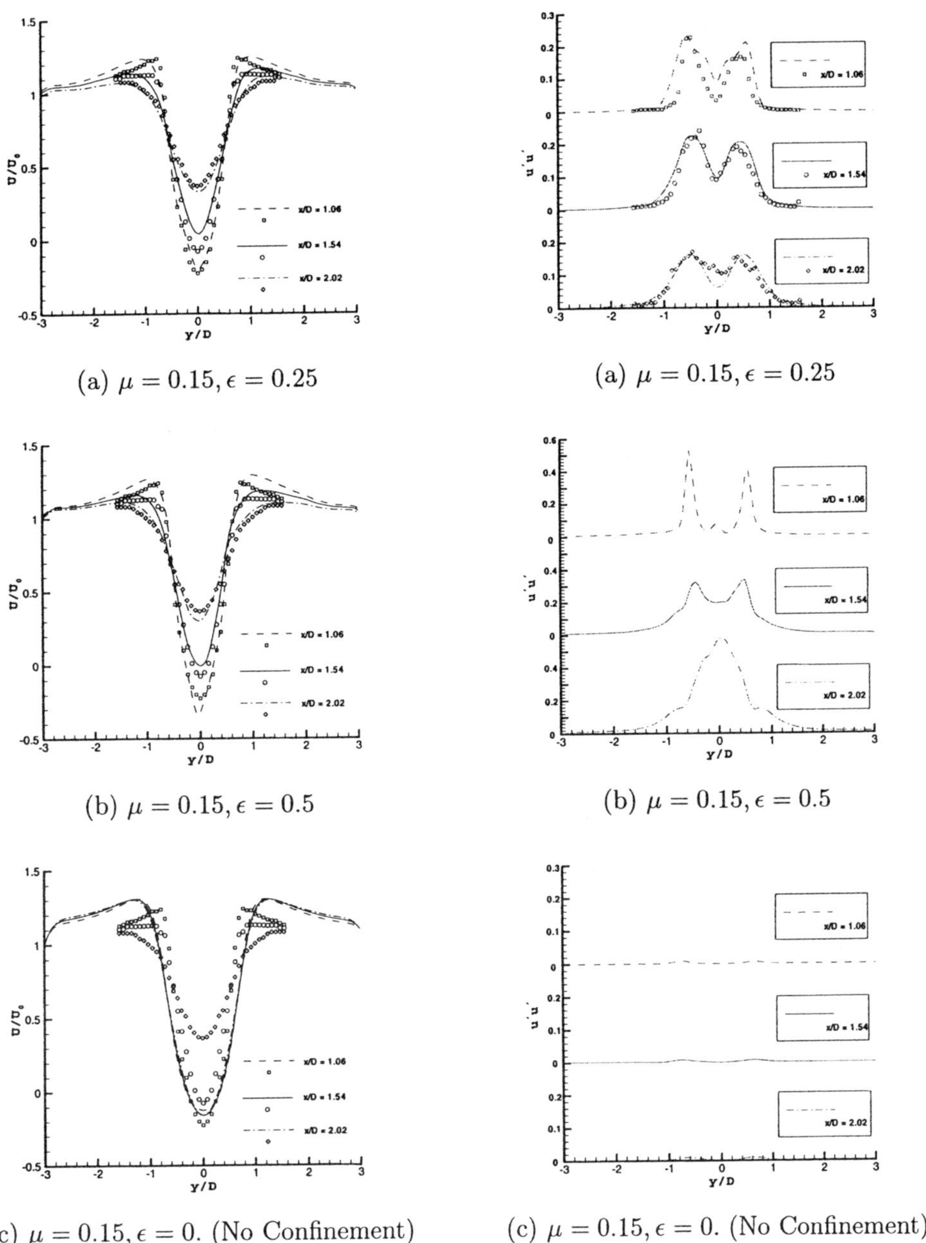

(a) $\mu = 0.15, \epsilon = 0.25$

(a) $\mu = 0.15, \epsilon = 0.25$

(b) $\mu = 0.15, \epsilon = 0.5$

(b) $\mu = 0.15, \epsilon = 0.5$

(c) $\mu = 0.15, \epsilon = 0.$ (No Confinement)

(c) $\mu = 0.15, \epsilon = 0.$ (No Confinement)

**Figure 1** Mean Streamwise Velocity Profiles

**Figure 2** Streamwise RMS Fluctuations

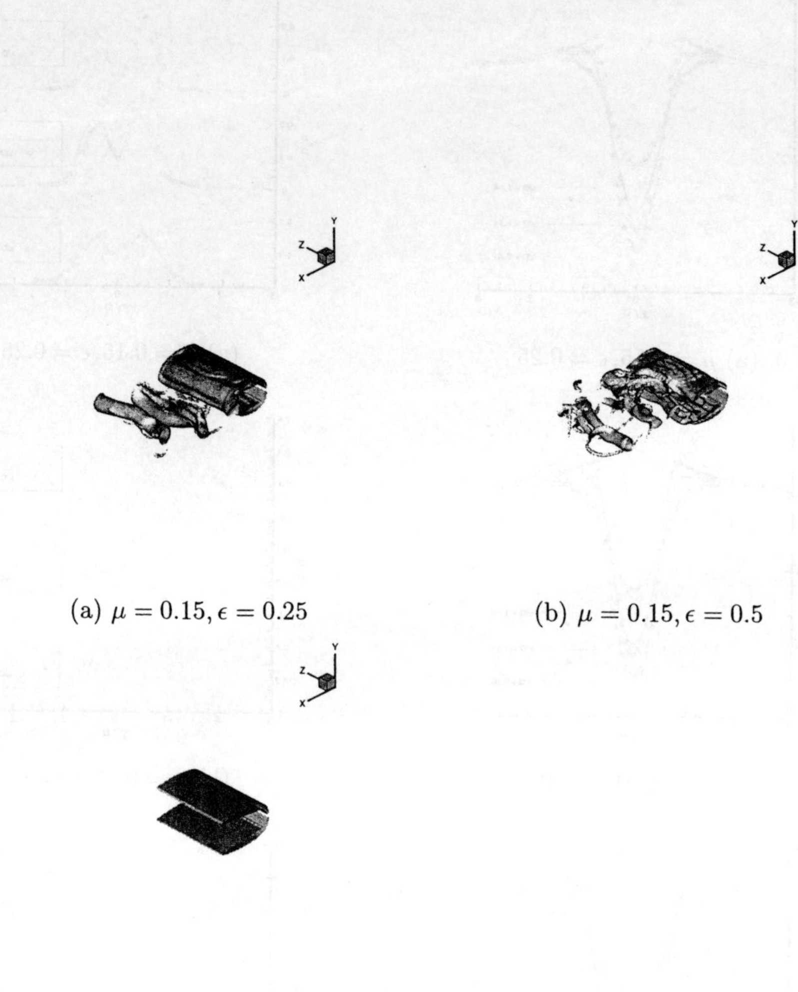

(a) $\mu = 0.15, \epsilon = 0.25$

(b) $\mu = 0.15, \epsilon = 0.5$

(c) $\mu = 0.15, \epsilon = 0.0$ (No Confinement)

**Figure 3** Isosurfaces of Vorticity Magnitude (Isosurface Level $= 0.15$)

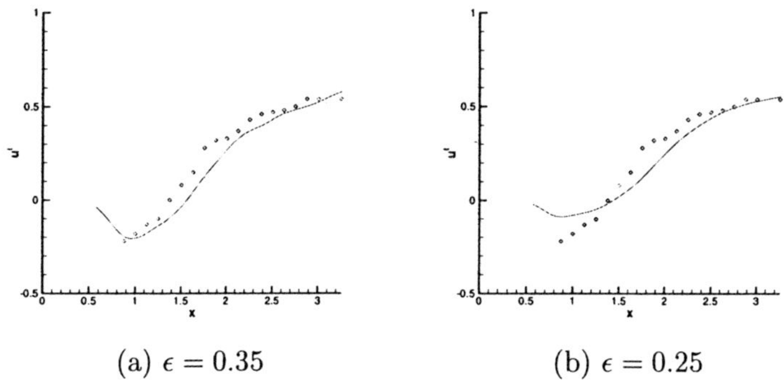

(a) $\epsilon = 0.35$        (b) $\epsilon = 0.25$

**Figure 4** Comparison of time-averaged streamwise velocity along a streamwise line. Symbols denote experimental data.

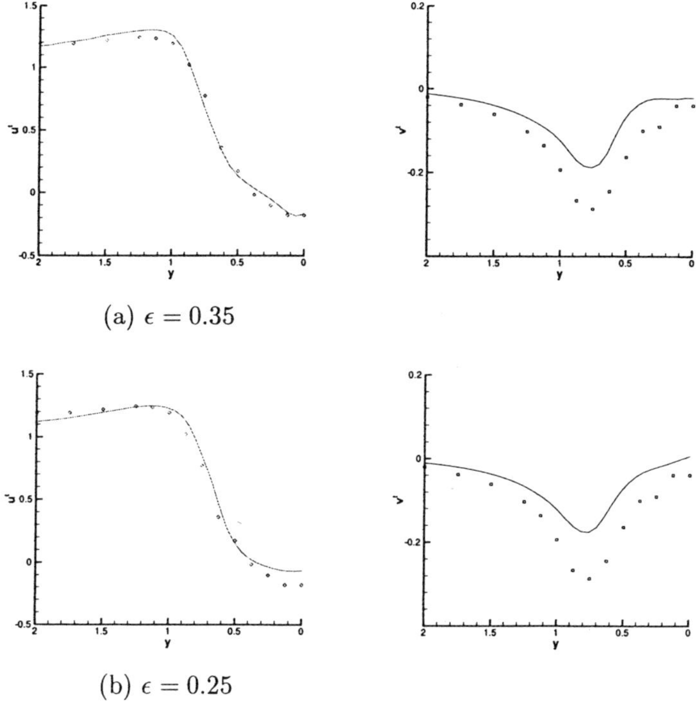

(a) $\epsilon = 0.35$

(b) $\epsilon = 0.25$

**Figure 5** Comparison of time-averaged velocity profiles at x = 1. Symbols are experimental data.

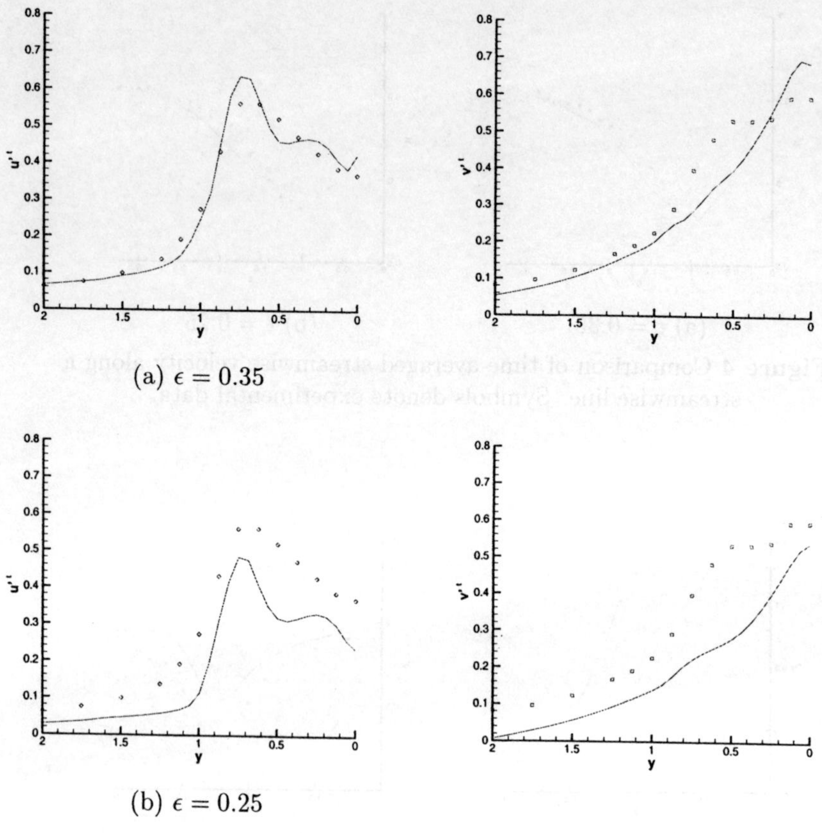



**Figure 6** Comparison of root mean square velocity fluctuation profiles at $x = 1$. Symbols are experimental data.

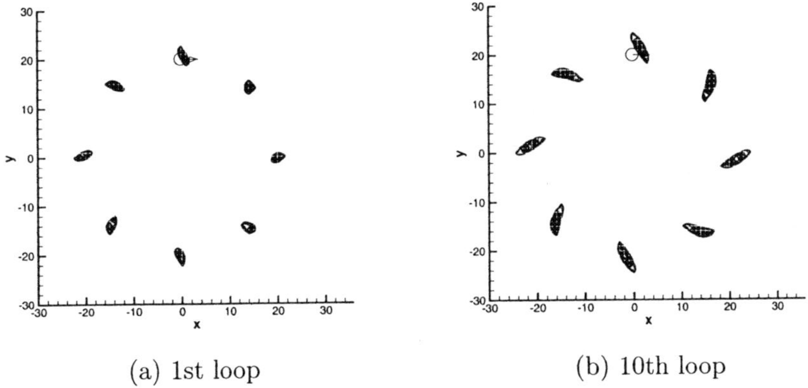

(a) 1st loop            (b) 10th loop

**Figure 7** Scalar convection

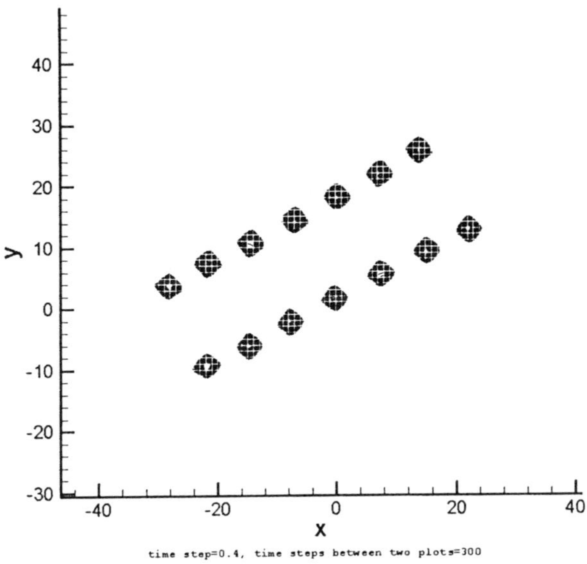



**Figure 8** Vorticity contour plots of self-induced flow by two point vortices (15 cells apart) with opposite sign

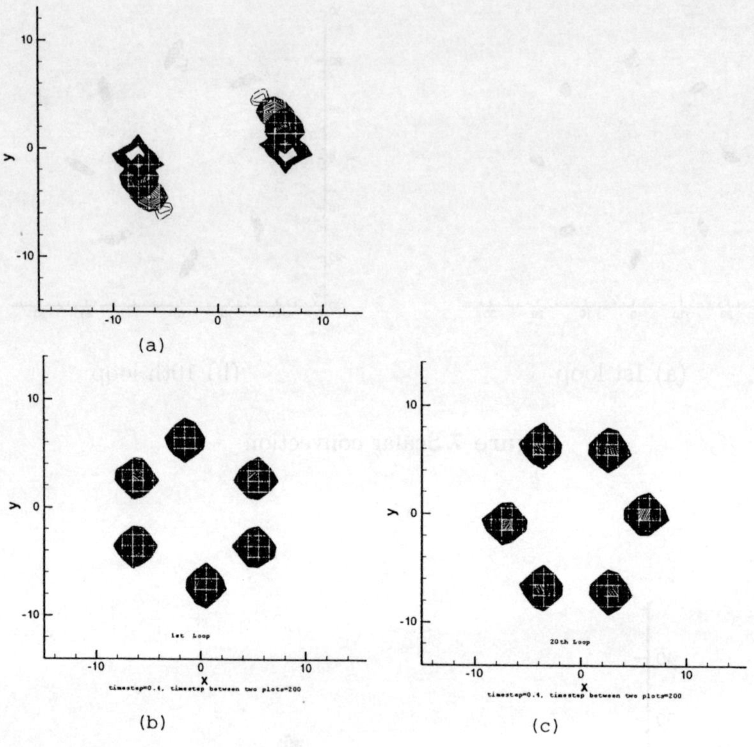



**Figure 9** Vorticity contour plots of self-induced flow by two vortices (14 cells apart) with same sign. (a) without Vorticity Confinement; (b)with Vorticity Confinement, 1st loop; (c)with Vorticity Confinement, after 20th loops.

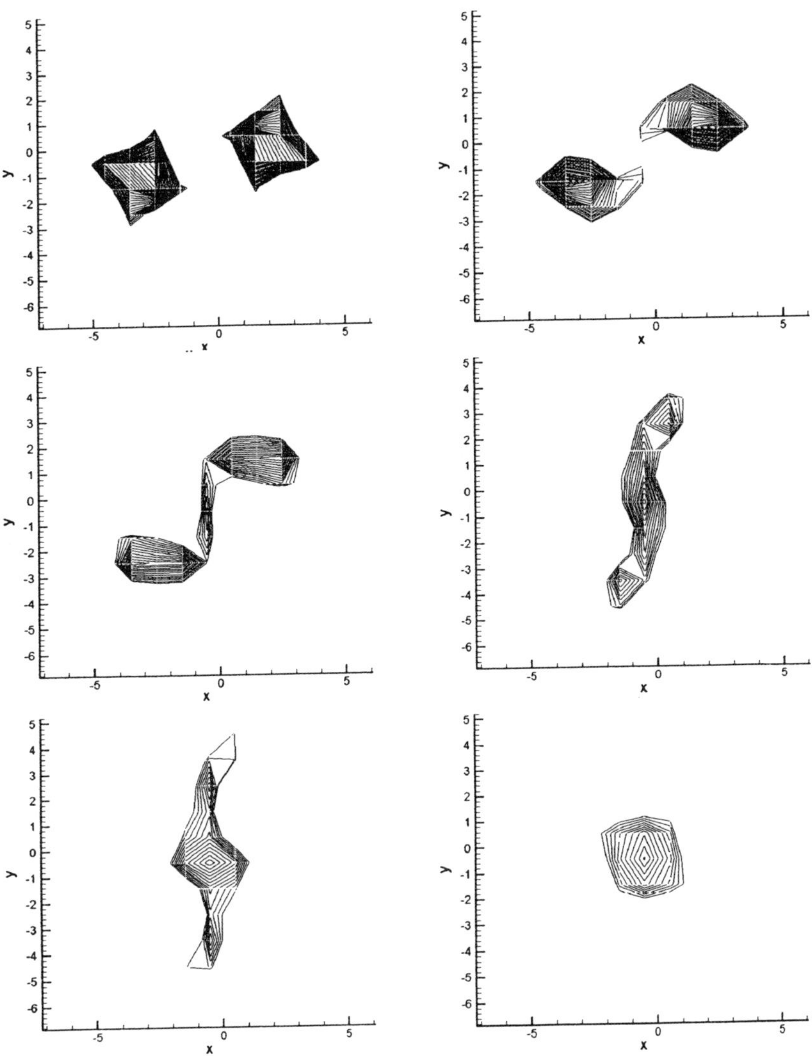

**Figure 10** Vorticity contour plots of self-induced flow by two vortices (5 cells apart) with same sign, with Vorticity Confinement

Figure 10 Vortices capture slices of semianalytical flow by two small vortices apart with equal area, with Vortex 5. Contour event.

# Applications in Aeronautics

# OPTIMIZATION OF PROPELLERS USING A VORTEX MODEL

Jean-Jacques Chattot

University of California Davis, Davis, CA 95616

## Abstract

An approach to the optimization of propellers is presented that is based on a helicoidal vortex model and the rigorous calculation of the induced velocities to maximize the thrust for a given output torque of the engine. The mathematical model is presented for inviscid flow and for the viscous correction. The design of a two-bladed propeller at low advance ratio is carried out and the distributions of circulation, induced velocities, and chord are obtained. Comparison with published data indicates that the method is efficient and flexible and produces detailed and accurate results.

## 1  Introduction

The aerodynamics of propellers is a challenging topic, even today, despite the advances in computing power. The inviscid model has been described in landmark papers and books by Prandtl and Betz [1], Glauert [2], Goldstein [3], Lock [4] and Theodorsen [5]. For large Reynolds numbers the model is expected to remain accurate and give a correct representation of the blade/vortex sheet interaction, even though the flow contraction behind the blades is generally neglected. This helicoidal structure is the most important aspect of propeller flow as it influences the blade operation through the induced velocities in the streamline and azimuthal direction, which in turn determine the local angle of attack of the blade element, its lift and induced drag.

The vortex sheet is a free surface and its location and strength are part of the solution. However, capturing the vortex sheet beyond 2 or 3 turns of the helix with an Euler or Navier-Stokes solver is beyond the current capabilities of modern computational fluid dynamics (CFD), rendering the simulation inaccurate for analysis or design purposes. The present work is based on the Goldstein model with a helicoidal vortex structure depending on the advance ratio $adv = \frac{V}{\Omega R}$, the power coefficient $P_\tau = \frac{2P}{\rho \Omega^3 R^5}$ and the root location $y_0$. This is an improvement on the method of Adkins and Liebeck [6] who combine the actuator disk and rotor blade element equations, the so-called the blade element theory (or vortex theory, see [7]).

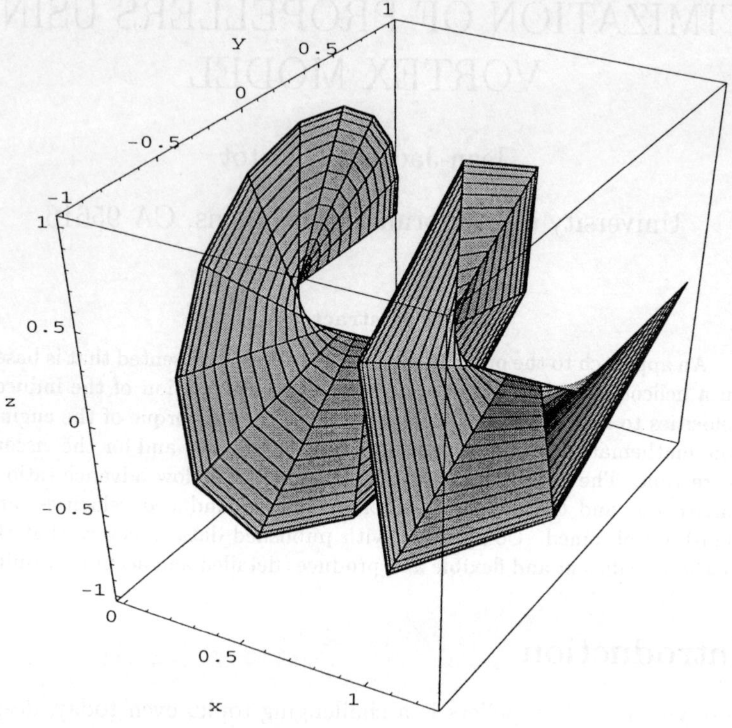

Figure 1: Partial view of the vortex sheet behind a propeller blade

Once the vortex structure is defined, the Biot-Savart law is used to obtain the influence coefficients relating the vortex strength $\frac{d\Gamma}{dy}$ to the induced axial and azimuthal (downwash) velocities $u(y)$ and $w(y)$. The Goldstein model combined with the Biot-Savart law removes some of the empirism associated with the actuator disk/annulus flow model. There remains, however, the assumption of a frozen vortex structure. In reference [3], Goldstein finds a particular analytic solution for the flow past a regular screw surface starting from the root at $y_0 = 0$. Here we make use of the computer resources available today to evaluate the influence coefficients of the vortex sheet on the blade and to use a minimization procedure to find the solution $\Gamma(y)$ as well as the geometry of the optimum blade. Optimization proceeds along the line of a previous paper [8]. A viscous correction is added to the model using strip theory and two-dimensional viscous profile data. This is assumed to be acceptable since the objective is to design a large Reynolds number propeller with best possible performance and a realistic lift coefficient distribution.

# 2 Helicoidal vortex structure

The minimum energy loss condition of Betz [1] requires the vortex sheet to be a regular screw surface. The pitch of the screw varies with the average axial velocity, as estimated by the actuator disk momentum theorem, from $u_b$ at the blade plane to $u_T = 2u_b$ in the Trefftz plane, far downstream. This flow acceleration is accompanied by a contraction of the stream tube which will be neglected. The theory predicts that the thrust of the propeller is

$$T = -2\pi\rho R^2(V + u_b)u_b$$

and the power delivered by the propeller

$$P = 2\pi\rho R^2(V + u_b)^2 u_b$$

Dimensionless quantities are introduced such as the thrust coefficient acting opposite to the drag (negative drag)

$$C_D = -\frac{2T}{\rho V^2 R^2}$$

the lift coefficient (in the case of a wing)

$$C_L = \frac{2L}{\rho V^2 R^2}$$

the torque coefficient

$$C_\tau = \frac{2\tau}{\rho V^2 R^3} = \frac{P_\tau}{adv^2}$$

the power coefficient $P_\tau$ , given above, and all the coordinates, velocities, circulation and chord distribution as

$$x = R\tilde{x}, \ y = R\tilde{y}, \ z = R\tilde{z}, \ u = V\tilde{u}, \ v = V\tilde{v}, \ w = V\tilde{w}, \ \Gamma = VR\tilde{\Gamma}, \ c = R\tilde{c}.$$

Unless otherwise indicated, the dimensionless variables are used in the rest of the paper and the tilde is dropped for simplicity. The induced velocities on the blade are given by the Biot-Savart law

$$u(y') = \int_{y_0}^1 \int_0^\infty \frac{1}{4\pi} \frac{d\Gamma(y)}{dy} \frac{\left(y'-y_v\right)\frac{dz_v}{dx} + z_v\frac{dy_v}{dx}}{[x^2+(y'-y_v)^2+z_v^2]^{\frac{3}{2}}} dxdy$$

$$w(y') = -\int_{y_0}^1 \int_0^\infty \frac{1}{4\pi} \frac{d\Gamma(y)}{dy} \frac{\left(y'-y_v\right)+x\frac{dy_v}{dx}}{[x^2+(y'-y_v)^2+z_v^2]^{\frac{3}{2}}} dxdy$$

where the equation of the vortex sheet is given by

$$\begin{cases} y_v = y\cos\left(\frac{x}{adv}\right) \\ z_v = y\sin\left(\frac{x}{adv}\right) \end{cases} \quad x \geq 0, \ y_0 \leq y \leq 1$$

Two turns of the vortex sheet shed by one blade is shown in figure 1.

The increase in induced axial velocity from $u_b$ to $u_T$ is reflected in the stretching in the construction of the helix. This is accomplished with a variable $adv$ in the above equations, that increases from $adv(1 + u_b)$ to $adv(1 + 2u_b)$ on three turns of the screw. Numerical tests have shown that little difference in results occur when the stretching is done on one or five turns.

The discrete vortex sheet is constructed as a vortex lattice. In the $x$-direction, a variable step allows for the clustering of points near the blade where the integrand in the Biot-Savart formulae is large, until a sufficient distance $L$ from the rotor is reached, typically 20 radii. Beyond this position, an asymptotic formula is used. In the $y$-direction, a cosine distribution of control points is used as is done for the lifting line problem:

$$x_i = x_{i-1} + dx_i \ , \ dx_i = sdx_{i-1} \ , \ i = 1, ..., ix$$

$$y'_j = y_0 + (1 - y_0)\cos\theta_j \ , \ \theta_j = \frac{j-1}{jx-1}\pi \ , \ j = 1, ..., jx$$

with the following choice of parameters

$$x_0 = 0 \,, \; dx_0 = 1.0\,10^{-6} \,, \; s = 1.001 \,, \; ix = 10,001, \; jx = 101$$

Integration points are placed between the control points to avoid the singularity in the integrand as in [8]:

$$y_k = y_0 + (1 - y_0)\cos(\theta_k + \tfrac{\Delta\theta}{2}) \,, \quad \Delta\theta = \tfrac{\pi}{jx-1} \,, \quad k = 1, ..., jx - 1$$

The induced velocities are given by

$$w_k = \sum_{j=1}^{jx-1} (\Gamma_{j+1} - \Gamma_j) a_{j,k}$$
$$u_k = \sum_{j=1}^{jx-1} (\Gamma_{j+1} - \Gamma_j) b_{j,k}$$

where

$$a_{j,k} = -\sum_{i=2}^{ix} \frac{1}{4\pi} \frac{\left(y_k' - \left(\frac{y_{v\,i-1}+y_{v\,i}}{2}\right)_j\right)(x_i - x_{i-1}) + \left(\frac{x_{i-1}+x_i}{2}\right)(y_{v\,i} - y_{v\,i-1})_j}{\left[\left(\frac{x_{i-1}+x_i}{2}\right)_j^2 + \left(y_k' - \left(\frac{y_{v\,i-1}+y_{v\,i}}{2}\right)_j\right)^2 + \left(\frac{z_{v\,i-1}+z_{v\,i}}{2}\right)_j^2\right]^{\frac{3}{2}}}$$

$$b_{j,k} = \sum_{i=2}^{ix} \frac{1}{4\pi} \frac{\left(y_k' - \left(\frac{y_{v\,i-1}+y_{v\,i}}{2}\right)_j\right)(z_{vi} - z_{v\,i-1})_j + \left(\frac{z_{v\,i-1}+z_{v\,i}}{2}\right)_j (y_{v\,i} - y_{v\,i-1})_j}{\left[\left(\frac{x_{i-1}+x_i}{2}\right)_j^2 + \left(y_k' - \left(\frac{y_{v\,i-1}+y_{v\,i}}{2}\right)_j\right)^2 + \left(\frac{z_{v\,i-1}+z_{v\,i}}{2}\right)_j^2\right]^{\frac{3}{2}}}$$

The asymptotic remainders are found to be

$$\Delta a_{j,k} = \frac{1}{4\pi} \frac{y_k' - 2y_j \cos\left(\frac{L}{adv}\right)}{2L^2} + O\left(\frac{adv}{L^3}\right)$$
$$\Delta b_{j,k} = -\frac{1}{4\pi} \frac{y_j^2}{adv\, L^2} + O\left(\frac{1}{L^3}\right)$$

# 3 Inviscid model

In dimensional form, for an element $dy$ of blade in inviscid flow, according to the Kutta-Joukowski Lift theorem, the force $d\vec{F}(y)$ is acting perpendicular to the local incoming flow $\vec{q}(y)$ and equals $d\vec{F}(y) = \rho\,\vec{q}(y) \wedge \Gamma(y)\,\vec{j}\,dy$.

In dimensionless form, the elementary contributions to drag (negative), lift and torque coefficients, for a two bladed rotor, are respectively

$$\begin{cases} C_D = -4\int_{y_0}^1 \Gamma(y)\left(\frac{y}{adv} + w(y)\right)dy \\ C_L = 4\int_{y_0}^1 \Gamma(y)\left(1 + u(y)\right)dy \\ C_\tau = 4\int_{y_0}^1 \Gamma(y)\left(1 + u(y)\right)y\,dy \end{cases}$$

Note that for an infinite advance ratio and $y_0 = -1$ the above formulae reduce to the lifting line results for a wing.

Discrete analogs are now defined:

$$\begin{cases} C_D = -4\sum_{k=2}^{jx-1} \Gamma_k\left(\frac{y_k'}{adv} + w_k\right)(y_k - y_{k-1}) \\ C_L = 4\sum_{k=2}^{jx-1} \Gamma_k\left(1 + u_k\right)(y_k - y_{k-1}) \\ C_\tau = 4\sum_{k=2}^{jx-1} \Gamma_k\left(1 + u_k\right)y_k'(y_k - y_{k-1}) \end{cases}$$

Minimization of the induced drag amounts to maximizing the thrust. The linear non-homogeneous system for the $\Gamma_j$'s, where the rotation term $\frac{y_j'}{adv}$ plays the role of a source term, can be solved by relaxation.

With $jx = 101$ and $\omega = 1.8$ the solution converges in a few hundreds iterations. The solution obtained corresponds to a certain torque. In order to prescribe the torque, a constraint must be added. Define the cost functional to be $Cost = C_D + \lambda C_\tau$. $\lambda$ is the Lagrange multiplier that will control the torque in the solution.

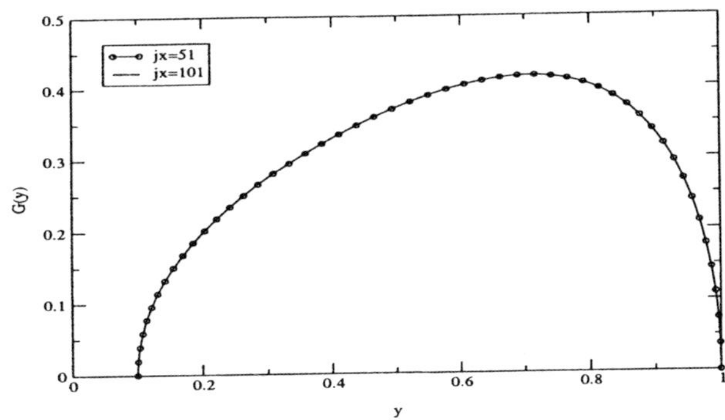

Figure 2: Optimum distribution of circulation ($adv = .1$, $P_T = .01$, $y_0 = .1$)

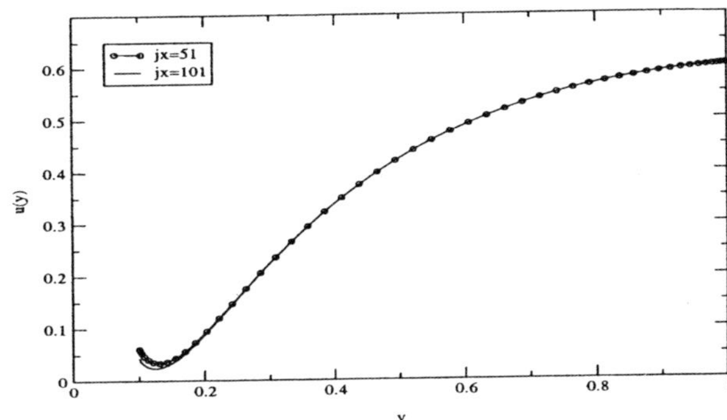

Figure 3: Optimum distribution of axial velocity ($adv = .1$, $P_T = .01$, $y_0 = .1$)

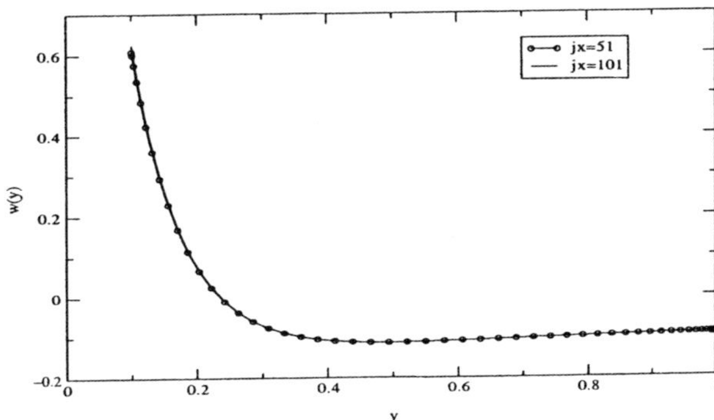

Figure 4: Optimum distribution of downwash ($adv = .1$, $P_T = .01$, $y_0 = .1$)

In order to find the value of the Lagrange multiplier that will correspond to the desired value of the torque coefficient, say $C_{\tau target}$, let's note that $C_D$ and $C_\tau$ can be decomposed into a linear and a bilinear forms in terms of the $\Gamma_j$'s

$$C_D = C_{D1} + C_{D2} = -4 \sum_{k=2}^{jx-1} \Gamma_k \frac{y'_k}{adv}(y_k - y_{k-1}) - 4 \sum_{k=2}^{jx-1} \Gamma_k w_k (y_k - y_{k-1})$$
$$C_\tau = C_{\tau 1} + C_{\tau 2} = 4 \sum_{k=2}^{jx-1} \Gamma_k y'_k (y_k - y_{k-1}) + 4 \sum_{k=2}^{jx-1} \Gamma_k u_k y'_k (y_k - y_{k-1})$$

$C_{D1}$ and $C_{\tau 1}$ are homogeneous of degree one. $C_{D2}$ and $C_{\tau 2}$ are homogeneous of degree two. Using the properties of homogeneous forms, the summation of the minimization equations multiplied each by the corresponding $\Gamma_j$ results in the identity

$$\sum_{j=1}^{jx} \left( \frac{\partial C_D}{\partial \Gamma_j} + \lambda \frac{\partial C_\tau}{\partial \Gamma_j} \right) \Gamma_j = C_{D1} + 2C_{D2} + \lambda (C_{\tau 1} + 2C_{\tau 2}) = 0$$

If we assume that the optimum distributions of circulation for different $C_{\tau target}$ vary approximately by a multiplication factor, then knowing the solution $\Gamma_j$ for a particular value of $\lambda$ (say $\lambda = 0$) allows to find a new value of the Lagrange multiplier as follows. Change the $\Gamma_j$'s to new values $\kappa \Gamma_j$. In order to satisfy the constraint one must have $C_{\tau target} = \kappa C_{\tau 1} + \kappa^2 C_{\tau 2}$. Solving for $\kappa$ yields

$$\kappa = \frac{\sqrt{C_{\tau 1}^2 + 4C_{\tau 2} C_{\tau target}} - C_{\tau 1}}{2C_{\tau 2}}$$

The new estimate for $\lambda$ is obtained from the above identity as $\lambda = -\frac{C_{D1} + 2\kappa C_{D2}}{C_{\tau 1} + 2\kappa C_{\tau 2}}$
This procedure is repeated 2 or 3 times to produce the desired solution.

The optimum distributions of circulation and induced velocities for $adv = 0.1$, $P_\tau = 0.01$ and $y_0 = 0.1$ are shown in figures 2-4. Also shown is the effect of mesh refinement from $jx = 51$ to $jx = 101$. The fine mesh result is an accurate solution to the integro-differential equation.

# 4　Viscous correction

The viscous correction accounts for second order effects. It is assumed that the Reynolds number is large and that the boundary layer is attached. This is reasonable in a design procedure since the chord distribution can be chosen such that the lift coefficient remains well below the maximum lift coefficient in any section of the blade. Let $C_l(y)$ and $C_d(y)$ be the usual lift and drag coefficients of the two dimensional blade section. The dimensional form of the elementary force on the blade section $(y, y + dy)$ is now

$$d\vec{F}(y) = \rho \vec{q}(y) \wedge \Gamma(y) \vec{j} \, dy + \tfrac{1}{2} \rho \| \vec{q}(y) \| \vec{q}(y) C_d(y) c(y) dy$$

where $\vec{q}$ is the local incoming flow velocity in the $(x, z)$-plane. Projection of the force on the $x$-axis and the $z$-axis yields the dimensionless force and moment coefficients for a two-bladed propeller:

$$\begin{cases}
C_D = -4 \int_{y_0}^{1} \Gamma(y) \left( \frac{y}{adv} + w(y) \right) dy \\
\quad + 2 \int_{y_0}^{1} \sqrt{(1+u(y))^2 + \left( \frac{y}{adv} + w(y) \right)^2} \, (1+u(y)) \, C_d(y) c(y) dy \\
C_L = 4 \int_{y_0}^{1} \Gamma(y) \left( 1+u(y) \right) dy \\
\quad + 2 \int_{y_0}^{1} \sqrt{(1+u(y))^2 + \left( \frac{y}{adv} + w(y) \right)^2} \left( \frac{y}{adv} + w(y) \right) C_d(y) c(y) dy \\
C_\tau = 4 \int_{y_0}^{1} \Gamma(y) \left( 1+u(y) \right) y \, dy \\
\quad + 2 \int_{y_0}^{1} \sqrt{(1+u(y))^2 + \left( \frac{y}{adv} + w(y) \right)^2} \left( \frac{y}{adv} + w(y) \right) C_d(y) c(y) y \, dy
\end{cases}$$

The propeller efficiency is $\eta = -\frac{VT}{\tau\Omega} = -adv \frac{C_D}{C_\tau}$.

Note that the chord distribution $c(y)$ is now explicitly in the formulation. It can be prescribed once for all or it can be part of the design process. At any rate, an initial chord distribution is needed. This system is nonlinear due to the viscous correction. To simplify the iteration process, only the dependency of $C_d$ on $\Gamma$ in the viscous terms will be used in the minimization process, all the other terms being frozen from the previous iteration. However, the above equations will be satisfied at convergence.

In order to obtain a minimization equation with a linear contribution in $\Gamma$ from the viscous terms, we approximate the profile polar by piecewise parabolic curves as $C_d = (C_{d0})_m + (C_{d1})_m C_l + (C_{d2})_m C_l^2$. The relation between the lift coefficient $C_l$ and the circulation $\Gamma$ is $C_l = \frac{2\Gamma}{\| \vec{q} \| c}$

Let $C_{Di} = C_{D1} + C_{D2}$ and $C_{\tau i} = C_{\tau 1} + C_{\tau 2}$ ($i$ for inviscid or induced) and the new cost functional be $Cost = C_{Di} + C_{Dv} + \lambda \left( C_{\tau i} + C_{\tau v} \right)$ where the discrete analogs of the viscous terms are

$$\begin{cases}
C_{Dv} = 2 \sum_{k=2}^{jx-1} q_k \left( 1 + u_k \right) \\
\quad \times \left( (C_{d0})_{m(k)} + (C_{d1})_{m(k)} C_{lk} + (C_{d2})_{m(k)} C_{lk}^2 \right) c_k (y_k - y_{k-1})
\end{cases}$$

$$\begin{cases}
C_{\tau v} = 2 \sum_{k=2}^{jx-1} q_k \left( \frac{y_k'}{adv} + w_k \right) \\
\quad \times \left( (C_{d0})_{m(k)} + (C_{d1})_{m(k)} C_{lk} + (C_{d2})_{m(k)} C_{lk}^2 \right) c_k y_k' (y_k - y_{k-1})
\end{cases}$$

and $q_k = \sqrt{(1+u_k)^2 + \left( \frac{y_k'}{adv} + w_k \right)^2}$ is the dimensionless incoming velocity.

In turn, we can decompose the viscous contributions into 3 terms, one independent of $\Gamma$, one linear and one quadratic function of $\Gamma$ as $C_{Dv} = C_{Dv0} + C_{Dv1} + C_{Dv2}$ and $C_{\tau v} = C_{\tau v0} + C_{\tau v1} + C_{\tau v2}$.

The update of the Lagrange multiplier proceeds along similar lines as in the inviscid case. Now $\kappa$ and $\lambda$ are defined as

$$\begin{cases}
\kappa = \frac{\sqrt{(C_{\tau 1}+C_{\tau v1})^2 + 4(C_{\tau 2}+C_{\tau v2})(C_{\tau target}-C_{\tau 0})} - (C_{\tau 1}+C_{\tau v1})}{2(C_{\tau 2}+C_{\tau v2})} \\
\lambda = -\frac{C_{D1} + 2\kappa C_{D2} + C_{Dv1} + 2\kappa C_{Dv2}}{C_{\tau 1} + 2\kappa C_{\tau 2} + C_{\tau v1} + 2\kappa C_{\tau v2}}
\end{cases}$$

In inviscid flow theory, the chord distribution as well as other geometrical characteristics do not appear in the optimization. Hence, there is an infinite number of blade geometries that will produce the optimum distribution of $\Gamma$. On the other hand, the viscous problem is in all generality unmanageable. However, using a single known profile freezes the relative thickness and camber and reduces the problem to finding the optimum chord and twist distributions, given the profile polar at the appropriate Reynolds number. Again, the flow is assumed

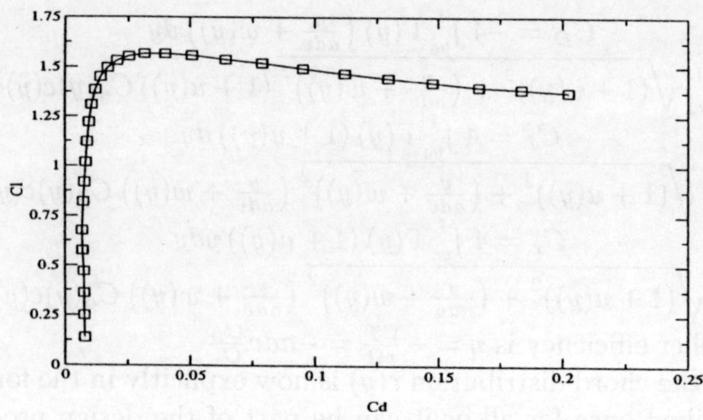

Figure 5: Calculated polar for the NACA 4415 at $R_e = 10^6$ (reference [9])

to be attached in design conditions.

The design procedure consists in specifying an initial chord distribution, say constant chord $c$, large enough to produce a first solution with a lift coefficient below $(C_l)_{max}$. Then a new chord distribution is calculated as $c(y) = \dfrac{2\Gamma(y)}{\|\vec{q}(y)\|(C_l)_{opt}}$, where the value $(C_l)_{opt}$ for minimum drag corresponds to maximum $\dfrac{C_l}{C_d}$. It was found that 5 design cycles were sufficient to obtain a converged solution.

# 5   Test case

The design of a propeller for $adv = 0.223$, $P_\tau = 0.01$ and $y_0 = 0.174$ is compared with that of Adkins and Liebeck [6]. The blade section is the NACA 4415 at Reynolds number $10^6$. The polar characteristics are calculated using Drela's XFOIL code [9]. The angle of attack is swept from -3 degrees to 25 degrees by steps of 1 degree. It is found that $(C_l)_{opt} = 1.2$ for $\alpha = 6.8$ degrees. The data are shown in figure 5.

The distributions of $\Gamma$, $u$ and $w$ are shown in figure 6. The agreement is fair. It is clear that enforcing zero values for the induced velocities at the tip, as is done in reference [6], is not correct. Surprisingly the efficiencies compare well: $\eta_{A\&L} = 0.86996$, $\eta_{present\,method} = 0.8695$. This coincidence is fortuitous since the models are different and the Adkins and Liebeck calculation relies only on 6 radial control points.

The geometry of the blade is determined by the chord distribution. The leading and trailing edges locations versus $y$ are shown in figure 7, where the quarter-chord has been placed along the $y$-axis.

The propeller design obtained at $adv = 0.223$ is tested for a range of advance ratios, for the fixed power output $P_\tau = 0.01$ used in the design. The results are presented in figure 8. At $adv = 0.18$, the maximum lift coefficient is almost reached at the root section where $C_l = 1.563$. At $adv = 0.35$ the root section bears no lift. At $adv = 0.4$, half of the blade is functioning as a windmill.

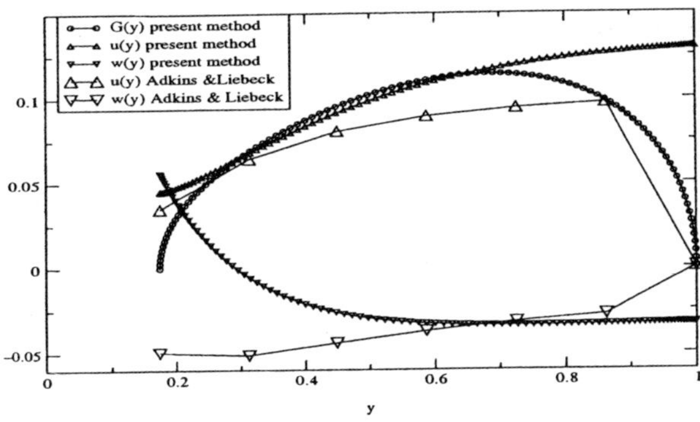

Figure 6: Comparison of the optimum design with results of reference [6]

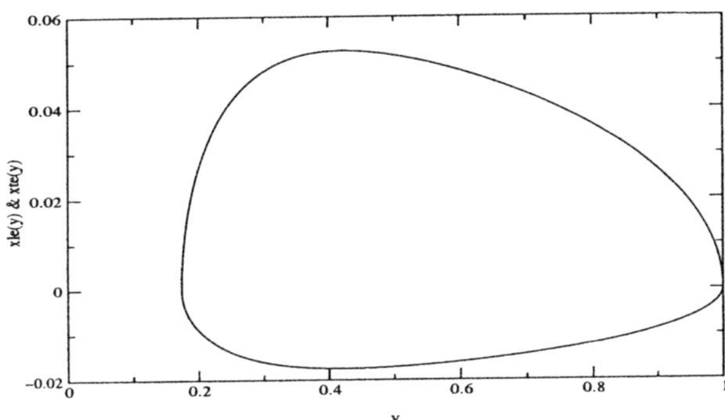

Figure 7: Chord distribution for the optimum design

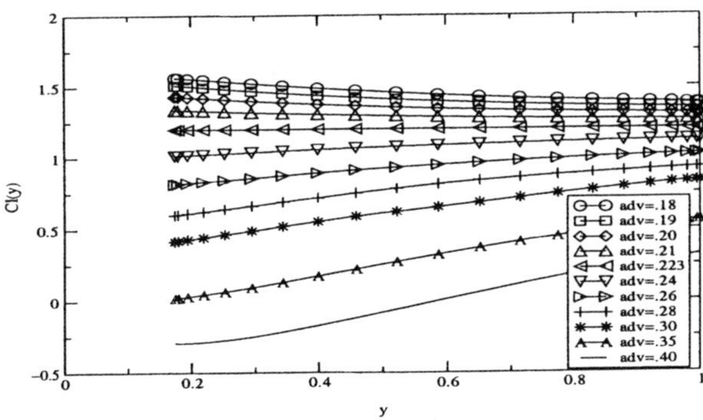

Figure 8: Lift coefficient distribution for a range of advance ratios

# 6   Conclusion

An improved model for the optimization of propellers, using a vortex structure and the Biot-Savart law for the flow interaction, has been established, based on the maximization of the thrust for a given torque. The inviscid model is corrected with two-dimensional viscous data from experiments or computation, using strip theory. However, interaction between the various blade sections is allowed with this model, in a way similar to the Prandtl lifting line theory. The results indicate that the method is flexible and efficient. The lifting line results are recovered accurately for large advance ratios. The design of a blade at low advance ratio compares well with previous results. The optimum geometry is also tested at off-design advance ratios, up to stall, using an analysis code based on the same formulation. This development allows to study the efficiency of the propeller in different working conditions. More details about the formulation can be found in [10].

# References


[1] Prandtl, L. and Betz, A., "Vier Abhandlungen zur Hydrodynamik und Aerodynamik," Gottingen Nachr., Gottingen, Selbstverlag des Kaiser Wilhelminstituts fur Stromungsforschung, 1927.

[2] Glauert, H., "Airplane Propellers," in Aerodynamic Theory, edited by W. F. Durand, Dover, Vol IV, 1963, pp. 169-269.

[3] Goldstein, S., "On the Vortex Theory of Screw Propellers," Proceedings of the Royal Society of London, Series A, Vol. 123, 1929, pp. 440-465.

[4] Lock, C., "Application of Goldstein's Airscrew Theory to Design," Aeronautical Research Committee, RM 1377, Nov. 1930.

[5] Theodorsen, T., Theory of Propellers, McGraw-Hill, 1948.

[6] Adkins, C. and Liebeck, R., "Design of Optimum Propellers," Journal of Propulsion and Power, Sept.-Oct. 1994, Vol 10, No 5, pp. 676-682.

[7] Wilson,R. and Lissaman, P., "Applied Aerodynamics of Wind Power Machines," Oregon State University, NSF-RA-N-74-113, PB 238 595, Corvallis, OR, May 1974.

[8] Chattot, J.-J., "Optimization in Applied Aerodynamics," Computational Fluid Dynamics Journal, Vol. 9, No. 3, Oct. 2000.

[9] Drela, M., and Giles, M.B., "Viscous-Inviscid Analysis of Transonic and Low Reynolds Number Airfoils," AIAA Journal, Vol. 25, No. 10, 1987, pp. 1347-1355.

[10] Chattot, J.-J., "Optimization of Propellers Using Helicoidal Vortex Model", CFD Journal, Vol. 10, No. 4, January 2002.

# A NONITERATIVE METHOD FOR BOUNDARY-LAYER EQUATIONS

## I. TWO-DIMENSIONAL LAMINAR FLOWS

**Tuncer Cebeci[*] and Jian Ping Shao[†]**

[*] The Boeing Company
Long Beach, CA90840, USA
email: tuncer.cebeci@boeing.com

[†] The Boeing Company
Long Beach, CA90840, USA
email: jian.p.shao@boeing.com

**Key words:** Boundary-Layer Equations, Laminar Flow, Iterative Method, Noniterative Method, Nonlinear Parabolic Partial-Differential Equation.

**Abstract.** *A noniterative method for nonlinear parabolic partial-differential equations is described and applied to boundary-layer equations for two-dimensional incompressible laminar flows. Comparison of calculated results indicates that the accuracy of this method is comparable to those obtained with the iterative method.*

## 1   INTRODUCTION

It is well known that the boundary-layer equations for laminar and turbulent flows are parabolic and nonlinear. Their solutions can be obtained in their nonlinear form by using a shooting method or in linearized form by a finite-difference method [1]. The latter choice is more common and practical, especially in turbulent flows since a finite-difference method is more efficient than an integration method such as the shooting method.

It is also well known that the solutions of the boundary-layer equations in differential form are more time consuming than those based on integral form since in the latter case the solutions are obtained for ordinary differential equations. Even though significant advances have been made in the development of efficient and accurate numerical methods to solve the differential form of the boundary-layer equations, the computer times associated with them are still considerably more than those of integral methods.

386

The "edge" the integral methods have over the differential methods becomes even more significant in methods which employ inviscid/viscous interaction techniques where the boundary-layer equations are solved several times for a given pressure distribution by making several sweeps on the body [2]. In each sweep, the solutions at each streamwise location are obtained iteratively since the equations have been linearized. This procedure is not only applied on every streamwise station but is also used in each sweep. It is clear that significant savings in computer time can result if the solutions are not iterated at each streamwise location without sacrificing accuracy.

In this paper, we address this need and describe a new numerical method that allows computer savings of 75% of a modern finite-difference method such as that described in [1]. The new method is based on an noniterative scheme and its accuracy is comparable to that resulting from an iterative scheme. In this paper, we describe this method for two-dimensional laminar flows. To the authors' knowledge, this is the first paper that describes the solution of the nonlinear parabolic partial-differential equations with a noniterative scheme. As such, it is a useful and efficient tool that can lead to considerable computer savings in many engineering problems.

## 2  BOUNDARY-LAYER EQUATIONS

The boundary-layer equations and their boundary conditions for two-dimensional laminar flows are well known and can be written as

$$\frac{\partial u}{\partial x} + \frac{\partial v}{\partial y} = 0 \tag{1}$$

$$u\frac{\partial u}{\partial x} + v\frac{\partial u}{\partial y} = u_e\frac{du_e}{dx} + v\frac{\partial^2 u}{\partial y^2} \tag{2}$$

$$y = 0, \ u = 0, \ v = 0; \ \ y \to \infty, \ u \to u_e(x) \tag{3},(4)$$

As discussed in [1], it is more efficient to solve the boundary-layer equations in transformed variables. A convenient one is the Falkner-Skan transformation in which a similarity parameter $\eta$ is defined by

$$\eta = \sqrt{\frac{u_e}{vx}}\,y \tag{5a}$$

and a dimensionless stream function $f(x, \eta)$ by

$$f(x,\eta) = \frac{\psi(x, y)}{\sqrt{u_e \nu x}} \tag{5b}$$

Here $\psi(x,y)$ is a dimensional stream function that satisfies equation (1),

$$u = \frac{\partial \psi}{\partial y}, \qquad v = -\frac{\partial \psi}{\partial x} \tag{6}$$

With this transformation, Eqs. (1) to (3) can be written as

$$f''' + \frac{m+1}{2} ff'' + m\left[1 - (f')^2\right] = x\left(f'\frac{\partial f'}{\partial x} - f''\frac{\partial f}{\partial x}\right) \tag{7}$$

$$\eta = 0, \ f = f' = 0; \qquad \eta = \eta_e, \ f' = 1 \tag{8}$$

Here m is a dimensionless pressure-gradient parameter

$$m = \frac{x}{u_e}\frac{du_e}{dx}. \tag{9}$$

## 3  ITERATIVE METHOD

Boundary-layer equations being parabolic in nature can be solved by using several numerical methods; of these, finite difference methods are at present the most common with Crank-Nicolson [3] and Keller's box [4] methods being the most popular ones.  The latter has several advantages over the former and is considered here.

According to Keller's method, Eq. (7) is first  expressed as a first-order system by introducing new variables $u(x,\eta)$, and $v(x,\eta)$

$$f' = u \tag{10a}$$

$$u' = v \tag{10b}$$

so that Eq. (7) becomes

$$v' + \frac{m+1}{2} fv + m\left(1 - u^2\right) = x(u \frac{\partial u}{\partial x} - v \frac{\partial f}{\partial x}). \tag{10c}$$

The boundary conditions now become

$$\eta = 0, \ u = 0, \quad f = 0; \qquad \qquad \eta = \eta_e, \ u = 1 \tag{11}$$

## 3.1  Iterative Method

To solve the system given by Eqs. (10) and (11) with Keller's box method, which is an iterative method, we consider the net rectangle shown in Fig. 1

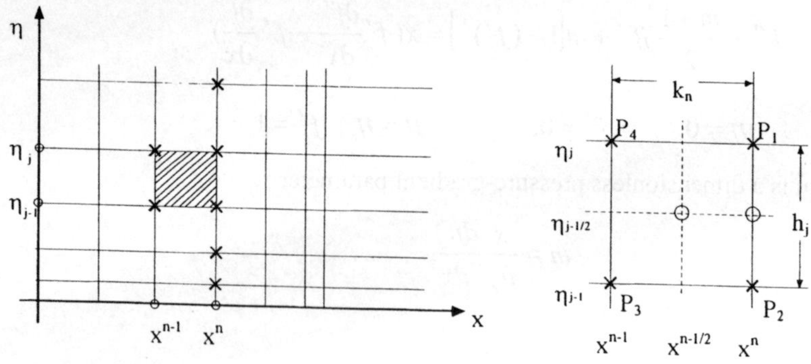


Figure 1

$$x^0 = 0, \quad x^n = x^{n-1} + k_n, \quad n = 1,2,...,N$$

$$\eta_0 = 0, \quad \eta_j = \eta_{j-1} + h_j, \quad j = 1,2,...,J; \qquad \eta_J = \eta_e \tag{12}$$

and write difference equations that are to approximate Eqs. (10). We write the finite difference approximations of Eq. (10a,b) at $(x^n, \eta_{j-1/2})$ using centered-difference derivatives.

$$\frac{f_j^n - f_{j-1}^n}{h_j} - \frac{u_j^n + u_{j-1}^n}{2} = 0 \tag{13a}$$

$$\frac{u_j^n - u_{j-1}^n}{h_j} - \frac{v_j^n + v_{j-1}^n}{2} = 0 \tag{13b}$$

Similarly, Eq. (10c) is approximated at the midpoint $(x^{n-1/2}, \eta_{j-1/2})$ of the net rectangle $P_1P_2P_3P_4$,

$$\frac{1}{2}\left[\frac{v_j^n - v_{j-1}^n}{h_j} + \frac{v_j^{n-1} - v_{j-1}^{n-1}}{h_j}\right]$$

$$+ \frac{m^{n-1/2} + 1}{4}\left[(fv)_{j-1/2}^n + (fv)_{j-1/2}^{n-1}\right]$$

$$+ m^{n-1/2}\left(1 - \frac{\left((u^2)_{j-1/2}^n + (u^2)_{j-1/2}^{n-1}\right)}{2}\right) \tag{13c}$$

$$= x^{n-1/2}\left[u_{j-1/2}^{n-1/2}\frac{u_{j-1/2}^n - u_{j-1/2}^{n-1}}{k_n} - v_{j-1/2}^{n-1/2}\frac{f_{j-1/2}^n - f_{j-1/2}^{n-1}}{k_n}\right]$$

The boundary conditions, Eq. (11) become

$$u_0^n = 0; \quad f_0^n = 0; \quad u_J^n = 1 \tag{14}$$

If we assume $f_j^{n-1}$, $u_j^{n-1}$, $v_j^{n-1}$ to be known for $0 \le j \le J$, then Eqs. (13) and (14) form a system of $3J+3$ equations for the solution of $3J+3$ unknowns $f_j^n$, $u_j^n$, $v_j^n$, $0 \le j \le J$. To solve this nonlinear system, we use Newton's method and solve the resulting linear system with the block-elimination method described in detail in [1].

### 3.2  Noniterative Method

To solve the system given by Eqs. (10) and (11) with the noniterative method, we first write Eq. (10c) at $(x^{n-1/2}, \eta)$ as

390

$$\frac{(v')^n + (v')^{n-1}}{2}$$

$$+ \frac{m^{n-1/2} + 1}{4} \left( f^n v^{n-1} + f^{n-1} v^n \right)$$

$$+ m^{n-1/2} \left( 1 - u^n u^{n-1} \right)$$

$$= \frac{x^{n-1/2}}{2} \left( u^n \frac{\partial u^{n-1}}{\partial x} + u^{n-1} \frac{\partial u^n}{\partial x} - v^n \frac{\partial f^{n-1}}{\partial x} - v^{n-1} \frac{\partial f^n}{\partial x} \right) \tag{15}$$

Since

$$f^n = f^{n-1/2} + \frac{\partial f^{n-1/2}}{\partial x} \frac{k_n}{2} + \frac{1}{2!} \frac{\partial^2 f^{n-1/2}}{\partial x^2} \left( \frac{k_n}{2} \right)^2 + \mathrm{o}(k_n^3) \tag{16a}$$

and

$$f^{n-1} = f^{n-1/2} - \frac{\partial f^{n-1/2}}{\partial x} \frac{k_n}{2} + \frac{1}{2!} \frac{\partial^2 f^{n-1/2}}{\partial x^2} \left( \frac{k_n}{2} \right)^2 + \mathrm{o}(k_n^3) \tag{16b}$$

we can write

$$(fv)^{n-1/2} = \frac{\left( f^n v^{n-1} + f^{n-1} v^n \right)}{2} + \mathrm{o}(k_n^2). \tag{16c}$$

Similarly, we can represent the other terms in Eq. (15) with second order approximations,

$$(u_x)_{j-1/2}^{n-1} = a_1 u_{j-1/2}^{n-3} + a_2 u_{j-1/2}^{n-2} + a_3 u_{j-1/2}^{n-1}$$

$$(u_x)_{j-1/2}^{n} = \tilde{a}_1 u_{j-1/2}^{n-2} + \tilde{a}_2 u_{j-1/2}^{n-1} + \tilde{a}_3 u_{j-1/2}^{n} \tag{17}$$

where

$$a_1 = \frac{x_{n-1} - x_{n-2}}{(x_{n-2} - x_{n-3})(x_{n-1} - x_{n-3})}$$

$$a_2 = -\frac{x_{n-1} - x_{n-3}}{(x_{n-2} - x_{n-3})(x_{n-1} - x_{n-2})}$$

$$a_3 = \frac{2x_{n-1} - x_{n-2} - x_{n-3}}{(x_{n-1} - x_{n-2})(x_{n-1} - x_{n-3})}$$

$$\tilde{a}_1 = \frac{x_n - x_{n-1}}{(x_{n-1} - x_{n-2})(x_n - x_{n-2})}$$

$$\tilde{a}_2 = -\frac{x_n - x_{n-2}}{(x_{n-1} - x_{n-2})(x_n - x_{n-1})}$$

$$\tilde{a}_3 = \frac{2x_n - x_{n-1} - x_{n-2}}{(x_n - x_{n-1})(x_n - x_{n-2})}$$

$$(18)$$

Eq. (15) can then be written as

$$
\begin{aligned}
\frac{(v)_j^n - (v)_{j-1}^n}{h_j} & \\
+ \frac{(v)_j^{n-1} - (v)_{j-1}^{n-1}}{h_j} & \\
+ \frac{m^n + 1}{2}\left(f_{j-1/2}^n v_{j-1/2}^{n-1} + v_{j-1/2}^n f_{j-1/2}^{n-1}\right) & \\
+ 2m^n\left(1 - u_{j-1/2}^n u_{j-1/2}^{n-1}\right) & \\
= x^{n-1/2}\left[
\begin{array}{l}
u_{j-1/2}^n\left(a_1 u_{j-1/2}^{n-3} + a_2 u_{j-1/2}^{n-2} + a_3 u_{j-1/2}^{n-1}\right) \\
+ u_{j-1/2}^{n-1}\left(\tilde{a}_1 u_{j-1/2}^{n-2} + \tilde{a}_2 u_{j-1/2}^{n-1} + \tilde{a}_3 u_{j-1/2}^n\right) \\
- v_{j-1/2}^n\left(a_1 f_{j-1/2}^{n-3} + a_2 f_{j-1/2}^{n-2} + a_3 f_{j-1/2}^{n-1}\right) \\
- v_{j-1/2}^{n-1}\left(\tilde{a}_1 f_{j-1/2}^{n-2} + \tilde{a}_2 f_{j-1/2}^{n-1} + \tilde{a}_3 f_{j-1/2}^n\right)
\end{array}
\right]
\end{aligned}
$$

$$(19)$$

If we assume $f_j^n$, $u_j^n$, $v_j^n$ to be unknowns for $0 \le j \le J$, then Eqs. (13a, b) and (19) form a linear algebraic system of $3J+3$ equations that can be solved with the same procedure used for the iterative method [1].

## 4   ACCURACY OF THE NONITERATIVE METHOD

To evaluate the accuracy of the noniterative method, we consider Howarth's flow for which the inviscid velocity distribution is given by

$$u_e(x) = 1 - \frac{1}{8}x$$

and perform calculations with fixed uniform spacing in the x-direction with $k = 1/16$ and 1/32.

This flow has a separation at x=0.960 which is generally obtained by extrapolation. For calculations to proceed as close to this location, much finer spacings are needed around x $\approx$ 0.95. So we expect the calculations to break down with course spacing earlier than those with fine spacing.

Tables 1 to 2 show the calculated wall-shear parameter for k=1/16 and 1/32, respectively. A choice of $k = 1/16$ yields 16 x-stations, the last being at x=0.9375. This is a rather severe test case because, since the iterative method uses Newton's method, a course grid will increase the number of iterations.

Table 1 shows the results for k=1/16. Since the noniterative method requires that solutions to previous three x-stations are known, the comparison of the calculated results in Table 1 begin at the fourth x-station, x= 0.25. Also, this table contains additional x-stations between x= 0.9375 and x = 0.96 with $\Delta x = 0.005$, due to flow separation.

As can be seen, the solutions obtained from the noniterative method agree well with those obtained from the iterative method, especially for values of x away from the separation location. As we approach the separation location, say x $\approx$ 0.90, the difference between the two solutions increases gradually. Within a plotting accuracy shown in Fig. 2, the agreement is very good, including the location of flow separation point. In addition, the comparison shows that while the iterative method requires a total of 49 iterations, the noniterative method requires only 16 iterations, leading to an efficiency of approximately 70%.

Table 2 and Fig. 3 show the results for k=1/32 with conclusions essentially the same as those obtained with k = 1/16. Again, both results agree very well until x $\approx$ 0.90 and begin to differ with increasing x. The efficiency of the noniterative method (83 iterations vs 31 iterations) is again good, but this time it is around 60%.

Table 1. Comparison of calculated results for $k=1/16$

| $\Delta x=1/16$ | Iterative Method | | Noniterative Method | | $\left(f_w''\right)_{iter}-\left(f_w''\right)_{non}$ | Relative |
|---|---|---|---|---|---|---|
| x | $\left(f_w''\right)_{iter}$ | Iter | $\left(f_w''\right)_{non}$ | Iter | Diff. | Diff. |
| 0.25 | 2.80E-01 | 3 | 2.80E-01 | 1 | 1.00E-04 | 3.57E-04 |
| 0.3125 | 2.66E-01 | 3 | 2.66E-01 | 1 | 0.00E+00 | 0.00E+00 |
| 0.375 | 2.50E-01 | 3 | 2.50E-01 | 1 | 1.00E-04 | 3.99E-04 |
| 0.4375 | 2.34E-01 | 3 | 2.34E-01 | 1 | 0.00E+00 | 0.00E+00 |
| 0.5 | 2.18E-01 | 3 | 2.18E-01 | 1 | 1.00E-04 | 4.60E-04 |
| 0.5625 | 2.00E-01 | 3 | 2.00E-01 | 1 | 0.00E+00 | 0.00E+00 |
| 0.625 | 1.81E-01 | 3 | 1.80E-01 | 1 | 2.00E-04 | 1.11E-03 |
| 0.6875 | 1.60E-01 | 3 | 1.60E-01 | 1 | 2.00E-04 | 1.25E-03 |
| 0.75 | 1.37E-01 | 3 | 1.37E-01 | 1 | 3.00E-04 | 2.19E-03 |
| 0.8125 | 1.11E-01 | 3 | 1.11E-01 | 1 | 5.00E-04 | 4.50E-03 |
| 0.875 | 8.02E-02 | 3 | 7.95E-02 | 1 | 7.10E-04 | 8.85E-03 |
| 0.9375 | 3.95E-02 | 4 | 3.88E-02 | 1 | 6.30E-04 | 1.60E-02 |
| 0.9425 | 3.42E-02 | 3 | 3.25E-02 | 1 | 1.68E-03 | 4.92E-02 |
| 0.9475 | 2.83E-02 | 3 | 2.66E-02 | 1 | 1.75E-03 | 6.18E-02 |
| 0.9525 | 2.18E-02 | 3 | 1.93E-02 | 1 | 2.48E-03 | 1.14E-01 |
| 0.9575 | 1.32E-02 | 3 | 9.44E-03 | 1 | 3.74E-03 | 2.84E-01 |
| | Total Iterations | 49 | | 16 | | |

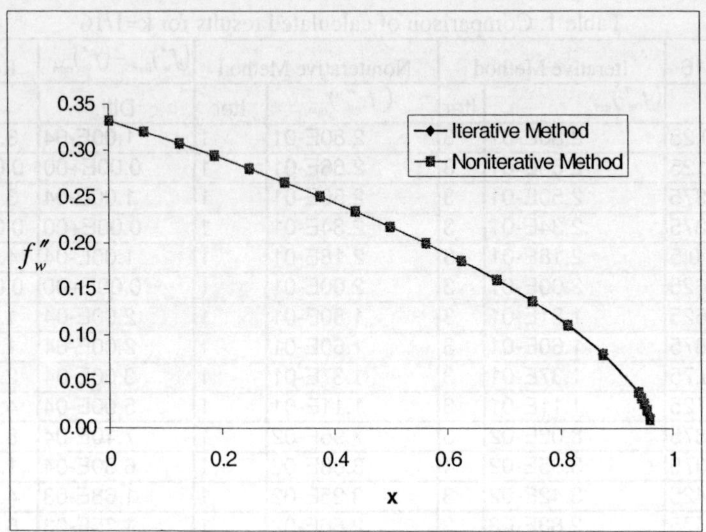


Figure 2.  Comparison of calculated wall shear values with both methods for k =1/16

## Table 2. Comparison of calculated results for k=1/32

| $\Delta x = 1/32$ | Iterative Method | | Noniterative Method | | $\left\|\left(f_w''\right)_{iter} - \left(f_w''\right)_{non}\right\|$ | Relative |
|---|---|---|---|---|---|---|
| x | $\left(f_w''\right)_{iter}$ | Iter | $\left(f_w''\right)_{non}$ | Iter | Diff. | Diff. |
| 0.125 | 3.07E-01 | 2 | 3.07E-01 | 1 | 0.00E+00 | 0.00E+00 |
| 0.1563 | 3.01E-01 | 2 | 3.01E-01 | 1 | 0.00E+00 | 0.00E+00 |
| 0.1875 | 2.94E-01 | 2 | 2.94E-01 | 1 | 0.00E+00 | 0.00E+00 |
| 0.2188 | 2.87E-01 | 2 | 2.87E-01 | 1 | 0.00E+00 | 0.00E+00 |
| 0.25 | 2.80E-01 | 2 | 2.80E-01 | 1 | -1.00E-04 | -3.57E-04 |
| 0.2813 | 2.73E-01 | 2 | 2.73E-01 | 1 | -1.00E-04 | -3.67E-04 |
| 0.3125 | 2.66E-01 | 2 | 2.66E-01 | 1 | 0.00E+00 | 0.00E+00 |
| 0.3438 | 2.58E-01 | 2 | 2.58E-01 | 1 | 0.00E+00 | 0.00E+00 |
| 0.375 | 2.50E-01 | 2 | 2.50E-01 | 1 | 0.00E+00 | 0.00E+00 |
| 0.4063 | 2.43E-01 | 2 | 2.43E-01 | 1 | 0.00E+00 | 0.00E+00 |
| 0.4375 | 2.35E-01 | 2 | 2.35E-01 | 1 | 0.00E+00 | 0.00E+00 |
| 0.4688 | 2.26E-01 | 2 | 2.26E-01 | 1 | 0.00E+00 | 0.00E+00 |
| 0.5 | 2.18E-01 | 3 | 2.18E-01 | 1 | 1.00E-04 | 4.60E-04 |
| 0.5313 | 2.09E-01 | 3 | 2.09E-01 | 1 | 0.00E+00 | 0.00E+00 |
| 0.5625 | 2.00E-01 | 3 | 2.00E-01 | 1 | 1.00E-04 | 5.01E-04 |
| 0.5938 | 1.90E-01 | 3 | 1.90E-01 | 1 | 0.00E+00 | 0.00E+00 |
| 0.625 | 1.81E-01 | 3 | 1.81E-01 | 1 | 1.00E-04 | 5.54E-04 |
| 0.6563 | 1.70E-01 | 3 | 1.70E-01 | 1 | 1.00E-04 | 5.87E-04 |
| 0.6875 | 1.60E-01 | 3 | 1.60E-01 | 1 | 1.00E-04 | 6.26E-04 |
| 0.7188 | 1.49E-01 | 3 | 1.49E-01 | 1 | 2.00E-04 | 1.35E-03 |
| 0.75 | 1.37E-01 | 3 | 1.37E-01 | 1 | 2.00E-04 | 1.46E-03 |
| 0.7813 | 1.24E-01 | 3 | 1.24E-01 | 1 | 2.00E-04 | 1.61E-03 |
| 0.8125 | 1.11E-01 | 3 | 1.11E-01 | 1 | 3.00E-04 | 2.70E-03 |
| 0.8438 | 9.65E-02 | 3 | 9.61E-02 | 1 | 3.50E-04 | 3.63E-03 |
| 0.875 | 8.02E-02 | 3 | 7.98E-02 | 1 | 4.90E-04 | 6.11E-03 |
| 0.9063 | 6.12E-02 | 3 | 6.05E-02 | 1 | 6.90E-04 | 1.13E-02 |
| 0.9375 | 3.83E-02 | 4 | 3.73E-02 | 1 | 1.00E-03 | 2.61E-02 |
| 0.9425 | 3.31E-02 | 3 | 3.14E-02 | 1 | 1.67E-03 | 5.05E-02 |
| 0.9475 | 2.71E-02 | 3 | 2.52E-02 | 1 | 1.91E-03 | 7.05E-02 |
| 0.9525 | 2.04E-02 | 3 | 1.79E-02 | 1 | 2.58E-03 | 1.26E-01 |
| 0.9575 | 1.12E-02 | 4 | 6.88E-03 | 1 | 4.30E-03 | 3.84E-01 |
| Total iterations | | 83 | | 31 | | |

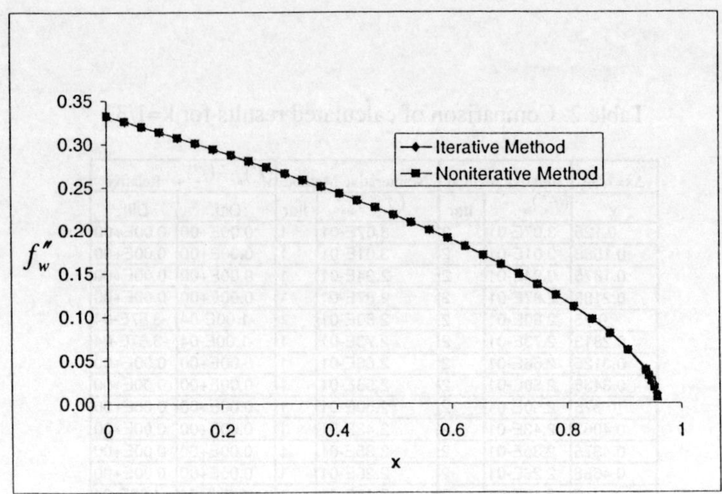

Figure 3. Comparison of calculated wall shear values with both methods for k =1/32.

## SUMMARY

The application of the noniterative method to two-dimensional laminar flows shows that its accuracy is sufficient for most engineering problems and savings in computer time are substantial.  The method is currently being applied to two-dimensional laminar and turbulent flows with and without separation.

## REFERENCES

[1] Tuncer Cebeci and Jean Cousteix, *Modeling and Computation of Boundary-Layer Flows*, Horizons Publishing, Long Beach, CA and Springer-Varlag, Heidelberg, 1998.

[2] Tuncer Cebeci, *An Engineering Approach to the Calculation of Aerodynamic Flows,*  Springer-Varlag, Heidelberg, Horizons Publishing, Long Beach, CA. 1999.

[3] Crank, J. and Nicholson, *A Practical Method for Numerical Evaluation of Solutions of Partial Differential Equations of the Heat-Conduction Type,* Proceedings of the Cambridge Philosophical Society, Vol. 43, pp. 50-67, January. 1947.

[4] Keller, H. B, Numerical Methods for Two-Point Boundary-Value Problems, Blaisdell Publishing Co., 1968.

# A NONITERATIVE METHOD FOR BOUNDARY-LAYER EQUATIONS

## II. TWO-DIMENSIONAL LAMINAR AND TURBULENT FLOWS

**Tuncer Cebeci**[*], **Jian Ping Shao**[†]

• The Boeing Company
Long Beach, CA90840, USA
email: tuncer.cebeci@boeing.com

[†] The Boeing Company
Long Beach, CA90840, USA
email: jian.p.shao@boeing.com

**Key words:** Boundary-Layer Equation, Laminar Flow, Turbulent Flow, Iterative Method, Noniterative Method, Nonlinear Parabolic Partial-Differential Equation.

**Abstract.** *A noniterative method for nonlinear parabolic partial-differential equations is described and applied to boundary-layer equations for two-dimensional laminar and turbulent flows. Comparison of calculated results indicates that the accuracy of this method is comparable to those obtained with an iterative method.*

## 1   INTRODUCTION

It is well known that the boundary-layer equations for laminar and turbulent flows are parabolic and nonlinear. Their solutions can be obtained in their nonlinear form by using a shooting method or in linearized form by a finite-difference method [1]. The latter choice is more common and practical, especially in turbulent flows since a finite-difference method is more efficient than an integration method such as a shooting method.

It is also well known that the solutions of the boundary-layer equations in differential form are more time consuming than those based on integral form since in the latter case the solutions are obtained for ordinary differential equations. Even though significant advances have been made in the development of efficient and accurate numerical methods to solve the differential form of the boundary-layer equations, the computer times associated with them are still considerably more than those of integral methods.

The "edge" the integral methods have over the differential methods becomes even more significant in methods which employ inviscid/viscous interaction techniques where the boundary-layer equations are solved several times for a given pressure distribution by making several sweeps on the body. In each sweep, the solutions at each streamwise location are obtained iteratively since the equations have been linearized. This procedure is not only applied on every streamwise station but it is also used in each sweep. It is clear that significant savings in computer time can result if the solutions are not iterated at each streamwise location without sacrificing accuracy.

In the present paper, we address this need and describe a new numerical method that will allow computer savings of 75% of a modern finite-difference method such as that described in [2]. The method is based on a noniterative scheme and its accuracy is comparable to that resulting from an iterative scheme. In this paper, we describe this method for two-dimensional laminar and turbulent flows and compare its solutions with an iterative scheme such as that described in [2].

To the authors' knowledge, this is the first paper that describes the solution of the nonlinear parabolic partial-differential equations with a noniterative scheme. As such, it is a useful and efficient tool leading to considerable computer savings and will find applications in many engineering problems that require the solution of nonlinear parabolic partial differential equations.

## 2  BOUNDARY-LAYER EQUATIONS

The boundary-layer equations and their boundary conditions for two-dimensional laminar and turbulent flows are well known. With an eddy viscosity ($\varepsilon_m$) concept,

$$\varepsilon_m = \frac{(-\overline{u'v'})}{\dfrac{\partial u}{\partial y}} \tag{1}$$

they can be written as

$$\frac{\partial u}{\partial x} + \frac{\partial v}{\partial y} = 0 \tag{2}$$

$$u\frac{\partial u}{\partial x}+v\frac{\partial u}{\partial y}=u_e\frac{du_e}{dx}+\frac{\partial}{\partial y}\left(b\frac{\partial u}{\partial y}\right) \tag{3}$$

$$y=0,\ u=0,\ v=0\ ;\ y\rightarrow\infty,\ u\rightarrow u_e(x) \tag{4}$$

where

$$b=\varepsilon_m+v.$$

As discussed in [2], it is more efficient to solve the boundary-layer equations in transformed variables. A convenient one is the Falkner-Skan transformation in which a similarity parameter $\eta$ is defined by

$$\eta=\sqrt{\frac{u_e}{vx}}\,y \tag{5a}$$

and a dimensionless stream function $f(x,\eta)$ by

$$f(x,\eta)=\frac{\psi(x,y)}{\sqrt{u_e vx}} \tag{5b}$$

Here $\psi(x,y)$ is a dimensional stream function that satisfies equation (2),

$$u=\frac{\partial\psi}{\partial y},\qquad v=-\frac{\partial\psi}{\partial x} \tag{6}$$

With this transformation, Eqs. (2) to (4) can be written as

$$(bf'')'+\frac{m+1}{2}ff''+m\left[1-(f')^2\right]=x\left(f'\frac{\partial f'}{\partial x}-f''\frac{\partial f}{\partial x}\right) \tag{7}$$

$$\eta=0,\ f=f'=0;\qquad \eta=\eta_e,\ f'=1 \tag{8}$$

Here m is a dimensionless pressure-gradient parameter

$$m = \frac{x}{u_e}\frac{du_e}{dx} \tag{9}$$

## 3  ITERATIVE METHOD

Boundary-layer equations being parabolic in nature can be solved by using several numerical methods; of these, finite difference methods are at present the most common with Crank-Nicolson [2] and Keller's box [3] methods being the most popular ones. The latter has several advantages over the former and is considered here.

According to Keller's method, Eq. (7) is first expressed as a first-order system by introducing new variables $u(x,\eta)$, and $v(x,\eta)$

$$f' = u \tag{10a}$$

$$u' = v \tag{10b}$$

so that Eq. (7) becomes

$$(bv)' + \frac{m+1}{2}fv + m\left(1 - u^2\right) = x(u\frac{\partial u}{\partial x} - v\frac{\partial f}{\partial x}) \tag{10c}$$

The boundary conditions now become

$$\eta = 0,\; u = 0,\quad f = 0; \qquad\qquad \eta = \eta_e,\; u = 1 \tag{11}$$

### 3.1  Iterative Method

To solve the system given by Eqs. (10) and (11) with Keller's box method, which is an iterative method, we consider the net rectangle shown in Fig. 1

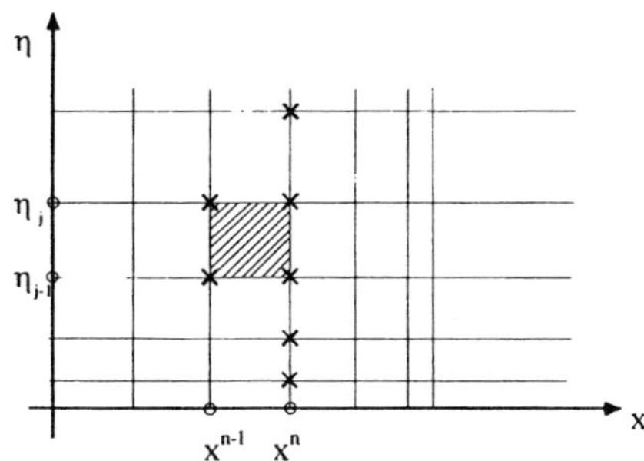 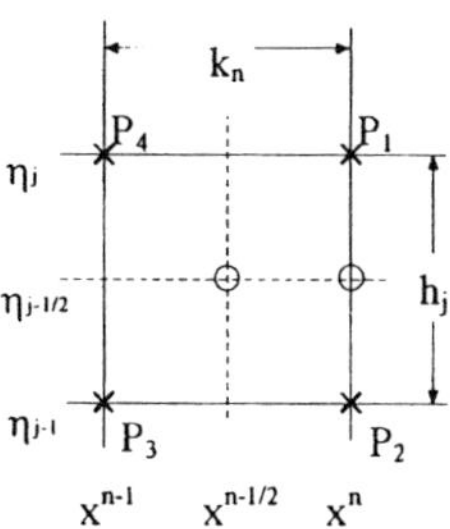

Figure 1

$$x^0 = 0, \quad x^n = x^{n-1} + k_n, \quad n = 1,2,\ldots,N$$
$$\eta_0 = 0, \quad \eta_j = \eta_{j-1} + h_j, \quad j = 1,2,\ldots,J; \qquad \eta_J = \eta_e \tag{12}$$

and write difference equations that are to approximate Eqs. (10). We write the finite difference approximations of Eq. (10a,b) at $(x^n, \eta_{j-1/2})$ using centered-difference derivatives.

$$\frac{f_j^n - f_{j-1}^n}{h_j} - \frac{u_j^n + u_{j-1}^n}{2} = 0 \tag{13a}$$

$$\frac{u_j^n - u_{j-1}^n}{h_j} - \frac{v_j^n + v_{j-1}^n}{2} = 0 \tag{13b}$$

Similarly, Eq. (10c) is approximated at the midpoint $(x^{n-1/2}, \eta_{j-1/2})$ of the net rectangle $P_1 P_2 P_3 P_4$,

$$\frac{1}{2}\left[\frac{(bv)_j^n-(bv)_{j-1}^n}{h_j}+\frac{(bv)_j^{n-1}-(bv)_{j-1}^{n-1}}{h_j}\right]$$

$$+\frac{m^{n-1/2}+1}{4}\left[(fv)_{j-1/2}^n+(fv)_{j-1/2}^{n-1}\right]$$

$$+m^{n-1/2}\left(1-\frac{\left((u^2)_{j-1/2}^n+(u^2)_{j-1/2}^{n-1}\right)}{2}\right) \tag{13c}$$

$$=x^{n-1/2}\left[u_{j-1/2}^{n-1/2}\frac{u_{j-1/2}^n-u_{j-1/2}^{n-1}}{k_n}-v_{j-1/2}^{n-1/2}\frac{f_{j-1/2}^n-f_{j-1/2}^{n-1}}{k_n}\right]$$

The boundary conditions, Eq. (11) become

$$u_0^n=0;\quad f_0^n=0;\quad u_J^n=1 \tag{14}$$

If we assume $f_j^{n-1}$, $u_j^{n-1}$, $v_j^{n-1}$ to be known for $0\le j\le J$, then Eqs. (13) and (14) form a system of $3J+3$ equations for the solution of $3J+3$ unknowns $f_j^n$, $u_j^n$, $v_j^n$, $0\le j\le J$. To solve this nonlinear system, we use Newton's method and solve the resulting linear system with the block-elimination method described in detail in [2].

## 3.2  Noniterative Method

To solve the system given by Eqs. (10) and (11) with the noniterative method, we first write Eq. (10c) at $(x^{n-1/2},\eta)$ as

$$\frac{\left((bv)'\right)^{n} + \left((bv)'\right)^{n-1}}{2}$$
$$+ \frac{m^{n-1/2}+1}{4}\left(f^{n}v^{n-1} + f^{n-1}v^{n}\right)$$
$$+ m^{n-1/2}\left(1 - u^{n}u^{n-1}\right)$$
$$= \frac{x^{n-1/2}}{2}\left(u^{n}\frac{\partial u^{n-1}}{\partial x} + u^{n-1}\frac{\partial u^{n}}{\partial x} - v^{n}\frac{\partial f^{n-1}}{\partial x} - v^{n-1}\frac{\partial f^{n}}{\partial x}\right) \tag{15}$$

Since

$$f^{n} = f^{n-1/2} + \frac{\partial f^{n-1/2}}{\partial x}\frac{k_{n}}{2} + \frac{1}{2!}\frac{\partial^{2} f^{n-1/2}}{\partial x^{2}}\left(\frac{k_{n}}{2}\right)^{2} + o(k_{n}^{3}) \tag{16a}$$

and

$$f^{n-1} = f^{n-1/2} - \frac{\partial f^{n-1/2}}{\partial x}\frac{k_{n}}{2} + \frac{1}{2!}\frac{\partial^{2} f^{n-1/2}}{\partial x^{2}}\left(\frac{k_{n}}{2}\right)^{2} + o(k_{n}^{3}) \tag{16b}$$

we can write

$$(fv)^{n-1/2} = \frac{\left(f^{n}v^{n-1} + f^{n-1}v^{n}\right)}{2} + o(k_{n}^{2}). \tag{16c}$$

Similarly, we can represent the other terms in Eq. (15) with second order approximations,

$$(u_{x})_{j-1/2}^{n-1} = a_{1}u_{j-1/2}^{n-3} + a_{2}u_{j-1/2}^{n-2} + a_{3}u_{j-1/2}^{n-1}$$
$$(u_{x})_{j-1/2}^{n} = \tilde{a}_{1}u_{j-1/2}^{n-2} + \tilde{a}_{2}u_{j-1/2}^{n-1} + \tilde{a}_{3}u_{j-1/2}^{n} \tag{17}$$

where

$$a_1 = \frac{x_{n-1} - x_{n-2}}{(x_{n-2} - x_{n-3})(x_{n-1} - x_{n-3})} \qquad a_2 = -\frac{x_{n-1} - x_{n-3}}{(x_{n-2} - x_{n-3})(x_{n-1} - x_{n-2})}$$

$$a_3 = \frac{2x_{n-1} - x_{n-2} - x_{n-3}}{(x_{n-1} - x_{n-2})(x_{n-1} - x_{n-3})} \qquad \tilde{a}_1 = \frac{x_n - x_{n-1}}{(x_{n-1} - x_{n-2})(x_n - x_{n-2})}$$

$$\tilde{a}_2 = -\frac{x_n - x_{n-2}}{(x_{n-1} - x_{n-2})(x_n - x_{n-1})} \qquad \tilde{a}_3 = \frac{2x_n - x_{n-1} - x_{n-2}}{(x_n - x_{n-1})(x_n - x_{n-2})} \tag{18}$$

Eq. (15) can then be written as

$$\frac{(bv)_j^n - (bv)_{j-1}^n}{h_j}$$

$$+ \frac{(bv)_j^{n-1} - (bv)_{j-1}^{n-1}}{h_j}$$

$$+ \frac{m^n + 1}{2}\left(f_{j-1/2}^n v_{j-1/2}^{n-1} + v_{j-1/2}^n f_{j-1/2}^{n-1}\right) \tag{19}$$

$$+ 2m^n\left(1 - u_{j-1/2}^n u_{j-1/2}^{n-1}\right)$$

$$= x^{n-1/2}\begin{bmatrix} u_{j-1/2}^n\left(a_1 u_{j-1/2}^{n-3} + a_2 u_{j-1/2}^{n-2} + a_3 u_{j-1/2}^{n-1}\right) \\ + u_{j-1/2}^{n-1}\left(\tilde{a}_1 u_{j-1/2}^{n-2} + \tilde{a}_2 u_{j-1/2}^{n-1} + \tilde{a}_3 u_{j-1/2}^n\right) \\ - v_{j-1/2}^n\left(a_1 f_{j-1/2}^{n-3} + a_2 f_{j-1/2}^{n-2} + a_3 f_{j-1/2}^{n-1}\right) \\ - v_{j-1/2}^{n-1}\left(\tilde{a}_1 f_{j-1/2}^{n-2} + \tilde{a}_2 f_{j-1/2}^{n-1} + \tilde{a}_3 f_{j-1/2}^n\right) \end{bmatrix}$$

If we assume $f_j^n$, $u_j^n$, $v_j^n$ to be unknowns for $0 \le j \le J$, then Eqs. (13a, b) and (19) form a linear algebraic system of $3J+3$ equations that can be solved with the same procedure used for the iterative method [2].


## 4  ACCURACY OF THE NONITERATIVE METHOD

To evaluate the accuracy of the noniterative method, we have performed calculations with both methods for laminar and turbulent flows as discussed below.

## 4.1 Laminar Flow

We consider Howarth's flow for which the inviscid velocity distribution is given by

$$u_e(x) = 1 - \frac{1}{8}x$$

and perform calculations with fixed uniform spacing in the x-direction with $k = 1/16$ and 1/32.

This flow has a separation at x=0.960 which is generally obtained by extrapolation. For calculations to proceed as close to this location, much finer spacings are needed around $x \approx 0.95$. So we expect the calculations to break down with course spacing earlier than those with fine spacing.

Tables 1 to 2 show the calculated wall-shear parameter for k=1/16 and 1/32, respectively. A choice of $k = 1/16$ yields 16 x-stations, the last being at x=0.9375. This is a rather severe test case because, since the iterative method uses Newton's method, a course grid will increase the number of iterations.

Table 1 shows the results for k=1/16. Since the noniterative method requires that solutions to previous three x-stations are known, the comparison of the calculated results in Table 1 begin at the fourth x-station, x= 0.25. Also, this table contains additional x-stations between x= 0.9375 and x = 0.96 with $\Delta x = 0.005$, due to flow separation.

As can be seen, the solutions obtained from the noniterative method agree well with those obtained from the iterative method, especially for values of x away from the separation location. As we approach the separation location, say $x \approx 0.90$, the difference between the two solutions increases gradually. Within a plotting accuracy shown in Fig. 2, the agreement is very good, including the location of flow separation point. In addition, the comparison shows that while the iterative method requires a total of 49 iterations, the noniterative method requires only 16 iterations, leading to an efficiency of $\approx 70\%$

Table 2 and Fig. 3 show the results for k=1/32 with conclusions essentially the same as those obtained with $k = 1/16$. Again, both results agree very well until x $\approx 0.90$ and begin to differ with increasing x. The efficiency of the noniterative method (83 iterations vs 31 iterations) is again good, but this time it is around 60%.

Table 1. Comparison of calculated results for k=1/16

| $\Delta x=1/16$ | Iterative Method | | Noniterative Method | | $\left(f_w''\right)_{iter} - \left(f_w''\right)_{non}$ | Relative |
|---|---|---|---|---|---|---|
| x | $\left(f_w''\right)_{iter}$ | Iter | $\left(f_w''\right)_{non}$ | Iter | Diff. | Diff. |
| 0.25 | 2.80E-01 | 3 | 2.80E-01 | 1 | 1.00E-04 | 3.57E-04 |
| 0.3125 | 2.66E-01 | 3 | 2.66E-01 | 1 | 0.00E+00 | 0.00E+00 |
| 0.375 | 2.50E-01 | 3 | 2.50E-01 | 1 | 1.00E-04 | 3.99E-04 |
| 0.4375 | 2.34E-01 | 3 | 2.34E-01 | 1 | 0.00E+00 | 0.00E+00 |
| 0.5 | 2.18E-01 | 3 | 2.18E-01 | 1 | 1.00E-04 | 4.60E-04 |
| 0.5625 | 2.00E-01 | 3 | 2.00E-01 | 1 | 0.00E+00 | 0.00E+00 |
| 0.625 | 1.81E-01 | 3 | 1.80E-01 | 1 | 2.00E-04 | 1.11E-03 |
| 0.6875 | 1.60E-01 | 3 | 1.60E-01 | 1 | 2.00E-04 | 1.25E-03 |
| 0.75 | 1.37E-01 | 3 | 1.37E-01 | 1 | 3.00E-04 | 2.19E-03 |
| 0.8125 | 1.11E-01 | 3 | 1.11E-01 | 1 | 5.00E-04 | 4.50E-03 |
| 0.875 | 8.02E-02 | 3 | 7.95E-02 | 1 | 7.10E-04 | 8.85E-03 |
| 0.9375 | 3.95E-02 | 4 | 3.88E-02 | 1 | 6.30E-04 | 1.60E-02 |
| 0.9425 | 3.42E-02 | 3 | 3.25E-02 | 1 | 1.68E-03 | 4.92E-02 |
| 0.9475 | 2.83E-02 | 3 | 2.66E-02 | 1 | 1.75E-03 | 6.18E-02 |
| 0.9525 | 2.18E-02 | 3 | 1.93E-02 | 1 | 2.48E-03 | 1.14E-01 |
| 0.9575 | 1.32E-02 | 3 | 9.44E-03 | 1 | 3.74E-03 | 2.84E-01 |
| | Total Iteration | 49 | | 16 | | |

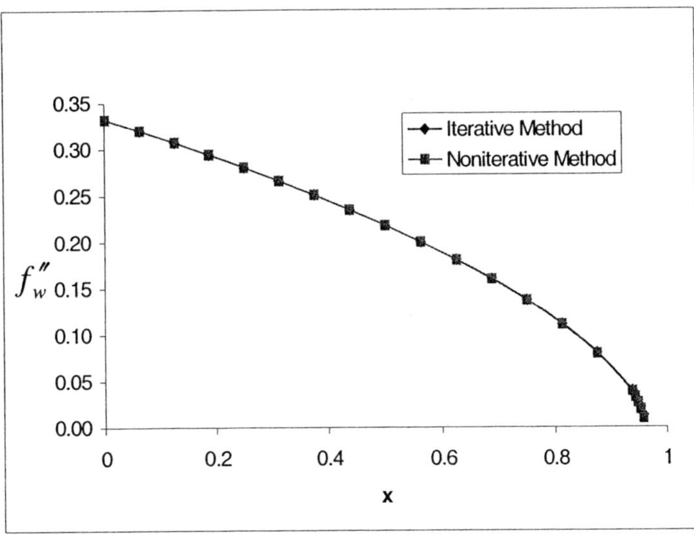

Figure 2. Comparison of calculated wall shear values with both methods for k =1/16

## Table 2. Comparison of calculated results for k=1/16

| $\Delta x = 1/32$ | Iterative Method | | Noniterative Method | | $(f_w'')_{iter} - (f_w'')_{non}$ | Relative |
| --- | --- | --- | --- | --- | --- | --- |
| x | $(f_w'')_{iter}$ | Iter | $(f_w'')_{non}$ | Iter | Diff. | Diff. |
| 0.125 | 3.07E-01 | 2 | 3.07E-01 | 1 | 0.00E+00 | 0.00E+00 |
| 0.1563 | 3.01E-01 | 2 | 3.01E-01 | 1 | 0.00E+00 | 0.00E+00 |
| 0.1875 | 2.94E-01 | 2 | 2.94E-01 | 1 | 0.00E+00 | 0.00E+00 |
| 0.2188 | 2.87E-01 | 2 | 2.87E-01 | 1 | 0.00E+00 | 0.00E+00 |
| 0.25 | 2.80E-01 | 2 | 2.80E-01 | 1 | -1.00E-04 | -3.57E-04 |
| 0.2813 | 2.73E-01 | 2 | 2.73E-01 | 1 | -1.00E-04 | -3.67E-04 |
| 0.3125 | 2.66E-01 | 2 | 2.66E-01 | 1 | 0.00E+00 | 0.00E+00 |
| 0.3438 | 2.58E-01 | 2 | 2.58E-01 | 1 | 0.00E+00 | 0.00E+00 |
| 0.375 | 2.50E-01 | 2 | 2.50E-01 | 1 | 0.00E+00 | 0.00E+00 |
| 0.4063 | 2.43E-01 | 2 | 2.43E-01 | 1 | 0.00E+00 | 0.00E+00 |
| 0.4375 | 2.35E-01 | 2 | 2.35E-01 | 1 | 0.00E+00 | 0.00E+00 |
| 0.4688 | 2.26E-01 | 2 | 2.26E-01 | 1 | 0.00E+00 | 0.00E+00 |
| 0.5 | 2.18E-01 | 3 | 2.18E-01 | 1 | 1.00E-04 | 4.60E-04 |
| 0.5313 | 2.09E-01 | 3 | 2.09E-01 | 1 | 0.00E+00 | 0.00E+00 |
| 0.5625 | 2.00E-01 | 3 | 2.00E-01 | 1 | 1.00E-04 | 5.01E-04 |
| 0.5938 | 1.90E-01 | 3 | 1.90E-01 | 1 | 0.00E+00 | 0.00E+00 |
| 0.625 | 1.81E-01 | 3 | 1.81E-01 | 1 | 1.00E-04 | 5.54E-04 |
| 0.6563 | 1.70E-01 | 3 | 1.70E-01 | 1 | 1.00E-04 | 5.87E-04 |
| 0.6875 | 1.60E-01 | 3 | 1.60E-01 | 1 | 1.00E-04 | 6.26E-04 |
| 0.7188 | 1.49E-01 | 3 | 1.49E-01 | 1 | 2.00E-04 | 1.35E-03 |
| 0.75 | 1.37E-01 | 3 | 1.37E-01 | 1 | 2.00E-04 | 1.46E-03 |
| 0.7813 | 1.24E-01 | 3 | 1.24E-01 | 1 | 2.00E-04 | 1.61E-03 |
| 0.8125 | 1.11E-01 | 3 | 1.11E-01 | 1 | 3.00E-04 | 2.70E-03 |
| 0.8438 | 9.65E-02 | 3 | 9.61E-02 | 1 | 3.50E-04 | 3.63E-03 |
| 0.875 | 8.02E-02 | 3 | 7.98E-02 | 1 | 4.90E-04 | 6.11E-03 |
| 0.9063 | 6.12E-02 | 3 | 6.05E-02 | 1 | 6.90E-04 | 1.13E-02 |
| 0.9375 | 3.83E-02 | 4 | 3.73E-02 | 1 | 1.00E-03 | 2.61E-02 |
| 0.9425 | 3.31E-02 | 3 | 3.14E-02 | 1 | 1.67E-03 | 5.05E-02 |
| 0.9475 | 2.71E-02 | 3 | 2.52E-02 | 1 | 1.91E-03 | 7.05E-02 |
| 0.9525 | 2.04E-02 | 3 | 1.79E-02 | 1 | 2.58E-03 | 1.26E-01 |
| 0.9575 | 1.12E-02 | 4 | 6.88E-03 | 1 | 4.30E-03 | 3.84E-01 |
| | Total | 83 | | 31 | | |

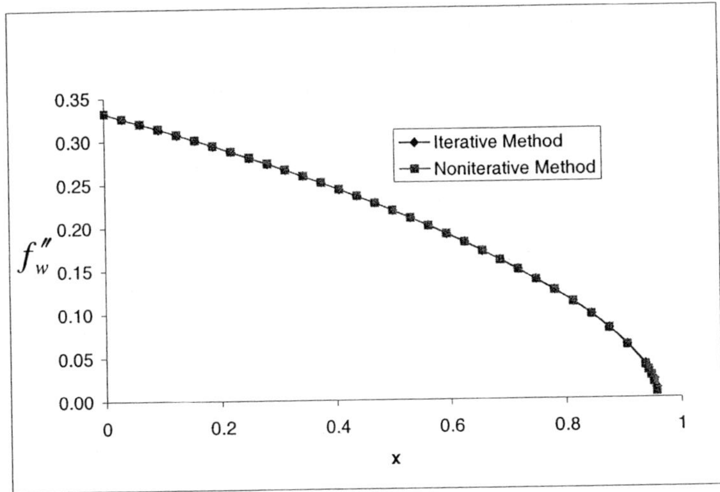



Figure 3. Comparison of calculated wall shear values with both methods for k =1/32.

## 4.2 Turbulent Flow

To compare the predictions of the noniterative method with the iterative method, we have considered a turbulent flow with zero pressure gradient with eddy viscousity formulation given by the Cebeci-Smith model[3]. This formulation treats a turbulent boundary layer as a composite layer with inner and outer regions and uses separate eddy-viscosity formulas in each region, expressing $b$ in Eq. (7) as a function of $f''$ and $f$. As a result, there are several choices for writing finite-difference approximations to $(bf'')'$ in the noniterative method. Here we assume that b is given with values at $x_{n-1}$, thus making this term known prior to performing calculations at $x = x_n$. Other choices are being investigated and will be reported later.

Table 3 and Fig. 4 show the results for a turbulent flow on a flat plate with unit Raynolds number corresponding to $10^6$. The calculations with the iterative method started as laminar at $x=0$ and transition location was specified at $x=0.2$ and were continued with $k = 0.1$ in [0, 1] and $k=1$ in [1,10]. The calculations with the noniterative method started at the fourth x-station, x= 0.40. As can be seen, the predictions of the noniterative method are in very good agreement with those obtained with the iterative method. The efficiency of the noniterative method is similar to the efficiency in laminar flows: whereas the iterative method requires 41 iterations, the noniterative method requires only 16.

Table 4 and Fig. 5 present the results similar to those in Table 3 and Fig. 4 except $k$ in [1,10] is 2.5. As can be seen, the predictions of the noniterative method differ from those predicted with the iterative method. The reason for this is clearly the way $b$ is calculated. Other choices for handling the $b$-term are in progress and will be reported later.

Table 3. Comparison of calculated results for turbulent flow on a flat plate

| x | Iterative Method $\left(f_w''\right)_{iter}$ | Iter | Noniterative Method $\left(f_w''\right)_{non}$ | Iter | $\left(f_w''\right)_{iter}-\left(f_w''\right)_{non}$ Diff. | Relative Diff. |
|---|---|---|---|---|---|---|
| 0.4 | 1.27E+00 | 5 | 1.33E+00 | 1 | -5.50E-02 | -4.32E-02 |
| 0.5 | 1.38E+00 | 2 | 1.39E+00 | 1 | -9.00E-03 | -6.54E-03 |
| 0.6 | 1.43E+00 | 3 | 1.46E+00 | 1 | -2.10E-02 | -1.46E-02 |
| 0.7 | 1.51E+00 | 2 | 1.52E+00 | 1 | -2.00E-03 | -1.32E-03 |
| 0.8 | 1.56E+00 | 2 | 1.58E+00 | 1 | -1.50E-02 | -9.62E-03 |
| 0.9 | 1.63E+00 | 2 | 1.63E+00 | 1 | -2.00E-03 | -1.23E-03 |
| 1 | 1.68E+00 | 2 | 1.69E+00 | 1 | -1.30E-02 | -7.75E-03 |
| 2 | 2.04E+00 | 2 | 2.08E+00 | 1 | -3.40E-02 | -1.67E-02 |
| 3 | 2.40E+00 | 6 | 2.43E+00 | 1 | -2.80E-02 | -1.17E-02 |
| 4 | 2.65E+00 | 2 | 2.67E+00 | 1 | -1.60E-02 | -6.03E-03 |
| 5 | 2.88E+00 | 2 | 2.93E+00 | 1 | -5.30E-02 | -1.84E-02 |
| 6 | 3.12E+00 | 3 | 3.10E+00 | 1 | 2.00E-02 | 6.42E-03 |
| 7 | 3.29E+00 | 2 | 3.29E+00 | 1 | 3.00E-03 | 9.11E-04 |
| 8 | 3.45E+00 | 2 | 3.44E+00 | 1 | 7.00E-03 | 2.03E-03 |
| 9 | 3.61E+00 | 2 | 3.60E+00 | 1 | 1.40E-02 | 3.88E-03 |
| 10 | 3.75E+00 | 2 | 3.76E+00 | 1 | -1.30E-02 | -3.47E-03 |
| | Total Iterations | 41 | | 16 | | |

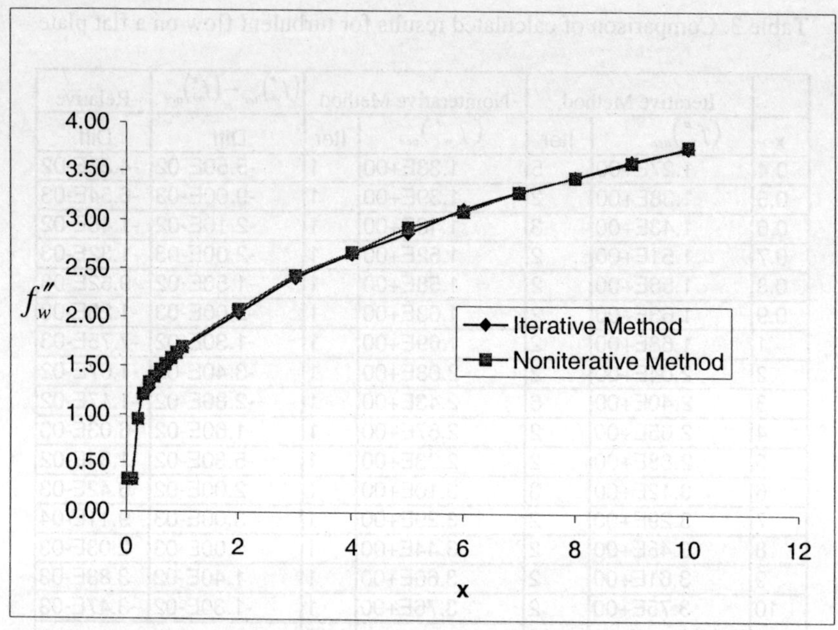

Figure 4. Comparison of calculated wall shear values for turbulent flow on a flat plate.

Table 4. Comparison of calculated results for turbulent flow on a flat plate

| x | Iterative Method $(f_w'')_{iter}$ | Iter | Noniterative Method $(f_w'')_{non}$ | Iter | $(f_w'')_{iter} - (f_w'')_{non}$ Diff. | Relative Diff. |
|---|---|---|---|---|---|---|
| 0.4 | 1.24E+00 | 1 | 1.24E+00 | 1 | 4.00E-03 | 3.22E-03 |
| 0.5 | 1.32E+00 | 2 | 1.33E+00 | 1 | -7.00E-03 | -5.29E-03 |
| 0.6 | 1.40E+00 | 2 | 1.40E+00 | 1 | -1.00E-03 | -7.15E-04 |
| 0.7 | 1.47E+00 | 2 | 1.48E+00 | 1 | -2.00E-03 | -1.36E-03 |
| 0.8 | 1.53E+00 | 2 | 1.55E+00 | 1 | -1.30E-02 | -8.47E-03 |
| 0.9 | 1.60E+00 | 1 | 1.60E+00 | 1 | 0.00E+00 | 0.00E+00 |
| 1 | 1.66E+00 | 2 | 1.66E+00 | 1 | 2.00E-03 | 1.20E-03 |
| 2.5 | 2.17E+00 | 2 | 2.17E+00 | 1 | -2.00E-03 | -9.21E-04 |
| 5 | 2.90E+00 | 1 | 2.87E+00 | 1 | 2.20E-02 | 7.60E-03 |
| 7.5 | 3.33E+00 | 2 | 3.34E+00 | 1 | -1.30E-02 | -3.91E-03 |
| 10 | 3.70E+00 | 2 | 3.83E+00 | 1 | -1.29E-01 | -3.48E-02 |
| | Total Iterations | 19 | | 11 | | |

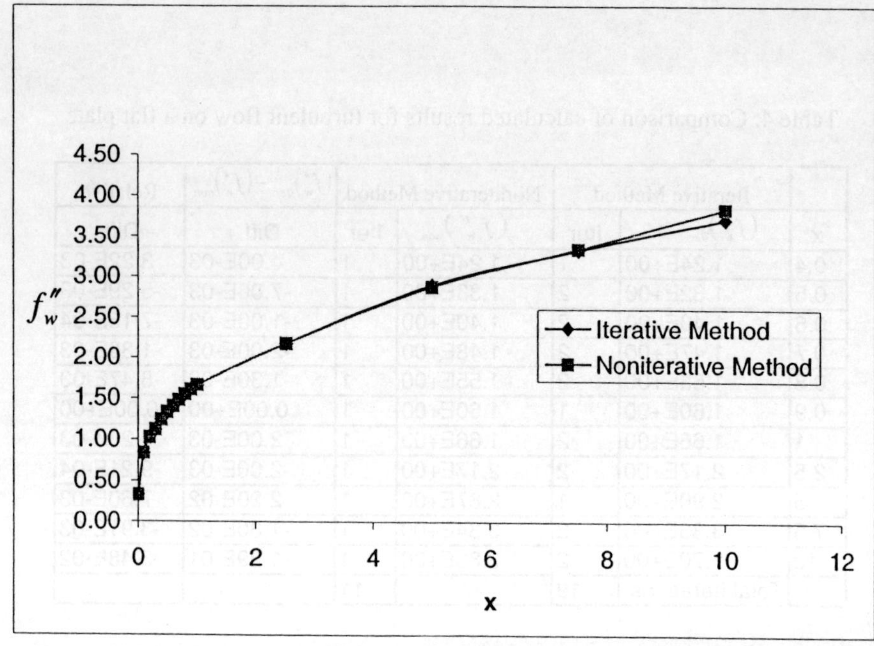



Figure 5. Comparison of calculated wall shear values for turbulent flow on a flat plate

## REFERENCES


[1]  Keller, H. B, *Numerical Methods for Two-Point Boundary-Value Problems*, Blaisdell Publishing Co., 1968.

[2]  Tuncer Cebeci and Jean Cousteix, *Modeling and Computation of Boundary-Layer Flows*, Horizons Publishing, Long Beach, CA and Springer-Verlag, Heidelberg, 1998.

[3]  Tuncer Cebeci, *An Engineering Approach to the Calculation of Aerodynamic Flows,* Horizons Publishing, Long Beach, CA and Springer-Verlag, Heidelberg, 1999.

# Parallelized Flow Analysis over High-Lift Airfoils Using Incompressible/Compressible Navier-Stokes Equations

Chang Sung Kim,[*] Chongam Kim,[†] and Oh Hyun Rho[‡]
*Seoul National University, Seoul 151-742, Republic of Korea*

## Abstract

Parallel computations of viscous turbulent flows over high-lift airfoils are performed using unsteady, incompressible and compressible Reynolds-averaged Navier-Stokes equations with three two-equation turbulence models (the standard $k - \varepsilon$, $k - \omega$, and $k - \omega$ SST model). Both the incompressible and compressible flow solvers are developed and demonstrated to investigate the compressibility effects. The compressible code involves an upwind-differenced scheme for the convective terms and a lower-upper symmetric Gauss-Seidel scheme for temporal integration. The incompressible code with pseudo-compressibility method also adopts the same schemes as the compressible code. Both codes are parallel-processed using the message passing interface programming method and show good parallel speedups. The compressible and incompressible codes are validated by predicting the flow around the RAE 2822 transonic airfoil and NACA 4412 airfoil, respectively. In addition, both the incompressible and compressible code using the Chimera overlapping grid scheme are used to compute the flow over the NLR 7301 airfoil with flap and the NASA GAW-1 high-lift airfoil. Compressibility effects on surface pressure coefficients, velocity profiles, and skin friction coefficients are numerically simulated.

## 1    Introduction

As computational power advances, more accurate and efficient solutions to the compressible and incompressible Navier-Stokes equations have become of major interest in the field of computational fluid dynamics (CFD). For the purpose of code validation, the flow over a high-lift airfoil involving multiple elements is still a most attractive topic to CFD researchers in which massive flow-separation, confluent boundary layer, unsteadiness, compressibility effects at highly loaded leading-edge slat, and transition from laminar to turbulence might be all included as shown in Fig. 1. Regardless of these difficulties originated from high-lift aerodynamics [1], many researchers have performed remarkable studies on flow analysis by adopting a suitable turbulence model [2-6] and even on high-lift airfoil design and optimization [7-10]. Also for the intrinsic geometric complexity of multielement airfoil and wing, various grid-generation techniques such as unstructured grid, multi-block grid, patched grid, and

---

[*] Post Doctorial Researcher, Institute of Advanced Machinery and Design.
[†] Assistant Professor, Dep't of Aerospace Engineering.
[‡] Professor, Dep't of Aerospace Engineering.

overlapping grid topologies have been applied to high-lift aerodynamic computations. Kwak *et al.* [2] reported that the use of the Chimera overlaid grid approach is much easier than using a patched grid scheme for solving the flows over multielement airfoils, while both grid approaches are capable of producing fairly accurate solutions. Anderson and Bonhaus [4] presented an implicit Navier-Stokes solution algorithm for the computation of turbulent flow on unstructured grids. Godin *et al.* [5] provided a detailed comparison of one- and two-equation turbulence models in context of two-dimensional high-lift aerodynamic flows on multi-block grids. In addition, parallel-processing implementation can be an issue of interest to efficiently obtain the flow solution on the computational grid generated by one of those grid-generation techniques.[6]

Kwak *et al.* [7, 8] presented a method of optimizing the geometry and position of high-lift system elements, which uses an incompressible Navier-Stokes flow solver (INS2D), a Chimera overlaid grid system (PEGSUS), and a constrained numerical optimizer (DOT). Besnard and Schmitz [9] also presented a two-dimensional aircraft high-lift system design and optimization method, which uses a gradient-based local optimizer and can be easily extended to three dimensions. The representation of airfoils by general shape functions as well as element positioning such as deflection angle, gap, and overlap is considered. Jameson *et al.* [10] derived and presented an adjoint-based Navier-Stokes design and optimization method for two-dimensional multielement high-lift configurations. The compressible Reynolds-Averaged Navier-Stokes (RANS) equations are used as a flow model together with the Spalart-Allmaras turbulence model to account for high-Reynolds number effects. Using a viscous continuous adjoint formulation, the necessary aerodynamic gradient information is obtained with large computational savings over traditional finite difference methods.

In general, the flows over high-lift airfoils can be assumed to be incompressible because the freestream Mach number is approximately 0.1 to 0.4 and chord Reynolds numbers are usually 1 to 40 million [1]. Thus, an incompressible Navier-Stokes flow solver, INS2D/3D developed by Kwak *et al.* [11, 12] has been used for efficient computations of the flows over high-lift airfoils. Rogers *et al.* [2] efficiently simulated the two-dimensional incompressible viscous flow over single and multielement airfoils using this INS2D code. An incompressible flow solver has the advantage of requiring less computing time over a compressible code, because the equation of energy transport is generally decoupled in the incompressible formulation, and compressible flow solvers may have a slow convergence at these low freestream Mach numbers.

However, it may not be computationally cheaper to use an incompressible flow solver over a compressible one to study poststall, unsteady airfoil flows because of the difficult nature of solving the unsteady incompressible Navier-Stokes equations. Also, compressibility effects can be significant in some cases of multielement airfoils with a highly loaded element near the leading edge. The experimental results by Valarezo *et al.* [13] show the effect of Reynolds and Mach numbers on flow over several multielement configurations. These results show very little difference in the lift coefficient vs. angle-of-attack curves between Mach numbers of 0.15-0.20 for most configurations. For a Mach number of 0.26, the compressible effects become significant, as there is a significant drop in maximum lift and stall angle compared with the lower Mach numbers. There will be conditions where a freestream Mach number of 0.2 will cause a small localized region of sonic flow at a slat leading edge, which could lead to a dramatic compressible effects such as the formation of a shock. Therefore, the compressible computations are necessary for these flow regimes.

For these reasons, Kim *et al.* [6] developed a parallelized compressible Navier-Stokes flow solver with Chimera overlaid grids and compared its results with those computed from a parallelized incompressible Navier-Stokes flow solver to carefully study the effect of compressibility. They reported that compressibility effects might be significant as angle of attack increases even at the same freestream Mach number. They also proposed a PPCG (Parallel Processing for Chimera Grid) algorithm to efficiently compute the flow over complex geometries such as multielement airfoils. Both codes are parallel-processed using a message passing interface (MPI) programming method [14, 15] on a Cray T3E parallel machine. Both the compressible parallel code and incompressible parallel code with different two-equation turbulence models are examined via numerous computations.

Various turbulence models such as algebraic models, one-equation models, and two-equation models have been developed for the engineering computations of various turbulent flows. Compared to the Direct Numerical Simulation (DNS) and Large Eddy Simulation (LES) approaches, turbulence modeling uses the minimum amount of complexity while capturing the essence of turbulence under the Boussinesq eddy-viscosity assumption. Since the 1960's, turbulence models based upon the equation for the turbulence kinetic energy have become the cornerstone of the turbulence modeling research. The simplest of all turbulence models based on the Boussinesq eddy-viscosity approximation are known as algebraic models. Unlike one- and two-equation models, algebraic models do not have the transport equation of turbulent kinetic energy. One- and two-equation turbulence models also retain the Boussinesq eddy-viscosity approximation, but differ in one important respect. One-equation models are not incomplete as they relate the turbulence length scale to some typical flow dimension. By contrast, two-equation models provide an equation for the turbulence length scale or its equivalent and are thus complete. For second-order closure turbulence models, sufficient computer resources became available to permit serious development of these models. However, because of the large number of equations and complexity involved in second-order closure models, they have thus far found their way into a relatively small number of applications compared to algebraic and two-equation models. [16]

It is, however, still difficult to accurately predict viscous, turbulent flows around multielement airfoils due to confluent boundary layers and massive flow separation at a high angle of attack close to the stall angle. Many studies adopting one-equation models were performed for the computations of such flows with a varying degree of success. [2, 3, 5] One-equation models may not be preferable for separated flows and wakes. Recently, two-equation turbulence models have been used for the turbulent flows over multi-element airfoils. [5, 6] In the present work, three two-equation turbulence models of the $k - \omega$ of Wilcox [16, 17], the standard $k - \varepsilon$ [16, 19], and the $k - \omega$ SST of Menter [18-19] are carefully assessed through computing the flows over single and multi-element airfoils. Several $k - \varepsilon$ models involving different wall functions for the accurate calculation of the inner boundary layer have been developed and have shown good results. Unlike the $k - \varepsilon$ models, the $k - \omega$ model does not need a wall function to resolve a viscous sub-layer, which is quite suitable in parallel-processed programming. On the other hand, the $k - \varepsilon$ model is insensitive of the freestream turbulence level in the outer boundary layer while the $k - \omega$ model is highly sensitive to freestream turbulence values. [18] Combining the advantages of the two models, the $k - \omega$ shear stress transport (SST) model is found to have the property of the original $k - \omega$ model in the near wall region and the $k - \varepsilon$ model in the outer of boundary layer. [19]

Both the compressible parallel code and incompressible parallel code with different two-equation turbulence models are examined via numerous computations. The compressible code is validated by computing the flow over the RAE2822 transonic airfoil and shows a good property to predict surface pressure coefficients and velocity profiles in the boundary layers. The incompressible code is validated through computations of the flow over the NACA 4412 airfoil, for a wide range of angles of attack from zero to maximum lift. In particular, computations of the flow over the NLR7301 airfoil with flap, and the NASA GAW-1 high-lift airfoil with a leading-edge slat and a trailing-edge flap are performed using both the incompressible and compressible code to study the compressibility effects on the flow near the high-loaded leading edge.

## 2　Numerical Background

### 2.1　Governing Equations

The governing equations are the two-dimensional, unsteady, compressible Navier-Stokes equations and are written in conservation law form as

$$\frac{\partial \rho}{\partial t} + \frac{\partial}{\partial x_k}\left(\rho u_k\right) = 0,$$

$$\frac{\partial}{\partial t}\left(\rho u_i\right) + \frac{\partial}{\partial x_j}\left(\rho u_i u_j\right) = -\frac{\partial p}{\partial x_i} + \frac{\partial \hat{\tau}_{ij}}{\partial x_j},$$

(1)

$$\frac{\partial \rho E}{\partial t} + \frac{\partial}{\partial x_j}\left(\rho E u_j\right) = -\frac{\partial p u_j}{\partial x_j} + \frac{\partial}{\partial x_j}\left[u_i \hat{\tau}_{ij} - q_j\right],$$

where $E$ is the total energy and $\hat{\tau}_{ij}$ are the sum of molecular and Reynolds stresses defined as

$$\hat{\tau}_{ij} = 2\mu\left(S_{ij} - S_{kk}\delta_{ij}/3\right) + \tau_{ij},$$

$$\tau_{ij} = 2\mu_T\left(S_{ij} - S_{kk}\delta_{ij}/3\right) - 2\rho k\delta_{ij}/3, \qquad S_{ij} = \frac{1}{2}\left(\frac{\partial u_i}{\partial x_j} + \frac{\partial u_j}{\partial x_i}\right),$$

and $q_j$ is the total heat-flux rate defined as

$$q_j = -\left(\frac{\gamma}{\gamma-1}\right)\left(\frac{\mu}{\Pr} + \frac{\mu_T}{\Pr_T}\right)\frac{\partial T}{\partial x_j}.$$

The perfect gas equation of state is introduced to complete the set of compressible equations as

$$p = \rho(\gamma - 1)\left[E - \frac{1}{2}u_i u_i\right] \tag{2}$$

For incompressible problems, the energy equation does not need to included unless heat transfer is a matter of concern.

## 2.2 Turbulence Models

In the present work, the standard $k - \varepsilon$ model [16, 19] without wall functions, the $k - \omega$ model [16, 17], and the $k - \omega$ SST model [18, 19] are evaluated to predict the viscous turbulent flows over high-lift airfoils. The $k - \omega$ SST model combining the $k - \omega$ model with the $k - \varepsilon$ model is written as follows.

$$\frac{\partial \rho k}{\partial t} + \frac{\partial}{\partial x_j}\left(\rho k u_j\right) = \tau_{ij}\frac{\partial u_i}{\partial x_j} - \beta^* \rho \omega k + \frac{\partial}{\partial x_j}\left[\left(\mu + \sigma_k \mu_T\right)\frac{\partial k}{\partial x_j}\right],$$

$$\frac{\partial \rho \omega}{\partial t} + \frac{\partial}{\partial x_j}\left(\rho \omega u_j\right) = \frac{\gamma}{\nu_T}\tau_{ij}\frac{\partial u_i}{\partial x_j} - \beta \rho \omega^2 + \frac{\partial}{\partial x_j}\left[\left(\mu + \sigma_\omega \mu_T\right)\frac{\partial \omega}{\partial x_j}\right]$$
$$+ 2(1 - F_1)\rho \sigma_{\omega_2}\frac{1}{\omega}\frac{\partial k}{\partial x_j}\frac{\partial \omega}{\partial x_j} \tag{3}$$

The combined constants are calculated in the following relation.

420

$$\Phi = F_1 \Phi_1 + (1 - F_1) \Phi_2 \, ,$$

where $\Phi_1$ represent the constants of the $k - \omega$ model except for $\sigma_{k_1}$ and $\gamma_1$,

$$\sigma_{k_1} = 0.85, \quad \sigma_{\omega_1} = 0.5, \quad \beta_1 = 0.075, \quad \beta^* = 0.09 \, ,$$
$$\kappa = 0.41, \quad \gamma_1 = \beta_1 / \beta^* - \sigma_{\omega_1} \kappa^2 / \sqrt{\beta^*}$$

and $\Phi_2$ represent the constants of the $k - \varepsilon$ model as

$$\sigma_{k_2} = 1.0, \quad \sigma_{\omega_2} = 0.856, \quad \beta_2 = 0.0828, \quad \beta^* = 0.09 \, ,$$
$$\kappa = 0.41, \quad \gamma_2 = \beta_2 / \beta^* - \sigma_{\omega_2} \kappa^2 / \sqrt{\beta^*}$$

$F_1$ is set to be zero for the $k - \varepsilon$ model and one for the $k - \omega$ model. For the $k - \omega$ SST model, $F_1$ is defined as follows.

$$F_1 = \tanh\left\{ \left( \min\left[ \max\left[ \frac{\sqrt{k}}{\beta^* \omega y}; \frac{500\nu}{y^2 \omega} \right]; \frac{4\rho\sigma_{\omega_2} k}{CD_{kw} y^2} \right] \right)^4 \right\} \, ,$$

where $y$ is the distance to the nearest surface and $CD_{k\omega}$ is the positive portion of the cross-diffusion term.

$$CD_{k\omega} = \max\left[ 2\rho\sigma_{\omega_2} \frac{1}{\omega} \frac{\partial k}{\partial x_j} \frac{\partial \omega}{\partial x_j}; 10^{-20} \right] .$$

The eddy viscosity is defined to account for the transport of the principal turbulent shear stress as follows.

$$\mu_T = \rho k / \max(\omega; \Omega F_2 / 0.31) \, ,$$

where $\Omega$ is the absolute value of the vorticity and

$$F_2 = \tanh\left\{ \left( \max\left[ \frac{2\sqrt{k}}{\beta^* \omega y}; \frac{500\nu}{y^2 \omega} \right] \right)^2 \right\} .$$

## 2.3 Numerical Approach

The governing equations are discretized with a finite-volume method. For the compressible code, convective terms are upwind-differenced based on the Roe's Flux Difference Splitting (FDS) scheme [20]. A MUSCL (Monotone Upstream Centered Scheme for Conservation Laws) approach [21] using a third order interpolation is used to obtain a higher order of spatial accuracy. The third order of spatial accuracy is kept in all calculations. For a temporal integration, Lower-Upper Symmetric Gauss-Seidel (LU-SGS) scheme by Yoon *et al.* [22, 23] is adopted. The incompressible code using the following pseudo-compressibility relation has the same spatial and temporal schemes as the compressible code.

$$\frac{\partial p}{\partial \tau} = -\beta \nabla \cdot u, \tag{4}$$

where $\tau$ denotes pseudo time and $\beta$ is a pseudo-compressibility parameter. Numerical experiments must be performed to determine a value of $\beta$ yielding the best convergence. [11, 12]

Because a multi-element airfoil generally has a complex geometry, it is very difficult to generate a high quality grid system in a single block. Thus, patched grids, overlapping grids or unstructured grids are popularly adopted. In the present study, the overlapping grid approach known as the Chimera grid scheme [24] is used for an efficient grid generation. Sub-grids around flap and slat are overlapped on the main grid around the basic airfoil. Information is exchanged between sub-grid and main grid domain through bilinear interpolation, which is found to be robust and easy to implement.

Wall boundary conditions are applied explicitly with no slip condition. The pressure is extrapolated from the interior points and the other variables are specified from the free-stream values at the inflow, while the pressure is specified and the other variables are extrapolated from the interior points at the outflow.

For the angles of attack near maximum lift, steady-state computations using local time stepping method do not converge completely. In these cases unsteady computations must be performed in a time-accurate manner. Thus, the dual time stepping method is adopted both in the incompressible and compressible code.

### 2.4 Parallel Processing Algorithm for Chimera Grid

The search routines for fringe cells and donor cells are pre-processed. A fringe cell is to receive the flow information from four donor cells in an overlapped grid domain as shown in Fig. 2. A Parallel Process for Chimera Grid (PPCG algorithm, hereafter) is implemented as follows.

Step 1.  Divide the main grid and sub-grids into sub-domains considering load balance.
Step 2.  Search each sub-domain for fringe cells and donor cells.
Step 3.  MPI_send and MPI_receive the flow information of donor cells from each

Step 4.    processor to the processor 0.

Step 4.   Calculate the conservative variables of fringe cells through bilinear interpolation in the processor 0.

Step 5.   MPI_send and MPI_receive the flow information of fringe cells from the processor 0 to each processor.

Step 6.   Repeat the process from Step 3 to Step 5 until converged solutions are obtained.

The PPCG algorithm is found to be simple and efficient to implement in the parallel computations for the flows over multi-element airfoils.

# 3   Results and Discussion

Both the compressible code and incompressible code were parallel-processed using MPI (Message Passing Interface) programming method. The same grid points were distributed in all sub-domains for load balance. Figure 3 shows the parallel speedup of the compressible and incompressible code using 1 to 24 processors. The parallel speedup was defined by the ratio of computing times according to the number of processors. Both codes show a good parallel speedup on the Cray T3E parallel machine.

## 3.1   RAE 2822 Transonic Airfoil

The flow over the RAE 2822 transonic airfoil with the maximum thickness of 12.1% of chord and a sharp trailing edge was computed to validate the compressible code at the Mach number of 0.73 and the Reynolds number of 6.5 million. A $129 \times 65$ hyperbolic O-grid was used with the spacing at the wall of $1 \times 10^{-5}$ based on unit chord length. Also a $241 \times 81$ O-grid with the wall spacing of $5 \times 10^{-6}$ chord was used to test the sensitivity to grid resolution. The outer boundary was extended to 25 chords.

Computed results from the standard $k - \varepsilon$ model, the $k - \omega$ model, and the $k - \omega$ SST model were compared with the experimental data presented by Cook et al. [25] at the angle of attack of 2.79 degrees. Comparison between computed and measured lift, drag and pitching moment coefficients are shown in Table 1.

Table 1. Comparison of computed and measured load coefficients
( RAE 2822, M=0.73, Re=6.5million, $\alpha$ =2.79 deg.)

| | $k-\varepsilon$ | $k-\omega$ | $k-\omega$ SST | Exp. |
|---|---|---|---|---|
| $C_L$ | 0.7957 | 0.8486 | 0.7910 | 0.803 |
| $C_D(p)$ | 0.01299 | 0.01446 | 0.01228 | - |
| $C_D(v)$ | 0.00733 | 0.00591 | 0.00519 | - |
| $C_D$ | 0.02032 | 0.02037 | 0.01747 | 0.0168 |
| $C_M$ | -0.09406 | -0.10518 | -0.09245 | -0.099 |

$C_D(p)$ : pressure drag, $C_D(v)$ : skin friction drag

All the three turbulence models had little sensitivity to grid resolution in the sense that the computed results from the coarse and fine mesh were almost the same. Computed surface pressure coefficients using the fine mesh are compared with experimental data in Fig. 4. All models show a slight difference in shock wave position compared with experimental data. The $k-\omega$ model shows a difference in shock wave position compared with the other two turbulence models. Figure 5 shows velocity profiles from the upper surface boundary layer at the stations of $x/c = 0.574, 0.650, 0.750$, and $0.900$ using the fine mesh. The $k-\omega$ SST model yields a better agreement with experimental velocity profiles in the downstream region after the shock wave, while the $k-\omega$ model produces a better performance in the immediate vicinity of the shock wave. The convergence histories of the compressible code using three turbulence models are shown in Fig. 6. The maximum iteration count was 10000 and the maximum residual was less than $10^{-4}$ in about 2000 iterations. For the coarse mesh of $129 \times 65$ size, computing time using 24 processors takes about 40 seconds for 2000 iterations on the Cray T3E parallel machine.

## 3.2    NACA 4412 Airfoil with Separation

The flow over the NACA 4412 airfoil was computed to validate the incompressible code at the Reynolds number of 1.52 million. A $245 \times 65$ hyperbolic O-grid was used with the wall spacing of $1 \times 10^{-5}$ chord and the outer boundary was extended to 25 chords.

Computed results from the standard $k-\varepsilon$ model, the $k-\omega$ model, and $k-\omega$ SST model were compared with the experimental data presented by Coles and Wadcock [26] at the angle of attack of 13.87 degrees for which a steady trailing edge separation occurs. Trip-strips were employed in the experiment on the suction and pressure sides at $x/c = 0.025$ and $0.103$, respectively. The production term in the transport equation for the turbulent kinetic energy was set to zero upstream of these locations to represent transition effects. Computed surface pressure coefficients with and without transition were compared with experimental data in

424

Fig. 7. Transition has a slight effect on the surface pressure coefficients from the $k - \omega$ SST model while having little effect on those from the standard $k - \varepsilon$ or the $k - \omega$ model. The transition effect can be seen more clearly by comparing the velocity profiles from the upper surface boundary layer at the stations of $x/c = 0.620, 0.731, 0.842,$ and $0.897$ in Fig. 8. The $k - \omega$ SST model shows a fairly good agreement with experimental velocity profiles in the downstream adverse pressure gradient region while the standard $k - \varepsilon$ model and the $k - \omega$ model yield quite a difference.

Convergence characteristics were sensitive to the value of $\beta$. It was found that the incompressible code converged well for the $\beta$ of 20 ~ 200 through numerical tests. The convergence histories of the code for the three turbulence models are shown in Fig. 9. It can be seen that transition slightly delay convergence characteristics. The maximum iteration count was 10000 and the maximum residual of the continuity equation was less than $10^{-4}$ in about 3000 iterations. For the $245 \times 65$ mesh, computing time on the Cray T3E using 24 processors is about 72 seconds for 3000 iterations.

Figure 10 shows lift coefficients at the wide range of angles of attack from zero to maximum lift. The $k - \omega$ SST model with transition gives a closer agreement in the lift, especially in the prediction of stall. For the high angles of attack over 13.87 deg, steady-state computations overestimate the flow separation massively and do not converge completely. In these cases unsteady computations with the dual time stepping method were performed in a time-accurate manner. For 16 deg, lift coefficients vs. time are shown in Fig. 11. While the standard $k - \varepsilon$ model converges to a steady-state solution, the $k - \omega$ model and the $k - \omega$ SST model show a periodic unsteadiness. For the $k - \omega$ and $k - \omega$ SST model with transition, the leading-edge laminar bubble is periodically shedding downstream to the trailing edge with different modes. Unsteady computations were performed with the non-dimensional time step of 0.05 until the non-dimensional time reached 800. For each physical time step, 30 to 40 subiterations were required until both the maximum divergence of velocity and the maximum residual were less than $10^{-4}$.


### 3.3  Two-Element Airfoil

Compressible and incompressible computations for the flow over the NLR7301 with a 32%$c$ flap were performed at the Mach number of 0.185 and the Reynolds number of 2.51 million. The flap was positioned with the deflection angle of 20 deg, the overlap of 5.3%$c$ and the gap of 2.6%$c$ in the experiment by Berg [27]. A $249 \times 81$ O-grid for the basic airfoil and a $125 \times 41$ O-grid for the flap was used for the Chimera grid scheme with the wall spacing on the order of $10^{-6}$ chord. The outer boundary was extended to about 30 chords.

Figure 12 shows the parallel speedup of the incompressible and compressible code adopting the PPCG algorithm using 5 to 20 processors. Because the main grid had 4 times larger grid size than the sub-grid, the processors were distributed as evenly as possible, proportional to the grid size for load balance. When 5 processors were used, 4 processors were given for the main grid and 1 processor for the sub-grid. And when 20 processors were used, 16 processors were given for the main grid and 4 processors for the sub-grid. Both the incompressible and compressible codes show a good parallel speedup on the Cray T3E parallel machine.

Computed surface pressure coefficients are compared with the experimental data at the angle of attack of 6.0 deg in Fig. 13. The local peak Mach number near the leading edge is about 0.54 at this angle of attack. The incompressible and compressible results are similar, even in the suction peak at the leading edge of the basic airfoil. All computations using the standard $k - \varepsilon$, the $k - \omega$ model and the $k - \omega$ SST model give a fairly good agreement with the experimental data. Computed skin friction coefficients are compared with the experimental data in Fig. 14. It is found that the flow is attached downstream to the trailing edge.

The convergence histories of the incompressible and compressible code with three turbulence models are shown in Fig. 15. For the incompressible computations, the maximum iteration count was 10000 and the maximum residual was less than $10^{-4}$ in about 3000 iterations. Computing time is about 160 seconds for 3000 iterations on the Cray T3E using 20 processors. For the compressible computations, converged solutions are obtained in about 3000 iterations. Computing time on the Cray T3E using 20 processors is about 240 seconds for 3000 iterations.

For the critical angle of attack of 13.1 degrees where no flow separation observed downstream to the trailing edge except for a very tiny laminar separation bubble near the leading edge in the experiment, both the steady-state incompressible and compressible computations predicted massive flow separation and did not converge completely. For this reason, computations were performed in a time-accurate, unsteady manner using the dual time stepping method. Figure 16 shows lift coefficients vs. time at this angle of attack. Incompressible computations were performed with the non-dimensional time step of 0.01 and converged in 4000 physical time steps. Computing time using 20 processors was about 6100 seconds. Compressible computations were performed with the non-dimensional time step of 0.1 and converged in 2000 physical time steps. Computing time using 20 processors was about 6900 seconds. In both cases, 5 to 40 subiterations were required for each time step during the course of calculations. In the cases of time-accurate computations, it was found that the incompressible code had little advantage in terms of computing time over the compressible code.

Computed surface pressure coefficients are compared with the experimental data at 13.1 deg in Fig. 17. A discrepancy between the incompressible and compressible results is shown in the suction peak near the leading edge of the basic airfoil. For the freestream Mach number of 0.185, compressibility effects near the suction peak become significant because the local Mach number near the leading edge can be as large as 0.72. Figure 18 shows the velocity profiles at $x/c = 0.05$, 0.94 on the basic airfoil and at $x_F/c = 0.119$, 0.219 on the upper flap surface. The incompressible and compressible results are similar except at the 5 % chord position. This is more evident by comparing the velocity profiles at 5 % chord position at the angles of attack of 6 and 13.1 deg shown in Fig. 19. The discrepancy between the incompressible and compressible velocity profiles is considerable at 13.1 deg. Compressibility effects can be significant as angle of attack increases even at the same freestream Mach number.

### 3.4 Three-Element Airfoil

Also like in two-element airfoil case, both the parallelized incompressible and compressible code were successfully implemented to computations for the flow over the $17\%c$ thick NASA GAW-1 airfoil with a $15\%c$ leading-edge slat and a $29\%c$ trailing-edge flap. The slat was positioned with the deflection angle of 42 deg, the overlap of $1.5\%c$ and the gap of $1.5\%c$ while the flap was positioned with the deflection angle of 30 deg, the overlap of $0.0\%c$ and the gap of $2.5\%c$. The experiment was carried out at the Mach number of 0.15 and the Reynolds number of 0.62 million by Braden *et al.* [28]. A $249 \times 81$ O-grid for the basic airfoil, a $125 \times 41$ O-grid for slat and a $125 \times 41$ O-grid for flap was used for the Chimera grid scheme with the wall spacing on the order of $10^{-5}$ chord. The outer boundary was extended to about 30 chords.

Figure 20 shows the parallel speedup of the incompressible and compressible code using from 6 to 24 processors. Because the main grid had 4 times larger grid size than the sub-grids, the processors were distributed proportional to the grid size for load balance. For example, 4 processors were given for the main grid, 1 processor for the slat and flap grid, respectively.

Computed surface pressure coefficients are compared with the experimental data at the angle of attack of 12.0 deg in Fig. 21. For both incompressible and compressible computations, a discrepancy between the computational and experimental results is found on the suction surface of the slat and flap due to the ambiguity in flap and slat position. The three turbulence models give slight differences in surface pressure coefficients at this angle of attack. The convergence history of the three-element airfoil case is similar with those of the two-element airfoil case in Fig. 15. For the incompressible computations, the maximum iteration count was 10000 and the maximum residual was less than $10^{-4}$ in about 3000 iterations. Computing time using 24 processors is about 180 seconds for 3000 iterations on the Cray T3E. For the compressible computations, converged solutions are obtained in about 3000 iterations. Computing time using 24 processors is about 290 seconds for 3000 iterations.

### 3.5 Application to High-Lift Design Optimization

The CFANS2D flow solver [6] and CSAAV2D sensitivity code [29] are used for high-lift design optimization according to the design procedure in Fig. 22. Optimization is performed using the Broydon-Fletcher-Goldfarb-Shanno (BFGS) variable metric method supported by the DOT (Design Optimization Tool) commercial software [30]. The essence of high-lift design optimization is undoubtedly to improve $Cl_{max}$ by modifying the geometric shape on each element and varying angle of attack as well. To validate the present design topology based on unsteady flow analysis and sensitivity analysis, the NLR 7301 airfoil with flap is tested as a baseline model at a Mach number of 0.185, a Reynolds number of $2.51 \times 10^6$ and an angle of attack of 13.1 deg. Among total 44 design variables, a flow design variable of angle of attack and 43 geometric variables are given: 20 design variables on surfaces of the basic airfoil and flap, respectively, using Hicks-Henne functions [31], and 3 geometric changes of flap deflection angle, overlap, and gap. For the accurate prediction of the flowfield very close to stall during the one-dimensional search, flow computations are performed in a time-accurate, unsteady manner using the dual time stepping method.

The objective is determined to maximize the lift coefficient at floating angle of attack. Thus, as already mentioned, angle of attack is also included as a design variable. The initial and designed surface pressure coefficients are compared in Fig. 23. With an extension of stall angle from 13.1 to 14.17 deg, the lift coefficient increases from 3.2982 to 3.8399 after 29 calls of the flow solver and 4 calls of the sensitivity analysis code. Figure 24 shows an improvement of $Cl_{\max}$ by avoiding a massive-separated flow over a high-lift airfoil using the present design optimization tool composed of the unsteady CFANS2D solver and the CSAAV2D sensitivity code.


## 4  Concluding Remarks

Accurate and efficient compressible/incompressible computations were performed for the viscous turbulent flows over high-lift airfoils using two-equation turbulence models under a parallel computing environment. The parallelized compressible and incompressible flow solvers on Chimera grid showed fairly good parallel speedups. The three turbulence models of the standard $k - \varepsilon$, $k - \omega$, and $k - \omega$ SST were evaluated by comparing the surface pressure coefficients, skin friction coefficients and velocity profiles in the boundary layers with the experimental data. Overall, the $k - \omega$ SST model shows a better performance compared to the other two-equation models in predicting the adverse pressure gradient region on the suction surface of high-lift airfoils. In particular, the flow over the NLR7301 airfoil with a flap was carefully investigated using both the incompressible and compressible code with two-equation turbulence models. It was found that the compressibility effect is significant when the leading edge is highly loaded even at the low freestream Mach number. Both the parallelized incompressible and compressible code adopting the PPCG algorithm can be efficient computational tools for multi-element airfoil design. Actually the CFANS2D flow solver and CSAAV2D sensitivity code are successfully applied to a high-lift design optimization.

428

## 5  Acknowledgement

This paper is dedicated to Dr. Dochan Kwak in commemoration of his sixtieth anniversary as a tribute of respect to him.

## 6  References


[1]  Brune, G. W., and McMasters, J. H., "Computational Aerodynamics Applied to High-Lift System," Edited by P. A. Henne, *Progress in Astronautics and Aeronautics*, Vol. 125, 1990, pp. 389 - 493.

[2]  Rogers, S. E., Wiltberger, N. L., and Kwak, D., "Efficient Simulation of Incompressible Viscous Flow over Single and Multielement Airfoils," *Journal of Aircraft*, Vol.30, No.5, Sep.-Oct. 1993, pp. 736-743.

[3]  Shima, E., Egami, K., and Amano, K., "Navier-Stokes Computation of a High Lift System Using Spalart-Allmaras Turbulence Model," AIAA Paper 94-0162, Reno, NV, Jan. 1994.

[4]  Anderson, W. K., and Bonhaus, D. L., "An Implicit Upwind Algorithm for Computing Turbulent Flows on Unstructured Grids," *Computers and Fluids*, Vol. 23, No.1, 1994, pp. 1-21.

[5]  Godin, P., Zingg, D. W., and Nelson, T. E., "High-Lift Aerodynamic Computations with One- and Two-Equation Turbulence Models," *AIAA Journal*, Vol. 35, No. 2, Feb. 1997, pp. 237-243.

[6]  Kim, C. S., Kim, C., and Rho, O. H., "Parallel Computations of High-Lift Airfoil Flows Using Two-Equation Turbulence Models," *AIAA Journal* Vol.38, No.8, August 2000, pp.1360-1368.

[7]  Eyi, S., Lee, K. D., Rogers, S. E., and Kwak, D., "High-Lift Design Optimization Using the Navier-Stokes Equations," AIAA Paper 95-0477, Reno, NV, Jan. 1995.

[8]  Eyi, S., Chand, K. K., Lee, K. D., Rogers, S. E., and Kwak, D., "Multi-Element High-Lift Design Using the Navier-Stokes Equations," AIAA Paper 96-1943, New Orleans, LA, June 1996.

[9]  Besnard, E., Schmitz, A., Boscher, E., Garcia, N., and Cebeci, T., "Two-Dimensional Aircraft High Lift System Design and Optimization," AIAA Paper 98-0123, 1998.

[10]  Kim, S., Alonso, J. J., and Jameson, Antony, "Two-dimensional High-Lift Aerodynamic Optimization Using the Continuous Adjoint Method," AIAA Paper 2000-4741, 8th AIAA/USAF/NASA/ISSMO Symposium on Multidisciplinary Analysis and Optimization, Long Beach, CA, Sep. 2000.

[11]  Kwak, D., Chang, J. L., Shanks, S. P., and Chakravarthy, S., "A Three-Dimensional Incompressible Navier-Stokes Flow Solver Using Primitive Variables," *AIAA Journal*, Vol. 24, 1986, pp.390-396.

[12]  Rogers, S. E., and Kwak, D., "Upwind Differencing Scheme for the Time Accurate Incompressible Navier-Stokes Equations," *AIAA Journal*, Vol. 28, No. 2, Feb. 1990, pp. 253-262.

[13] Valarezo, W. O., Dominik, C. J., and McGhee, R. J., "Reynolds and Mach Number Effects on Multielement Airfoils," *Proceedings of the Fifth Symposium on Numerical and Physical Aspects of Aerodynamic Flows*, California State Univ., Long Beach, CA, Jan. 1992.

[14] MPI Forum, "MPI-A Message Passing Interface Standard," *Journal of Supercomputer Applications*, Vol.8, 1994.

[15] MPI-2: Extensions to the Message Passing Interface, Published by the University of Tenesse, Knoxville, TN, 1996.

[16] Wilcox, D. C., "Turbulence Modeling for CFD," DCW Industries, Inc., 5354 Palm Drive, La Cañada, California, 1993, pp. 73-170.

[17] Wilcox, D. C., "Comparison of Two-Equation Turbulence Models for Boundary Layers with Pressure Gradient," *AIAA Journal*, Vol. 31, No. 8, 1993, pp. 1414-1421.

[18] Menter, F. R., "Influence of Freestream Values on the $k - \omega$ Turbulence Model Predictions," *AIAA Journal*, Vol. 30, No. 6, August 1992, pp. 1651-1659.

[19] Menter, F. R., "Two-Equation Eddy-Viscosity turbulence Models for Engineering Applications," *AIAA Journal*, Vol. 32, No. 8, August 1994, pp. 1598-1605.

[20] Roe, P. L., "Approximate Riemann Solvers, Parameter Vectors and Difference Schemes," *Journal of Computational Physics*, Vol.43, 1983, pp.357-372.

[21] Hwang, S. W., "Numerical Analysis of Unsteady Supersonic Flow over Double Cavity," Ph.D. Dissertation, Seoul Nat'l Univ., Seoul, Korea, 1996.

[22] Yoon, S., and Jameson, A., "Lower-Upper Symmetric-Gauss-Seidel Method for the Euler and Navier-Stokes Equations, " *AIAA Journal*, Vol. 26, 1988, pp. 1025-1026.

[23] Yoon, S., and Kwak, D., "Three-Dimensional Incompressible Navier-Stokes Solver Using Lower-Upper Symmetric-Gauss-Seidel Algorithm," *AIAA Journal*, Vol. 29, June 1991, pp. 874-875.

[24] Steger, J. L., Doughty, F. C., and Beneck, J. A., "A Chimera Grid Scheme," Advances in Grid Generation, FED, Vol.5, ASME, Edited by Ghia, K. N., New York, 1983, pp. 59-69.

[25] Cook, P. H., McDonald, M. A., and Firmin, M. C. P., "Aerofoil RAE 2822 – Pressure Distributions, and Boundary Layer and Wake Measurements," AGARD AR 138, May 1979, A6-1 to A6-77.

[26] Coles, D., and Wadcock, A. J., "Flying-Hot-Wire Study of Flow Pat an NACA 4412 Airfoil at Maximum Lift," *AIAA Journal*, Vol. 17, April 1979, pp. 321-329.

[27] Berg, B. v. d., "Boundary Layer Measurements On a Two-Dimensional Wing With Flap," NLR TR 79009 U, Jan. 1997.

[28] Braden, J. A., Whipkey, R. R., Jones, G. S., and Lilley, D. E., "Experimental Study of the Separating Confluent Boundary-Layer," NASA Contractor Report 3655.

[29] Kim, C. S., Kim, C., and Rho, O. H., "Sensitivity Analysis for the Navier-Stokes Equations with Two-Equation Turbulence Models," *AIAA Journal*, Vol. 39, No. 5, 2001, pp. 838-845.

[30] DOT Users Manual, Version 4.00, VMA Engineering, Vanderplaats, Miura & Associate, Inc., 1993.

[31] Hicks, R. M., and Henne, P. A., "Wing Design by Numerical Optimization," *Journal of Aircraft*, Vol.15, No.7, 1978, pp. 407-412.

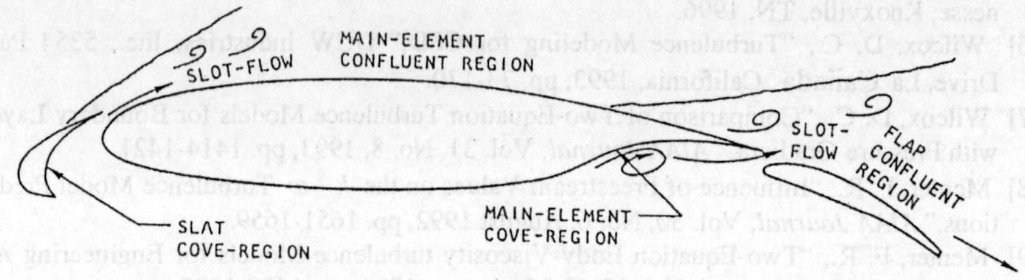



Fig. 1 Region exhibiting confluent boundary layer flow on high-lift airfoil. [28]

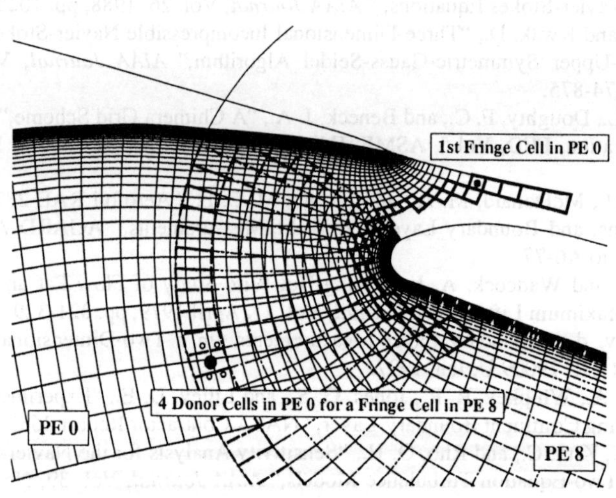



Fig. 2  Parallel implementation on a Chimera overlaid grid system.

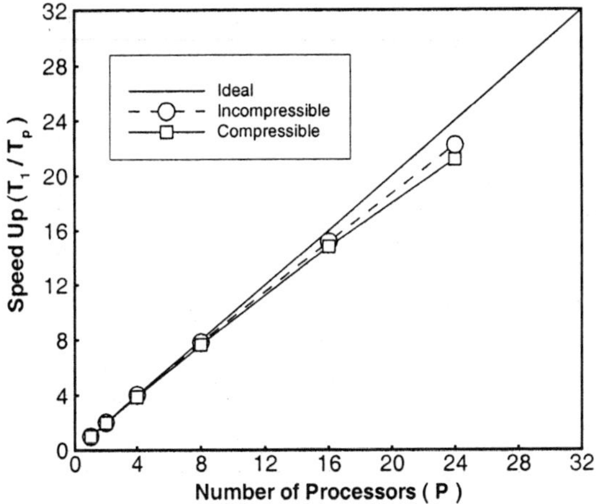

Fig. 3  Parallel speed up for single airfoil case using the incompressible and compressible code.

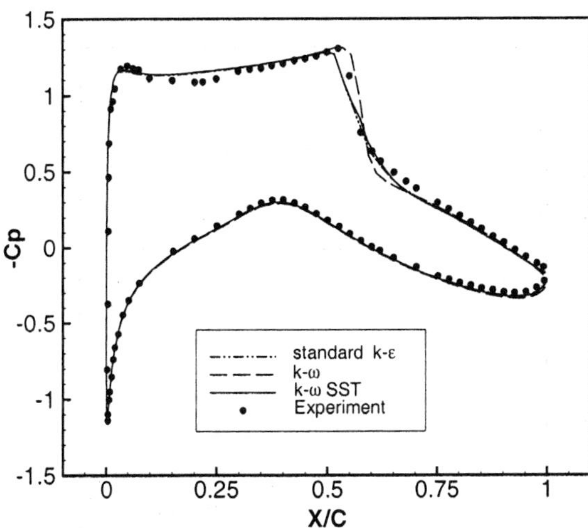

Fig. 4  Surface pressure coefficients over the RAE 2822 airfoil at $\alpha$ =2.79 deg, M=0.73, and Re=6.5 million.

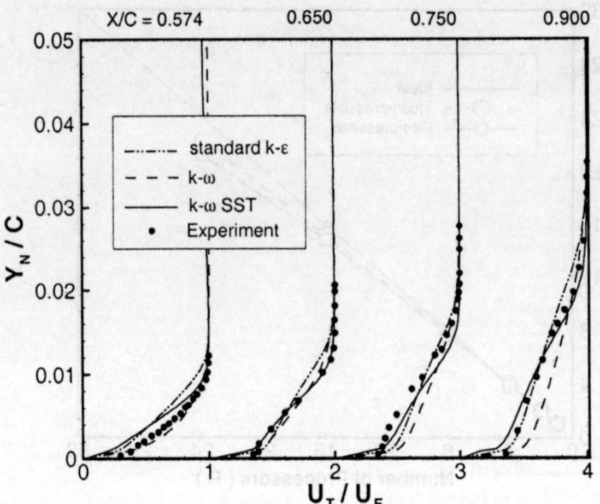



Fig. 5  Velocity profiles on the upper surface of the RAE 2822 airfoil at $\alpha$ =2.79 deg.

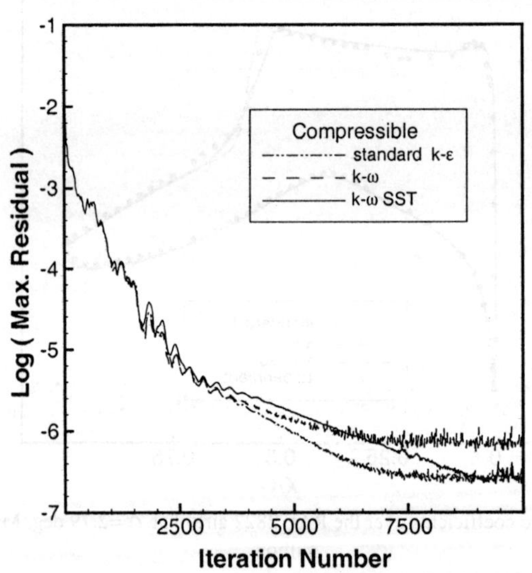



Fig. 6  Convergence history for the compressible flow over the RAE 2822 airfoil.

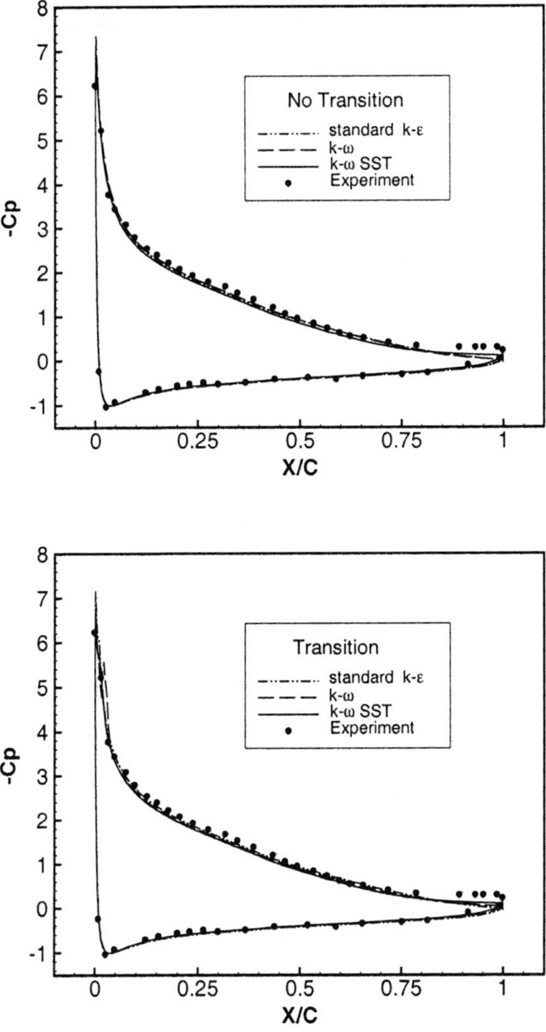

Fig. 7  Surface pressure coefficients over a NACA 4412 airfoil at $\alpha$ =13.87 deg and Re=1.52 million.

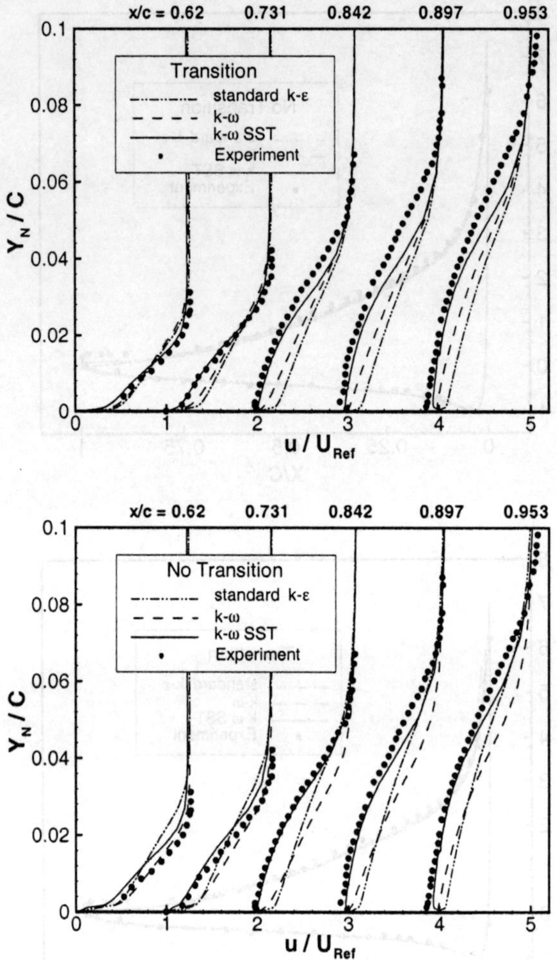

Fig. 8  Velocity profiles on the upper surface of a NACA 4412 airfoil at $\alpha$ =13.87 deg.

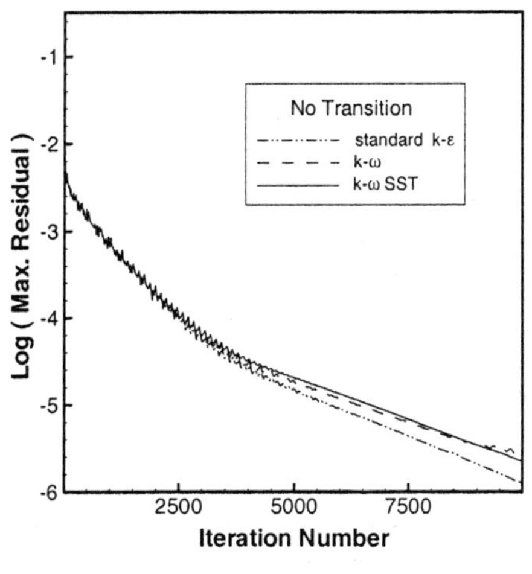

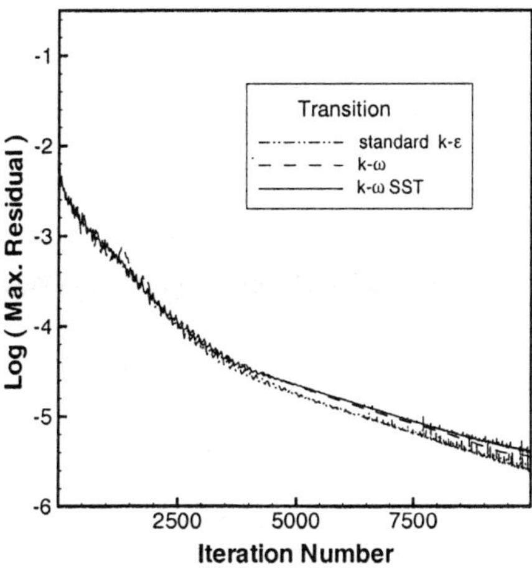

Fig. 9  Convergence history for the incompressible flow over a NACA 4412 airfoil.

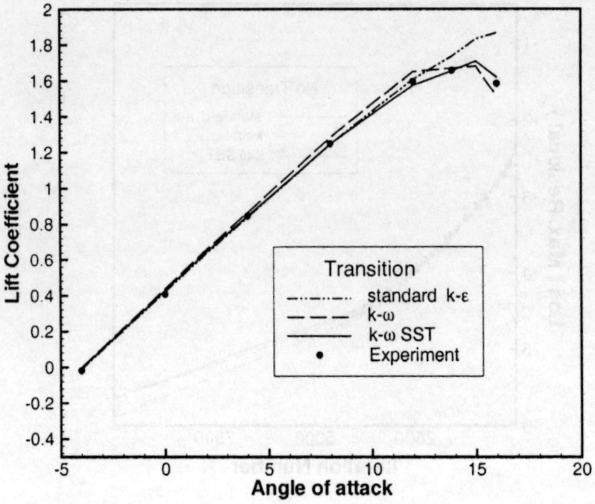

Fig. 10  Lift coefficient vs. angle of attack for a NACA 4412 airfoil.

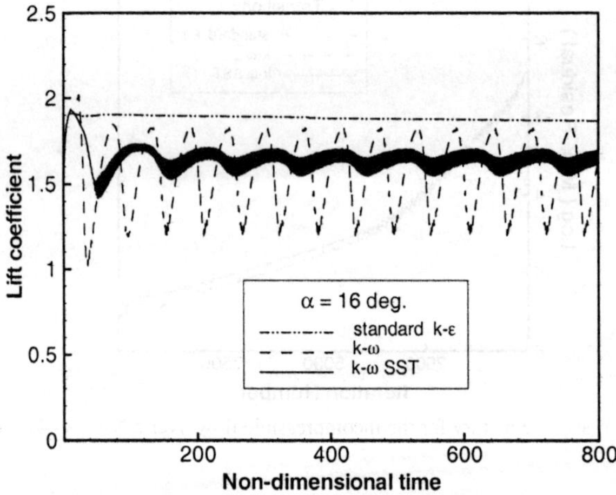

Fig. 11  Lift coefficient vs. time for a NACA 4412 airfoil at $\alpha = 16$ deg.

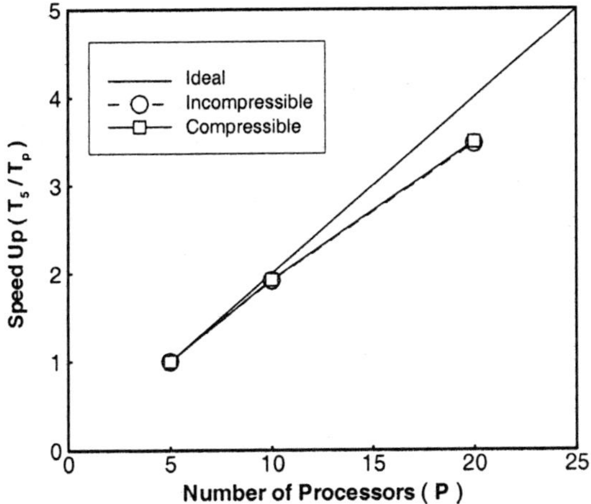



Fig. 12  Parallel speed up for two-element airfoil case using Chimera grid scheme

438

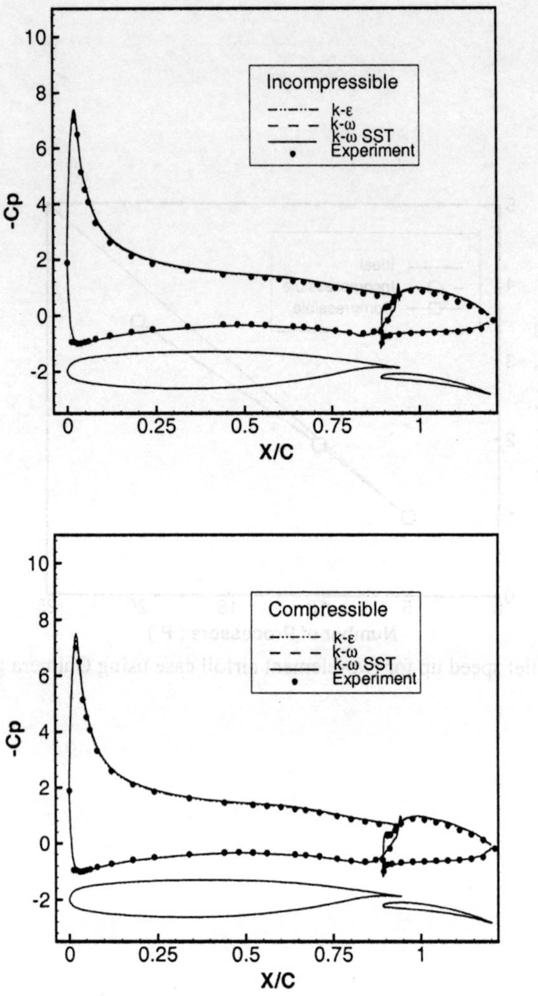

Fig. 13  Surface pressure coefficients over the NLR 7301 airfoil with flap at $\alpha$ =6.0 deg, M=0.185, and Re=2.51 million.

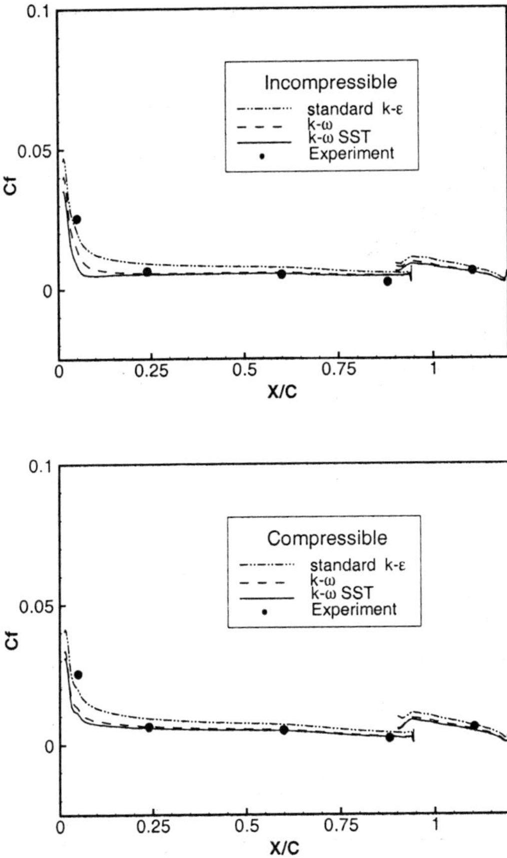



Fig. 14  Skin friction coefficients on the upper surface of the NLR 7301 airfoil with flap at $\alpha = 6.0$ deg.

440

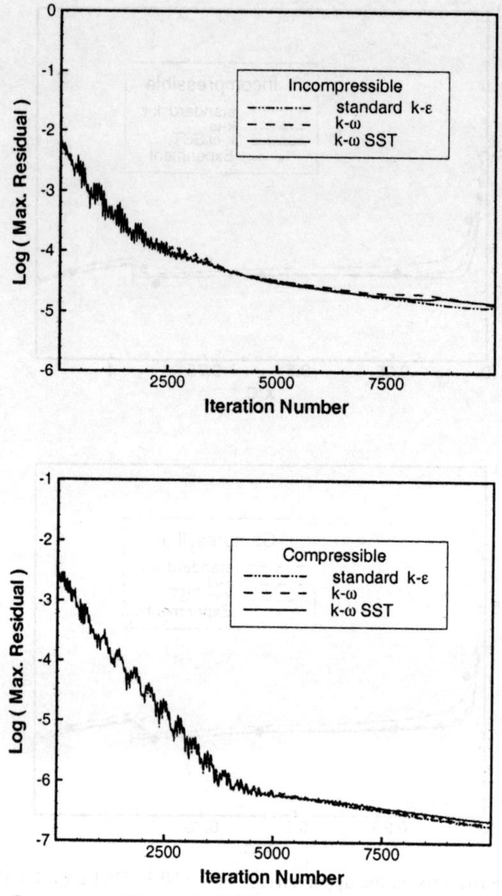

Fig. 15 Convergence history for the flow over the NLR 7301 airfoil with flap.

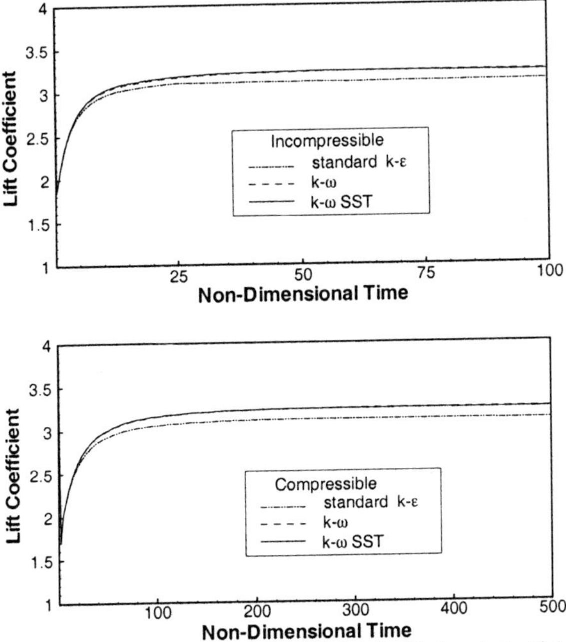

\Fig. 16  Lift coefficient vs. time for the NLR 7301 airfoil with flap at $\alpha$ =13.1 deg in unsteady computations.

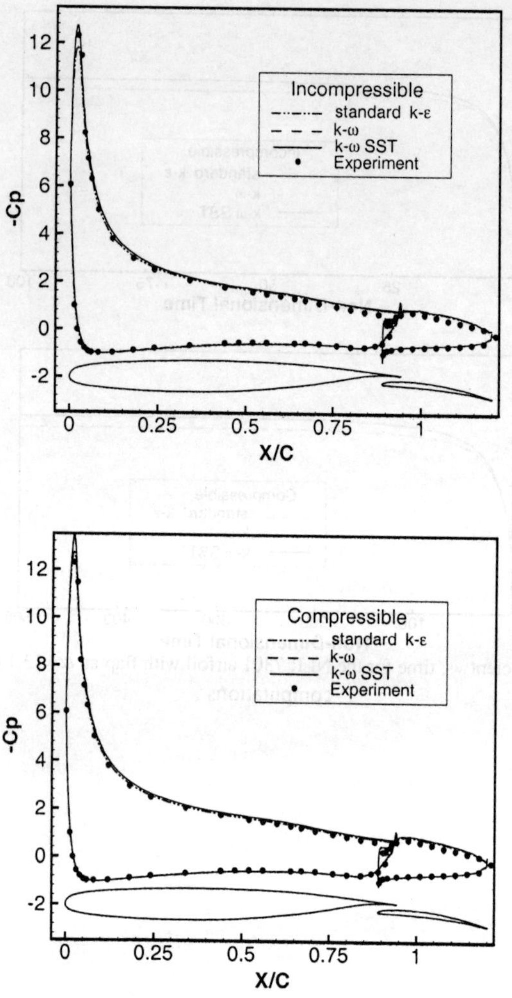

Fig. 17 Surface pressure coefficients over the NLR 7301 airfoil with flap at $\alpha$ =13.1 deg by unsteady computations.

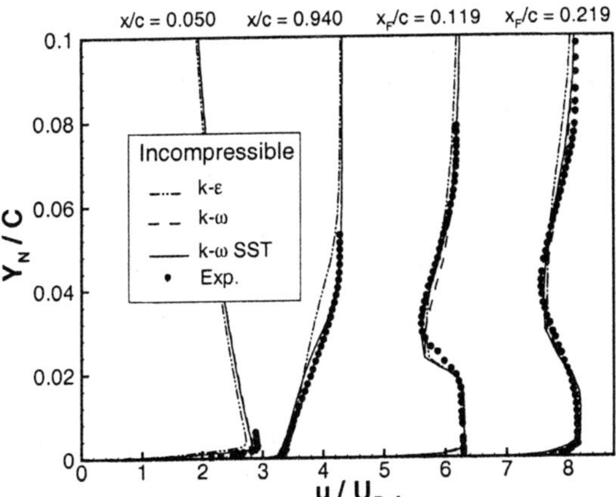



Fig. 18 Velocity profiles on the upper surfaces of the NLR 7301 airfoil with flap at $\alpha$ =13.1 deg.

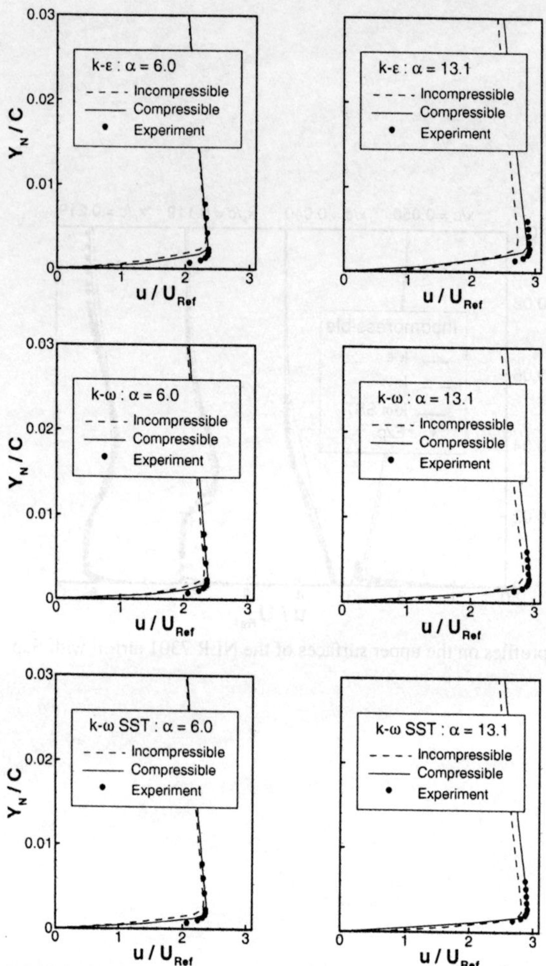

Fig. 19  Compressibility effects on velocity profiles at 5% chord for the NLR 7301 airfoil with flap.

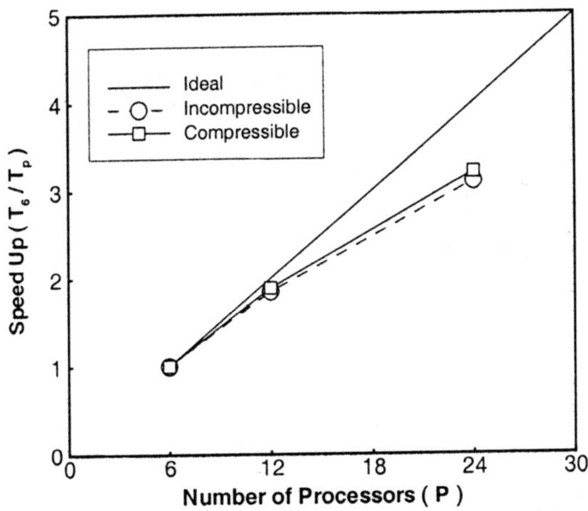



Fig. 20  Parallel speed up for three-element airfoil case using Chimera grid scheme.

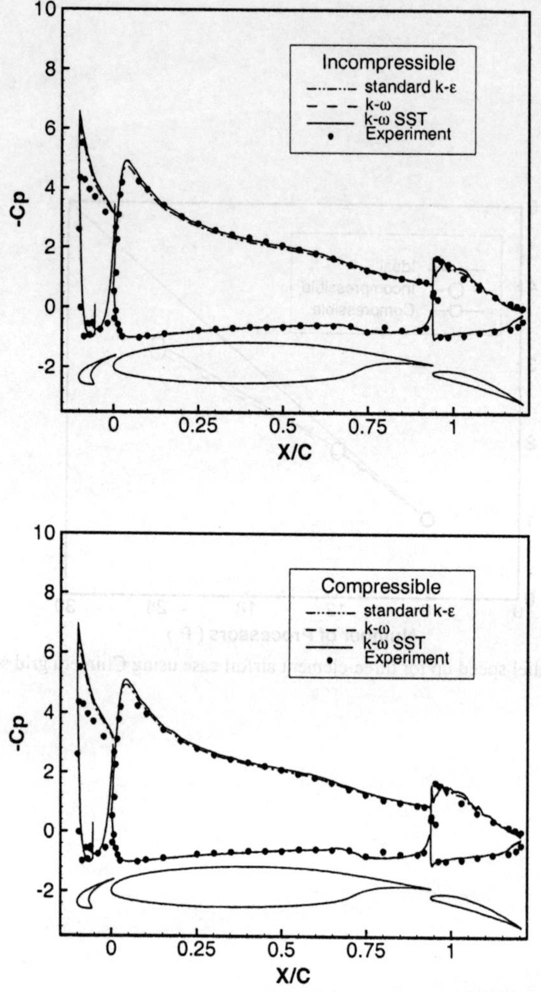

Fig. 21  Surface pressure coefficients over the GAW-1 high-lift airfoil at $\alpha$ =12 deg, M=0.15, and Re=0.62 million.

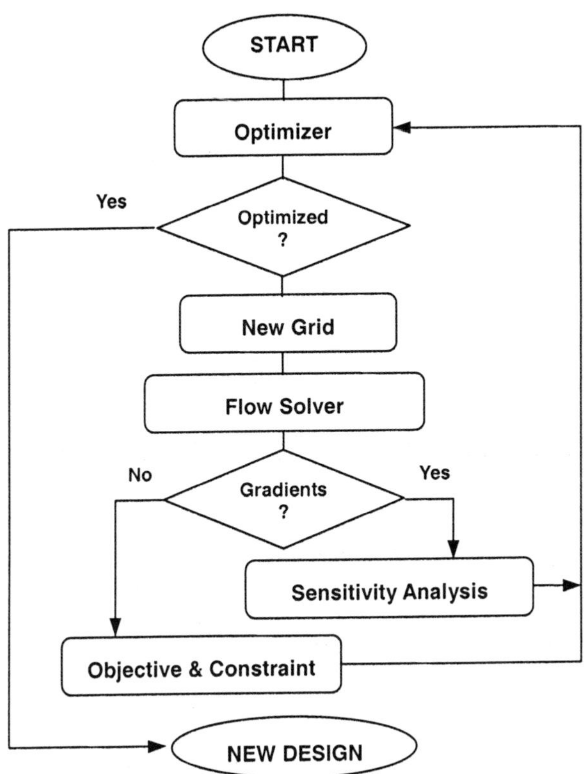



Fig. 22 Flow chart of a gradient-based design optimization

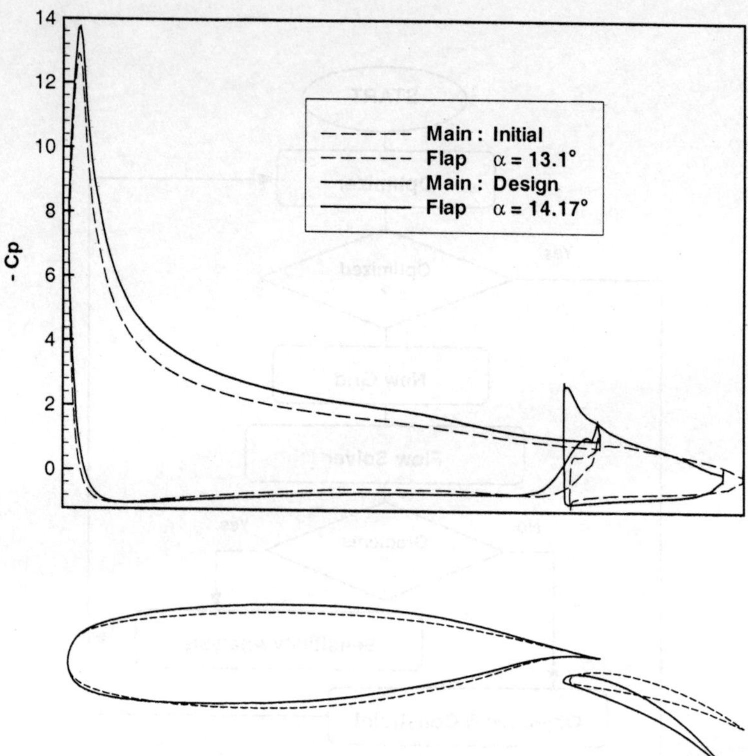



Fig. 23 Improvement of $Cl_{max}$ for multielement airfoil.

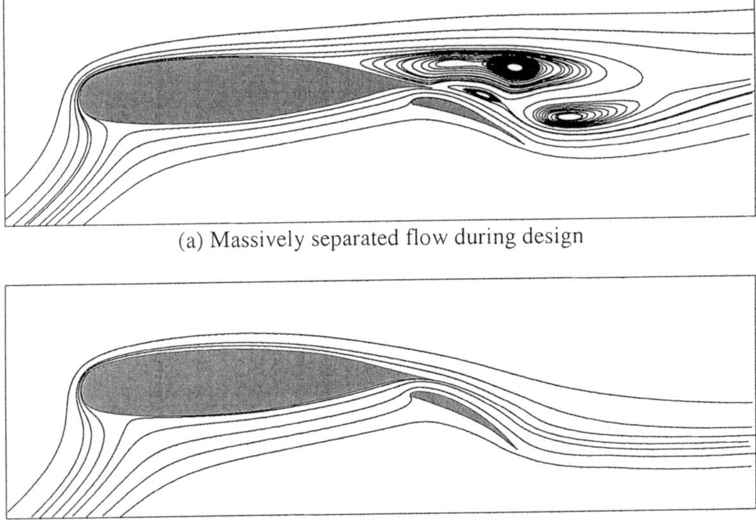

(a) Massively separated flow during design

(b) Attached flow after design

Fig. 24 Comparison of streamlines over high-lift airfoil with flap.

# Ice Effects on Aerodynamic Performance of Airfoils

K. D. Lee, J. Shim, and J. Chung
Aeronautical and Astronautical Engineering
University of Illinois
Urbana, Illinois 61801

The effect of ice shape on aerodynamic performance of ice-contaminated airfoils is investigated by comparing several different ice shapes. The computational method used is based on Navier-Stokes flow physics and validated with experimental data. The effect of turbulence modeling is first studied for different ice shapes to select appropriate models for the iced airfoil study. Results show that ice shapes can be classified into a few geometric categories based on their common characteristics. Aerodynamic performance of iced airfoils is strongly dependent on several key ice shape features such as ice ridges or horns, and the geometry of the suction side of an iced airfoil is more sensitive in determining overall aerodynamic characteristics.

## INTRODUCTION

Three different engineering research tools – flight test, wind tunnel, and CFD – are being used in icing research, and each has different merits and shortcomings. While flight test can provide real-world database, it should pay for its limited accessibility and controllability of test conditions. Wind-tunnel data are often quite different from flight test data due to the Reynolds number effect, wall effect, turbulence level, etc. CFD provides advantages in both running time and engineering time, but also possesses other shortcomings resulted from turbulence modeling, computational grid, transition modeling, surface roughness effect, etc. However, some recent studies show promising results in icing simulation.[1-3] Turbulence modeling is one of the most important factors which determine the acceptability of CFD results. Recently the present authors studied the turbulence modeling effect in iced airfoil simulation and reported the behavior of turbulence models for different types of ice geometry.[4] Because massive or local flow separation is a general flow feature in iced airfoil flows, algebraic turbulence models are not adequate. Therefore, one- and two-equation turbulence models are considered in this study. Among turbulence models tested in the previous study, the one-equation Spalart-Allmaras model and the two-equation Shear Stress Transport model are selected because they perform better comparatively. A systematic investigation is also performed to evaluate the ice geometry effect using various types of ice shapes. The geometry of ice shapes are obtained from IRT (Icing Research Tunnel) test data at the NASA Glenn Research Center.[5] Computational grids are generated using SMG2D-ICE[6] (Structured Multi-block Grid generation for ICEd airfoils), a grid generation code developed by the authors for iced airfoil flow analysis. Due to irregular surface roughness, a boundary-orthogonal grid technique is adopted to improve grid quality near the surface. The WIND[7] code of the NPARC Alliance is used as flow solver. WIND is a general-purpose Navier-Stokes code and has been validated in various applications.

## GEOMETRY AND SURFACE MODELING

### Iced Airfoil Geometry

Among extensive data of ice geometry obtained from IRT, ice shapes accreted on the NLF0414 and GLC305 airfoils are selected for this study. The airfoils with accreted ice shapes are shown in Figure 1. Ice shapes on the NLF0414 airfoil are used for the turbulence modeling effect study, because experimental aerodynamic data are available.[8] Ice shapes on the GLC305 airfoil are used for the geometry effect study. Among various ice shapes, six different ice shapes are selected considering overall patterns of ice accretion. Each ice shape represents a different icing condition in the IRT test. Rime ice around the leading edge is usually formed at lower temperature when the droplet freezes on the impact. In contrast, glaze ice is formed at higher temperature near freezing when the water droplet runs back on the upper surface before freezing. The rime ice is less irregular in shape and more streamlined than the glaze ice. The glaze ice has more aerodynamic penalties due to its less aerodynamic shape and subsequently a larger separation zone behind the horn.

### Surface Modeling

While maintaining basic geometric characteristics, ice geometry from test data should be locally smoothed through an appropriate surface modeling process to reduce the computational cost. The sensitivity of the surface modeling and ice shape smoothing in CFD simulation was studied by Chung *et al.*[1] They reported that ice shape smoothing with 50% of original control points are acceptable without losing basic characteristics. The surface modeling in the present study is performed using SmaggIce[9] (Surface Modeling and Grid Generation for ICEd airfoils), a software system being developed at NASA Glenn Research Center.

## GRID GENERATION

The irregularity and roughness of ice geometry make it very difficult to generate quality grids around iced airfoils using existing CFD tools. Ice shapes often accompany many concave and convex corners and cavities. Especially, the glaze ice shape includes large prominent horns on the suction side of the airfoil. Conventional grid generation around these shapes often results in poor quality grids. Thus, grid generation issues for iced airfoil geometry were studied and implemented in SMG2D-ICE.[6] In this code, the grid around an iced airfoil is generated with transfinite interpolation. The grid-orthogonality condition is imposed on the airfoil surface, and the grid is smoothed using a Laplace smoother.

### Surface Orthogonality

The surface orthogonalization technique by Tysell and Hedman[10] was adopted to handle the irregularity and roughness of the iced airfoil surface. In this technique, a specified length of the surface-orthogonal grid layer is added at each surface point into the surface normal direction. Then a smoothing process is used to account for convex and concave corners while maintaining grid orthogonality at the surface. Figure 2 shows a grid around an ice shape with surface orthogonality.

### Overset Grid

To reduce the computational cost in flow calculation, an overset grid with inner and outer blocks was used to avoid unnecessary extension of near-filed grid lines to the outer far-field region. The number of grid points and boundary locations of the outer grid is fixed regardless of the ice shape. The flow quantities at the overlapped interface are interpolated through the chimera boundary condition of the WIND code. The far field of the outer grid is located at 20 chord lengths away from the airfoil surface. Figure 3 shows an example of the grid around an ice shape. The normal wall spacing at the airfoil surface is $1.3 \times 10^{-5}$ of the chord length.

## TURBULENCE MODELING

Flows including massive separation of multiple length-scales and unsteadiness must be treated as complex turbulent flow. The complicated and irregular geometry of ice also requires accurate turbulent flow modeling. In those cases, algebraic turbulence modeling is not adequate to describe turbulent flow phenomena such as transport properties, boundary layer profiles, etc. Many efforts were given to overcome the weakness of algebraic models, resulting in one- and two-equation turbulence models. These models solve one or two partial differential equations for transport properties such as turbulent kinetic energy and turbulent length scale. The present study tested two different turbulence models – Spalart-Allmaras (SA) one-equation model[11] and Menter's k-$\omega$ Shear Stress Transport (SST) model.[12] The SA model was developed with the objective to accurately predict the turbulent phenomena in separated regions and multiple shear layers. It has similarities with other one-equation models, but it is not derived from two-equation models and includes a destruction term that depends on the distance to the wall. The SA model has some numerical advantages such that it is forgiving in terms of near-wall resolution and stiffness. It also provides relatively smooth laminar-turbulent transition, but only at points specified by the users.

In two-equation turbulence models, two partial differential equations are solved for the turbulent transport properties. The SST model was developed from the original k-$\omega$ model of Wilcox. It is identical to the k-$\omega$ model in the inner region of the boundary layer but changes gradually to the standard k-$\varepsilon$ model towards the outer edge of the boundary layer. Compared to the original k-$\omega$ model, it is known to have the ability to account for the transport of turbulent shear stress in boundary layers with adverse pressure gradients. In author's previous study, the Baldwind-Barth's one-equation model and the Chien's k-ε model were also tested and compared with the experimental data. But for the ice shapes selected for this study, the SA and SST models have shown better results in predictions of iced airfoil characteristics. Thus, only the SA and SST models are used in the present study.

## DISCUSSION OF RESULTS

All computations are made at the flow condition of M=0.28, Re=6M, p=13.46 psi, and T=482.07°R. Focus is given to the geometry effect. Figures 4 and 5 compare the predicted pressure distributions around the iced NLF0414 airfoil with experimental data. Two types of ice shape are considered – rime ice (R1) and glaze ice (G1). Both SA and SST turbulence models produced results in a good agreement with experiment, except the glaze ice at high angles of attack (Figure 5-c). This means that a more accurate turbulence model is needed when the flow involves massive flow separations such as the flow around the glaze ice.

In iced airfoils, the aerodynamic penalty on drag is usually paid more attention than the lift, since it is often more prominent and severely affects the actual flight at stall. Thus it is important to accurately predict how a certain type of ice shape can affect the drag. Tables 1 and 2 show a pattern of drag rise with the angle of attack selected from the author's previous work with typical rime and glaze ice shapes (R1 and G1 on the NLF0414 airfoil).

Table 1.  Drag increase due to rime ice accretion (R1)

| $\alpha$ | Clean | Iced | Increase (%) |
|---|---|---|---|
| 0 | 0.0113047 | 0.0149507 | 32.25 |
| 2 | 0.0140069 | 0.0191857 | 36.97 |
| 4 | 0.0189420 | 0.0272152 | 43.67 |
| 6 | 0.0244065 | 0.0392255 | 60.71 |
| 8 | 0.0312773 | 0.0601952 | 92.45 |

Table 2.  Drag increase due to glaze ice accretion (G1)

| $\alpha$ | Clean | Iced | Increase (%) |
|---|---|---|---|
| 1 | 0.0124817 | 0.0367728 | 194.61 |
| 3 | 0.0161261 | 0.0550917 | 241.63 |
| 5 | 0.0214611 | 0.0769682 | 258.64 |
| 7 | 0.0277991 | 0.1090730 | 292.36 |

Figures 6 and 7 show the lift and drag coefficients versus angle of attack for ice shapes of Group A in the GLC305 airfoil, respectively. Results from both SA and SST models are shown. The overall lift and drag characteristics predicted by both turbulent models are approximately the same. The only thing noticeably different is that the SST model is predicting the stall to occur at a slightly lower angle of attack than the SA model. The predicted maximum lift coefficient values are also smaller with the SST model. The ice geometry A3 shows almost no lift loss compared to the clean airfoil with both turbulence models. In both lift and drag, the SA model is predicting some aerodynamics penalties in ice geometry A3, but SST is not capturing the difference. Thus, the SA model may be showing somewhat better performance for the rime ice case, which is consistent with our previous observation.[4]  The ice shape A3 does not have any protuberant ice ridge on its surface and only have some ice accreted on the lower leading-edge surface. In contrast, A1 and A2 have a slightly sharp ice horn on the upper leading-edge surface. The lower leading-edge surface roughness is bigger in A1 than A2. However, the aerodynamic penalties for these two ice shapes are almost identical, although A1 is the worse case. This indicates that prominent ice shape factors like ice ridge and horn are more important, in determining overall aerodynamic characteristics, than slight changes in surface roughness.

In contrast to Group A, ice shapes of Group B have small or big ice ridges, on both upper and lower leading-edge surfaces. The predicted lift and drag characteristics for these ice shapes are shown in Figures 8 and 9. In this case, all three shapes are showing some noticeable aerodynamic penalties. The worst ice shape in this case is B3, which has the biggest ice horns on both upper and lower leading-edge surfaces. The ice shape B3 has more aerodynamic penalties in drag than in lift. The lift coefficient of B3 is not much smaller than that of other two cases, but the drag increase is obvious. The observation indicates that the ice geometry with big upper and lower horns produces a large stagnation region especially at high angles of attack. The presence of the high-pressure region is thought to be responsible for the noticeable drag penalty. From simple geometric consideration, it may be speculated that the aerodynamic penalty would be bigger for B1 than B2. As Figures 8 and 9 show, however, the actual lift loss and drag rise is bigger for B2. One noticeable difference between these two ice shapes is that B2 has a sharper upper horn than B1. In analysis with Group A, it is noticed that lower-surface roughness is not as much critical for flow characteristics as upper-surface ice ridge. Thus, we can conjecture that the critical ice shape factor on the suction side is playing a more dominant role in determining overall aerodynamic characteristics.

Throughout Figures 6-9, there are some common patterns in iced airfoil characteristics. At lower angles of attack, all of the ice shapes showed similar results with relatively small deviations from the clean airfoil. The lift slope is slightly smaller in iced airfoils and almost constant regardless of ice shape. Also, it is almost independent of ice geometry although it may change for rougher ice shapes. All ice shapes show finite aerodynamic performance degradations resulting in stall at lower angles of attack, lower maximum lift coefficient, and increased drag. Ice shape A3 has the least aerodynamic penalty and B3 has the most.

The drag penalty due to ice increases with the angle of attack, and it is even more drastic for the glaze ice. All ice shapes considered in the present study follow this characteristic pattern with minor variations. Tables 3 and 4 show the quantitative comparison of drag for all the ice shapes at zero and stall angles of attack. Again, we can see that the ice shape A3 is showing the smallest value of drag rise and B3 the largest value.

Table 3. Comparison of drag rise ($\alpha = 0°$)

| Ice geometry | Cd | Increase(%) |
|---|---|---|
| clean | 0.0101915 | 0.0 |
| A1 | 0.0170832 | 67.62 |
| A2 | 0.0127854 | 25.45 |
| A3 | 0.0104893 | 2.92 |
| B1 | 0.0112298 | 10.19 |
| B2 | 0.0123856 | 21.53 |
| B3 | 0.0267231 | 162.21 |

Table 4. Comparison of drag rise ($\alpha = 8°$)

| Ice geometry | Cd | Increase(%) |
|---|---|---|
| clean | 0.0186465 | 0.0 |
| A1 | 0.0786233 | 321.65 |
| A2 | 0.0712284 | 281.99 |
| A3 | 0.0264516 | 41.86 |
| B1 | 0.0734005 | 293.64 |
| B2 | 0.0888354 | 376.42 |
| B3 | 0.1062750 | 469.95 |

The Cp distributions for ice shapes of Group A are compared in Figure 10. At low angles of attack ($\alpha = 0°$ and 4°), the Cp distribution is almost the same as that of the clean airfoil, except in the front portion of the airfoil. The effect of ice accretion is confined to local regions at low angles of attack. But at a higher, near-stall angle of attack, the Cp distribution begins to change noticeably. At the angle of attack of 8°, which is just before the stall angle of attack, the Cp distribution looks much different from that of the clean airfoil. Ice accretion decreases the pressure at the lower surface and increases at the upper surface, making the lift smaller. Also, high expansion disappears on the upper leading-edge surface, due to the flow separation behind the upper ice ridge. This flow separation changes the flow pattern on the upper surface completely and eventually plays a critical role in creating a bigger aerodynamic penalty. The pressure drop and its wiggly distribution are also eminent on the lower surface due to the surface roughness. But, the Cp distribution around ice shape A3 is not much different from that of the clean airfoil even at near-stall angle of attack (Figure 10-c). Figure 11 shows the Cp distributions for ice shapes of Group B. One noticeable difference from Group A results is that there are higher spikes in Cp curves right after the leading edge on both the upper and lower surfaces. These are caused by separated flows behind the leading-edge ice horns. This is more obvious in B3 than in B2, which is consistent with the lift and drag characteristics discussed.

Figures 12 and 13 show the typical stream traces around ice shapes A1 and B3, which contain a big separation region and small local separation bubbles near the rough ice surface. Large protuberant horns in the glaze ice make the separation bigger and reduce the stall angle of attack, making the flow unsteady at a relatively smaller angle of attack compared with the rime ice.

## CONCLUDING REMARKS

Several different ice shapes were selected and tested to investigate critical factors that dominate the degradation of aerodynamic performance in iced airfoils. In analysis of airfoils with ice accretion, the flow solution varies significantly with the choice of turbulence modeling. The geometry effect on aerodynamic performance of ice-contaminated airfoils was thus studied by comparing different ice shapes with selected turbulence models. The results show that aerodynamic performance of an ice-contaminated airfoil is more sensitive to some critical ice shape feature such as ice horn and less dependent on local surface roughness. Also, it is conjectured that the geometric factor on the suction side of the airfoil is more dominant in determining overall aerodynamic characteristics.

## ACKNOWLEDGEMENTS

This research was supported by a grant from the NASA Glenn Research Center with Dr. Yung Choo as technical administrator. The authors would like to thank to Dr. Choo for providing the ice geometry and experimental data and for sharing his expertise on icing aerodynamics.

## REFERENCES

1. Chung, J., Reehorst, A., Choo, Y., and Potapczuk, M., "Effect of Airfoil Ice Shape Smoothing on the Aerodynamic Performance," AIAA Paper 98-3242, 34th AIAA/ASME/SAE/ASEE Joint Propulsion Conference & Exhibit, Cleveland, OH, July 1998.

2. Chung, J., Choo, Y., Reehorst, A., Potapczuk, M., and Slater, J., "Navier-Stokes Analysis of the Flowfield Characteristics of an Ice Contaminated Aircraft Wing," AIAA Paper 99-0375, 37th Aerospace Sciences Meeting & Exhibit, Reno, NV, January 1999.

3. Chung, J. and Addy, H. Jr., "A Numerical Evaluation of Icing Effects on a Natural Laminar Flow Airfoil," AIAA Paper 2000-0096, 38th Aerospace Sciences Meeting & Exhibit, Reno, NV, January 2000.

4. Shim, J., Chung, J., and Lee, K., "A Comparison of Turbulence Modeling in Flow Analysis of Iced Airfoils," AIAA Paper 2000-3920, 18th Applied Aerodynamics Conferences and Exhibit, Denver, CO, August 2000.

5. Wright, W. and Rutkowski, A., "Validation Results for LEWICE 2.0," NASA CR-1999-208690, January 1999.

6. Shim, J., Chung, J., and Lee, K., "A Grid Generation Strategy for CFD Analysis of Iced Airfoil Flows," 7th International Conference on Numerical Grid Generation in Computational Field Simulations, Whistler, Canada, September 2000.

7. Bush, R., Power G., and Towne, C., "WIND: The Production Flow Solver of the NPARC Alliance," AIAA Paper 98-0935, 36th Aerospace Sciences Meeting & Exhibit, Reno, NV, January 1998.

8.  Addy, H. Jr. and Chung, J., "A Wind Tunnel Study of Icing on a Natural Laminar Flow Airfoil," AIAA Paper 2000-0095, 38th Aerospace Sciences Meeting & Exhibit, Reno, NV, January 2000.

9.  Vickerman, M. B., Choo, Y. K., Braun, D. C., Baez, and M., Gnepp, S., "SmaggIce : Surface Modeling and Grid Generation for Iced Airfoils-Phase 1 Results," AIAA Paper 2000-0235, 38th Aerospace Sciences Meeting & Exhibit, Reno, NV, January 2000.

10. Tysell, L. G. and Hedman, S. G., "Toward a General Three-Dimensional Grid Generation System," ICAS-88-4.7.4, 16th Congress of the International Council of the Aeronautical Sciences, Jerusalem, Israel, August 1988.

11. Spalart, P. R. and Allmaras, S. R., "A One-Equation Turbulence Model for Aerodynamic Flows," AIAA Paper 92-0439, 30th Aerospace Sciences Meeting & Exhibit, Reno, NV, January 1992.

12. Menter, F. R., "Zonal Two Equation k-$\omega$ Turbulence Models for Aerodynamic Flows," AIAA Paper 93-2906, 24th AIAA Fluid Dynamics Conference, Orlando, FL, July 1993.

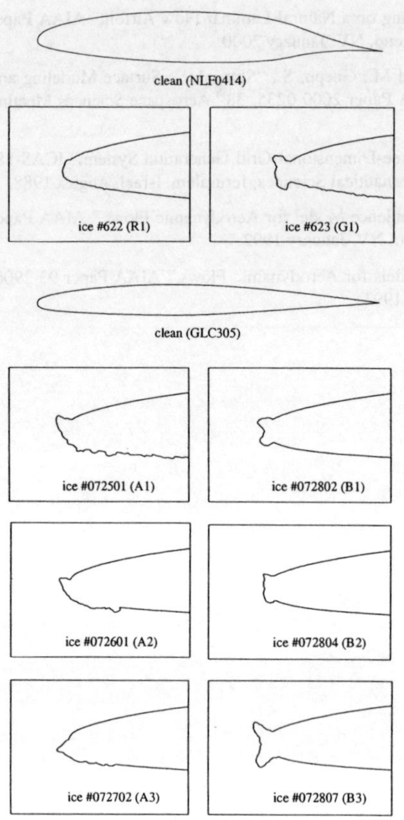



Figure 1. Airfoil geometry with various ice accretion

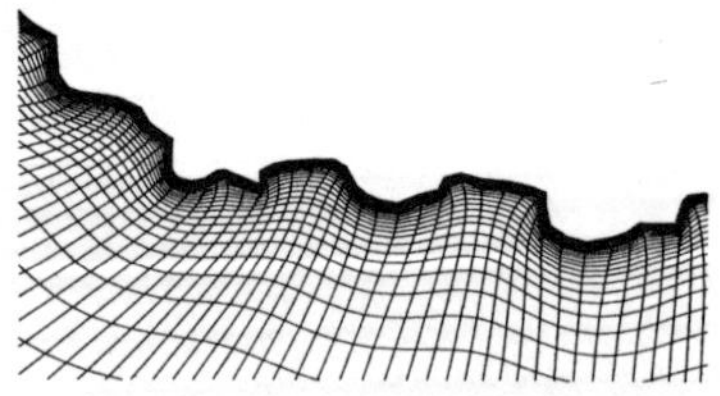

Figure 2. Grid distribution near surface

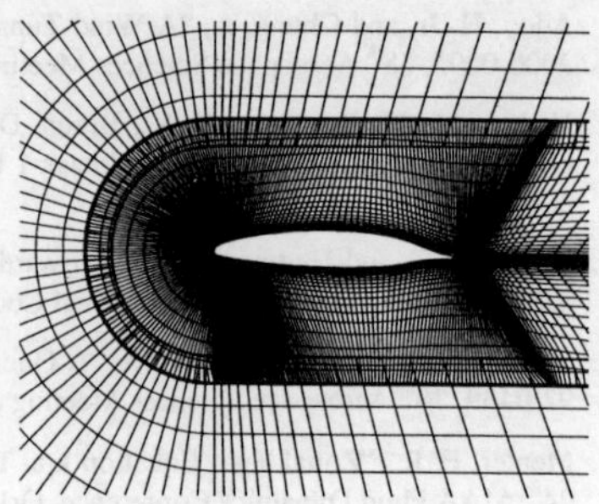

Figure 3. Two-zone overset grid system

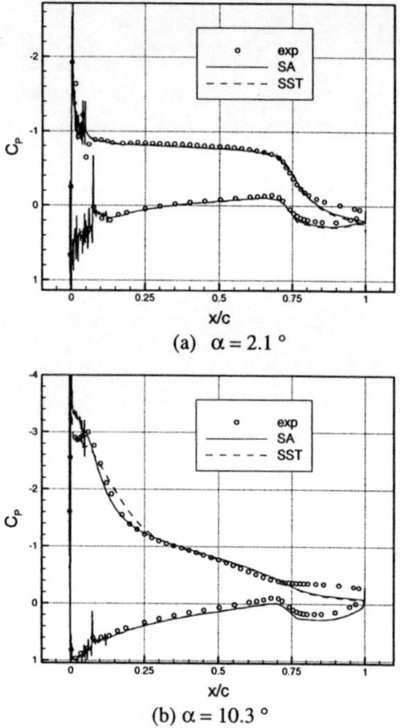



(a) $\alpha = 2.1\,^\circ$

(b) $\alpha = 10.3\,^\circ$

Figure 4. Cp distribution (Ice R1 on NLF0414)

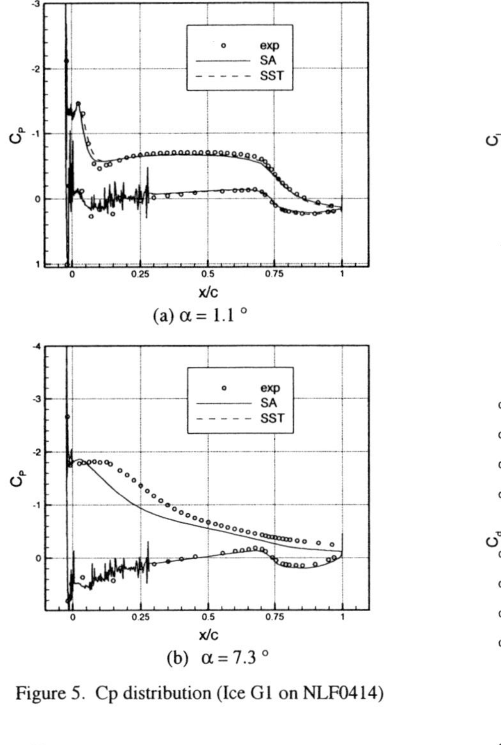

(a) $\alpha = 1.1\,^\circ$

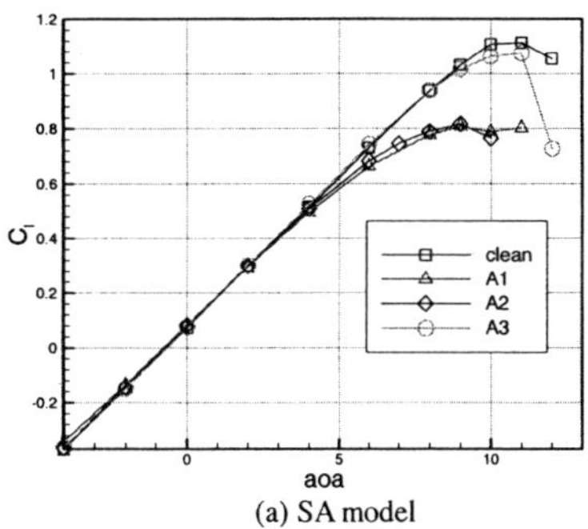

(b) $\alpha = 7.3\,^\circ$

Figure 5. Cp distribution (Ice G1 on NLF0414)

(a) SA model

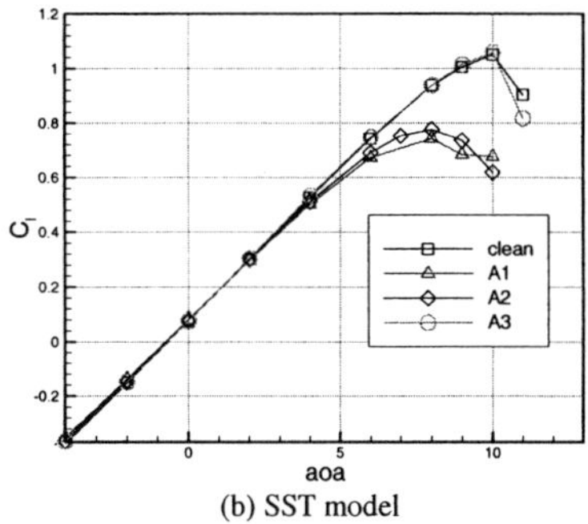

(b) SST model

Figure 6. Ice effect on lift – Group A

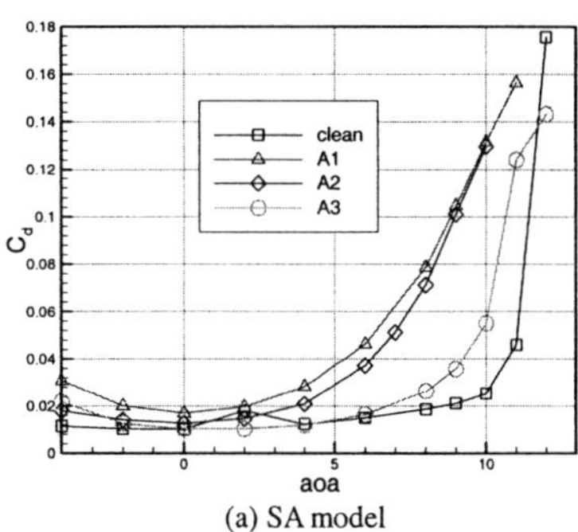

(a) SA model

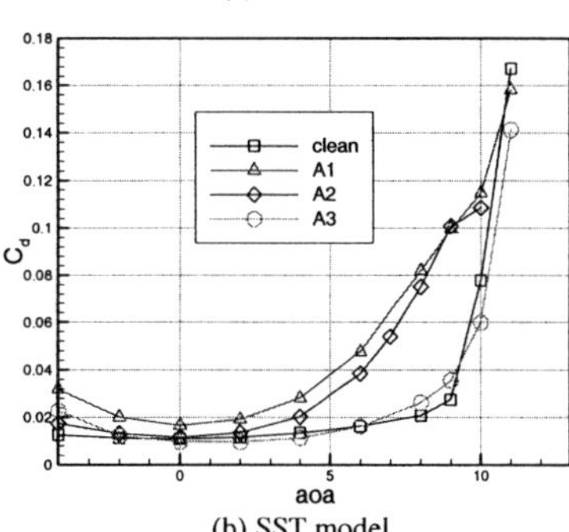

(b) SST model

Figure 7. Ice effect on drag – Group A

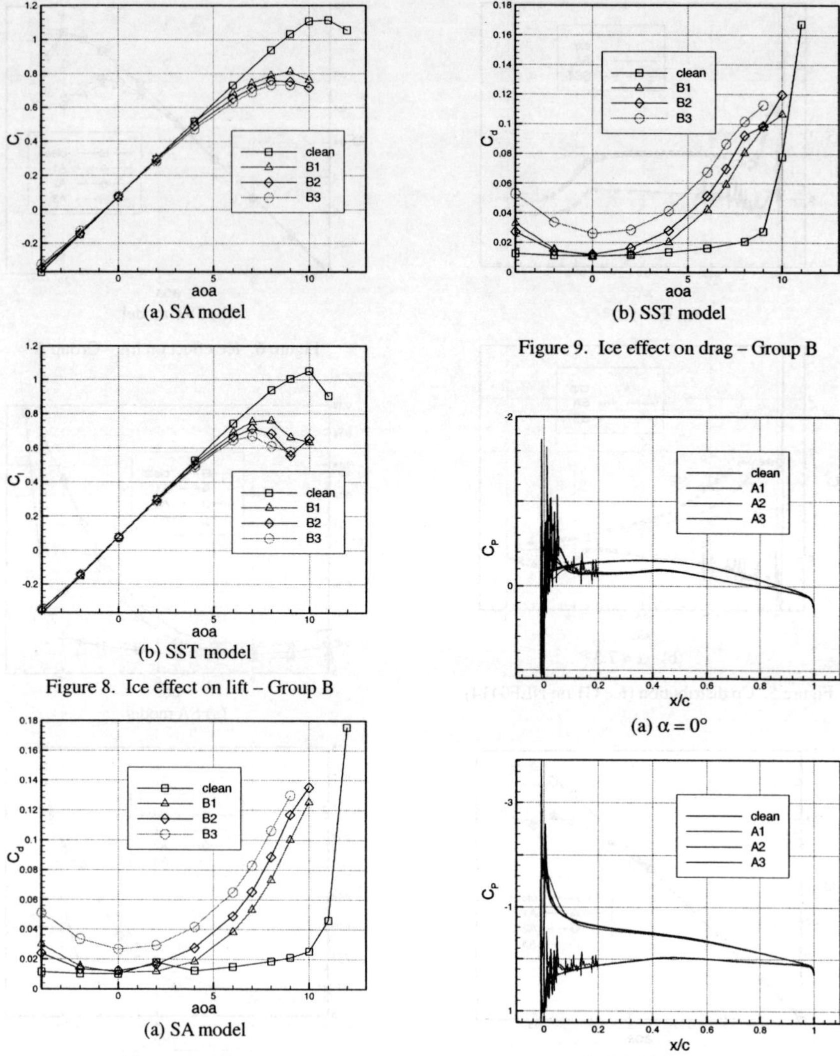

(a) SA model

(b) SST model

Figure 9.  Ice effect on drag – Group B

(b) SST model

Figure 8.  Ice effect on lift – Group B

(a) α = 0°

(a) SA model

(b) α = 4 °

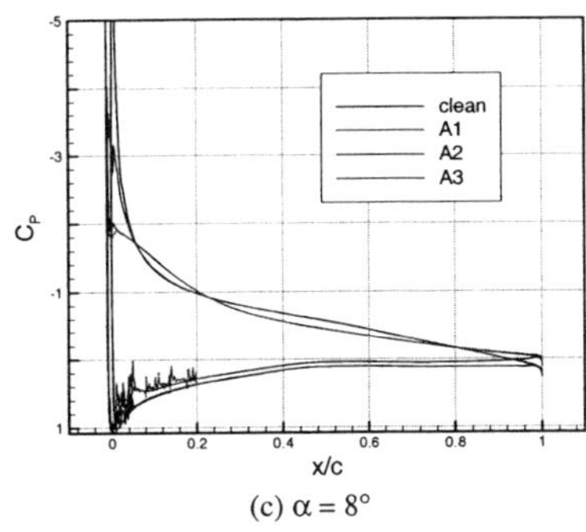

(c) $\alpha = 8°$

Figure 10. Cp distribution – Group A

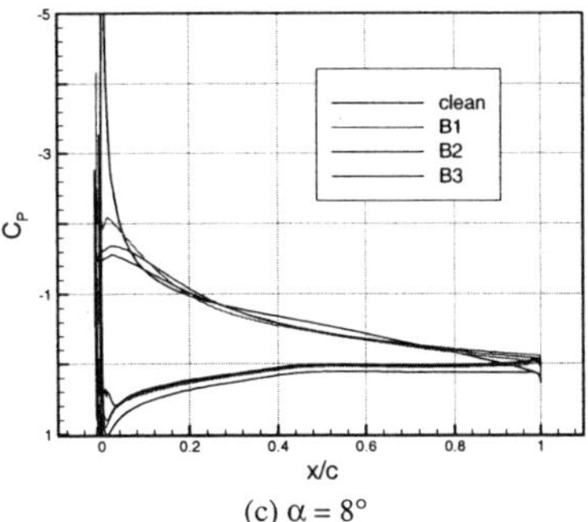

(c) $\alpha = 8°$

Figure 11. Cp distribution – Group B

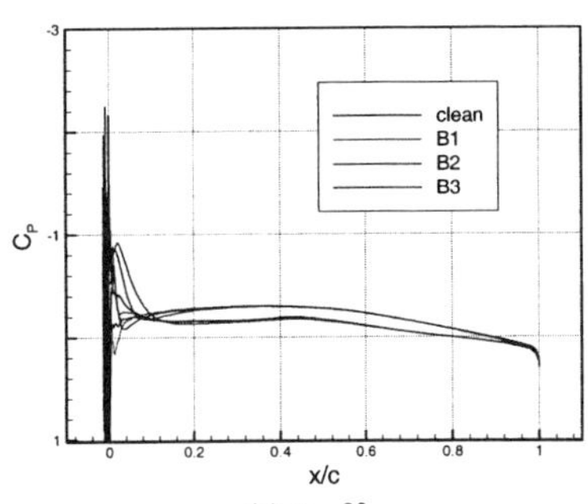

(a) $\alpha = 0°$

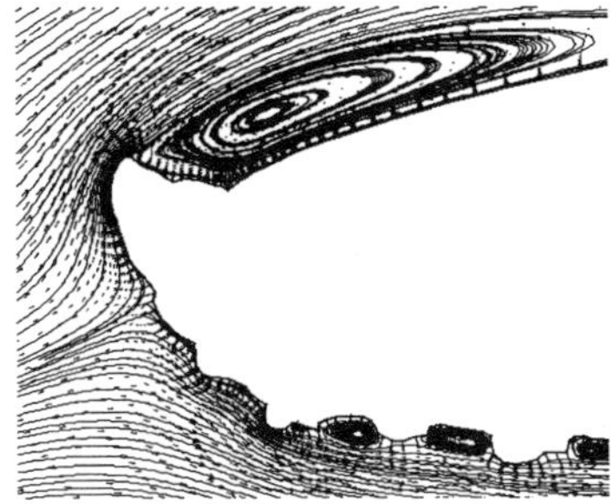

Figure 12. Stream traces (A1, $\alpha = 4°$)

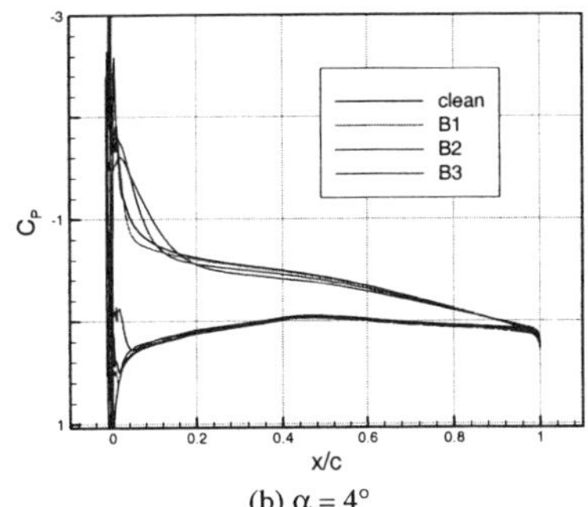

(b) $\alpha = 4°$

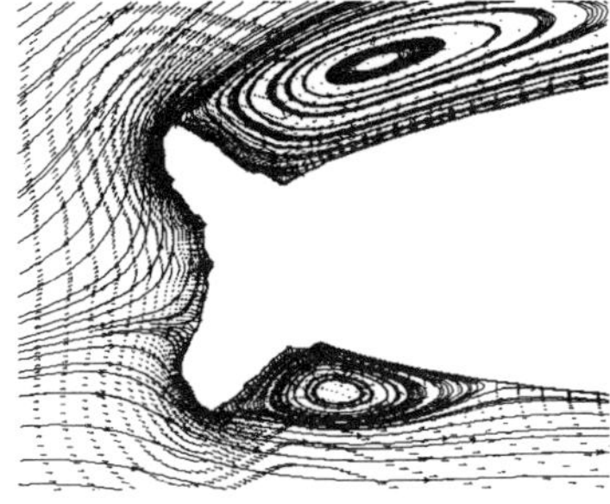

Figure 13. Stream traces (B3, $\alpha = 4°$)

# Applications Beyond Aeronautics

# FLOW IN THE STENOTIC CAROTID BIFURCATION

Stanley A. Berger and Vitaliy L. Rayz
Department of Mechanical Engineering
University of California
Berkeley, CA 94720

**Abstract.** Blood flow in a severely stenotic carotid bifurcation has been analyzed. A realistic three-dimensional model of an atherosclerotic carotid bifurcation was obtained using MR images. An unstructured grid was then generated on the domain and the full Navier-Stokes equations were solved numerically. Steady flows at different Reynolds numbers as well as an unsteady pulsatile flow have been calculated. Non-Newtonian behavior of blood was considered in some simulations. The results obtained are found to be in good qualitative agreement with simulations carried out earlier for two-dimensional flows.

## 1 Introduction

Numerical simulations of the blood flow in arteries can contribute to elucidating the influence of atherosclerotic plaques on the flow and vice versa (Berger and Jou, 2000; Stroud, 2000). The amount of blood passing through a severely occluded artery is often significantly reduced. Under certain conditions the plaque in a vessel can fracture and rupture. The occurrence of this in an artery supplying blood to the heart or brain may lead to a heart attack or stroke, respectively. Thus modeling flow in such vessels is particularly important. We have concentrated our studies on the carotid artery, one of the principal arteries supplying blood to the head and neck, particularly on the carotid bifurcation, where the common carotid artery, arising from the aorta, divides into two daughter vessels, a larger one, the internal carotid, supplying blood to the brain, and a smaller vessel, the external carotid, supplying blood to the facial tissues. Such bifurcations are especially prone to the development of plaque. Modeling of severely stenotic blood vessels presents challenges due to extremely irregular geometries. Typically, stenotic vessels have been modeled as cylindrical pipes with constrictions. A two-dimensional model of a realistic severely stenotic carotid bifurcation was analyzed by Stroud et al. (2002). Their simulations showed that morphological details of a particular artery have important effects on the blood flow. However, some features of the flow could not be fully captured in a two-dimensional model since the flow has important three-dimensional aspects. The objective of the current work is to continue these investigations in fully three-dimensional realistic geometries.

## 2 Three-Dimensional Geometry

### 2.1 Grid Generation

A three-dimensional model of a severely stenotic carotid bifurcation has been created from MR images of an atherosclerotic plaque obtained at the San Francisco VA Medical Center. The contours of the images were transferred to the CFD Research Corporation Geometry Module and used as the boundaries for the three-dimensional model. An unstructured tetrahedral grid was then generated on the domain to be used for the numerical solution of the governing equations. The three-dimensional grid is shown in Figure 1. First a boundary surface grid is created. The software then automatically generates a 3D grid starting from the surface and propagating inside the domain. The density of the grid is controlled by specifying the size of a minimum and maximum allowable cell face on the surface. An increase of the grid density allows for better accuracy of the solution at the cost of significantly increased calculation time. At some point a further increase of the grid density has very little influence on the solution accuracy while dramatically increasing the time. It is necessary to find a compromise, so that a sufficiently accurate solution can be obtained in a reasonable time.

## 2.2 Three models with different grids

To obtain an optimum grid three models with different grid densities were created. The "coarse grid" model contains 32,203 tetrahedral cells, the "fine grid" has 60,781 cells, and the "ultra-fine grid" has 99,062 cells. These grids are shown in Figure 2. The density of the grid was increased by reducing the size of a maximum allowable element on the surface of the domain.

# 3 Finite-Volume Flow Solver

## 3.1 Discretization of the Governing Equations

To a very good approximation blood flowing in large arteries can be modeled as a Newtonian fluid, with a constant dynamic viscosity of 0.0035 N.sec/m$^2$ and mass density of 1060 kg/m$^3$. We have made the assumption of Newtonian behavior in most of our studies, but have also considered the possible effects of non-Newtonian behavior. The walls of the artery are assumed to be rigid. The Navier-Stokes equations for unsteady, incompressible flow are:

$$\nabla \cdot \mathbf{u} = 0$$
$$\rho \left( \frac{\partial \mathbf{u}}{\partial t} + \mathbf{u} \cdot \nabla \mathbf{u} \right) = -\nabla p + \nabla \cdot \boldsymbol{\tau}$$

where $\mathbf{u}$ is velocity vector, p is pressure and $\boldsymbol{\tau}$ is the viscous stress tensor. For a Newtonian fluid the momentum equation becomes:

$$\rho \left( \frac{\partial \mathbf{u}}{\partial t} + \mathbf{u} \cdot \nabla \mathbf{u} \right) = -\nabla p + \mu \nabla^2 \mathbf{u} \, ,$$

where $\mu$ is the dynamic viscosity. The finite-volume CFD-ACE(U) package was used for discretization and solution of the equations. A first-order accurate Euler scheme is used for time discretization. A blending of the second-order central difference scheme and the first-order upwind scheme is used in space discretization. The equations are numerically integrated over each of the computational cells. The average value of a variable in a computational cell is given by its value at the cell center and calculated using the values at the neighboring cells. For a flow property $\phi$ at the center of a cell P this results in the following equation:

$$a_P \phi_P = \sum_{nb} a_{nb} \phi_{nb} + S_\phi$$

where the subscript "nb" denotes values at neighboring cells and $S_\phi$ is a source term. The Semi-Implicit Method for Pressure-Linked Equations Consistent (SIMPLEC) algorithm is used in CFD-ACE(U) to solve for the pressure field. The algorithm, an enhancement of the well known SIMPLE algorithm, can be summarized as follows:

1. The discretized momentum equations are solved for an initial estimate of the velocity, $\mathbf{u}^*$, using a guessed (or last iterate) pressure field p*. In general, the velocity components obtained in this solution will not satisfy the continuity equation.

2. It is assumed that the actual velocity and pressure are equal to the current velocity and pressure plus unknown corrections:

$$\mathbf{u} = \mathbf{u}^* + \mathbf{u}', p = p^* + p'$$

3. An expression relating the velocity and pressure corrections is obtained using the discretized momentum equations.

4. The continuity equation is discretized as $\sum \rho u^* A + \sum \rho u' A = 0$

5. The pressure correction is found by solving $a_P p'_P = \sum a_{nb} p'_{nb} + S_m$ , where the mass source $S_m$ is evaluated as $S_m = \sum \rho u^* A$.

6. The velocity field is corrected by solving the momentum equations with the corrected pressure field.

This procedure is repeated until convergence is obtained.

## 3.2 Boundary and Initial Conditions

The boundary conditions are no-slip and no penetration at the walls. Additional boundary conditions are uniform velocity at the inlet and fixed pressure at both outlets. The flow becomes fully developed upstream of the bifurcation. It is assumed that the velocity vector has only a stream-wise component at the inlet. For steady flow simulations the magnitude of the inlet velocity was dictated by the desired Reynolds number based on the common carotid diameter. The initial values for velocity components and pressure were set equal to zero. For the unsteady flow simulations the inlet velocity values were obtained from the available volume flow rate data. A steady-state solution was used as the initial condition for the unsteady simulations.

## 3.3 Convergence criteria

The number of iterations necessary to obtain convergence is determined by the reduction of the residuals. Residuals for each variable - pressure and three velocity components in our case - are defined as the sum of the absolute values of the residuals for that variable at all the computational cells. Generally, a five order of magnitude reduction is taken as the indication that the solution has converged. To facilitate convergence for the simulations of the flow at high Reynolds numbers it was necessary to change the relaxation parameters.

## 3.4 Visualization of the Results

The pressure and velocity fields obtained by solving the Navier-Stokes equations can be visualized for an arbitrary cross-section using CFDRC post-processing software. In Cartesian coordinates the model is oriented along the z-axis, so this direction can be considered as the "axial" or "stream-wise" direction of the flow. Hence a cross-section showing this "stream-wise" velocity component should be chosen perpendicular to the X-Y plane. It was found that a good view at the stenotic part of the internal carotid artery could be obtained at the cross-section shown in Figure 3. The three-dimensional nature of the flow can be appreciated by plotting the flow streamlines.

## 4 Steady Flow

## 4.1 Comparison of the three grid density models

Numerical solutions of the governing equations have been obtained for steady laminar Newtonian flow at different Reynolds numbers. The Reynolds number was based on the inlet velocity and the average diameter of the common carotid artery (8mm). A steady flow solution was obtained for each of the grid density models. The quality of the grids was evaluated by comparing the velocity fields obtained for each grid. "Stream-wise" velocity distributions obtained for the three grids for steady flow at a Reynolds number of 200 are shown in Figure 4. Significant differences can be seen between the coarse grid and the fine grid results, while the differences between the fine grid and the ultra-fine grid results are much less pronounced. Similar conclusions can be made by comparing the velocity fields obtained for the flow at a Reynolds number of 600 (Figure 5). Since the ultra-fine grid model requires considerably more calculation time while yielding results very similar to those obtained for the fine grid model, the fine grid model was considered sufficient for the current analysis and all subsequent simulations were performed for this grid density only.

## 4.2 Steady Velocity Distribution

The "stream-wise" velocity distribution and a few three-dimensional streamlines obtained for steady flow at a Reynolds number of 200 are shown in Figure 6. There is a large recirculation region in the carotid sinus (the large widening at the entrance of the internal carotid artery), upstream of the stenosis. A high velocity jet develops in the narrowest stenotic part of the internal carotid. Another large recirculation region is formed downstream of the stenosis. Steady solutions obtained for the flow at Reynolds numbers of 400 and 600 show a similar velocity field with a stronger jet propagating along the inner internal carotid wall and a larger recirculation zone downstream of the stenosis. The results obtained are in a good agreement with those obtained by Stroud et al. (2001).

## 5 Pulsatile Flow

### 5.1 Physiological Pulse Initial and Boundary Conditions

A physiological waveform was used in the unsteady flow simulations. A physiologically realistic volume flow rate versus time curve for a single heart pulse is shown in Figure 7. The pulse consists of a short systolic part (left ventricular contraction) with a rapid increase and decrease of the flow rate, and a longer diastolic part (left ventricular relaxation) where the flow changes are much smaller. A time step of 0.01 second was used for the systolic part of the pulse, while for the smoother diastolic part of the pulse the time step was increased to 0.02 second to save computational time. The magnitude of the inlet velocity was calculated at each time step using the volume flow rate and the approximate cross-section area of the common carotid at the inlet. To specify the time-dependent boundary conditions it was assumed that only the stream-wise velocity component is present at the inlet and its magnitude at each time step is equal to the inlet velocity value calculated from the waveform. The number of iterations necessary to reach convergence were carried out at each time step, and the values thus obtained were used as the initial condition for the next time step.

### 5.2 Unsteady Velocity Distribution

The "stream-wise" velocity field is shown in Figure 8 and the instantaneous streamlines in Figure 9. In the systolic part of the pulse a rapidly growing high-speed jet is formed at the stenosis in the internal carotid artery, and a large recirculation zone can be observed downstream of the plaque. As the flow accelerates the jet extends along the inner internal carotid wall. There is also a recirculation zone in the carotid sinus. During diastole the jet diminishes and the flow in the carotid sinus is more stagnant. The streamlines at systole change dramatically with each time step. At diastole the streamlines are still fairly unsteady but the changes are not that abrupt. It is interesting that a streamline passing through the internal carotid artery can then switch to the external carotid and back.

## 6 Non-Newtonian Flow

### 6.1 Power Law

It was recently shown (Lowe et al., 1997; Sloop and Garber, 1997) that atheroscletic disease could cause blood to exhibit some non-Newtonian properties. The non-Newtonian properties of blood can influence the flow in low shear regions, such as the recirculation zone downstream of the stenotic plaque. Steady and transient simulations were made to assess non-Newtonian effects in the flow. The viscosity is calculated from the following power law:

$$\mu = K\mu_0 \exp\left(a_1 T - a_2 T^2\right) \cdot \max(\dot{\gamma}, \dot{\gamma}_0)]^3, \text{ where } c = n - 1 + a_3 ln(\dot{\gamma}) + a_4 T$$

where $\gamma$ is the shear rate, T is the fluid temperature, $n = 0.8$, $K = 1$, $\mu_0 = 0.0035$ (SI), and the cutoff shear rate is allowed to vary but will generally taken to be zero. Because blood viscosity does not depend strongly on temperature in the physiological range, the constant coefficients $a_i$ are taken to be zero.

## 6.2 Non-Newtonian velocity distribution

Both steady and unsteady simulations were performed for the non-Newtonian flow. Comparison between a Newtonian and non-Newtonian steady flow at a Reynolds number of 200 is shown in Figure 10. Similar velocity fields can be observed in both cases, but the magnitudes of the velocities in the non-Newtonian case are smaller. The jet for the non-Newtonian flow is less pronounced but the recirculation area downstream of the plaque is increased. This is consistent with the findings of Stroud et al. (2001) for two-dimensional stenotic flow in the same geometry.

## 6.3 Unsteady Non-Newtonian Flow

The systolic phase of the physiological waveform has been simulated using the above non-Newtonian viscosity. Similar to the steady case, non-Newtonian properties have a damping effect on the development of the flow in systole, so the velocity field resembles the one obtained in the Newtonian flow simulations but the magnitudes of the velocity and pressure are smaller.

## Acknowledgement

The authors wish to acknowledge the support of NIH Grant HL 61823.

## References

Berger, S.A. and Jou, L-D., 2000, Flows in Stenotic Vessels, *Annual Review of Fluid Mechanics*, Vol. 32, pp. 347-382.

Stroud J.S., Berger S.A., and Saloner D., 2002, "Numerical analysis of flow through a severely stenotic carotid artery bifurcation," *Journal of Biomechanical Engineering* (ASME),(in press, to appear in Feb. 2002).

Stroud, J.S., 2000, *Numerical Simulation of Blood Flow in the Stenotic Artery Bifurcation*," Ph.D. Thesis.

Lowe, G.D.O., Lee, A.J., Rumley, A., Price, J.F., and Fowkes, F.G.R., 1997, "Blood viscosity and risk of cardiovascular events: the Edinburgh Artery Study", *British Journal of Haematology*, Vol. 96, pp. 168-173.

Sloop, G.D. and Garber, D.W., 1997, "The effects of low-density lipoprotein and high-density lipoprotein on blood viscosity correlate with their association with risk of atherosclerosis in humans", *Clinical Science*, Vol. 92: 473-479.

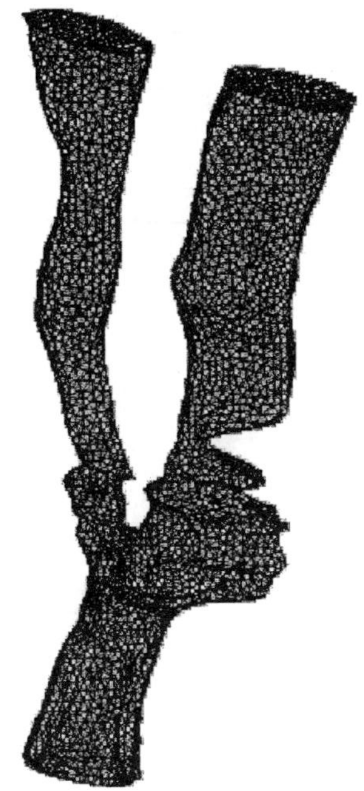

Figure 1: Three-Dimensional Model of Stenotic Carotid Bifurcation.

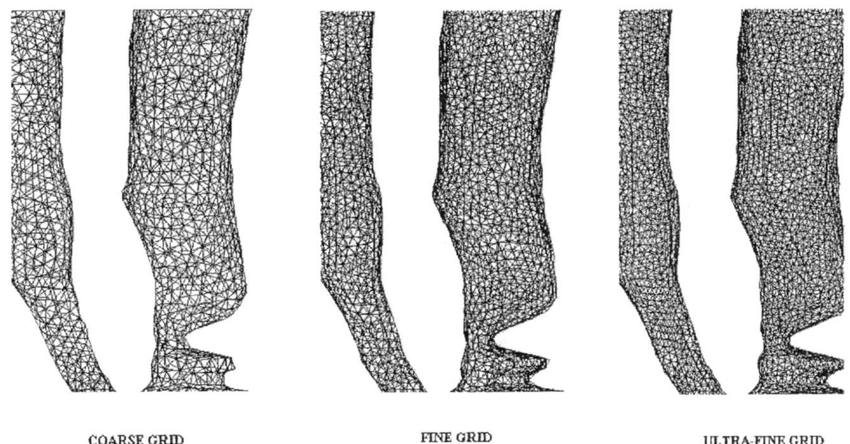


Figure 2: Three-dimensional grids of various densities.

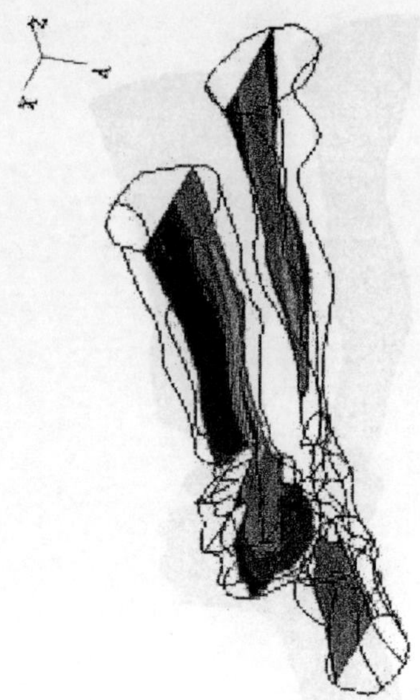

Figure 3: Cross-section plane.

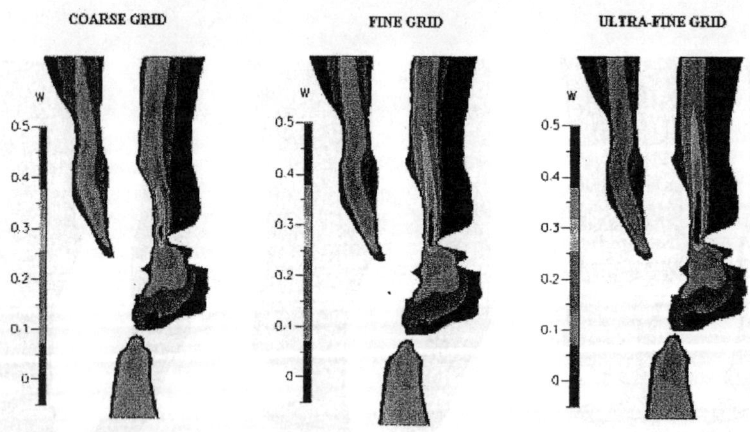



Figure 4: Velocity Distribution for Steady Laminar Newtonian Flow at Reynolds number of 200 for various grid densities.

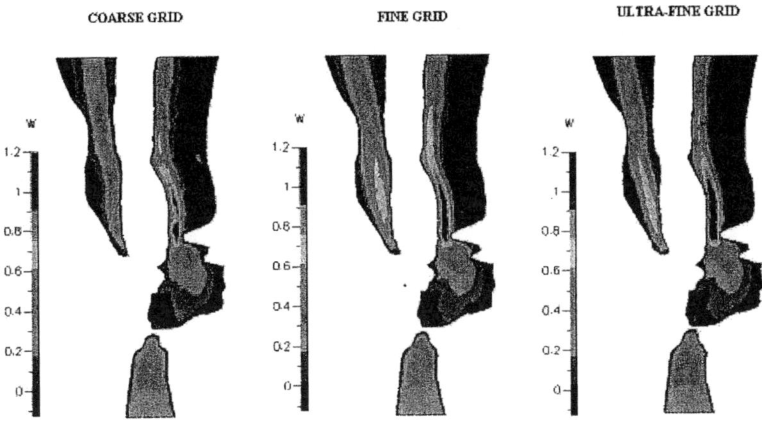



Figure 5: Velocity Distribution for Steady Laminar Newtonian Flow at Reynolds number of 600 for various grid densities.

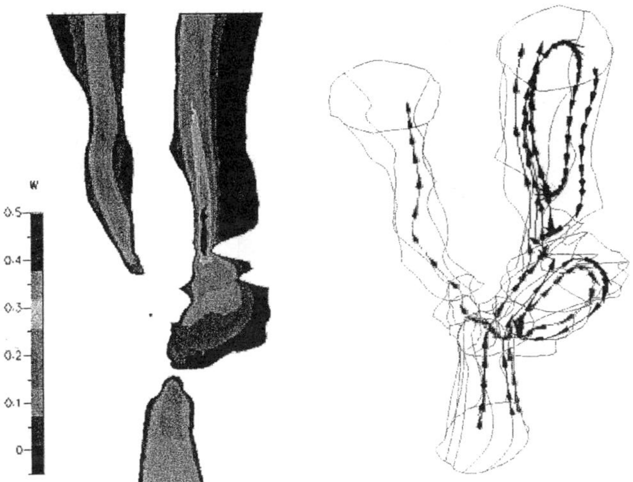



Figure 6: Steady Newtonian flow at Reynolds number of 200. "Stream-wise" velocity distribution and 3D streamlines.

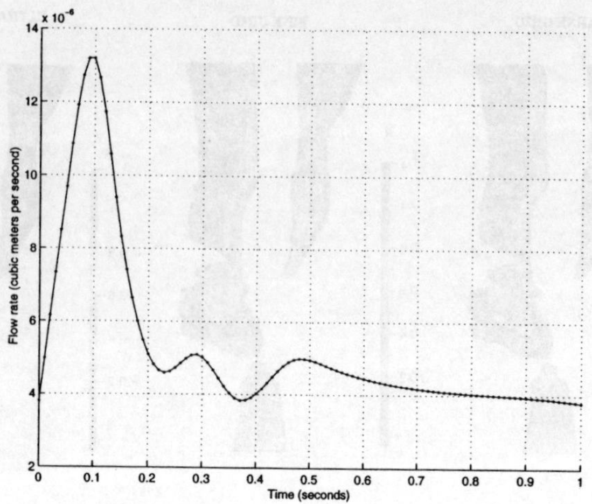

Figure 7: Physiological Waveform Used for Pulsatile Flow Simulations.

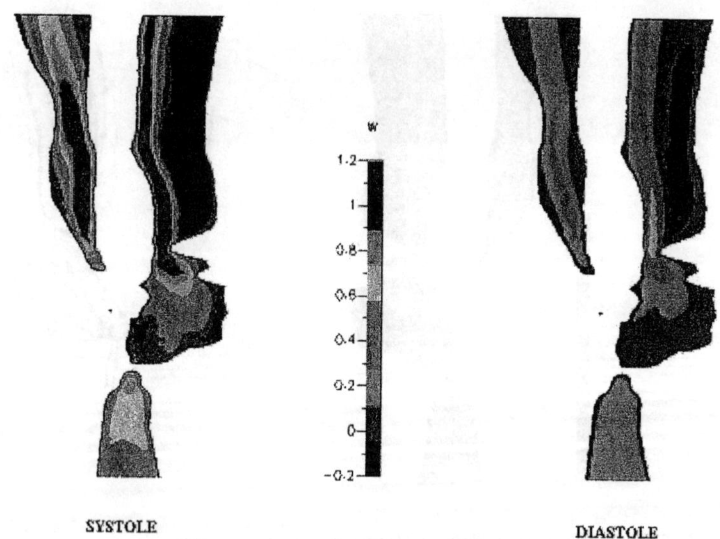



Figure 8: "Stream-wise" Velocity Distribution at Peak Systole and Diastole for Unsteady Newtonian Flow.

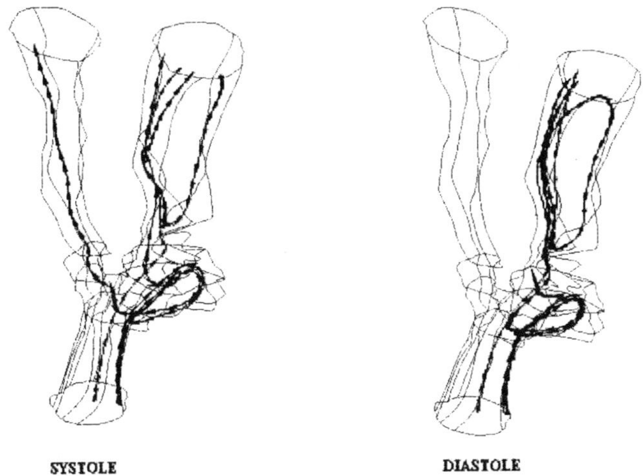



Figure 9: Instantaneous Flow Streamlines at Peak Systole and Diastole for Unsteady Newtonian Flow.

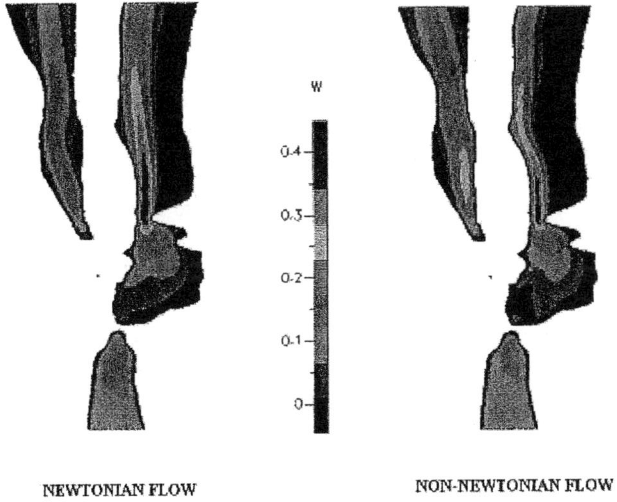



Figure 10: Steady Newtonian and non-Newtonian flows at Reynolds number of 200. "Streamwise" velocity distribution.

# Direct Simulation of the Vaporization of Droplet Arrays

M. T. Parra* ,H. A. Dwyer, ***, and F. Castro*
Department of Mechanical and Aeronautical Engineering
University of California, Davis
Davis, California 95616

*Departamento de Ingeniería Energética y Fluidomecánica
University of Valladolid, Spain

## Summary

The objective of this work is to determine the behavior of a group of droplets interacting at an intermediate Reynolds flow. The algorithm CHIMERA has been used to accomplish the direct simulation of the vaporization of water droplets.

Results for an isolated drop have been compared with available results to validate the model. We have analyzed the global transport coefficient for groups of droplets in order to determine the agreement with experimental correlations that are commonly used in classic droplet vaporization models.

## 1. Introduction

There are different many applications of spray technology to problems in heat exchange, power and materials processing, and the present work focuses on a water spray problem. A water spray barrier is an effective flame extinction mechanism. When the water droplets contact the hot gas mixture, they absorb energy from the gas mixture changing from liquid to steam, which means an increase of around 1700 times its volume. Hence the concentration of oxygen decreases on the flame surroundings and the reaction rate is reduced. The slow vaporization rate of water provides a large droplet lifetime.

Detailed numerical and experimental vaporization studies of single droplet and droplet groups behavior are available. These studies are necessary to develop an understanding of the physical processes of evaporation and mixing. Due to relative motion between the droplet and the ambient gas, boundary layers and wakes appear on the flow. Therefore, every droplet is under the influence of the neighboring droplets and its behavior will differ from the one of the isolated droplet.

In a spray, there are many droplets with a range of concentrations. It is not possible to treat the detailed interaction between the gas phase and the disperse phase. There are different formulations of the spray equations such as two-continua formulation, discrete particle or the probabilistic formulations. Both of them consider the interaction between the phases as source terms for drag, droplet break-up, heating and vaporization models. These approximate models utilize as inputs transport coefficients obtained from experimental or numerical computation correlations. So that, drag coefficient, Nusselt and Sherwood numbers are evaluated according the range of variation of the Reynolds, Prandtl and Schmidt numbers.

The nature of direct simulation research provides additional data for further verification of numerical codes. The main goal of the present work is to make the direct simulation of a group of droplets interchanging mass, momentum and energy throughout the gas. The results will be used to test the agreement between the transport coefficients and the correlations developed for conventional vaporization model of sprays simulations.

## 2. Numerical model

The correct performance of a direct simulation requires a high spatial resolution. The use of an overset grid provides substantial advantages such as the use of orthogonal meshes to fit the spherical droplets. Another feature is that the technique is a natural domain decomposition scheme in case parallel processing computations are considered.

The overset technique used is the CHIMERA grid scheme which uses separate meshes in different regions of a domain, [1]. For the simulations presented, the Chimera scheme produces a large major mesh for the gas phase, N droplet meshes and N surrounding droplet meshes that capture the viscous and thermal boundary layer.

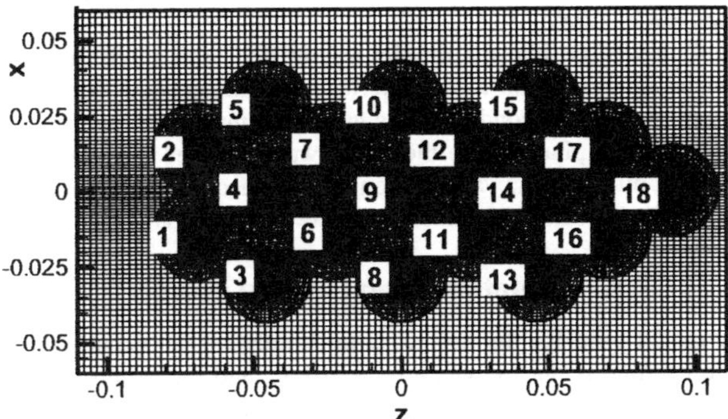

Figure 1. Mesh for the droplet array at a separation center to center of 3,5 D

The domain is 1,2 × 0,5 × 2,2 mm with and spatial resolution of 81 × 31 × 101 grid points. Every droplet has a diameter of 75 microns. They are surrounded by a spherical mesh of 21 × 31 × 21 nodes with a diameter 4 times the droplet diameter. The droplets have a separation center to center around 3,5 times the droplet diameter.

The equations solved are the Navier-Stokes equations for a three dimensional, viscous and low Mach number flow on a time dependent mesh to adapt the minor meshes to the droplet size. It is expected that the droplets will have different vaporization rates.

$$\frac{d}{dt}\iiint_v \rho\,dv + \iint_A \rho\left(\vec{V} - \vec{V}_b\right)d\vec{A} = 0 \tag{1}$$

$$\iiint_v \rho\frac{\partial u}{\partial t}dv + \iiint_v \rho\left(\vec{V} - \vec{V}_b\right)\vec{\nabla}u\,dv = \iint_A P\vec{i}\,d\vec{A} + \iint_A \vec{i}\,\overline{\overline{\tau}}d\vec{A} \tag{2}$$

$$\iiint_v \rho\frac{\partial v}{\partial t}dv + \iiint_v \rho\left(\vec{V} - \vec{V}_b\right)\vec{\nabla}v\,dv = \iint_A P\vec{j}\,d\vec{A} + \iint_A \vec{j}\,\overline{\overline{\tau}}d\vec{A} \tag{3}$$

$$\iiint_v \rho\frac{\partial w}{\partial t}dv + \iiint_v \rho\left(\vec{V} - \vec{V}_b\right)\vec{\nabla}w\,dv = \iint_A P\vec{k}\,d\vec{A} + \iint_A \vec{k}\,\overline{\overline{\tau}}d\vec{A} \tag{4}$$

$$\iiint_v \rho c_p \frac{\partial T}{\partial t} dv + \iiint_v \rho c_p \left(\vec{V} - \vec{V}_b\right)\vec{\nabla}T dv =$$

$$= \iint_A k\vec{\nabla}T d\vec{A} - \iiint_v \rho \sum_{i=1}^{kk} Y_i c_{p,i} \vec{V}_i \vec{\nabla}T dv - \iiint_v \sum_{i=1}^{kk} h_i \dot{\omega}_i W_i dv \qquad (5)$$

$$\iiint_v \rho \frac{\partial Y_i}{\partial t} dv + \iiint_v \rho \left(\vec{V} - \vec{V}_b\right)\vec{\nabla}Y_i dv = \iint_A \rho Y_i \vec{V}_i d\vec{A} + \iiint_v \dot{\omega}_i W_i dv \qquad (6)$$

The gas properties and transport coefficients [2] are obtained from the library of subroutines CHEMKIN, whereas the liquid properties are calculated according to [3]. The solutions were obtained from a predictor/corrector algorithm developed by Dwyer [4]. It uses a second order accuracy scheme without the use of artificial diffusion, and the temporal resolution is based on the diffusive time scale. More details of the resolution of the equations as well as different applications can be found on [5 and 6].

The internal circulation into the droplet was treated with an enhanced one-dimensional model, and the species concentration at the interface was determined from the thermodynamic equilibrium condition. The inflow condition was uniform temperature and pressure conditions of 1000 K and 1 bar in the gas phase while the initial water droplet temperature was 300 K. The simulations were carried out until the vaporization rate reached a quasi-steady state. The criterion used is that the last droplet heat flux does not change the third decimal.

## 3. Validation case: Isolated droplet

To validate the model used for the droplet array simulation, an isolated droplet was studied with incompressible conditions, gas temperature = 301 K, and the results were compared to the available data. Some numerical calculations of the flow past the isolated droplet have been made for Reynolds number of 1, 20, 50 and 100. The agreement between these calculated flow patterns and experimental results were good.

Figure 2 shows the distribution of the dimensionless pressure at surface of droplet versus the angle from the stagnation point at Reynolds number equal to 100. The solid line is the numerical result for the water droplet in air proposed by Woo, [7]. It is noticed that the minimum pressure occurs at the side of the droplet, since it is evident that the speed is maximum there. It is possible to evaluate the pressure and the viscous contribution to the drag coefficient from the local pressure and the shear forces at the surface of the droplet. These values are evaluated according the equations 7.

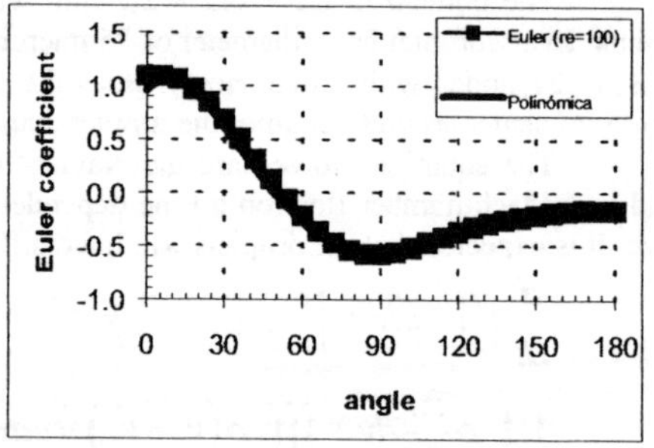


Figure 2. Surface distribution of $(P_s-P_\infty)/(\rho V^2_\infty/2)$ @ Re = 100

$$C_{DP} = \frac{\iint_A \left(-P\vec{n}\right)d\vec{A}}{\rho V_\infty A_P/2} \qquad C_{DV} = \frac{\iint_A \left(\bar{\tau}\vec{s}\right)d\vec{A}}{\rho V_\infty A_P/2} \qquad (7)$$

Table 1 presents the pressure and viscous drag coefficients at Reynolds numbers of 50 and 100 as well as the error with reference to the data for water droplets proposed on [7]. The total drag coefficient is the sum of both.

*Table 1. Drag coefficients for water drops in air under incompressible conditions*

|         | Reynolds = 50 | | | Reynolds = 100 | | |
|---------|---------|--------|---------|---------|--------|---------|
|         | Ref. [7] | Model | Error | Ref. [7] | Model | Error |
| $CD_P$  | 0,64 | 0,7443 | 16,30 % | 0,49 | 0,5629 | 14,88 % |
| $CD_V$  | 0,92 | 0,9222 | 0,24 % | 0,59 | 0,5829 | - 1,20 % |
| $CD_T$  | 1,56 | 1,6665 | 6,83 % | 1,08 | 1,1458 | 6,09 % |

The local Nusselt number is the local heat flux on the interface expressed by unit of area normalized with a reference heat flux based on the conditions of free stream and stagnation point:

$$Nu = \frac{q}{k(T_\infty - T_s)/D} \qquad (8)$$

Figure 3 show the distribution of the local Nusselt number versus the angle to the stagnation point at different Reynolds numbers. The largest Nusselt number is located at the stagnation point where the temperature gradient is the highest. At Reynolds unity, there is no flow separation and the Nusselt number decreases slightly from front to rear. At Reynolds number of 50, there is flow re-circulation that provides a slow increase of the Nusselt number on the rear of the droplet. At Reynolds 100, the minimum Nusselt number is clearly located at 138°, which means after the flow separation.

Another kind of validation is based on the length of the standing eddies that appear at the rear of most bodies due to the flow separation, [8]. Dots on figure 4 present the Chimera results. The region of closed streamlines behind a sphere forms, according to experimental results, at about Reynolds number of 24. As Reynolds number increases, the recirculating wake ground up. The errors are 7.6 % and 15 % for Reynolds number of 100 and 50 respectively. Figure 5 shows the streamlines on the rear of the droplet for different Reynolds numbers ranging from 1 to

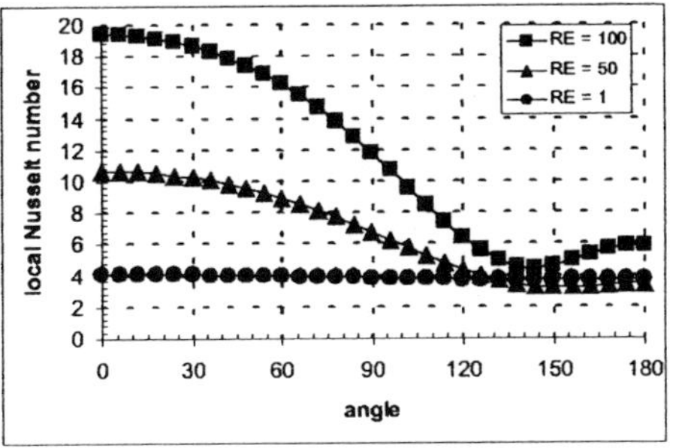


Figure 3. Surface distribution of Nusselt number, @ $T_\infty = 301$ K

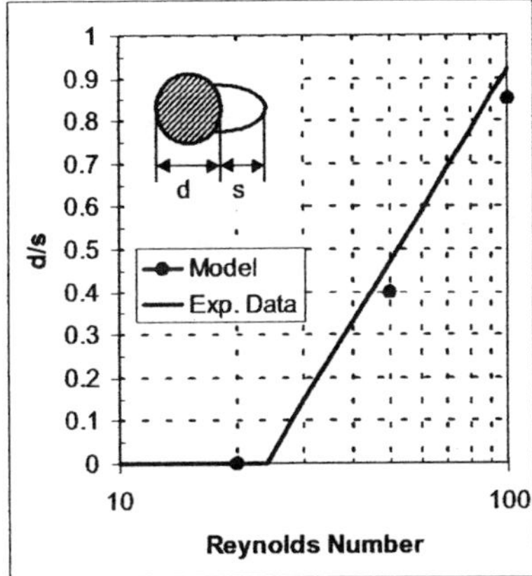


Figure 4. Region of closed streamlines behind the droplet.

478

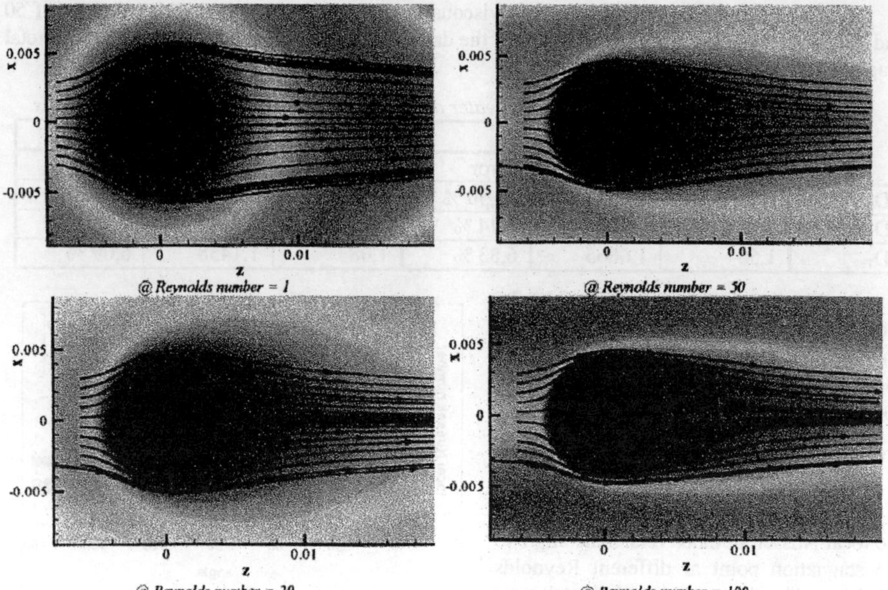

@ Reynolds number = 1

@ Reynolds number = 50

@ Reynolds number = 20

@ Reynolds number = 100

Figure 5. *Streamlines for a droplet at different Reynolds number*

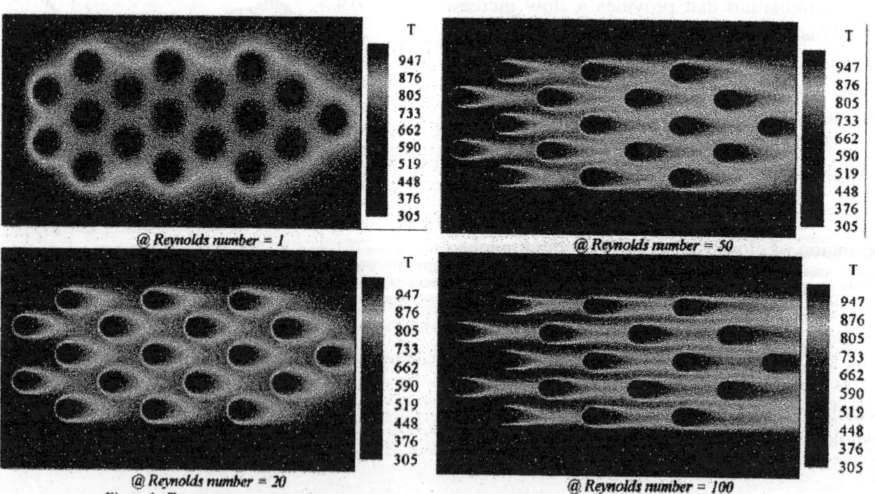



Figure 6. *Temperature contours for water droplet array in the symmetry plane at different Reynolds number*

## 4.  Vaporization of a droplet array

A droplet array involves a few interacting droplets that interchange mass, momentum and energy throughout the gas at high temperature. A decrease in droplet spacing leads to an increase in vapor concentration and a depletion of gas temperature. Therefore there is a tendency to decrease the heat and mass transfer rates. The simulations have been carried on a group of droplets with a center to center separation around 3,5 times the droplet diameter. See figure 6 for the temperature contours at different Reynolds numbers. The transport coefficients present lower values than the corresponding with isolated droplet.

Figure 7 show the distribution of the local Nusselt number on different droplets located on the centered line of the array at a free stream Reynolds number of 100. It is clear that the rear droplets exhibit lower heat transport than the front droplet. This difference between consecutive droplets is less important for droplets on the rear of the array. Note: The droplet locations can be found from figure (1).

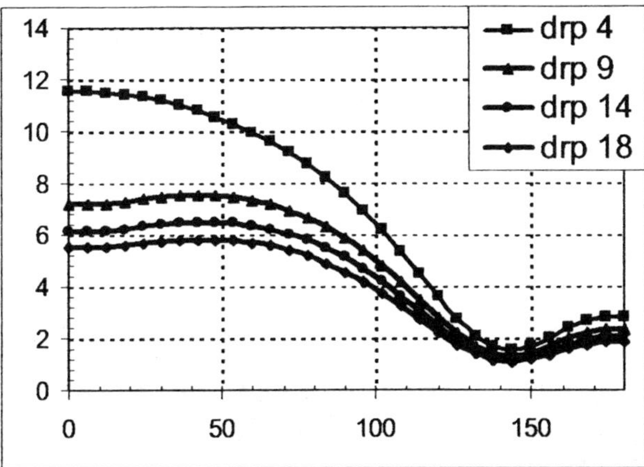


Figure 7. Distribution of the local Nusselt number versus the
angle to the stagnation point, @ Re = 100 and $T_\infty$ = 1000 K

The global Nusselt number of each droplet has been obtained from averages on the surface area.

$$Nu = \frac{\iint_A q\bar{n}\, d\bar{A}}{\pi k D\left(T_\infty - T_s\right)} \tag{9}$$

The influence of the Reynolds number on the global Nusselt number is shown on figure 8. The Nusselt number is represented for the droplets of the centered line of the array. Downstream droplets have lower values than that of the lead droplet due to each droplet in on the wake of the preceding one. The decrease of the Nusselt number on the rear region is important for intermediate Reynolds numbers characterized by a large wake. Lower Reynolds numbers do not have strong wakes and therefore downstream droplets suffer less effect of the droplets ahead.

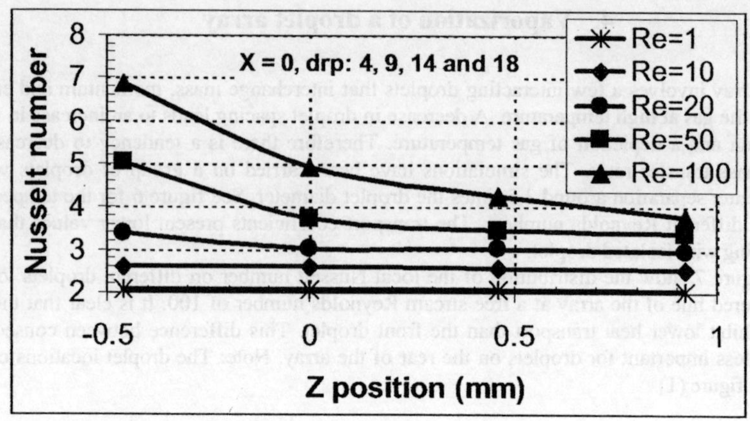



Figure 8. Global Nusselt number of the centered line droplets at different Reynolds number

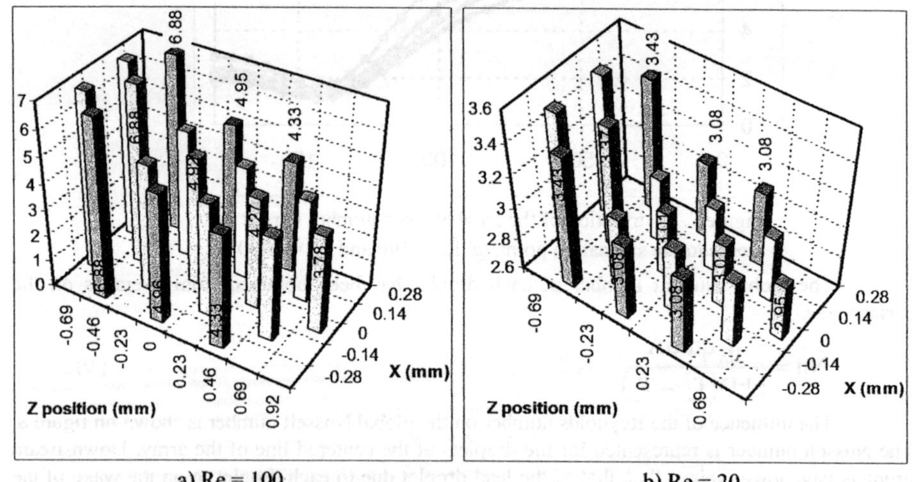



a) Re = 100                                    b) Re = 20

Figure 9. Global Nusselt number versus droplet location

In figure 9a, the global Nusselt number at Reynolds number of 100 show that the decrease on the Nusselt number is significant on the main flow direction. The interaction between droplets side by side is less important. The internal droplet has a slightly lower coefficient than the external droplets. At lower Reynolds number, see picture 9b for Re = 20, the interaction of parallel droplets is more important than the interaction on the flow direction, due to the boundary layer is thicker and the recirculation region is shorter. Vaporization model

In a conventional spray simulation, it is common the use of momentum, heat and mass transfer models to simulate the interaction between the disperse phase and the gas flow.

$$F_D = C_D \frac{1}{2} \rho V_r^2 A_P \tag{10}$$

A classic droplet vaporization model is the Spalding one, [9, 10 and 11], which was initially used in the KIVA code, as well as the PHOENICS code.

$$\frac{dm_{H2O}}{dt} = +\pi D \, \rho D_m \, ShB_M \tag{11}$$

$$\dot{Q} = \pi D K \, Nu \left( T_G - T_{H2O} \right) \tag{12}$$

That involves the necessity of correlations for the drag coefficient as well as the Nusselt and Sherwood numbers.

An important problem is the estimation of the fluid properties for correlations. It is common to use an averaging rule for weighting the surface properties heavier than the free stream properties. Abramzon and Sirignano recommend properties were evaluated at conditions that are 1/3 the free stream conditions and 2/3 of the droplet surface conditions, [9].

Chiang [12 and 13] proposed the following results for isolated droplets at Reynolds lower than 160 and higher than 10:

$$C_{D,iso} = \left(1 + B_H\right)^{-0,27} \frac{24,432}{Re^{0,721}} \tag{13}$$

$$Nu_{iso} = 1,275 \, Re^{0,438} \, Pr^{0,619} \left(1 + B_H\right)^{-0,678} \tag{14}$$

Reynold and Prandtl numbers are based on the 1/3 average gas-film properties except the free stream density, which is used for the Reynolds number.

Chiang and Sirignano [12 and 13] analyzed groups of two or three droplets where the lead droplet behaves according to the following correlations where *Dist* is the spacing between droplet centers made non-dimensional with the droplet radius and ranging from 2,5 to 32. *Siz* is the ratio of the droplet diameters ranging from 0,17 to 2,0.

$$C_{D,1} = C_{D,iso} \, 0,877 \, Re^{0,003} \left(1 + B_H\right)^{-0,04} Dist^{0,048} Siz^{-0,098} \tag{15}$$

$$Nu_1 = Nu_{iso} \, 1,245 \, Re^{-0,073} \, Pr^{0,15} \left(1 + B_H\right)^{-0,122} Dist^{0,013} Siz^{-0,056} \tag{16}$$

The down stream droplet follows the correlations:

$$C_{D,2} = C_{D,iso} \, 0,549 \, Re^{-0,098} \left(1 + B_H\right)^{0,132} Dist^{0,275} Siz^{0,521} \tag{17}$$

$$Nu_2 = Nu_{iso} \, 0,528 \, Re^{-0,146} \, Pr^{-0,768} \left(1 + B_H\right)^{0,356} Dist^{0,262} Siz^{0,147} \tag{18}$$

482

Chiang and Sirignano found out that there is a slight difference between the second and the third droplet coefficients. Figure 10 present the transport coefficients at a Reynolds number ranging from 10 to 100 produced by the correlations (plotted in solid lines) and the direct simulation carried out with the Chimera code (dot symbols). The subscript –iso refers to the isolated droplet, -1 to the front droplet and –2 to the second droplet.

There is a reasonable agreement between the Nusselt numbers obtained from the Chimera code and those provided by the Chiang and Sirignano correlations. Nusselt number for the lead droplet is slightly lower than that of the correlation. The heat flux for the droplets of the rear region are fairly well predicted.

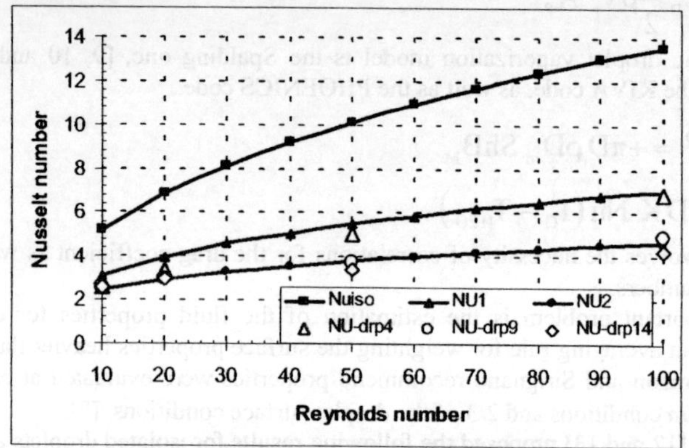


Figure 10. Nusselt number provided by the correlations and the direct
simulations for droplets separated 3,5 D

With reference to the drag coefficient, the front droplet behaves in a similar way to an isolated droplet. Results from the Chimera code tend supply a slightly bigger drag coefficient, such as it happens with the validation test, figure 11.

It is possible to conclude that the direct simulation is a reasonable procedure to validate the use of a correlation in a certain range of variation of the parameters in order to use it in a spray model.

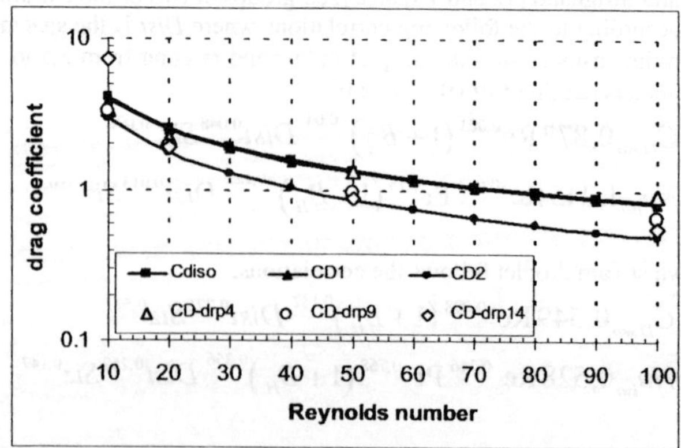


Figure 11. Drag coefficients provided by the correlations and
the direct simulations for droplets separated 3,5 D

## 5. CONCLUSIONS

Groups of droplets have been modeled with the use of an overset grid to accomplish the direct simulation of a high-temperature gas flow past a group of water droplets. Results for an isolated drop have been contrasted with available results in order to validate the model. It has been analyzed the interaction between the droplets at different Reynolds numbers. Finally, the transport coefficients have been compared with experimental correlations. It seems the direct simulation results match well with the experimental results.

## References


[1] Shahcheraghi N.: *Numerical Study of Three Dimensional Incompressible Flow and Heat Transfer Over a Stationary and Moving and Rotating Sphere in a Pipe*. Ph. D. Dissertation of the University of California, Davis (1996).

[2] Kee R. J., Dixon-Lewis G., Warnatz J., Coltrin M. E. and Miller J. A.: *A Fortran Computer Code Package for the Evaluation of Gas-Phase, Multicomponent Transport Properties*. Sandia National Laboratories Rep. SAND 86-8246 (1986).

[3] Reid R. C., Prausnitz J. M. and Poling B. E.; Bubbles: *The Properties of Gases and Liquids*. McGraw-Hill (1987).

[4] Dwyer H. A.: *Calculations of droplets dynamics in high temperature enviroments*. Prog. Energy Combust. Sci. 15:135 (1989).

[5] Dwyer H. A., DeBus K. and Shahcheraghi N.: *The Use of Overset Meshes in Particle and Porous Media Three-Dimensional Flows*. International Journal for Numerical Methods in Fluids 31:393-406 (1999).

[6] Dwyer H. A., Stapf P. and Maly R.: *Unsteady Vaporization and Ignition of a Three-Dimensional Droplet Array*. Combustion and Flame 121:181-194 (2000).

[7] Clift R., Grace J. R. and Weber M. E.: *Bubbles, Drops and Particles*, 1$^{st}$ edn. Academic Press, New York (1978).

[8] Batchelor, G. K: *An introduction to fluid dynamics*. Cambridge University Press. (2000)

[9] Curtis E. W., Uludogan A. and Reitz R. D.: *A new high pressure droplet vaporization model for diesel engine modeling*. Society of Automotive Engineers, 952431. (1995)

[10] Reitz R. D.: *Modeling atomization processes in high-pressure vaporizing sprays*. Atomisation and Spray Technology, Vol. 3, pp 309-337. (1987).

[11] Fueyo N., Hamill I. and Zhang Q., *The Gentra user guide*, CHAM Report TR/211. (1992).

[12] Sirignano W. A.: *Fluid dynamics of sprays--1992 Freeman scholar lecture*. Journal of Fluids Engineering, Vol. 115, pp.345-378. (1993).

[13] Sirignano W. A.: *Fluid Dynamics and Transport of Droplets and Sprays*. Cambridge (1999).

# Incompressible Simulations on the Flowfield
# Around a Square Cylinder at Ground Proximity

Dong-Ho Lee[‡], Jaeho Hwang[*]
Seoul National University, Seoul 151-742, Korea

## Abstract

This paper presents unsteady computations on the flow field around a square cylinder with a gap between the body and the ground plane. Two-dimensional unsteady, incompressible Navier-Stokes codes are developed for the computation of the viscous turbulent flows. The incompressible code with pseudo-compressibility and dual-time stepping method adopts a conventional upwind differencing scheme for the convective terms and DP-SGS scheme for efficient temporal integration. The grid system around a two-dimensional square body is efficiently generated by adopting Chimera hole concept using IB array. By computing the flow around a square cylinder without ground effect, three two-equation turbulence models are evaluated and the developed code is validated. The results show a good agreement with experimental values and other computational results. Critical gap height at which the formation of Kàrman vortex streets is interrupted, is demonstrated by the present numerical investigations.

## I. INTRODUCTION

In designing structurally safe large-scale civil structures or small-scale high-speed electronic parts, the dynamic structural behavior in response to unsteady wind loadings should be investigated. Since many objects in real world are bluff bodies, flow separates, reattaches, and forms an unsteady vortex in the wake region behind the bodies. These complex flow phenomena would be more complicate if a gap between a bluff body and another wall-which is generally ground for a civil structure or a ground vehicle-is introduced. The present work, in order to investigate the gap effect on the wind loading, is focused on the development of the time-accurate flow solver and its application to the various configurations. In general, flow around civil-structures can be assumed incompressible because the free stream Mach number is

‡ Professor, Head, School of Mechanical and Aerospace Engineering, Senior Member AIAA.

* Post Doctor, School of Mechanical and Aerospace Engineering

less than 0.1 and Reynolds numbers are usually 0.01 to 0.1 million. Thus, incompressible Navier-Stokes equations are used for efficient computation.

So far, many turbulence models such as algebraic models, one-equation models, and two-equation models have been developed for accurate computations of various turbulent flows. But, it seems to be impossible to accurately predict engineering flows to the fine details of turbulence, even with the most advanced computers. From the engineering point of view, unpractical direct numerical simulation(DNS) or large-eddy Simulation(LES) is not considered for the current application. The several $k - \varepsilon$ models involving different wall functions for the accurate calculation of the inner boundary layer has been developed and showed good results. Unlike the $k - \varepsilon$ model, the $k - \omega$ model shows a good property in the sublayer without wall functions. Thus, the $k - \omega$ model has superiority to other models for its simplicity, especially in parallel programming. On the other hand, the $k - \varepsilon$ model has the advantage of the freestream independence in the outer boundary layer while the $k - \omega$ model is highly sensitive to freestream values. Menter's base line model (BSL) was developed to achieve these two desired features in the sublayer and the outer boundary layer. This BSL model is similar to the original $k - \omega$ model, but has no freestream dependency. The $k - \omega$ shear stress transport (SST) model has the modified eddy viscosity to account for the transport of turbulent shear stress. The $k - \omega$ SST model was found to have the property of the original $k - \omega$ model in the near wall region, and the advantage of the free-stream independence of $k - \varepsilon$ model in the outer of boundary layer.

In the present study, the incompressible codes with different turbulence models are examined via numerous computations. The codes are parallel-processed using MPI(Message Passing Interface) programming method. The incompressible code is validated by predicting the flow around a square cylinder without ground effect. After the validation, flowfields around a square cylinder at ground proximity on six gap heights(G/D = 0.1, 0.2, 0.3, 0.4, 0.5, 1.0) are numerically investigated and discussed.

## II. NUMERICAL PROCEDURE

### Governing equations
The governing equations are two-dimensional unsteady incompressible Navier-Stokes equations, and are written as

$$\frac{\partial u_k}{\partial x_k} = 0$$

$$\frac{\partial u_i}{\partial t} + \frac{\partial}{\partial x_j}\left(u_j u_i\right) = -\frac{\partial \hat{p}}{\partial x_i} + \frac{\partial}{\partial x_j}\left[\left(\mu + \mu_t\right)\left(\frac{\partial u_i}{\partial x_j} + \frac{\partial u_j}{\partial x_i}\right)\right] \tag{1}$$

## Turbulence Models

The original $k-\omega$ model of Wilcox[1,2], the standard $k-\varepsilon$ model without wall functions, and Menter's $k-\omega$ SST model[3,4] are tested here to predict the viscous turbulent flow around a bluff body.

$k-\omega$ *model:*

$$\frac{D\rho k}{Dt} = \tau_{ij}\frac{\partial u_i}{\partial x_j} - \beta^* \rho \omega k + \frac{\partial}{\partial x_j}\left[\left(\mu + \sigma_{k_1}\mu_T\right)\frac{\partial k}{\partial x_j}\right]$$

$$\frac{D\rho\omega}{Dt} = \frac{\gamma_1}{\nu_T}\tau_{ij}\frac{\partial u_i}{\partial x_j} - \beta_1\rho\omega^2 + \frac{\partial}{\partial x_j}\left[\left(\mu + \sigma_{\omega_1}\mu_T\right)\frac{\partial \omega}{\partial x_j}\right] \tag{2}$$

where

$$\sigma_{k_1} = 0.5, \quad \sigma_{\omega_1} = 0.5, \quad \beta_1 = 0.075, \quad \beta^* = 0.09$$

$$\kappa = 0.41, \quad \gamma_1 = \beta_1 / \beta^* - \sigma_{\omega_1}\kappa^2 / \sqrt{\beta^*}$$

$k-\varepsilon$ *model:*

$$\frac{D\rho k}{Dt} = \tau_{ij}\frac{\partial u_i}{\partial x_j} - \beta^* \rho \omega k + \frac{\partial}{\partial x_j}\left[\left(\mu + \sigma_{k_2}\mu_T\right)\frac{\partial k}{\partial x_j}\right]$$

$$\frac{D\rho\omega}{Dt} = \frac{\gamma_2}{\nu_T}\tau_{ij}\frac{\partial u_i}{\partial x_j} - \beta_2\rho\omega^2 + \frac{\partial}{\partial x_j}\left[\left(\mu + \sigma_{\omega_2}\mu_T\right)\frac{\partial \omega}{\partial x_j}\right] + 2\rho\sigma_{\omega_2}\frac{1}{\omega}\frac{\partial k}{\partial x_j}\frac{\partial \omega}{\partial x_j} \tag{3}$$

where

$$\sigma_{k_2} = 1.0, \quad \sigma_{\omega_2} = 0.856, \quad \beta_2 = 0.0828, \quad \beta^* = 0.09$$

$$\kappa = 0.41, \quad \gamma_2 = \beta_2 / \beta^* - \sigma_{\omega_2}\kappa^2 / \sqrt{\beta^*}$$

$k-\omega$ *SST model:*

$$\frac{D\rho k}{Dt} = \tau_{ij}\frac{\partial u_i}{\partial x_j} - \beta^* \rho\omega k + \frac{\partial}{\partial x_j}\left[(\mu + \sigma_k \mu_T)\frac{\partial k}{\partial x_j}\right]$$

$$\frac{D\rho\omega}{Dt} = \frac{\gamma}{\nu_T}\tau_{ij}\frac{\partial u_i}{\partial x_j} - \beta\rho\omega^2 + \frac{\partial}{\partial x_j}\left[(\mu + \sigma_\omega \mu_T)\frac{\partial\omega}{\partial x_j}\right] + 2(1-F_1)\rho\sigma_{\omega_2}\frac{1}{\omega}\frac{\partial k}{\partial x_j}\frac{\partial\omega}{\partial x_j} \quad (4)$$

The combined constants are calculated in the following relation

$$\Phi = F_1\Phi_1 + (1-F_1)\Phi_2$$

where $\Phi_1$ and $\Phi_2$ represent the constants of equations (2) and (3) except for $\sigma_{k_1} = 0.85$. $F_1$ is defined as follows.

$$F_1 = tanh(arg_1^4)$$

$$arg_1 = min\left[max\left(\frac{\sqrt{k}}{\beta^*\omega y}; \frac{500\nu}{y^2\omega}\right); \frac{4\rho\sigma_{\omega_2}k}{CD_{kw}y^2}\right]$$

where $y$ is the distance to the nearest surface and $CD_{k\omega}$ is the positive portion of the cross-diffusion term

$$CD_{k\omega} = max\left(2\rho\sigma_{\omega_2}\frac{1}{\omega}\frac{\partial k}{\partial x_j}\frac{\partial\omega}{\partial x_j}; 10^{-20}\right).$$

The eddy viscosity is defined as

$$\nu_T = 0.31k / max(0.31\omega; \Omega F_2)$$

where $\Omega$ is the absolute value of the vorticity and

$$F_2 = tanh\left(arg_2^2\right)$$

$$arg_2 = max\left(2\frac{\sqrt{k}}{\beta^*\omega y}; \frac{500\nu}{y^2\omega}\right)$$

For the $k-\varepsilon$ model and the original $k-\omega$ model, the eddy viscosity is defined as

$$\nu_T = k / \omega.$$

## Numerical Approach

The governing equations are solved with a finite-volume method. In the incompressible code with pseudo-compressibility and dual-time stepping method, convective terms are upwind-differenced based on Roe's Flux Difference Splitting (FDS) scheme[5]. A MUSCL (Monotone Upstream Centered Scheme for Conservation Laws) approach using the third order interpolation is used to obtain a higher order of spatial accuracy. The third order of spatial accuracy is kept in all calculations. For a temporal integration, Lee's DP-SGS[11] scheme was adopted which is a modified form of Yoon's LU-SGS[6] for MPP with domain decomposition. In the solver developed for the current work, the mean flow equations and the turbulence model equations are fully coupled for fast convergence and numerical stability.

Grid systems are clustered to the wall until $\Delta y_1^+$ is set to be the order of one. Wall boundary conditions are applied explicitly with no slip condition. The pressure is extrapolated from the interior points and the other variables are specified from the freestream values at the inflow, while the pressure is specified and the other variables are extrapolated from the interior points at outflow.

## Grid Treatment

As shown in Fig. 1, single block h-type grid system is used. In the figure, because square cylinder is located in the domain, it is difficult to generate a single block grid around square cylinder. On a sequential solver, multi-block grid system is proper choice for cost-efficient computation but a load balancing problem arises in parallel solver using multi-block grid system.

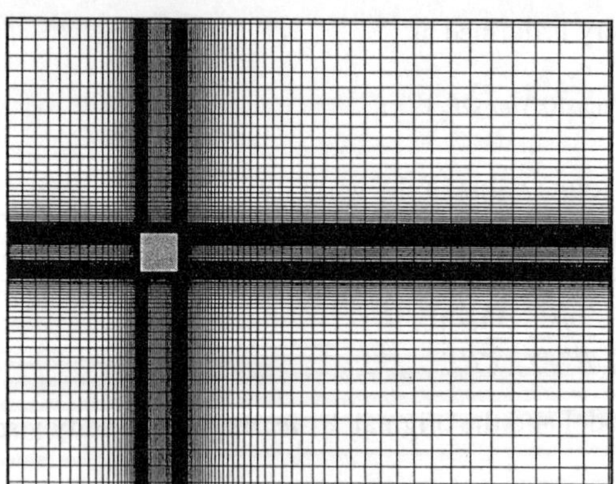

**Fig. 1 H-type Grid System**

Another choice is the unstructured grid system. But a square cylinder is geometrically simple, unstructured grid system isn't cost-effective. And the data structures of unstructured grid system is so complex, communication patterns between processors on MPP are also complex. Another difficulty on the unstructured grid system is its incapability of resolving wall vicinity.

For these reasons, the solver developed in this study adopts single block structured grid system with blanked cells by using IB array. This approach could be treated as a kind of Chimera[9] grid technique. In the domain, the cells in the bodies are marked as zero and the outer cells are marked as one.(see fig. 2) The only additional memory requirement is the array denotes which cells are blanked and minimal modifications of solution algorithm are needed. In the computation, the cells marked as zero are excluded automatically.

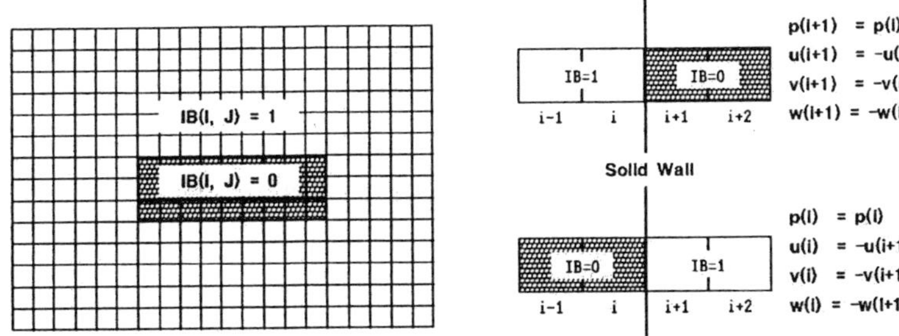


**Fig. 2 Application of IB Array**

As shown in fig. 2, at the cell interface where IB value is changed from zero to one or from one to zero, wall boundary conditions are applied during the calculation of fluxes. For simplicity, the mean flow boundary conditions are appeared in this figure. And time integration algorithm in LU-SGS is modified like below,

$$\left(LD^{-1}U\right)\Delta Q^{n} = -\Delta\tau R\left(Q^{n}\right)$$
$$Q^{*} = -\Delta\tau L^{-1}R\left(Q^{n}\right)\mathbf{IB}$$
$$Q^{**} = DQ^{*}\mathbf{IB}$$
$$\Delta Q = U^{-1}Q^{**}$$

(5)

At the cell where IB is zero, $\Delta Q$ is unchanged and excluded automatically in the solution processes. So no special treatments except for marking IB array are needed.

| Case, T/B Model | $\Delta t$ | $\Delta x$ | $C_L$ | $C_D$ | St |
|---|---|---|---|---|---|
| 1. $k-\varepsilon$ | 0.025 | 0.005 | 0.99 | 1.83$\pm$0.04 | 0.137 |
| 2. $k-\omega$ | 0.025 | 0.005 | 1.96 | 2.12$\pm$0.18 | 0.138 |
| 3. $k-\omega\,SST$ | 0.05 | 0.005 | 1.68 | 2.04$\pm$0.09 | 0.136 |
| 4. $k-\omega\,SST$ | 0.025 | 0.005 | 1.69 | 2.06$\pm$0.10 | 0.138 |
| 5. $k-\varepsilon$ [12] | 0.02 | 0.035 | 0.56 | 1.75$\pm$0.004 | 0.138 |
| 6. RNG $k-\varepsilon$ [12] | 0.02 | 0.035 | 2.08 | 2.12$\pm$0.23 | 0.133 |
| 7. Low Re $k-\varepsilon$ [12] | 0.05 | 0.015 | 1.13 | 2.01$\pm$0.018 | 0.138 |
| 8. Low Re $k-\varepsilon$ [12] | 0.02 | 0.0025 | 1.55 | 2.08$\pm$0.036 | 0.134 |
| 9. Low Re $k-\varepsilon$ [12] | 0.02 | 0.0012 | 1.61 | 2.10$\pm$0.04 | 0.134 |
| 10. LES [13] | 0.001 | 0.022 | 1.6 | 2.09$\pm$0.13 | 0.132 |
| 11. Exp. [12] | | | | 2.14$\pm$0.09 | 0.134 |

Table. 1 Force coefficients with various turbulence models

## III. RESULTS AND DISCUSSIONS

### Code Validation

For the validation of the current code, calculation of unsteady turbulent flowfields around two dimensional square cylinder without ground effect at Re = 22,000 is performed because experimental data and numerical results are widely available for this problem. In the computation, 129x117 H-type grid system as shown in fig. 1 is used and $\Delta y_1^+$ is set to be less than 6, Time step is set to 0.025 and 0.05, and sufficient sub-iterations to assess the accuracy of the results are performed during the computation. The cases for the validation are shown in Table. 1 denoted by 1, 2, 3, 4 respectively.

As shown in fig. 3-a), periodicity of aerodynamic force is achieved after 5-6 periods from the initial condition. This sinusoidal variation represents the periodic vortex formation on the surface and its shedding to the wake region. In figure 3-b), comparison between the adopted three turbulence models - $k-\varepsilon$ model, $k-\omega$ model and $k-\omega\,SST$ - is shown. It can be argued that the computed frequencies is not so different for three turbulence models unlike the amplitude of the variation and the absolute values.

Table. 1 lists the current computational results and other computational and experimental results by Lee[12] and Murakami[13] with respect to the aerodynamic force coefficient and Strouhal number(St $= fD/U$). Strouhal number is acquired through FFT of the sinusoidal aerodynamic force histories. For the case 1, computations using $k - \varepsilon$ model do not predict the force coefficients properly. More sub-iterations are required to satisfy the given convergence criteria between the time marching when $k - \varepsilon$ model is used, which could be explained by instability of the solution at the wake region because the cross diffusion term at the extremely large-aspect-ratio cell in the far down-stream induced the instability.

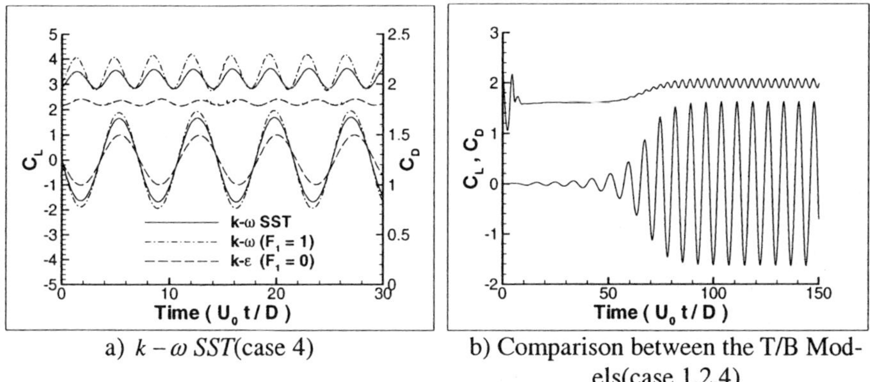

a) $k - \omega$ $SST$(case 4)      b) Comparison between the T/B Models(case 1,2,4)

**Fig. 3 History of force coefficients**

During the computations, it is experienced that $k - \omega$ SST and $k - \omega$ model is numerically more stable than $k - \varepsilon$ model, as suggested by Menter[4]. On the other hand, the cases of $k - \omega$ SST model show appropriate coincidence with experimental and LES results on both the aerodynamic force coefficients and the Strouhal number, and moreover they show satisfactory results with less clustered surface grid than the cases of $Low\ Re\ k - \varepsilon$ model. Fig. 4 depicts an instantaneous contour of turbulent kinetic energy for the case 4. It clearly shows the periodic vortex shedding to the wake region.

492

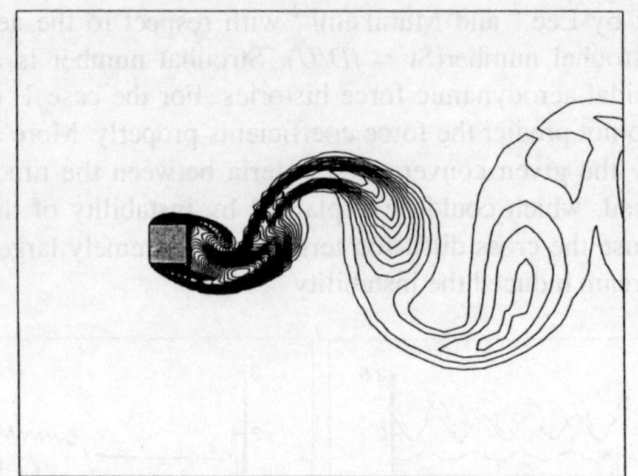

**Fig. 4 $k$ contours at maximum lift(case 4, $\Delta k$ = 0.005)**

## Gap Effects

The flow past 2-D square cylinder at ground proximity exhibits very different flow characteristics from the case where the body is placed in a free, uniform flow condition. The wake of a bluff body becomes completely random or steady when the gap between the bluff body and the plane wall is small, particularly smaller than a critical gap height.

The onset of vortex formation and regular vortex shedding occurs above a critical gap height, which depends on many parameters such as the Reynolds number, the width-to-height ratio, the blockage ratio, the level of free stream turbulence intensity and the on-coming wall boundary layer thickness.[14-18] The existence of wall boundary layer affects the flow under the body by inhibiting normal velocities and growing boundary layer along the surface.

In the case of a circular cylinder, Bearman and Zdravkovich[15] showed that the regular vortex shedding persisted down to a gap height of 0.3 times diameter of the cylinder. Taniguchi et al.[16] found that the critical gap height and the boundary layer thickness are correlated with each other, resulting in the critical gap height ranging 0.3 for thin boundary layer ($\delta/D < 0.4$) to 0.8 at $\delta/D \approx 1$ from measurements of drag, lift forces and flow visualization.

Only a few studies were carried out for a square cylinder placed near a wall. Kenjo et al.[14] investigated the influence of wall with a two-component Laser-Doppler Velocimeter (LDV) at Re = 23,000 and $\delta/D = 1.5$ with varying G/D. They presented that the vortex shedding was suppressed at G/D = 0.25 and the Strouhal number increases with decreasing gap height from 0.133 at free-standing condition to 0.164 at G/D =

0.3. Durao et al.[17] found that vortex shedding was absent for G/D < 0.35, whereas a predominant frequency was detected at G/D = 0.375, which for this and larger gap heights had a value of St = 0.133. Bosch et al.[18] carried out experiments at Re = 22,000 and for a constant boundary layer thickness $\delta$/D = 0.8, finding that the complete vortex shedding did not occur at G/D = 0.25 but perfect shedding at G/D = 0.5. They found that the critical gap height was determined to be in the range of G/D = 0.35 ~ 0.5 and there was no sharp transition from purely non-periodic flow to purely periodic flow.

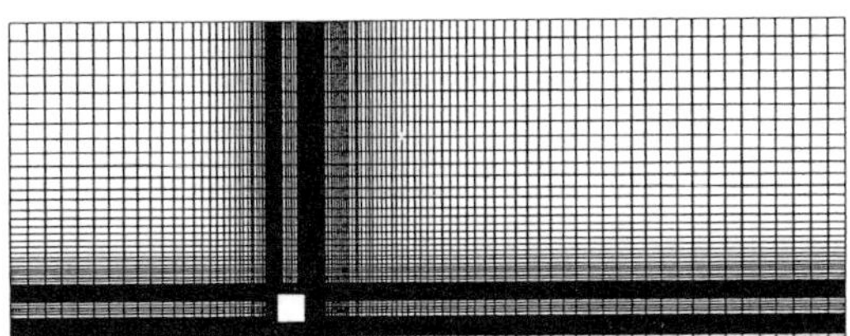

**Fig. 5 H-type Grid System(G/D = 1.0, 178x113)**

The present study deals with the numerical investigation of the flowfields around a square cylinder at ground proximity on the six gap heights(G/D = 0.1, 0.2, 0.3, 0.4, 0.5, 1.0). The objective of the current study is the correct computation of the critical gap height and the aerodynamic force histories- St and $C_D$ by using 2-equation turbulence models. Through the validation in the current study, we compute the flowfield at Re = 22,000 with the $k - \omega$ SST model. Time step is set to $\Delta t$ = 0.02 with sufficient subiterations and the grid resolution is set to $\Delta x$ = 0.005 for the computations.

The computational conditions for the 6 cases dealing the gap effect are the same except the change of the number of grid points along the y-axis(see fig.5).

Aerodynamic load histories for G/D = 0.3,0.4,0.5,1.0 are shown in fig. 6. As for the case without ground effect, the flowfield is characterized by vortex street in the wake region, originated by the alternate separation at the front corners(see fig.4). For the cases of G/D = 0.4,0.5,1.0, the load histories show periodic behavior, which can be explained by similar manner. However, for the cases of G/D being less than 0.4, the periodic behavior is not observed. By this result, we can expect that the critical gap height would be between 0.3 and 0.4. And, this value falls in the range of the experimental researches[14,17,18]

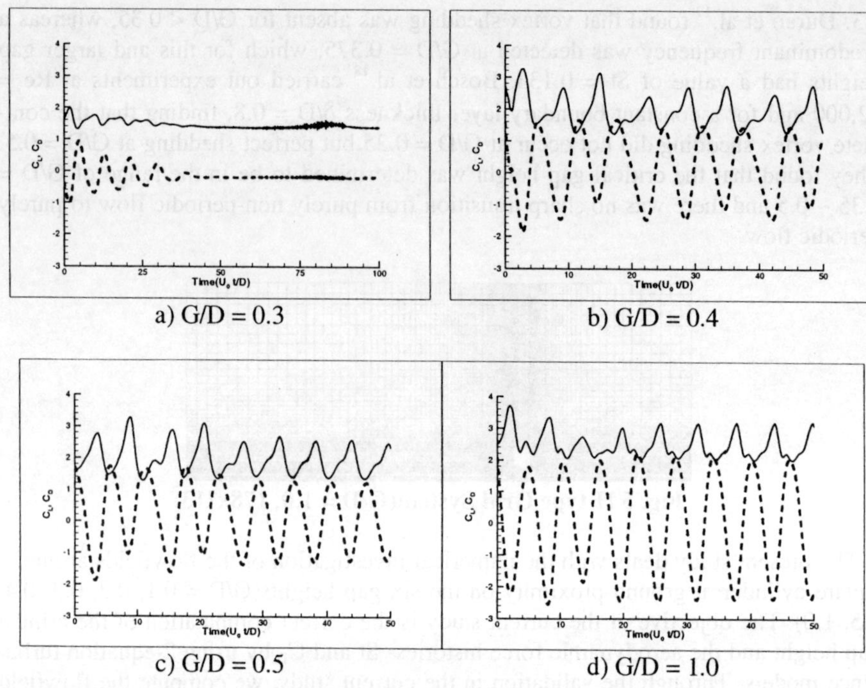

**Fig. 6 Aerodynamic Load Histories on various G/D**

As shown clearly in fig. 7, the periodicity comes from the shedding of vortex for the cases above G/D = 0.3, but the shedding is not originated by the alternate separation at the front edges as is the case of the cube in the freestream. It is driven by up-washes from the ground, and the upwashes are controlled by the separation bubble near the ground plane. Thus, the characteristics of the boundary layer might be the main factor influencing the shedding processes. If one wants to control the shedding process, the most easy way to achieve it might be the governing of the flow at the gap exit region or just behind the gap exit.

a) G/D = 0.3

b) G/D = 0.4

c) G/D = 0.5

d) G/D = 1.0

**Fig. 7 Instantaneous *k*-contour on Various G/D**

In fig.8 and 9, shedding frequency and drag coefficient as a function of G/D are shown as quantitative comparisons with results from Kenjo et al.[14] Compared with the others' results, current numerical approach seems to fail in accurate predictions both of the Strouhal number and of the drag coeffcient, though the current study does not yield unphysical results. From the St as function of G/D as shown in fig. 8, it can be argued that around G/D = 0.5 there may exist maximum St and this fact is reproduced by an experiment by our team as shown in fig.10. The details about the experiment can be found in reference[19]. This possibility will be investigated by more systematic approach in the near future.

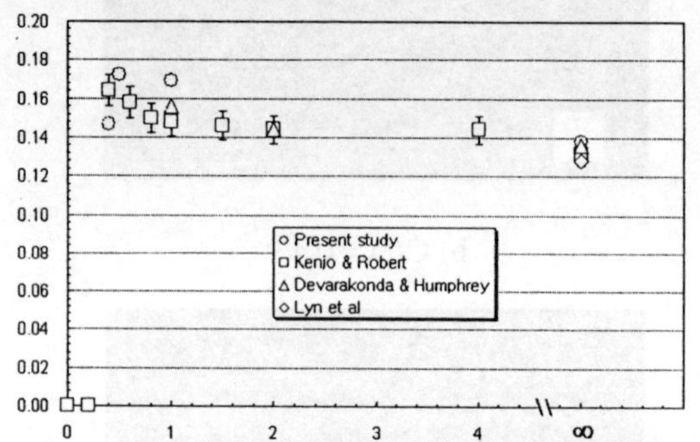



**Fig. 8 Shedding frequency(St) as a function of G/D**

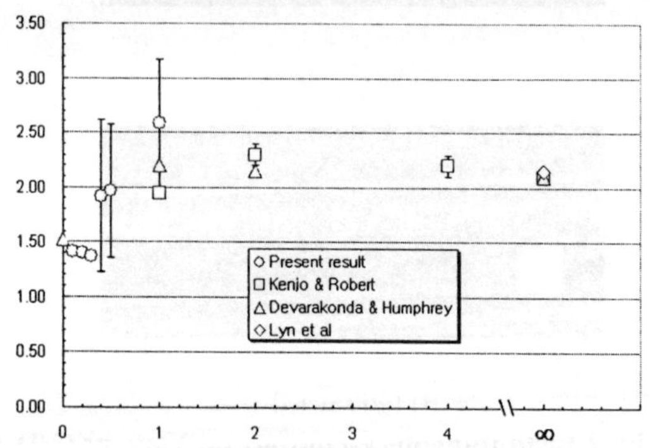



**Fig. 9 Drag coefficient($C_D$) as a function of G/D**

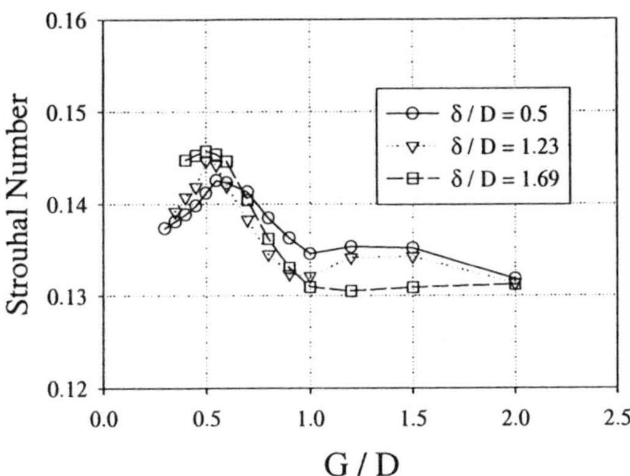


**Fig. 10 Strouhal number distribution for different boundary layer thickness(experiment[19])**

The flowfield around a bluff body like a cube placed on the ground is defined by stagnation, separation, reattachment, circulation, and vortex street. At the front face, a standing vortex is formed just beneath a stagnation point and the flow separates at the upper edge and reattaches on the upper surface. And, to the downstream, recirculation zone and vortex street are formed. If the gap effect is taken into account, two changes in flow structures can be mentioned compared with the case without gap effect. Firstly, the standing vortex at the front face, becomes much more weaker, and shows time dependent behavior, which may results in bigger change in aerodynamic drag. Secondly, the bound vortex at the recirculation zone is weakened by the streamwise momentum transfer at the bottom of the square cylinder. So, if the gap is taken into account, the fluctuation of drag force can be magnified as is depicted in fig.9. And, this possibility also will be examined in the near future.

### III. CONCLUDING REMARKS

The incompressible Naiver-Stokes code were developed and applied to the turbulent viscous flow around bluff bodies at ground proximity using two-equation turbulence models. The codes are validated by predicting the flow around a square cylinder without ground effect. Through the validation, it is pointed out that $k - \omega$ SST model is superior to the other models in predictions and in numerical stability.

By using $k - \omega$ SST model, the flowfields around a square cylinder at ground proximity on the six gap heights(G/D = 0.1, 0.2, 0.3, 0.4, 0.5, 1.0) are numerically investigated and discussed. The critical gap height(G/D) where the vortex shedding is suppressed by the presence of the ground plane, can be simulated using the practical 2-equation turbulence model, and it is pointed that the separation bubble near the ground plane and the upwashes are the main factors influencing the shedding process. It is argued that there might exist maximum value around G/D = 0.5 from the current computations. In order to predict Strouhal number and drag coefficient more accurately, more systematic study with refined numerical approach is required.

## ACKNOWLEDGEMENT

The authors appreciate the financial supports from the Government of Korea. This research is a partial result of 'Aerodynamic Design & Analyses of High-Speed Railway System'. And, this research is partly supported by iDOT ERC and BK MECHA SNU.

## NOMENCLATURE

| | |
|---|---|
| D | cylinder diameter |
| f | frequency, Hz. |
| G | gap height. |
| G/D | non-dimensioned gap height. |
| Re | Reynolds number based on D |
| St | Strouhal number. |
| $U_0$ | freestream velocity |
| $\delta$ | boundary layer thickness. |

# <u>REFERENCES</u>


[1] Wilcox, D. C., "Reassessment of the Scale-Determining Equation for Advanced Turbulence Models," *AIAA Journal*, Vol. 26, No. 11, 1988, pp. 1299-1310.

[2] Wilcox, D. C., "Simulation of Transition with a Two-Equation Turbulence Model," AIAA Journal, Vol. 32, No. 2, Feb. 1994, pp. 247-255.

[3] Menter, F. R., "Influence of Freestream Values on the $k - \omega$ Turbulence Model Predictions," *AIAA Journal*, Vol. 30, No. 6, August 1992, pp. 1651-1659.

[4] Menter, F. R., "Two-Equation Eddy-Viscosity turbulence Models for Engineering Applications," *AIAA Journal*, Vol. 32, No. 8, August 1994, pp. 1598-1605.

[5] Roe, P. L., "Approximate Riemann Solvers, Parameter Vectors and Difference Schemes," Journal of Computational Physics, Vol.43, 1983, pp.357-372.

[6] Yoon, S. and Kwak, D., "Three-Dimensional Incompressible Navier-Stokes Solver Using Lower-Upper Symmetric-Gauss-Seidel Algorithm," *AIAA Journal*, Vol. 29, June 1991, pp. 874-875.

[7] Rogers, S. E. and Kwak, D., "Upwind Differencing Scheme for the Time Accurate Incompressible Navier-Stokes Equations," *AIAA Journal*, Vol. 28, No. 2, February 1990, pp. 253-262.

[8] Rogers, S. E., Wiltberger, N. L., Kwak, D., "Efficient Simulation of Incompressible Viscous Flow over Single and Multielement Airfoils," *Journal of Aircraft*, Vol.30, No.5, Sept-Oct. 1993, pp736-743.

[9] Steger, J. L., Doughty, F. C., and Beneck, J. A., "A Chimera Grid Scheme," Advances in Grid Generation, FED, Vol.5, ASME, Edited by Ghia, K. N., New York, 1983, pp. 59-69.

[10] Wright, M. J., Candler, G. V., Prampolini, M., "Data-Parallel Lower-Upper Relaxation Method for the Navier-Stokes Equations", *AIAA Journal*, Vol. 34, No. 7, 1996, pp. 1371~1377

[11] Lee, B., Lee, D., "Data Parallel Symmetric Gauss-Seidel Algorithm for Efficient Distributed Computing Using Massively Parallel Supercomputers", AIAA-97-2138, 1997

[12] Sangsan Lee, Unsteady aerodynamic force prediction on a square cylinder using turbulence models, J. Wind Eng. Ind. Aerodyn. 67&68 (1997), pp.79-90

[13] S. Murakami, A. Mochida, On turbulent vortex shedding flow past 2D square cylinder predicted by CFD, J. Wind Eng. Ind. Aerodyn. 54&55 (1995), pp. 191-211

[14] Kenjo C. Q. Wu and Robert J. Marinuzzi, "Experimental Study of the Turbulent Wake Flow behind a Square Cylinder Near a Wall," ASME, FEDSM 97-3151, June, 1997

[15] Bearman, P. W. and Zdravkovich, M. M., "Flow around a Circular Cylinder Near a Plane Boundary," Journal of Fluid Mechanics, Vol. 89, part 1, pp.33-47, 1978

[16] Taniguchi, S. and Miyahoshi, K., "Fluctuating Fluid Forced Acting on a Circular Cylinder and Interference with a Plane Wall," Experiments in Fluids, Vol. 9, pp. 197-204, 1990

[17] Durao, D. E. F., Gouveia, P. S. T. and Pereira, J. C. F., "Velocity Characteristics of the Flow around a Square Section Cylinder Placed near a Channel Wall," Experiments in Fluids, Vol. 11, pp. 341-350, 1991

[18] Bosch, G., Kappler, M. and Rodi, W., "Experiments on the Flow Past a Square Cylinder Placed near a Wall," Experimental Thermal and Fluid Science, Vol. 13, pp. 292-305, 1996

[19] Y. Park, Y. Cho, and D. Lee, Vortex Shedding around a Square Cylinder near a Wall, AIAA-2001-0450

# NUMERICAL SIMULATION OF AN UNDERWATER HOT LAUNCH EXHAUST PLUME

W. G. SZYMCZAK
Naval Research Laboratory
Washington DC, USA  20375-5350

T. HSIEH, W. P. CHEPREN, A. E. BERGER and T. F. ZIEN
Naval Surface Warfare Center Dahlgren Division
Dahlgren, VA , USA  22448-5100

## Abstract

Simulations of an underwater missile hot launch exhausting plume from a Concentric Canister Launcher (CCL) using a computer code, BUB2D, based on a generalized formulation of hydrodynamic free surface problems, are described.  In this paper the missile motion is not considered.  Three cases were calculated: (1) a prescribed pressure history at the CCL exit, (2) considering the CCL as a chamber with prescribed rocket motor exhaust at one end, and (3) an interacting model, which consists of a gas dynamic model in the CCL chamber as a subsystem and the gas plume and water above the CCL device as another subsystem with the two subsystems interacting.  The computed results are described.

## Introduction

A new system for an underwater hot launch using a Concentric Canister Launcher (CCL)[1] device is under consideration. A sketch of the CCL is shown in Fig. 1.  Two concentric circular hollow cylinders with a hemispherical plenum covering the lower end of the outer cylinder constitute the basic structure.  The missile is sitting in the center of the inner cylinder.  After the ignition of missile motor, the exhaust gas will turn around the bottom of plenum, move up through the uptake and shoot out of the exit into the water above to form a gas plume. The flow field resulting from the interaction of the missile, the hot exhaust plume, and the surrounding water is very important in determining the success of the proposed launch system.  In this paper, we will first consider the case without the missile motion to study the behavior of the exhausting plume in the flow domain between the launcher exit and the free surface of air above.

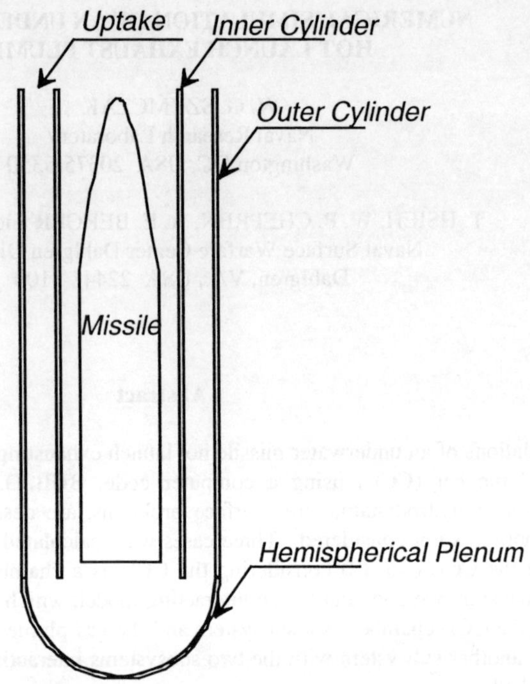


Figure 1. Sketch of Concentric Canister Launcher

A numerical algorithm for hydrodynamic free boundary problems for incompressible fluid flows has been developed at NSWC to simulate flow fields involving water and gas bubbles in the computational domain including physically based treatment of fluid on fluid collisions. This algorithm solves the conservation of mass and momentum subject to constraints on density and pressure on an "Eulerian" grid. The computational method uses three separate steps to advance the flow field. The first step employs a second order Godunov method for the approximation of the nonlinear convection terms. A variational inequality is solved for the redistribution of density so that the density constraint is satisfied and mass is conserved. This step is coupled with a momentum redistribution, which ensures that energy cannot be increased by the density redistribution. Finally, the pressure is used as a Lagrange multiplier to project the velocity onto a divergence free subspace. The pressure projection and redistribution steps use a tri-linear finite element discretization for the three-dimensional problem. The resulting computer code is called BUB3D. For the exhaust plume problem application, an axisymmetric version of the computer code, called BUB2D, was extended to treat the CCL device by incorporating modeling of the gas source from the rocket motor.

Three examples will be discussed in this paper: (1) given pressure history at the CCL exit, (2) considering the CCL as a chamber with prescribed rocket exhaust at one end, and (3) an interacting model, which consists of a gas dynamic model in the CCL chamber as a subsystem and the gas plume and water above the CCL device as another subsystem with the two subsystems interacting.

## Numerical Method

The computational model used for both the two- and three-dimensional codes is based on a generalized formulation of hydrodynamics as described in both Rogers et al.[2] and Szymczak et al.[3,4] This formulation uses a fixed spatial domain $\Omega$, where the density $\rho$, velocity $\mathbf{u}$, and the pressure p are governed by the mass and momentum conservation equations

$$\rho_t + \nabla \cdot (\rho\mathbf{u}) = 0 \tag{1}$$

$$(\rho\mathbf{u})_t + \nabla \cdot (\rho\mathbf{uu}) = -\rho\tilde{g}\mathbf{k} - \nabla p \tag{2}$$

subject to the constraint

$$\rho \le \rho_0 \tag{3}$$

where $\rho_0$ is the constant density of the incompressible liquid. In (2) $-\mathbf{k}$ is the unit vector in the direction of the gravitational force, and $\tilde{g} = \tilde{g}(t)$ is an acceleration term, which can be used to model vertical rigid body motion. (When the body is fixed $\tilde{g}(t) = g$, the gravitational constant.) In regions where $\rho = \rho_0$ equation (1) becomes the usual divergence free condition for incompressible flow. We define the time varying "liquid" domain $\mathbf{D}(t)$ by

$$\mathbf{D}(t) = \{\mathbf{x} : \rho(\mathbf{x},t) = \rho_0\} . \tag{4}$$

The non-liquid domain is defined using

$$\Omega - \mathbf{D}(t) = \mathbf{A}(t) \cup \mathbf{B}(t) \cup \mathbf{C}(t) \tag{5}$$

where the regions $\mathbf{A}$, $\mathbf{B}$, and $\mathbf{C}$ are disjoint. Within these regions the pressure is assumed to be uniform; that is

$$p(\mathbf{x},t) = \begin{cases} p_A & \mathbf{x} \in \mathbf{A}(t) \\ p_B(t) & \mathbf{x} \in \mathbf{B}(t) \\ p_C & \mathbf{x} \in \mathbf{C}(t) \end{cases} . \tag{6}$$

In the above $p_A$ represents the constant ambient "air" pressure. The "bubble" pressure, $p_B(t)$, is usually determined using an adiabatic gas assumption

$$p_B(t) = c / (V_B(t))^\gamma \tag{7}$$

where $c$ is constant, $\gamma$ is the (constant) ratio of the heat capacities at constant pressure and at constant volume of the bubble gases, and $V_B(t)$ is the bubble volume which can be determined using

$$V_B(t) = \int_{B(t)} \left(1 - \frac{\rho(\mathbf{x},t)}{\rho_0}\right) dx \ . \tag{8}$$

Finally, $p_C$ is the "cavitation" pressure, which is usually set to the vapor pressure of the liquid at some specified temperature. When cavitation is to be modeled an additional constraint is imposed on the pressure, namely that $p(\mathbf{x},t) \geq p_C$.[4,6] The adiabatic model (7) will be refined when the effects of condensation are considered.

Assume that the density and velocity, $\rho^n, \mathbf{u}^n$ at time step $n$ are known together with the pressure gradient at the previous half step, $\nabla p^{n-1/2}$. This solution is evolved from time $t = t^n \rightarrow t^n + \tau \equiv t^{n+1}$ using the following three step time split procedure.

## Convection

The solution is first advanced $(\rho^n, \mathbf{u}^n) \rightarrow (\tilde{\rho}, \tilde{\mathbf{u}})$ by "solving" the conservation laws (1), (2) without including the term $\nabla p$ on the right-hand side of (2) and without regard to the constraint (3). This step is fully discretized using a formally second-order Godunov-type method, which uses slope limiting in space and explicit predictor-corrector time stepping following the method proposed by Davis.[14] Although the pressure gradient is not explicitly added to the momentum here it is used within the predictor step of the Godunov method. Further details of this step may be found in Szymczak et al.[4] for axially symmetric flow problems and in Szymczak et al.[10] for three-dimensional problems.

## Redistribution of Density and Momenta

Next, the density and momenta are redistributed $(\tilde{\rho}, \tilde{\mathbf{u}}) \rightarrow (\overline{\rho}, \overline{\mathbf{u}})$ so that the constraint (3) is satisfied, the global conservation of mass and momenta are maintained, and the energy is non-increasing. The density is redistributed using an approximate solution to the obstacle problem

$$\nabla^2 H = \begin{cases} \rho_0 - \tilde{\rho} & \text{if} \quad H > 0 \\ 0 & \text{if} \quad H = 0 \end{cases} \tag{9a}$$

and setting

$$\overline{\rho} = \tilde{\rho} + \nabla^2 H \tag{9b}$$

These equations have been derived by considering a Boltzmann formulation for modeling inelastic fluid collisions and are discussed in greater detail in Rogers et al.[2,8] and Szymczak et al.[4] These references also contain details of the numerical discretization of Eq. (9) and the solution procedure which employs a constrained direction preconditioned conjugate gradient method. The momenta redistributions are determined as solutions of two (or three for three-dimensional problems) elliptic self-adjoint problems

$$\overline{\rho \mathbf{u}} = \tilde{\rho}\tilde{\mathbf{u}} + \nabla^2(H\overline{\mathbf{u}}) \tag{10}$$

Discretizations of Eq. (10) yield systems of linear equations with diagonally dominant matrices, which are efficiently solved using a diagonally preconditioned conjugate gradient method.

This step has been shown to be important to the overall accuracy and stability of the algorithm.[5] As mentioned in the introduction it is a major distinguishing feature between this algorithm and other VOF approaches,[11,13] which would simply truncate the density values, thereby violating mass conservation.

### Pressure Projection

After the redistribution step, the density $\rho^{n+1}$ at the next time level is given by $\bar{\rho}$, and the new non-liquid region is then determined along with the pressure in each of its connected subsets. In the computational space the new liquid region, $\mathbf{D}^{n+1} = \mathbf{D}(t^{n+1})$, is defined to be the collection of grid cells $C_l$ such that

$$\rho_l^{n+1} \geq (1 - \varepsilon_\rho)\rho_0 \tag{11}$$

where $\rho_l^{n+1}$ is the density in cell $C_l$. Through numerical experimentation we have found the results to be relatively insensitive to the choice of $\varepsilon_\rho$. Small values of $\varepsilon_\rho$ will cause cells with only slightly less density than the liquid to be treated as regions with uniform pressure, while larger values will cause more of the "spray" (where $0 < \rho < \rho_0$) to be treated as a variable density incompressible region. For the results in this paper we have selected $\varepsilon_\rho = 0.04$.

In the non-liquid region $\mathbf{u}^{n+1} = \bar{\mathbf{u}}$. However, $\bar{\mathbf{u}}$ is not consistent with Eq. (1) in the new liquid region. In this region the velocity is corrected, $\bar{\mathbf{u}} \to \mathbf{u}^{n+1}$, using the gradient of the new pressure, $\nabla p^{n+1/2}$, where the pressure $p^{n+1/2}$ is the solution of

$$\tau \nabla \bullet \left(\frac{\nabla p}{\rho^{n+1}}\right) = \nabla \bullet \bar{\mathbf{u}} \text{ in } \mathbf{D}^{n+1} \tag{12}$$

The new velocity, given by

$$\mathbf{u}^{n+1} = \bar{\mathbf{u}} - \tau \frac{\nabla p^{n+1/2}}{\rho^{n+1}} \tag{13}$$

is divergence free in $\mathbf{D}^{n+1}$ and thus is consistent with (1). Equations (12) and (13) define a projection $\mathbf{u}^{n+1} = P(\bar{\mathbf{u}})$ onto the space of divergence free velocities. The equation (12) is discretized using a finite element method with bilinear (in 2D) or tri-linear (in 3D) elements. This spatial discretization produces an approximate projection which has been discussed in Almgren et al.[15] The linear system resulting from the discretization of Eq. (12) is solved using an incomplete Cholesky preconditioned conjugate gradient method.

506

To determine the pressure uniquely using Eq. (12), boundary conditions must be specified. On those portions of the boundary of $\mathbf{D}^{n+1}$ that correspond to "wall" boundaries of $\Omega$ a Neumann condition is specified, namely

$$\frac{\partial p}{\partial \mathbf{n}} = \frac{\rho^{n+1}}{\tau} \bar{\mathbf{u}} \bullet \mathbf{n} \tag{14}$$

Note that this condition, together with Eq. (13), implies that $\mathbf{u}^{n+1} \bullet \mathbf{n} = 0$ along wall boundaries. Since the pressure is assumed to be continuous Dirichlet conditions for the pressure along the non-liquid regions are determined according to Eq. (2-6). In particular, we specify

$$p(\mathbf{x},t) = \begin{cases} p_A & \mathbf{x} \in \partial\mathbf{A}^{n+1} \cap \partial\mathbf{D}^{n+1} \\ p_B^{n+1} & \mathbf{x} \in \partial\mathbf{B}^{n+1} \cap \partial\mathbf{D}^{n+1} \\ p_C & \mathbf{x} \in \partial\mathbf{C}^{n+1} \cap \partial\mathbf{D}^{n+1} \end{cases} \tag{15}$$

along those boundaries common to the non-liquid regions. Hydrostatic pressure Dirichlet conditions are set on other portions of the boundary of $\Omega$ to model "cutoff" boundaries.

## Numerical Results

Descriptions of the computational grid, BUB2D CCL model geometry, the interaction models with the exhaust pressures and numerical results are presented in this section.

## Computational Grid

A portion of the computational grid used for the simulations is shown in Fig. 2. The grid uses axially symmetric $(r,z)$ coordinates and the image shown in Fig. 2 was reflected about $r = 0$ for visualization purposes only. The grid is comprised of a region of $50 \times 80$ uniform square cells of size 0.02 m in the region bounded by $0 \leq r \leq 1$ and $-7.7 \leq z \leq -6.1$. The top of the launch platform is located at $z = -7.62$ and the surface of the water is $z = 0$. The platform and cylindrical missile are outlined with dark lines and the cylinder is indicated by diagonal lines. Outside of the uniform region, the grid spacing is stretched radially using 70 additional cells to $r = 10$ with the restriction that the maximum spacing is 0.2 m. Similarly, stretched cells are used to extend the computational domain downward to $z = -14.62$ and upward to $z = 8$. The entire grid contains $120 \times 211$ cells.

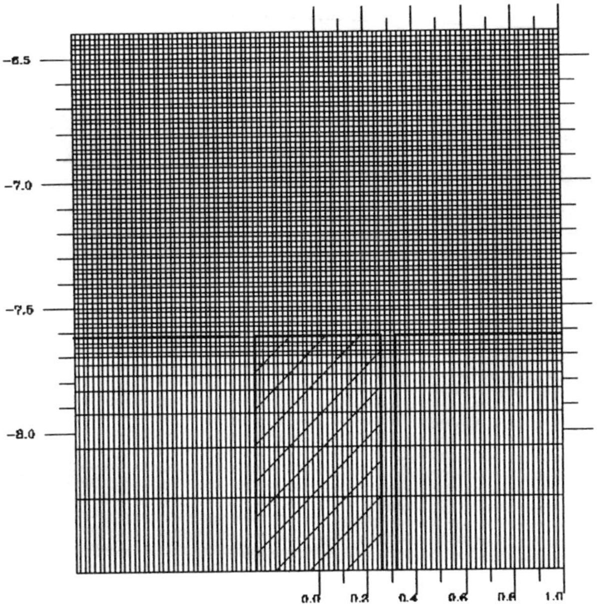

Fig. 2  Computational grid near the CCL exit.

## Given CCL exit pressure (P1)

The first example to demonstrate the capability of the numerical code is to impose a given pressure history for the plume which is adjacent to the CCL and has the same pressure as the CCL exit pressure at all time.  To begin with, a small plume (bubble) with the pressure of the initial pressure given by the CCL exit pressure history curve, see Fig. 3, is assume to exist.  With a selected time interval $\Delta t$, a new plume pressure is obtained from the CCL exit pressure history curve.  The new updated plume pressure is used to determine the hydrodynamics including new plume volume.  This procedure is repeated over the entire CCL exit pressure history of 2.5 seconds.  The calculated plume shapes at six different times are shown in Fig. 4.

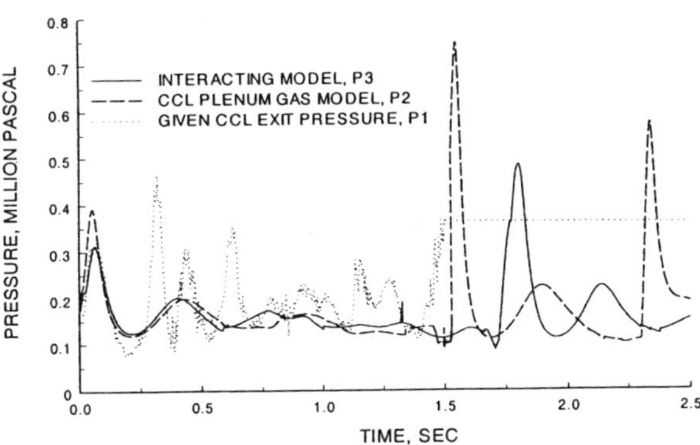


Fig. 3 Pressure histories at CCL exit

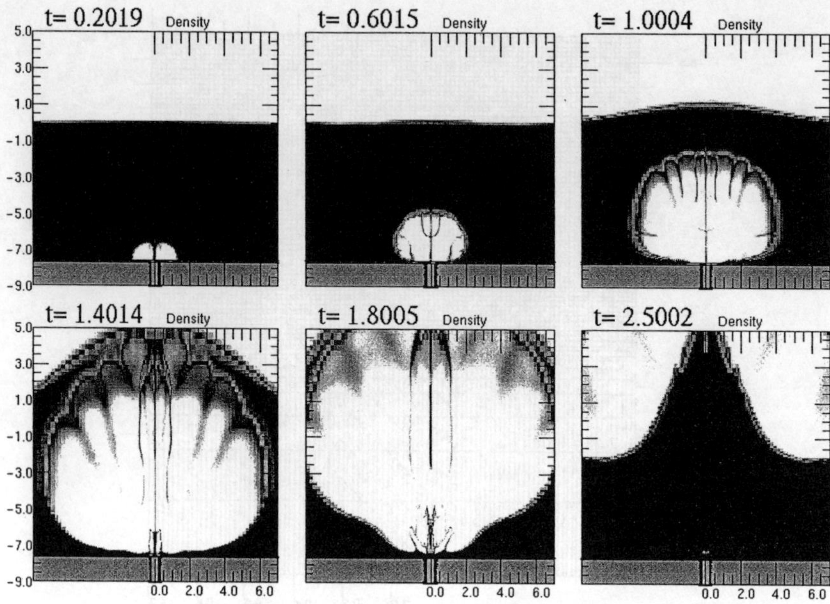

Fig. 4.  Plume shape for given CCL exit pressure, P1

## Model rocket motor exhaust gas into CCL chamber and plume

The second example is to treat the CCL plenum and uptake as a chamber connected to the plume.  A split step approach is used to treat influx of gas from a source such as a rocket motor into the chamber.  Instead of explicitly specifying the pressure, the gas in the plume and chamber is modeled as an ideal gas.  After a complete step from time $t$ to time $t + \tau$ of the basic generalized hydrodynamics algorithm (but before gas sources are treated) the plume adjacent to the CCL (note there may be more than one plume component, each of which have a different state) and the chamber has a volume $V_0$, temperature $T_0$, number of moles $N_0$, and its pressure $P_0$ satisfies the ideal gas $P_0 = N_0 R T_0 / V_0$ , where $R$ is the gas constant 8.31451 J/(mol K).  The heat capacity $C_v$ of all the gases present in the computation is assumed to be the same, and so the exponent $\gamma$ in the adiabatic relationship $PV^\gamma = constant$ is $\gamma = 1 + \left( R / C_v \right)$.  This adiabatic relationship is used in the basic generalized hydrodynamics algorithm to obtain $T_0$ and $P_0$.  An inflow rate $r_i$ (mol/s) of gas at a temperature $T_i$ from the rocket motor into the chamber and plume adjacent to the CCL during the time interval $[t, t + \tau]$ is obtained from a given table containing missile rocket motor exhaust pressure, $P_i$, temperature, $T_i$, and nozzle throat velocity, $V_i$, data as a function of time as shown in Fig. 5.  The new thermodynamic state ($P$, $V$, $N$, $T$) of the plume at time $t + \tau$, accounting for the source gas, is then $V = V_0$, $N = N_0 + r_i \tau$, $T = \left( N_0 T_0 + r_i \tau T_i \right) / \left( N_0 + r_i \tau \right)$, and $P = NRT / V$.  The calculated plume shapes at 6 different times are shown in Fig. 6.

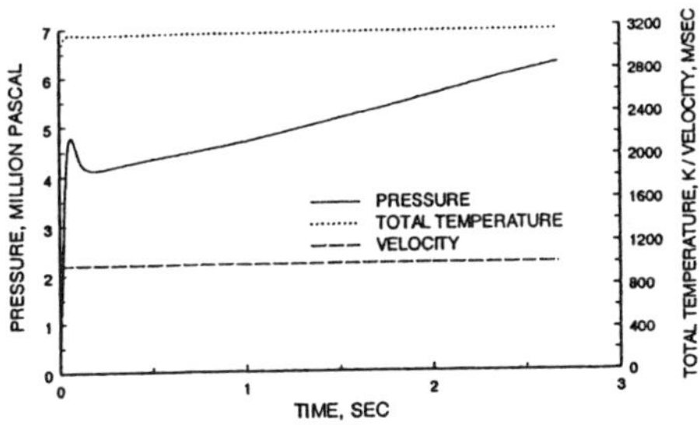

Fig. 5  Characteristics of missile rocket motor exhaust data (CCL inflow).

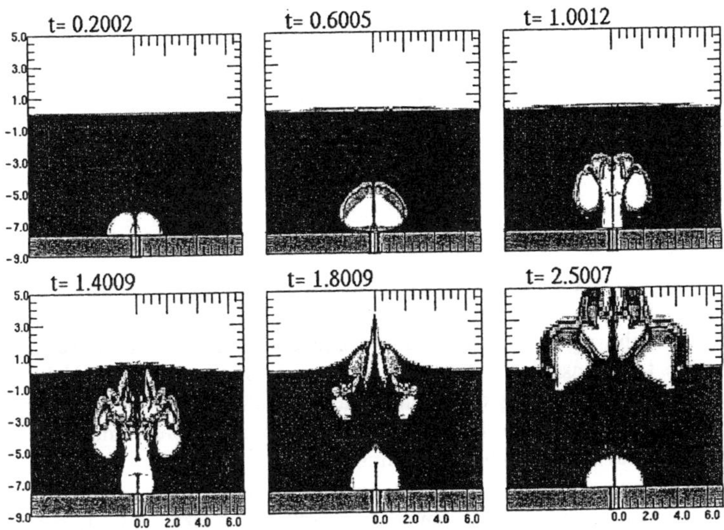

Fig. 6 Plume shapes using passive CCL plenum gas model, P2.

## CCL and plume interacting model

The third approach is to consider the CCL and the plume as two separate subsystems and let them interact as shown in Fig. 7.

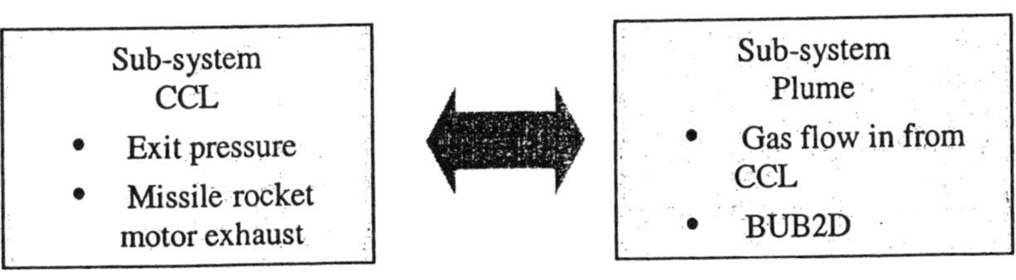



Figure 7.  CCL and plume Interacting Model

510

Flow in the Concentric Canister Launcher was modeled using a lumped parameter model to simulate the plenum and a one-dimensional CFD model to simulate the uptake. The plenum model was developed from the integral forms of the mass and energy conservation equations. A table containing missile rocket motor exhaust pressure, chamber temperature and nozzle throat velocity data as a function of time was used to simulate mass flow into the plenum as was used in example 2. In the plenum, the conservation equations were solved to determine temperature and density. Pressure obtained from the ideal gas equation of state was used to calculate a velocity at the interface between the plenum and the uptake. The uptake was simulated using the split coefficient matrix method to discretize the nonconservative form of the one-dimensional Euler equations. The equations were transformed as described in Anderson et al[16] so that one-sided finite differencing could be applied to account for signal propagation in the characteristic directions. The transformed equations are shown below

$$
\begin{bmatrix} \dfrac{\partial P}{\partial t} \\ \dfrac{\partial u}{\partial t} \\ \dfrac{\partial T}{\partial t} \end{bmatrix} + A^{+} \begin{bmatrix} \dfrac{\partial P}{\partial x} \\ \dfrac{\partial u}{\partial t} \\ \dfrac{\partial T}{\partial x} \end{bmatrix} + A^{-} \begin{bmatrix} \dfrac{\partial P}{\partial x} \\ \dfrac{\partial u}{\partial t} \\ \dfrac{\partial T}{\partial x} \end{bmatrix} = \begin{bmatrix} 0 \\ F_{friction} \\ F_{dissipation} \end{bmatrix} \tag{16}
$$

The $A^{+}$ and $A^{-}$ arrays were derived from the original coefficient matrix. Expressions to account for wall friction, viscous dissipation and heat transfer at the uptake walls were incorporated into the right hand side of the equation. The spatial derivatives were discretized such that backward differencing was used for terms multiplied by the $A^{+}$ matrix and forward differencing was used for terms multiplied by the $A^{-}$ matrix. The CFD model communicated with the plenum and the domain representing the seawater through a set of characteristic boundary conditions. =

The procedures to carry out the computation from time t to t+$\Delta$t are as follows: (1) Use intermediate plume pressure as boundary conditions at exit of CCL model. (2) Solve CCL model, Equation (16), to determine velocity, temperature, and density of gas at CCL exit. (3) Include gas source from CCL model to update the gas pressure of the plume adjacent to the CCL, similar to example 2. (4) Use updated plume pressure to determine hydrodynamics including new plume volume and intermediate pressure. This completes a cycle of calculation and return to step 1 with a new time step. The calculated plume shapes at 6 different time are shown in Fig. 8.

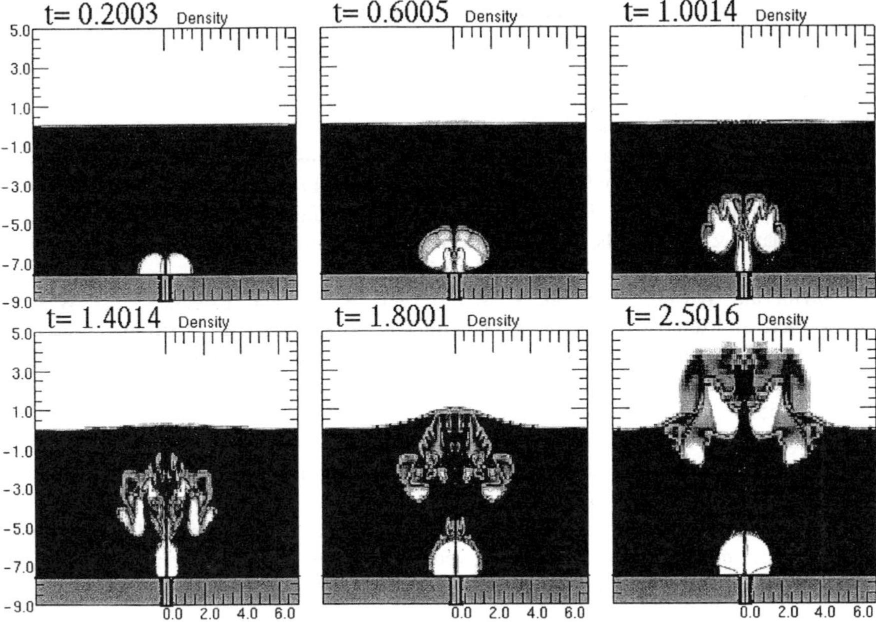

Figure 8.  Plume shape using CCL-plume interacting model.

## Concluding Remark

An axisymmetric two-dimensional code, base on a generalized formulation of incompressible hydrodynamics of free boundary problems, has been successfully applied to calculate plume shapes of an underwater hot launched missile from a Concentric Canister Launcher.  Three examples of calculated plume shapes for (1) given CCL exit pressure, (2) using CCL plenum pressure with rocket motor exhaust data, and (3) using CCL and plume interacting model were described.  The third approach is considered to be the best model.  The accuracy of the predicted plume shapes can not be evaluated at the present time for lack of experimental data to compare with.  To treat the complete problem it will, of course, be necessary to account for the motion of the missile and its interaction with the flow field.

## Acknowledgement

Partial support of this work is from the Seed and Venture Program of NSWCDD monitored by Larry C. Triola.

## References


1. Busic, John F., Anderson, Leon P., Jr., Vendettuoli, Derrick N., "One-Dimensional Computational Fluid Dynamic Analysis of Experimental Data for the Scaled Concentric Canister Launcher Restrained Firing Tests," NSWCDD/TR-99/68, May 1999

2. Rogers, J. C. W., Szymczak, W. G., Berger, A. E. and Solomon, J. M., "Numerical Solution of Hydrodynamic Free Boundary Problems," *Free Boundary Problems,* (Proceedings of the Conference on Free Boundary Problems – Numerical Methods and Optimal Control, July 9-19, 1989, Oberwolfach, West Germany), K.-H. Hoffmann & J. Sprekels, eds., Birkhäuser Verlag (International Series of Numerical Mathematics **95**), Basel, 1990, pp. 241-266.

3. Szymczak, W. G., Solomon, J. M., Berger, A. E., and Rogers, J. C. W., "A Numerical Method Based on a Generalized Formulation of Hydrodynamic Free Surface Problems," paper no. AIAA-91-1541-CP in *Proceedings of the AIAA 10th Computational Fluid Dynamics Conference*, Honolulu, Hawaii, 24-27 Jun 1991, pp. 155-165.

4. Szymczak, W. G., Rogers, J. C. W., Solomon, J. M., and Berger, A. E., "A Numerical Algorithm for Hydrodynamic Free Boundary Problems," *J. Comput. Phys.*, Vol. 106, 1993, pp. 319-336.

5. Szymczak, W. G., "Energy Losses in Non-Classical Free Surface Flows," *Bubble Dynamics and Interface Phenomena,* (ed. J. R. Blake, J. M. Boulton-Stone & N. H. Thomas), Dordrecht, The Netherlands: Kluwer Academic Publishers, 1994, pp. 413-420.

6. Szymczak, W. G., Rogers, J. C. W., Solomon, J. M., and Berger, A. E., "Numerical Simulations of Cavitation Inception," *Proceedings of CAV '95, The International Symposium on Cavitation, May, 1995, Deauville, France,* 1995, pp. 399-406.

7. Szymczak, W. G. and Solomon, J. M., "Computations and Experiments of Shallow Depth Explosion Plumes," NSWCDD/TR-94/156, 1996, Naval Surface Warfare Center Dahlgren Division, Dahlgren, VA.

8. Rogers, J. C. W. and Szymczak, W. G., "Computations of Violent Surface Motions: Comparisons with Theory and Experiment," *Phil. Trans. R. Soc. Lond.* A, Vol. 355, 1997, pp. 1-15.

9. Szymczak, W. G., "Computations and Experiments of Pressure Loadings from Underwater Explosion Bubbles," accepted for publication in *Proceedings of the 1997 ASME Pressure Vessels & Piping Conference*, Orlando, Florida, 27-31 July, 1997,

10. Szymczak, W. G., Solomon, J. M., and Berger, A. E., "BUB3D–Description of Mathematical Model, Algorithms, and User's Guide," NSWCDD/TR-95/42, (to appear) Naval Surface Warfare Center Dahlgren Division, Dahlgren VA.

11. Hirt, C. W. and Nichols, B. D., "Volume of Fluid (VOF) Method for the Dynamics of Free Boundaries," *J. of Comput. Phys.*, Vol. 39, 1981, pp. 201-225.

12. Kothe, D. E. and Mjolsness, R. C., "RIPPLE: A New Model for Incompressible Flows with Free Surfaces," *AIAA Journal*, Vol. 30, 1992, pp. 2694-2700.

13. Kan, K.-K., and Stuhmiller, J. H., "Simulation of the Collapse of an Underwater Explosion Bubble Under a Circular Plate," to appear.

14. Davis, S. F., "Simplified Second Order Godunov-Type Methods," *SIAM J. Sci. Stat. Comput.*, Vol. 9, 1988, pp. 445-473.

15. Almgren, A. S., Bell, J. B., & Szymczak, W. G., "A Numerical Method for the Incompressible Navier-Stokes Equations Based on an Approximate Projection," *SIAM J. Sci. Stat. Comput.*, Vol. 17, 1996, pp. 358-369.
16. Anderson, D.A., Tannehill, J.C., Pletcher, R.H., <u>Computational Fluid Mechanics And Heat Transfer</u>, Hemisphere Publishing Corporation, New York, 1984, pp. 288-302.

# Multiphase and Cavitating Flows

# AUSM-family Schemes for Multiphase Flows at All Speeds

Meng-Sing Liou[1] and Jack R. Edwards[2]

[1] Turbomachinery and Propulsion Systems Division, NASA Glenn Research Center at Lewis Field, Cleveland, OH 44135, USA
[2] Department of Mechanical and Aerospace Engineering, North Carolina State University, Raleigh, NC 27695

**Abstract.** In this paper, we will show the ideas and procedure how the AUSM-family schemes are extended for solving two phase flows equations. Specifically, we shall consider three mathematical models for describing multiphase flows, the single phase model with homogeneous phase transition, two phase mixture model, and two phase gas-solid model. Details of the numerical formulation and issues involved are discussed and the effectiveness of the method are demonstrated for several practical examples.

## 1 Introduction

Two phase flows can be found in broad situations in nature, biological systems, and industry devices and they often involve diverse and complex mechanisms. While physical models may be specific to certain situations, the mathematical formulation and numerical treatment for solving the governing equations can be general. Based on the continuum mechanics, we shall treat the fluid as a mixture consisting of two interacting phases (or materials) occupying the same region in space at any given moment. Hence, we will require information concerning each individual phase as needed in a single phase, but also the interactions between them. These interaction terms, however, pose additional numerical challenges. For example, the mathematical equations are not guaranteed to be hyperbolic in time for all conditions. Moreover, due to disparate differences in time scales, fluid compressibility and nonlinearity become acute, further complicating the numerical procedures.

This paper summarizes some recent developments in developing a robust, accurate (low-diffusion) and efficient upwind scheme for multiphase flows, specifically the so-called AUSM-family schemes including AUSM+ [1,2], AUSMDV [3], and LDFSS [4]. They have been proven to be accurate, simple, robust, and easy to extend to other types of conservation laws, thus providing an attractive alternative to the existing schemes; a summary is given recently in [5]. Several attempts have been made over the last ten years, e.g., Refs. [6]-[7] to improve the original scheme [1] and in general, some successes have been achieved. Now various versions of the AUSM-family schemes have been incorporated into both research and commercial codes worldwide.

Prior efforts in the construction of AUSM-type algorithms [2,4] have assumed that the fluid behaves as an ideal gas or a mixture thereof. This paper details recent extensions of AUSM-type schemes that are valid for generalized state equations, which may describe single-phase liquid, gas, or supercritical fluid behavior of a given substance. Another thrust of this paper is to provide extensions suitable for solving the general multiphase flow problem for arbitrary flow speeds and arbitrary levels of compressibility. Thus, attention will be focused on three situations (physical modelings): a single-component real fluid governed by a generalized equation of state, a two-phase "mixture" model, and a separated two-phase gas-solid flow model. Where appropriate, the schemes will be analyzed on a term-by-term basis to illustrate various scaling issues.

Due to the dramatic differences in compressibility among fluids in the different states and possible large differences in the flow speed, the "preconditioned" forms of the flux-splitting methods are utilized in the extensions. The concept of time-derivative preconditioning, for example [8–11], is now widely used to enable standard compressible gas codes to operate at very low Mach numbers, and more recently, there have been several efforts designed to extend its applicability to supercritical fluids [12], flows governed by generalized equations of state [13], and some two-phase flows [14,15]. The process of rendering some two-phase flow equation systems hyperbolic in time can also be viewed as a type of preconditioning, and examples abound in the multiphase literature, [16] and references cited within.

One concern with the use of time-derivative preconditioning is its lack of robustness in many situations (low-speed reactive flows, for one), which may sometimes necessitate the use of implicit methods for time-advancement. Another concern is the need to specify, in general, a "reference velocity" to avoid numerical difficulties in stagnation regions. This quantity is problem-dependent, and as yet, no firm guidelines are available for its choice. Even with these concerns, the framework offered by time-derivative preconditioning offers a strong potential as a basis for developing the extensions mentioned above. Numerical examples will be shown to confirm the effectiveness of the numerical schemes proposed.

## 2 Governing Equations for Multiphase Flows

A set of equations of general conservation laws is considered:

$$\mathbf{Q}_t^{(k)} + \mathrm{div}(\mathbf{F}^{(k)} + \mathbf{F}_v^{(k)}) = \mathbf{S}^{(k)}. \tag{1}$$

The conservative variables are given in $\mathbf{Q}^{(k)}$ where the superscript $k(= 1, 2, ...)$ is introduced to include multifluid models, often adopted for describing multiphase flows [16,17]. The inviscid and viscous fluxes are denoted respectively by $\mathbf{F}^{(k)}$ and $\mathbf{F}_v^{(k)}$, whose definitions are rather standard and hence omitted herein. However, the source terms are dependent upon the physical problems studied. For multiphase flows, they can contain terms describing interfacial balances of mass, momentum, and energy transfers due to phase differences/changes. We include this option because examples will be given later in the paper.

The discretization of viscous terms is rather standard and is generally done with centered schemes. On the other hand, treatment of the source terms varies considerably and this subject is not so easy because it is quite problem-dependent and the terms can be extremely complicated. See for example the ones involved in the fluidized bed [18]. The subject is beyond the scope of this paper and will not be dealt with here. We shall restrict ourselves instead to the numerical representation of inviscid fluxes, which has been a subject of intensive effort by many researchers over the past three decades.

## 2.1  Time-Derivative Preconditioning

During the 90s, a great deal of interest has been focused upon the development of a (local) preconditioning method to improve the convergence rate and robustness of solutions in the low Mach number regime. This is accomplished by premultiplying the time-derivative term with a conditioning matrix $\Gamma$,

$$\Gamma\tilde{\mathbf{Q}}_t^{(k)} + \mathrm{div}(\mathbf{F}^{(k)} + \mathbf{F}_{(v)}^{(k)}) = \mathbf{S}^{(k)}, \tag{2}$$

where $\tilde{\mathbf{Q}}^{(k)}$ can take different forms, such as $(p^{(k)} \text{ or } \rho^{(k)}, V^{(k)}, T^{(k)})^{\mathrm{T}}$. Its details will be dealt with later as specific multiphase models are presented. Several forms of the local preconditioning matrix $\Gamma$ have been proposed in the literature, e.g., Refs. [8–11].

To consider a time-dependent problem, the physical time derivative term is kept in a dual-time formulation,

$$\Gamma\tilde{\mathbf{Q}}_t^{(k)} + \mathbf{Q}_t^{(k)} + \mathrm{div}(\mathbf{F}^{(k)} + \mathbf{F}_{(v)}^{(k)}) = \mathbf{S}^{(k)}. \tag{3}$$

A three point backward discretization is applied to the time-accurate $\mathbf{Q}_t^{(k)}$ and implicit subiteration employed for the preconditioned term $\Gamma\tilde{\mathbf{Q}}_t^{(k)}$. Typically, 20-30 subiterations are required to maintain accuracy, such as mass conservation.

The utility of time-derivative preconditioning in the solution of the real fluid system described above lies in its ability to provide a smooth transition between nearly incompressible conditions (such as liquid phase or low Mach number vapor or supercritical fluid phase flows) and strongly compressible conditions (such as two-phase flows or high-speed vapor phase flows). As discussed previously, modifications to AUSM-type discretization are required to extend their range of applicability to flows at all speeds. These modifications depend on the choice of preconditioner, through the use of the eigenvalues of the preconditioned system. As in the previous work, the real fluid extension utilizes the preconditioner of Weiss and Smith [11], which may be expressed as a rank-one perturbation of the Jacobian matrix $\partial\mathbf{Q}/\partial\tilde{\mathbf{Q}}$. The time-derivative term in the real fluid Euler system is replaced by

$$\Gamma \equiv (\frac{\partial\mathbf{Q}}{\partial\tilde{\mathbf{Q}}} + \Theta uv^T), \tag{4}$$

where for example,

$$u = [1, u, H]^T \tag{5}$$

$$v = [p_\rho, 0, p_T]^T, \tag{6}$$

$$\Theta = \frac{1}{a^2}\left(\frac{1}{M_*^2} - 1\right) = \frac{1}{V_*^2} - \frac{1}{a^2}. \tag{7}$$

At low speeds, this preconditioner essentially replaces the physical thermodynamic derivative $\rho_p$ with $1/V_*^2$, rescaling the eigenvalues of the Euler system so that the condition number remains bounded. This allows uniform convergence in both low-speed and high-speed flows. The eigenvalues of $\Gamma^{-1}A$ ($A = \partial \mathbf{F}/\partial \tilde{\mathbf{Q}}$) are $u, u' \pm a'$. We see that the eigenvalue associated with the linear field is still $u$, but those associated with the nonlinear fields have been scaled to be $u' \pm a'$. For most common situations, they have these forms,

$$u' \pm a' = \frac{1 + M_*^2}{2}\left(u \pm a\frac{\sqrt{(1 - M_*^2)^2 M^2 + 4M_*^2}}{1 + M_*^2}\right). \tag{8}$$

## 3　Numerical Flux: AUSM-Family

In this section, we will look at the algorithm involved in extending the AUSM-family schemes to multiphase flows. Also, we will present some new developments about the pressure flux. For more details and other numerical properties, the reader should consult with the cited references.

As a first step common in the AUSM schemes, we explicitly split the inviscid flux (written in one space-dimension without loss of generality) into two parts:

$$\mathbf{F} = \mathbf{F}^{(c)} + \mathbf{P} = \dot{m}\psi + \mathbf{P}, \tag{9}$$

where

$$\dot{m} = \rho V, \quad \psi = (1, u, H)^T, \quad \mathbf{P} = (0, p, 0)^T. \tag{10}$$

The first term in $\mathbf{F}$ is the *convective* flux $\mathbf{F}^{(c)}$, indicating the convection of $\psi$ by the mass flux $\dot{m}$ and the second term is the *pressure* flux $\mathbf{P}$, containing nothing but the pressure. It is noted that the conservation of the total enthalpy is guaranteed if $H$, instead of total internal energy ($E$), is contained in $\psi$. Halt and Agarwal[19] split $H$ into $E$ and $p/\rho$ and put them respectively in the $\psi$ and $\mathbf{P}$ terms and called it the WPS scheme.

We note that in a three-dimensional control volume, the flux along the surface normal formally looks identical to Eqs. (9-10). Hence, at each control surface, the mass flux is treated in an one-dimensional fashion.

Making distinction of these two fluxes gives rise to one of unique features of the AUSM-family schemes. Since the mass flux appears in all equations, its effects will be felt by all the variables. Hence, it is desirable to observe this fact at the discrete level as well when devising a new scheme. Significant benefits can be derived as well. For example, the numerical dissipation term is scalar even for the system of equations; it is just as easy to add more conservation equations insofar as the numerical flux is concerned.

It is possible to write a numerical flux, mimicking the expression at the continuum level, in terms of a common mass flux in the following general upwind form.

$$\mathbf{f}_{1/2} = \dot{m}_{1/2}\,\psi_{L/R} + \mathbf{p}_{1/2} = \dot{m}_{1/2}^{+}\psi_L + \dot{m}_{1/2}^{-}\psi_R + \mathbf{p}_{1/2}. \tag{11}$$

Here the contributions of $\psi_L$ and $\psi_R$ are weighted by the split masses $(\dot{m}_{1/2}^{+}, \dot{m}_{1/2}^{-})$, which must follow the consistency requirement,

$$\dot{m}_{1/2} = \dot{m}_{1/2}^{+} + \dot{m}_{1/2}^{-}. \tag{12}$$

This fact is automatically satisfied by the first element of $\mathbf{f}$. One can rewrite Eq. (11), using Eq. (12), as

$$\mathbf{f}_{1/2} = \frac{1}{2}\dot{m}_{1/2}\,(\psi_L + \psi_R) - \frac{1}{2}\mathcal{D}_m(\psi_R - \psi_L) + \mathbf{p}_{1/2}, \tag{13}$$

The dissipation term, $\mathcal{D}_m$, is

$$\mathcal{D}_m = \dot{m}_{1/2}^{+} - \dot{m}_{1/2}^{-}. \tag{14}$$

The subscripts "L" and "R" are understood to mean the cell centers on either side of the interface at which the normal vector is assumed to point from "L" to "R".

The quantities $(\dot{m}_{1/2}^{+}, \dot{m}_{1/2}^{-})$ are required to satisfy these conditions,

$$(\dot{m}_{1/2}^{+}) \geq 0, \quad (\dot{m}_{1/2}^{-}) \leq 0, \tag{15}$$

so that they provide proper upwinding, thus ensuring stability. In most of the AUSM-family schemes, these two variables are mutually exclusive, i.e.,

$$(\dot{m}_{1/2}^{+})\,(\dot{m}_{1/2}^{-}) \equiv 0, \tag{16}$$

It must be noted that the flux expressed in the form of Eq. (11) implies that the numerical dissipation is of the scalar, rather than the matrix form, because the same factors $\dot{m}_{1/2}^{\pm}$ are applied throughout for all conservation equations. The flux difference splitting schemes are known to belong to the category of matrix dissipation. On the other hand, the category of scalar dissipation encompasses several existing schemes other than the AUSM-family schemes, such as central differencing with artificial damping, the Van Leer flux vector scheme, the HLLE scheme. Indeed, there are several attractive properties associated with the scalar form of dissipation. From the algorithmic viewpoint, it offers simplicity, efficiency, and generality allowing for an easy extension to other systems of equations.

The 1991 AUSM scheme has served well by laying out the basis for further developments. One of the important developments is the concept of common speed of sound, which makes an accurate resolution of contact and shock discontinuities possible for both steady and unsteady flows.

In the following sections, we shall describe separately the details of mass and pressure fluxes, $\dot{m}_{1/2}$ and $p_{1/2}$.

## 3.1   Mass Flux

In what follows, we will give some basic formulas used to define the mass flux of two variants in the AUSM-family. To facilitate the discussion, we first define the following split functions.

$$\begin{aligned}
\mathcal{M}^{\pm}_{(1)}(M) &= \tfrac{1}{2}(M \pm |M|), \\
\mathcal{M}^{\pm}_{(2)}(M) &= \pm\tfrac{1}{4}(M \pm 1)^2
\end{aligned} \tag{17}$$

$$\mathcal{M}^{\pm}_{(4)}(M) = \begin{cases} \mathcal{M}^{\pm}_{(1)}(M), & \text{if } |M| \geq 1, \\ \mathcal{M}^{\pm}_{(2)}(1 \mp 2\mathcal{M}^{\mp}_{(2)}), & \text{otherwise,} \end{cases} \tag{18}$$

and

$$\mathcal{P}_{(5)}{}^{\pm}(M) = \begin{cases} \frac{1}{M}\mathcal{M}^{\pm}_{(1)}(M), & \text{if } |M| \geq 1, \\ \mathcal{M}^{\pm}_{(2)}[(\pm 2 - M) \mp 3M\mathcal{M}^{\mp}_{(2)}], & \text{otherwise.} \end{cases} \tag{19}$$

The numerals in the subscript of $\mathcal{M}^{\pm}_{(1)}$, $\mathcal{M}^{\pm}_{(2)}$, $\mathcal{M}^{\pm}_{(4)}$, and $\mathcal{P}^{\pm}_{(5)}$ indicate the degree of polynomials.

As it has become known during last decade that for low speed flows, the upwind scheme produces excessive numerical dissipation. To eradicate this phenomenon, we propose the concept of numerical speed of sound [20] which scales with the flow speed, thereby diminishing with Mach number. The numerical speed of sound is easily expressed in terms of a scaling function as follows,

$$\tilde{a}_{1/2} = f(\bar{M}; M_*)a_{1/2}, \quad a_{1/2} = (a_L + a_R)/2 \ \text{ for example,} \tag{20}$$

where the scaling factor may be of this form,

$$f(\bar{M}; M_*) = \frac{\sqrt{(1 - M_*^2)^2 \bar{M}^2 + 4M_*^2}}{1 + M_*^2}, \quad \bar{M}^2 = (M_L^2 + M_R^2)/2 \tag{21}$$

with the reference Mach number,

$$M_*^2 = \min(1, \max(\bar{M}^2, M_{co}^2)). \tag{22}$$

Some other suitably chosen forms may be appropriate, as will be detailed in the section on multiphase flows modeling. The cutoff parameter $M_{co}$ is introduced to prevent a singularity at stagnation point. It is a user-specified parameter and $M_{co}^2 = 10^{-4}$ has been used. Details can be found in Ref. [20].

A major consequence of using the numerical speed of sound is that the condition number of the preconditioned system becomes,

$$\kappa_\Gamma = \frac{|u| + \tilde{a}}{|u|} \to O(1), \quad \text{while } \tilde{a} \to 0, \quad \text{as } |u| \to 0. \tag{23}$$

That is, the condition number remains order of unity at low speeds. Further, the numerical dissipation based on this new speed of sound now scales with the local speed $|u|$, instead of the local speed of sound $a$. As a result, the accuracy can be restored as it is applied to low Mach flows, as evident in Fig. 1.

**Fig. 1.** Pressure contours for the shuttle external tank problem for $M_\infty = 0.01$. a): using the standard AUSM$^+$ at N=6400 time steps; and b): using numerical speed of sound at N=1000 time steps.

Using the numerical speed of sound $\tilde{a}_{1/2}$ to define

$$M_L = u_L/\tilde{a}_{1/2} \quad M_R = u_R/\tilde{a}_{1/2}. \tag{24}$$

Noticing that for simplicity we have dropped from $M_{L/R}$ the "tilde" sign used to accent the application of $\tilde{a}$. We now show two variants of the AUSM schemes about how the interface Mach number and eventually the mass flux is defined. They all have the common structure, namely

$$M_{1/2} = \mathcal{M}_{1/2} + M_p. \tag{25}$$

But each has a slightly-varying form for the two terms, $\mathcal{M}_{1/2}$ and $M_p$, where $M_p$ is the pressure diffusion term introduced to enhance calculations of low Mach number and multiphase flows. Let us begin by denoting

$$\Delta p = p_R - p_L, \tag{26}$$

and

$$\Delta \mathcal{M} = \left[ \mathcal{M}^+_{(4)}(M_L) - \mathcal{M}^+_{(1)}(M_L) - \mathcal{M}^-_{(4)}(M_R) + \mathcal{M}^-_{(1)}(M_R) \right] \geq 0. \tag{27}$$

Below we give expression of $\mathcal{M}_{1/2}$ and $M_p$ for AUSM$^+$ and LDFSS.

1. AUSM$^+$:

$$\mathcal{M}_{1/2} = \mathcal{M}^+_{(4)}(M_L) + \mathcal{M}^-_{(4)}(M_R), \tag{28}$$

and

$$M_p = -(1 - M^2_*)\Delta\mathcal{M}\frac{\Delta p}{(\rho_L + \rho_R)V^2_*}, \tag{29}$$

2. LDFSS:

$$M^+_{1/2} = \mathcal{M}^+_{(4),L} - \frac{1}{2}\Delta\mathcal{M}(1 + \frac{[\Delta p - \delta|\Delta p|]}{2\rho_L V^2_*}) \tag{30}$$

$$M^-_{1/2} = \mathcal{M}^-_{(4),R} + \frac{1}{2}\Delta\mathcal{M}(1 - \frac{[\Delta p + \delta|\Delta p|]}{2\rho_R V^2_*}), \tag{31}$$

where $\delta$ is a suitably-chosen scaling constant. Then,

$$\mathcal{M}_{1/2} = M^+_{1/2} + M^-_{1/2}. \tag{32}$$

There are several key similarities and differences among these two forms. First, both are designed to scale properly in the incompressible limit, meaning that the physical speed of sound, wherever it appears, must be replaced by some quantity that scales as the velocity magnitude as the local Mach number approaches zero. The preconditioned system provides two natural quantities, $\tilde{a}_{1/2}$ and $M^2_*$, that can effect this transition. In particular, $M^2_* \to 1$ as the local Mach number approaches or exceeds unity (the compressible limit) and $M^2_* \to \max(\bar{M}^2, M^2_{co})$ in the incompressible limit. Similarly, $\tilde{a}_{1/2} \to a_{1/2}$ as the local Mach number approaches or exceeds unity, and $\tilde{a}_{1/2} \to a_{1/2}\sqrt{\bar{M}^2 + 4M^2_*}$ in the incompressible limit. Both also reduce to a form of advective upwinding of the flux, based on the sign of an average velocity at the interface. The precise form of this average velocity differs between the schemes, but the key trait that this component of the flux vanishes as the velocity vanishes is present in both.

Both schemes also incorporate in the interface flux a "pressure diffusion" term proportional to:

$$\frac{1}{2}\Delta\mathcal{M}\frac{\Delta p}{\rho V^2_*} \tag{33}$$

For AUSM$^+$, this term is scaled by $1 - M^2_*$, meaning that its effects will vanish as the local Mach number approaches the sonic speed. LDFSS retains this component at all speeds, as it assists in ensuring a monotone capturing of non-grid aligned discontinuities at higher speeds. The pressure diffusion mechanism is found in other schemes, such as Roe's method and is known to cause the "carbuncle" often noted in calculations of supersonic flows past blunt bodies.[21,22] However, its inclusion is extremely beneficial for very low Mach number flows, as it provides pressure-velocity coupling necessary for stability and accuracy.

While AUSM$^+$ under most circumstances will not admit a "carbuncle" solution, LDFSS can, for the reasons mentioned above. To counteract this, an additional dissipative pressure diffusion mechanism scaled by $\delta$ was introduced, as shown in Eq. (31), first in [4] and further refined in [23]. This construct has been shown to suppress, if not entirely eliminate, the "carbuncle" response. It should be noted that the forms for $\tilde{a}_{1/2}$ and $M_*$ allow the schemes to respond in a manner that is independent of the thermodynamic state of the fluid in the incompressible limit. Furthermore, for conditions approaching a stationary contact discontinuity ($u_L = u_R = 0$, $p_L = p_R$), the numerical convective flux for all of the schemes vanishes as it should.

Now the mass flux is immediately determined by upwinding:

$$\overset{\bullet}{m} = \tilde{a}_{1/2} M_{1/2} \begin{cases} \rho_L \psi_L, & \text{if } M_{1/2} > 0, \\ \rho_R \psi_R, & \text{otherwise,} \end{cases} \tag{34}$$

## 3.2 Pressure Flux

Lastly, we show the definition of the pressure flux. We see in

$$\mathbf{P}_{1/2} = p_{1/2}(0, 1, 0)^T, \tag{35}$$

that all one needs is simply the definition of $p_{1/2}$.

In all the AUSM-family schemes, a general Van Leer/Liou - type interface pressure is used as a starting point,

$$p_{1/2} = \mathcal{P}^+_{(m),L} p_L + \mathcal{P}^-_{(m),R} p_R \tag{36}$$

where $m = 1, 3,$ or $5$ corresponds to the polynomial choices given above.

As simple as it may seem, there apparently are enough opportunities to further enhance the AUSM-family. As shown in Refs. [24,25], it needs modification to be suitable for general fluids at all speeds. The latest forms of the pressure splitting alleviates some of the ambiguities of the earlier attempts and is described next.

A recent effort in this regard has been reported [5], with an aim at further improving the scheme's robustness and accuracy. The complaint voiced commonly about the AUSM$^+$ concerns the overshoots resulting from strong shock-shock interactions. In the AUSMDV[3] scheme, the convective part of the momentum flux consists of blending the flux-difference and flux-vector procedures (thus denoted with DV). A notable advantage of this strategy is that it gives smooth shock profiles, e.g., in shock-shock interactions. The notion was adopted in the AUSM$^+$-W[6], but its effectiveness has never been extensively tested. This blended flux can be recast and the extra terms can be reassigned to the pressure flux, giving rise to a new pressure flux containing a term proportional to the velocity difference. This interpretation, however, has a sound connection to the characteristic equations,

$$dp \pm \rho a \, du = 0, \quad \text{along} \quad \frac{dx}{dt} = u \pm a. \tag{37}$$

An integrated form for the interface pressure for $|u| < a$ is

$$p_{1/2} = \frac{1}{2}[(p_L + p_R) - \rho_{1/2} a_{1/2}(u_R - u_L)]. \tag{38}$$

Hence, we can beef up the pressure flux, Eq. (36) by including the velocity difference term,

$$p_{1/2} = \mathcal{P}^+_{(5)}(M_L)p_L + \mathcal{P}^-_{(5)}(M_R)p_R \\ - \mathcal{P}^+_{(5)}(M_L)\mathcal{P}^-_{(5)}(M_R)\rho_{1/2}\tilde{a}^2_{1/2}(M_R - M_L),$$

(39)

where we may use $\rho_{1/2} = (\rho_L + \rho_R)/2$.

The coefficient involving $\mathcal{P}^+_{(5)}(M_L)$ and $\mathcal{P}^-_{(5)}(M_R)$ is introduced to automatically transition between supersonic and subsonic conditions. The pressure now is explicitly coupled with the velocity field by the addition of the velocity difference term.

As $M$ tends to zero, the pressure flux reduces to a form similar to Eq. (38), but with a much smaller dissipation coefficient $\tilde{a}$, rather that $a$,

$$p_{1/2} = \frac{1}{2}[(p_L + p_R) - \frac{1}{2}\rho_{1/2}\tilde{a}_{1/2}(u_R - u_L)].$$

(40)

To denote this new version first appeared in Ref. [5]), we use the suffix "u" (for velocity diffusion) and call it AUSM$^+$-u. Thus, when $M_p$ is also included in Eq. (25), the scheme reads AUSM$^+$-up. Or if the numerical speed of sound is also activated, the version becomes AUSM$^+$-au. Note that if AUSM$^+$-au is used, then the coefficient in Eq. (39) is scaled however, with the numerical speed of sound $\tilde{a}$. This is just what we want for low Mach number flows since $\tilde{a} = O(u)$ as $|u| \to 0$ and hence the coefficient is scaled by the magnitude of local velocity.

Another formulation given in Ref. [25] is explained below. Rewriting Eq. (36) as,

$$p_{1/2} = \frac{1}{2}(\mathcal{P}^+_{(m),L} - \mathcal{P}^-_{(m),R})(p_L - p_R) + \frac{1}{2}(p_L + p_R)(\mathcal{P}^+_{(m),L} + \mathcal{P}^-_{(m),R})$$
$$= \frac{1}{2}(p_L + p_R) + \frac{1}{2}(\mathcal{P}^+_{(m),L} - \mathcal{P}^-_{(m),R})(p_L - p_R)$$
$$+ \frac{1}{2}(p_L + p_R)(\mathcal{P}^+_{(m),L} + \mathcal{P}^-_{(m),R} - 1)$$

(41)

The first term on the right-hand side is a simple average and is considered to be appropriate for low speed flow. The second term is considered to be a dissipative term and is $O(\Delta p)$ since its coefficient $\mathcal{P}^+_{(m),L} - \mathcal{P}^-_{(m),R}$ is of order unity. However, the third term remains of order unity because terms in both brackets are of order unity. This becomes excessively dissipative at low speeds, as shown in [25], and is unacceptable in the incompressible limit. This major difficulty can be rectified by replacing the cell-average pressure $\frac{1}{2}(p_L + p_R)$ in this term by $\rho_{1/2}V_*^2$ or $\rho_{1/2}\tilde{a}^2_{1/2}$. In either case, the term now scales as the velocity magnitude in the incompressible limit and is independent of the thermodynamic state of the fluid. The second choice above is similar to that given above by Liou [5].

The final form of the pressure splitting is thus

$$p_{1/2} = \frac{1}{2}(p_L + p_R) + \frac{1}{2}(\mathcal{P}^+_{(m),L} - \mathcal{P}^-_{(m),R})(p_L - p_R) \\ + \rho_{1/2}V_*^2(\mathcal{P}^+_{(m),L} + \mathcal{P}^-_{(m),R} - 1).$$

(42)

It should be noted that this new splitting will be slightly more diffusive for compressible gas calculations than the standard splitting techniques. The real-fluid and multiphase flow calculations presented later all use this version of the pressure splitting.

The new pressure flux $AUSM^+$-up scheme completely removes the overshoots, yielding results as good as that by the Godunov scheme, suggesting that the new pressure flux formula is a worthy replacement of the old one and will be used again in all the following tests.

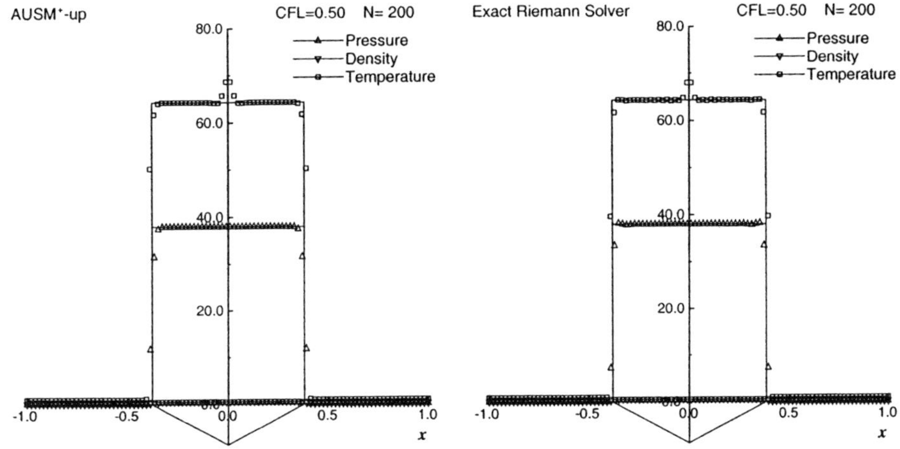


**Fig. 2.** Colliding shock problem

The second problem concerns a shock moving slowly against a flow, as studied in Ref. [26]. Figures 3 shows the strength of linear and nonlinear waves by the $AUSM^+$-u ($AUSM^+$-up is indistinguishable and not shown), and Godunov schemes. It confirms in the figure that the Godunov schemes produces a noticeable long wave trailing the shock, while the AUSM scheme performs well.

Shown in Fig. 4 is a blunt body problem, often used by Radespiel[27]; this problem has several features to study. The grid is clearly for the viscous calculations, but is tested here for the Euler solutions. First, the solution is free of carbuncle phenomena, consistent with the conjecture given in Ref. [28] since there is no explicit pressure diffusion term in the mass flux. Secondly, the pressure contours are smooth, not only near the wall, shown in the blow-up view near the stagnation region, but also near the sonic line region.

## 4   Multiphase Flow Modelings

Recently, the AUSM-family has been extended to the multiphase flow calculations, e.g., in Refs. [29,24,30]. Paillère et al.[30] solved a system of two-fluid models with interfacial source terms included. Several features that are different from the usual equations for aerodynamic flows add complexity significantly.

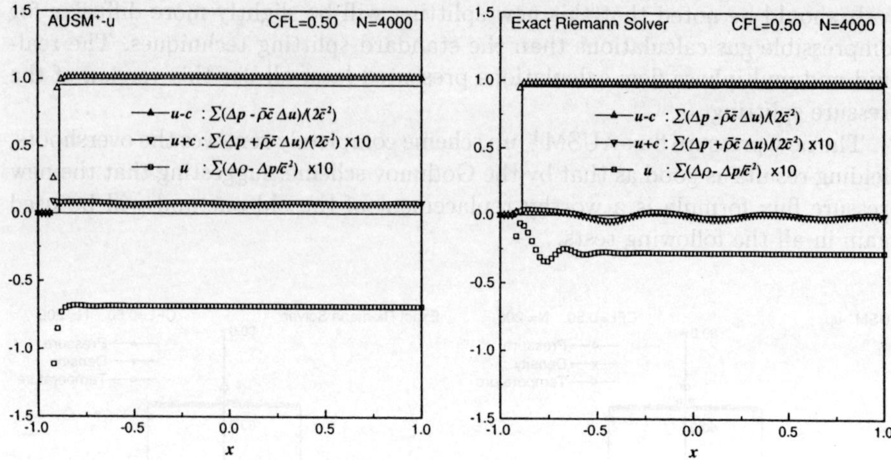

**Fig. 3.** Slowly moving shock problem.

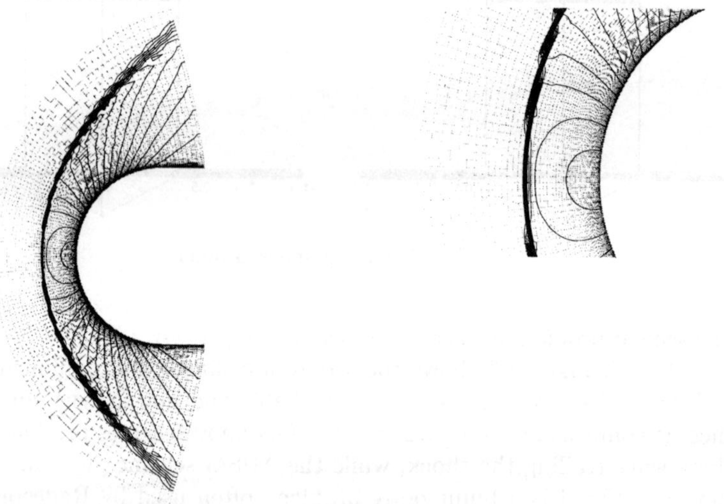

**Fig. 4.** Blunt body problem, $M_\infty = 10$.

That is, the system is no longer in conservative form because of the presence of the source terms and the system is not guaranteed to be hyperbolic because it admits complex eigenvalues.

Another recent accomplishment of using the AUSM scheme has been reported in Ref. [18] by De Wilde et al. for an extremely complicated set of equations and flow patterns involving solid particles and gas phase in an industrial scale riser. Due to the inelastic particle-particle collisions, flow instability is triggered and a periodic slugging flow pattern was obtained by them.

These extensions differ in detail, but not in concept, from those outlined in the remainder of the paper, giving another indication of flexibility of the

AUSM-family. In what follows, we shall describe in detail about the models for multiphase flows, from the simplest to complicated multi-fluid forms.

## 4.1 Single Fluid Formulation - homogeneous Equilibrium Two-Phase Flow Model

The thermodynamic state of a single-phase real fluid is defined by the relations $p = p(\rho, T)$ and $h = h(\rho, T)$. In the present work, we utilize the Peng-Robinson equation of state [31], a cubic formulation similar to the Van der Waals equation but generally much more accurate in the liquid phase. Similar equation can be used to describe the thermodynamic relationship of a given fluid, such as the one by Sanchez and Lacombe [32] which will be used in this paper as well. The Peng-Robinson equation is given by

$$p \qquad = Z(\tilde{\rho}, T)\tilde{\rho}RT, \tag{43}$$

$$Z(\tilde{\rho}, T) = \frac{1}{1 - b\tilde{\rho}} - \frac{a(T)}{RT} \frac{\tilde{\rho}}{(1 + 2b\tilde{\rho} - (b\tilde{\rho})^2)} \tag{44}$$

where $\tilde{\rho}$ is the molar density $\rho/M_w$, $M_w$ is the molecular weight, and $R$ is the universal gas constant.

Typical isotherms for the Peng-Robinson equation are plotted in Fig. 5 on a pressure-density diagram. Clearly indicated is the vapor regime, where pressure varies nearly linearly with density, and the liquid regime, where large pressure changes are required to induce a density change. For a given pressure and temperature, the solution of Eq. (44) returns one or three values of the compressibility factor $Z$ or density, the former corresponding to the single-phase region (either liquid or vapor) and the latter corresponding to the two-phase region, where vapor and liquid may co-exist. The corresponding densities for a pressure within the two-phase region are shown as points A-C. A and C represent saturated vapor and liquid states, while B is physically meaningless. For a particular temperature, the "allowable" two-phase region is bounded by the pressure values at D and E, which are local extrema. The loci of these pressure values for temperatures between the triple and critical points define liquid and vapor spinodal curves, dividing the two-phase region into metastable vapor, unstable, and metastable liquid regions. At a particular pressure between the liquid and vapor spinodal points, the system is in equilibrium, with the vapor and liquid fugacities attaining equal values. This pressure is known as the vapor pressure $p$vap and is calculated as a function of temperature by iterating on the equation

$$f(Z_v, T, p\text{vap}) = f(Z_l, T, p\text{vap}), \tag{45}$$

where the fugacity $f$ is given by

$$\ln \frac{f}{p} = Z - 1 - \ln(Z - B) - \frac{A}{2\sqrt{2}B} \ln\left(\frac{Z + (1 + \sqrt{2})B}{Z + (1 - \sqrt{2})B}\right), \tag{46}$$

where

$$A = \frac{a(T)p}{R^2 T^2}, \quad B = \frac{bp}{RT} \tag{47}$$

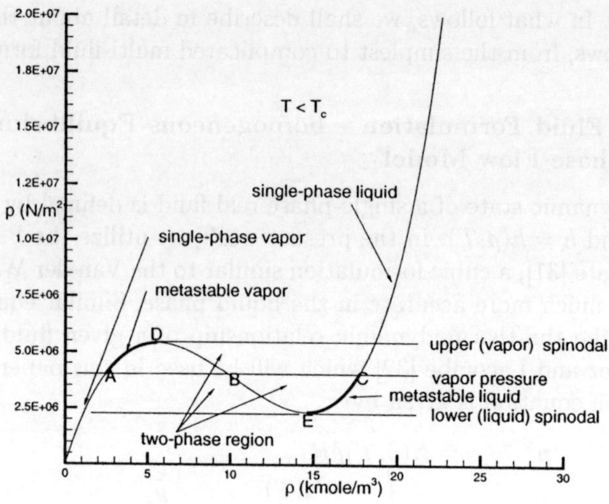



**Fig. 5.** Pressure vs. molar density (isotherm below critical temperature).

The spinodal pressure and density values (points D and E in Fig. 5) can be obtained analytically by solving the quartic equation $\frac{\partial p}{\partial \rho}|_T = 0$, discarding two meaningless roots that occur outside the range of validity of the Peng-Robinson equation. The spinodal pressure values bound the actual vapor pressure, and an appropriate linear combination can be used as an initial guess for the iteration described above.

The Peng-Robinson equation (and similar ones) gives no useful information in the unstable parts of the two-phase region. For densities between the spinodal values, it can be shown that the acoustic eigenvalues are complex, meaning that the Euler system is not hyperbolic in time and that conventional time-marching procedures for integrating the equations are ill posed. It is also of note that the liquid spinodal pressure may be negative for high molecular-weight liquids at lower temperatures, implying that the simulated expansion of a liquid might produce reasonable densities, but unphysical pressures, in the metastable region.

One means of avoiding these difficulties starts with the introduction of a void-fraction formalism for the two-phase region and the assumption of thermodynamic and kinematic equilibrium between the phases. For an equilibrium two-phase flow, the vapor pressure $p_{vap}(T)$ is directly related to the temperature through the Clausius-Clapeyron equation, and the density and temperature are independent variables. Given updated values for the density and temperature at a grid point as determined from a time-integration method, the following procedure is performed:

1. Determine the vapor pressure at that temperature, either through reference to a curve-fit or by the iterative procedure described above, and establish

the saturation densities $\rho_l(T)$ and $\rho_v(T)$ and the saturation enthalpies $h_l(T)$ and $h_v(T)$ using (44), and the specific-enthalpy equation

$$h(\rho, T) = h_I(T) + \frac{1}{M_w}[RT(Z - 1) + \frac{T\frac{da(T)}{dT} - a(T)}{2\sqrt{2}b}\ln(\frac{Z + (1 + \sqrt{2})B}{Z + (1 - \sqrt{2})B})]. \tag{48}$$

2. If the density is not between the saturation values or the temperature is greater than the critical value, then the single phase description given by the Peng-Robinson equation will be used to determine the pressure and enthalpy.

3. If the fluid density is within the saturation limits, the equilibrium equation of state for the homogeneous mixture of liquid and vapor is given by

$$p = p_{\mathrm{vap}}(T) \tag{49}$$

$$\rho h(\rho, T) = \rho_v(T)\alpha_v(\rho, T)h_v(T) + \rho_l(T)\alpha_l(\rho, T)h_l(T) \tag{50}$$

$$\alpha_v(\rho, T) = \frac{\rho - \rho_l(T)}{\rho_v(T) - \rho_l(T)} \tag{51}$$

$$\alpha_l(\rho, T) = 1 - \alpha_v \tag{52}$$

where the subscripts $l$ and $v$ refer to liquid and vapor, and $\alpha$ is the voidage. The vapor pressure $p_v$ and the associated saturation-state enthalpies and densities can be calculated directly from the equation of state, thus defining the "vapor dome" in a self-consistent manner.

In this description, the saturation-state values are strict functions of temperature; density dependence is introduced through the void fractions $\alpha$ and latent-heat effects arise through the change in departure enthalpy between the saturation states. The thermodynamic derivatives $p_\rho$, $p_T$, $(\rho h)_\rho$, and $(\rho h)_T$ needed in the time-integration method and in the sound speed definition can be computed by straightforward differentiation of the expressions above. These are discontinuous at phase transition points, leading to dramatic changes in the effective "sound speed" in the two-phase region. Figure 6 plots $a^2$ as a function of molar density for both the Peng-Robinson equation and the Peng-Robinson equation augmented by the equilibrium two-phase flow model Eqs. (49-52). The fluid is octane at a temperature of 350 K. As shown, the equilibrium two-phase description preserves a real value for the "sound speed", while the basic Peng-Robinson equation results in negative values for $a^2$. Also shown is a theoretical result for the sound speed in a homogeneous two-phase mixture of liquid and vapor [33]:

$$\frac{1}{\rho a^2} = \frac{\alpha_v}{\rho_v a_v^2(T)} + \frac{\alpha_l}{\rho_l a_l^2(T)} \tag{53}$$

where $a_{v,l}^2(T)$ may be obtained from

$$a^2 = \frac{(\rho h)_T p_\rho - p_T((\rho h)_\rho - h)}{(\rho h)_T - p_T}, \tag{54}$$

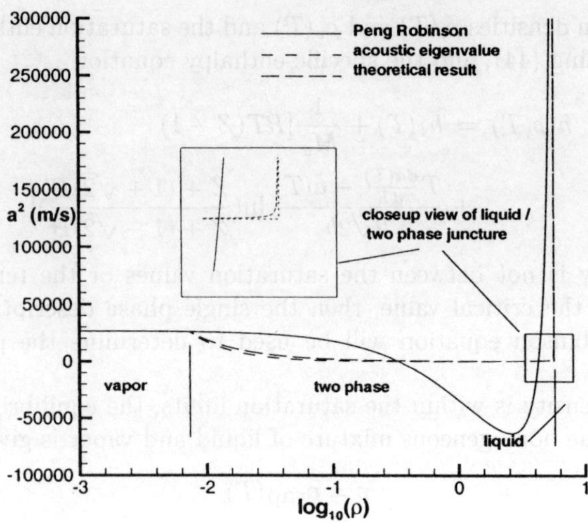



**Fig. 6.** $a^2$ vs. molar density (octane at 350 K).

and are evaluated at the saturation states $\rho_{v,l}(T)$. The eigenvalue calculation for $a^2$ agrees reasonably well with the theoretical estimate except near single-phase / two-phase junctures, where the latter blends smoothly with the saturation-state values and the former exhibits a jump discontinuity. The theoretical expression for the sound speed is numerically more robust and is used in all calculations presented herein. Both expressions result in very small values (on the order of meters per second) for the "sound speed" near the liquid phase / two-phase interface, meaning that a shift to a locally "supersonic" flow condition during a phase transition is a distinct possibility.

The above formulation neglects velocity-slip effects, with the velocity actually solved for being a mass-weighted average velocity. This system is hyperbolic in character and is similar to the Euler system in structure but admits such multiphase features as cavitation zones and condensation shocks. A key element is the use of density and temperature as the "working" thermodynamic variables, particularly in contrast with the low-speed formulation described earlier, which utilizes pressure and temperature as the "working" thermodynamic variables. This choice is driven by the equilibrium closure for the two-phase region, in which pressure and temperature are not independent variables.

The Euler system that describes this flow is identical to that of single phase fluid. In the homogeneous equilibrium model, it is necessary that $\tilde{\mathbf{Q}}$ contain $\rho$, rather than the pressure, as in the two-phase region, $p = p(T)$ only.

The eigenvalues of $\mathbf{\Gamma}^{-1}\frac{\partial \mathbf{F}}{\partial \tilde{\mathbf{Q}}}$ are

$$u, \quad \frac{1}{2}[(1 + M_*^2)u \pm \sqrt{u^2(1 - M_*^2)^2 + 4V_*^2}]. \tag{55}$$

In the limit of an incompressible flow, the acoustic eigenvalues revert to $\frac{1}{2}[u \pm \sqrt{u^2 + 4V_*^2}]$, which now scale as the reference velocity $V_*$. AUSM-family extensions for this real-fluid/two-phase flow model are very similar to the standard compressible flow formulation, with the exception being the use of the generalized state equation to define the sound speed, pressure, and enthalpy. Since $p \neq p(\rho)$ in the two-phase region, it is necessary to solve for the density as a working variable, rather than the pressure.

## 4.2    Compressible two-phase mixture model

The second fluid system describes a mixture of two fluid phases, each of which can consist of multiple components. This model has advantages over the homogeneous equilibrium model considered previously, as non-equilibrium phase transitions are allowed. A common pressure is assumed to hold for both phases, and in the simplest variant of this model, the velocities of each phase are assumed to be same as the mass-averaged velocity of the mixture. Mass, momentum, and energy exchange due to phasic velocity-slip can be handled by suitable drift flux models in more refined variants. Considering, for simplicity, a one-dimensional flow with each phase having only one component, the Euler system is given again by Eq. (1) with the vectors defined as

$$\mathbf{Q} = \begin{bmatrix} \rho Y_v \\ \rho \\ \rho u \\ \rho H + p \end{bmatrix}, \quad \mathbf{F} = \begin{bmatrix} \rho Y_v u \\ \rho u \\ \rho u^2 + p \\ \rho u H \end{bmatrix}, \tag{56}$$

In this, $Y_v$ is the mass fraction of vapor, which is is related to the vapor voidage $\alpha_v$ by

$$Y_v = \frac{\rho_v(p, T)\alpha_v}{\rho}, \tag{57}$$

with the bulk density $\rho$ defined by

$$\frac{1}{\rho} = \frac{Y_v}{\rho_v(p, T)} + \frac{Y_l}{\rho_l(p, T)}, \tag{58}$$

$$Y_l = 1 - Y_v \tag{59}$$

The mixture enthalpy $h$ is given by

$$h = Y_v h_v(p, T) + Y_l h_l(p, T) \tag{60}$$

In this model, each phase can be described by its own state equation: $\rho_{v,l} = \rho_{v,l}(p, T)$ and $h_{v,l} = h_{v,l}(p, T)$.

The preconditioner for this system has

$$\boldsymbol{u}^T = [Y_v, 1, u, H] \tag{61}$$

$$\boldsymbol{v}^T = [0, 1, 0, 0] \tag{62}$$

534

but otherwise is of the same structure as Eq. (4). For this system, a natural choice for $\tilde{\mathbf{Q}}$ is $[Y_v, p, u, T]^T$. The eigenvalues of $\boldsymbol{\Gamma}^{-1}\frac{\partial \mathbf{F}}{\partial \mathbf{Q}}$ are the same as Eq. (55), but with

$$a^2 = \frac{\rho h_T}{\rho(\rho_p h_T - \rho_T h_p) + \rho_T} \qquad (63)$$

AUSM-family extensions for this two-phase flow model simply replace the vector of advected variables $\psi$ with $(Y_v, 1, u, H)^T$ and utilize the appropriate "mixture" definitions for the sound speed, density, and enthalpy. Here, it is best to solve for the pressure, rather than the density, as the pressure will be nearly constant across phase interfaces.

## 4.3  Separated two-phase gas-solid flow model

The third fluid system considered in this work is a hyperbolic model for two-phase gas-solid flows developed by Gidaspow, et al. [34] In contrast to the previous models, this model solves separate momentum equations for each of the two phases. The system is rendered hyperbolic in time by concentrating the gas- and solids-phase pressure effects in the gas- and solids-phase momentum equations, respectively. This gas-solid system is simpler than separated two-phase models for compressible flows in that both the solids and gas densities are assumed constant and the energy equations are eliminated. Compressibility effects are thus absent from the gas-phase, but by virtue of solids compaction / dilution, significant solids compressibility effects are present. The Euler system is composed of these vectors:

$$\mathbf{Q} = \begin{bmatrix} \rho_g \alpha_g \\ \rho_g \alpha_g u_g \\ \rho_s \alpha_s \\ \rho_s \alpha_s u_s \end{bmatrix}, \quad \mathbf{F} = \begin{bmatrix} \rho_g \alpha_g u_g \\ \rho_g \alpha_g u_g^2 + p_g \\ \rho_s \alpha_s u_s \\ \rho_s \alpha_s u_s^2 + p_s \end{bmatrix},$$

$$\mathbf{S} = \begin{bmatrix} 0 \\ \rho_g g + C(u_g - u_s) \\ 0 \\ (\rho_s - \rho_g)\alpha_s g - C(u_g - u_s) \\ 0 \end{bmatrix}, \qquad (64)$$

In this, $\alpha_g$ is the gas-phase void fraction, $\alpha_s = 1 - \alpha_g$ is the solids-phase void fraction, $u_s, u_g, v_s, v_g$ are the solids and gas-phase velocities, and $g$ is acceleration due to gravity. The gas-phase intrinsic density $\rho_g$ and the solids-phase intrinsic density $\rho_s$ are assumed constant. The function $C$ scales the momentum exchange terms and is defined as follows:

$$C = 150\left(\frac{\alpha_s}{\alpha_g}\right)^2 \frac{\mu}{d_p^2} + 1.75\rho_g\left(\frac{\alpha_s}{\alpha_g}\right)\frac{|\mathbf{V}_g - \mathbf{V}_s|}{d_p}, \quad \alpha_g < 0.8 \qquad (65)$$

and

$$C = 0.75 C_D \frac{\alpha_s \rho_g |\mathbf{V}_g - \mathbf{V}_s| \alpha_g^{-2.65}}{d_p}, \quad \alpha_g > 0.8 \qquad (66)$$

In the preceding expressions, $\mu$ is the gas viscosity (assumed constant), $d_p$ is the particle diameter, and $C_D$ is the drag coefficient. The form for the solids pressure is taken from Boivin, et al. [35]:

$$p_s = \rho_s C_s [\alpha_s + 2\alpha_{s,\mathrm{max}} \ln(1 - \frac{\alpha_s}{\alpha_{s,\mathrm{max}}}) - \frac{\alpha_s \alpha_{s,\mathrm{max}}}{\alpha_s - \alpha_{s,\mathrm{max}}}], \qquad (67)$$

where $\alpha_{s,\mathrm{max}}$ is the solids volume fraction at maximum compaction (taken as 0.64) and $C_s$ is a scaling constant (taken as 0.02 for most calculations). A solids "sound speed" can be defined through the relation

$$\rho_s a_s^2 = \frac{\partial p_s}{\partial \alpha_s} = \rho_s C_s \left(\frac{\alpha_s}{\alpha_s - \alpha_{s,\mathrm{max}}}\right)^2 \qquad (68)$$

The solids sound speed ranges from zero at zero solids voidage to infinity at maximum compaction.

This system can be preconditioned by a Weiss-Smith - type construction [25], with $\tilde{\mathbf{Q}}$ chosen as $[p_g, u_g, \alpha_s, u_s]$:

$$\boldsymbol{\Gamma} = \begin{bmatrix} \frac{\rho_g}{\rho \beta_g^2} & 0 & -\rho_g & 0 \\ \frac{\rho_g u_g}{\rho \beta_g^2} & \rho_g \alpha_g & -\rho_g u_g & 0 \\ 0 & 0 & \rho_s(1 + \frac{\rho_s a_s^2}{\rho}\theta_s) & 0 \\ 0 & 0 & \rho_s u_s(1 + \frac{\rho_s a_s^2}{\rho}\theta_s) & \rho_s \alpha_s \end{bmatrix} \qquad (69)$$

In this, $\beta_g$ is a reference velocity for the gas phase, $\theta_s$ is a scaling parameter, defined as $\theta_s = \frac{1}{\beta_s^2} - \frac{1}{a_s^2}$, and $\beta_s$ is a reference velocity for the solids phase. The quantities $\tilde{\rho}$ and $\bar{\rho}$ are reference densities. Appropriate choices [25] for these quantities are

$$\beta_g^2 = \max(u_g^2 + v_g^2, U_{\infty,g}^2), \qquad (70)$$

$$\beta_s^2 = \min(a_s^2, \max(u_s^2 + v_s^2, U_{\infty,s}^2)), \qquad (71)$$

$$\rho = \rho_g \alpha_g + \rho_s \alpha_s \qquad (72)$$

with $U_{\infty,g}$ and $U_{\infty,s}$ both set to user-specified constants. The forms for $\beta_g$ and $\beta_s$ are consistent with the assumptions of an incompressible gas phase and a "compressible" solids phase in the sense that the solids phase can become more dense or dilute as conditions change. The preconditioning strategy for the solids phase replaces the solids sound speed by a quantity proportional to $\beta_s$ as the degree of compaction increases. In contrast, a "supersonic flow" situation may prevail for dilute gas-solid flows, as the solids phase velocity may be much higher than the solids sound speed. The scaling by the bulk density $\rho$ implies that the numerical formulations for each phase (discussed later) will be coupled through the eigenvalues.

The eigenvalues of $\boldsymbol{\Gamma}^{-1}\mathbf{A}$, where $A = \frac{\partial \mathbf{F}}{\partial \tilde{\mathbf{Q}}}$ are: for the gas phase,

$$\{u_g, \frac{1}{2}(u_g \pm \sqrt{u_g^2 + 4V_{*,g}^2})\}, \qquad (73)$$

and for the solid phase,

$$\{u_s, \frac{1}{2}[(1 + M_{*,s}^2) \pm \sqrt{u_s^2(1 - M_{*,s}^2)^2 + 4V_{*,s}^2}]\}, \tag{74}$$

where

$$V_{*,g}^2 = \frac{\rho}{\rho_g}\beta_g^2 \tag{75}$$

$$\frac{1}{M_{*,s}^2} = 1 + \frac{\rho_s}{\rho}\frac{1 - M_s^2}{M_s^2} \tag{76}$$

$$M_s^2 = \frac{\beta_s^2}{a_s^2} \tag{77}$$

$$V_{*,s}^2 = a_s^2 M_{*,s}^2 \tag{78}$$

It is of note that the eigenvalues corresponding to the gas phase are identical with those obtained for Chorin's artificial compressibility method, with an effective reference velocity of $\sqrt{\frac{\rho}{\rho_g}\beta_g^2}$. The eigenvalues corresponding to the solids phase are identical to those outlined above for the other two fluid systems, except with a specially-defined reference Mach number given above. AUSM-family discretization of Gidaspow's two-fluid model are applied to each phase separately. This is possible as pressure and convective effects are segregated to separate phases in Gidaspow's model. Further details may be found in Ref. [25].

## 5   Applications

The test cases below illustrate the effectiveness of the schemes outlined above for solving two phase flows. Results the AUSM$^+$ and LDFSS variants are shown.

### 5.1   Supercritical ethylene injection into nitrogen - real fluid model with equilibrium phase transitions

Figure 7 presents results from supercritical ethylene injection into gaseous nitrogen under conditions identical to those considered by Wu, et al. [36] Upon expansion, the ethylene transitions from a supercritical fluid to a two-phase liquid-vapor mixture to a gas. Liquid-phase voidage contours are shown in the figure and indicate the effects of increasing chamber pressure in reducing the spreading rate of the jet.

At all chamber pressures, the fluid is choked within the nozzle by virtue of the lower speed of sound in the two-phase mixture. As the chamber pressure increases, the liquid content also increases but is nowhere larger than 5% by volume. Though not shown, a Mach disk is present for the chamber pressure of 0.41 MPa but disappears as the chamber pressure is raised. The Peng-Robinson [31] generalized state equation, supplemented by the homogeneous equilibrium model discussed earlier, is used in this calculation, and the LDFSS scheme is employed.

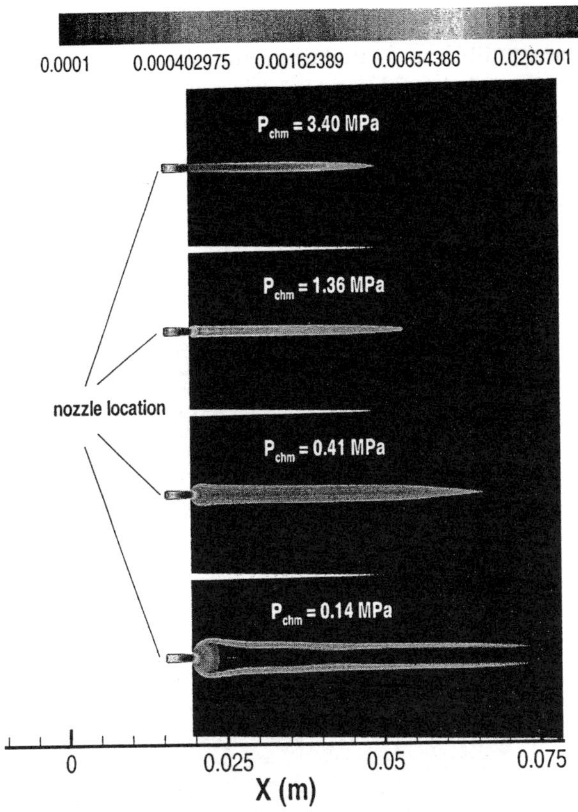



**Fig. 7.** Effect of chamber pressure on liquid-phase volume fraction: injection of super-critical ethylene into nitrogen.

## 5.2 Cavitating water flow over a hemisphere/cylinder geometry - real fluid model with equilibrium phase transitions and two-phase mixture model

Another example of multiphase flows involves a water flow over a hemispherical cylinder. The flow can undergo cavitation if the pressure difference ( cavitation number, $K = 2(p_\infty - p_v)/\rho_\infty U_\infty^2$) is low enough.

Figures 8 and 9 correspond to cavitating water flow over a hemisphere/cylinder geometry. Two flow models are used, with the first being a real fluid/homogeneous equilibrium model based on the Sanchez-Lacombe state equation. The second employs the two-phase mixture model, with liquid water described by an extended Tait's equation [37] and water vapor described by the ideal gas law. Vapor generation is described by a rate law similar to that proposed by Kunz, et al. [14] but recalibrated in Neaves and Edwards [38]. The LDFSS scheme is again used. The results are parameterized by a cavitation number $K = \frac{p_\infty - p_v}{1/2\rho_\infty U_\infty^2}$, with lower values resulting in are vapor generation. Figure 8 compares density contours for the equilibrium formulation and the finite-rate model. As shown, both formulations result in a sharp capturing of the cavitation bubble interface but differ in their predictions of the collapse of the cavity in the "wake" region. Figure 9 shows that the finite rate model, if calibrated carefully, can yield predictions superior to the equilibrium model.

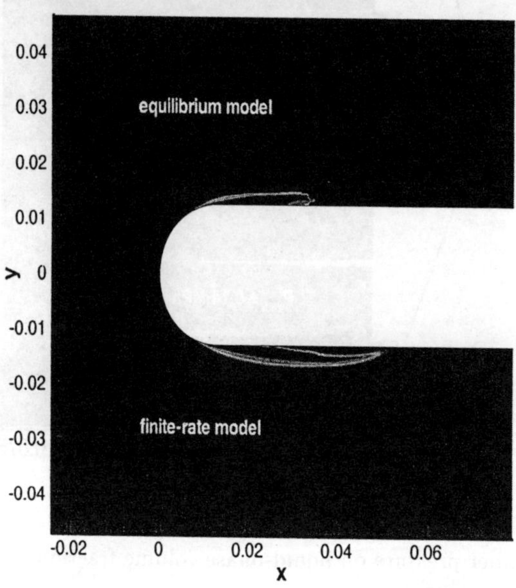


**Fig. 8.** Density contours: liquid water flow over a hemisphere/cylinder geometry.

## 5.3   Aerated liquid injector simulations - two-phase mixture model

Figure 10 presents snapshots of gas volume fraction within an aerated liquid injector. In this configuration, a central tube injects gas into a co-flowing liquid, forming bubbles that later merge into a gaseous core. This gaseous core pushes the liquid toward the walls of the injector. The injector configuration is a 2-D idealization of that considered by Lin, et al. [39], and the conditions correspond to a liquid flow rate of 0.38 l/min and a gas-to-liquid mass ratio of 0.08%. Snapshots at 5 millisecond intervals illustrate (from bottom to top of the figure) the initial formation and growth of the gas bubble and the eventual formation of a core-annulus structure in the straight nozzle passageway. In this calculation, the Tait's equation of state is used for liquid water and the ideal gas law is used for gaseous nitrogen. The AUSM$^+$ scheme is used in this calculation.

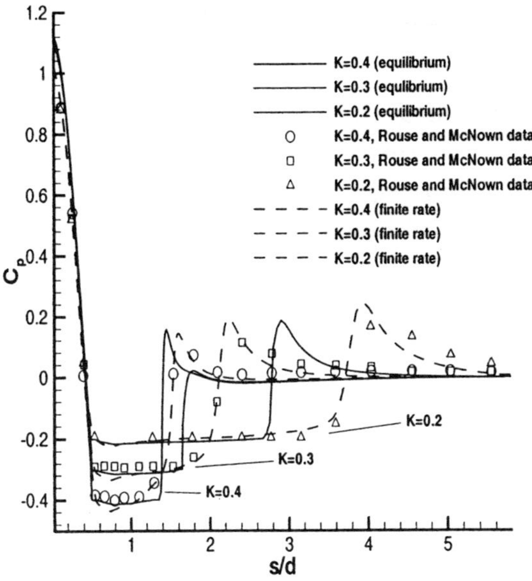


**Fig. 9.** Surface pressure distributions: liquid water flow over a hemisphere/cylinder geometry.

## 5.4 Gas-solid fluidized bed simulations - separated two-phase gas-solid model

In Fig. 11, we display the time-dependent response of a minimally-fluidized bed of solid particles to gas injection along the centerline. The hyperbolic two-phase flow equation system discussed previously is used, along with the AUSM$^+$ scheme. The sequence of events shown in Fig. 11 traces the initial formation of a bubble in the fluidized bed in response to jet injection. This bubble grows larger, and displaced solids are forced toward the bottom of the tank, where they act to impede the jet at later times. Bubbles formed later are thus smaller and the entire process is more chaotic. The numerical scheme (extended to second order) sharply captures the gas-solid interfaces and displays a physically consistent response, including a "spouting" behavior at later times. Solids mass loss is less than 0.2% over one second of elapsed time. The figure also highlights the effect of the choice of $m$ (degree of polynomial) in the pressure splitting, indicating nearly identical results.

## 6   Concluding Remarks

Simple modifications for extending the AUSM-family schemes, specifically AUSM$^+$ and LDFSS, for calculations of real fluids at all speeds and at all states of com-

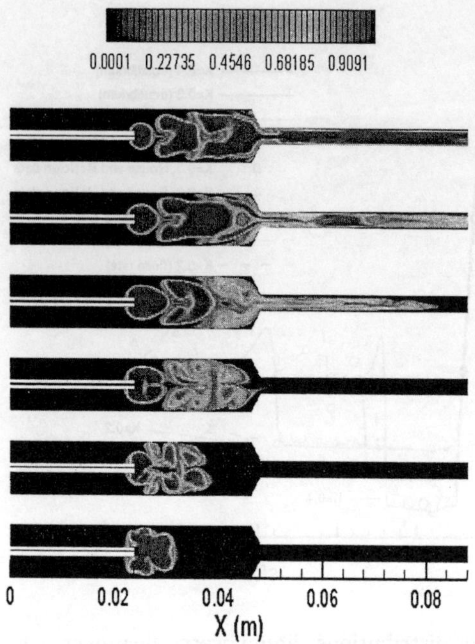

**Fig. 10.** Gas volume fraction evolution for time-dependent 2-D flow within an aerated liquid injector.

pressibility have been outlined in this paper. Three approaches for describing multiphase flows have been presented. The first model is based on the real fluid equation of state (such as those by Peng and Robinson, Sanchez and Lacombe) and applies thermodynamic equilibrium for treating liquid-vapor phase transitions. The other two models are nonequilibrium models which solve each phase separately via transport equations, with velocity and/or pressure equilibrium between phases. Detailed inter-phase interactions are problem dependent and can have rather complicated mathematical forms. Numerical results have shown that the modifications based on the AUSM-family schemes are effective in simulating incompressible liquid and compressible vapor responses as well as multiphase flow phenomena involved in realistic industrial devices, such as the appearance of cavitation bubbles, vapor-liquid condensation shocks, bubble formation in an aerated injector, and complicate dynamics of gas-solid interfaces.

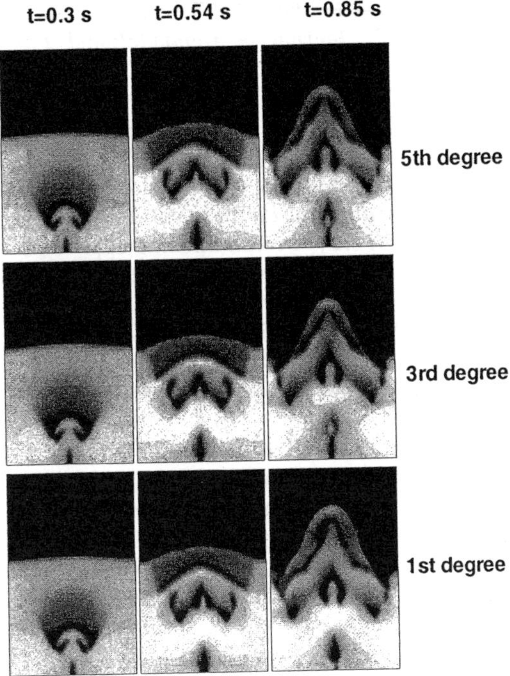


**Fig. 11.** Solids voidage contours: response of fluidized bed to gas injection.

# References


1. Liou, M.-S. and Steffen, Jr., C. J., "A New Flux Splitting Scheme," *Journal of Computational Physics*, Vol. 107, 1993, pp. 23–39, Also NASA TM 104404, May 1991.
2. Liou, M.-S., "A Sequel to AUSM: AUSM$^+$," *Journal of Computational Physics*, Vol. 129, 1996, pp. 364–382, Also NASA TM 106524, March 1994.
3. Wada, Y. and Liou, M.-S., "An accurate and robust flux splitting scheme for shock and contact discontinuities," *SIAM Journal on Scientific and Statistical Computing*, Vol. 18, 1997, pp. 633–657.
4. Edwards, J. R., "A low-diffusion flux-splitting scheme for Navier-Stokes calculations," *Computers & Fluids*, Vol. 26, 1997, pp. 635–659.
5. Liou, M.-S., "Ten Years in the Making–AUSM-Family," AIAA Paper 2001-2521-CP, 15th AIAA CFD Conference, June 11-14 2001.
6. Liou, M.-S., "Progress Towards an Improved CFD Method: AUSM$^+$," AIAA Paper 95-1701-CP, 12th AIAA CFD Conference, 1995.
7. Moschetta, J.-M. and Gressier, J., "The sonic point glitch problem: a numerical solution," *Lecture Notes in Physics, Volume 515*, 1998, pp. 403–408.

8. Turkel, E., "Preconditioned Methods for Solving Incompressible and Low Speed Compressible Equations," *Journal of Computational Physics*, Vol. 72, 1987, pp. 277–298.

9. van Leer, B., Lee, W. T., and Roe, P. L., "Characteristic time stepping or local preconditioning of the Euler equations," AIAA Paper 91-1552, AIAA 10th CFD Conference, June 1991.

10. Choi, Y. H. and Merkle, C. L., "The Application of Preconditioning in Viscous Flows," *Journal of Computational Physics*, Vol. 105, 1993, pp. 207–223.

11. Weiss, J. M. and Smith, W. A., "Preconditioning applied to variable and constant density time-accurate flows on unstructured meshes," AIAA Paper 94-2209, June 1994.

12. Oefelein, J. and Yang, V., "Simulation and Analysis of Supercritical Multiphase Combustion Processes," AIAA Paper 96-2880, 1996.

13. Merkle, C., Sullivan, J., Buelow, P., and Venkateswaran, S., "Computation of Flows with Arbitrary Equations of State,," *AIAA Journal*, Vol. 36, 1998, pp. 515–521.

14. et al., K., "A Preconditioned Navier-Stokes Method for Two-Phase Flows with Application to Cavitation Prediction," AIAA Paper 99-3329, 1999.

15. et al., V. A., "A Hybrid Unstructured Mesh Solver for Multi-Fluid Mixtures," AIAA Paper 99-3330, 1999.

16. Toumi, I., Kumbaro, A., and Paillère, H., "Approximate Riemann Solvers and Flux Vector Splitting Schemes for two-phase flow," VKI Lecture Series 1999-03, 30th Computational Fluid Dynamics, von Karmann Institute, Belgium, 1999.

17. Stewart, H. B. and Wendroff, B., "Two-phase flow: models and methods," *Journal of Computational Physics*, Vol. 56, 1984, pp. 363–409.

18. De Wilde, J., Vierendeels, J., and Dick, E., "An extension of the precondi-tioned AUSM to two-phase flows for the 3D calculations of circulating flu-idized beds," *Third Intern. Symposium on Computational Technologies for Fluid/Thermal/Chemical Systems with Applications, ASME-PVP*, 2001.

19. Halt, D. W. and Agarwal, R. K., "A novel algorithm for the solution of compressible Euler equations in wave/particle split (WPS) form," Open forum paper, 11th AIAA CFD Conference, 1993.

20. Liou, M.-S. and Edwards, J. R., "Numerical Speed of Sound and Its Application to Schemes for All Speeds," AIAA Paper 99-3268-CP, 14th AIAA CFD Conference, 1999.

21. Peery, K. M. and Imlay, S. T., "Blunt-Body Flow Simulations," AIAA Paper 88-2904, AIAA/SAE/ASME/ASEE 24th Joint Propulsion Conference, 1988.

22. Quirk, J. J., "A Contribution to the Great Riemann Solver Debate," *Int. J. Numer. Methods Fluids*, Vol. 18, 1994, pp. 555–574, Also ICASE Report 92-64, 1992.

23. Edwards, J. and Liou, M.-S., "Low-diffusion flux-splitting methods for flows at all speeds," *AIAA Journal*, Vol. 36, 1998, pp. 1610–1617.

24. Edwards, J. R., Franklin, R. K., and Liou, M.-S., "Low-diffusion flux-splitting methods for real fluid flows at all speeds," *AIAA Journal*, Vol. 38, 2000, pp. 1624–1633.

25. Edwards, J. and Mao, D., "Development of Low-Diffusion Flux-Splitting Methods for Dense Gas-Solid Flows,," AIAA Paper 2001-2649-CP, 2001.

26. Roberts, T. W., "The Behavior of Flux Difference Splitting Schemes near Slowly Moving Shock Waves," *Journal of Computational Physics*, Vol. 90, 1990, pp. 141–160.

27. Radespiel, R., "Efficient computation of compressible flows," *Lecture Notes in Physics*, Vol. 515, 1998, pp. 237–253.

28. Liou, M.-S., "Mass Flux Schemes and Connection to Shock Instability," *Journal of Computational Physics*, Vol. 160, 2000, pp. 623–648.

29. Liou, M.-S. and Edwards, J. R., "AUSM schemes and extensions for low Mach and multiphase flows," VKI lecture series 1999-03, VKI, Belgium, 1999.

30. Paillère, H., Core, C., and Garcia, J., "On the Extension of the AUSM$^+$ Scheme to Compressible Two-Fluid Models," to appear (2002).

31. Peng, D.-Y. and Robinson, D. C., "A New Two-Constant Equation of State," *Ind. Eng. Chem. Fundam.*, Vol. 15, No. 1, 1976, pp. 59–64.

32. Sanchez, I. C. and Lacombe, R. H., "An Elementary Molecular Theory of Classical Fluids: Pure Fluids," *Journal of Physical Chemistry*, Vol. 80, No. 21, 1976, pp. 2352–2362.

33. Brennen, C. E., *Cavitation and Bubble Dynamics, Oxford Engineering Science Series 44*, Oxford University Press, New York, 1995.

34. Gidaspow, D., *Multiphase Flow and Fluidization*, Academic Press, 1994.

35. Boivin, S., Cayre, F., and Herard, J. M., "A Finite Volume Scheme to Compute Incompressible Gas-Solid Two-Phase Flows," AIAA Paper 2000-2665, 2000.

36. Wu, P.-K., Nejad, T. H. C. A. S., and Carter, C. D., "Injection of Supercritical Ethylene into Nitrogen," *Journal of Propulsion and Power*, Vol. 12, 1996, pp. 770–777.

37. Cocchi, R. S. J. P. and Butler, P. B., "Numerical Study of Cavitation in the Wake of a Hypervelocity Underwater Projectile," *Journal of Propulsion and Power*, Vol. 15, 1999, pp. 513–521.

38. Neaves, M. D. and Edwards, J. R., "Hypervelocity Underwater Projectile Calculations Using A Preconditioned Density-Based Algorithm," AIAA Paper (submitted), 2003.

39. Lin, K. C., Kennedy, P. J., and Jackson, T. A., "Structures of Internal Flow and the Corresponding Spray for Aerated-Liquid Injectors," AIAA Paper 2001-3569, 2001.

# Modeling and Computation of Unsteady Cavitating Flows based on Bubble Dynamics

G.H. Schnerr

## 1 Abstract

The aim of this research is to provide a physical complete and numerical efficient simulation method to predict developed cavitation in hydrodynamic turbomachinery as well as in micro fluid dynamic applications, e.g. in high pressure injection nozzles of combustion engines. Cavitating two-phase flows are always very unstable, highly unsteady, 3-D and turbulent. To separate and to understand the inertia controlled cavitation dynamics and its interaction with viscous effects like boundary layers and separation, we start with Euler simulations by neglecting the viscosity of the fluid. Next we introduce the single-phase turbulence $k - \omega$ model of Wilcox [1] without modifications with respect to dispersed structures of bubbly liquids, which overestimates viscous effects in the transitional regime between the vapor and liquid phase and tends to suppress typical cavitation instabilities. Consequently our third approach consists of modifications of the single-phase Wilcox model to account for the strong nonlinear variation of the turbulent viscosity $\mu_t$, depending in the local void fraction $\alpha$.

The key issue of all numerical methods for simulation of cavitating flows is the treatment of the sudden density change of the fluid, in cold water up to 40.000:1, embedded in a global incompressible liquid flow. Here the two-phase fluid is modeled as dispersed mixture of an incompressible liquid and tiny vapor bubbles which grow or collapse, accordingly to the local static pressure and their convective transport. Therefore, the standard VOF method for capturing distinct interfaces without phase transition, e.g. free surface flow or single bubbles, is extended to include phase transition of dispersed mixtures. For simulation of bubble dynamics we apply the Rayleigh equation, which is completed by an energy balance to account for thermal effects, if hot water or if technical fluids others than water, e.g. refrigerants, with high vapor densities are considered.

By using our CFD tool CAVKA we present examples of cavitating flow around hydrofoils and through single hole injection nozzles. Comparing Euler and single-phase turbulence simulations, results based on the inviscid approach are closer to experiments, which indicates an interesting option to reduce computational time.

## 2  Introduction

Cavitation occurs in flows of liquids when the local static pressure falls below the vapor pressure, resulting in regions of vapor. It can occur in a broad variety of technical devices, where it usually degrades their performance, accompanied with intense erosion of structural components and very loud noise, see e.g. the review by Arndt [2]. It is interesting to note that cavitation can also be found in nature, e.g. snapping shrimps produce a loud snapping sound by an extremely fast closure of their snappers, which produce a high speed collapsing jet to kill prey animals ([3]).

Progress in understanding of cavitation phenomena is of great importance and interest for industry. In the past, several models were developed to simulate cavitation ([4], [5]), which do in general not model the complicated and highly transient bubble growth and collapse. This process is responsible for the cavitation damage and therefore, the location where the bubbles collapse is of great interest and the need for a cavitation model rises that also includes effects based on bubble dynamics. Since hydrofoils make up so many different types of machines - pumps, turbines, propellers - the study of as how cavitation affects hydrofoil performance is of special interest.

The type of cavitation most frequently observed at hydrofoils with a well-rounded leading edge is the so-called traveling bubble cavitation. From experiments it can be seen that for such foils the cavitation region is made up of individual bubbles rather than by a large vapor-filled cavity. These bubbles originate from small cavitation nuclei (particles, air bubbles), which are already existing in the bulk flow. The nuclei reach the low pressure region (suction peak) and grow to vapor bubbles while they are convected downstream, supposed the static pressure is sufficiently below the vapor pressure. Then, the bubbles are swept in the region of higher pressure and finally collapse. Many experimantal studies, see e.g. [6], [7], [8], point out the importance of the nuclei content of the liquid for the inception and development of cavitation. The nuclei content affects the tensile strength of the liquid and therefore is responsible if or if not cavitation occurs at given conditions.

At the beginning clouds form a dispersed sheet attached to the solid surface, which seems to be steady at first sight. Due to the entrainment of bubbles into the wake the sheet becomes unstable, causing large parts of the cavity to break off. The resulting dynamics forms by interaction between the dynamics of individual bubbles and the motion of the liquid fluid. During collapse single bubbles near solid walls create extremely high pressure pulses which seem to responsible for erosion and destruction, and global (hydrodynamic) pressure waves, formed during the collapse of bubble clouds ([9]). The aim of the inviscid approach is to improve understanding of the global pressure dynamics, which is not controlled by the viscosity of the fluid.

We reproduce recently performed experiments of Keller and Arndt (Technical University Munich and University of Michigan) [7]. Despite the fact that the present model is not yet complete, comparison of experiment and inviscid flow simulations shows surprisingly good agreement of the global structure of the unsteady cavitation dynamics around an inclined NACA 0015 hydrofoil. The time dependent resolution of the simulation of the unsteady cavitation dynamics allows interesting insight and analysis of the interaction of the surrounding unsteady hydrodynamic pressure field and the dynamics of bubble clouds, and of the strong impact of individual boundary conditions on strength and extension of cavitation regimes.

Due to very high inlet pressures of typically up to 1000-2000 bar in injection nozzles of Diesel engines cavitation cannot be avoided. In addition, the pressure upstream and behind the nozzle fluctuates with high frequencies of the order of 1-100 kHz. From experiments it is well known that spray characteristics changes, if cavitation is present [10]. Advanced partial evaporation and sudden collapse of bubbles within the jet indicate an interesting potential to improve the spray quality by optimization of the interaction of cavitation and the jet flow. Because of the limited access to the micro scale problem of $100\mu m - 1mm$ by experiments, simulations are most important. Therefore, in the second chapter we show systematic tests and parameter variations for identification of the main model parameters and their sensitivity in the operating regime.

It is well known, that single-phase turbulence models tend to suppress natural instabilities of partial cavitatig flows with their typical re-entry jets and bubble cloud separation. Several empirical modifications are proposed in the literature ([1],[11], [12]). Here we present first results of modifications of the turbulent viscosity $\mu_t$, dependent on the local vapor content $\alpha$. These investigations are certainly not yet complete. However, the modified turbulence model provides better agreement with experiments, as compared with single-phase turbulence models.

# 3 Modeling

## 3.1 Physical modeling

Our cavitation model bases on the following physical assumptions: Cavitation is modeled as the growth and collapse process of vapor bubbles. The bubbles originate from nuclei, which already exist in the bulk flow and grow or collapse depending on the surrounding conditions (pressure and temperature). It is assumed, that the slip between the vapor bubbles and the liquid can be neglected. The growth of vapor bubbles due to pseudo or gaseous cavitation is not yet taken into account. Cavitation is assumed to be dominated by heterogeneous nucleation, hence a homogeneous nucleation theory, known e.g. from modeling of condensing, flows is not employed. The interested reader may refer to see [13].

The perhaps most important feature of the model is, that it resolves the interior dispersed structure of bubble clouds. From the numerical point of view, the cavitation model simulates the production (bubble growth), destruction (bubble collapse) and convection of the vapor phase. But due to the underlying modeling of the nuclei content, it is possible to reconstruct from the vapor content of the computational cell, the number of bubbles that are currently in the computational cell and their radii. Therefore, a certain value of the vapor fraction $\alpha$ directly corresponds to a certain bubble radius $R$. Radius and vapor fraction are related by Eq. 5.

## 3.2 Numerical modeling

The formation of cavitation leads to a two-phase flow with phase transition. Entering or leaving the cavitation region, the mixture density varies from that of the pure liquid to a much smaller value or vice versa.

To overcome problems due to a discontinuous density distribution, a special numerical treatment is required. Front tracking methods (level set, marker particles, surface-fitting) were not found to be suitable, since they require the presence of distinct interfaces to be tracked. In this context, the interface is simply the bubble wall, that separates the vapor from the liquid phase. Because of the huge number of bubbles (typically 1000 per $cm^3$ liquid) and consequently a huge number of interfaces to be tracked, application of interface tracking methods would result in enormous CPU time and storage requirements. Instead, we prefer the usage of a front capturing method, namely the Volume-of-Fluid technique, proposed by Nichols & Hirt [14]. For a detailed description of our CFD tool CAVKA see [15], [26], [28].

## 3.3 Modified Volume-of-Fluid method for simulation of developed cavitation

The Volume-of-Fluid (VOF) method tracks the motion of a certain fluid volume through the computational domain, irrespective whether the volume contains pure liquid, pure vapor or a mixture of vapor bubbles and liquid. Within the scope of the VOF approach, the two-phase flow is treated as a homogeneous mixture and hence only one set of equations is used for description. The VOF method requires in addition to the continuity and the momentum equations (which are coupled by a SIMPLE algorithm) the solution of a transport equation for the cell vapor fraction $\alpha$, which is defined as the ratio of the vapor (gas) volume to the cell volume, see Eq. 5:

$$\frac{\partial \alpha}{\partial t} + \frac{\partial(\alpha u)}{\partial x} + \frac{\partial(\alpha v)}{\partial y} = 0, \tag{1}$$

where $u$ and $v$ are Cartesian components of the velocity vector $\mathbf{u}$. The equations of motion are closed with the constitutive relations for the density and dynamic viscosity:

$$\varrho = \alpha \varrho_v + (1 - \alpha)\varrho_l, \tag{2}$$

$$\mu = \alpha\mu_v + (1 - \alpha)\mu_l. \tag{3}$$

Here the subscripts $l$ and $v$ stand for the properties of pure liquid and pure vapor which are assumed to be constant. The equations derived are general and describe the motion of two fluids with an interface between them. A more detailed discussion of the governing equations without phase transition can be found for example in the work of Ubbink [16]. For simulation of cavitation in fuel injection nozzles, including simulation of the direct interaction of cavitation dynamics with the external jet flow behind the nozzle exit, this set of constitutive equations is extended by a second fluid component "gas", see chapter 4.2. As suggested by Spalding [17], the continuity equation is used in its non-conservative form:

$$\nabla \cdot \mathbf{u} = -\frac{1}{\varrho}\frac{d\varrho}{dt}. \tag{4}$$

The usage of the volume fluxes rather than mass fluxes (conservative form) accounts for the numerical advantage, that the volume fluxes are continuous at the interfaces and thus simplifies the solution of the pressure correction equation. For standard VOF applications, i.e. both fluids are assumed to be incompressible and no phase transition takes place, the rhs. of Eq. 4 reduces to zero, meaning that the flow field is divergence-free. Care has to be taken for discretization of the transport equation of the volume fraction Eq. 1. In order to avoid smearing of the interface, special methods are used to derive the cell face values for the void fraction $\alpha$. For that reason, the CICSAM scheme (Compressive Interface Capturing on Arbitrary Meshes) as proposed by Ubbink [16] was implemented and the code was verified by several test cases (convection tests, sloshing, dam breaking). Note that the CICSAM scheme is employed, if there exists a discrete sharp interface to be tracked, e.g. the motion of the free surface, see Sauer [15]. In the case of cavitation, the vapor bubbles and hence the vapor fraction are homogenously distributed in the computational cell. A schematic sketch of the distribution of the gaseous phase for a standard VOF application and for cavitation is presented in Fig. 1 to explain this difference.

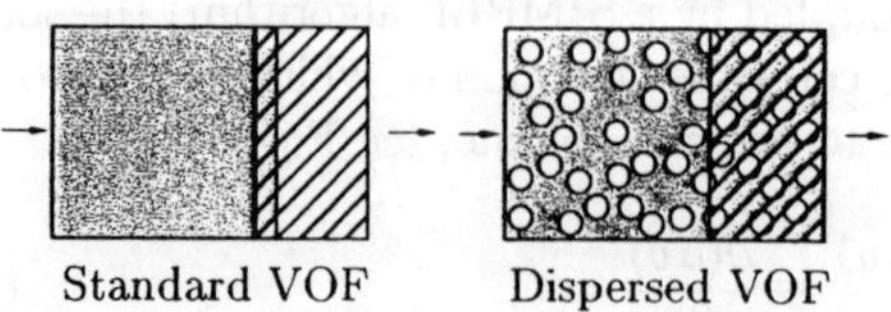



**Fig. 1.** Distribution of the gaseous phase in a computational cell. Left: Standard VOF application, non-homogeneous distribution. Right: Cavitation, homogeneous distribution of the gaseous phase within the liquid. The marked volume leaves the cell due to convection.

In contrast to standard VOF applications, cavitation leads to a poly-dispersed two-phase flow including phase transition. The bubbles grow and collapse and hence change the vapor fraction in a computational cell, in addition to the convective transport. The standard VOF method does account for convective transport, but not for phase transition. With respect to cavitation, the void fraction $\alpha$ may be reformulated as follows:

$$\alpha = \frac{V_v}{V_{cell}} = \frac{N_{bubbles} \cdot \frac{4}{3}\pi R^3}{V_v + V_l} = \frac{n_0 V_l \cdot \frac{4}{3}\pi R^3}{n_0 V_l \cdot \frac{4}{3}\pi R^3 + V_l} = \frac{n_0 \cdot \frac{4}{3}\pi R^3}{1 + n_0 \cdot \frac{4}{3}\pi R^3}, \quad (5)$$

where $V_{cell}$ is the volume of the computational cell, $V_v$ and $V_l$ are the volumes occupied by vapor and liquid, respectively and $N_{bubbles}$ is the number of bubbles in the computational cell. As a consequence of the bubble growth process, the velocity field is no longer divergence-free (Eq. 6) and the void fraction transport equation has to be extended by a vapor production source term:

$$\nabla \cdot \mathbf{u} = -\frac{\varrho_v - \varrho_l}{\alpha\varrho_v + (1-\alpha)\varrho_l}\frac{d\alpha}{dt}, \quad (6)$$

$$\frac{\partial\alpha}{\partial t} + \frac{\partial(\alpha u)}{\partial x} + \frac{\partial(\alpha v)}{\partial y} = \left(\frac{n_0}{1 + n_0 \cdot \frac{4}{3}\pi R^3}\right)\frac{d}{dt}\left(\frac{4}{3}\pi R^3\right). \quad (7)$$

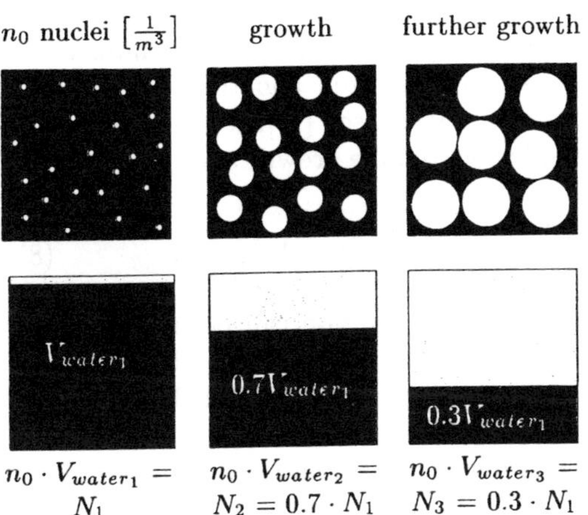


**Fig. 2.** Number of bubbles depending on the water fraction in a computational cell. Left: Initial simulation. Middle: Growth of the nuclei. Right: Further growth of the vapor bubbles.

The vapor production is taken into account by the source term on the right hand side of Eq. 7. The change of the cell vapor fraction does now depend on the number of bubbles per cell volume (rhs:1st term) times the volume change of a single bubble (rhs:2nd term) and on the convective transport. The parameter $n_0$ is defined as the bubble concentration per unit volume of pure liquid. Using this definition, the number of nuclei is explicitly coupled to the water volume in a cell. The physical background of this more formal aspect is, that relating the number of bubbles to the water volume (rather than to the volume of the mixture) guarantees the conservation of the number of bubbles. If the nuclei grow, the vapor fraction rises and hence the water fraction decreases. Therefore, the number of bubbles in the cell does also decrease. In Fig. 2 the growth of nuclei in a cell is schematically depicted. For the initial situtation, Fig. 2 left, the cell contains a large water fraction

550

and hence a large number of nuclei. If the nuclei grow, Fig. 2 middle, they displace water and at the same time other bubbles. Therefore, the gain of vapor fraction due to nuclei growth is reduced by a loss of vapor fraction due to bubbles that are displaced out of the cell. Further growth, Fig. 2 right, leads to a further reduction of the number of bubbles. Since every vapor bubble originates from a nuclei, the definition of $n_o$ also holds for vapor bubbles. A given volume of water $V_l$ always contains $N = n_o \cdot V_l$ bubbles, the bubbles can either be nuclei of radius $R = R_0$ or vapor bubbles with arbitrary larger radius. The interested reader may refer to Sauer [15] for further explanation of modeling the nuclei content of the flow.

Note that for the discretization of the convection terms of Eq. 7 the UP-WIND differencing scheme of Ferziger and Perić [18] is employed. This is consistent with the homogeneous assumption. A mixture of liquid and bubbles leaves the cell, as shown in Fig. 1. The composition of the fluid volume that is convected out of the cell has the same composition than the fluid that is currently in that cell. This physical aspect is numerically taken into account by using UPWIND discretization.

To complete the derivation of the numerical method, a relation to model the bubble growth is needed. Under the assumptions that bubble-bubble interactions and bubble coalescence can be neglected and that the bubbles remain spherical, the Rayleigh-Plesset equation, see [19], together with the energy equation is well suited to model the bubble growth and collapse process:

$$R\frac{d^2 R}{dt^2} + \frac{3}{2}\left(\frac{dR}{dt}\right)^2 = \frac{p(R) - p_\infty}{\rho_l} - \frac{2\sigma}{\rho_l R} - 4\frac{\mu}{\rho_l R}\frac{dR}{dt}. \tag{8}$$

Note that in our model the Rayleigh-Plesset equation can be treated as an ordinary differential equation, even when dealing with 2-D or 3-D flow problems, and can e.g. be solved by a Runge - Kutta method. For details of the solution procedure we refer to the work of Lee et al. [20].

If the system pressure is sufficiently low and the pressure difference $p(R) - p_\infty$ is large, the Rayleigh relation (9) may be considered as an adequate description for the so-called inertia controlled bubble growth:

$$\dot{R} = \sqrt{\frac{2}{3}\frac{p(R) - p_\infty}{\varrho_l}}, \tag{9}$$

where $p(R)$ is the pressure in the liquid at the bubble boundary and $p_\infty$ is the pressure in the liquid at a large distance from the bubble. Within the scope of this model, $p(R)$ is set equal to the vapor pressure $p_{sat}$ and $p_\infty$ to the ambient cell pressure. The Rayleigh relation was used to obtain the results presented in the following sections. To simulate cavitation in liquids others than cold water, i.e. if thermal effects are important (organic fluids, hot water), the model is extended by a simplified equation for the mixture enthalpy $h$, to account for the temperature change of the mixture caused by cavitation. For details and results see [21]. The advantage of the present modeling is the option for simultaneous application of both methods, the standard front capturing method without phase transition and the modified VOF technique to track dispersed voids. Numerical simulations of unsteady cloud cavitation including free surface flow effects are interesting examples for simultaneous application of these two methods.

## 4  Results

### 4.1  Inviscid Flow

**Unsteady cavitating nozzle flow** First the cavitation model is applied to simulate steady and unsteady cavitating nozzle flows. The geometry and boundary conditions of the 2-D plane nozzle are depicted in Fig. 3. The working fluid is cold water at a constant temperature of T = 293.15K. It is well known that under these conditions thermal effects are quantitatively not important. Both phases are assumed to inviscid, compressibility is only taken into account for mixtures if $0 < \alpha < 1$. The nuclei concentration $n_0$ is set to $10^8$ nuclei/m$^3$water, the vapor fraction at the nozzle inlet $a_0 = 10^{-5}$, which corresponds to a nucleus radius of $R_0 = 30\mu$m. The diffuser part downstream of the throat forces all bubbles to collapse inside the nozzle, which prevents additional complications at the outflow boundary. As boundary conditions of the reference case we prescribe the velocity at the inlet $U_{ref} = 10$m/s and a constant value of the normalized static pressure at the exit $\sigma_{exit} = 5.45$, the reference length $L_{ref} = 0.1$m. To reduce the computational time, we assume symmetry with respect to the nozzle axis. Therefore, only the flow in the lower half of the symmetric nozzle is calculated. The reference grid consists of $58 \times 16$ cells.

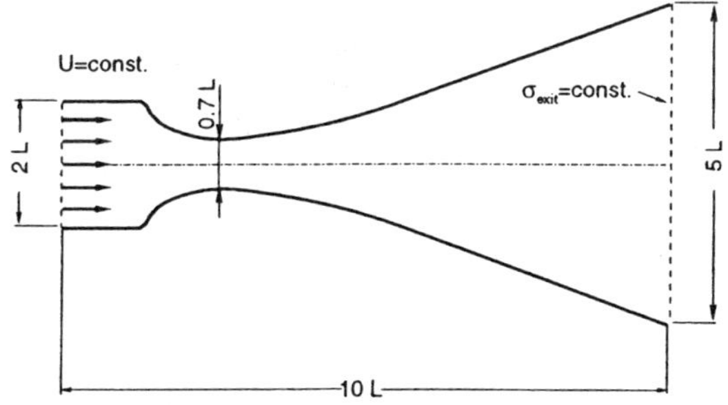


**Fig. 3.**

Geometrical setup of the investigated nozzle,

$$\sigma_{exit} = \frac{p_{exit} - p_v}{\frac{1}{2}\rho_l U_{ref}^2}.$$

*Discretization effect* Extensive numerical tests have been performed to identify the limit between the steady and unsteady cavitating flow regime. Herein we concentrate on the more relevant unsteady cavitating flows. Figure 4 shows the time dependent total vapor fraction $\alpha_{total}$ (the integral value of the instantaneous local vapor fraction $\alpha$ in the entire computational domain) dependent on the numerical time step, based on the single-phase CFL numbers. Obviously, for CFL< 0.1 the solutions are in good agreement, including the resolution of the secondary peak which represents the instantaneous formation of a bubble cloud. For CFL> 0.1 this important detail is no more resolved. Figure 4 shows also clearly that after initialisation periodicity establishes after the second cycle, the corresponding oscillation frequency is $f = 9.49$Hz.

To check the influence of the spatial discretization we compare results obtained with double resolution, i.e. with $115 \times 31$ cells, with the reference case. Results for the instantaneous and time averaged $\sigma$-distribution along the lower nozzle wall are depicted in Fig. 5. It is interesting to note that the existing differences vary depending on the location in the nozzle. In the throat region the deviations are much lower than at positions downstream. Quantitatively the relative difference shown here disappears immediately for stream lines apart from the nozzle wall, especially along the center line.

*Scaling effect* An important aspect of the design of hydraulic machinery concerns scaling effects, i.e. effects which cannot be scaled up or down accordingly to classical similarity laws. From extensive experiments of Keller [22] it is well known, that the development of cavitation depends on the length and velocity scale of the flow field, even if the normalized parameters are kept constant. Reduction of the reference length by one order of magnitude reproduces in tendency what is known from experiments of Keller. In the larger nozzle locally more vapor is formed and the shedding frequency is lower, see Fig. 6.

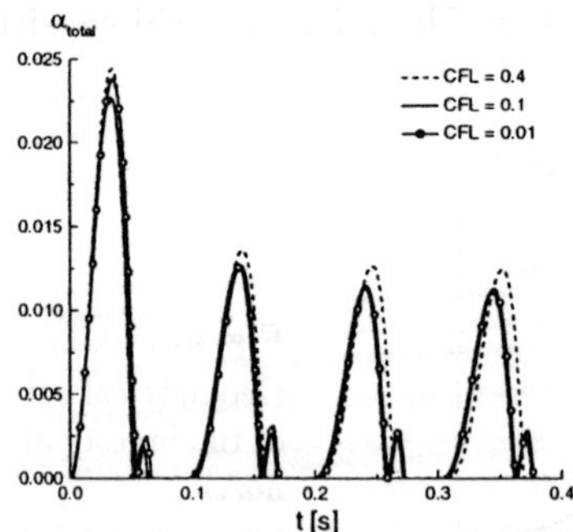

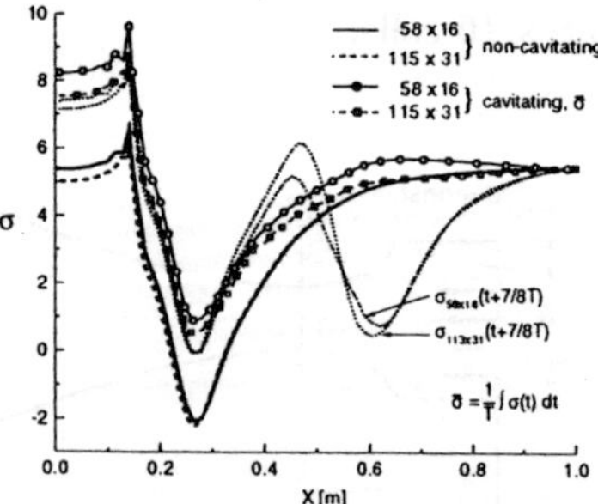

Fig. 4. Influence of CFL number. Total vapor fraction in the nozzle vs. time. $\sigma_{exit} = 5.45$, $U_{ref} = 10$m/s, $L_{ref} = 0.1$m, $n_{0,ref} = 10^8$nuclei/m$^3$water, $R_{0,ref} = 3 \cdot 10^{-5}$m, water at 293.15K.

Fig. 5. Mesh effect. Comparison of the $\sigma$-distributions at the nozzle wall for different meshes. Dotted: instantaneous $\sigma$-distribution. $\sigma_{exit} = 5.45$, $U_{ref} = 10$m/s, $L_{ref} = 0.1$m, $n_{0,ref} = 10^8$nuclei/m$^3$water, $R_{0,ref} = 3 \cdot 10^{-5}$m, water at 293.15K.

Note that in Fig. 6 the total vapor fraction $\alpha_{total}$ is plotted against the normalized time $t^*$ and reduction of the length scale of one order increases the oscillation frequency from $f = 9.49$Hz to $f = 83.52$Hz. From the simplified bubble dynamics Eq. 9 it follows immediately that the present model cannot reveal velocity scale effects.

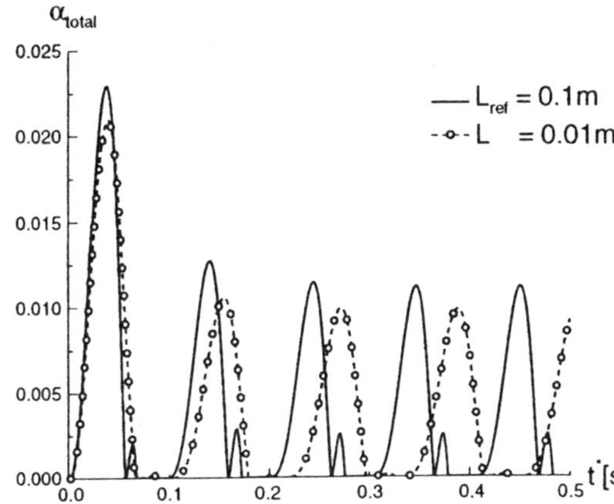


**Fig. 6.** Length scale effect. Total vapor fraction in the nozzle vs. time. $t^* = L_{ref}/L \cdot t$, $\sigma_{exit} = 5.45$, $U_{ref} = 10$m/s, $n_{0,ref} = 10^8$nuclei/m$^3$water, $R_{0,ref} = 3 \cdot 10^{-5}$m, water at 293.15K.

*Variation of cavitation number $\sigma_{ref}$* In the following we investigate the influence of the cavitation number $\sigma_{exit}$ in the smaller nozzle with $L_{ref} = 0.01$m. Above $\sigma_{exit} = 6.87$ no cavitation occurs. If the exit pressure starts to decrease we observe small attached vapor sheets with low amplitude oscillations. With further decrease of $\sigma_{exit}$ the unsteady vapor region growth with increasing frequency toward the nozzle exit (Fig. 7). Figure 8 shows the variation of the frequency and of the maximum of the local vapor fraction $\alpha_{max}$ against the exit boundary condition $\sigma_{exit}$. Near $\sigma_{exit} = 2$ the frequency reaches a maximum of about 100Hz, then it starts to decrease before the flow turns into steady supercavitating state.

Up to now the constant pressure boundary condition was prescribed at the nozzle exit. Figure 9 shows a simplified model to demonstrate the influence of this assumption on the pressure gradient and on the cavitation number after instantaneous formation of a certain vapor volume in the throat region. If the pressure is kept constant at the nozzle exit the displacement of the local vapor volume forces the pressure to increase ahead which tends to weaken and to diminish the vaporization itself. If $\sigma_{ref}$ is fixed at the nozzle inlet the opposite tendency occurs and cavitation will be intensified. In practice the pressure will be affected at both sides, i.e. inflow and outflow of the test section will simultaneously be affected.

**Unsteady cavitating flow over NACA 0015 hydrofoil** The layout of the 2-D plane test section used by Keller & Arndt [7] for investigation of

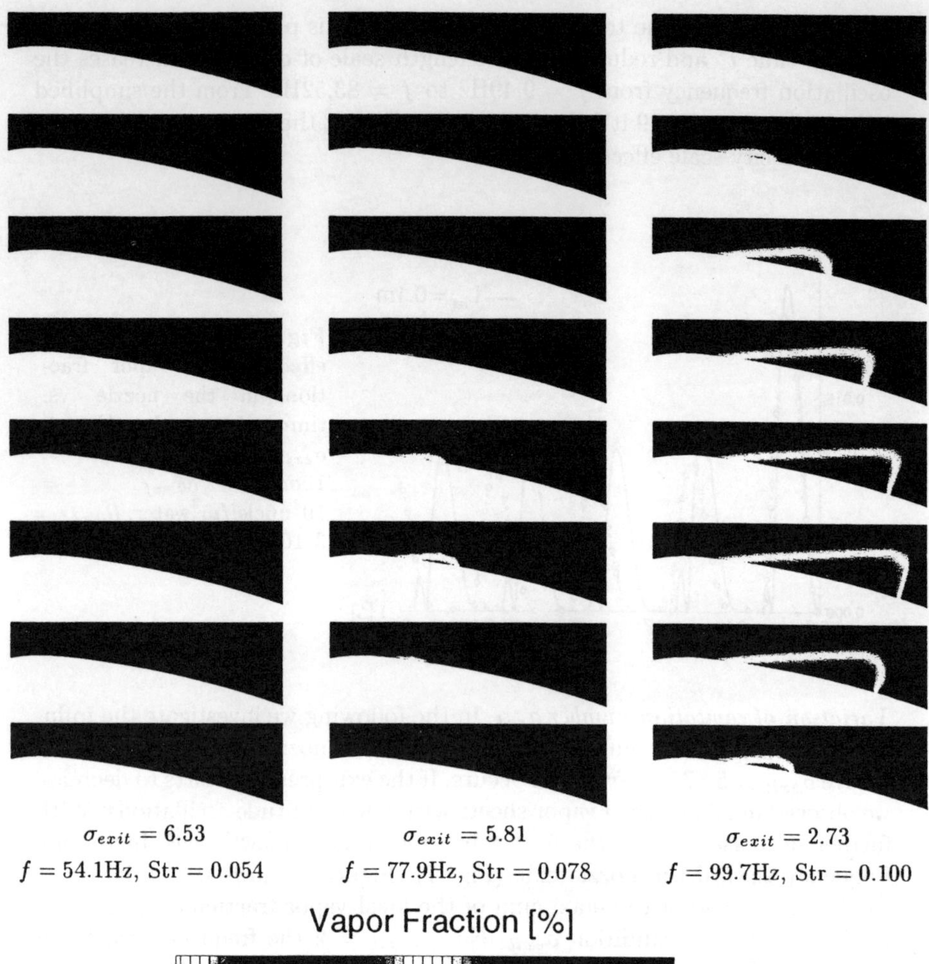

$\sigma_{exit} = 6.53$  $\quad\quad\quad$  $\sigma_{exit} = 5.81$  $\quad\quad\quad$  $\sigma_{exit} = 2.73$

$f = 54.1\text{Hz}, \text{Str} = 0.054$  $\quad$  $f = 77.9\text{Hz}, \text{Str} = 0.078$  $\quad$  $f = 99.7\text{Hz}, \text{Str} = 0.100$

**Vapor Fraction [%]**

100　　80　　60　　40　　20　　0

**Fig. 7.** Variation of cavitation number $\sigma_{exit}$ at nozzle exit. Vapor fraction distribution, time increment $\Delta t = T_{period}/8$. $U_{ref} = 10\text{m/s}$, $L = 0.01\text{m}$, $n_{0,ref} = 10^8\,\text{nuclei/m}^3\text{water}$, $R_{0,ref} = 3.0 \cdot 10^{-5}\text{m}$, water at 293.15K.

the unsteady cavitating flow over a NACA 0015 hydrofoil is shown in Fig. 10. For comparison with our numerical simulation the experiment defined by an inlet velocity U=12m/s, $\sigma_{ref} = 1.0$ and an angle of attack $\alpha_A = 6^o$ was chosen, because under these conditions in good approximation a 2-D periodic unsteady cavitating flow was observed. Figure 10 also includes different modifications concerning the numerical pressure boundary condition, which are investigated in the following chapter. Figure 11 depicts one cycle of the observed cavitation, the frequency $f \sim 16\text{Hz}$, the flow direction is from left to

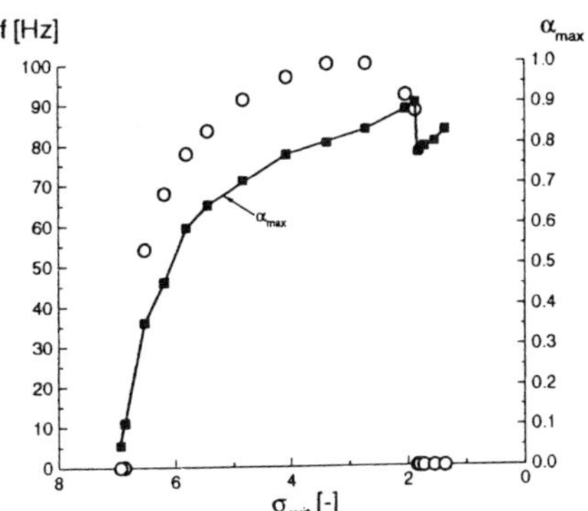

**Fig. 8.** Effect of cavitation number $\sigma_{exit}$ at nozzle exit on the frequency and the maximum vapor fraction. $U_{ref} = 10\text{m/s}$, $L = 0.01\text{m}$, $n_{0,ref} = 10^8 \text{nuclei/m}^3\text{water}$, $R_{0,ref} = 3.0 \cdot 10^{-5}\text{m}$, water at 293.15K.

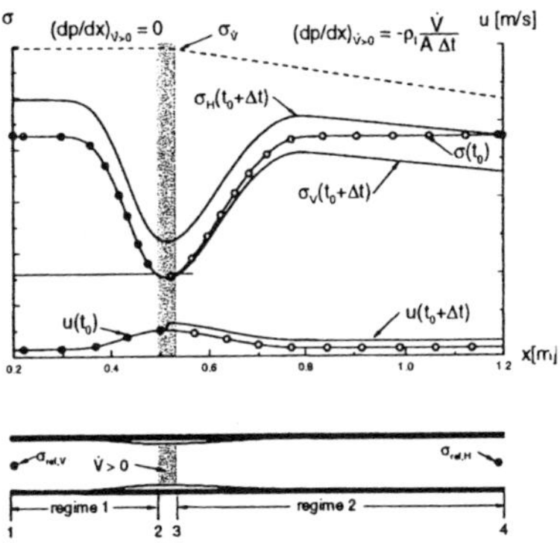

**Fig. 9.** Inviscid planar channel flow with volume source. Distribution of the pressure and the velocity along the axis before initiating the volume source ($t = t_0$) and thereafter, depending on the position of the reference pressure point. Dashed line: additional pressure gradient owing to volume source.

right, the line of sight is on the suction surface, the time marches from the upper left to the lower right picture.

Picture 1: Cavitation starts at the leading edge of the foil and an attached cavity grows. A second cavitation area exists which originates from the previous cavitation cycle. Pic. 2: The first cavity grows further and the second one is convected downstream. The picture emplies that there is a vortex existent, meaning that the water flows upwards in the middle of the foil, passes the second cavitation area and then flows downward at the trailing edge. Here, the numerical simluation yields a vortex at the upper side of the foil, see Fig. 12. The vortex rotates clockwise, the experiment (Pic. 3) also implies this direction. Pics. 4 and 5: The formation of a jet is visible which re-enters the cavity. The secondary cavitation region collapses. The primary cavity still grows and reaches its maximum length, Pic. 7. The re-entrant jet reaches the leading edge and breaks the cavity off. The first part collapses, the second one is convected downstream and the cavitation cycle restarts again.

*Variation of location of $\sigma_{ref}$ - without free surface* The numerical simulation bases on the following additional assumptions: The fluid is cold water at 295.15K with a nuclei concentration $n_0 = 10^8 \frac{nuclei}{m^3 water}$ and a nuclei radius $R_0 = 3 \cdot 10^{-5}\text{m}$ in order to match the experimental conditions. The inlet velocity U is set to 12m/s, the fluid is still assumed to be inviscid. The cavitation number is set constant to $\sigma_{ref} = 1.0$ at the reference point, located 1.5 chord length upstream of the leading edge, see Fig. 10 left.

The series of instantaneous pictures presented in Fig. 12 shows the vapor fraction distribution $\alpha$ during one cycle of the periodic formation and destruction of the vapor phase. The computed frequency is $f \approx 11\text{Hz}$, the time increment is $\Delta t \approx 0.011\text{s}$. The vapor cavity starts growing at the leading edge and grows until a re-entrant jets forms and breaks off a part of the cavity (Pics. 1-3). The first part starts to collapse (Pic. 4) and the second part of the cavity is swept downstream. A secondary vapor region forms (Pic. 6), merges with the already existing vapor cloud and finally collapses. The growth and collapse of the vapor phase do significantly alter the pressure distribution and thus change the lift and drag of the hydrofoil. The single-phase calculation yields a lift of $L = 7300N$ per unit meter span and a drag of $D \sim 0$ (inviscid calculation), compared to a time averaged lift of $L_c = 4200N$ and drag of $D_c = 750N$ under cavitating conditions. Due to cavitation, the lift of the hydrofoil has been substantially decreased (drag increased), which is in agreement with experimental observations of Keller & Arndt [7].

Figure 12 also illustrates the important unsteady effect on the lift which varies to about 100% within one cycle and may even become negative after the collapse of the dispersed vapor cloud near the trailing edge. Frequency and quantitative variation of the lift up to 100% are in good agreement with the experiment.

Since the nuclei content of the flow is approximated by an average radius $R_0$ and an average nuclei concentration $n_0$, parameter variations have been performed to investigate the effect of the nuclei content on the time averaged lift and drag. The simulations show, that lift and drag are rather insensitive to a variation of the number of nuclei, although the time dependent vapor fraction distribution is quite different for the $n_0$ variation performed. The common fact of all simulations is, that there is one large vortex created, which is convected downstream at the suction side of the hydrofoil. This vortex controls the pressure distribution and hence affects lift and drag. The amount of vapor that is present during one cycle is not so significant for lift and drag.

To demonstrate the importance of the a priori not well defined boundary conditions, especially of the static pressure, we compare this result, where the reference point of the constant static pressure, i.e. constant value of $\sigma_{ref} = 1.0$, is located ahead of the hydrofoil in the entrance region, with the opposite situation, where the reference point is downstream near the bottom of the test section, see Fig. 10. Now the cavitating flow remains steady, only a very small attached vapor sheet develops at the suction side, which causes a lift decrease of about 16% compared with the single-phase flow.

*Interaction of free surface and cavitation dynamics* To avoid this very strong dependence on the assumption of the reference location with identical absolute values of the static pressure, the cavitation problem is coupled with a free surface flow, where the static pressure is constant above the liquid surface, i.e. independent of the location of the reference point. At first, a steady state cavitating flow develops with a slight increase of the lift of about 2% compared with single-phase flow. Figure 13 depicts unsteady cavitating flows with separating vapor clouds similar to Fig. 12. For equal values of $\sigma_{ref}$ the instantaneous displacement of the free surface reduces the cavitation intensity, i.e. the volume fraction $\alpha$ and the frequency of the shedded vapor clouds. The significant effects on the lift (Table 1), only dependent on the location of the reference point and for identical values of the static pressure at this point follow directly from the cavitation dynamics inside the test section, i.e. from the interaction of the local vapor production, the convective transport and the resulting reaction of the velocity and pressure field. It emphasizes the importance of appropriate boundary conditions and of the inclusion of the dynamical interaction with the entire test circuit.

**Table 1.** Time averaged lift and drag (per unit meter span) depending on the cavitation number $\sigma$, **reference point above free surface.**

| $\sigma$ | lift [N] | drag [N] | $f$ [Hz] | Str $= \frac{f \cdot l_c}{U}$ |
|---|---|---|---|---|
| 2.5 * | 7200 | 40 | steady | steady |
| 2.5 ** | 3600 | 50 | steady | steady |
| 0.97 *** | 3665 | 150 | steady | steady |
| 0.87 | 3160 | 480 | 5.4 | 0.0585 |
| 0.77 | 3080 | 550 | 5.75 | 0.062 |
| 0.67 | 2660 | 580 | 6.0 | 0.065 |
| 0.57 | 2750 | 680 | $\sim 9.0$ | 0.0975 |

   *    noncavitating **without** free surface
  **   noncavitating **with** free surface
***cavitating **with** free surface and
       equal static pressure at the tip of
       the hydrofoil as in *.

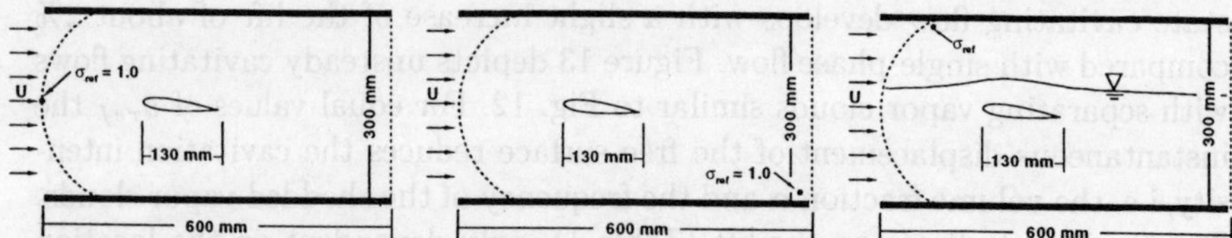

**Fig. 10.** Geometry and boundary conditions for simulation of cavitating flows. NACA 0015 hydrofoil, $U = 12\text{m/s}$, $l_c = 0.13\text{m}$, $\alpha_A = 6^\circ$. Left: $\sigma$-reference location upstream of hydrofoil. Middle: $\sigma$-reference location downstream. Right: $\sigma$-reference location upstream, above the free surface.

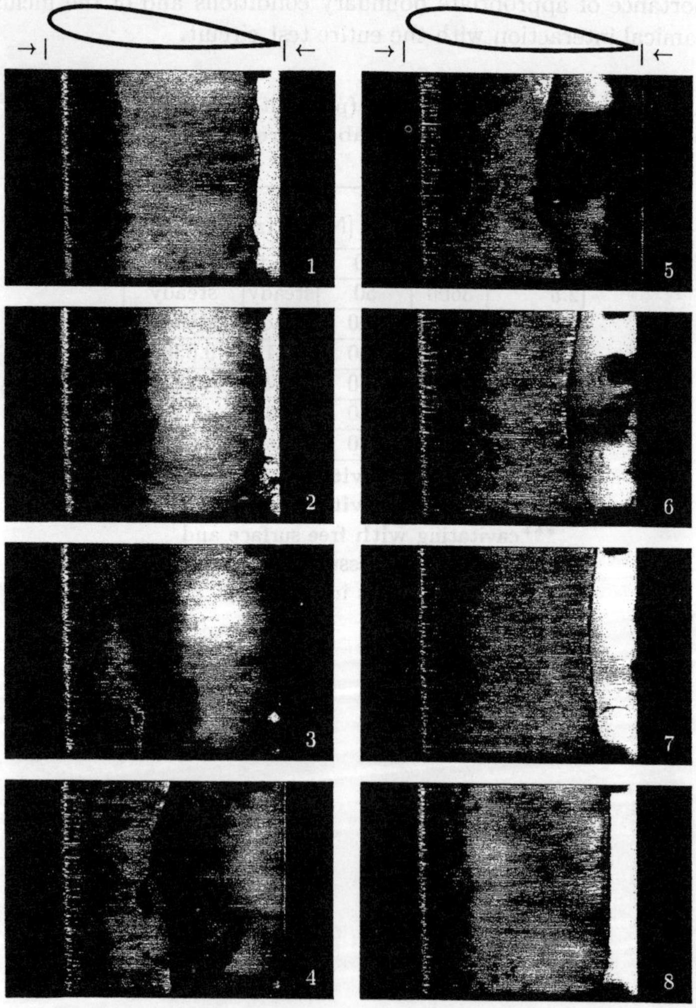

**Fig. 11.** Cavitation at a NACA 0015 hydrofoil, flow from left, view from above on suction side. Time increment of the photographs: $\Delta T/8$. $f_{\text{exp}} \sim 16\text{Hz}$, $\sigma_{ref} = 1.0$, $U = 12\text{m/s}$, $L_c = 0.13m$, cold water; experiment of Keller and Arndt [7].

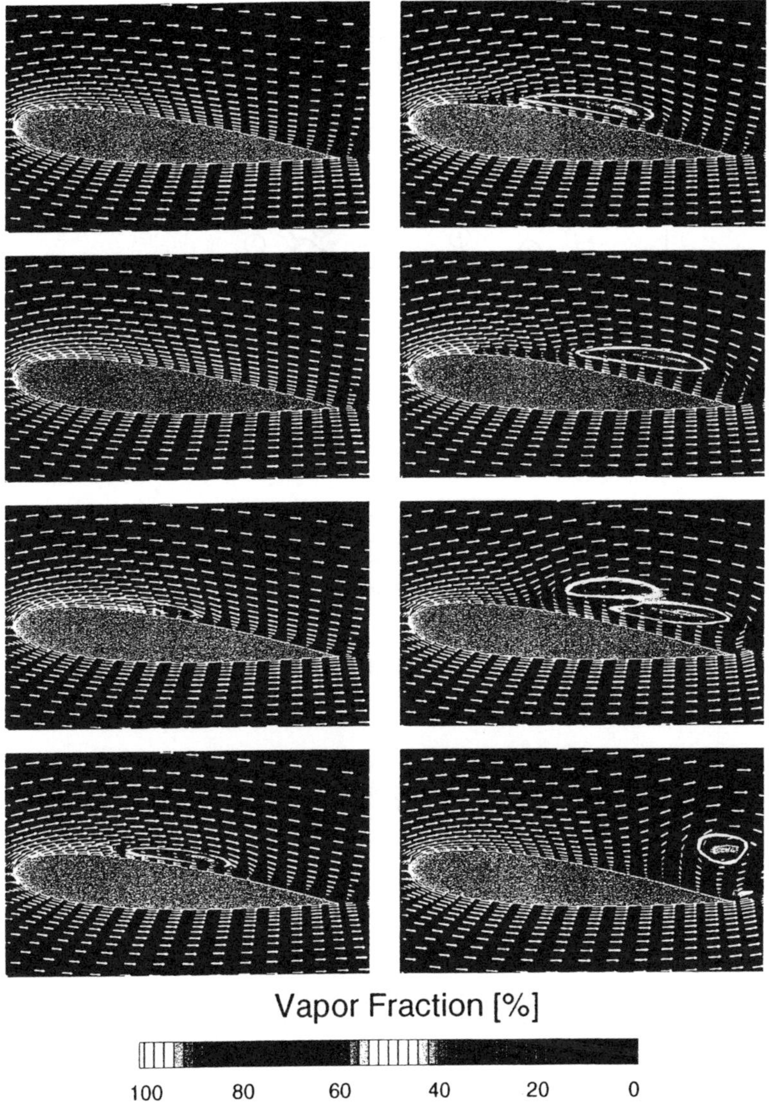



**Fig. 12.** One cycle of the periodic unsteady cavitating flow over a NACA 0015 hydrofoil. Vapor fraction distribution and velocity vectors. Time increment $\Delta t = T/8$, $f \approx 11\text{Hz}$, $L_c = 0.13\text{m}$, $\sigma_{ref} = 1.0$, $U = 12\text{m/s}$, $n_{0,ref} = 10^8 \text{nuclei/m}^3$, $R_{0,ref} = 3.0 \cdot 10^{-5}\text{m}$, water 293.15K, **reference point ahead hydrofoil.**

560

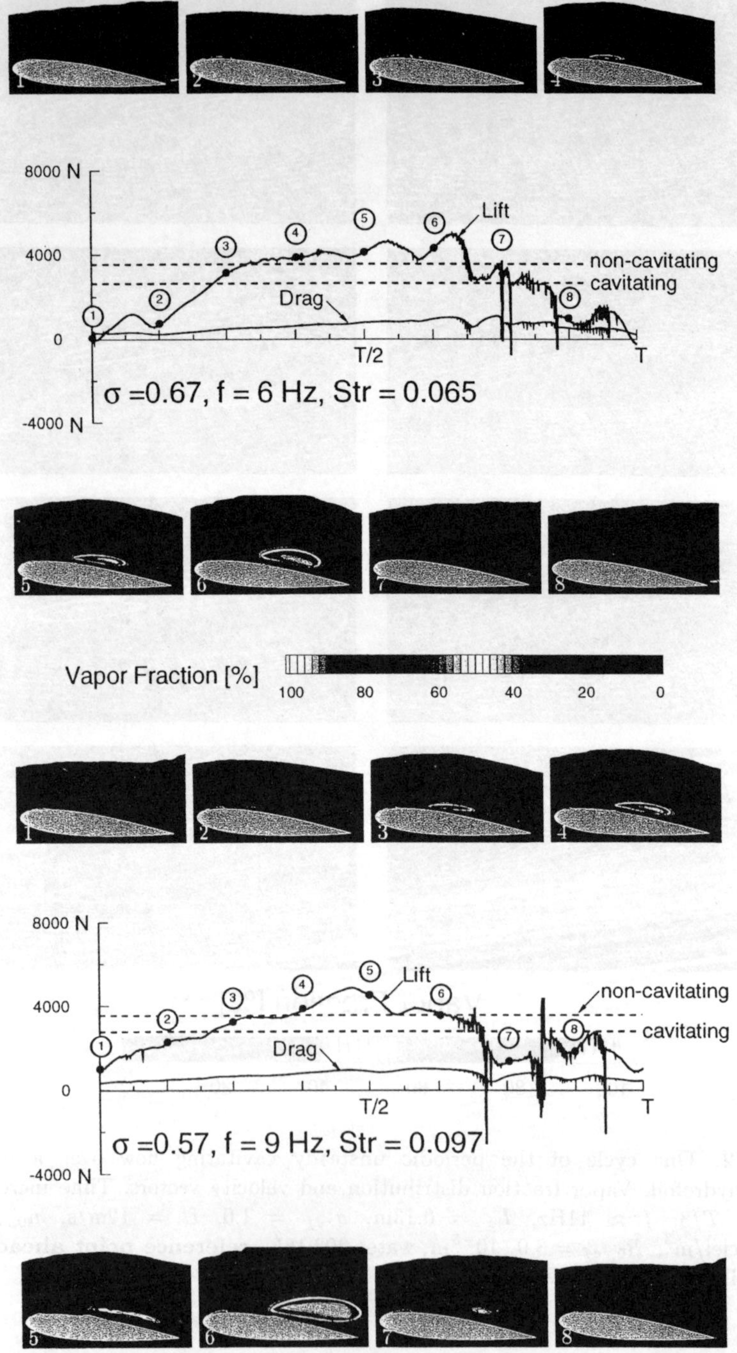


**Fig. 13.** Unsteady lift, drag and vapor distribution for one cavitation cycle. Time increment: $\Delta t = T/8$, $l_c = 0.13$m, $n_0 = 10^8$ nuclei/m$^3$ water, $R_0 = 3.0 \cdot 10^{-5}$m, water at 293.15K, **reference point above free surface.**

## 4.2 Turbulent Cavitating Flows

### Single-phase $k - \omega$ model of Wilcox without modification

*Cavitation in high pressure fuel injection nozzles* From experiments it is well know, that cavitation in injection nozzles of Diesel engines has a strong impact on spray angle and momentum of the jet, which are important for control and efficiency of primary break up and atomization of the liquid fuel and the combustion process. Further important details, e.g. dispersion and droplet size, dependent on the cavitation intensity, are not yet know, but of great interest and therefore in focus of many research projects dealing with this subject ([10],[23],[24], [25] ). Due to the very short time scale of the flow of $1 - 10\mu s$ and the micro scale of the nozzles of $100\mu m - 1mm$, experiments and detailed quantitative data are very rare. Therefore, accurate physical modeling together with efficient numerical simulation methods are most important and promising to improve understanding of this complex unsteady turbulent two-phase multi-component micro fluid dynamic problem. Here the viscous version of our simulation tool CAVKA is applied. Because the Reynolds number is of the order of 30.000 we make use of the single-phase $k - \omega$ turbulence model of Wilcox [1] with wall fuctions, but without modification in the two-phase regime.

Here the cavitation model parameters, nuclei concentration $n_0$ and nuclei size $R_0$, are unknown and it is important to realize, that quantitative verification of these submicron data by experiments in the real micro scale flow, including thermodynamic nonequilibrium is not to be expected in future. Therefore, we have performed systematic tests with parameter variations to identify necessary nuclei sizes $R_0$ and concentrations $n_0$, which lead to cavitation structures as observed in injection nozzle experiments. This fact emphasizes an important difference with respect to classical hydraulic cavitation in macro scale, as presented in the previous chapter. There representative data of $n_0$ and $R_0$ are known from direct measurements. It is interesting to note, that the required nuclei size is of the order of $10^{-7}m$, but the required concentration is about $10^{14} nuclei/m^3$ water, i.e. 6-7 orders higher as compared with hydraulic cavitation. Such high concentrations are not typical for pre-existing heterogeneous nuclei. In conclusion, the formation of homogeneous vapor nuclei cannot be excluded. This enlightens another difficult but quite interesting physical aspect of this engineering problem.

Due to the high sensitivity of cavitation on the imposed initial and boundary conditions, simulations with restriction to the internal nozzle flow, i.e. without direct interaction with the external jet flow are qualitatively and quantitatively incorrect. Simulation of the external jet flowing into the surrounding air requires inclusion of a second non-condensable fluid component (gas or air), for details see [26]. To close the system of conservation equations for unsteady turbulent flows, for the mixture density $\varrho$ and the mixture viscosity $\mu$ we introduce the following extended set of constitutive equations:

$$\varrho = \alpha_l \varrho_l + \alpha_v \varrho_v + \alpha_g \varrho_g, \tag{10}$$

$$\mu = \alpha_l \mu_l + \alpha_v \mu_v + \alpha_g \mu_g, \tag{11}$$

where the subscripts $_l$, $_v$ and $_g$ stand for the values of pure liquid, pure vapor and pure gas (air), $\alpha_i$ indicates the volume fraction of the fluid component "i" with the compatibility condition:

$$\alpha_l + \alpha_v + \alpha_g = 1. \tag{12}$$

As before the properties $\varrho$ and $\mu$ of pure liquid, pure vapor and of pure gas are assumed to be constant. Turbulence modeling is carried out, $\mu_{eff}$ is therefore used instead of the molecular viscosity $\mu$:

$$\mu_{eff} = \mu + \mu_t, \tag{13}$$

where $\mu_t$ represents the turbulence viscosity which is modeled using the Wilcox $k - \omega$ model. For details of the turbulence modeling see [26].

For the volume fractions $\alpha_i$ we derive the following transport equations:

$$\frac{\partial \alpha_v}{\partial t} + \nabla \cdot (\alpha_v \mathbf{c}) = \frac{(1 - \alpha_g)\varrho_l + \alpha_g \varrho_g}{\varrho} \frac{d\alpha_v}{dt}, \tag{14}$$

$$\frac{\partial \alpha_l}{\partial t} + \nabla \cdot (\alpha_l \mathbf{c}) = -\frac{(1 - \alpha_g)\varrho_v + \alpha_g \varrho_g}{\varrho} \frac{d\alpha_v}{dt}, \tag{15}$$

$$\frac{\partial \alpha_g}{\partial t} + \nabla \cdot (\alpha_g \mathbf{c}) = \alpha_g \frac{\varrho_l - \varrho_v}{\varrho} \frac{d\alpha_v}{dt}. \tag{16}$$

Multiplication of the volume fraction $\alpha_i$ with the density $\rho_i$ yields the mass conservation for each fluid component

$$\frac{\partial(\varrho_v \alpha_v)}{\partial t} + \nabla \cdot (\varrho_v \alpha_v \mathbf{c}) = \varrho_v \frac{(1 - \alpha_g)\varrho_l + \alpha_g \varrho_g}{\varrho} \frac{d\alpha_v}{dt}, \tag{17}$$

$$\frac{\partial(\varrho_l \alpha_l)}{\partial t} + \nabla \cdot (\varrho_l \alpha_l \mathbf{c}) = -\varrho_l \frac{(1 - \alpha_g)\varrho_v + \alpha_g \varrho_g}{\varrho} \frac{d\alpha_v}{dt}, \tag{18}$$

$$\frac{\partial(\varrho_g \alpha_g)}{\partial t} + \nabla \cdot (\varrho_g \alpha_g \mathbf{c}) = \varrho_g \alpha_g \frac{\varrho_l - \varrho_v}{\varrho} \frac{d\alpha_v}{dt}. \tag{19}$$

Summation of Eqs. 17, 18 and 19 represents the continuity equation.

*Scaling of model nozzle* Here the 2-D plane experimental test cases of Roosen et al. [27] are recalculated, to validate our numerical scheme and to study the physical effects concerning the interaction between the cavitating flow and the external jet of the injection nozzle. The fluid used in the experiments was tap water. The "bore hole" of the nozzle consists of a rectangular-shaped channel of 0.2mm × 0.28mm × 1mm (*width × height × length*). The calculated flow regime includes simultaneous treatment of internal and external flow. Figure 14 shows the nozzle geometry and boundary conditions. To reduce the computational time, we assume a 2-D flow and symmetry with respect to the nozzle axis. A computational mesh of 95 × 13 nodes for the nozzle block and 79 × 49 for the outflow block is used for the spatial discretization. Stationary liquid in the nozzle and a sharp liquid interface with the air behind the nozzle exit are assumed as initial conditions. Both inflow and outflow boundaries are modeled as constant pressure surfaces. At the inlet, the turbulence kinetic energy $k$ is set to be $6 \times 10^{-4} u_{in}^2$. The value of the specific dissipation rate $\omega$ is selected using the length scale equation, see Ferziger et al. [18] and Yuan et al. [26]. On the wall, the boundary conditions are the impermeability and no-slip for the velocity, and the normal gradient of the pressure is assumed to be zero. Wall functions based on the law of wall are used as boundary conditions for turbulence modeling.

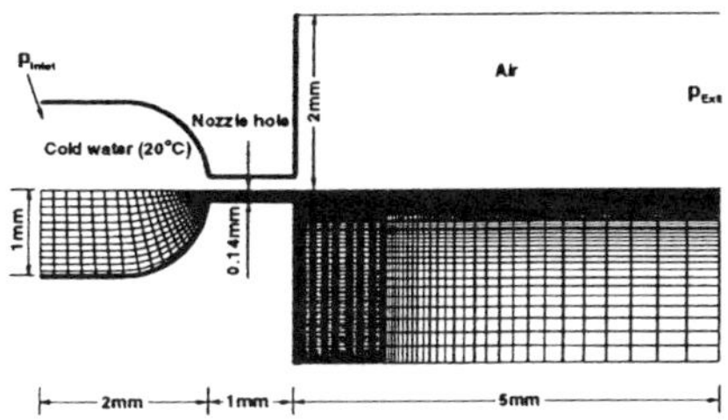


**Fig. 14.** Computational mesh and boundary conditions of the 2-D plane model injection nozzle.

Inception of cavitation in a liquid requires the presence of nuclei. Our parameter tests have shown that the numerical solutions using the nuclei concentration of the order of $n_0 = 1 \times 10^{14} \frac{\text{nuclei}}{m^3 \text{water}}$ agree very well with experimental observations of Roosen et al. and the nuclei radius $R_0 = 0.3\mu$m is in the insensitive range. In this study these values will be used, which correspond to an initial vapor fraction $\alpha_{v0} = 0.0017\%$.

*Moderate and high injection pressure and steady cavitating flow* The calculation is first carried out for the case with $p_{Injection} = 80$bar and $p_{Exit} = 21$bar. The computed result yields a length of the cavitation region of about $200\mu$m, which agrees well with the experimental observation of Roosen et al. [27], see Figs. 15, 16. Chaves et al. [10] have observed, that after a small increase of

564

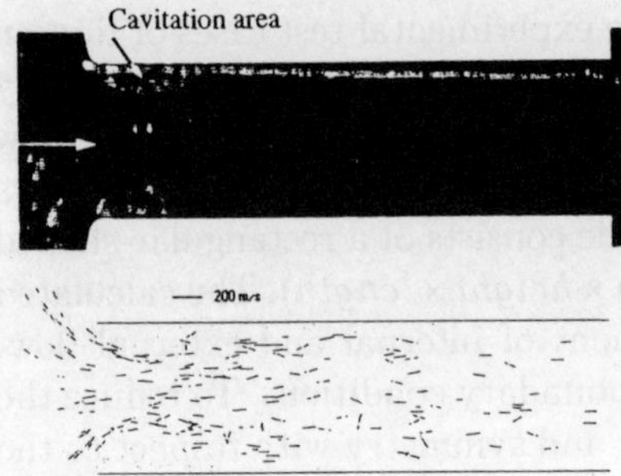


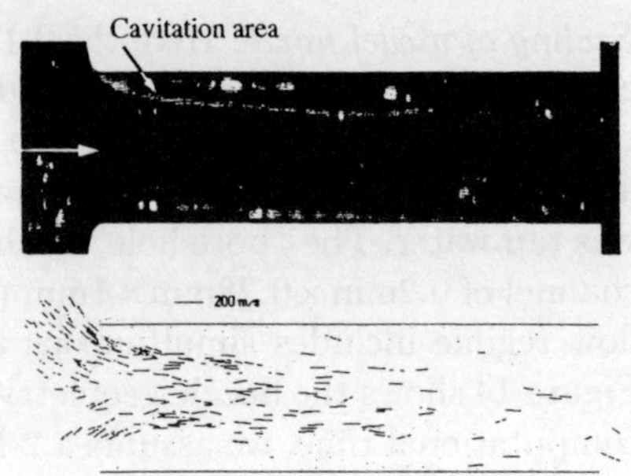


**Fig. 15.** Experimental density gradient and velocity distribution in the nozzle hole for $p_{Injection}$ = 80bar and $p_{Exit}$ = 21bar, flow from left to right, from Roosen et al. [27].

**Fig. 17.** Experimental density gradient and velocity distribution in the nozzle hole for $p_{Injection}$ = 80bar and $p_{Exit}$ = 11bar, flow from left to right, from Roosen et al. [27].

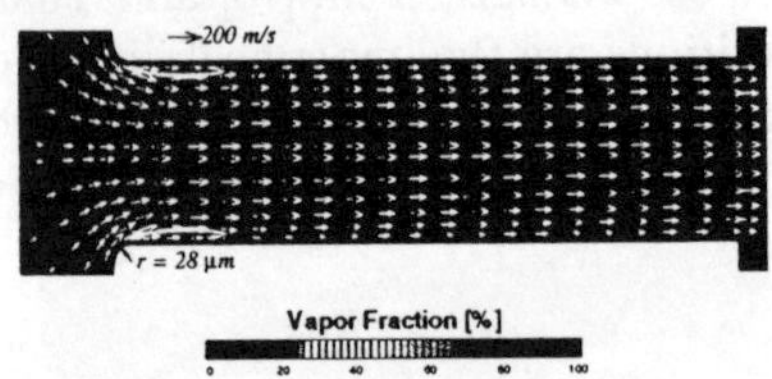


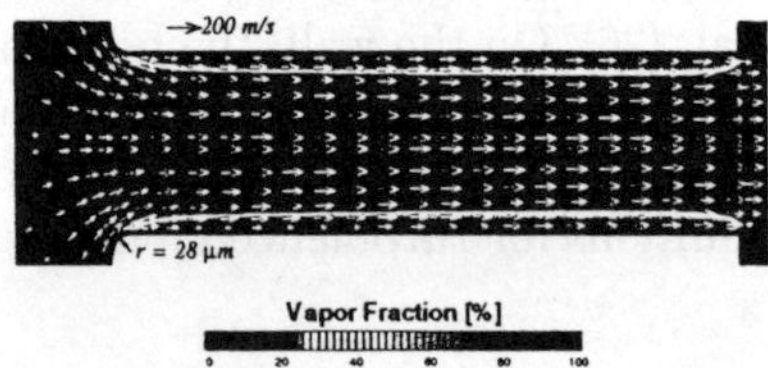


**Fig. 16.** Computed steady vapor fraction distribution in the nozzle hole. $p_{Inlet}$ = 80bar, $p_{Exit}$ = 21bar = const., $Re = \frac{\bar{u}_{Inlet} \cdot H}{\nu} \approx 2.78 \times 10^4$.

**Fig. 18.** Computed steady vapor fraction distribution in the nozzle hole. $p_{Inlet}$ = 80bar, $p_{Exit}$ = 11bar = const., $Re = \frac{\bar{u}_{Inlet} \cdot H}{\nu} \approx 2.78 \times 10^4$.

the injection pressure above the pressure at which cavitation first appears at the nozzle lip, the cavitation regime reaches the nozzle exit. To confirm their observation, another test case with $p_{Injection}$ = 80bar and $p_{Exit}$ = 11bar is performed, which demonstrates that under these conditions the cavitation extends toward the nozzle exit, see Figs. 17, 18. It is obvious, that in this supercavitating situation inclusion of the outflow is necessary, because some of the cavitation bubbles collapse outside.

Cavitation has its own length scale which is not possible to be scaled up, see Chaves et al. [10] and Yuan & Schnerr [28]. This length scale is determined by the characteristic collapse time of a given cavity and the flow velocity. If the collapse of a bubble or cavity occurs downstream of the nozzle exit, i.e. in the external jet, this is called supercavitation. Supercavitation occurs, if the collapse time of the vapor cavities is longer than the transit time of the fluid through the nozzle. It can be expected that higher pressure differences cause stronger supercavitation. For demonstration an example with a realistic pressure of $p_{Injection}$ = 800bar and $p_{Exit}$ = 11bar is calculated. Figure 19

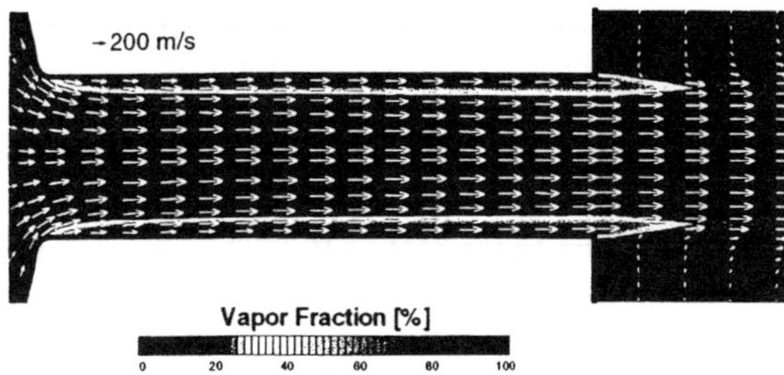



**Fig. 19.** Computed steady vapor fraction distribution. $p_{Inlet} = 800$bar, $p_{Exit} = 11$bar = const., $Re = \frac{\bar{u}_{Inlet} \cdot H}{\nu} \approx 8.82 \times 10^4$.

shows the computed steady vapor distribution.

*High frequency periodic inlet pressure excitation* Chaves et al. [10] have claimed that supercavitation is an inherent unsteady process that can only occur at high flow velocities. So far, this unsteadiness has not yet been captured, because of the constant outflow boundary conditions and the single-phase turbulence model. The collapse process is strongly influenced by the boundary conditions, the nozzle geometry and the main axial flow. The single-phase turbulence model overestimates the viscous effects in two-phase regions.

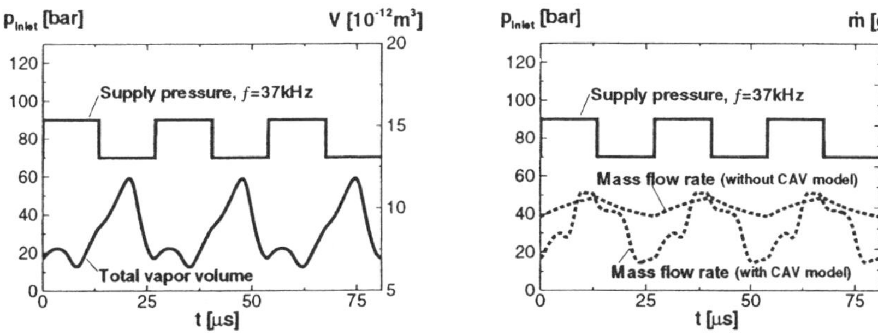



**Fig. 20.** Integrated total vapor volume within and outside the nozzle (top) and mass flow rate at the nozzle exit (bottom) depending on the rectangular inlet pressure pulse. Back pressure $p_{Exit} = 11$bar = const., $Re = \frac{\bar{u}_{Inlet} \cdot H}{\nu} \approx 2.78 \times 10^4$.

In reality the pressure upstream and behind the nozzle fluctuates with high frequencies. The reason for the upstream pressure fluctuation is the well known "water hammer" effect in the needle chamber. In order to understand the effects caused by the fluctuation of the upstream pressure further, calculations for the case with a rectangular inlet pressure pulse are performed. The pressure pulse and the calculated results are shown in Figs. 20-22.

The integrated total vapor volume in the nozzle shown in Fig. 20 indicates that the cavitation process in the injection nozzle is very complex. As a difference to the unsteady partial cavitation, see Yuan and Schnerr [28], here the total vapor volume reveals two peaks during one inlet pressure period. Also the vapor distribution, shown in Fig. 21 confirms, that the cavitation reaches the nozzle exit two times within the same period. The physical reason is the inflow of air from the nozzle outlet. The velocity field shows that the cavitating flow separates instantaneously in the nozzle. In this situation the external pressure forces the gas (air) to flow into the cavitated region (hydraulic flip), first observed by Bergwerk [29].

The calculated mass flow rates shown in Fig. 20 demonstrate that the discharge of the nozzle strongly depends on the cavitation intensity. The variation of the computed mass flow rate, with and without cavitation model, is of the order of 10% of the averaged mass flow rate, i.e. cavitation reduces the discharge significantly. In addition, it is recognized from Fig. 20 that the cavitation process enhances the mass flow rate fluctuation of $\pm 15\%$, which intensifies jet atomization, see Fig. 22.

**Modified $k - \omega$ model for simulation of cavitating flows** The intention is to enhance the single-phase $k - \omega$ model with empirical modifications to provide a two-equation turbulence model for quantitative correct simulation of steady and unsteady partial cavitating flows. From the actual literature modifications are known with respect to

- turbulent viscosity $\mu_t$ [11]
- local compressibility [1]
- pressure fluctuations [12].

From preliminary tests we found that inclusion of pressure fluctuations has no significant influence on developed or partial cavitation. This is what is to be expected, except if dealing with incipient cavitation. Because the effect of the first two modifications is similar to some extend, results of turbulent viscosity modifications are shown here and compared with Euler and single-phase $k - \omega$ simulations.

*Modification of turbulent viscosity $\mu_t$* In the two-phase regime all fluid properties vary accordingly to the local void fraction $\alpha$ between the threshold values of the pure liquid and the vapor phase. In the original model the turbulent viscosity $\mu_t$ varies explicitly proportional to the fluid density. To account for the typical very large density variation of 40.000 : 1 (water 293.15K), the linear dependence on the mixtures density is replaced by $f(\varrho)$ with the free parameter N:

$$\mu_{turb} = f(\varrho) \cdot \frac{k}{\omega} \tag{20}$$

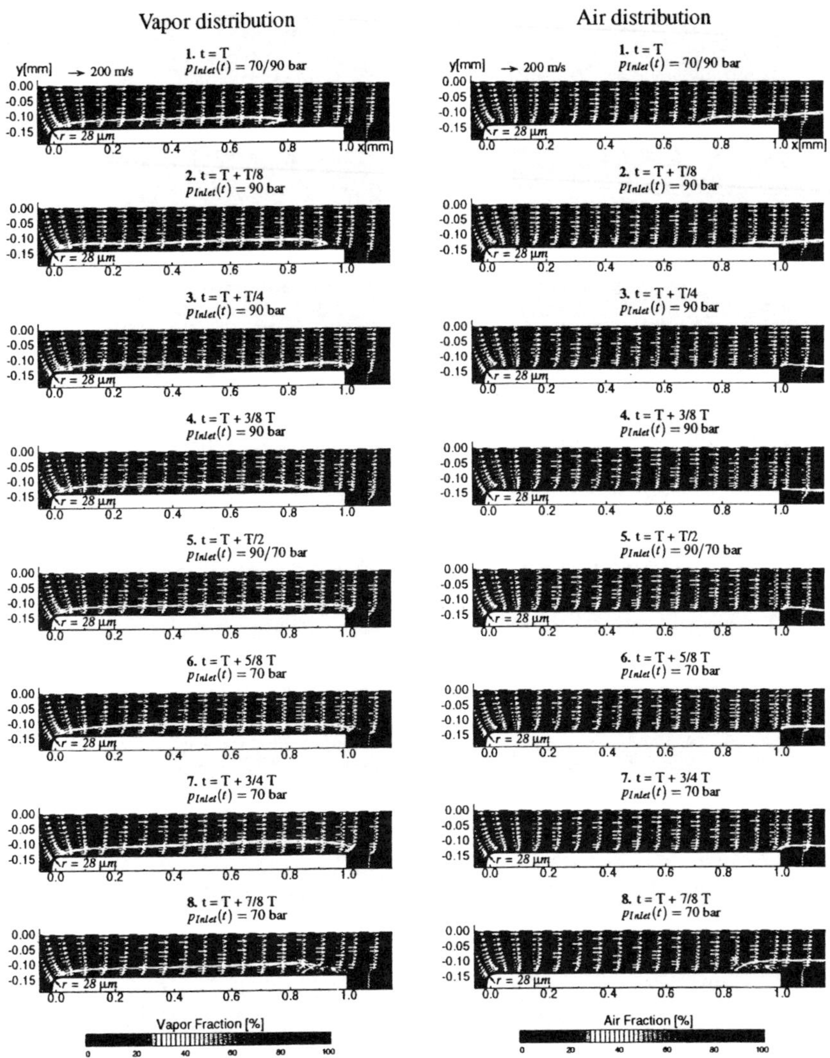



**Fig. 21.** Unsteady supercavitation: vapor and air distribution of one cycle. Periodic unsteady inlet pressure $p_{Inlet} = 80 \pm 10\mathrm{bar}$, $f=37\mathrm{kHz}$, back pressure $p_{Exit} = 11\mathrm{bar} = \mathrm{const.}$; $Re = \frac{\bar{u}_{Inlet} \cdot H}{\nu} \approx 2.78 \times 10^4$.

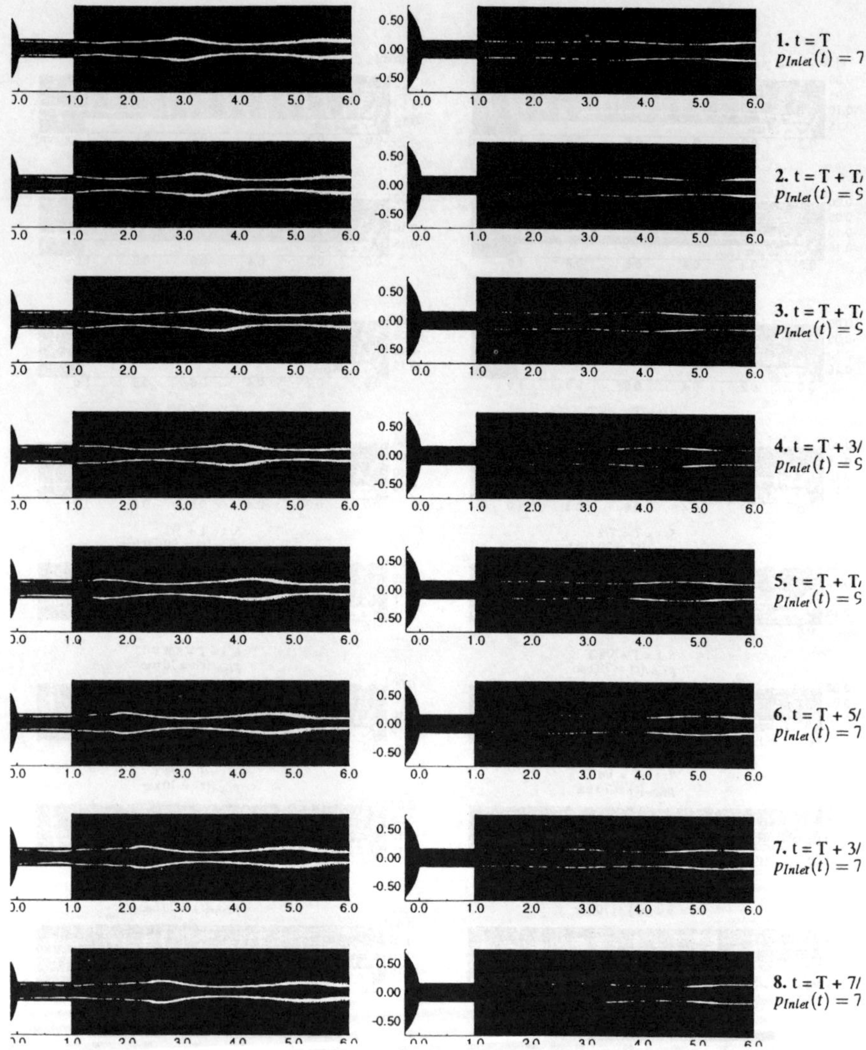


**Fig. 22.** Unsteady jet flow nearby the nozzle exit. Periodic unsteady inlet pressure $p_{Inlet} = 80 \pm 10\text{bar}$, $f=37\text{kHz}$, back pressure $p_{Exit} =11\text{bar}$. Left: computed with cavitation model, $Re = \frac{\bar{u}_{Inlet} \cdot H}{\nu} \approx 2.78 \times 10^4$; right: without cavitation model, $Re = \frac{\bar{u}_{Inlet} \cdot H}{\nu} \approx 3.0 \times 10^4$.

$$f(\varrho) = \varrho_{vap} + \left(\frac{\varrho_{vap} - \varrho}{\varrho_{vap} - \varrho_{liq}}\right)^N \cdot (\varrho_{liq} - \varrho_{vap}). \tag{21}$$

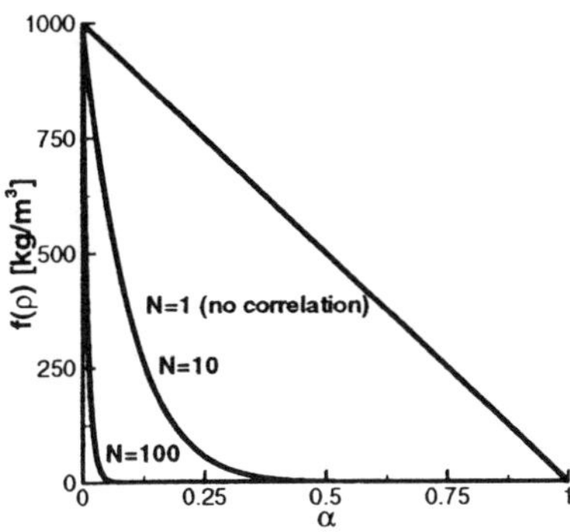


**Fig. 23.** Turbulent viscosity dependence of $\alpha$

In the pure liquid and vapor phase $\mu_t$ remains unchanged. In the two-phase regime and especially for higher values of N $\mu_t$ decreases suddenly and approximates the lower value of the pure vapor phase, see Fig. 23. Figures 24, 25 and 26 demonstrate the effect by the vapor fraction $\alpha$. Figure 24 starts with the inviscid model, we observe the typical instability with formation of a re-entry jet and separating bubbly clouds, $f = 5.2Hz$. If the Euler equation is replaced by the Reynolds averaged Navier-Stokes equation, closed by the single-phase $k - \omega$ model, the global instability of the dispersed vapor sheet disappears (Fig. 25). Due to the strong gradients in the collapse region there is still a slight local unsteadiness with $f = 7.7Hz$. However cloud separation does not exist. Modification of the turbulent viscosity $\mu_t$ with N=1000 re-establishes the global unsteadiness, similar to the inviscid case with $f = 6.6Hz$ This demonstrates the potential of the method. Experiments are necessary and under preparation to calibrate the parameter N quantitatively.

The time dependent development of the total vapor volume $V[dm^3]$ depicts the typical difference between the quasi steady result Fig. 28 with the maximum mean vapor production rate of all three cases and the global unsteady flows of Figs. 27 and 29, with the typical instantaneous disappearance of the vapor production at the beginning of each cycle. Obviously after about 5 cycles periodicity establishes, independent of the model used.

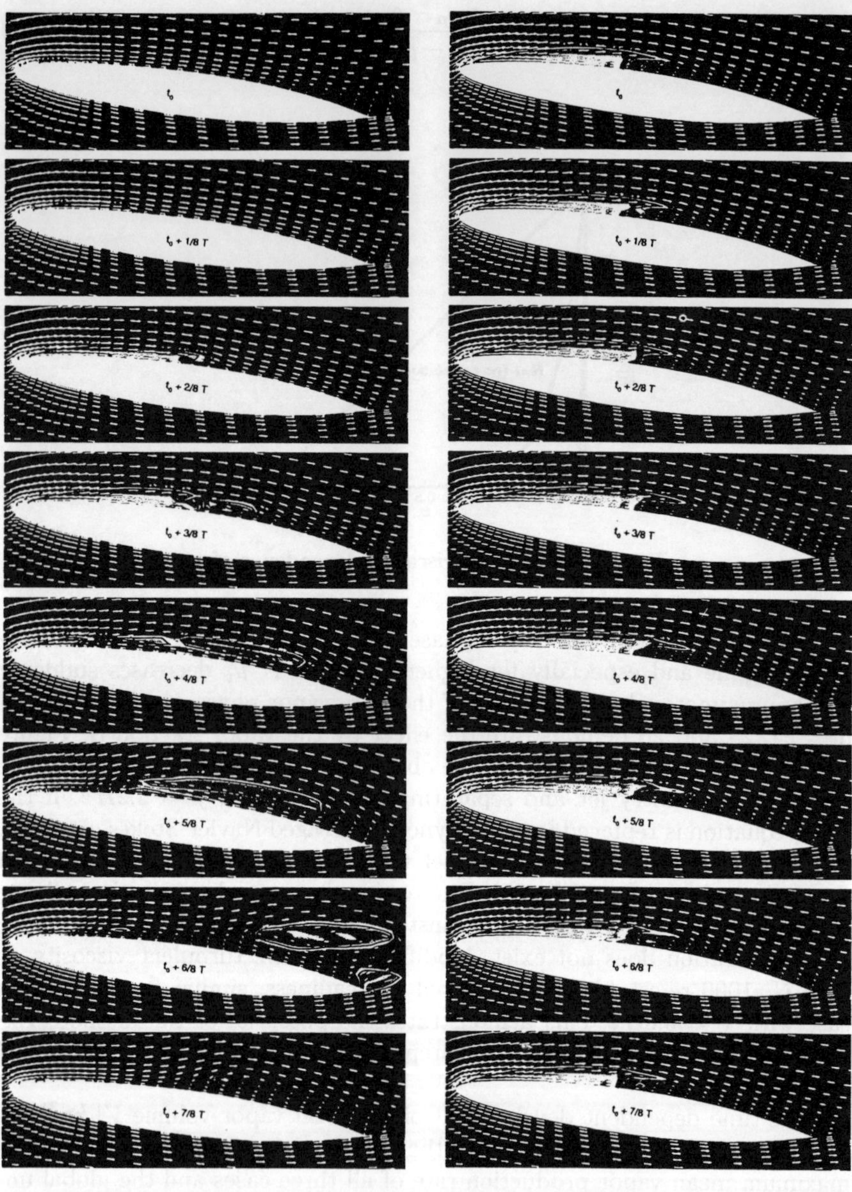

**Fig. 24.** Unsteady cavitating Euler flow. $\sigma_{ref} = 0.8$, f=5.2 Hz, $u_{inlet} = 6m/s, p_{exit} = 9528.6Pa, n_0 = 10^8 nuclei/m^3, R_0 = 3x10^{-6}m, l_c = 0.1m, Re_c = 6x10^5$.

**Fig. 25.** Unsteady cavitating flow with single-phase $k - \omega$ model of Wilcox. $\sigma_{ref} = 0.8$, f=7.7 Hz, $u_{inlet} = 6m/s, p_{exit} = 9528.6Pa, n_0 = 10^8 nuclei/m^3, R_0 = 3x10^{-6}m, l_c = 0.1m, Re_c = 6x10^5$.

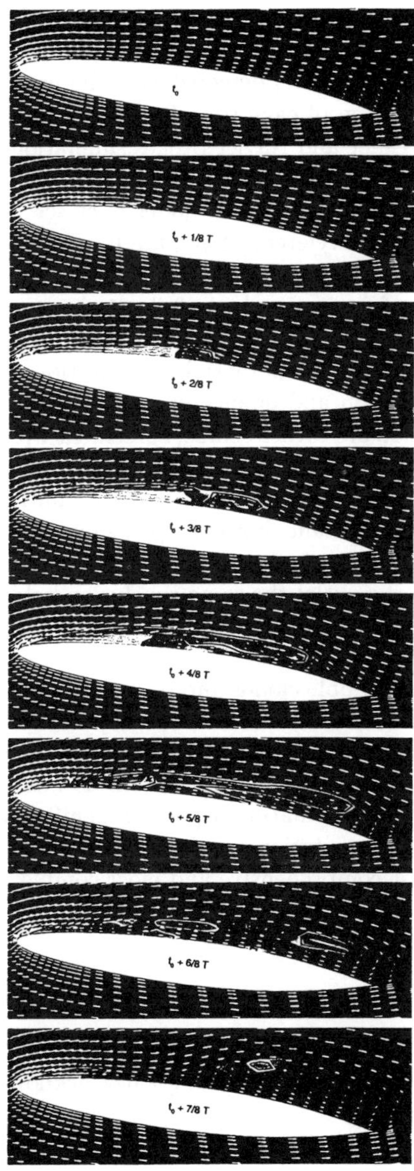


**Fig. 26.** Unsteady cavitating flow with turbulent viscosity modification, N=1000. $\sigma_{ref} = 0.8$, f=6.6 Hz, $u_{inlet} = 6m/s, p_{exit} = 9528.6Pa, n_0 = 10^8 nuclei/m^3, R_0 = 3x10^{-6}m, l_c = 0.1m, Re_c = 6x10^5$.

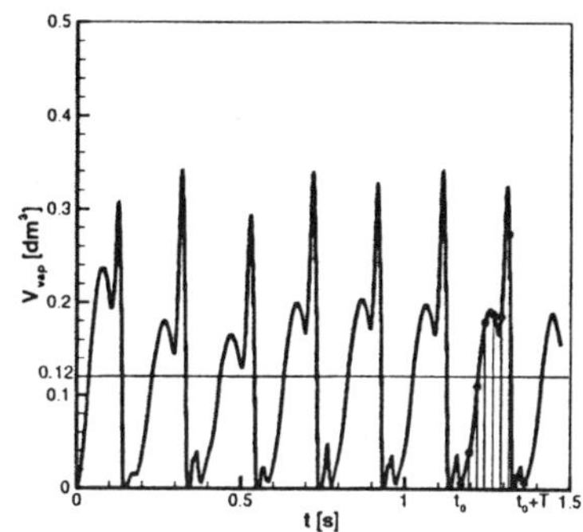

**Fig. 27.** Vapor volume in Euler flow. $\sigma_{ref} = 0.8$, f=5.2 Hz, see Fig. 24.

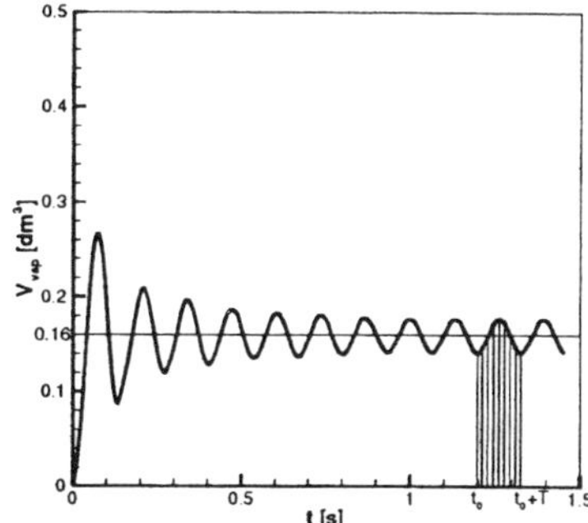

**Fig. 28.** Vapor volume with single-phase $k-\omega$ model of Wilcox. $\sigma_{ref} = 0.8$, f=7.7 Hz, see Fig 25.

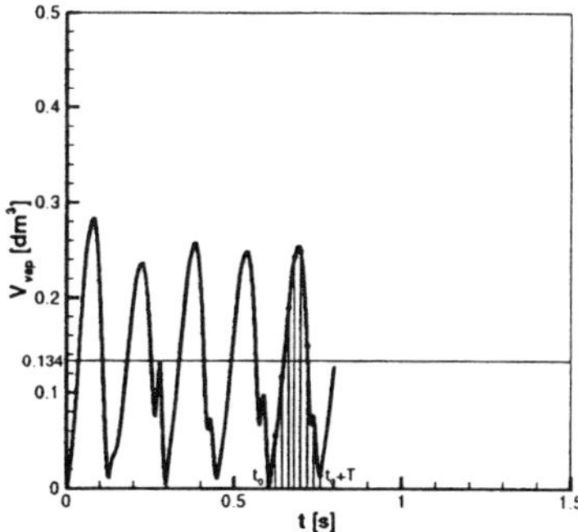

**Fig. 29.** Vapor volume with turbulent viscosity modification, N=1000. $\sigma_{ref} = 0.8$, f=6.6 Hz, see Fig. 26.

# 5 Conclusions

Application of our dispersed cavitation model to different test cases with developed cavitation in internal and external flows demonstrates the capability of our CFD tool CAVKA for simulation of cavitating flows, especially in the unsteady regime. Comparison of Euler and Navier-Stokes simulations with experiments emphasizes the necessity of further tests and development of appropriate models. Fortunately the simpler inviscid model is closer to experiments, mainly because the single-phase reference flows are free of separation.

In tendency lift and drag dynamics of cavitating hydrofoils are in good agreement with experiments. Application of our model to the injection nozzle problem demonstrates interesting details and the potential of cavitation as a fluid dynamic tool to control the formation and quality of sprays. Again, despite the fact, that the model assumption are rather coarse, agreement with experiments is surprisingly good and promising for further developments.

Due to the natural stiffness of the two-phase problem appropriate initial and boundary conditions are extremely important. For example, simulation of the injection nozzle problem without appropriate unsteady pressure fluctuations at the inlet are not realistic. At the outlet the same conclusion holds for the linkage of the higher pressure outside with the bubble collapse regime inside the nozzle.

Because simulation of sheet formation, bubble clouds and re-entry jets are natural capabilities of our dispersed cavitation model without any need to define artificial closure conditions, extension to supercavitation for simulation of external hydrodynamic flows around high speed moving bodies is possible without any model change. Due to the fascinating aspect of significant viscous drag reduction, simulation of supercavitation is subject of many research projects around the world. A most comprehensive and most complete documentation of the state of the art in this field can be found in [30].

**Acknowledgement**
The support of the Deutsche Forschungsgemeinschaft (DFG) Bonn by grant SCHN 352/16-2 during performing this research is greatly acknowledged.

# References

1. D.C. Wilcox: *Turbulence Modeling for CFD*, 2nd edn. (DCW Industries, California, 1998)
2. R.E. Arndt: Ann. Rev. Fluid Mech. **13**, 273-328 (1981)
3. M. Versluis, B. Schmitz, A. von der Heydt, D. Lohse: Science **289**, 2214-2117 (2000)

4. G.H. Schnerr, S. Adam, K. Lanzenberger, R. Schulz: 'Multiphase flows: Condensation and cavitation problems'. In: *Computational Fluid Dynamics REVIEW* ed. by M. Hafez and K. Oshima (John Wiley & Sons, Ltd., 1995)

5. J.L. Reboud,Y. Delannoy: 'Two-phase flow modeling of unsteady cavitation'. In: *Proc. 2nd Int. Symp. on Cavitation, Tokyo, Japan, 1994*

6. B. Gindroz, M.L. Billet: J. of Fluids Engineering, 87-97 (1976)

7. A. Keller, R.A. Arndt: Private communication (1999)

8. C.Y. Li, S.L. Ceccio: ASME-FED. **190**, 283-290 (1994)

9. C.E. Brennen: Multiphase Science and Technology. **10**, 303-321 (1998)

10. H. Chaves, M. Knapp, A. Kubitzek, F. Obermeier, T. Schneider: Journal of engines, SAE 1995 Transactions, Section 3, **104**, 645-657 (1995)

11. O. Coutier-Delgosha, R. Fortes-Patella, J.L. Reboud: 'Evaluation of the turbulence model influence on the simulations of unsteady cavitation'. In: *Proc. of ASME FEDSM'01, New Orleans, Lousiana, 2001*

12. M.M. Athavale, H.Y. Li, Y. Jiang, A.K. Singhal: 'Application of the full cavitation model to pumps and inducers'. In: *Proc. of ISROMAC-8, Honolulu, 2000*

13. G. Winkler, M. Heiler, G.H. Schnerr:'Homogeneous/heterogeneous condensation dynamics in turbulent low pressure turbine flows'. In: *Proc. Int. Conf. SYMCOM99, Cieplne Maszyny Przeplywowe, Turbomachinery, No. 115, Lodz, Polen, pp. 445-456, 1999*

14. B.D. Nichols, C.W. Hirt: Journal of Comp. Physics, **12**, 343-448 (1973)

15. J. Sauer: 2000. Instationär kavitierende Strömungen - Ein neues Modell, basierend auf Front Capturing (VoF) und Blasendynamik. Ph.D. thesis, University of Karlsruhe, Germany (2000)

16. O. Ubbink: Numerical prediction of two fluid systems with sharp interfaces. Ph.D. thesis, University of London, London (1997)

17. D.B. Spalding: CHAM report 910/2 (1974)

18. J.H. Ferziger, M. Perić: *Computational methods for fluid dynamics*, (Springer, Berlin Heidelberg, 1996)

19. M.S. Plesset, A. Prosperetti: Annual Review of Fluid Mechanics. **9**, 145-185 (1977)

20. H.S. Lee, H. Merte: Journal of Heat Mass Transfer. **39**, 12, 2427-2447 (1996)

21. J. Sauer, G.H. Schnerr: 'Unsteady Cavitating Flow - A New Cavitation Model Based on a Modified Front Capturing Method and Bubble Dynamics'. In: *ASMEFEDSM2000-11095, 2000*

22. A. Keller, R.A. Arndt: 'Cavitation scale effects - a presentation of its visual appearance and empirically found relations'. In: *Proc. of NCTC50 International Conference on Propeller Cavitation, April 2000, Newcastle upon Tyne*

23. J. Walther, J.K. Schaller, R. Wirth. C. Tropea: 'Characterization of Cavitating Flow Field in Transparent Diesel Injection Nozzles using Fluorescent Particle Image Velocimetry (FPIV)'. In: *Proc. ILASS - 16th Annual Conference on Liquid Atomization and Spray Systems, Darmstadt, Germany, September 11-13, 2000 (eds. Tropea, C., Heukelbach., K.) (2000) I.8 1- I.8.6*

24. A. Alajbegovic, D.A. Drew, R.T. Lahey: Chem. Eng. Comm. **174**, 85-133 (1999)

25. A. Alajbegovic, G. Meister: 'Three-phase Cavitating Flows in High Pressure Swirls Injectors'. In: *Proc. ICMF- 2001 - 4th International Conference on Multiphase Flow, New Orleans, USA, May 27-June 1, 2001 (ed. E.E. Michaelides). Tulane University (2001) CD-ROM*

26. W. Yuan, J. Sauer, G.H. Schnerr: Mécanique & Industries. **2**, 383-394 (2001)

27. P. Roosen, O. Unruh, M. Behmann M.: 'Untersuchung und Modellierung des transienten Verhaltens von Kavitationserscheinungen bei ein- und mehrkomponentigen Kraftstoffen in schnell durchströmten Düsen'. In: *Report of the Institute for Technical Thermodynamics, RWTH Aachen, Germany, March 1996*

28. W. Yuan, G.H. Schnerr: 'Cavitation in injection nozzles - Effect of injection pressure fluctuations'. In: *Proc. 4th International Symposium on Cavitation, California Institute of Technology, Pasadena, CA, USA, June 20-23, 2001*

29. W. Bergwerk W.: 'Flow pattern in Diesel nozzle spray holes'. In: *Proceedings of the Institute on Mechanical Engineers, Vol. 173, No. 25, pp. 655-660, 1959*

30. Proc. HSH 2002 - High Speed Hydrodynamics - Cheboksary Washington, Russia June 16-23, 2002 (eds. Cherny, G.G., Tulin, M.P., Terentiev, A.G., Serebryakov, V.V.). Cortana Corp. USA (2002)

# Special Topics

# Advances in Hydrodynamic Stability Theory: the Wave / Vortex Eigenmodes' Interaction

Oleg S. Ryzhov
Department of Mechanical & Aeronautical Engineering
University of California, Davis

## Abstract

Recent studies suggest that the wave / vortex eigenmodes' coupling occurs in the linear stage of disturbance development under the influence of centrifugal forces. The boundary layers of two different types provide pertinent examples. The first one relates to a three-dimensional motion with crossflow on a flat plate. The curvature of streamsurfaces naturally warping in the crossflow direction maintains centrifugal forces. The two-dimensional boundary layer on a concave cylindrical surface illustrates the second example. Here centrifugal forces are supported by the fixed curvature of a solid surface bending in the streamwise direction. In both cases centrifugal forces are balanced out by the normal-to-wall pressure gradient. The inclusion of the vortex eigenmodes brought about by centrifugal forces makes asymptotic models self-consistent, with the Cauchy problems well posed in the limit of large Reynolds numbers. An inference of conceptual importance from the asymptotic models is that the boundary layers under consideration suffer absolute instability in the streamwise direction which leads to earlier transition.

## 1  Introduction

It is becoming clear that most outstanding results in hydrodynamic stability theory are restricted to two-dimensional boundary layers on flat (or convex) surfaces. An analogous statement applies to wind-tunnel tests conducted under carefully controlled conditions with fairly weak artificial disturbances. In most cases a harmonically vibrating ribbon is used as an external source to excite TS waves of various frequencies and then trace their spatial evolution downstream. Even in this simplest environment the measurements of the amplitude of enhancing and decaying oscillations appears to be not a simple matter and there are still continuous efforts aimed at refining existing experimental data and improving the shape of the neutral stability curve (Ross et al. 1970; Klingmann et al. 1993). Typically, the linear stage of the TS wave amplification in the two-dimensional Blasius boundary layer extends over a few hundred wavelengths if the amplitude of the vibrating ribbon is small enough.

Numerous observations show that the disturbance field becomes essentially three-dimensional prior to transition. In general, three-dimensionality develops at approximately the same location where nonlinear effects come into operation. The three-dimensional wave patterns unraveled by Klebanoff et al. (1962) consist of alternating "peaks" and "valleys" which are associated with elongated domains of maximum and minimum pulsation intensity, respectively. The process as a whole is driven by convectively unstable disturbances sweeping downstream of a region where they were given birth. The convective instability is a dominant feature of the Blasius boundary layer response to external perturbations. The development of three-dimensionality in the form of peaks and valleys heralds the onset of transition due to convective instability. It is worth mentioning that the three-dimensional nonlinear stage of transition in the Blasius boundary layer is very short and extends over a characteristic distance of only ten TS wavelengths.

The transition scenario dramatically changes if the boundary layer is three-dimensional owing to the presence of crossflow in the base steady state. Gray (1952) was the first to observe the formation of seemingly steady vortices developing in the streamwise direction on a swept wing under real flight conditions. In refined experiments the three-dimensional boundary layer is created on a swept flat plate owing to a displacement body on top. Here nonlinear interactions dominate transition to turbulence nearly from a chordwise position where artificially introduced disturbances are grown to experimentally detectable size. They can terminate, depending on the test environments, in stationary streamwise vortices and coexisting travelling waves. The former of these modes derives from crossflow intrinsic to the three-dimensional boundary layer. Until recently the latter mode was widely believed to be governed by a mechanism of the TS type typical of the two-dimensional boundary layer. Most experimentalists agreed that the measured frequencies and phase velocities of travelling waves and wavelengths of stationary crossflow vortices were fairly well predicted by hydrodynamic stability theory. However they could not reach a good consensus of opinion on the group velocities and amplitude amplification rates. These discrepancies are fixed in Kachanov (1996) and Reed, Saric & Arnal (1996) whereas Bippes (1999) obviates to bring to light different views.

The latest theoretical findings (Lingwood 1997; Ryzhov & Terent'ev 1998) show that the interaction of TS waves and crossflow vortical disturbances occur even in the linear stage of disturbance amplification. Most importantly, the specific linear interactions give rise to absolute instability in the streamwise direction of the three-dimensional boundary layer. Asymptotically, absolute instability takes place in any direction, nevertheless it is not of the general type discussed by Huerre & Monkewitz (1990). Thus, interactions of the unstable TS and crossflow eigenmodes appear to be at the heart of the transition onset on swept wings and turbine blades due to the new mechanism of absolute instability in any particular direction. This suggests the grounds (among many others) that can explain why the nonlinear stage of pulsations in the initially two-dimensional Blasius boundary layer evolves so rapidly over a very short distance.

Strong streamwise disturbances are known to emerge also in a two-dimensional boundary layer on a concave surface in the form of the Goertler vortices. As a rule they are recorded as stationary elongated structures (Saric 1994). For this reason the time-dependence was usually omitted in theoretical studies. Recently Timoshin (1990) and Denier, Hall & Seddougui (1991) identified five different asymptotic regimes in the limit of large Goertler numbers by including time-dependent velocity fields into mathematical treatment. Starting from one of these regimes, Ruban (1990) made an attempt to calculate the wave-packet propagation in the boundary layer on a curved surface. However the system of governing equations underlying his model has an inherent deficiency due to the ill-posedness of the Cauchy problem and Choudhari, Hall & Streett (1994) call into question the results from the approximate model. Important similarities between the crossflow vortical disturbances in the three-dimensional boundary layer and the Goertler vortices in the two-dimensional flow on a concave cylindrical surface were clarified by Ryzhov & Terent'ev (1997).

The results set forth below are based on the well-known triple-deck theory on the assumption that the Reynolds number is large enough. As usual, an asymptotic study is restricted to the leading-order terms. However this simplest approach turns out to be insufficient as applied to wave/vortex interactions provoking absolute instability in the streamwise direction because the eigenmodes with different frequencies and wavenumbers that are involved in the process become inextricably coupled in the linear stage of amplification. The terms induced by TS waves are of lowest order in magnitude and form the leading-order approximation. The second-order terms arise from accounting for centrifugal forces maintained by strong surface curvature in the crossflow direction of the three-dimensional boundary layer. The fourth-order terms come into play when considering the impact of the curvature of a given cylindrical surface on the spectral parameters of the Goertler vortices. At a glance, the inclusion in analysis of the higher-order terms makes little sense because they are too small to cause any change of conceptual significance in the base disturbance field. However it is in the range of the wave/vortex mode interactions that the higher-order effects build up to magnitudes comparable with the leading-order contributions. The broadening of the TS pulsation spectra derives in both cases from the eigenmode coupling and results in the side bands of frequencies and wavenumbers featuring the interaction range. A small parameter depending on the Reynolds number enters the dispersion relation typical of the eigenmode coupling regime. In either case, a particular value of this parameter determines the time/space scaling of oscillations responsible for absolute instability in the streamwise direction. A strong singularity appearing in the dispersion relation is at the bottom of the remarkable mathematical analogy between the two types of vortical disturbances (Ryzhov & Terent'ev 1997). The singularity underlies the mechanism driving wave packets upstream, against the oncoming boundary layer.

## 2   Normal-to-wall pressure gradient

We start with the system of Navier-Stokes equations for an incompressible fluid which can symbolically be written as

$$N. - S. \text{ Eqs.} \tag{2.1}$$

Depending on the problem, the base steady motion is supposed to be either a general three-dimensional boundary layer on a flat plate or a two-dimensional boundary layer on a concave cylindrical surface with generators orthogonal to the direction of the oncoming uniform stream. As has been mentioned, a displacement body on top is used in experimental set-ups to create the three-dimensional flow pattern with desired properties. An extended triple-deck theory provides the foundation for a study of lower-branch instabilities arising from the interactions of TS waves and vortical eigenmodes. In the three-dimensional boundary layer these are the time-dependent crossflow disturbances, whereas in the two-dimensional boundary layer on a concave surface they take the form of unsteady Goertler vortices. In either case, the wave/vortex coupling results in a new eigenmode of upstream advancing pulsation systems ( Ryzhov & Terent'ev 1997, 1998 ).

Let the velocity $U_\infty^*$ of the oncoming external flow and the distance $L^*$ from the wing leading edge be utilized as the reference time and length. The velocity components and the pressure are nondimensionalized, respectively, with $U_\infty^*$ and $\rho_\infty^* U_\infty^{*2}$ where $\rho_\infty^*$ denotes the constant density of an incompressible fluid. We choose a special system of units adopted in the classical triple-deck theory to scale both the independent variables and desired functions in terms of a small parameter (Stewartson 1969; Messiter 1970)

$$\varepsilon = \mathrm{Re}^{-\frac{1}{8}} \tag{2.2}$$

Here the Reynolds number $\mathrm{Re} = \left( U_\infty^* \rho_\infty^* L^* \right) / \mu_\infty^*$ based on $U_\infty^*$, $L^*$, $\rho_\infty^*$ and the fluid viscosity $\mu_\infty^*$ is supposed to be large.

Let us consider first most of the boundary layer to furnish the clues to extending the triple deck by including higher-order terms. To this end, we introduce the nondimensional time $\bar{t}$ and Cartesian coordinates $\bar{x}$, $\bar{z}$ in a plane tangent to the body surface as

$$\bar{t} = \varepsilon^2 t, \ \bar{x} = 1 + \varepsilon^3 x, \ \bar{z} = \bar{z}_0 + \varepsilon^3 z \tag{2.3a, b, c}$$

and put the normal-to-wall distance equal to the standard Prandtl variable

$$\bar{y} = \varepsilon^4 y_2 \tag{2.4}$$

The further analysis depends on the type of the base flow.

## 2.1 Three-dimensional boundary layer on a flat plate.

The nondimensional components $\overline{u}$, $\overline{v}$, $\overline{w}$ of the velocity vector and the excess pressure $\overline{p}$ are

$$\overline{u} = U_0(y_2) + \varepsilon u_{21}(t, x, y_2, z) + \varepsilon^2 u_{22}(t, x, y_2, z) + \dots \tag{2.5a}$$

$$\overline{v} = \varepsilon^2 v_{21}(t, x, y_2, z) + \varepsilon^3 v_{22}(t, x, y_2, z) + \dots \tag{2.5b}$$

$$\overline{w} = W_0(y_2) + \varepsilon w_{21}(t, x, y_2, z) + \varepsilon^2 w_{22}(t, x, y_2, z) + \dots \tag{2.5c}$$

$$\overline{p} = \varepsilon^2 p_{21}(t, x, y_2, z) + \varepsilon^3 p_{22}(t, x, y_2, z) + \dots \tag{2.5d}$$

respectively, to the second-order accuracy. Here $U_0(y_2)$ and $W_0(y_2)$ prescribe the velocity profile in an initially unperturbed boundary layer in the vicinity of $\overline{x} = 1$, $\overline{z} = \overline{z}_0$. We consider $U_0$ to be in the direction of the oncoming inviscid stream and $W_0$ aligned with crossflow. Notice that both components of velocity in a plane tangent to the solid surface have similar representations in terms of $\varepsilon$, whereas the normal-to-wall component is $O(\varepsilon^2)$. Substitution of (2.3), (2.4) and the asymptotic expansion (2.5a-d) into (2.1) yields in the first approximation

$$\frac{\partial u_{21}}{\partial x} + \frac{\partial v_{21}}{\partial y_2} + \frac{\partial w_{21}}{\partial z} = 0 \tag{2.6a}$$

$$U_0 \frac{\partial u_{21}}{\partial x} + v_{21} \frac{dU_0}{dy_2} + W_0 \frac{\partial u_{21}}{\partial z} = 0 \tag{2.6b}$$

$$U_0 \frac{\partial w_{21}}{\partial x} + v_{21} \frac{dW_0}{dy_2} + W_0 \frac{\partial w_{21}}{\partial z} = 0 \tag{2.6c}$$

and the condition

$$\frac{\partial p_{21}}{\partial y_2} = 0 \tag{2.7}$$

Thus, the self-induced pressure does not vary to leading order across the bulk of the boundary layer referred to as the main deck (Stewartson 1979; Messiter 1970). The velocity field comes from (2.6a-c) in an elegant form

$$u_{21} = A(t,x,z)\frac{dU_0}{dy_2} \tag{2.8a}$$

$$v_{21} = -\frac{\partial A}{\partial x}U_0(y_2) - \frac{\partial A}{\partial z}W_0(y_2) \tag{2.8b}$$

$$w_{21} = A(t,x,z)\frac{dW_0}{dy_2} \tag{2.8c}$$

where an arbitrary function $-A(t,x,z)$ may be treated as the instantaneous displacement thickness.

Integration of the full system of equations for the second-order functions presents a tedious task. However our goal is to investigate the impact which can bring about alterations of conceptual importance associated with crossflow rather than to calculate small corrections to the leading-order effects. For this reason we may restrict ourselves to the relation

$$U_0 \frac{\partial v_{21}}{\partial x} + W_0 \frac{\partial v_{21}}{\partial z} = -\frac{\partial p_{22}}{\partial y_2} \tag{2.9}$$

determining the normal-to-wall pressure gradient in terms of the first-order transverse component $v_{21}$ of velocity. It is important that the remaining equations for the second-order functions do not affect $\partial p_{22}/\partial y_2$ at all. Substitution of (2.8b) for $v_{21}$ in (2.9) leads to

$$\frac{\partial p_{22}}{\partial y_2} = U_0^2 \frac{\partial^2 A}{\partial x^2} + 2U_0 W_0 \frac{\partial^2 A}{\partial x \partial z} + W_0^2 \frac{\partial^2 A}{\partial z^2} \tag{2.10}$$

The first term on the right-hand side of (2.10) is characteristic of the two-dimensional velocity field and evidently has nothing to do with crossflow. The second term derives from the streamsurface deformation caused in the boundary layer by both components of velocity. The third term traces its origin solely to crossflow. In what follows the first two terms on the right-hand side of (2.10) are omitted. In consequence the normal-to-wall pressure gradient becomes

$$\frac{\partial p_{22}}{\partial y_2} \sim \frac{\partial p_{22}^{(cr)}}{\partial y_2} = W_0^2(y_2)\frac{\partial^2 A}{\partial z^2} \qquad (2.11)$$

This expression would asymptotically hold in the limit as $\varepsilon \to 0$. It has a simple physical meaning. The function $-A(t,x,z)$ prescribes the instantaneous displacement thickness and thereby defines the deviation of a streamsurface from its original position in the unperturbed boundary layer. The second derivative $\partial^2 A/\partial z^2$ is proportional to the streamsurface curvature in the crossflow direction. Hence the normal-to-wall pressure gradient across most of the boundary layer is supported by centrifugal forces due to streamsurface warping at a right angle to the mainstream.

## 2.2  Two-dimensional boundary layer on a curved surface.

Let us turn to three-dimensional disturbances evolving in an initially steady two-dimensional boundary layer adjacent to a concave cylindrical surface with generators orthogonal to the oncoming external stream. The triple-deck scaling of independent variables given in (2.3) and (2.4) in terms of the small parameter $\varepsilon$ remains valid, however, the higher-order asymptotic sequences for nondimensional components of velocity and the excess pressure should be introduced in this case to capture the curvature effects giving rise to centrifugal forces. More specifically, the asymptotic expansions read

$$\bar{u} = U_0(y_2) + \varepsilon u_{21} + \varepsilon^2 u_{22} + \varepsilon^3 u_{23} + \varepsilon^4 u_{24} + \dots \qquad (2.12a)$$

$$\bar{v} = \varepsilon^2 v_{21} + \varepsilon^3 v_{22} + \varepsilon^4 v_{23} + \varepsilon^5 v_{24} + \dots \qquad (2.12b)$$

$$\bar{w} = \varepsilon^2 w_{21} + \varepsilon^3 w_{22} + \varepsilon^4 w_{23} + \varepsilon^5 w_{24} + \dots \qquad (2.12c)$$

$$\bar{p} = \varepsilon^2 p_{21} + \varepsilon^3 p_{22} + \varepsilon^4 p_{23} + \varepsilon^5 p_{24} + \dots \qquad (2.12d)$$

Terms containing $\ln \varepsilon$ along with powers in $\varepsilon$ may also enter (2.12a-d). They are forced by $\ln y_2$ featuring the behavior of the main deck solution as $y_2 \to 0$. Logarithmic terms have to be included, for instance, when taking into account the influence of the non-parallel velocity field in a thickening boundary layer even if the Blasius flow over a flat plate is under consideration. Smith (1979) suggested the corresponding extension

$$\bar{u} = U_0(y_2) + \varepsilon u_{21} + \varepsilon^2 u_{22} + \varepsilon^3 u_{23} + \varepsilon^4 \ln \varepsilon u_{24L} + \varepsilon^4 u_{24} + \dots$$

$$\bar{v} = \varepsilon^2 v_{21} + \varepsilon^3 v_{22} + \varepsilon^4 v_{23} + \varepsilon^5 \ln \varepsilon v_{24L} + \varepsilon^5 v_{24} + \ldots$$

$$\bar{w} = \varepsilon^2 w_{21} + \varepsilon^3 w_{22} + \varepsilon^4 w_{23} + \varepsilon^5 \ln \varepsilon w_{24L} + \varepsilon^5 w_{24} + \ldots$$

$$\bar{p} = \varepsilon^2 p_{21} + \varepsilon^3 p_{22} + \varepsilon^4 p_{23} + \varepsilon^5 \ln \varepsilon p_{24L} + \varepsilon^5 p_{24} + \ldots$$

of asymptotic expansions for evaluating higher-order corrections to the non-parallel flow stability. Since the curvature and non-parallel flow effects can be examined independently, the logarithmic terms are omitted from (2.12a-d).

The wall curvature first affects the terms labeled 23 which contain the contribution from the base unperturbed flow. The self-induced pressure deriving from the curvature effects is included in the functions marked by the next index 24. Notice that, unlike (2.5a-d), both components $\bar{v}$ and $\bar{w}$ are supposed in (2.12a-d) to be of equal order in magnitude. In spite of this difference we may apply the solution in (2.8a, b) to the first-order velocities $u_{21}$ and $v_{21}$ to obtain

$$u_{21} = A(t,x,z)\frac{dU_0}{dy_2}, \quad v_{21} = -\frac{\partial A}{\partial x} U_0(y_2) \tag{2.13a, b}$$

with allowance made for the fact that $W_0(y_2) = 0$ identically. The self-induced pressure

$$p_{21} = p_{21}(t,x,z) \tag{2.14}$$

is in accord with (2.7). The lateral component of velocity comes in the first approximation from an equation

$$U_0(y_2)\frac{\partial w_{21}}{\partial x} + \frac{\partial p_{21}}{\partial z} = 0 \tag{2.15}$$

showing that $w_{21}$ is driven by the spanwise pressure gradient.

Integration of the full system of equations for the higher-order functions entering (2.12a-d) presents a formidable task. However, we pursue well-defined goals and focus on investigating the impact which can bring about alterations of conceptual importance associated with the surface bending. Calculation of small corrections to the leading-order effects falls far beyond the scope of the present study. Asymptotically, a contribution to the normal-to-wall pressure gradient $\partial p_{24}/\partial y_2$ from eigenmodes affected by the surface curvature $\kappa$ takes the form

$$\frac{\partial p_{24}}{\partial y_2} - \frac{\partial p_{24}^{(\kappa)}}{\partial y_2} = -2\kappa u_{21} U_0 \tag{2.16}$$

Contrary to (2.9), it is determined by the streamwise component $u_{21}$ of perturbed velocity. On account of (2.13a) we have

$$\frac{\partial p_{24}^{(\kappa)}}{\partial y_2} = -\kappa A(t, x, z)\frac{dU_0^2}{dy_2} \tag{2.17}$$

in place of (2.11). It should be emphasized that $p_{24}^{(\kappa)}$ gives only a part of the self-induced pressure striking balance with centrifugal forces, the other contributions to $p_{24}$ from the higher-order terms entering the asymptotic expansions in powers of $\varepsilon$ are omitted, they do not bring new physics.

Let us summarize the results obtained up to this point. According to (2.7) and (2.14) the self-induced pressure holds constant to leading order across the main part of the boundary layer regardless of its initial state. In the three-dimensional flow over a flat plate the normal-to-wall pressure gradient arises from the second-order terms whereas in the two-dimensional velocity field the fourth-order corrections are required to develop the pressure variations in the direction orthogonal to the curved surface. In both cases the higher-order normal-to-wall pressure gradient is given birth by the first-order velocity field where the streamsurface curvature plays a decisive role in providing support to centrifugal forces. However, there are three important distinctions. First, the curvature of streamsurfaces naturally originates due to their warping in the crossflow direction within the three-dimensional boundary layer over a flat plate but keeps a fixed value characteristic of the solid cylindrical surface bending in the direction of the two-dimensional oncoming flow. Second, comparison of (2.9) and (2.16) shows that the normal-to-wall pressure gradient is based on the transverse velocity of the three-dimensional boundary layer past a plane and on the streamwise velocity of the two-dimensional boundary layer adjacent to a curved surface with generators at right angle to the direction of the oncoming stream. Third, unlike (2.11), the normal-to-wall pressure gradient in (2.17) varies linearly with the instantaneous displacement thickness $-A(t, x, z)$. A dependence on the second derivative $\partial^2 A/\partial x^2$ is included in (2.10) and in view of the above comments may be omitted in the limit as $\varepsilon \to 0$.

## 3  Viscous near-wall sublayer

No curvature effects need to be accounted for to within the accuracy adopted when deriving asymptotic equations controlling the velocity field in the thin viscous near-wall sublayer referred to as the lower deck. Here the normal-to-wall distance is scaled and normalized through

$$\overline{y} = \varepsilon^5 \tau_w^{-\frac{3}{4}} y_3 \tag{3.1}$$

586

where $\tau_w$ designates the skin friction (Stewartson 1969; Messiter 1970). To comply with (3.1), the nondimensional time and spatial coordinates in a plane tangent to the solid surface should be transformed by means of

$$t \to \tau_w^{-\frac{3}{2}} t \, , \; x \to \tau_w^{-\frac{5}{4}} x \, , \; z \to \tau_w^{-\frac{5}{4}} z \qquad (3.2\text{a, b, c})$$

The corresponding components $\overline{u}$, $\overline{v}$, $\overline{w}$ of the velocity vector and the self-induced pressure $\overline{p}$ are sought in the form of asymptotic sequences

$$\overline{u} = \varepsilon \tau_w^{\frac{1}{4}} u_3 \left( t, x, y_3, z \right) + \dots \qquad (3.3\text{a})$$

$$\overline{v} = \varepsilon^3 \tau_w^{\frac{3}{4}} v_3 \left( t, x, y_3, z \right) + \dots \qquad (3.3\text{b})$$

$$\overline{w} = \varepsilon \tau_w^{\frac{1}{4}} w_3 \left( t, x, y_3, z \right) + \dots \qquad (3.3\text{c})$$

$$\overline{p} = \varepsilon^2 \tau_w^{\frac{1}{2}} p_3 \left( t, x, y_3, z \right) + \dots \qquad (3.3\text{d})$$

for the boundary layers of both types under consideration.

The first-order system of Prandtl equations

$$\frac{\partial u_3}{\partial x} + \frac{\partial v_3}{\partial y_3} + \frac{\partial w_3}{\partial z} = 0 \qquad (3.4\text{a})$$

$$\frac{\partial u_3}{\partial t} + u_3 \frac{\partial u_3}{\partial x} + v_3 \frac{\partial u_3}{\partial y_3} + w_3 \frac{\partial u_3}{\partial z} = -\frac{\partial p_3}{\partial x} + \frac{\partial^2 u_3}{\partial y_3^2} \qquad (3.4\text{b})$$

$$\frac{\partial w_3}{\partial t} + u_3 \frac{\partial w_3}{\partial x} + v_3 \frac{\partial w_3}{\partial y_3} + w_3 \frac{\partial w_3}{\partial z} = -\frac{\partial p_3}{\partial z} + \frac{\partial^2 w_3}{\partial y_3^2} \qquad (3.4\text{c})$$

for the unsteady boundary-layer motion comes into operation as a result of substitution of (2.3), (3.1) and asymptotic expansions (3.3a-d) into (2.1). Owing to the additional affine transformation (3.2) the skin friction $\tau_w$ drops out of (3.4a-c) supplemented with the condition that

$$\frac{\partial p_3}{\partial y_3} = 0 \qquad (3.5)$$

The system of linearized boundary-layer equations which belongs to the second-order approximation is homogeneous and therefore can be incorporated into the first-order approximation. The error introduced by omitting the curvature-related terms in original Navier-Stokes equations falls beyond the scope of the asymptotic approach under examination. Thus, in the study of the disturbance patterns in the viscous near-wall sublayer we may restrict ourselves to the first-order functions in the asymptotic sequences (3.3a-d). As is typical of the triple-deck theory, the disturbance pattern obeys the Prandtl equations where both components $\partial p_3/\partial x$ and

$\partial p_3/\partial z$ of the self-induced pressure gradient in a plane tangent to the solid surface are not known in advance and have to be determined simultaneously with the velocity field (Stewartson 1969; Messiter 1970).

The limit conditions at the upper reaches $y_3 \to \infty$ of the viscous near-wall sublayer derive from the first-order matching of asymptotic expansions which specify the velocity fields in the bulk of the boundary layer and the thin near-wall sublayer. In the initially three-dimensional motion including crossflow the limit conditions read

$$u_3 - \tau_x y_3 \to \tau_x A(t,x,z), \quad w_3 - \tau_z y_3 \to \tau_z A(t,x,z) \quad \text{as } y_3 \to \infty \tag{3.6a, b}$$

Here $\tau_x$ and $\tau_z$ denote the normalized skin friction components such that $\tau_x^2 + \tau_z^2 = 1$. In the case of the two-dimensional flow on a curved surface the limit conditions are

$$u_3 - y_3 \to A(t,x,z) + O\left(y_3^{-1}\right), \quad w_3 \to O\left(y_3^{-1}\right) \quad \text{as } y_3 \to \infty \tag{3.7a, b}$$

in a crude approximation nevertheless sufficient for our purposes. If necessary, the $O\left(y_3^{-1}\right)$-terms can be cast in an explicit form.

The no-slip conditions

$$u_3 = v_3 = w_3 = 0 \text{ at } y_3 = 0 \tag{3.8}$$

for the boundary layers of both types under examination are homogeneous provided that no local humps or dents occur on a smooth surface.

The above formulation is not complete: along with the components $u_3$, $v_3$, $w_3$ of velocity it contains two additional unknown functions, the self-induced pressure $p_3$ and the instantaneous displacement thickness $-A$, both independent of $y_3$. The interaction law is needed to connect them and make the problem closed.

## 4 Interaction law

As the first step, the matching of the self-induced pressures in the limit as $y_2 \to 0$ and $y_3 \to \infty$ yields

$$p_3 = p_2(t, x, 0, z) \tag{4.1}$$

with $p_2$ being defined by the asymptotic expansion (2.12d). Thus, the limit process is used at this point to develop a composite model including the most important higher-order terms. However, the leading-order term $p_{21}$ in (2.12d) remains still to be found. As usual, it comes from the matching of a solution for most of the boundary layer with predominantly inviscid flow in the upper deck.

The normal-to-wall distance

$$\bar{y} = \varepsilon^3 \tau_w^{-\frac{5}{4}} y_1 \tag{4.2}$$

in the potential region of the upper deck has the same characteristic size as both reference lengths in a plane tangent to the solid surface. The first-order potential function $\varphi_1(t, x, y_1, z)$ obeys the Laplace equation

$$\frac{\partial^2 \varphi_1}{\partial x^2} + \frac{\partial^2 \varphi_1}{\partial y_1^2} + \frac{\partial^2 \varphi_1}{\partial z^2} = 0 \tag{4.3}$$

a solution of which defines the velocity components

$$u_1 = \frac{\partial \varphi_1}{\partial x}, \; v_1 = \frac{\partial \varphi_1}{\partial y_1}, \; w_1 = \frac{\partial \varphi_1}{\partial x} \tag{4.4a, b, c}$$

entering the asymptotic expansions

$$\bar{u} = 1 + \varepsilon^2 \tau_w^{\frac{1}{2}} u_1(t, x, y_1, z) + \dots \tag{4.5a}$$

$$\bar{v} = \varepsilon^2 \tau_w^{\frac{1}{2}} v_1(t, x, y_1, z) + \dots \tag{4.5b}$$

$$\bar{w} = \varepsilon^2 \tau_w^{\frac{1}{2}} w_1(t, x, y_1, z) + \dots \tag{4.5c}$$

$$\bar{p} = \varepsilon^2 \tau_w^{\frac{1}{2}} p_1(t, x, y_1, z) + \dots \tag{4.5d}$$

The boundary condition

$$\frac{\partial \varphi_1}{\partial y_1} = -\frac{\partial A}{\partial x} \quad \text{at } y_1 = 0 \tag{4.6}$$

originates from the matching of the transverse velocities in the limit as $y_1 \to 0$ and $y_2 \to \infty$. With this expression in hand we can write down a desired solution to (4.3) in the form

$$\frac{\partial \varphi_1}{\partial x} = \frac{1}{2\pi} \int_{-\infty}^{\infty} d\xi \int_{-\infty}^{\infty} \frac{\partial^2 A(t,\xi,\zeta)/\partial\xi^2}{\left[(x-\xi)^2 + y_1^2 + (z-\zeta)^2\right]^{\frac{1}{2}}} d\zeta \tag{4.7}$$

Then the Bernoulli integral

$$p_1 = -\frac{\partial \varphi_1}{\partial x} \tag{4.8}$$

provides the excess pressure induced by the instantaneous displacement thickness.

We are now in a position to make the second step and derive the interaction law. It follows from the first-order matching of the self-induced pressures in an intermediate domain where $y_1 \to 0$ and $y_2 \to \infty$ that $p_{21}(t,x,z) = p_1(t,x,0,z)$ whence

$$p_{21} = -\frac{1}{2\pi} \int_{-\infty}^{\infty} d\xi \int_{-\infty}^{\infty} \frac{\partial^2 A(t,\xi,\zeta)/\partial\xi^2}{\left[(x-\xi)^2 + (z-\zeta)^2\right]^{\frac{1}{2}}} d\zeta \tag{4.9}$$

A few comments are due on matching higher-order terms. From (2.11), the second-order normal-to-wall pressure gradient induced by crossflow vortices exerts no influence upon the external inviscid flow field because $W_0(y_2)$ exponentially vanishes to zero when approaching the upper reaches of the three-dimensional boundary layer. The matching becomes trivial in this case. On the contrary, the fourth-order matching of the normal-to-wall pressure gradient intrinsic to the Goertler vortices on a curved cylindrical surface requires caution. Higher-order terms associated with the surface curvature appear also in the outer asymptotic expansions. Those terms can be shown to be consistent with (2.17), so the matching of solutions for the upper and main decks is easily achievable. We leave aside pure technical details of the higher-order limit process.

Combining (4.1) with (2.11) and (4.9) with due account for the affine transformation (3.2) results in (Ryzhov & Terent'ev 1998)

590

$$p_3 = -\frac{1}{2\pi} \int_{-\infty}^{\infty} d\xi \int_{-\infty}^{\infty} \frac{\partial^2 A(t,\xi,\zeta)/\partial\xi^2}{\left[(x-\xi)^2 + (z-\zeta)^2\right]^{\frac{1}{2}}} d\zeta - \varepsilon D_{cr} \frac{\partial^2 A}{\partial z^2} \qquad (4.10a)$$

$$D_{cr} = \tau_w^{\frac{5}{4}} D_{(zz)} \qquad (4.10b)$$

for the initially three-dimensional boundary layer on a plane. Here the coefficient

$$D_{(zz)} = \int_0^{\infty} W_0^2(y_2)\, dy_2 \qquad (4.11)$$

conforms with the above comment that disturbances which are generated by crossflow vortices would be absent in the external flow. As applied to (4.1) combined with (2.17) and (4.9), an analogous procedure leads to

$$p_3 = -\frac{1}{2\pi} \int_{-\infty}^{\infty} d\xi \int_{-\infty}^{\infty} \frac{\partial^2 A(t,\xi,\zeta)/\partial\xi^2}{\left[(x-\xi)^2 + (z-\zeta)^2\right]^{\frac{1}{2}}} d\zeta + \varepsilon^3 D_G \kappa A \qquad (4.12a)$$

$$D_G = \tau_w^{-\frac{5}{4}} \qquad (4.12b)$$

In general, the interaction law provides a relationship to connect the excess pressure built up in the upper predominantly inviscid velocity field with the instantaneous displacement thickness generated in the lower essentially viscous sublayer. Both the excess pressure and the displacement thickness are brought about by pulsation systems propagating through the two regions and interacting with each other. The relationships cast in (4.10a, b) and (4.12a, b) have an analogous meaning and specify the interaction laws as applied to the boundary layers created in dissimilar environments. However, these relationships point also to the interaction of eigenmodes of different physical nature. The first terms on the right-hand sides of (4.10a) and (4.12a) are given birth by the TS waves whereas the second terms trace their origins to eigenmodes driven by vortical disturbance patterns elongated in the streamwise direction. These are more properly referred to as the crossflow vortices and Goertler vortices, respectively. Centrifugal forces are at the bottom of the vortical eigenmodes. As (4.10a) and (4.12a) show, centrifugal forces are maintained by the naturally emerging curvature $\partial^2 A/\partial z^2$ of streamsurfaces in the first case and the fixed curvature $\kappa$ of the solid surface in the second case. Thus, the interaction laws (4.10a, b) and (4.12a, b) describe the wave/vortex eigenmode coupling. It is just this coupling which triggers specific instabilities evolving in the boundary layers in the streamwise direction. The instability of such a kind does not arise in the Blasius flow.

## 5   Dispersion relation

Let us simplify the problem by linearizing the Prandtl equations (3.4a-c) as well as the limit conditions in the form of (3.6a, b) or (3.7a, b) and the boundary conditions (3.8) about an initially unperturbed flow

$$u = \tau_x y \,, \; w = \tau_z y \,, \; v = p = A = 0 \qquad\qquad (5.1\text{a, b, c})$$

with the subscript 3 omitted from the designation of both the normal-to-wall distance and desired functions. In the two-dimensional case $\tau_x = 1$ and $\tau_z$ becomes identically zero. Setting

$$(u - \tau_x y, v, w - \tau_z y, p, A) = \sigma\left(\tau_x \hat{u}, \hat{v}, \tau_z \hat{w}, \hat{p}, \hat{A}\right) \qquad\qquad (5.2)$$

where $\sigma \to 0$, and then

$$\left(\hat{u}, \hat{v}, \hat{w}, \hat{p}, \hat{A}\right) = e^{\omega t + i(kx + mz)}\left[\overline{u}_c\left(y\right), \overline{v}_c\left(y\right), \overline{w}_c\left(y\right), \overline{p}_c, \overline{A}_c\right] \qquad\qquad (5.3)$$

leaves us with infinitesimal disturbances governed by a system of ordinary linear differential equations for the complex amplitudes $\overline{u}_c$, $\overline{v}_c$, $\overline{w}_c$ which involve $\overline{p}_c$ and $\overline{A}_c$ as parameters. Details of handling this system can be found in many papers (see for example Ryzhov & Terent'ev 1998). Since the no-slip conditions (3.8) are homogeneous the final formulation reduces to a problem in eigenvalues. It has a nontrivial solution if the frequency $\omega$ and both wavenumbers $k$ and $m$ satisfy the dispersion relation

$$\Phi(\Omega) = Q(k, m) \qquad\qquad (5.4)$$

In both cases the left-hand side has a standard form

$$\Phi(\Omega) = \frac{dAi(\Omega)}{dY} I^{-1}(\Omega) \qquad\qquad (5.5)$$

in terms of the first derivative $dAi(\Omega)/dY$ and an improper integral

$$I(\Omega) = \int_\Omega^\infty Ai(Y)\,dY \qquad\qquad (5.6)$$

of the Airy function $Ai(\Omega)$. The argument $\Omega$ of the left-hand side depends on the type of the base motion. For the three-dimensional boundary layer containing crossflow,

$$\Omega = i^{-\frac{2}{3}}\omega K^{-\frac{2}{3}} \tag{5.7}$$

with the reduced wavenumber $K$ being given by

$$K = k\tau_x + m\tau_z \tag{5.8}$$

For the two-dimensional boundary layer on a curved cylindrical surface

$$\Omega = i^{-\frac{2}{3}}\omega k^{-\frac{2}{3}} \tag{5.9}$$

regardless of the lateral wavenumber $m$.

The right-hand side in (5.4) also varies with the type of the base motion. For the three-dimensional boundary layer with crossflow,

$$Q = i^{\frac{1}{3}}\frac{k^2 + m^2}{K^{\frac{5}{3}}}\left[\frac{k^2}{\sqrt{k^2 + m^2}} + \varepsilon D_{cr}m^2\right] \tag{5.10}$$

as shown by Ryzhov & Terent'ev (1998) in accord with the interaction law (4.10a, b). For the two-dimensional boundary layer on a concave surface the right-hand side becomes

$$Q = i^{\frac{1}{3}}\frac{k^2 + m^2}{k^{\frac{5}{3}}}\left[\frac{k^2}{\sqrt{k^2 + m^2}} + \varepsilon^3 D_G\kappa\right] \tag{5.11}$$

with the correction term in square brackets remaining constant no matter what the values of both wavenumbers are in the exponent entering (5.3).

The Blasius flow takes place on a flat plate submerged in a uniform oncoming stream experiencing no pressure gradient from outside. Then $\tau_z = D_{(zz)} = \kappa = 0$ and (5.10) as well as (5.11) reduce to

$$Q = ik^{\frac{1}{3}}\sqrt{k^2 + m^2} \tag{5.12}$$

This is the dispersion relation typical of an asymptotic description of TS waves within the framework of a classical triple-deck theory. The dispersion relation (5.12) holds also in the TS spectral range irrespective of the boundary layer provided that $\varepsilon \to 0$.

# 6 Spectral side bands

The first-order correction term in square brackets of (5.10) and the third-order term in square brackets of (5.11) are intrinsic to vortical eigenmodes emerging in more complicated base motions analyzed in the present study. A comparison of (5.10) and (5.11) with (5.12) lend support to the statement that the wave/vortex mode coupling features the disturbance patterns in these motions inevitably coming into being in real-life environments. To shed more light on the mode coupling let us treat each type of boundary layers separately.

## 6.1 Flat plate in three-dimensional flow.

On the strength of (4.10), centrifugal forces are supported by the naturally emerging curvature of streamsurfaces strongly warping in the crossflow direction and giving rise to vortical oscillations in the range

$$\omega = (\varepsilon D_{cr})^{-\frac{2}{7}} \omega_{cr} = \mathrm{Re}^{\frac{1}{28}} D_{cr}^{-\frac{2}{7}} \omega_{cr} \approx 1.17\omega_{cr} \tag{6.1a}$$

$$k = (\varepsilon D_{cr})^{-\frac{1}{7}} k_{cr} = \mathrm{Re}^{\frac{1}{56}} D_{cr}^{-\frac{1}{7}} k_{cr} \approx 1.26 k_{cr} \tag{6.1b}$$

$$m = (\varepsilon D_{cr})^{-\frac{3}{7}} m_{cr} = \mathrm{Re}^{\frac{3}{56}} D_{cr}^{-\frac{3}{7}} m_{cr} \approx 2.02 m_{cr} \tag{6.1c}$$

typical of transition conditions with $\mathrm{Re} \approx 5 \times 10^5$. In this spectral range both terms in square brackets of (5.10) become of the same order in magnitude and equally affect the disturbance pattern. As a matter of fact, oscillating crossflow vortices blend into the TS spectral range even though asymptotically, as $\varepsilon \to 0$, they make up a separate side band of higher frequencies and larger streamwise wavenumbers.

One more reason to consider the TS wave and crossflow vortical eigenmodes to be inextricably coupled comes from a purely mathematical reasoning (Ryzhov & Terent'ev 1997). Let us omit for a while the correction term in (5.10), put $m = ck^3$, $c > 0$ in line with (6.1b,c) and pass to a limit $k \to \infty$. Then we have

$$\Phi(\Omega) \sim i^{\frac{1}{3}} c^{-\frac{2}{3}} \tau_z^{-\frac{5}{3}} \tag{6.2}$$

to leading order, with the right-hand side being independent of $k$ and $m$. However, the magnitude of the frequency ensuing from (5.7) varies with $k$ as

$$\omega = i^{\frac{2}{3}} \Omega (c\tau_\tau)^{\frac{2}{3}} k^2 \tag{6.3}$$

An important conclusion can be drawn from a comparison of (6.2) with the dispersion relation (5.12) where the spanwise wavenumber $m = 0$. The amplitude of these two-dimensional TS waves remains constant if the streamwise wavenumber $k = k_* = 1.0005$. Hence it follows that a constant $c = c_* = k_*^{-2} \tau_z^{-\frac{5}{2}} = 0.999 \tau_z^{-\frac{5}{2}}$ determines the asymptotics of neutral oscillations. The frequency $\omega = \omega_*$ of neutral oscillations is fixed by (6.3). Thus, the curve $m = c_* k^3$ is an asymptote to separate exponentially enhancing eigenmodes given by $c < c_*$ from stable TS waves with $c > c_*$. The amplitude amplification rate of growing eigenmodes with $c < c_*$ inherent in the three-dimensional boundary layer would be estimated from (6.3) by $\Re(\omega) \sim k^2 \to \infty$ in the limit as $k \to \infty$. The occurrence of singularity that arises from discarding the small correction term in (5.10) is of conceptual importance since the Cauchy problem turns to be ill posed in the framework of a linear version of the classical triple-deck theory. Inclusion in the interaction law (4.10a, b) of the centrifugal force effects induced by the streamsurface curvature in the crossflow direction renders the extended asymptotic theory free from strong singularities of this kind.

## 6.2     Curved cylindrical surface in two-dimensional stream.

In view of (4.12a, b), centrifugal forces are maintained by the fixed curvature of a solid surface bending in the streamwise direction and provoking unsteady Goertler vortices in the range

$$\omega = \varepsilon^{\frac{6}{7}} (\kappa D_G)^{\frac{2}{7}} \omega_G = \mathrm{Re}^{-\frac{3}{28}} (\kappa D_G)^{\frac{2}{7}} \omega_G \approx 0.25 \omega_G \tag{6.4a}$$

$$k = \varepsilon^{\frac{9}{7}} (\kappa D_G)^{\frac{3}{7}} k_G = \mathrm{Re}^{-\frac{9}{56}} (\kappa D_G)^{\frac{3}{7}} k_G \approx 0.12 k_G \tag{6.4b}$$

$$m = \varepsilon^{-\frac{3}{7}} (\kappa D_G)^{-\frac{1}{7}} m_G = \mathrm{Re}^{\frac{3}{56}} (\kappa D_G)^{-\frac{1}{7}} m_G \approx 2.02 m_G \tag{6.4c}$$

of transition approximately specified by $\mathrm{Re} \approx 5 \times 10^5$. In this spectral range the two terms in square brackets of (5.11) have the same order of magnitude thereby exerting an equal impact on the disturbance field. In accordance with (6.4a-c), oscillating Goertler vortices do not blend into the TS spectral range, rather they form a separate side band of lower frequencies and smaller streamwise wavenumbers even under real transition conditions and not just in the limit $\varepsilon \to 0$. Notice that the spacing between neighboring crossflow and Goertler vortices is identical by virtue of (6.1c) and (6.4c).

    Let us consider pure mathematical aspects of the problem at hand. Related to the Goertler range of frequencies and wavenumbers the dispersion relation reads

$$\Phi(\Omega_G) = i^{\frac{1}{3}} \frac{\mathrm{Re}^{-\frac{3}{7}}(\kappa D_G)^{\frac{8}{7}} k_G^2 + m_G^2}{k_G^{\frac{5}{3}}} \left[ \frac{k_G^2}{\sqrt{\mathrm{Re}^{-\frac{3}{7}}(\kappa D_G)^{\frac{8}{7}} k_G^2 + m_G^2}} + 1 \right] \tag{6.5}$$

where $\Omega_G = i^{-\frac{2}{3}} \omega_G k_G^{-\frac{2}{3}} = \Omega$ remains invariant under the affine transformation (6.4a-c). In order to gain insight into the asymptotic behavior of the first wave/vortex eigenmode we apply an expression

$$\Phi(\Omega_G) \sim -\Omega_G - \Omega_G^{-\frac{1}{2}} + \dots \quad \text{as } |\Omega_G| \to \infty \tag{6.6}$$

in two limits $k_G \to \pm\infty$ and $k_G \to \pm 0$ with Re however large but fixed. It follows in both cases that

$$\omega_G \sim -e^{i\frac{\pi}{2}} \frac{\mathrm{Re}^{-\frac{3}{7}}(\kappa D_G)^{\frac{8}{7}} k_G^2 + m_G^2}{k_G} \left[ \frac{k_G^2}{\sqrt{\mathrm{Re}^{-\frac{3}{7}}(\kappa D_G)^{\frac{8}{7}} k_G^2 + m_G^2}} + 1 \right]$$

$$+ e^{-i\frac{\pi}{4}} \frac{k_G^{\frac{3}{2}}}{\sqrt{\mathrm{Re}^{-\frac{3}{7}}(\kappa D_G)^{\frac{8}{7}} k_G^2 + m_G^2}} \left[ \frac{k_G^2}{\sqrt{\mathrm{Re}^{-\frac{3}{7}}(\kappa D_G)^{\frac{8}{7}} k_G^2 + m_G^2}} + 1 \right]^{-\frac{1}{2}} + \dots \tag{6.7}$$

If $k_G \to \pm\infty$ and $m_G = m_{G0} = const$, (6.7) converges to

$$\Re[\omega_G(k_G)] \to \frac{\sqrt{2}}{2} \mathrm{Re}^{\frac{3}{28}}(\kappa D_G)^{-\frac{2}{7}} \tag{6.8}$$

This limit determines the spectral range of TS waves. Coming back from $\omega_G$ and $k_G$ to $\omega$ and $k$ by means of (6.4a, b), the amplitude amplification rate takes on a well-known constant value

$$\Re[\omega(k)] \to \frac{\sqrt{2}}{2} \text{ as } k \to \pm\infty \tag{6.9}$$

peculiar to the classical triple-deck theory.

With $O\left(\mathrm{Re}^{-\frac{3}{7}}\right)$-terms omitted, (6.5) reduces to the form

$$Q = i^{\frac{1}{3}} \frac{m_{G0}^2}{k_G^{\frac{5}{3}}} \left( \frac{k_G^2}{|m_{G0}|} + 1 \right) \tag{6.10}$$

exploited by Ruban (1990) in his treatment of the unsteady Goertler vortices. The amplitude amplification rate that corresponds to (6.10) stems from (6.7) on neglect of $O\left(\mathrm{Re}^{-\frac{3}{7}}\right)$-terms. Within this simplified approach we would arrive at

$$\Re[\omega_G(k_G)] \to \frac{\sqrt{2}}{2}\left(\frac{|k_G|}{|m_{G0}|}\right)^{\frac{1}{2}} \text{ as } k_G \to \pm\infty.\tag{6.11}$$

It is just the unrealistically large growth rate which makes the Cauchy problem ill posed. An improper account of the TS eigenmode results in a conceptual flaw of the mathematical model. Thus, the coupling of the TS wave and Goertler vortical eigenmodes proves to be decisive in putting forward the extended asymptotic theory.

In the second limit $k_G \to \pm 0$, (6.7) yields

$$\Re[\omega_G(k_G)] \to \frac{\sqrt{2}}{2}\frac{|k_G|^{\frac{3}{2}}}{|m_{G0}|}\tag{6.12}$$

showing that at the classical triple-deck scale the amplitude amplification rate

$$\Re[\omega(k)] \to \frac{\sqrt{2}}{2}\mathrm{Re}^{\frac{3}{16}}(\kappa D_G)^{-\frac{1}{2}}\frac{|k|^{\frac{3}{2}}}{|m_0|} \text{ as } k \to 0\tag{6.13}$$

indefinitely increases with $\mathrm{Re} \to \infty$. It might be useful to point out that the integral term responsible in the interaction law (4.12a, b) for the pressure variations induced by travelling waves becomes negligible to leading order for small $k$. The excess pressure comes solely from the curvature effects in the main deck.

## 7    Conclusion: absolute instability in the streamwise direction

A far-reaching mathematical analogy between the disturbance patterns in the boundary layers under scrutiny derives from singularities, identical in type, that are involved in the dispersion relations specified by their right-hand sides (5.10) and (5.11). In the case of (5.10) the singularity ensues from the factor $K^{5/3}$ entering the denominator. This factor is analogous to $k^{5/3}$ in the denominator of (5.11). On the contrary, no singularity is present in the dispersion relation (5.12) controlling three-dimensional TS waves in the Blasius boundary layer. This sharp distinction in the form of dispersion relations brings new physics into instabilities evolving in more complicated flows due to the wave/vortex interaction.

An exhaustive analysis of disturbances in an initially three-dimensional boundary layer with crossflow is available in Ryzhov & Terent'ev (1998). As it has been demonstrated, a passage of $K$ through zero gives rise to an eigenmode with peculiar properties previously unknown in the theory of hydrodynamic stability (Lin 1955; Drazin & Reid 1981). A typical map of this eigenmode onto the complex $\omega_{cr}$-plane shown in Figure 1 for $\mathrm{Re}^{-\frac{1}{8}} D_{cr} = 0.01$, $\tau_x = 2\sqrt{2}/3$, $\tau_z = 1/3$ and a moderate value $m_{cr} = m_{cr0} = 0.45$ of the crossflow wavenumber includes two separate branches, asymmetrically located in the lower and upper half-planes. In turn, each of the branches consists of two segments (lobes) asymptotically tending to straight lines parallel to the imaginary axis. More precisely, the behaviour of both lobes in the limits $k_{cr} \to \pm\infty$ and $K_{cr} = \left( \mathrm{Re}^{-\frac{1}{28}} D_{cr}^{\frac{2}{7}} k_{cr} \tau_x + m_{cr} \tau_z \right) \to \pm 0$ with $\mathrm{Re}$ however large but fixed follows from (6.6) and reads (cf. (6.7))

$$
\omega_{cr} \sim -e^{i\frac{\pi}{2}} \frac{\mathrm{Re}^{-\frac{1}{14}} D_{cr}^{\frac{4}{7}} k_{cr}^2 + m_{cr}^2}{K_{cr}} \left[ \frac{k_{cr}^2}{\sqrt{\mathrm{Re}^{-\frac{1}{14}} D_{cr}^{\frac{4}{7}} k_{cr}^2 + m_{cr}^2}} + m_{cr}^2 \right]
$$
$$
+ e^{-i\frac{\pi}{4}} \frac{K_{cr}^{\frac{3}{2}}}{\sqrt{\mathrm{Re}^{-\frac{1}{14}} D_{cr}^{\frac{4}{7}} k_{cr}^2 + m_{cr}^2}} \left[ \frac{k_{cr}^2}{\sqrt{\mathrm{Re}^{-\frac{1}{14}} D_{cr}^{\frac{4}{7}} k_{cr}^2 + m_{cr}^2}} + m_{cr}^2 \right]^{-\frac{1}{2}} + \ldots
\tag{7.1}
$$

Hence, the amplitude growth rate is positive along either of the two asymptotes. The formation of a thin loop that combines the lobes of the lower branch into a single curve is a distinctive feature of the plot in Figure 1. The tip of the loop determines the magnitude of a small positive peak of $\Re(\omega_{cr})$. This tip, labeled $d$, belongs to the left-hand lobe with the asymptote positioned closer to the imaginary axis. As it was established by Ryzhov & Terent'ev (1998), the derivative

$$
\frac{d\Im(\omega_{cr})}{dk_{cr}} > 0
\tag{7.2}
$$

at all points of this lobe provided that the crossflow wavenumber $m_{cr}$ is supposed to be constant and equal to $m_{cr0}$.

The group velocity $V_{gr}$ of the modulated signal generated by any $\max[\Re(\omega_{cr})]$ is (Landau & Lifshitz 1959)

$$
V_{gr}(k_{cr}) = -\frac{d\Im(\omega_{cr})}{dk_{cr}}
\tag{7.3}
$$

The right-hand segment of the lower branch has in its upper portion two maxima of $\Re(\omega_{cr})$ denoted by $g$ and $l$. The group velocity at both maxima, $\max_g[\Re(\omega_{cr})]$ and $\max_l[\Re(\omega_{cr})]$, is positive. So, they bring about the wave packets sweeping downstream in accord with the conventional scenario of convective instability typical of the Blasius boundary layer on a flat plate. In fact, these wave packets merge into a single highly modulated disturbance dominated by $\max_g[\Re(\omega_{cr})]$.

The group velocity corresponding to $\max_d[\Re(\omega_{cr})]$ at the tip of the loop turns out to be negative. We may conclude herefrom that in the three-dimensional boundary layer with crossflow, the wave packets are capable of advancing upstream, against the oncoming outer flow. When penetrating the region in front of a perturbing agency, they may provoke earlier transition. The term "absolute instability in the streamwise direction" coined by Lingwood (1997) and Ryzhov & Terent'ev (1998) reflects the essence of the matter and serves to distinguish between disturbances moving against the oncoming flow and oscillations initiating the absolute instability proper (Huerre & Menkewitz 1990). Notice that the group velocity defined by (7.3) has nothing to do with the phase velocity

$$c_{cr} = -\frac{\omega_{cr}(k_{cr})}{ik_{cr}} \tag{7.4}$$

of monochromatic wave trains. The absolute instability in the streamwise direction is caused by the modulated signals rather than harmonic Tollmien-Schlichting waves. The Blasius boundary layer, while experiencing convective instability, does not suffer absolute instabilities of any kind.

When $m_{cr0}$ increases to greater values the loop in the shape of the lower branch fades away. An example in Figure 2 illustrates the same conditions $\mathrm{Re}^{-\frac{1}{3}} D_{cr} = 0.01$, $\tau_x = 2\sqrt{2}/3$, $\tau_z = 1/3$ in the boundary layer, but the direction of disturbance propagation is fixed by the crossflow wavenumber $m_{cr0} = 1.25$. The wave packet capable of advancing upstream ceases to exist as a separate modulated disturbance. This does not mean that the three-dimensional boundary layer suffers no absolute instability in the streamwise direction if $m_{cr0}$ becomes large enough. Instead, continuous contributions from the monotonic variations of $\Re(\omega_{cr})$ along the left-hand lobes of both branches can affect the wave system to a substantial degree and induce amplifying pulsations upstream. According to Ryzhov & Terent'ev (1998), an indivisible wave packet is mostly driven by $\max_g[\Re(\omega_{cr})]$ and moves as a whole downstream. However, the tail part of this exponentially growing disturbance penetrates the region in front of a perturbing agency.

It is unnecessary to elaborate upon detailed arguments that the two-dimensional boundary layer on a concave cylindrical surface is also susceptible to absolute instability in the streamwise direction. The occurrence of upstream advancing pulsations immediately follows from the remarkable mathematical analogy outlined at the beginning of this section. On the other hand, (6.8) and (6.12) directly point to the existence of two lobes making up both branches of a special eigenmode similar to that discussed above in connection with the three-dimensional boundary layer with crossflow. However, in the case at hand, the branches located in the lower and upper half-planes of the complex $\omega_G$-plane are mirror images of each other about the real axis for the denominator of an expression on the right-hand side of (6.5) vanishes with $k_G = 0$ and any $m_G \neq 0$.

The lower branch of the special eigenmode in the complex $\omega_G$-plane is displayed in Figure 3 for $\mathrm{Re}^{-\frac{3}{4}}(\kappa D_G)^{\frac{5}{8}} = 0.02$ and a moderate value $m_G = m_{G0} = 0.5$ of the lateral wavenumber. The two-lobe shape of a curve in the plot agrees with the above general statement. As in the case of crossflow vortices illustrated in Figure 1, the lobes are connected together by a tiny loop. The tip of the loop with a small positive peak of $\Re(\omega_G)$ controls the formation of a modulated signal advancing upstream, against the oncoming two-dimensional flow. The wave/vortex interaction maintained by centrifugal forces results in absolute instability in the streamwise direction of the boundary layer on a concave cylindrical surface. However the ratio $\max_d[\Re(\omega_G)]/\max_g[\Re(\omega_G)]$ characteristic of the dispersion curve in Figure 3 is much less than the corresponding ratio $\max_d[\Re(\omega_{cr})]/\max_g[\Re(\omega_{cr})]$ found from Figure 1 for crossflow vortices. When evaluating the upstream and downstream moving wave packets, their relative amplitudes would be reasonably expected on these grounds to differ dramatically in the two cases under comparison. Indeed, it would even be impossible to accurately evaluate the disturbance advancing upstream on a concave surface of a given curvature without a special technique used to filter out short-scaled large-sized oscillation cycles making up the downstream propagating signal intrinsic to convective instability.

The long-scaled small-sized pulsations are seen in Figure 4 ahead of a ribbon operating in the pulse mode at $x_{cr} = 0$ under conditions specified in Figure 3. Notice that the amplitude of the wave packet behind the ribbon would be many orders greater if no filtering has been utilized in the computation. In comparison to the crossflow oscillations, the upstream influence triggered by the spiral-type Görtler vortices is a much weaker effect. Nevertheless, the laser velocity measurements by Mangalam, Dagenhart, Hepner & Meyers (1985) clearly indicate small spanwise variations at or just ahead of the beginning of the concave zone of a body in their experimental set-up. This observation is at variance with the conventional stability results according to which no streamwise vortices can start developing on a flat portion of the surface located in front of a concave insert.

The loop does not appear in the shape of the dispersion curves drawn with $m_{G0}$ increasing beyond a certain value. Figure 5 provides an example. As in the case of the three-dimensional boundary layer with crossflow, there emerges a single modulated signal which propagates downstream as a whole. However, continuous contributions from the monotonic variations of $\Re(\omega_G)$ along the left-hand lobes of both symmetrically positioned branches of the special eigenmode strongly affect the disturbance pattern. The oscillation cycles bringing up the rear of the signal penetrate upstream and may cause the initially two-dimensional boundary layer on a concave cylindrical surface to break down ahead of a perturbing source. These oscillations capable of advancing against the oncoming stream become at the heart of absolute instability in the streamwise direction.

## 8  Acknowledgements

The author wishes to express his sincere thanks to Professor M. Hafez for discussions and encouraging comments.

This work was sponsored by the Air Force Office of Scientific Research, Air Force Materials Command, USAF, under Grant number F49620-00-1-0066 DEF. The US Government is authorized to reproduce and distribute reprints for governmental purposes notwithstanding any copyright notation thereon. The views and conclusions contained herein are those of the author and should not be interpreted as necessarily representing the official policies or endorsements, either expressed or implied, of the Air Force Office of Scientific Research or the US Government.

## 9  References

[1]  Bippes, H. **1999** Basic experiments on transition in three-dimensional boundary layers dominated by crossflow instability. *Progr. Aero. Sci.* <u>35</u>, pp.363-412.

[2]  Choudhari, M., Hall, P. & Streett, C. **1994** On spatial evolution of long-wavelength Goertler vortices governed by a viscous-inviscid interaction. Part I: The linear case. *Q. J. Mech. Appl. Math* <u>47</u>, pp. 207-229.

[3]  Denier, J.P., Hall, P. & Seddougui, S.O. **1991** On the receptivity problem for Goertler vortices: vortex motions induced by wall roughness. *Phil. Trans. R. Soc. London* <u>A335</u>, pp. 51-85.

[4]  Drazin, P.G. & Reid, W.H. **1981** Hydrodynamic stability. Cambridge University Press.

[5]   Gray, W. E. **1952** The effect of wing sweep on laminar flows. RAE TM Aero. 255.

[6]   Huerre, P.S., Monkewitz, P.A. **1990** Local and global instabilities in spatially developing flows. *Ann. Rev. Fluid Mech.* 22, pp. 473-537.

[7]   Kachanov, Yu.S. **1996** Experimental studies of three-dimensional instability of boundary layers. *AIAA Paper 96-1778.*

[8]   Klebanoff, P.S., Tidstrom, K.D. & Sargent, L.M. **1962** The three-dimensional nature of boundary layer instability. *J. Fluid Mech.* 12, pp. 1-34.

[9]   Klingman, B.G.B., Boiko, A.V., Westlin, K.J.A., Kozlov, V.V. & Alfredsson, P.H. **1993** Experiments on the stability of Tollmien-Schlichting waves. *Europ. J. Mech. B / Fluids* 12, pp. 493-514.

[10]  Landau, L.D. & Lifshitz, E.M. **1959** Fluid mechanics. Pergamon Press.

[11]  Lin, C.C. 1955 The theory of hydrodynamic stability. Cambridge University Press.

[12]  Lingwood, R.J. **1997** On the impulse response for swept boundary layer flows. *J. Fluid Mech.* 344, pp. 317-344.

[13]  Mangalam, S.M., Dagenhart, J.R., Hepner, T.E. & Meyers, J.F. **1985** *AIAA Paper 85-0491.*

[14]  Messiter, A.F. **1970** Boundary-layer flow near the trailing edge of a flat plate. SIAM *J. Appl. Maths.* 18, pp. 241-257.

[15]  Reed, H.L., Saric, W.C. & Arnal, D. **1996** Linear stability theory applied to boundary layers. *Ann. Rev. Fluid Mech.* 28 , pp. 389-428.

[16]  Ross, J.A, Barnes, F.H., Burnes, J.G. & Ross, M.A.S. **1970** The flat plate boundary layer. Part 3. Comparison of theory with experiment. *J. Fluid Mech.* 43, pp. 819-832.

[17]  Ruban, A. I. **1990** Propagation of wave packets in the boundary on a curved surface. *Fluid Dyn.* 25, pp. 213-221.

[18]  Ryzhov, O.S. & Terent'ev, E.D. **1997** A composite asymptotic model for the wave motion in a steady three-dimensional subsonic boundary layer. *J. Fluid Mech.* 337, pp. 103-128.

[19]  Ryzhov, O.S. & Terent'ev, E.D. **1998** Streamwise absolute instability of a three-dimensional boundary layer at high Reynolds numbers. *J. Fluid Mech.* 373, pp.111-153.

[20]  Saric, W.C. **1994** Goertler vortices. *Ann. Rev. Fluid Mech.* 26, pp. 379-409.

[21]  Smith, F.T. **1979** On the non-parallel flow stability of the Blasius boundary layer. Proc. R. Soc. London A366, pp. 91-109.

[22]  Stewartson, K. **1969** On the flow near the trailing edge of a flat plate. II. *Mathematika* 16, pp. 106-121.

[23]  Timoshin, S.N. **1990** Asymptotic analysis of a spatially unstable Goertler vortex spectrum. *Fluid Dyn.* 25, pp. 18-25.

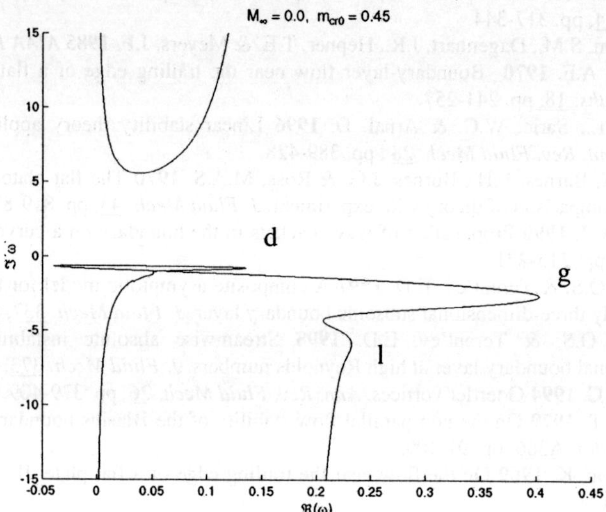


Figure 1. Two asymmetric branches of the special eigenmode in the plane of complex crossflow frequencies for $m_{cr0} = 0.45$.

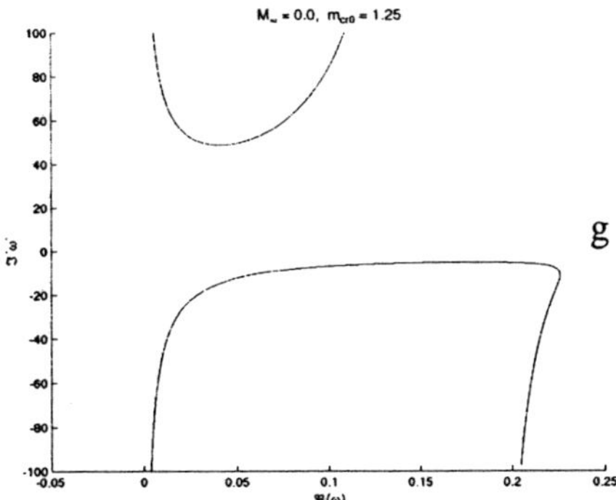

Figure 2. The wave/vortex eigenmode in the plane of complex crossflow frequencies for $m_{cr0} = 1.25$ .

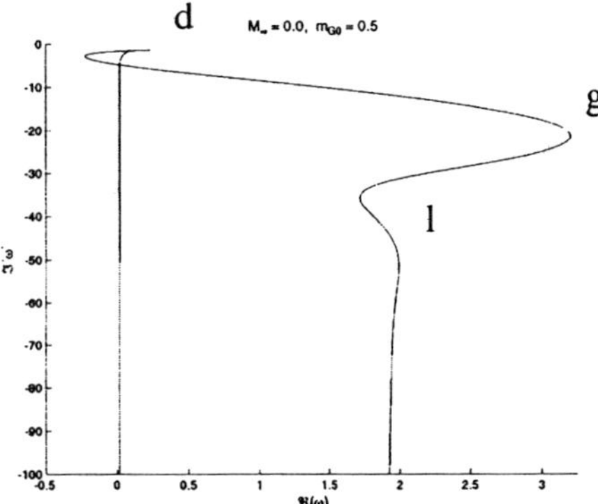

Figure 3. The lower branch of the special eigenmode in the plane of complex Görtler frequencies for $m_{G0} = 0.5$ .

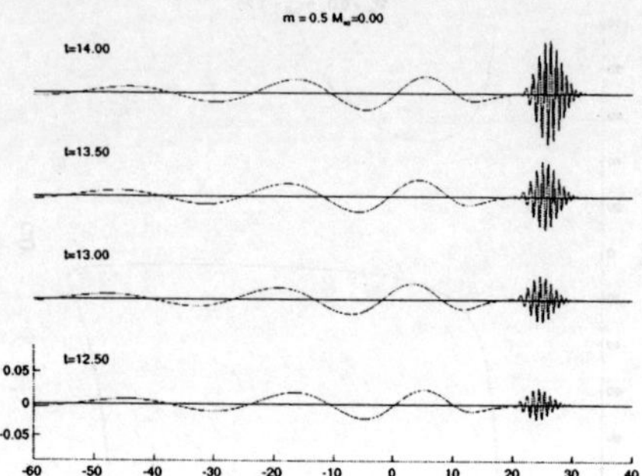

Figure 4. Wave packets induced by the wave/vortex interaction in the range of Görtler frequencies with $m_{G0} = 0.5$.

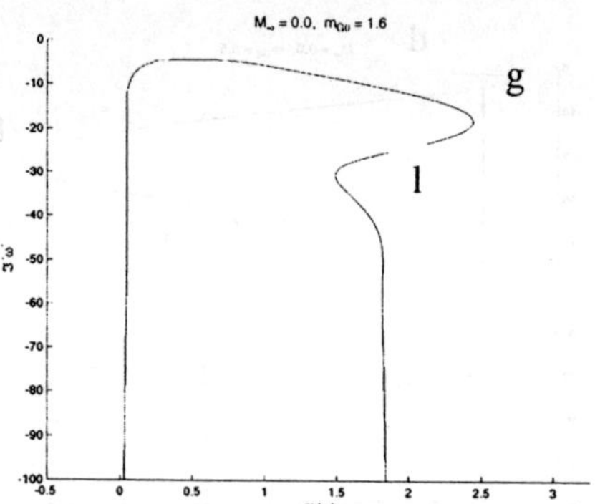

Figure 5. The lower branch of the wave/vortex eigenmode in the plane of complex Görtler frequencies for $m_{G0} = 1.6$.

# Evaluation of a Preconditioned Flow Solver for a Broad Range of Mach Number and Temperature Ratio

B. K. Lambert, L. K. Taylor, W. R. Briley [1]

## 1 Abstract

The convergence behavior and solution accuracy of a preconditioned characteristic–based viscous flow solver are evaluated for steady laminar compressible flow past a flate plate, over a range of Mach number and wall temperature ratio ($0.1 \leq M_\infty \leq 3.0$ , $0.1 \leq T_w/T_\infty \leq 10.0$). The flat–plate is useful as a test problem since easily computed similarity solutions for arbitrary Mach number and wall temperature ratio are available that themselves have been validated against experimental measurements. The Navier–Stokes solutions are evaluated for asymptotic convergence behavior, spatial discretization error, and are assessed for conformance to similarity and agreement with the similar solutions. The algorithm converges well for all Mach numbers tested when the wall temperature ratio is between 0.5 and 2.0, and a reduced time step gave solutions for temperatue ratios of 0.1 and 10.0. The validation comparisons indicate consistent and acceptable accuracy over a broad range of flow speeds and temperature ratios, subject to the theoretical thin–layer similarity assumption and to a lesser extent the grid resolution.

---

[1] ERC Computational Simulation and Design Center, Mississippi State University, Mississippi State,MS 39762-9627.

## 2 Introduction

Low speed flows of a compressible gas with local regions of high velocity, variable density or temperature occur in numerous engineering applications. Propeller, jet or rocket–powered vehicles designed for vertical take–off involve nearly stagnant far–field flow with embedded regions of localized high velocity and large temperature variations caused by hot exhaust. Low–speed flows with high heat transfer or energy release also involve large density changes. Examples include combustion and advanced space propulsion concepts wherein thrust is produced in convergent–divergent nozzles by lasers or other energy sources. Some examples of flow simulations for problems of this type are given in [1–5].

During the past twenty years, a number of researchers (e.g., [6–16]) have explored the use of preconditioned algorithms for compressible flows as a means of addressing compressible flows at both high and low Mach number using the same basic flow solver. Since the purpose of these preconditioned algorithms is to address a broad range of flows, it is beneficial to evaluate trends in performance and accuracy of the algorithm over a range of flow parameters, for the same physical problem.

The need to evaluate new or improved flow solvers on appropriate test problems is well recognized. Once the solution algorithm has been demonstrated and validated for relatively simple problems, it can then be applied to increasingly complex problems and geometries, within its range of applicability.

### 2.1 Objective of This Study

The objective of the present study is to evaluate a recent preconditioned high–resolution viscous flow solver [17] over a wide range of parameter values for both freestream Mach number $M_\infty$ and wall temperature $T_w$. This evaluation includes an assessment of both algorithm performance (rate of convergence to a steady solution) and validation of solution accuracy. The test problem is steady laminar compressible flow past a flat plate with either adiabatic wall or imposed constant temperature.

In general, the validation of solution accuracy over wide parameter ranges poses difficulties in that validation–quality experimental data may not be available, and multiple experimental facilities may be required to cover the desired parameter ranges. Consequently, algorithms are typically validated against experimental data for only one or two flow conditions on a given geometry. The laminar flat–plate boundary layer is especially useful as a test problem in this regard, since there exist analytical similarity solutions for arbitrary Mach number and wall temperature ratio that themselves have been validated against experimental measurements. These similarity solutions are easily evaluated for comparison with the present Navier–Stokes solutions.

The present validation using the flat–plate similarity solutions involves two considerations. First, the Navier–Stokes solution must be evaluated for spatial discretization error. Secondly, the Navier–Stokes solution must be assessed in regard to its conformance to the similarity property of the similarity solutions. Suppose that the numerical discretization error is small, and that the Navier–Stokes solution conforms to similarity and agrees with the similarity solutions. Then the Navier–Stokes solution can be regarded as validated to the degree that the similarity solution itself has been validated by experimental comparisons at specific parameter values. The advantage of this approach is that it is not limited to the small number of cases for which experimental data exists. On the other hand, judgements regarding validation comparisons must be made in the context of the similarity assumptions.

## 3 Preconditioned Algorithm

The preconditioned algorithm examined in the present work is that presented in [17]. It employs a preconditioned numerical flux related to Roe's flux–difference scheme [18] and is applied within an iterative implicit unsteady solution algorithm adapted in [17] from the incompressible flow solver in references [19–21]. The preconditioned primitive–variable flux–difference formulation of [17] is illustrated here for the following one–dimensional hyperbolic conservation law, expressed in primitive variables:

$$\frac{\partial}{\partial t} Q(q) + \frac{\partial}{\partial x} F\left[Q(q)\right] = 0 \tag{1}$$

Here, $Q$ is the solution vector for conservative variables, $q$ is a set of primitive variables, and $F$ is the flux vector.

For the compressible Euler equations, these equations are expressed in nondimensional form using the following reference quantities: density, $\rho_r$; velocity, $U_r$; temperature, $T_r$; pressure, $\rho_r U_r^2$; length, $L_r$; time, $L_r/U_r$; energy and enthalpy, $h_r$. This gives $Q = (\rho, \rho u, \rho e_t)^T$ and $F(Q) = (\rho u, \rho u^2 + p, \rho u h_t)^T$, where $\rho$ is density, $u$ is velocity, $p$ is pressure, $e_t$ is total energy, and $h_t$ is total enthalpy. These variables are related by $h_t = e_t + E_c p/\rho$, where $E_c$ is an Eckardt number defined by $E_c = U_r^2/h_r$. These equations are supplemented by an equation of state of the form $\rho = \rho(p, T)$. For a perfect gas with constant specific heat, $h_r \equiv C_p T_r$, and it follows that $E_c = (\gamma - 1)M_r^2$, where $M_r = u_r/c_r$ is a reference Mach number, $c_r^2 = \gamma R T_r$ is the reference sound speed, $\gamma = C_p/C_v$ is specific heat ratio, and $R$ is the gas constant. The state equation is then given by

$$p = \rho T/\gamma M_r^2 = \rho c^2/\gamma, \quad c^2 = T/M_r^2 \tag{2}$$

A preconditioning matrix $\Gamma_q$ is introduced into (1) as in

$$MT_q^{-1}\frac{\partial q}{\partial t} + \frac{\partial}{\partial x}F(q) = 0 \tag{3}$$

where $M \equiv \partial Q/\partial q$ is the Jacobian matrix for the change of variables. The following scheme

$$(MT_q^{-1})_i\,\frac{\partial q_i}{\partial t} + (dA/dV)_i\,(F_{i+\frac{1}{2}} - F_{i-\frac{1}{2}}) = 0 \tag{4}$$

then approximates (3) on a finite volume $dV$ with cell interface area $dA$. Here, $(MT_q^{-1})_i$ and $q_i$ are volume–averaged values. The primitive variable choice $q \equiv (\rho, u, p)^T$ gives the simplest eigensystem and was used for all present solutions. Implementations for other primitive variable choices are given in [17].

The interfacial fluxes are obtained from the following preconditioned characteristic–based flux–difference approximation:

$$F_{i+\frac{1}{2}} = \frac{1}{2}(F_L + F_R) - \frac{1}{2}\hat{M}\hat{T}_q^{-1}\left|\hat{T}_q\hat{a}\right|\Delta q \tag{5}$$

Here, $a = M^{-1}AM$ is the system matrix for $q$, and $\Delta q \equiv q_R - q_L$, where $q_L$ and $q_R$ denote the left and right state variables at the cell interface. The matrices in (5) are evaluated from averaged variables $\hat{q}(q_R, q_L)$, where

$$\hat{\rho} = \frac{1}{2}(\rho_L + \rho_R), \quad \hat{u} = \frac{1}{2}(u_L + u_R)$$
$$\hat{h}_t = \frac{1}{2}(h_{tL} + h_{tR}), \quad \hat{c}^2 = \frac{\hat{h}_t}{M_r^2} - \frac{\gamma - 1}{2}\hat{u}^2 \tag{6}$$

These flux evaluations require an eigensystem for the preconditioned system matrix $a_\Gamma \equiv \Gamma_q a$. Let $\Lambda_\Gamma$ denote the diagonal matrix of eigenvalues of $a_\Gamma$, and let $R_\Gamma$ denote a set of right eigenvectors of $a_\Gamma$. Then $a_\Gamma$ can be expressed as $a_\Gamma = R_\Gamma \Lambda_\Gamma R_\Gamma^{-1}$, and then $\left|\hat{a}_\Gamma\right| \equiv \hat{a}_\Gamma^{\,+} - \hat{a}_\Gamma^{\,-}$, where $\hat{a}_\Gamma^{\pm} = \hat{R}_\Gamma \hat{\Lambda}_\Gamma^{\pm} \hat{R}_\Gamma^{-1}$.

A constant preconditioning matrix $\hat{\Gamma}_q = diag\,[1, 1, \beta\,]$ is used here with the following definition:

$$\beta \equiv \begin{cases} M_\infty^2 & (M_\infty < 1) \\ 1 & (M_\infty \geq 1) \end{cases} \tag{7}$$

where $M_r$ is taken as the freestream Mach number $M_\infty = u_\infty/\sqrt{\gamma RT_\infty}$.

The eigenvalues for the preconditioned system can be written as

$$\hat{u}, \quad \frac{\hat{u}}{2}(1 + M_\infty^2) \pm \sqrt{\frac{\hat{u}^2}{4}(1 - M_\infty^2)^2 + \hat{T}} \tag{8}$$

These eigenvalues are well behaved as $M_\infty \to 0$, since the nondimensional, local, averaged variables $\hat{u}$ and $\hat{T}$ are both $O(1)$ quantities. This primitive–

variable numerical flux is related to Roe's flux–difference scheme and preserves contact discontinuities using primitive variables, with or without preconditioning. Nonsingular eigensystems for the three–dimensional Euler equations in general dynamic curvilinear coordinate systems are given in [17].

The preconditioned flow solver has the same basic structure as the incompressible Discretized Newton–Relaxation (DNR) scheme of Taylor and Whitfield [19–21] but is adapted for compressible flow with preconditioning and is implemented in primitive variables. It is comprised of an iterative implicit finite–volume scheme, conservative primitive–variable fluxes with third–order MUSCL extrapolation, numerically computed state–vector flux linearizations, and approximate–Newton iteration solved using LU/SGS relaxation. The LU/SGS relaxation is equivalent to a lower–upper approximate factorization (LU/AF) scheme (see, for example, [22]), and interpretation as an LU/AF scheme clarifies the need for preconditioning. The present computations employ structured grids expressed with curvilinear coordinates, and characteristic–variable inflow/outflow boundary conditions and preconditioning are applied for subsonic cases. Further details of the algorithm can be found in [17] and in the references contained therein.

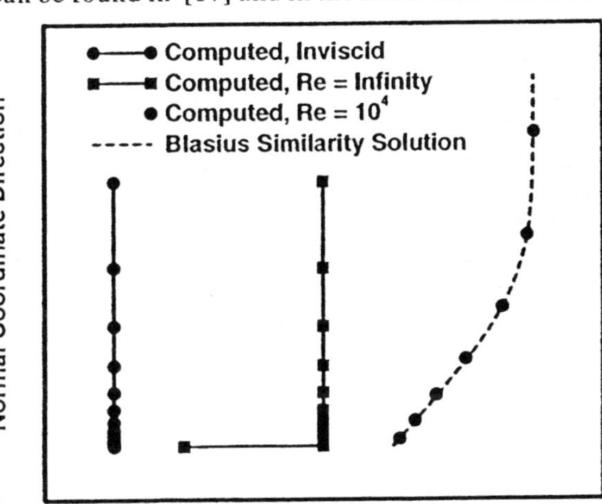


**Figure 1** Computed midchord velocity profiles for flow past a flat plate: Inviscid flow with slip and no–slip conditions, and viscous flow at Re=10$^4$ compared with the Blasius solution.

### .3.1 High–Resolution Inviscid Fluxes for Viscous Flows

There is some evidence (e.g., [23]) that characteristic–based inviscid fluxes that preserve slip–line discontinuities also improve the resolution of viscous flows containing thin shear layers. Consider the classical flat–plate boundary–layer solution in the limit of *infinite* Reynolds number. This limiting solution is a slip–line discontinuity, a uniform flow satisfying the no–slip condition on the wall. An inviscid flux that exactly captures this discontinuity with finite mesh spacing seems logically preferable to one that converges only asymptotically as mesh spacing approaches zero. The viscous terms then act to diffuse the discontinuity to an appropriate thickness $\delta \simeq O(\mathrm{Re}^{-1/2})$, in accordance with boundary layer theory.

This property is demonstrated for the present scheme in Figure NO TAG, where mid–chord velocity profiles are shown for a rectangular $(71 \times 21)$ grid with 50 uniform intervals along the plate and minimum wall spacing of $10^{-4}$ chord. Two of the solutions are inviscid solutions computed for zero viscosity. The first is a uniform flow with inviscid slip condition at the wall boundary; the second satisfies the no–slip condition and gives an inviscid slip–line discontinuity located at the wall. Both of these solutions reproduce the corresponding exact solution. The third solution is a viscous solution computed for $\mathrm{Re} = 10^4$, which agrees very well with the Blasius similarity solution, even with only six grid intervals in the boundary layer.

## 4 Validation Approach

The algorithm evaluation consists of two parts: algorithm performance and algorithm accuracy, both of which have been evaluated by running test cases for compressible viscous laminar flow over a flat plate. The test parameters of primary interest for the algorithm are time step and grid distribution, and for the physical problem they are wall temperature and free–stream Mach number.

The present metric for algorithm performance is the asymptotic convergence rate, measured for a range of Mach numbers at constant wall temperature, as well as for a constant Mach number over a range of wall temperatures. The algorithm accuracy is assessed by comparing computed flow solutions with results from theoretical similarity solutions for compressible laminar boundary–layer flow past a flat plate. The similarity solutions are easily evaluated and enable accuracy comparisons for a wide range of Mach numbers and wall temperature ratios. Schetz [24] outlines several similarity transformations that reduce the laminar compressible boundary layer equations to ordinary differential equations for flat plate flow. The particular similarity solution used here is that of Illingworth, as presented by White [25]. The Illingworth transformation was chosen for two reasons: 1) it employs a power–law viscosity relation which matches, to good accuracy, that in the Navier–Stokes flow solver, and 2) it is easily solved using commercially available mathematics software packages.

In addition to the assumptions of boundary layer theory (thin shear and thermal layers and asymptotically large Reynolds number), the similarity analysis assumes that the velocity and temperature profiles at each streamwise location have a similar shape when plotted in terms of appropriate similarity variables. It is not to be expected that the Navier–Stokes and similarity solutions would agree exactly, only that they will agree to within the limitations of the similarity assumptions. This issue has been explored in [26] by comparing solutions for $M_\infty = 0.1$ and $T_w/T_\infty = 2.0$ for $Re_L = 10^4$, $10^5$ and $10^6$, using a $142 \times 62$ grid with wall spacing of 0.00025. The velocity and temperature profiles from all of these solutions agree reasonably well with the similarity solution, except near the leading and trailing edges, and are thus suitable test cases for the present comparisons. The difference in skin friction coefficient at mid chord varies from about 8% at $Re_L = 10^4$ to 5% at $Re_L = 10^6$.

Grid resolution and mesh distribution both affect the accuracy of the computed solutions, whereas the Reynolds number affects how closely the Navier–Stokes and similarity solutions should agree. Since it was desirable to keep these factors constant throughout the validation study, most of the test cases were run for $Re_L = 10^4$ and with the same baseline grid. This baseline grid has 71 points in the axial direction (with 51 points on the plate itself) and 31 points in the direction normal to the plate, with spacing normal to the wall of 0.001. The influence of grid resolution on the solution accuracy is documented in the following Section. Reference [26] contains further results which indicate that the degree of mesh refinement and Reynolds number chosen for this study are adequate for most cases. The two cases for which the mesh refinement or Reynolds number were not adequate are also discussed in the following Section. A complete accounting of the present results can be found in [26].

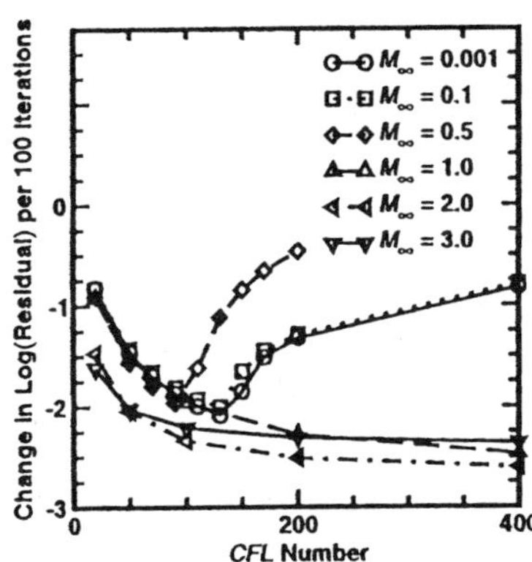

Figure 2  Asymptotic Convergence Rates for $T_w/T_\infty = 1.0$ and ($0.001 \leq M_\infty \leq 3.0$)

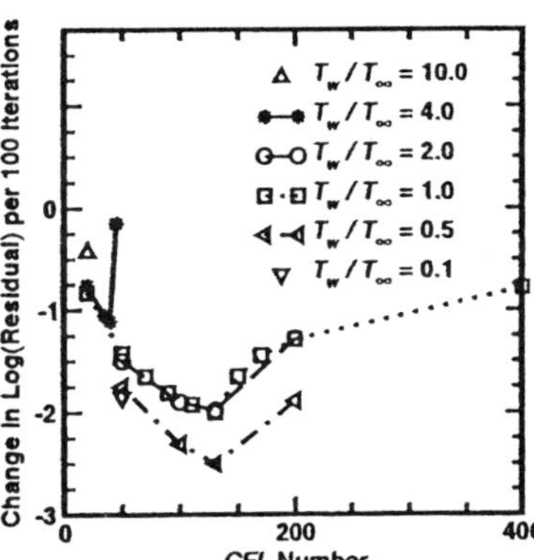

Figure 3  Asymptotic Convergence Rates for $M_\infty = 0.1$ and ($0.1 \leq T_w/T_\infty \leq 10.0$)

612

## 5 Computed Results

### 5.1 Algorithm Performance

The asymptotic convergence rate was determined empirically as a function of $CFL$ number for different Mach number and wall temperature values. A spatially varying time step was used and is defined by

$$\Delta t_{i,j} = CFL \left[ |\lambda_i|_{\max} + |\lambda_j|_{\max} \right]^{-1} \tag{9}$$

where $\lambda_i$ and $\lambda_j$ are the eigenvalues of the preconditioned flux Jacobian matrices for the $i$ and $j$ coordinate directions.

Figure 2 shows a plot of convergence rate as a function of $CFL$ number for $(0.001 \leq M_\infty \leq 3.0)$ with $T_w/T_\infty = 1.0$. The subsonic cases have an optimal $CFL$ value of 90–130, above which the convergence rate decreases, whereas the supersonic cases do not have this behavior. The maximum stable $CFL$ number is between 400 and 800 for the subsonic cases and about 900 for the sonic and supersonic cases.

Figure 3 shows convergence rate for $(0.1 \leq T_w/T_\infty \leq 10.0)$ with $M_\infty = 0.1$. The general behavior of convergence rate is similar to that in Figure 2 for $(0.5 \leq T_w/T_\infty \leq 2.0)$. Although the convergence rate for $T_w/T_\infty = 2.0$ is not significantly different from the baseline case of $T_w/T_\infty = 1.0$, the convergence is much faster for $T_w/T_\infty = 0.5$. Cases run at wall temperatures of 4.0 and 10.0 have a slower convergence rate and a much smaller maximum $CFL$ for convergence. The maximum $CFL$ is about 50 for $T_w/T_\infty = 4.0$ and 20 for $T_w/T_\infty = 10.0$. The cold–wall case $T_w/T_\infty = 0.1$ has a maximum $CFL$ number of about 50.

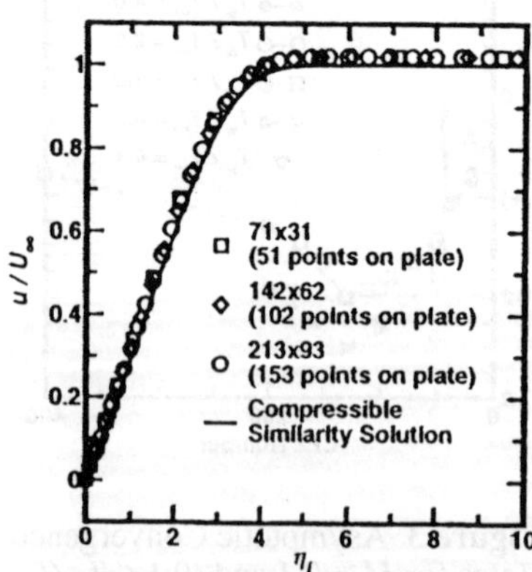

Figure 4 Computed Half–Chord U–Velocity Profiles for Three Grid Resolutions at $T_w/T_\infty=2.0$ and $M_\infty=0.5$

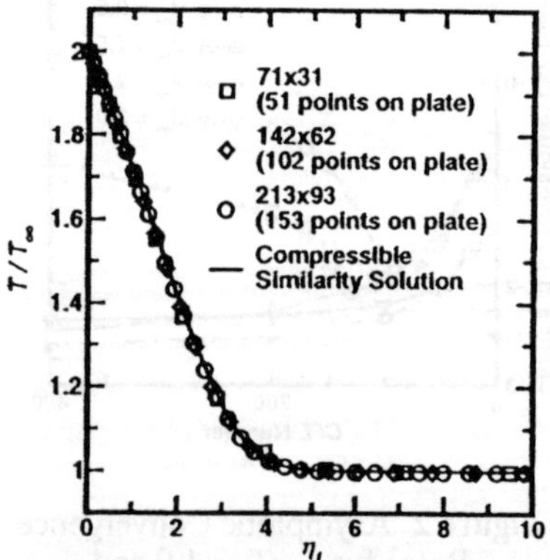

Figure 5 Computed Half–Chord Temperature Profiles for Three Grid Resolutions at $T_w/T_\infty=2.0$ and $M_\infty=0.5$

## 5.2 Assessment of Discretization Error

The algorithm accuracy is assessed by comparing computed solutions with the compressible similarity solution. The spatial discretization error is first investigated by comparing a solution computed on a baseline grid of $(71 \times 31)$ points with other solutions for $(142 \times 62)$ and $(213 \times 93)$ grids. The spacing adjacent to the wall was 0.001 for all grids.

Figures 4 and 5 show u–velocity and temperature profiles for free stream Mach number $M_\infty = 0.5$ and wall temperature $T_w/T_\infty = 2.0$ for each of these three levels of grid resolution. Although the trend is somewhat difficult to see on the scale of the plots, the computed profiles differ slightly from the similarity solution.

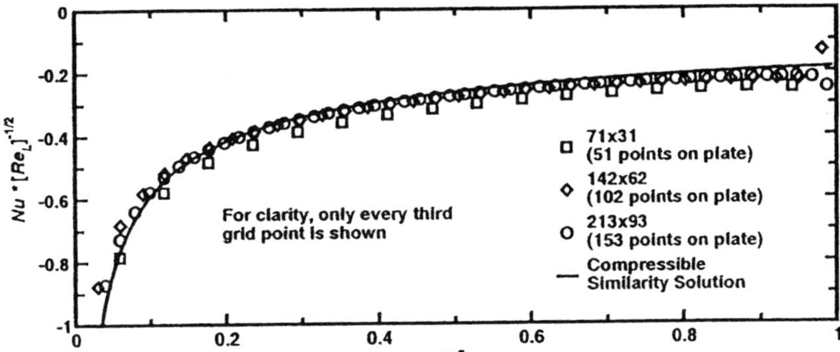



**Figure 6** Computed Nusselt Number for Three Grid Resolutions at $T_w/T_\infty$=0.5 and $M_\infty$=0.5

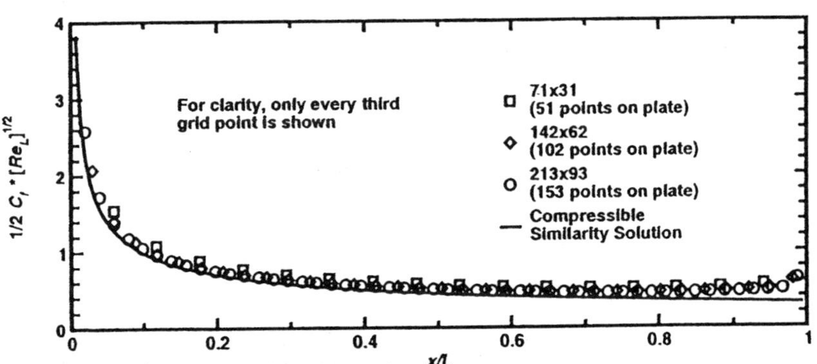



**Figure 7** Computed Skin Friction for Three Grid Resolutions at $T_w/T_\infty$=0.5 and $M_\infty$=0.5

Figures 6 and 7 show the skin friction and Nusselt number distributions along the plate. These quantities are proportional to the normal derivatives of u–velocity and temperature, respectively, and show the effect of grid resolution more clearly. Although the computed solution with highest resolu-

614

tion is essentially grid independent, the computed solutions converge towards a solution which again differs slightly from the theoretical result. This difference is due primarily to the relatively low Reynolds number of these cases (Re = $10^4$), as will be discussed in the next Section.

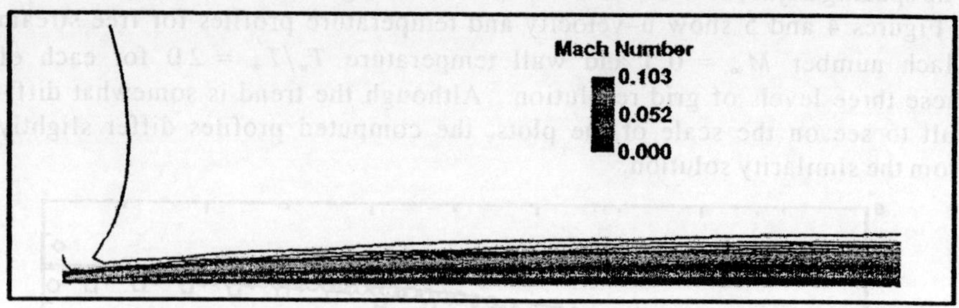

(8a). Mach Number Contours at $M_\infty$=0.1 and $T_w/T_\infty$=2.0.

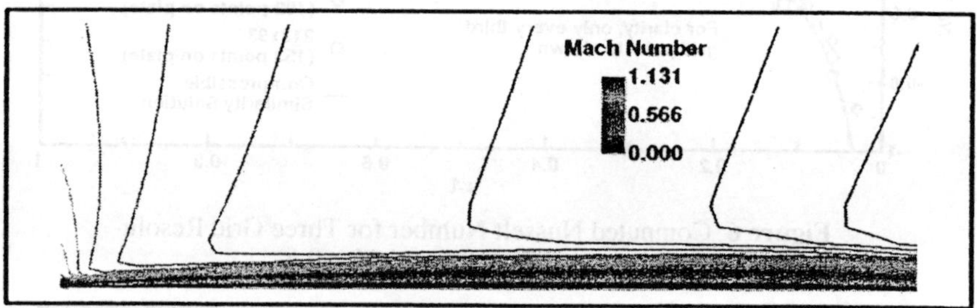

(8b). Mach Number Contours at $M_\infty$=1.0 and $T_w/T_\infty$=2.0.

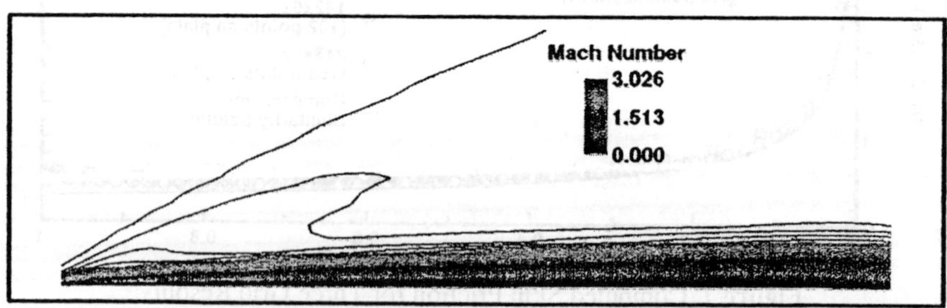

(8c). Mach Number Contours at $M_\infty$=3.0 and $T_w/T_\infty$=2.0.

**Figure 8** Mach Number Contour Plots for $T_w/T_\infty$=2.0 and for (0.1 $\leq M_\infty \leq$ 3.0)

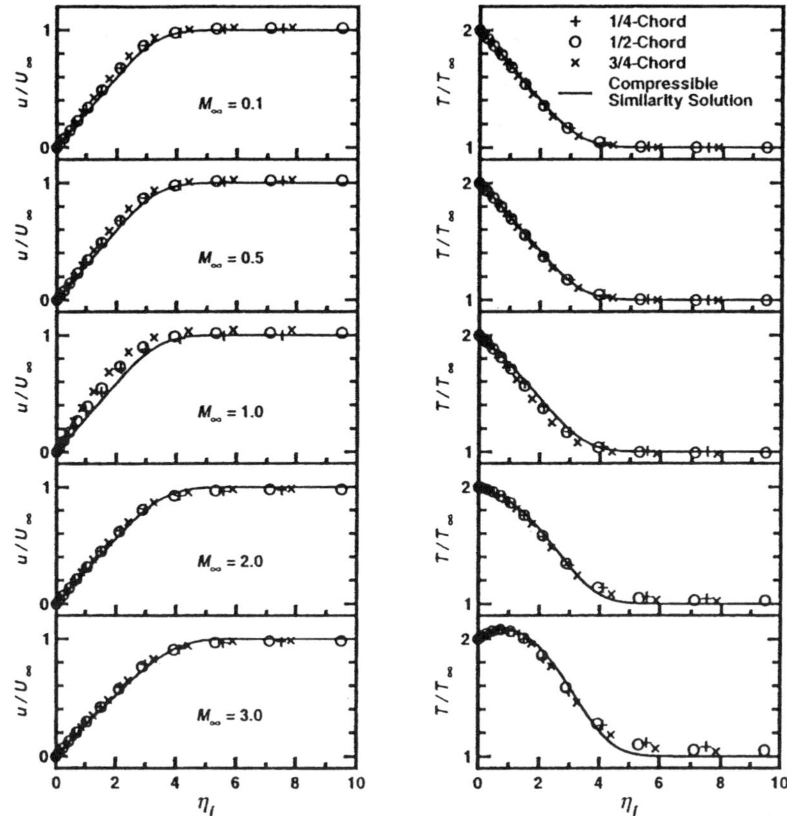

**Figure 9** Computed and Theoretical u–Velocity and Temperature Profiles for $T_w/T_\infty=2.0$ and for $(0.1 \leq M_\infty \leq 3.0)$

### 5.3 Comparisons for a Range of Mach number and Wall Temperature

Solution accuracy is examined here for two series of computed cases, first for a sequence of Mach numbers and then for a sequence of wall temperatures. The present results are representative of much more extensive tests reported in [26] and covering the following sequences: $M_\infty = 0.1, 0.5, 1.0, 2.0, 3.0$, each at $T_w/T_\infty = 0.5, 1.0, 2.0$, and $T_w/T_\infty = 0.1, 0.5, 1.0, 2.0, 10.0$, at $M_\infty = 0.1$.

*Mach Number Sequence* – Three computed Mach number contour plots are first shown in Figure 8 for $M_\infty = 0.1, 1.0, 3.0$ with $T_w/T_\infty = 2.0$. In all contour plots, the flow direction is left to right, and the plate leading and trailing edges are just inside the left and right edges of the figures, respectively. The leading–edge shock wave is visible in the $M_\infty = 3.0$ case.

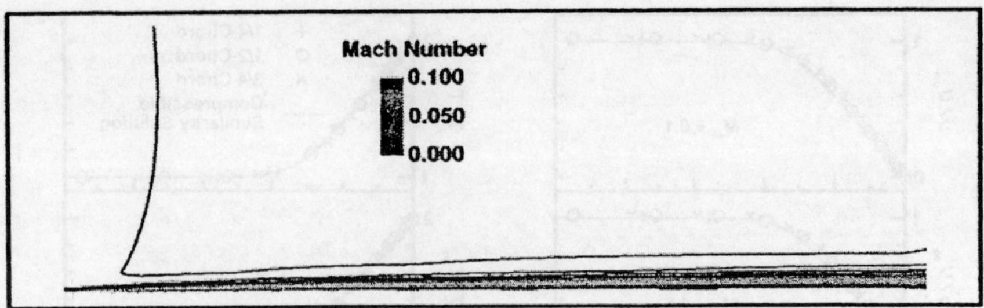

(10a). Mach Number Contours at $T_w/T_\infty$=0.1 and $M_\infty$=0.1

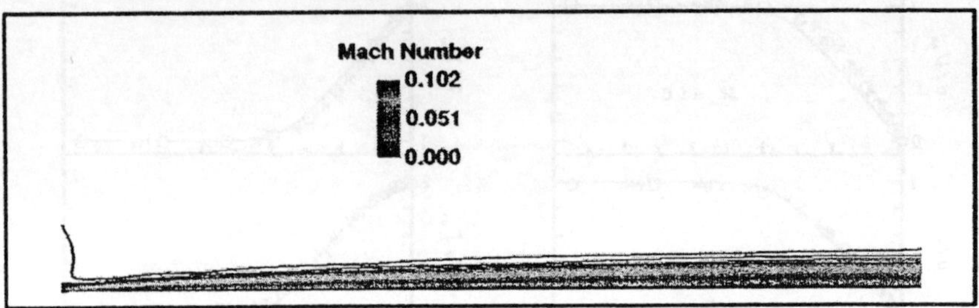

(10b). Mach Number Contours at $T_w/T_\infty$=1.0 and $M_\infty$=0.1

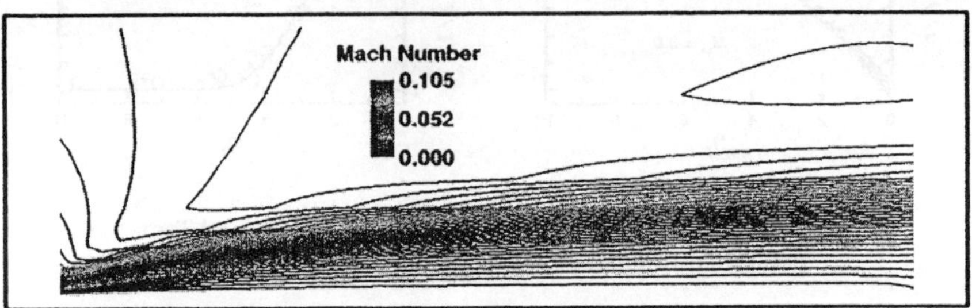

(10c). Mach Number Contours at $T_w/T_\infty$=10.0 and $M_\infty$=0.1

**Figure 10** Mach Number Contour Plots for $M_\infty$=0.1
and for ($0.1 \leq T_w/T_\infty \leq 10.0$)

In Figure 9, computed u–velocity and temperature profiles at three separate chord positions (0.25, 0.5 and 0.75 chord) are plotted in incompressible similarity variables $\eta \equiv y\,x^{-1}(\mathrm{Re}_x/2)^{0.5}$. This is a simple transformation for the compressible similarity solution and avoids a numerical integral of density on the much coarser Navier–Stokes grid. These are compared with the similarity solution, which applies at all three locations. There is good similarity among the computed solutions for the three chordal positions in all cases, and these solutions also agree well with the theoretical similarity solution. The degree of similarity in the computed Navier–Stokes solutions provides a means of assessing the validity of the similarity assumptions for these cases.

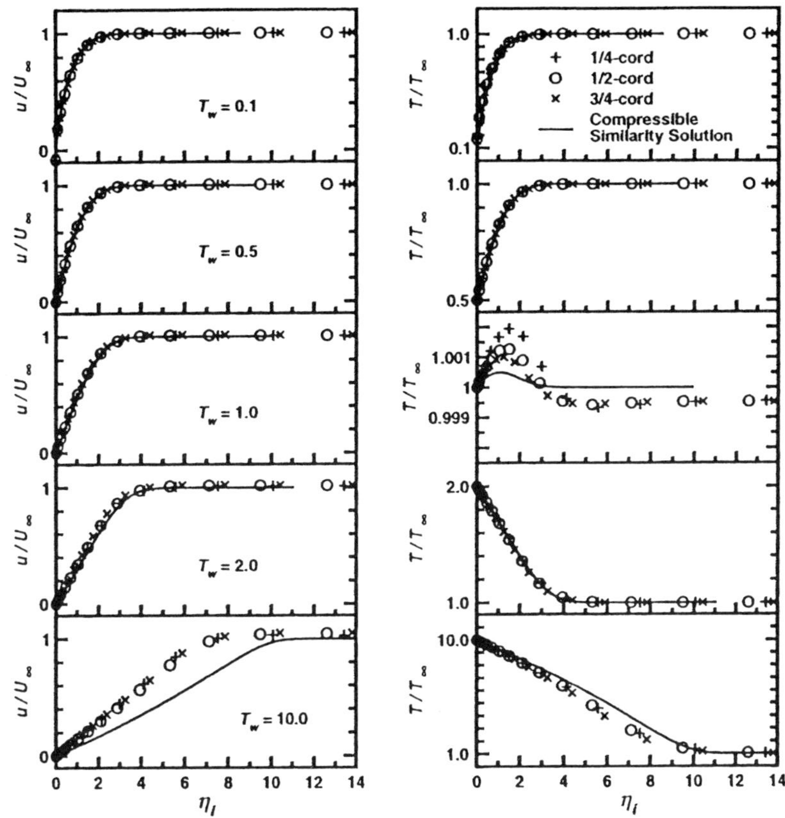

**Figure 11** Computed and Theoretical U–Velocity and Temperature Profiles for $M_\infty$=0.1 and for $(0.1 \leq T_w/T_\infty \leq 10.0)$. Note the Expanded Temperature Scale for $T_w/T_\infty = 1.0$.

Although there is generally good agreement, the computed velocity profiles for the subsonic and sonic cases lie above the similarity solution within the boundary layer, with a slight overshoot near the boundary–layer edge. The largest differences and departure from similarity occur for the $M_\infty = 1.0$ case. In the supersonic cases, the velocity profiles lie above the similarity solution within the boundary layer but have a slight undershoot external to the boundary layer. This undershoot is present in all of the supersonic cases [26] and is due to the leading–edge shock. The freestream velocity and Mach number are reduced behind the shock, causing the undershoot relative to the similarity solution, which does not account for the shock. Similarly, the supersonic temperature profiles are above the similarity solution, since the shock wave increases the freestream temperature. In all but one case, the overshoot or undershoot in flow variables outside the boundary layer disappears, and the variables attain their freestream values well inside the outer boundary. The one exception is in the $M_\infty = 1.0$ case, in which a weak normal shock forms downstream of the leading edge.

*Wall Temperature Sequence* – The second series of solutions has the wall–temperature sequence $T_w/T_\infty = 0.1, 0.5, 1.0, 2.0, 10.0$, at $M_\infty = 0.1$. The computed Mach number contour plots for $T_w/T_\infty = 0.1, 1.0, 10.0$ are shown in Figure 10. Note that the boundary layer is very thick for $T_w/T_\infty = 10.0$, and as a consequence, it is apparent that the baseline ($71 \times 31$) grid has insufficient resolution in the outer portion of the boundary layer in this case.

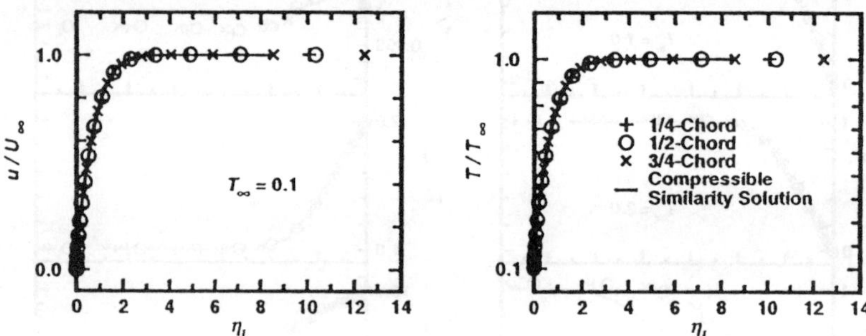


Figure 12 Improved Resolution of Boundary Condition for
$T_w/T_\infty$=0.1 and $M_\infty$=0.1

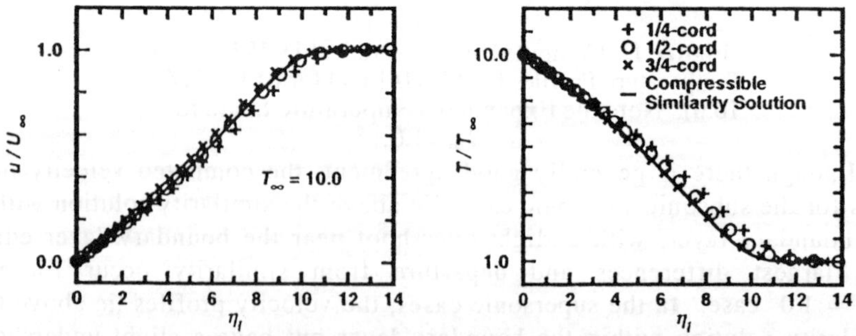


Figure 13 Improved Agreement at $T_w/T_\infty$=10.0, $M_\infty$=0.1
by Using Larger Reynolds Number

The computed u–velocity and temperature profiles for these cases are shown in Figure 11. There is both good similarity among the computed solutions and good agreement with the theoretical solution, with two notable exceptions. The first is that the solution for $T_w/T_\infty = 0.1$ is inaccurate near the wall boundary condition, and the second is that the solution for $T_w/T_\infty = 10.0$ differs substantially from the similarity solution.

The poor boundary resolution for $T_w/T_\infty = 0.1$ is associated with the thinner boundary layer, for which the wall spacing in the baseline grid is too large. This case was rerun using a smaller wall spacing, and the results in Figure 12 indicate that the wall boundary region has been much better resolved. The solution for $T_w/T_\infty = 10.0$ has a very thick thermal boundary layer, which disturbs the outer flow to the extent that the similarity assumptions begin to break down. Computed profiles for a thinner boundary layer at $Re_L = 10^6$ are given in Figure 13. These profiles are in much better agreement with the similarity solution at this higher Reynolds number, although there is a curious loss of similarity. The similarity can be improved to some extent by decreasing the wall spacing near the leading edge of the plate.

In all of the cases considered here, any slight overshoot or undershoot in flow variables near the boundary–layer edge disappears outside the boundary layer, and all variables reach their freestream values well inside the outer boundary.

## 6 Discussion

The influence of *CFL* on convergence rate differs for the subsonic and sonic/supersonic flow cases. Although the sonic and supersonic cases are insensitive to *CFL* number beyond a value of about 100, the subsonic cases converge rapidly in a narrow range of *CFL* values (cf., Figure 2). This difference in convergence behavior is thought to be a result of increased factorization errors in the subsonic case, since wave propagation occurs in both directions for subsonic flow.

Although there was some influence of wall temperature on convergence rate, of greater significance is that the maximum allowable *CFL* was significantly reduced for both heating and cooling cases (cf. Figure 3). The limitation on maximum *CFL* for the low–temperature case ($T_w/T_\infty = 0.1$) was found to be caused by insufficient wall spacing, as noted previously.

In the validation study, the agreement between computed and similarity solutions for subsonic cases was within expectations, given the grid resolution and Reynolds number used. The supersonic cases were also within expectations after accounting for the effect of the leading–edge shock wave.

The similarity solutions assume that conditions at the boundary–layer edge match the freestream conditions. This assumption becomes less valid for thicker boundary layers, since the outer flow must accelerate as it is diverted around the boundary layer, thereby exceeding the freestream velocity. This was especially apparent in the subsonic cases, and in the supersonic cases, the freestream flow outside the boundary layer is also altered by the leading–edge shock wave. Although it is theoretically possible to compensate for this leading–edge shock influence by recomputing the similarity solution using the Mach number from the Navier–Stokes solution at the boundary–layer edge, in practice this Mach number distribution is somewhat ambiguous in the Navier–Stokes solution.

The no–slip boundary condition was not accurately resolved for $T_w/T_\infty = 0.1$ in Figure 11. This was due to insufficient wall spacing in the baseline grid, which was based on viscosity for a wall temperature of 1.0. The colder wall reduces the viscosity, thus thinning the boundary layer and increasing the wall gradient. Reducing the wall spacing for cold walls compensates for this and also increases the maximum allowable $CFL$, although the smaller grid spacing reduces the convergence rate for the same $CFL$ number.

The final validation issue of note is the poor agreement in  9 for $M_\infty = 1.0$. In this case, a weak normal shock downstream of the leading edge causes an axial pressure gradient that violates the flat–plate similarity assumptions. The similarity solution is therefore not applicable.

## 7 Conclusions

The present study has addressed the convergence behavior and accuracy of a preconditioned compressible viscous flow solver for flat–plate flows over a broad range of Mach number and wall temperature. The preconditioned algorithm converges well for all Mach numbers tested with wall temperatures near the free stream value. Using a large $CFL$ number reduces the convergence rate for subsonic flows but not for supersonic flows. Although solutions were successfully computed for very high and low wall temperatures, the maximum allowable CFL number was reduced.

Comparisons with similarity solutions indicate that the preconditioned flow solver maintains a consistent and acceptable level of accuracy over a broad range of Mach number and wall temperature. Minor deviations from this pattern were observed for low Reynolds number with high wall temperature, or when the similarity assumptions break down due to leading–edge effects such as normal shocks or blockage due to thick boundary layers. There is also a variation in grid resolution or distribution requirements for some cases. In general, the preconditioned flow solver seems to be effective over a broad range of Mach number and wall temperature.

## 8 Acknowledgments

This work was sponsored in part by the National Science Foundation (Grant EEC–973–0365) and in part by NASA Ames Research Center (Grant NAG2–1232). This support is gratefully acknowledged.

# REFERENCES


[1]. D. M. Fricker, J. D. Holderman and S. P. Vanka, Calculations of Hot Gas Ingestion for a STOVL Aircraft Model, *Journal of Aircraft*, 31(1):236–242, 1994.

[2] C. L. Merkle, A. Gulati and Y.-H. Choi, The Effect of Strong Heat Addition on the Convergence of Implicit Schemes, *AIAA Journal*, 23(6):847–855, 1985.

[3] T.-S. Wang, Delta Clipper–Experimental In–Ground Effect on Base–Heating Environment, *Journal of Thermophysics and Heat Transfer*, 12(3):343–349, 1998.

[4] K. R. Roth, Comparison of Computation with Experiment for a Geometrically Simplified Powered–Lift Model, *Journal of Aircraft*, 34(2):160–167, 1997.

[5]. D. R. Poling, H. Rosenstein and G. Rajagopalan, Use of a Navier–Stokes Code in Understanding Tiltrotor Flowfields in Hover, *Journal of the American Helicopter Society*, 43(2):103–109, 1998.

[6]. Briley W. R., McDonald H., and S. J. Shamroth, A Low Mach Number Euler Formulation and Application to Time Iterative LBI Schemes. *AIAA J.*, 21(10):1467–1469, 1983.

[7]. Choi D., and C. L. Merkle, Application of Time–Iterative Schemes to Incompressible Flow. *AIAA J.*, 23(10):1518–1524, 1985.

[8]. Turkel, E., Preconditioned Methods for Solving the Incompressible and Low Speed Compressible Equations, *J. Comput. Phys.*, 72:277–298, 1987.

[9]. Merkle, C. L., and D. L. Choi, Computation of Low–Speed Flow with Heat Addition. *AIAA J.*, 25(6):831–838, 1987.

[10].Briley W. R., Buggeln R. C., and H. McDonald, Solution of the Incompressible Navier–Stokes Equations Using Artificial Compressibility. *Proc., 11th Int. Conf. on Numerical Methods in Fluid Mechanics*, Williamsburg, VA, New York, Springer–Verlag, 1989.

[11].Turkel, E., A Review of Preconditioning Methods for Fluid Dynamics, *Appl. Numer. Math.*, 12:257–284, 1993.

[12].Choi, Y. H., and C. L. Merkle, The Application of Preconditioning to Viscous Flows, *J. Comput. Phys.*, 105:207–223, 1993.

[13].Pletcher, R. H., and K.-H. Chen, On Solving the Compressible Navier–Stokes Equations for Unsteady Flows at Very Low Mach Number, *AIAA Paper No. 93–3368–CP*, 1993.

[14].Weiss, J. M., and W. A. Smith, Preconditioning Applied to Variable and Constant Density Flows, *AIAA J.*, 33:2050–2057, 1995.

[15].Edwards, J. R., and M–S Liou, Low–Diffusion Flux–Splitting Methods for Flows at All Speeds, *AIAA J.*, 36(9):1610–1617, 1998.

[16].Turkel, E., Preconditioning Techniques in Computational Fluid Dynamics, *Annu. Rev. Fluid Mech.*, 31:385–416, 1999.

[17] Briley, W. R., Taylor, L. K., and D. L. Whitfield: High–Resolution Viscous Flow Simulations at Arbitrary Mach Number, *MSU Report No. MSSU–COE–ERC–01–04*, Mar 2001.

[18] Roe, P. L., Approximate Riemann Solvers, Parameter Vector, and Difference Schemes. *J. Comput. Phys.*, 43:357, 1981.

[19] L. K. Taylor and D. L. Whitfield, "Unsteady Three–Dimensional Incompressible Euler and Navier–Stokes Solver for Stationary and Dynamic Grids," *AIAA Paper No. 91–1650*, 1991.

[20] D. L. Whitfield and L. K. Taylor, "Discretized Newton–Relaxation Solution of High Resolution Flux–Difference Split Schemes," *AIAA Paper No. 91–1539*, 1991.

[21] D. L. Whitfield and L. K. Taylor, "Numerical Solution of the Two–Dimensional Time–Dependent Incompressible Euler Equations," *MSU Report No. MSSU–EIRS–ERC–93–14*, 1994.

[22] A. Jameson, Successes and Challenges in Computational Aerodynamics, *Proc. 8th AIAA Computational Fluid Dynamics Conference, Honolulu, Hawaii*, pp. 1–35, 1987.

[23] Simpson, L. B., and D. L. Whitfield, Flux–Difference Split Algorithm for Unsteady Thin–Layer Navier–Stokes Solutions, *AIAA J.* 30(4):914–922, 1992.

[24] J. A. Schetz, *Boundary Layer Analysis*. Englewood Cliffs, 1993.

[25] F. M. White, *Viscous Fluid Flow*, 2nd Ed. Boston: Mc Graw Hill, 1991.

[26] Lambert, B. K., Accuracy and Performance Characteristics of a Modern CFD Algorithm Over a Range of Mach Numbers and Wall Temperatures," M. S. Thesis, Mississippi State University, May 2001.

# Evolution of topology in axi-symmetric and 3-d viscous flows

G. Scheuermann [1], W. Kollmann [2], X. Tricoche [1] and T. Wischgoll [1]

[1]Department of Computer Science, University of Kaiserslautern
D-67653 Kaiserslautern, Germany
[2] MAE Dept, University of California Davis
Davis, CA 95616, USA

Contact e-mail:scheuer,tricoche,wischgol@informatik.uni-kl.de, wkollmann@ucdavis.edu

**Abstract**

Topological methods are used to establish global and to extract local structure properties of vector fields in axi-symmetric and 3-d flows as function of time. The notion of topological skeleton is applied to the interpretation of vector fields generated numerically by the Navier-Stokes equations. The flows considered are swirling jets with super-critical swirl numbers that show low Reynolds number turbulence in the break-up region.

## 1 Introduction

A numerical study is conducted to investigate the temporal development of flow structures connected with the vortex breakdown in swirling free jets. There is experimental evidence (see Billant *et al.* [1]) that swirling jets exhibit a rich variety of flow phenomena. Simplified model equations assuming steady and rotationally symmetric motion above a conical stream surface show bi- and multi-stability (see [3], [4], [5], [6]) of several flow forms such as hysteresis and negative thrust phenomena. Direct numerical simulations at moderate Reynolds numbers allow relaxation of the model restrictions. The results show the development of the conical jet layer and its break up into three-dimensional vortex structures for moderate Reynolds numbers. The detection of these structures requires sophisticated tools, that are applicable to two and three dimensional, steady and unsteady flows. It is essential for the interpretation of these fields to reduce their properties to the bare minimum, hence the notion of the topological skeleton, introduced by Fomenko (see [7], [8]) for the classification of integrable Hamiltonian systems, is applied to the present case. The paper presents the theoretical aspects of vector field topology and numerical methods for their computation. The emphasis is on vector lines in 2-d sections of axi-symmetric and 3-d flows.

The organisation of the paper is as follows. First, the numerical method for the solution of the incompressible Navier-Stokes equations is briefly outlined. Then basic definitions and theorems for the topology of steady vector fields are discussed followed by the consideration of the topological skeleton and its temporal evolution. Numerical methods for the computation of the evolution of the vector field topology are developed and results for the swirling jet flows are presented.

## 2 Navier-Stokes solver

A hybrid spectral finite-difference method (developed in [9]) is used to solve the Navier-Stokes equations in cylindrical coordinates. The method will be outlined briefly, the details including the verification of its accuracy and convergence properties are available in [9] and [10]. The method takes advantage of the fact that all dependent variables $\varphi$ are periodic with respect to azimuthal variable $\theta$, hence is the discrete Fourier transform $\mathcal{F}_N$ (Fourier collocation method, Canuto *et al.*,[11]) applicable

$$\bar{\varphi}(k) = \mathcal{F}_N(\varphi) \equiv \frac{1}{N} \sum_{j=0}^{N-1} \varphi(\theta_j) \exp(-ik\theta_j), \quad k = -N/2, \cdots, N/2 - 1$$

$$\varphi(\theta_j) = \mathcal{F}_N^{-1}(\tilde{\varphi}) \equiv \sum_{k=-N/2}^{N/2-1} \tilde{\varphi}(k)\exp(ik\theta_j), \quad j=0,\cdots,N-1 \tag{1}$$

where $\mathcal{F}_N^{-1}$ denotes the inverse transform and the $\theta_j = 2\pi j/N$ are the discrete collocation points.

The dependent variables (see [9]) in the present formulation are the complex-valued Fourier modes ($k$ denotes the discrete azimuthal wavenumber) of:

The azimuthal velocity modes $\tilde{v}_\theta(r,k,z,t)$ for $0 \leq k \leq \frac{N}{2}$,

the azimuthal vorticity modes $\tilde{\Omega}_\theta(r,k,z,t)$ for $0 \leq k \leq \frac{N}{2}$,

the streamfunction modes $\tilde{\Psi}(r,k,z,t)$ for $0 \leq k \leq \frac{N}{2}$,

the pressure modes $\tilde{p}(r,k,z,t)$ for $1 \leq k \leq \frac{N}{2}$.

The azimuthal momentum balance provides the equation for the velocity modes $\tilde{v}_\theta(r,k,z,t)$

$$\frac{\partial \tilde{v}_\theta}{\partial t}(k) + \mathcal{F}_N[v_r\frac{\partial v_\theta}{\partial r} + v_z\frac{\partial v_\theta}{\partial z} + \frac{v_\theta}{r}(\frac{\partial v_\theta}{\partial\theta} + v_r)] = -i\frac{k}{r\rho}\tilde{p}(k)$$

$$+\frac{1}{Re}[\frac{\partial}{\partial r}(\frac{1}{r}\frac{\partial}{\partial r}(r\tilde{v}_\theta(k))) - \frac{k^2}{r^2}\tilde{v}_\theta(k) + \frac{\partial^2\tilde{v}_\theta}{\partial z^2}(k) + 2i\frac{k}{r^2}\tilde{v}_r(k)] \tag{2}$$

where the convective terms are evaluated using FFT, $Re$ denotes the Reynolds number. Application of the divergence operation to the momentum balances produces the equations for the pressure modes $\tilde{p}(r,k,z,t)$, which emerge in the form

$$\frac{1}{r}\frac{\partial}{\partial r}(r\frac{\partial\tilde{p}}{\partial r}) - \frac{k^2}{r^2}\tilde{p} + \frac{\partial^2\tilde{p}}{\partial z^2} =$$

$$-\rho\mathcal{F}_N[(\frac{\partial v_r}{\partial r})^2 + \frac{1}{r^2}(\frac{\partial v_\theta}{\partial\theta} + v_r)^2 + (\frac{\partial v_z}{\partial z})^2 + \frac{2}{r}\frac{\partial v_\theta}{\partial r}(\frac{\partial v_r}{\partial\theta} - v_\theta) + 2\frac{\partial v_r}{\partial z}\frac{\partial v_z}{\partial r} + \frac{2}{r}\frac{\partial v_z}{\partial\theta}\frac{\partial v_\theta}{\partial z}] \tag{3}$$

The transport equation for the azimuthal vorticity component $\tilde{\Omega}_\theta$ follows from its definition and the momentum balances. It appears in the form

$$\frac{\partial\tilde{\Omega}_\theta}{\partial t} = \frac{\partial}{\partial r}[\mathcal{F}_N(v_\theta\Omega_r) - \mathcal{F}_N(v_r\Omega_\theta)] + \frac{\partial}{\partial z}[\mathcal{F}_N(v_\theta\Omega_z) - \mathcal{F}_N(v_z\Omega_\theta)]$$

$$+\frac{1}{Re}[\frac{\partial}{\partial r}(\frac{1}{r}\frac{\partial}{\partial r}(r\tilde{\Omega}_\theta(k))) - \frac{k^2}{r^2}\tilde{\Omega}_\theta(k) + \frac{\partial^2\tilde{\Omega}_\theta}{\partial z^2}(k) + 2i\frac{k}{r^2}\tilde{\Omega}_r(k)] \tag{4}$$

It is shown in [9] that a complex-valued streamfunction $\tilde{\Psi}(r,k,z,t)$ exists for each wavenumber $k$ as a consequence of the Fourier transformed mass balance, which can be recast as the divergence of a 2-d vector being zero. This results in Poisson-Helmholtz type equations for $\tilde{\Psi}(r,k,z,t)$

$$\frac{\partial}{\partial r}\left(\frac{1}{r}\frac{\partial}{\partial r}(r\tilde{\Psi})\right) + \frac{\partial^2\tilde{\Psi}}{\partial z^2} = \tilde{\Omega}_\theta + ik\frac{\partial\langle\tilde{v}_\theta\rangle}{\partial z} \tag{5}$$

The velocity components can be recovered from

$$-r\tilde{v}_z = \frac{\partial}{\partial r}(r\tilde{\Psi}), \quad \tilde{v}_r = \frac{\partial\tilde{\Psi}}{\partial z} - ik\langle\tilde{v}_\theta\rangle \tag{6}$$

once $\tilde{v}_\theta$ and the streamfunctions $\tilde{\Psi}$ have been computed. The angular brackets are essential for the streamfunction formalism (5), (6), they are defined by

$$\langle\Phi\rangle(r,k,z,t) \equiv \frac{1}{r}\int_0^r dr'\,\Phi(r',k,z,t) \tag{7}$$

The presence of singular coefficients in (2) to (5) requires careful analysis of the pdes for $r \to 0$ ([12], [13]). The results (details in [9]) can be summarized as follows:

(I) Scalar modes $\tilde{\Phi}(r,k,z,t)$ and axial vector components satisfy $O(\tilde{\Phi}(r,k,z,t)) = r^k$ and are symmetric with respect to $r$ for $k$ even and skew-symmetric for $k$ odd.

(II) Radial and azimuthal vector components satisfy the kinematic conditions $\tilde{v}_r(0,k,z,t) + ik\tilde{v}_\theta(0,k,z,t) = 0$ and both modes are symmetric with respect to $r$ for $k = 1$, and $O(\tilde{v}_r(r,k,z,t)) = O(\tilde{v}_\theta(r,k,z,t)) = r$ for $k = 0$ and $O(\tilde{v}_r(r,k,z,t)) = O(\tilde{v}_\theta(r,k,z,t)) = r^{k-1}$ for $k > 1$, and they are skew-symmetric with respect to $r$ for $k$ even and symmetric for $k$ odd. The complex-valued streamfunctions can be shown to obey $O(\tilde{\Psi}(k)) = r^{k+1}$ and are skew-symmetric with respect to $r$ for $k$ even, symmetric for $k$ odd.

The conditions at the axis and the entrance, exit and outer boundaries are either Dirichlet or Neumann type or a combination of Dirichlet and Neumann conditions. Specifically, the axis conditions cannot be chosen arbitrarily but follow from the symmetry and smoothness properties at $r = 0$; the Fourier modes satisfy either homogenous Dirichlet or homogeneous Neumann conditions. At entrance and outer boundaries Dirichlet conditions are prescribed and convective boundary conditions are used at the exit.

The sectional domain $(r,z) \in [0,R] \times [0,L_z]$ is rectangular, it may be semi-infinite ($R = \infty$). It is mapped onto the unit quadrangle using separable and analytic functions. It is discretized with respect to radial and axial directions with uniform spacing in the image domain $[0,1] \times [0,1]$. Either central differences are used for first and second derivatives with explicit filters (Kennedy & Carpenter [14]) applied to the increments of the transport equations or upwind-biased differences (Li[15]) for the convective terms and central differences for the non-convective terms. The filter order is chosen such that it does not degrade the accuracy of the finite-difference schemes, no filters are applied for the upwind-biased schemes. The numerical integration required in (7) is done using a fourth order accurate difference method.

The time integrator is an explicit $4^{th}$-order state space Runge-Kutta method that requires minimal storage (see [9]). The Poisson/Helmholtz equations (2) and (5) are solved using LU-decomposition (LINPACK, see Dongarra et al.[16]) combined with deferred corrections to reduce the bandwidth of the coefficient matrix.

# 3 Topology of Steady Vector Fields

The study of the topology requires clear definitions and numerical methods for their computation. We define in the following the basic notions of vector field topology relevant for the present purpose. The description is similar to Fomenko et al. [7], [8], but the results in our examples are fundamentally different since our fields do not allow a Hamiltonian function. We start with some simple terms.

First, we clarify two definitions widely used in fluid mechanics: Streamlines and pathlines. Given an unsteady vector field $v(x,t)$ we compute the streamlines $\Phi(t^*)$ as solutions of the initial value problem

$$\frac{d\Phi}{dt^*} = v(\Phi(t^*),t) \tag{8}$$

with initial conditions determining the desired streamline. Streamlines are computed for the fixed time $t$ (instantaneous integral curves or snapshots) and the picture generated by them changes with time if the vector field is unsteady. On the other hand are the pathlines the solutions of

$$\frac{d\Phi}{dt} = v(\Phi(t),t) \tag{9}$$

plus initial conditions and it follows that pathlines are different from streamlines if the vector field is unsteady. The pathlines are the vector lines in the mathematical literature. We consider in the present chapter steady vector fields and instant snapshots of unsteady fields in the following section. The vector lines are streamlines and the line parameter $t$ is time if the field is steady, but for unsteady fields it is isomorphic to arclength.

**Definition 3.1** *A planar vector field is a map*

$$v : D \quad \rightarrow \quad \mathbb{R}^2 \,, \tag{10}$$
$$x \quad \mapsto \quad v(x)\,.$$

**Definition 3.2** *An* **integral curve through a point** $x \in \mathbb{R}^2$ *of a vector field* $v : D \to \mathbb{R}^2$ *is a map*

$$\alpha_x : \mathbb{R} \supset I \to D \,, \tag{11}$$

*where*

$$\alpha_x(0) \;=\; x_0 \,, \tag{12}$$
$$\dot{\alpha}_x(t) \;=\; v\,(\alpha(t)) \,, \qquad \forall\, t \in I \,. \tag{13}$$

Concerning the theorem on existence and uniqueness of integral curves, the *Lipschitz condition* has to be satisfied:

**Definition 3.3** *Let* $U \subset \mathbb{R}^2$ *be an open subset. A continous vector field* $v : \mathbb{R}^2 \to \mathbb{R}^2$ *satisfies a* **Lipschitz condition on** $U$ *provided there exists a real number* $K > 0$ *such that*

$$\|Dv(x) - Dv(y)\| \leq K\|x - y\| \tag{14}$$

*holds for all* $x, y \in U$ *and all* $t \in J$ *where* $Dv$ *denotes the differential of the vector field* $v$. *The constant* $K$ *is called* **Lipschitz constant**.

This allows to formulate the existence and uniqueness theorem:

**Theorem 3.4 (Existence and uniqueness of integral curves)** *Let* $v : D \to \mathbb{R}^2$ *be a vector field satisfying the Lipschitz condition on any open neighborhood around point* $x \in \mathbb{R}^2$. *Then there exists one and only one integral curve through any* $x_0 \in \mathbb{R}^2$.

**Proof:** See [20, pp. 66–68].

We also need some terms for the different directions of the vector field at the boundary.

**Definition 3.5** *Let* $D \subset \mathbb{R}^2$ *be a compact domain of a Lipschitz-continous vector field* $v : D \to \mathbb{R}^2$. *Let* $d \in \partial D$ *be a point on the boundary. We define the following three entities:*

*(1) The point* $d$ *is called an* **outflow point** *if every integral curve through* $d$ *ends at* $d$ *and there exists an integral curve through* $d$ *that does not contain another point of* $\partial D$. *The set of all outflow points is called* **outflow set**.

*(2) The point* $d$ *is called a* **inflow point** *if every integral curve through* $d$ *starts at* $d$ *and there exists an integral curve through* $d$ *that does not contain another point of* $\partial D$. *The set of all inflow points is called* **inflow set**.

*(3) The point* $d$ *is called a* **boundary flow point** *if there exists an integral curve* $\alpha_d : (-\epsilon, \epsilon) \to D$, $\epsilon > 0$, *through* $d$ *lying completely inside* $\partial D$. *The set of all boundary flow points is called* **boundary flow set**.

*(4) The point* $d$ *is called* **boundary saddle** *(switch point), if it does not belong to the other three types.*

If a vector field satisfies the Lipschitz condition for an open neighborhood of every point, integral curves can be extended over the whole time line if they do not reach the boundary at an outflow point.

**Lemma 3.6** *Let* $v : D \to \mathbb{R}^2$ *be Lipschitz continuous around each point. Then an integral curve can be defined up to* $+\infty$ *if it does not end at an outflow point. It can also be defined from* $-\infty$ *if it does not start at an inflow point.*

Since we want to keep the following notations simple, we define an integral curve at an inflow point as that point for the whole time before it enters the domain. Further, at an outflow point, we extend an integral curve by this point for all the time after it has left the domain. In this way, all integral curves are defined for all $t \in \mathbb{R}$. Topology is really concerned with the asymptotic behavior of integral curves, so we look at their asymptotic begin and end.

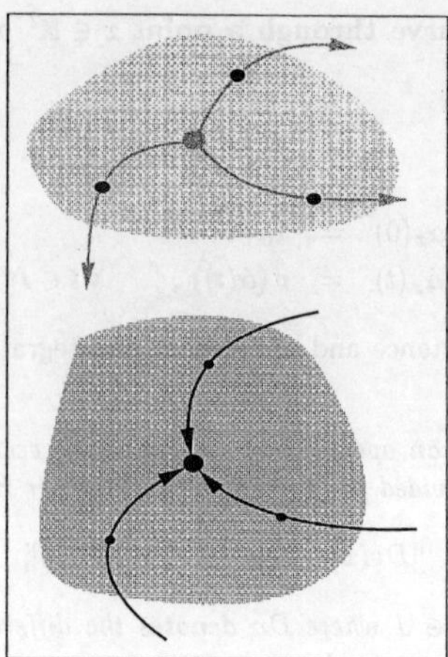

Figure 1: Around a source, every integral curve starts at the source; around a sink, every integral curve ends at the sink.

**Definition 3.7** *Let* $v : D \to \mathbb{R}^2$ *be a Lipschitz-continous vector field and* $c : \mathbb{R} \to \mathbb{R}^2$ *an integral curve. The set*

$$\{a \in \mathbb{R}^2 \,|\, \exists (t_n)_{n=0}^{\infty} \subset \mathbb{R}, t_n \to \infty, \lim_{n \to \infty} c(t_n) \to a\} \tag{15}$$

*is called* $\omega$-**limit set** *of* $c$. *The set*

$$\{a \in \mathbb{R}^2 \,|\, \exists (t_n)_{n=0}^{\infty} \subset \mathbb{R}, t_n \to -\infty, \lim_{n \to \infty} c(t_n) \to a\} \tag{16}$$

*is called* $\alpha$-**limit set** *of* $c$.

The most common types of limit sets are critical points.

**Definition 3.8** *A* **critical point** *of* $v : D \to \mathbb{R}^2$ *is a point* $p \in D$ *with* $v(p) = 0$.

Two cases are of interest: (1) sinks and (2) sources.

**Definition 3.9** *Let* $v : D \to \mathbb{R}^2$ *be a Lipschitz-continous vector field,* $z \in D$ *a critical point. If there is a neighborhood* $U \subset D$ *of* $z$ *such that all integral curves in* $U$ *have a* $\alpha$-*limit set consisting only of* $z$, *then* $z$ *is called a* **source**. *If there is a neighborhood* $U \in D$ *of* $z$ *such that all integral curves in* $U$ *have a* $\omega$-*limit set consisting only of* $z$, *then* $z$ *is called a* **sink**. *If there is a neighborhood* $U \in D$ *of* $z$ *such that there is a finite number of integral curves that start or end at* $z$ *and all other integral curves leave* $U$ *in both directions of time,* $z$ *is called a* **saddle**.

Figure 1 shows a source in the upper part. All integral curves around the critical point start at this critical point. In the lower part, a sink is shown. Every integral curve in the neighborhood ends at the sink. Another type of limit sets are closed integral curves (cycles).

**Definition 3.10** *Let* $v : D \to \mathbb{R}^2$ *be a Lipschitz continuous vector field. A* **closed streamline** $c : \mathbb{R} \to D$ *of* $v$ *is a streamline with a* $t_0 \in \mathbb{R}$ *with*

$$c(t + nt_0) = c(t) \forall n \in \mathbb{N}. \tag{17}$$

*(It is assumed that* $c$ *does not stay at one point all the time, i.e.* $c(t)$ *is not a critical point.)*

Figure 2 shows a typical example. Such a cycle is called structurally stable if, after small changes, the vector field still contains a closed streamline. Such stability arguments allow to concentrate on critical points, boundary regions and closed cycles as possible $\alpha$- and $\omega$-limit sets

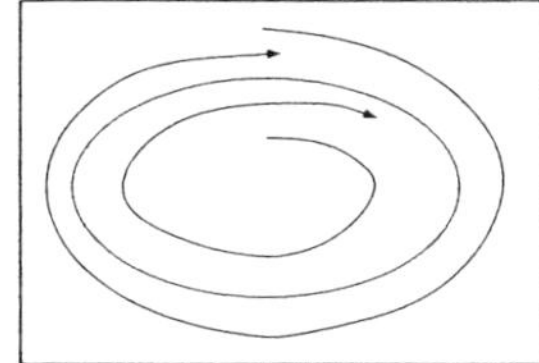

Figure 2: A limit cycle may attract streamlines in its neighborhood.

[18],[19]. For our analysis of a vector field, we have to determine the asymptotic behavior of all its integral curves. This is done by analyzing the union of all integral curves starting or ending at the same critical point, same cycle, or the same inflow/outflow region.

**Definition 3.11** *Let $v : D \to \mathbb{R}^2$ be a Lipschitz-continous vector field and $M \subset D$. The union of all integral curves of $v$ that converge to $M$ for $t \to -\infty$ is called $\alpha$-**basin** of $M$, denoted by $B_\alpha(M)$. The union of all integral curves of $v$ that converge to $M$ for $t \to \infty$ is called $\omega$-**basin of** $A$, denoted by $B_\omega(M)$.*

For sinks and sources, one can prove the following lemma:

**Lemma 3.12** *Let $a$ be a critical point of a Lipschitz continuous vector field $v : D \to \mathbb{R}^2$. If $a$ is a source then $B_\alpha(a)$ is an open subset of $D$. If $a$ is a sink then $B_\omega(a)$ is an open subset of $D$.*

**Proof:** See [19, pp. 181-182].

A similar result holds for cycles, i.e. a structurally stable cycle in 2D can be either a source and its $\alpha$-basin is an open subset of the domain or it is a sink and its $\omega$-basin is an open subset. A short comment on dimension prepares us for the final theorem of the section.

**Definition 3.13** *If a subset $M \subset \mathbb{R}^2$ allows a description as a pure n-dimensional manifold, we define the dimension of $M$ as $n$.*

**Remark 3.14** *We need the term dimension only for basins. The previous definition means that an open basin in $D$ is a 2-manifold, a basin consisting of a finite number of streamlines is a 1-manifold and a basin consisting of a finite number of points is a 0-manifold.*

Since we assume that we have only structurally stable limit sets, we obtain a decomposition of the domain into disjunct $\alpha$-basins and a decomposition into disjunct $\omega$-basins if we regard the connected component of the inflow set and the outflow set also as limit sets.

**Theorem 3.15** *Let $v : D \to \mathbb{R}^2$ be a Lipschitz continuous vector field with only structurally stable limit sets $L_i \subset D, i \in I$. Let $A_i \subset D$ denote the $\alpha$-basins of the limit sets and $\Omega_i \subset D$ denote the $\omega$-basins. Then:*

$$D = \bigcup_{i \in I} A_i \tag{18}$$

$$D = \bigcup_{i \in I} \Omega_i \tag{19}$$

*As **topology** of the vector field we define the set of limit sets $L_i$ and the connected components of the intersections $T_{ij} = A_i \cap \Omega_j$.*

Figure 3 shows the decomposition of a simple academic example flow into $\alpha$-basins, $\omega$-basins, and their intersections. Since the basins of sources, sinks, cycles, inflow, and outflow regions are all open subsets, the topology can be described by the one-dimensional basin intersections that form the boundaries of the open basin intersections. Because they separate the open basins, they are called **separatrices**. Since the boundary of an open set does not belong to the open set, the separatrices must start or end at saddles or boundary saddles. This fact is used to calculate the topology, i.e. we compute only the separatrices and obtain the other basin intersections as the resulting faces in the domain. All limit sets form the **topological skeleton** $S_v$ together with the separatrices.

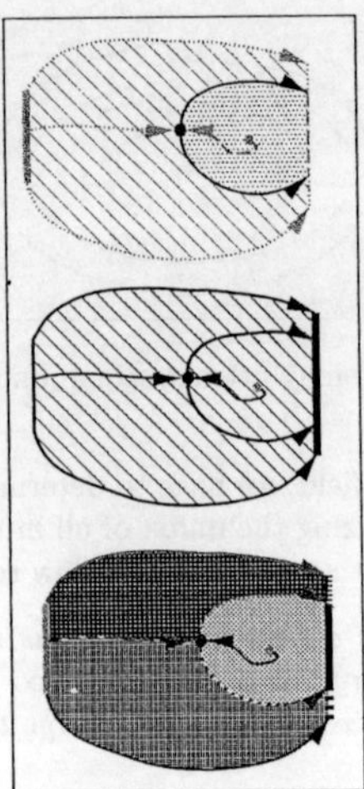

Figure 3: A simple vector field is first partioned into $\alpha$-basins and then into $\omega$-basins. The resulting different intersections of $\alpha$- and $\omega$-basins are shown in the bottom image. All $\alpha$-limit sets and $\alpha$-basins are colored green, all $\omega$-limit sets and $\omega$-basins are colored blue and the separatrices and saddles are colored red. The intersections are shown in different black patterns.

# 4   Evolution of Topology

We turn now to the analysis of unsteady vector fields. Our perspective is to look at the topology of the instantaneous integral curves (streamlines) and track their continuous evolution over time. The focus lies on the movement and substantial change of the limit sets. We do not consider the inflow and outflow regions here, but concentrate on critical points and the cycles. Again, the work is guided by results in the theory of dynamical systems. The main point is the search for structural changes of the vector field. As working definition, we use

**Definition 4.1** *Let* $v : D \times J \to \mathbb{R}^2$, $J \subset \mathbb{R}$ *interval, be a Lipschitz continuous vector field. A time* $s \in J$ *is called* **bifurcation time** *if the topological skeleton* $S(t)$ *at* $t \in (s - \delta, s)$ *is different from* $S(t)$ *at* $t \in (s, s + \delta)$, *where* $\delta$ *is so small that the Skeleton* $S(t)$ *does not change for* $t \in (s - \delta, s)$ *or for* $t \in (s, s + \delta)$.

## 4.1   Local Bifurcations

There are bifurcations that involve only the vicinity of a critical point. They are called **local bifurcations**. We have two such types: static fold bifurcation and Hopf bifurcation [17].

**Static Fold Bifurcation**   At the beginning, there are a saddle point and a sink, that are linked by a separatrix. If the attraction relation between both critical points along the separatrix weakens (i.e. the real positive eigenvalue of the Jacobian at the saddle point, corresponding to the considered separatrix, decreases and the negative real part of one of the Jacobian's eigenvalues at the sink increases), both critical points become closer and closer until they meet. At this point, an unstable critical point appears. As time goes on, this unstable structure vanishes: The new stable structure contains no critical point: This is a pairwise annihilation or static fold catastrophe (see figure 4). Inverting the direction of time, we get a pairwise creation. Similar transitions are obtained by inverting the direction of the flow (the sink becomes a source).

**Hopf Bifurcation.**   We start with a sink with spiraling behavior in its neighborhood. If the attracting effect of this sink weakens, the number of streamline rotations around this critical

Figure 4: Pairwise annihilation

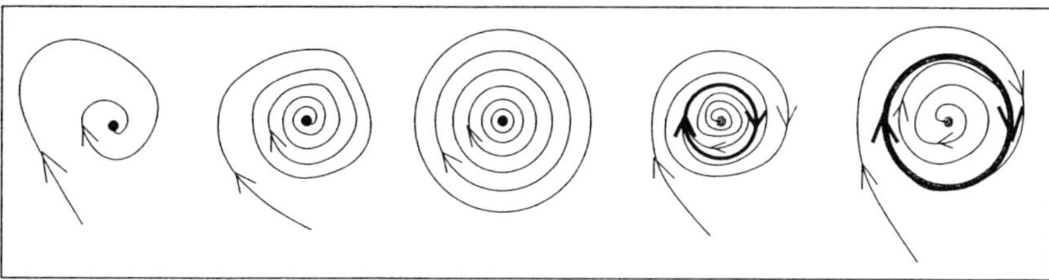

Figure 5: Hopf bifurcation

point increases (the convergence "slows down"). This corresponds mathematically to an increasing negative real part of the complex conjugate eigenvalues of the Jacobian matrix at the critical point. When the real part is zero, we get a center point, which is an unstable structure: A bifurcation has occurred. If the real part increases further, a new stable structure appears which consists of an attracting closed orbit typically moving away from the critical point. The critical point is transformed into a source. So, a sink has changed into a source with the emission of a cycle: This is a Hopf bifurcation. An illustration of this evolution is given in figure 5. Inverting the direction of time, we get the transition from a cycle containing a source into a sink over an instantaneous center. Similar transitions are obtained by inverting the direction of the flow (that is by replacing sources by sinks and vice versa).

## 4.2   Global Bifurcations

There are also three important global bifurcations in 2D, i.e. bifurcations that do not take place in an arbitrary small neighborhood of a point. The three types are periodic fold bifurcation, periodic blue sky in 2D bifurcation, and basin bifurcation. We describe them in detail.

**Periodic Fold Bifurcation.**   Figure 6 shows one structurally stable cycle enclosing another one in the left picture. In this situation without critical point in the ring between two cycles, one cycle is a source and the other cycle is a sink. If the two cycles drift towards each other, they eventually merge into an unstable cycle at the bifurcation time as can be seen in the middle of the figure. After this, both cycles have disappeared as in the right picture. Of course, a reversion of time is also possible, where two cycles around at least one critical point are created out of nothing.

**Periodic Blue Sky in 2D.**   In this type of bifurcation there are two different types of critical points involved: a saddle and a sink. Figure 7a shows the situation. As the attracting effect of the sink gets weaker and weaker, we see a homoclinic connection after some time where the saddle is connected to itself as shown in figure 7b. This results in a bifurcation: when this configuration breaks up again we find a limit cycle which simply appears out of the blue. The reason for the occurrence of the closed streamline is that the sink is totally unaffected by the whole event. Since there is an outflow to the critical point inside and to the saddle there must be a critical point or a closed streamline in this region according to the Poincaré-Bendixson theorem [18]. Because of the fact that there are only the two critical points a closed streamline emerged. This configuration is

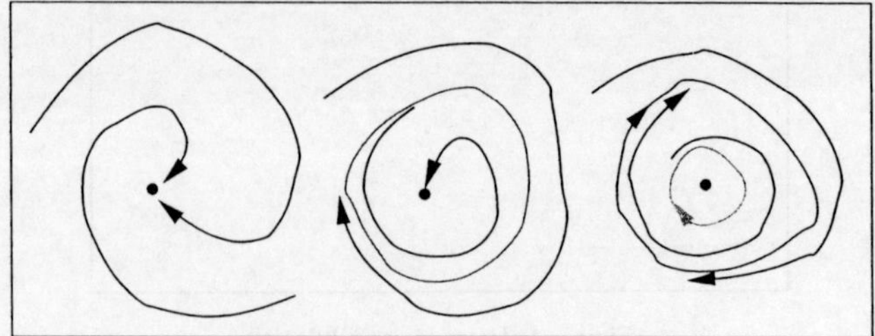

Figure 6: Periodic fold bifurcation

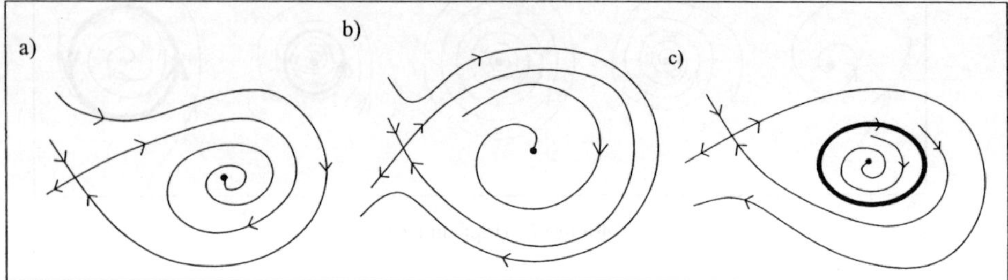



Figure 7: Periodic Blue Sky in 2D

shown in figure 7c. Other bifurcations of the same type can be constructed by replacing the sink with a source or by inverting the direction of time.

**Basin Bifurcation**   The remaining bifurcation that plays a role in our analysis does not involve critical points or cycles but only separatrices. It is called basin bifurcation, since only the basins and therefore the connectivity of the limit sets changes. It is characterized by a heteroclinic connection between two saddles at the bifurcation time. Figure 8 shows this event. The separatrices starting at the two saddles come closer and closer until they coincide at the bifurcation time. After this, the connection between the limit sets has changed — we have other basins than before.
The evolution of topology can now be defined.

**Definition 4.2** *Let $v : D \times J \to \mathbb{R}^2$, $J = [s_0, s_1] \subset \mathbb{R}$ an interval, be a Lipschitz continuous vector field. Let the vector field $v_t : D \to \mathbb{R}^2$, $v_t(x) = v(x, t)$ contain only structurally stable limit sets for all $t \in J$ except at a finite number of bifurcation times $t_1, \ldots, t_i, \ldots, t_m \in J$. The **evolution of topology** of $v$ is the topology at $s_0$, the maps of the movement of all limit sets, the maps of the movement of the separatrices, and the type and location of all bifurcations.*

We assume that our unsteady vector field is structurally stable, so that there takes place only one bifurcation at a bifurcation time. Especially, we do not consider combinations of bifurcations that interfer with each other. The evolution can be described by the movement of the limit sets, the movement of the separatrices and the bifurcations. The information about the bifurcations includes type, time, location, and the involved limit sets as well as separtrices.

# 5   Computation of Topology

For the topology, we need the limit sets and the basin boundaries as defined in section 3. Since we leave the boundary regions aside (for a clear analysis, see [22]), we are left with critical points, cycles, and the separatrices originating at saddles.

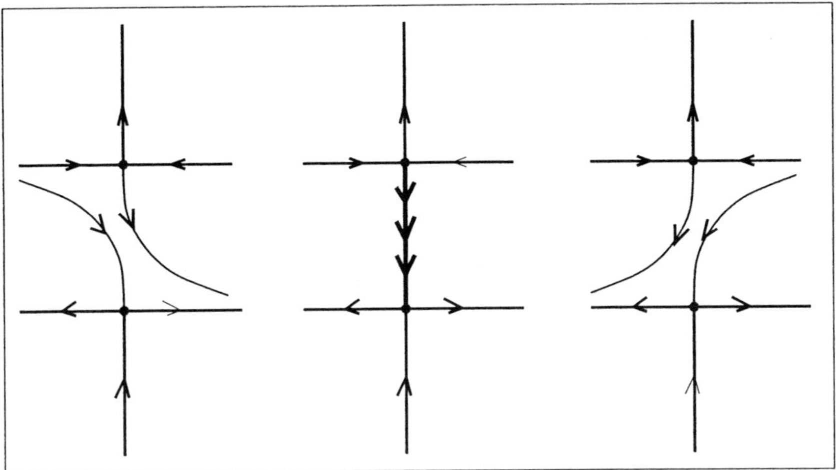

Figure 8: Basin bifurcation

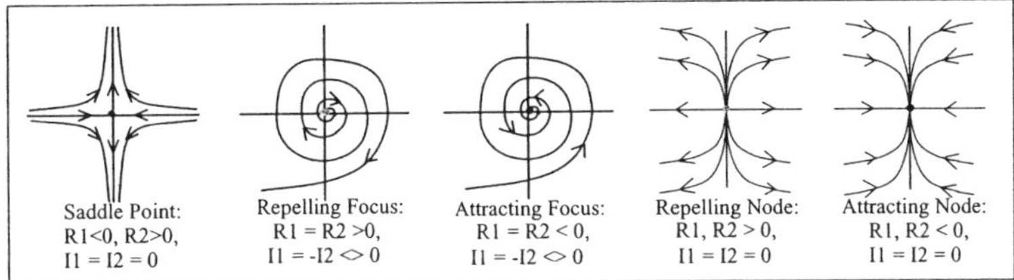


Figure 9: Common types of critical points.

## 5.1  Critical Points

A critical point is a zero of the field, so we have to do a zero search. This task depends heavily on the used interpolation scheme. In our case, we obtain a cartesian grid in a plane cut along the rotation axis of the cyclindric domain from the numerical method. Since we want to keep the computation effort, especially in the next section, at a manageable level, we triangulate the grid and use linear interpolation in each triangle. For the critical points, we can now solve a simple linear equation in each triangle. If the resulting zero is in the triangle, we have a critical point. Its type is easily found by calculating the (complex) eigenvalues of the Jacobian:

- Two negative real parts correspond to a sink.

- Two positive real parts correspond to a source.

- Mixed real parts correspond to a saddle.

A zero eigenvalue means a whole critical line of zeros in the triangle which we have excluded by our assumption of structurally stable vector fields. (We have also not found this case in practice.) Figure 9 shows the different cases. An analysis of critical points at the vertices of the grid is a little bit more difficult, but it can be done, see [23].

## 5.2  Closed Streamlines

As mentioned before, a closed streamline $d : \mathbb{R} \to \mathbb{R}^2$, $t \mapsto d(t)$ is a streamline of a vector field $v$ such that there is a $t_0 \in \mathbb{R}/\{0\}$ with $d(t + nt_0) = d(t)\,\forall n \in \mathbb{N}$ and $d$ not constant. In this subsection we present an algorithm that detects whether an arbitrary streamline $c$ converges to such a closed curve. This means that $c$ has $d$ as $\alpha$- or $\omega$-limit set depending on the orientation of integration. We do not assume any knowledge on the existence or location of the closed curve, so that the algorithm can detect closed streamlines. The principle of the algorithm works on any piecewise defined planar vector field where one can determine the topology inside the pieces. The basic idea of the algorithm is to determine a region of the vector field that is never left by a streamline. In case of a continuous vector field the Poincaré-Bendixson-Theorem ensures that this streamline approaches a closed streamline if no critical point exists in that region. A streamline approaching a closed streamline has to reenter the same cell. In this case we check if the cells were crossed by the streamline in the same order for the last two turns. This results in a *cell cycle* which identifies the above mentioned region. To examine if this cell cycle is left by the streamline we detect possible changes by checking the edges of the cells of the cell cycle. Therefore we identify points on each edge which we call *potential exits* where an outflow out of the cell cycle may occur in the vicinity. These points are identical with the vertices of the edge and points where the vector field is tangential to the edge.

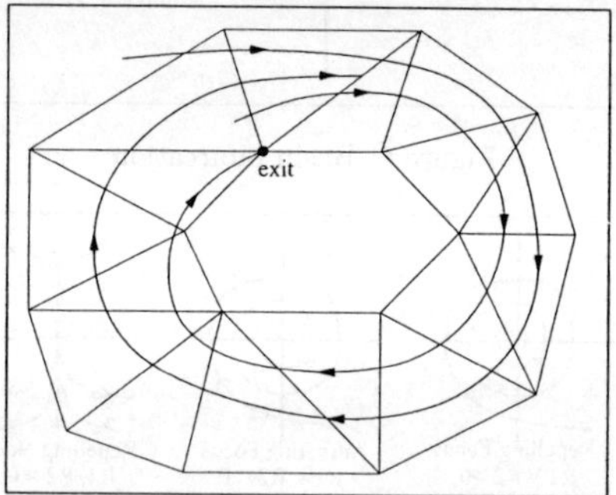


Figure 10: If a real exit can be reached, the streamline will leave the cell cycle.

We have to figure out if the actually investigated streamline will leave the cell cycle near such an exit. Therefore we integrate a streamline backwards from the potential exit to see if it leaves the cell cycle. If it does not leave after it crossed every cell of the cell cycle it converges to our streamline. We call this potential exit a *real exit* because the streamline will leave the cell cycle after a finite number of turns near that exit. Figure 10 displays an example for that case.

If the backward integrated streamline leaves the cell cycle, there will also be an entry point as shown in figure 11. A streamline starting at that point cannot be crossed by our actually investigated streamline. Consequently we cannot leave the cell cycle at this exit.

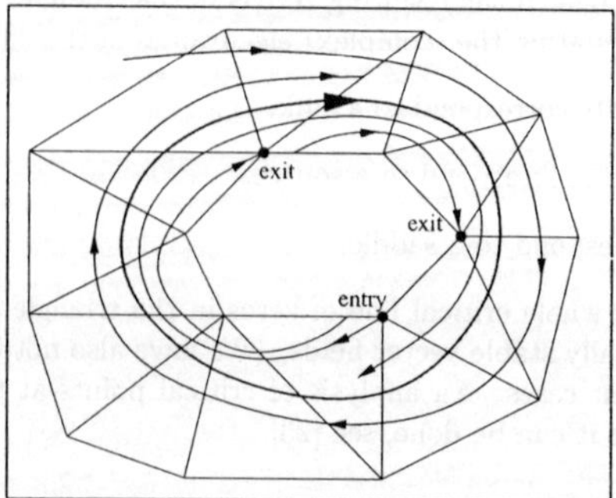


Figure 11: If no real exit can be reached, the streamline will approach a closed streamline.

If there is no real exit for the streamline, we have proven that the streamline will never leave the cell cycle. If there is no critical point inside the cell cycle the Poincaré-Bendixson-Theorem ensures that there exists a closed streamline in our cell cycle and the integral curve tends toward it. The exact position of the closed streamline can be found using the Poincaré map.

Let us assume that we have a two-dimensional continuous vector field containing one closed streamline. Then we can choose a point $P$ on the closed streamline and draw a *cross section S* which is a line segment nowhere parallel to the vector field. This line segment is then called a *Poincaré section*. If we start a streamline at an arbitrary point $x$ on $S$ and follow it until we cross the Poincaré section $S$ again, we get another point $R(x)$ on $S$. This results in the *Poincaré map* $R$. Figure 12 illustrates the situation. The left part shows the Poincaré section with the closed streamline in the middle, drawn with a thicker line. The right part displays the Poincaré map itself.

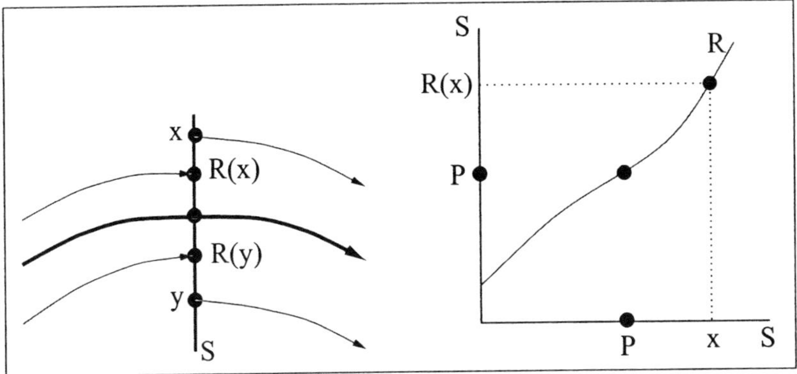


Figure 12: Poincaré section and Poincaré map.

$R$ maps the point on the closed streamline onto itself because the closed streamline will intersect the Poincaré section always at the same point. Consequently, it is sufficient to find the fix point to get a point on the closed streamline. If we find a cell cycle we can use the edge where we detected the cell cycle for the first time as a Poincaré section. To find the fix point of the Poincaré map we do a binary search: first we divide the edge into two parts at the mid point of the edge. Then we check which part gets intersected by the streamline starting at the intersection point after one turn. This part is subdivided again and we start another streamline. This continues until we are close enough at the fix point of the Poincaré map. If we start a streamline at this point we get the whole closed streamline after we crossed every cell of the cell cycle. This method terminates because we proved in the previous step where we detected the cell cycle that we converge to a closed streamline.

## 5.3  Separatrices

In section 3, it was explained that the separatrices can be found at the saddle points. Since we have only linear saddles in our vector field, we can use the eigenvectors of the Jacobian to find the separatrices in the triangle containing the saddle, see figure 13. We extend the separatrices by using an adaptive Runge-Kutta-Fehlberg integration [21]. We check for reaching the boundary, running into a critical point, and approaching a cycle as explained above. This completes our numerical analysis of the topology.

## 6  Computation of the Evolution

For the calculation of the evolution of the topology, we need an interpolation scheme over time. Since we want to keep the calculations simple, we do a linear interpolation between the stored time

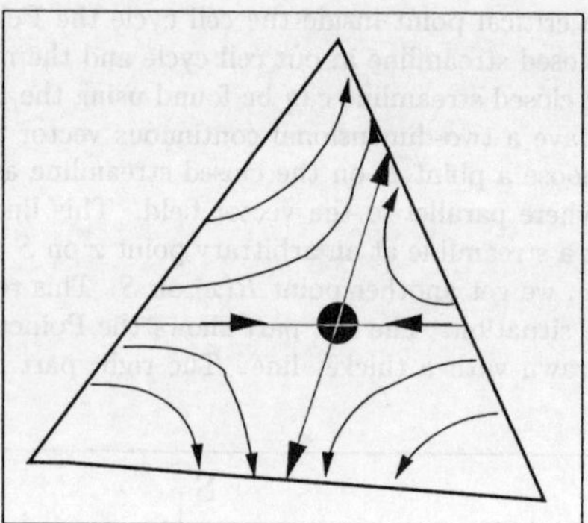

Figure 13: Separatrices for linear saddle.

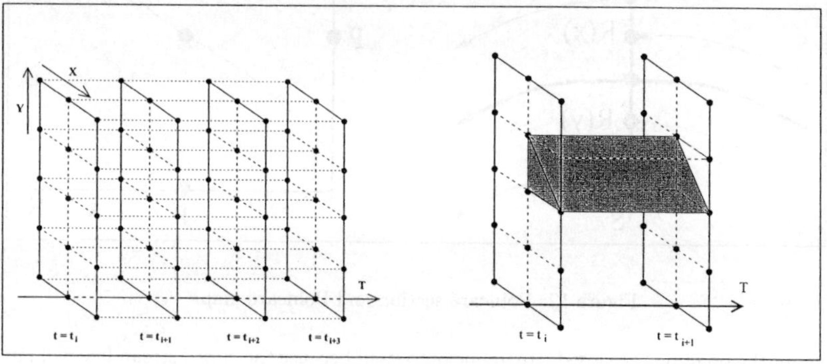

Figure 14: Space/Time grid

steps from the numerical simulation. The resulting grid is shown in figure 14.

**Definition 6.1** *Let $v_j : D \to \mathbb{R}^2$ be the given vector fields at time steps $t_0, \ldots, t_i, \ldots, t_m$. Then we define*

$$
\begin{aligned}
v : D \times [t_0, t_m] &\to \mathbb{R}^2 \\
(x, t) &\mapsto \frac{t_{i+1} - t}{t_{i+1} - t_i} v_i(x) + \frac{t - t_i}{t_{i+1} - t_i} \vec{a}_{i+1} v_{i+1}(x), \quad t \in [t_i, t_{i+1}].
\end{aligned}
\tag{20}
$$

We can now use the results from section 5 to obtain all critical points, cycles, and separatrices at $t_0$. We have to find now all the paths of critical points in all the prism cells, connect them and find the local bifurcations along the way.

## 6.1 Evolution of Critical Points

We locate and track critical points over time, using the interpolation scheme presented above. In particular, we seek the local bifurcations that may modify the topological structure: fold and Hopf bifurcations. To reduce the memory needs of our method, we do not process the whole grid at once: We consider only two consecutive time planes and handle the prism cells connecting them, forming a "time slice". More precisely, we first determine for each prism cell the possible entry and exit positions of singular points and then reconnect these pieces of information to track every singularity from one time plane to the next.

**Cellwise Analysis.** The interpolant has been chosen so that it contains at most one critical point for a fixed value of $t$ in a cell. This is due to the affin linear nature of its restriction to any time plane. If $t$ moves from $t_i$ to $t_{i+1}$, the singularity position describes a 3D curve. Yet, we are only interested in the curve sections that intersect the interior domain of the considered prism cell. A simple way to determine them is to consider the singularities lying on the side faces of the prism: Two triangular faces lying in $t = t_i$ and $t = t_{i+1}$ and 3 quadrilaterals connecting them. Finding the position of a singularity in each prism face requires the solution of a simple linear (resp.quadratic) system. If this is done, we sort the found (3D) positions in ascending time order and associate them pairwise. Indeed, since only one critical point can be present in a prism cell at a given time $t$, we know that a critical point must first leave the cell before it reenters it later. So, for each pair, we identify an entry and an exit position. We now need to pay attention to the singularity types to complete the topological information. The generic types of linear singularities are *source*, *sink* or *saddle* (omitting *center* points that correspond to a transition between source and sink). The transition from one type to another constitutes a bifurcation. Since a prism cell contains at most one critical point for a given $t$, we can assert that no fold bifurcation (involving two critical points) can occur inside it: *A fold bifurcation must occur at the cell boundary.* Consequently, we only seek Hopf bifurcations occurring between two consecutive entry and exit points. Practically, the determination of a singularity type is based upon the eigenvalues of the Jacobian at its position. Thus we compute the Jacobian matrix and its associated eigenvalues at each entry and exit point and check if they are the same. If this is the case, we can assert that no Hopf bifurcation has occurred since, otherwise, at least two bifurcations have taken place, which is impossible since our interpolant varies linearly over time. If the type has changed, a Hopf bifurcation is on the way. Because a Hopf bifurcation corresponds to a transition from a repelling to an attracting nature, the instantaneous nature of the intermediate singularity is a center and the trace of the Jacobian is zero at this point, which we use as criterion to determine the exact time location of the bifurcation. With the expression above, we obtain:

$$tr(J(t)) = tr\begin{pmatrix} a_{11}(t) & a_{12}(t) \\ a_{21}(t) & a_{22}(t) \end{pmatrix} = a_{11}(t) + a_{22}(t) = 0.$$

After straightforward calculus, we get finally

$$t = t_i - (t_{i+1} - t_i) \times \frac{a_{11}^i + a_{22}^i}{a_{11}^{i+1} - a_{11}^i + a_{22}^{i+1} - a_{22}^i}.$$

The corresponding position $(x(t), y(t), t)$ is the location of a Hopf bifurcation.

**Tracking between the Prisms.** At this stage, every cell knows the successive entry and exit positions of the critical point paths through its interior domain, as well as the possible presence of a Hopf bifurcation on the way. Yet, this information is scattered and must be reconnected to offer a global view of the topology evolution over time. A fundamental aspect of this task is to detect and identify the bifurcations that may take place on the faces. They are detected when connecting critical point path sections lying in neighbor cells as we show next.

We start in the time plane $\{t = t_i\}$. The restriction of the grid to this plane is a triangulation. In a boolean array, we indicate for each triangle if its corresponding prism must be further inspected or not. If it has to be inspected, we check the information collected during the cell analysis for a path section starting at an entry point located on the considered triangle face. This provides us with an exit point at which the tracking must be proceeded. This exit point can either lie on a side (quadrilateral) face of the prism or on the triangle face lying in $\{t = t_{i+1}\}$. In the latter case, we signify in the boolean array of the next time plane $\{t = t_{i+1}\}$ that this triangle must be further inspected. If the path leaves the cell at $t^* < t_{i+1}$ (side face), then it enters a neighboring cell at the same time. Back to a 2D representation, we get in time plane $\{t = t^*\}$ a critical point lying on the common edge of both neighboring triangle cells. Due to the discontinuity of the Jacobian matrix through this edge, the critical point may have another type when considered from the other side. This simple argument explains the possible existence of a bifurcation in such cases. Two situations may actually occur.

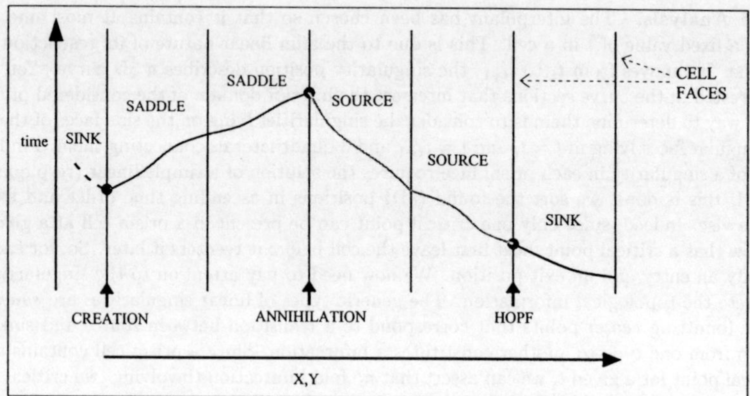



Figure 15: Possible cases when leaving a prism cell through a rectangular face

- There is a simple crossing of a critical point from one cell to another cell where no critical point was present so far. In this case, the type may change but, a saddle remains a saddle. A source may become a sink and vice versa. As mentioned previously, the transition source/sink or sink/source is a so-called Hopf bifurcation. We easily detect it by checking if the type of a source (resp. a sink) path changes after passing the face.

- The second situation corresponds to a merging of two coexisting critical points in neighboring cells on their common face (common edge in 2D). Now, an important property of the piecewise affin linear interpolant over a triangulation is that two neighboring triangles cannot contain two critical points with the same type which implies that only a saddle and a source or a saddle and a sink can merge that way. This type of bifurcation has been considered in section 4: It is a static fold bifurcation or its inverse.

When we leave the current cell through a side face, and a possible bifurcation has been detected, we identify this position. We proceed with tracking the path by asking the neighbor cell for the next exit position. We get three possible answers.

1. The exit position lies in the time plane $\{t = t_{i+1}\}$: We have reached the next time plane. The corresponding triangle face is marked `true` in the boolean array corresponding to the next time plane.

2. The "exit" position lies in the time plane $\{t = t_i\}$: If we were tracking the current path forward so far, we have found a pairwise annihilation. In this case, the corresponding triangle is set to `false` in the boolean array of $\{t = t_i\}$, indicating that this cell has been processed now.

3. The exit position lies on a side face for $t = t^{\#}$. If $t^{\#} < t^*$ and we were tracking the current path forward so far, we have found a pairwise annihilation and proceed with backward tracking. If $t^{\#} > t^*$ and we were tracking the current path backward so far, we have found a pairwise creation and proceed with forward tracking. In this case, we are back to the current situation where one enters a cell through a side face.

The possible cases are illustrated in figure 15.

## 6.2  Tracking the Cycles

Since there must be a critical point inside each closed streamline we use the critical point path containing a Hopf bifurcation as a starting point for our streamline algorithm from subsection 5.2 which detects the closed streamline if it exists. Therefore we follow the critical point path in

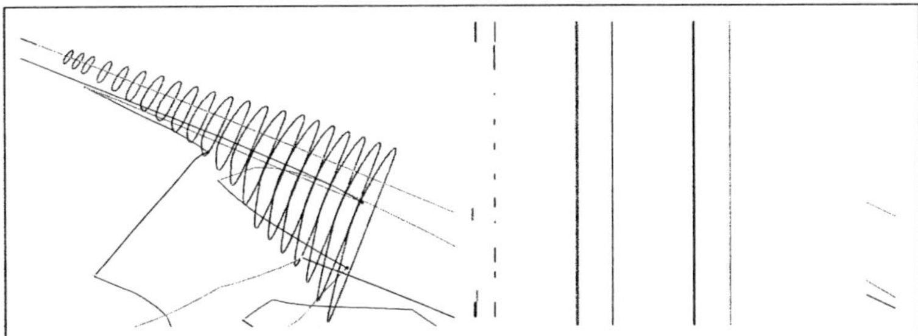

Figure 16: Closed streamlines visualized as wireframe and as a tube over time.

discrete steps in positive and negative directions starting at the bifurcation. After we have found the cell cycle containing the closed streamline we find the exact position using the Poincaré-map. Then we have to check if the closed streamline really surrounds the critical point. This is necessary because the streamline may have ran in another closed streamline in a totally different region of the flow. Obviously, closed streamlines surrounding the critical point occur only in one of the two temporal directions. This process continues until the cycles reach either another bifurcation which breaks them up or the border of the grid.

The left of figure 16 shows the result of this step, where we have found the closed streamlines at various timesteps. The closed streamlines are approximated by several line segments and the paths of the critical points are also shown using the same colors as in the original paper [24]. The Hopf bifurcation, where we started to detect the closed streamlines, is marked with a yellow sphere. In this example the life cycle of the closed streamline is started by a Hopf bifurcation and terminated by a Periodic Blue Sky in 2D bifurcation.

To visualize the evolution of closed streamlines, we construct tubes from the various closed streamlines similar to the pictures by Abraham and Shaw [17]. Therefore we construct surfaces consisting of triangles which connect the approximating line segments of the closed streamlines. The bifurcation point is connected to the tube using a parabolic surface approximated with triangles. The result is shown at the right of figure 16. We used an alpha value of 0.7 to get a better three dimensional impression and to prevent the tubes from hiding the paths of the critical points.

## 6.3  Separatrices

For the separatrices, we calculate each of them in smaller steps than the stored time steps from the simulation. We check for changes in the connectivity to find global bifurcations. It is possible to connect the instantaneous separatrices by surfaces visualizing their evolution, see [24]. Since these surfaces tend to cover the whole display, we do not show them in the final results. Right now, we do not have a method to find the basin bifurcations directly like the local bifurcations, so this is a good point for further research. During the calculation of the separatrices, we check for cycles all the time using the method from section 5. In this way, we hope to detect periodic fold bifurcations. If we find a cycle from one side, we go on with integral curve computation on an arbitrary point on the other side of the cycle to check for enclosed cycles. This allows to find all cycles for a structurally stable vector field. (Of course, we may miss some cycles due to our disregard of boundary saddles. Checks with random integral curves have not shown such missed cycles up to now.)

# 7  Results and discussion

The experiments of Billant *et al.* [1] document the breakdown of swirling water jets for Reynolds numbers in the range $300 < Re < 1200$ and swirl numbers $S$ in the range $[0, 1.42]$. The swirl number $S$ is defined by [1]

$$S \equiv \frac{2v_\theta(R/2, z_0)}{v_z(0, z_0)} \tag{21}$$

where $z_0 = 0.4D$, $D = 2R$ being the nozzle diameter. The Reynolds number is

$$Re \equiv \frac{v_z(0, z_0)D}{\nu} \tag{22}$$

The simulation of the 3d viscous flow calculates the values on a cylindrical grid. From this grid, we only take the values on a halfplan with the axis of the cylinder as one side and the radius as the other side. This results in an cartesian grid due to the set up of the direct numerical simulation. In figure 17, one can see the topology of the instantaneous integral curves after 20 timesteps. The figure does not show the whole grid but most of it. It is not difficult to see that the whole action takes place in a small area in the lower right corner. One can see vortical motion in action and might ask for a closer look at the critical points and cycles. For a better analysis, it may be mentioned that the program sets very small values to zero at the vertices to avoid a large number of critical points in the empty zone of the grid that may occur from numerical errors. If the values are very close to zero, the simulation method creates a large number of extremely weak critical points that do not have any physical significance, but result in a very disturbing visualization. In figure 18, we present a closer look at the topological structure of the flow. The flow does not contain a cycle right now, but one can see various blue sinks and green sources. All the curves are separatrices created by the red colored saddle points. The green part of the grid around the dominant structure in the middle is now used for an analysis of the topology evolution. We show this evolution in figure 19. The cone below shows the direction of time from time step 20 at the circle to time step 70 at the top. One can see clearly the paths of the saddles, sinks, and sources. All the small spheres stand for bifurcations. A static fold is shown by a purple sphere. A light blue sphere shows its inverse, i.e. a pairwise creation of a sink and a saddle or a source and a saddle. The yellow spheres denote Hopf bifurcations. As before, the transparent red surfaces show the evolution of cycles. They can either end/start at a Hopf bifurcation or at a blue sky in 2d bifurcation. As mentioned before, we have found no evidence for periodic fold bifurcations. Since we do not show the separatrices, one can also not see basin bifurcations.

We use the axisymmetric simulation to indicate some of the problems that we still have to solve. Figure 20 shows a large number of Hopf bifurcations along the trace of some of the sinks/sources. This is difficult for the analysis since it is nearly impossible to check for the cycles created (and destroyed) by these bifurcations. We present also a close-up in figure 21. Here, one can see that there are also sometimes larger number of static fold bifurcations that one would like to ignore since they are obviously not important for the behavior of the flow.

## 8   Conclusions

The analysis of the vector field topology based on instant vector lines, called streamlines, in plane sections of either axi-symmetric or three dimensional flows was developed. The vector fields were generated as solutions of the Navier-Stokes equations in cylindrical coordinates for swirling jet flows at moderate Reynolds numbers and supercritical swirl numbers, that produce recirculation and breakdown. The results for axi-symmetric flows develop a dense population of critical points and limit cycles as a result of flow normal to the plane considered. The temporal analysis using a sequence of planar ($r - z$ plane) snapshots of the early stage of a fully three dimensional swirling jet show the emergence of Hopf and blue sky bifurcations but no periodic fold bifurcation. The extension of the numerical tools for the topological analysis of vector fields to three dimension is currently under way and it can be expected that they will contribute to the understanding of the properties of turbulent fields. In particular the number of critical points in three dimensions is of interest, since it has been conjectured that it is much smaller than in two dimensional flows.

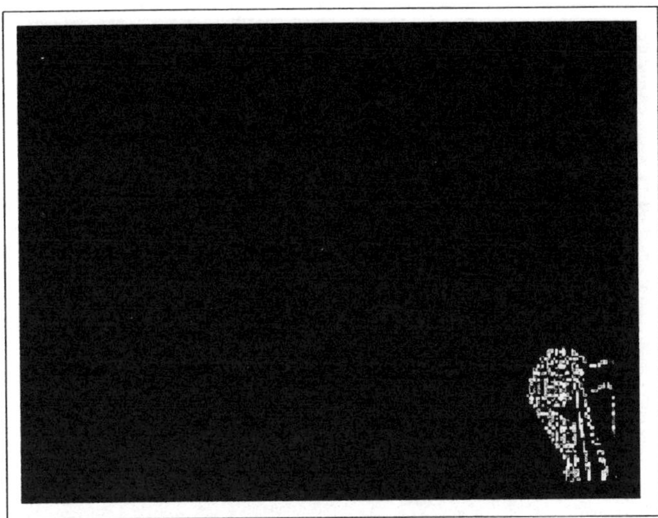

Figure 17: Topology of 3d viscous jet at time step 20.

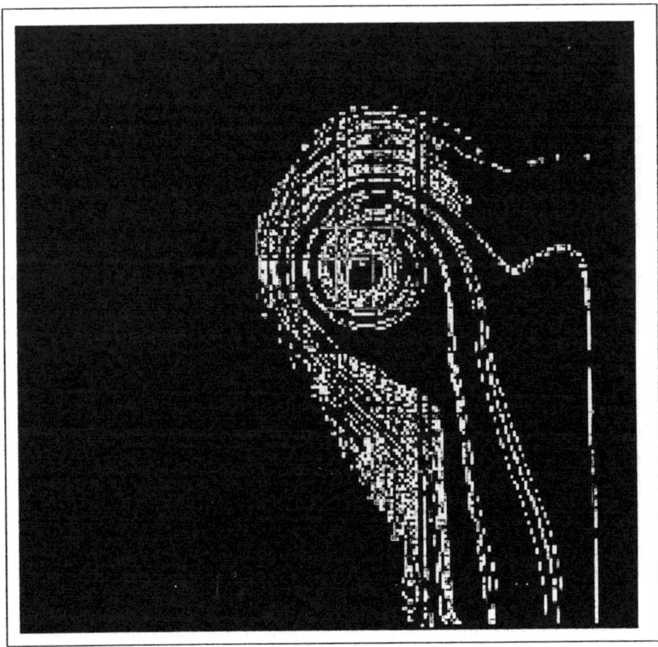

Figure 18: Interesting part of 3d viscous jet at time step 20. The green part of the grid is analyzed over time in the next figure.

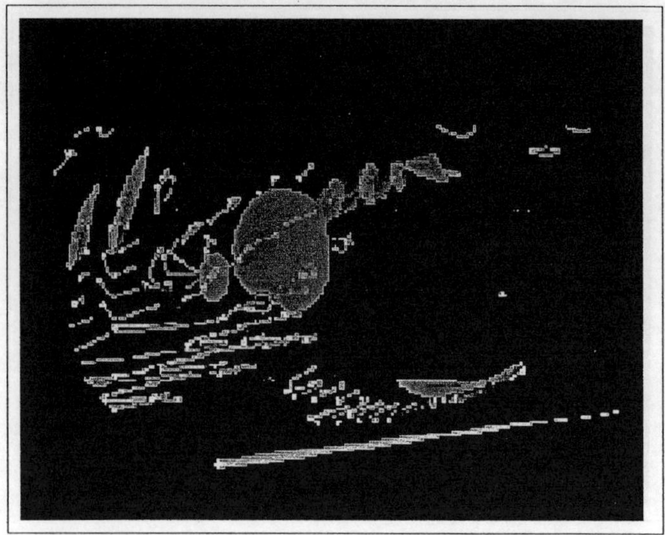

Figure 19: Evolution of the topology in the marked area over time.

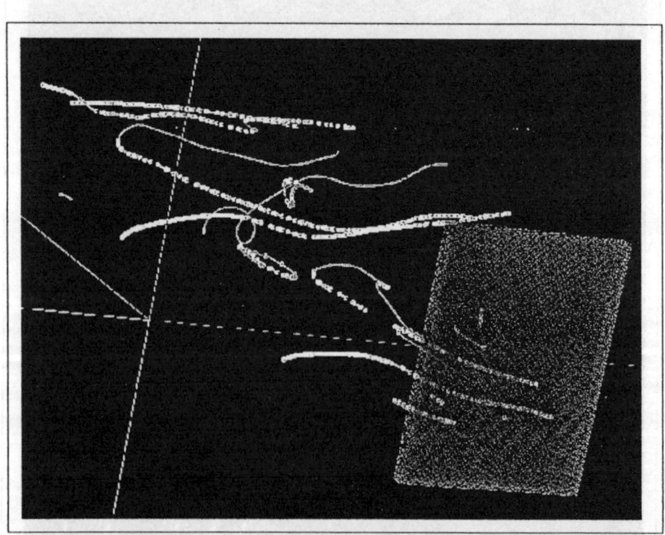

Figure 20: Cascade of Hopf bifurcations in the axisymmetric case.

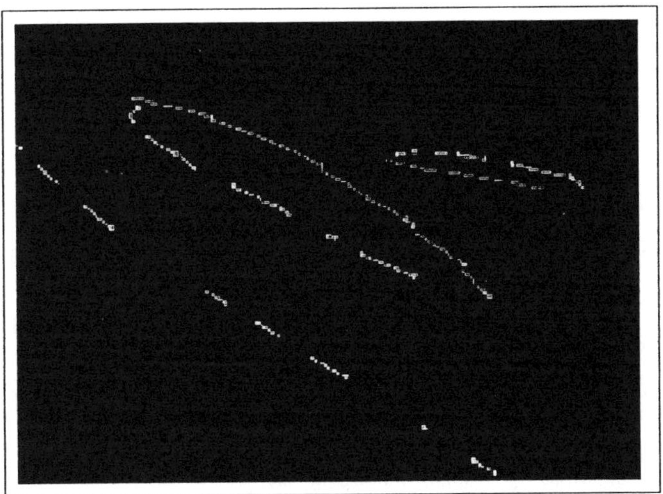

Figure 21: Zoom into some of the cascades.

# References

[1] P. Billant, J. Chomaz and P. Huerre (1999), "Experimental study of vortex breakdown in swirling jets", *J. Fluid Mech*, 376: 183-219.

[2] V. Shtern and F. Hussain (1996), "Hysteresis in swirling jets", J. Fluid Mech., **309**, 1-44

[3] R. Fernandez-Feria, J. Fernandez de la Mora and A. Barrero (1995), "Solution breakdown in a family of self-similar vortices", J. Fluid Mech., **307**, 77-94

[4] V. Shtern and F. Hussain (1998), "Instabilities of conical flows causing steady bifurcations", J. Fluid Mech., **366**, 33-85

[5] V. Shtern and F. Hussain (1999), "Collapse, symmetry breaking and hysteresis in swirling flows", in *Annu. Rev. Fluid Mech.*, vol. **31**, Annual Reviews, 537-566

[6] V. Shtern, F. Hussain and M. Herrada (2000), "New features or swirling jets", Physics Fluids **12**, 2868-2877

[7] A.T. Fomenko (1987), "The topology of surfaces of constant energy in integrable Hamiltonian systems and obstructions to integrability", Math. USSR Izv. **29**, 629-658.

[8] A.T. Fomenko (1991), "The Theory of Invariants of Multidimensional Integrable Hamiltonian Systems", Advances in Soviet Mathematics **6**, 1-35.

[9] W. Kollmann and J.Y. Roy (2000), "Hybrid Navier-Stokes solver in cylindrical coordinates I: Method", Computational Fluid Dynamics Journal **9**, 1-16.

[10] W. Kollmann and J.Y. Roy (2000), "Hybrid Navier-Stokes solver in cylindrical coordinates II: Validation", Computational Fluid Dynamics Journal **9**, 17-22.

[11] Canuto, C., Hussaini, M.Y. and Zang, T.A. (1988), "Spectral Methods in Fluid Dynamics", Springer-Verlag, Berlin.

[12] H.R. Lewis and P.M. Bellan (1990), "Physical constraints on the coefficients of Fourier expansions in cylindrical coordinates", *J. Math. Phys.* 31: 2592-2596.

[13] H. Eisen and W. Heinrichs and K. Witsch (1991), "Spectral collocation methods and polar coordinate singularities", J. Comput. Physics **96**, 241-257

[14] C.A. Kennedy and M.H. Carpenter (1994), "Several new numerical methods for compressible shear layer simulations", Appl. Num. Math. **14**, 397-433

[15] Li, Y. (1997), "Wavenumber Extended High-Order Upwind-Biased Finite Difference Schemes for Convective Scalar Transport", *J. Comput. Phys.* 133: 235-255.

[16] J.J. Dongarra and C.B. Moler and J.R. Bunch and G.W. Stewart (1979), "LINPACK user's guide", SIAM Philadelphia

[17] R. H. Abraham and C. D. Shaw (1982,1983,1985,1988), "Dynamics, the Geometry of Behavior I–IV", Aerial Press, Santa Cruz, CA.

[18] J. Guckenheimer and P. Holmes (1983), "Dynamical systems and bifurcation of vector fields", Springer-Verlag, New York, NY.

[19] M. W. Hirsch and S. Smale !974), "Differential equations, dynamical systems and linear algebra", Academic Press, New York, NY.

[20] S. Lang (1995), "Differential and Riemannian manifolds", Springer, New York, third edition.

[21] W. H. Press, S. A. Teukolsky, W. T. Vetterling, and B. P. Flannery !992), "Numerical Recipes in C", Cambridge University Press, Cambridge, UK, 2nd edition.

[22] G. Scheuermann, B. Hamann, K. I. Joy, and W. Kollmann (2000), "Visualizing local vector field topology", Journal of Electronic Imaging 9(4):356 – 367.

[23] X. Tricoche, G. Scheuermann, and H. Hagen (2000), "A topology simplification method for 2d vector fields", In IEEE Visualization 2000, IEEE Computer Society, Los Alamitos, CA, 359 – 366.

[24] X. Tricoche, G. Scheuermann, and H. Hagen (2001), "Topology-Based Visualisation Of Time-Dependent 2D Vector Fields", In Joint Eurographics and IEEE TCVG Symposium on Visualization 2001, Springer, Wien, 117 – 126.

644

# Data Compression for Incompressible Flow Solutions

Dohyung Lee[*†]

Key Words: Data Compression, Wavelets, Multiresolution, Numerical Solutions

## Abstract

Data compression for the incompressible flow solutions based on Supercompact multi-wavelets is presented. The proposed multiresolution method with supercompact wavelets offers high data compression for fluid simulation and experimental data including incompressible flow solutions. Supercompact wavelets provide advantageous benefits that it allows higher order accurate representation with compact support and therefore, it avoids unnecessary interaction with remotely located data such as across vortices. Thresholding for data compression is applied based on a covariance vector structure of multi-wavelets. Several numerical tests demonstrate large data compression ratios for the outputs of flow field simulation with various levels of fidelities.

## 1. Introduction

Large-scale simulations of fluid mechanics for integrated aero-vehicle design generate big output dataset. Depending on the level of fidelity, grid size, and unsteadiness consideration, the data set size increases dramatically. For instance, the required number of grid points substantially increases for the simulations with high fidelity simulations. Compared to low fidelity simulation such as the full potential equation solution, high fidelity simulations for the effects of viscosity, compressibility, turbulence and so on, require large number of grid points for the adequate resolution of the detailed physics.

Besides the grid size that directly reflects the output size, the unsteadiness


[*] Department of Mechanical Engineering, Hanyang University, 1271 Sa-1-dong, Ansan-si, Kyungki-do, 425-791, South Korea

[†] This work was supported by NASA Ames Research Center while the author was at NASA Ames Research Center as National Research Council Postdoctoral Fellow. This work was also supported by grant No. 2001-1-30400-023-3 from the Basic Research Program of the Korea Science & Engineering Foundation.

consideration also plays important role in determining the actual output size. To store the multi-time step solutions, the required storage significantly increases. In general, the data size for steady simulation of full-scale aircraft conducted at NASA Ames Research Center in 90's is less than 10 giga byte, but the unsteady solution storage requires over 100 giga byte, up to 1 tera byte size (See Table 1). In the course of actual aero-vehicle design, it's essential to construct an appropriate database that includes various flight conditions such as in Mach and Reynolds numbers, different fidelity levels in governing equations, and different grid size. In order to manage the huge database composed of individual big data sets, it is imperative to utilize a proper data management tool.

The solution data computed from large-scale simulations are sometimes too big for main memory, for local disks, and possibly even for a remote storage disk, creating tremendous processing time as well as technical difficulties in analyzing the data [24]. The excessive storage demands a corresponding huge I/O time as well as rendering time and transmission time between different computer systems. Sometimes, it is not possible to pull up a single big data set onto main memory in one piece, which means the simulation data set becomes useless unless a proper data management tool is proposed. In order to overcome the technical difficulties, various methods are being proposed and tested in computer science field, including progressive transmission technique [25].

Here, we propose a wavelet-based data compression method that is particularly tailored for computational fluid dynamics (CFD) simulations. Originally, the wavelet methodology was introduced in image processing field, for the purpose of data compression and feature extraction. The modulated version of wavelets and its application framework, multiresolution can be applied to CFD field for the specific needs such as data compression, feature extraction, convergence acceleration of solution algorithms, and grid refinement adaptation.

CFD or computational field simulation (CFS) data needs more careful treatment in applying the wavelet concept than photographic image data does. The typical photographic images and fingerprint images are unlike the numerical solutions of fluid dynamic equations. In general, the photographic images contain very little inherent smoothness, instead discontinuities almost everywhere. Meanwhile, computational solutions of fluid dynamics equations contain complex features in local areas that are embedded in an overall smooth solution. For instance, a high gradient viscous boundary layer lies under a smooth outer region; a strong vortex shed from a body is propagated with the flow speed in uniform flow; a strong discontinuous shock can be formed where otherwise the flow is smooth. The above features in CFD, in general, occur only in rather small spatial regions compared to the large overall size of the uniform field.

The wavelet bases should be amenable to the solution characteristics and constraints such as numerical accuracy and boundary condition. For solution characteristics, it is desirable that scaling function and wavelet bases have high order of accuracy since the data has smoothness in most of the areas. However, the number of support points substantially increases in most wavelet applications as higher order accurate method is used. In multi-dimension applications, the overall number of support points for the higher order schemes is significantly greater than to be handled in normal computations. The large support problem makes the numerical implementation very expensive, and sometimes, the implementation process is pretty annoying to consider all the combinations of support points, particularly near boundary. The big support might also introduce unnecessary interaction between the support points across the discontinuity or singular point. Wavelet representation across the discontinuity could be devastating for higher order schemes unless wavelet bases are not particularly suited. Therefore, compact support along with higher order accuracy, is strongly desired when developing wavelets for CFD.

Other than CFD solution characteristic, some other issues need to be addressed for wavelet application for CFD. While the image data is written on a regular spaced grid, the CFD or CFS data usually resides on a structured curvilinear grid or unstructured grid [28]. The other difficulty in CFD data is that CFD solutions consist of vectors instead of scalar values. Independent multiresolution application for each component of the vector solution may result in violation of important physical or implementation laws such as the conservation law of hyperbolic equations [29]. However, in this paper, only the CFD solution characteristic issue will be focused. Irregular spacing and the vector form of data issues remains to be studied.

In this paper, we consider compact schemes that have least support for higher order of accuracy. Before this work, several authors have attempted to develop wavelets that may have compact support with higher order of approximation and, if possible, additional beneficial features such as orthogonality and analytic (nonfractal) form. The most fundamental Haar wavelets (first order of approximation) were observed by Strang [8] to be compact, piecewise analytic and orthogonal. Daubechies [9] extended the order of approximation while retaining the orthogonality, but losing compactness. Geronimo, et. al [10] used multi-wavelets, but the results are similar to those found in [9]. Strang and Strela [11] extended the multi-wavelet work of Geronimo, et. al [10] for general orders of approximation. Donovan, et. al [12] used multiresolution analyses for the construction of multi-wavelets, which have arbitrary regularity and orthogonality, and are also symmetric and piecewise polynomial. Both Strang [11] and Donovan [12] wavelets have compact support to some degree but not to the level of the Haar compactness. As an early trial, Beam and Warming [13] extended Harten's interpolatory multiresolution to include Hermite interpolation, which leads to vector interpolation and multi-wavelets. These wavelets are nonfractal and accurate; however, they are not orthogonal (but biorthogonal) and are less compact than Haar wavelets.

Finally, based on the multi-wavelet basis of Alpert [14], Beam and Warming [15] constructed supercompact multi-wavelets which meet all the above conditions.

As far as supercompact support of the basis is concerned, the advantages of the multiresolution algorithm are two-fold. First, the application to functions defined on a finite interval does not require special treatment at the boundaries of the interval. Second, the application to functions which are only piecewise continuous between internal boundaries, can be efficiently implemented.

In this paper, Beam's supercompact wavelet is generalized to higher dimensions and demonstration of actual data compression for 3D fluid simulation data and its temporal performance in main solution operator are shown.

| *Data set name and Year* | *Number of vertices (Mil)* | *Number of time Steps* | *Size (GB)* |
|---|---|---|---|
| McDonnell Douglas F/A-18 `92 | 1.2 | 400 | 12.8 |
| Descending Delta Wing `93 | 0.9 | 1,800 | 64.8 |
| Bell-Boeing V-22 tilt rotor `93 | 1.3 | 1,450 | 140 |
| Bell-Boeing V-22 tilt rotor `96 | 5 | 1,450 | 300 |
| Bell-Boeing V-22 tilt rotor `97 | 10 | 1,450 | 600 |
| Neutron Star simulation (ASCI) | | | 1,000 |
| RLV Turbopump simulation `01 | 35 | 800 × (1–100) | 1,000 – 100,000 |

Table 1. Unsteady CFD Simulation data sets processed by NAS group in NASA Ames Research Center.

## 2. Harten's Interpolatory Multiresolution

A multiresolution analysis is the representation of a function using a series of scales. A discrete multiresolution analysis is decomposition into scales of a vector array which represents discrete values. A discrete multiresolution algorithm consists of a decomposition which generates the scale coefficients from the input array and a reconstruction which recovers the input array from the scale coefficients. The reconstruction is conservative in the sense that it reproduces the same input.

An interpolatory multiresolution algorithm due to Harten [16, 17, 3] can be briefly described as follows. At each step of the decomposition, there is a dyadic coarsening of the grid. During the coarsening the even-point values are projected and the odd-point values are decimated, i.e., discarded. For the reconstruction step, interpolation is used to predict the missing odd-point values at the finer level. Of course, in most cases, the interpolated values are not exact since there is an interpolation error. However, in each decomposition step the interpolation errors are saved before the odd-point values are decimated. Consequently, in each step of the reconstruction algorithm the interpolation errors are added to the interpolated values to obtain exact reconstruction. In an interpolatory multiresolution analysis, the interpolation errors are the scale coefficients.

The multiresolution transforms for the discrete data at level $p, v^{p'}$ s, are described algorithmically. The decomposition from finer level $p$ to coarser level $p - 1$ proceeds as

$$v^{p-1} = H_p^{p-1} v^p$$

$$d^p = G_p e^p = G_p (v^p - P_{p-1}^p v^{p-1}). \tag{1}$$

The vector at finer grid level, $v^p$, is decomposed into lower level representation, $v^{p-1}$, and scale coefficients, $d^p$, which are obtained from prediction errors, $e^p$. $H_p^{p-1}$ is a linear decimation operator that yields the discrete information at the resolution $p - 1$ from the discrete information at level $p$. The prediction operator, $P_{p-1}^p$, yields an approximation to the discrete information content at level $p$ from the discrete information content at level $p - 1$. The reconstruction process is

$$v^p = P^p_{p-1} v^{p-1} + E_p d^p .$$ (2)

The operator $G_p$ computes a prediction error in certain bases, and its inverse, $E_p$ simply computes a prediction error $e^p$ from scale coefficients $d^p$. Data compression occurs when some scale coefficients which fall below a certain threshold are replaced with zeros.

The basic relations between decimation and prediction operators, and between $E_p$ and $G_p$, are

$$H^{p-1}_p \cdot P^p_{p-1} = I_{p-1}$$ (3)

$$E_p \cdot G_p = I$$ (4)

which are consistency relations.

The choice of the linear decimation operator $H_p$ or prediction operator $G_p$ depends on the interpolation method and its order. Various forms of interpolation schemes (e.g. spline interpolation) are implemented to meet their own purposes. The lowest order interpolatory multiresolution is equivalent to the Haar wavelet, which decomposes discrete data into an average of two neighboring point values and its difference. The increase of interpolation order, in general, allows higher data compression ratio since it allows more accurate prediction from coarser level to finer level grid. The downside of the higher order representation is, however, that the interpolation scheme requires much bigger number of support points than the order of accuracy. The big support makes the numerical implementation expensive, and it may introduce unnecessary interaction with remotely located data (e.g. across a shock discontinuity or vortex). The problem becomes more serious in higher dimensions due to the substantial rise in the number of support points. For instance, if the number of support points in 1D is 3 times the order of accuracy, in 3D the factor of increase is 27. In the following section, a compact scheme is described that can significantly reduce the degree of support.

# 3. Supercompact Wavelets

## 3.1 Multi-Wavelets

In contrast to conventional wavelets which rely on only one wavelet, *Multiple wavelets* conduct decomposition and reconstruction processes using more than single mother wavelet. The motivation for multiple wavelets is that they offer fundamental advantages such as orthogonality, symmetry, short support, and a higher degree of accuracy [18]. Because of these beneficial characteristics, the multiple wavelets in general allow better data compression and feature extraction than a single wavelet. Beam and Warming's supercompact wavelets fall into this multi-wavelet category.

The use of multiple wavelets essentially entails additional pre- and post-processing for transforming given scalar data ( $u(x)_j$ ) to vector quantities ($\vec{\alpha}_j$). If the original given data has more than one dimension (vector, or matrix), the transformed data has one more additional dimension. The actual decomposition and reconstruction are performed on this transformed form ($\vec{\alpha}_j$) of data set.

## 3.2 Pre- and Post- Transformations

It is assumed that the discrete data on the interval (or element)-based grid system is given from a piecewise continuous scalar function $u(x)$ :

$$(\vec{u}_j)_m = u(x_m) \text{ for } x_m \in \vec{x}_j \tag{5}$$

$\vec{u}_j$'s are column vectors which represent the discrete data on interval $x_{j-1} \leq x \leq x_j$. $\vec{x}_j$ for the interval is composed of evenly distributed internal points. Note that the number of points on each interval determines the degree of polynomials $l$ and the order of accuracy ( $q = l + 1$ ).

The transformation from a vector $\vec{\alpha}_j$ to the discrete representation $\vec{u}_j$ is defined in matrix form:

$$\vec{u}_j = \mathbf{T}\,\vec{\alpha}_j \ . \tag{6}$$

When $l+1$ subinterval points are chosen based on the subinterval grid system, the polynomial coefficient $\vec{\alpha}$'s are the solution of the system of algebraic equations:

$$(\vec{u}_j)_m = \sum_{n=0}^{l} (\vec{\alpha}_j)_n \phi_n(x_m) \quad \text{for} \ \ x_m \in \vec{x}_j. \tag{7}$$

These equations are equivalent to those in matrix form (Eqn. (1)) and the transformation matrix becomes

$$\mathbf{T}_{mn} = \phi_n(x_m) \text{ for } x_m \in \vec{x}_j. \tag{8}$$

For example, third order accurate method ($l = 2$) can expand the Eqn. (6) to

$$\begin{bmatrix} u(x_j) \\ u(x_{j,1}) \\ u(x_{j,2}) \end{bmatrix} = \begin{bmatrix} \phi_0(x_j) & \phi_1(x_j) & \phi_2(x_j) \\ \phi_0(x_{j,1}) & \phi_1(x_{j,1}) & \phi_2(x_{j,1}) \\ \phi_0(x_{j,2}) & \phi_1(x_{j,2}) & \phi_2(x_{j,2}) \end{bmatrix} \begin{bmatrix} \alpha_0 \\ \alpha_1 \\ \alpha_2 \end{bmatrix}. \tag{9}$$

Various forms of orthogonal functions may be chosen for the basis functions $\phi_j$. Throughout this paper, orthonormal Legendre polynomials are used as basis functions. Then the transformation matrix becomes

$$\mathbf{T} = \begin{bmatrix} \sqrt{1/2} & -\sqrt{3/2} & \sqrt{5/2} \\ \sqrt{1/2} & 0 & -\sqrt{10}/4 \\ \sqrt{1/2} & \sqrt{3/2} & \sqrt{5/2} \end{bmatrix}. \tag{10}$$

The resulting $\vec{\alpha}_j \, (= \mathbf{T}^{-1} \cdot \vec{u}_j)$ can be interpreted as

652

$$\alpha_0 = \frac{\sqrt{2}}{6}\left[u(x_j) + 4u(x_{j,1}) + u(x_{j,2})\right] \approx \sqrt{2}\,\overline{u}_j\,, \qquad (11)$$

$$\alpha_1 = \frac{1}{\sqrt{6}}\left[-u(x_j) + u(x_{j,2})\right] \approx \frac{1}{\sqrt{6}}h\frac{d}{dx}u_j\,, \qquad (12)$$

$$\alpha_2 = \frac{\sqrt{10}}{15}\left[u(x_j) - 2u(x_{j,1}) + u(x_{j,2})\right] \approx \frac{\sqrt{10}}{15}\left(\frac{h}{2}\right)^2\frac{d^2}{dx^2}u_j\,. \qquad (13)$$

Equation (11 - 13) indicates that $\alpha_0, \alpha_1,$ and $\alpha_2$ are weighted discrete approximations to cell average, first derivative, and second derivative, respectively.

Even for the arbitrary order of moment, the one to one mapping is valid between $\vec{\alpha}_j$ elements and the quantities composed of average and various orders of difference approximations. $\vec{\alpha}_j$ can be redefined as

$$\vec{\alpha}_j = \left[\mathrm{w}_0 \cdot \overline{u},\ \mathrm{w}_1 \cdot u_x,\ \mathrm{w}_2 \cdot u_{xx},\ \mathrm{w}_3 \cdot u_{xxx},\ \cdots\right]^{\mathrm{T}}, \qquad (14)$$

where $\mathrm{w}_0$, $\mathrm{w}_1, \mathrm{w}_2$, and $\mathrm{w}_3$ are weighting factors.

3.3 Decomposition and Reconstruction

Supercompact wavelets retain the spatial compactness and orthogonality of Haar wavelets since the decomposition involves only two value interactions in both methods (Haar: scalar, Supercompact: vector). The higher level of accuracy of supercompact wavelets is attained by the vector data representation (higher order polynomials) shown in the previous subsection and also in the following accuracy analysis subsection.

We follow Harten's cell-average multiresolution framework, in which the decomposition is to compute (1) an average value and (2) the interpolation error ($e^p$)

or equivalently, a residual (scale coefficients, $d^p$).

Since the vector version of Haar wavelets is being implemented, information on only two adjacent intervals of a finer grid is needed for computing the average and error during coarsening. This illustrates why a small support is needed for a higher order technique in supercompact wavelets. Note that Harten's interpolatory multiresolution requires a large number of support points for higher order accurate representation. In general, the number of supporting points with interpolatory multiresolution is more than three times that of supercompact wavelets.

The one step supercompact decomposition is carried out by the following formulation [15]. The average vector for the coarser level grid is

$$\vec{\alpha}_j^{p-1} = \frac{1}{\sqrt{2}} (\mathbf{H}^0 \vec{\alpha}_{2j-1}^p + \mathbf{H}^1 \vec{\alpha}_{2j}^p). \tag{15}$$

The $\mathbf{H}^0$ and $\mathbf{H}^1$ are weighting coefficients for the averaging transformation based on orthonormal wavelet bases. For a residual we choose

$$\vec{r}_j^{p-1} = \frac{1}{\sqrt{2}} (\mathbf{G}^0 \vec{\alpha}_{2j-1}^p + \mathbf{G}^1 \vec{\alpha}_{2j}^p) \tag{16}$$

$\mathbf{G}^0$ and $\mathbf{G}^1$ are evaluated from $\mathbf{H}^0$ and $\mathbf{H}^1$ [15]. The coefficients $\vec{\alpha}_{2j-1}^p$, $\vec{\alpha}_{2j}^p$, $\vec{\alpha}_j^{p-1}$, and $\vec{r}_j^{p-1}$ represent odd and even interval vectors on the fine grid, and average and residual vectors on the coarse grid.

Equation (15) is based on the refinement equation, i.e. relationship of the polynomials between two grid scales.

$$\vec{\phi}_j^{p-1} = (\mathbf{H}^0 \vec{\phi}_{2j-1}^p + \mathbf{H}^1 \vec{\phi}_{2j}^p), \tag{17}$$

where $\vec{\phi}_j^{p-1} = [\phi_0(x), \phi_1(x), \cdots, \phi_l(x)]^{\mathrm{T}}$.

If we use the assumed orthogonality of $\phi_j$'s, the coefficients are simply

$$\mathbf{H}_{i+1,j+1}^0 = 2 \int_{-1}^0 \phi_j(2\xi + 1)\phi_i(\xi)d\xi , \quad i = 0,1,\cdots,l , \quad j = 0,1,\cdots,l \tag{18}$$

$$\mathbf{H}^1_{i+1,j+1} = 2\int_0^1 \phi_j(2\xi-1)\phi_i(\xi)d\xi \ , \ i=0,1,\cdots,l \ , \ j=0,1,\cdots,l \ . \tag{19}$$

The corresponding $\mathbf{G}$'s are given from known $\mathbf{H}^0$ and $\mathbf{H}^1$ [23,24,25].

The inverse transformation matrix for reconstruction is the transposed matrix due to the orthogonality of the multi-wavelets

$$\begin{bmatrix} \vec{\alpha}^p_{2j-1} \\ \vec{\alpha}^p_{2j} \end{bmatrix} = \sqrt{2}\begin{bmatrix} \mathbf{H}^0 & \mathbf{H}^1 \\ \mathbf{G}^0 & \mathbf{G}^1 \end{bmatrix}^{-1}\begin{bmatrix} \vec{\alpha}^{p-1}_j \\ \vec{r}^{p-1}_j \end{bmatrix}$$

$$= \sqrt{2}\begin{bmatrix} (\mathbf{H}^0)^{\mathrm{T}} & (\mathbf{G}^0)^{\mathrm{T}} \\ (\mathbf{H}^1)^{\mathrm{T}} & (\mathbf{G}^1)^{\mathrm{T}} \end{bmatrix}\begin{bmatrix} \vec{\alpha}^{p-1}_j \\ \vec{r}^{p-1}_j \end{bmatrix} . \tag{20}$$

## 3.4 Scaling functions and wavelets

Development of multiresolution algorithm does not require an explicit knowledge of wavelets. Even in application of multiresolution the exact analytic wavelet functions do not need to be given in the algorithm since only the discrete wavelet coefficients $(\mathbf{H},\mathbf{G})$ are required. However, the exact analytic multi-wavelet functions can be defined from the scaling functions by

$$\Psi^{p-1}_j = \mathbf{G}^0\phi^p_{2j-1} + \mathbf{G}^1\phi^p_{2j} \ ,$$

and

$$\Psi^p_j = \begin{bmatrix} \psi^k_{0,j} & \psi^k_{1,j} & \cdots\cdots & \psi^k_{l,j} \end{bmatrix} .$$

These orthonormal polynomials are the multiwavelets that form the basis of error expansion. For the third order scaling functions with Legendre polynomials are given in $\xi$ - space

$$\phi_0(\xi) = \frac{1}{\sqrt{2}}\,, \qquad\qquad\qquad -1 < \xi < 1\,,$$

$$\phi_1(\xi) = \sqrt{\frac{3}{2}}\,\xi\,, \qquad\qquad\qquad -1 < \xi < 1\,,$$

$$\phi_2(\xi) = \sqrt{\frac{5}{2}}\left(\frac{3}{2}\xi^2 - \frac{1}{2}\right), \qquad\qquad -1 < \xi < 1\,.$$

The corresponding multi-wavelet functions are defined on the interval $(-1,1]$

$$\psi_0(\xi) = \begin{cases} -\dfrac{1}{2}\sqrt{\dfrac{3}{2}}\left(3+16\xi+15\xi^2\right), & -1 < \xi \leq 0\,, \\[2ex] \dfrac{1}{2}\sqrt{\dfrac{3}{2}}\left(-3+16\xi-15\xi^2\right), & 0 < \xi \leq 1\,, \end{cases}$$

$$\psi_1(\xi) = \begin{cases} -\sqrt{\dfrac{1}{2}}\left(2+3\xi\right), & -1 < \xi \leq 0\,, \\[2ex] \sqrt{\dfrac{1}{2}}\left(2-3\xi\right), & 0 < \xi \leq 1\,, \end{cases}$$

$$\psi_2(\xi) = \begin{cases} -\sqrt{\dfrac{5}{2}}\left(1+6\xi+6\xi^2\right), & -1 < \xi \leq 0\,, \\[2ex] \sqrt{\dfrac{5}{2}}\left(1-6\xi+6\xi^2\right), & 0 < \xi \leq 1\,. \end{cases}$$

Multiscaling functions and multi-wavelets for the third order Legendre polynomials are plotted in Fig. 1.

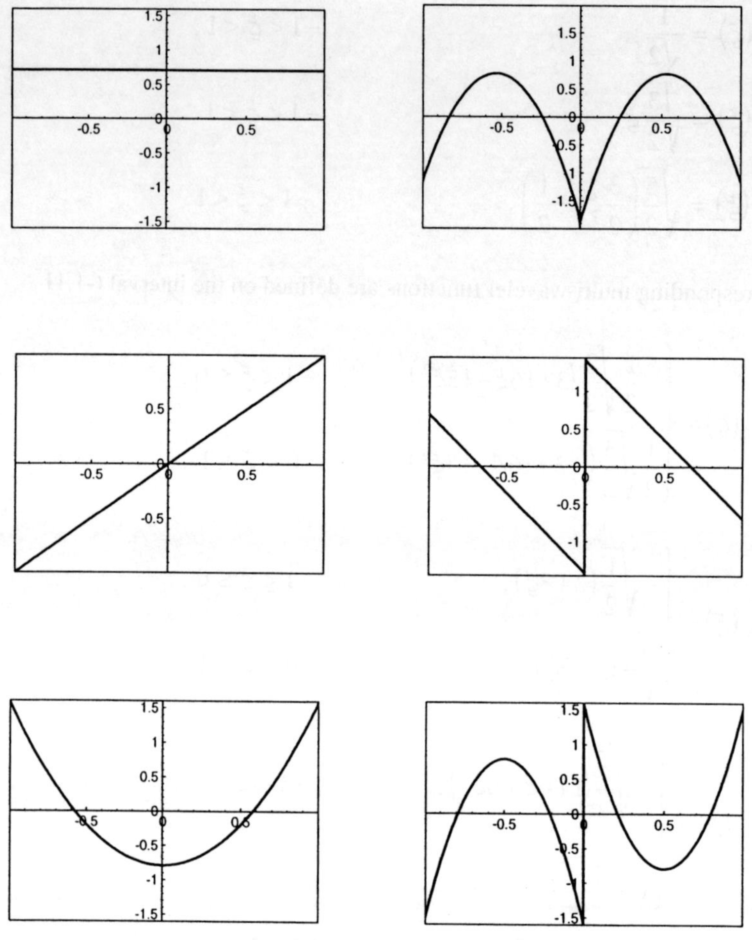

Fig. 1.  Scaling  function  (left-hand column) and  wavelets (right-hand column),  l=2.

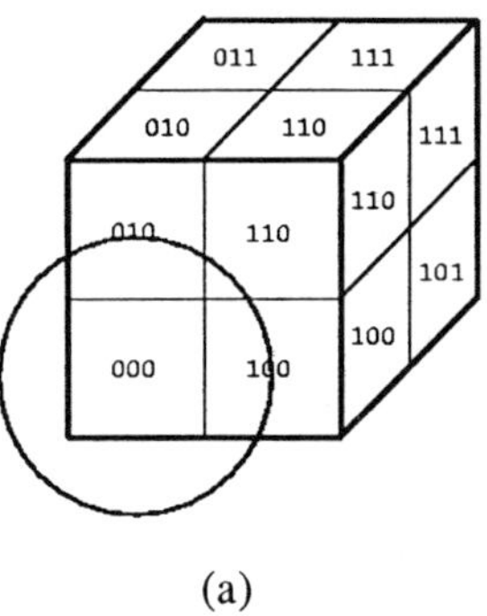



(a)

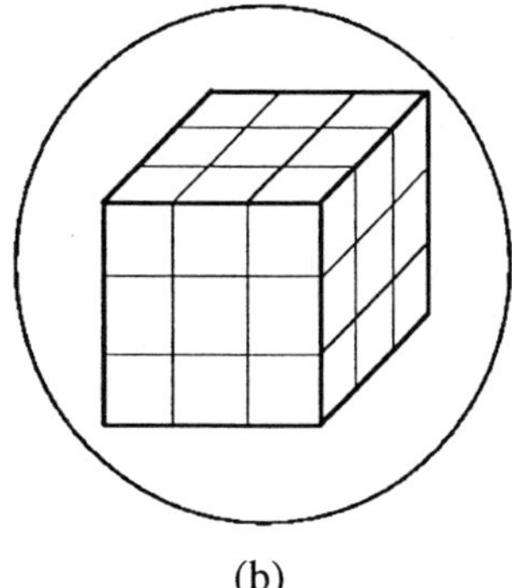

(b)

Fig.2. (a) one step multiresolution cell merging: eight sub cells merge into one cell during Decomposition; (b) fourth order cell configuration: the number of orders or determines the grid points for data.

## 4. Multi-dimensional Supercompact Wavelets

For supercompact wavelet generalization to multidimensions, some multiresolution methods are studied. One of the multiresolution methods for 3D data sets is to apply quasi -1D multiresolution by alternating the directions. Basic concept of this method is that, for one level multiresolution, the wavelet coefficients are decomposed three times by 1D decomposition operators along each three different direction. The advantage of the method is that the quasi-1D decomposition is not expensive compared to genuinely multidimensional multiresolution that includes expensive big matrix multiplication process.

However, supercompact wavelets are multi-wavelets that have more than one entry that can have the properties of average and various orders of derivatives. Since it is 3D application, the wavelets are given in the form of 3D tensor and higher order cross derivative quantities are included in the tensor entries. For the quasi-1D decomposition application, row or column vectors are extracted from the tensor form. When sweeping 1D decomposition along alternating direction, the same operators are applied to all extracted vectors regardless of the properties whether they are average and derivatives in the previous cross direction. It is not clear that a certain advantageous property such as orthogonality preservation can be guaranteed, which could lead to less data compression ratio. Meanwhile, data compression itself is not much of a problem since reconstruction process is reverse process of decomposition. Multiresolution with this method is still conservative since reconstruction without any thresholding guarantees that original solutions will be recovered. With adequate thresholding, data compression can be achieved.

In this study, a genuinely multidimensional multiresolution scheme is presented which relies on multidimensional scaling and wavelet functions. As in the 1D case, beneficial properties such as orthogonality are maintained even in the 3D generalization.

Compared to 1D case, in 3D more cells are involved on the finer grid during the merging process. One cell on the coarse grid corresponds to 8 sub cells on the fine grid (see Fig. 2). After one step of the decomposition is completed, the 8 solution vectors corresponding to each sub cell are replaced by one average (interpolation) vector and 7 residual (detail coefficient) vectors. Details of the transformation and decomposition and reconstruction processes are described in [23].

## 5. Boundary and Data Size Restriction

Interpolation based wavelets methods require modification in the vicinity of boundaries. In order to maintain the order of accuracy, the support points need to be replaced with the other points since some support points in the original form are located out of bound. Otherwise, the order of accuracy should be lowered. In contrast, supercompact multi-wavelets do not need any special modification in decomposition and reconstruction process near boundary since the weighting factor for each support points are the same. This is highly beneficial for the higher order methods in computational cost and coding complexity.

However, the application of multiresolution schemes has relied on data sets of which the grid size is the product of an order of accuracy and power of two along all the directions ($l \times 2^m$). In case the grid size is not ideal, the grid size for the multiresolution should be altered. If one shrinks the size of the problem, the overall data compression ratio drops since some parts of the solution remain untouched. When the size of the problem increases, extra memory is required for the extended problem and this also results in a reduction of the compression ratio. One has to choose an adequate method between shrinkage and expanding. It is suggested to use a dimension (power of two) that is closer to original dimension than the other choice.

The methods of extending the boundary in general include zero-padding, symmetrization, and smooth padding [30]. Smooth padding based on extrapolation is generally a proper choice for the outer boundary. In our numerical procedure, first or second order extrapolation is used as a smooth padding treatment.

However, when the problem dimension is too big, the difference between two possible choices (neighboring power of two numbers) substantially increases. Accordingly, the chance of efficiency loss in data compression becomes much greater. It is the biggest obstacle in wavelet application to the big size data set. One of practical remedies for this restriction is to divide the problem domain into small groups that independent multiresolution will be applied. The individual multiresolution schemes have flexibility in multiresolution dimensions (3D, 2D or 1D) and application size by varying order of accuracy and number of multiresolution levels. In other words, it is hybrid method that combines 3D, 2D and 1D versions of multiresolutions and different order of moments and multiresolutions levels. Since 2D version of the method is applied layer by layer and 1D version is processed in one line of data, most of the problem domain will be covered by the method.

The danger of the hybrid method is that degenerated version of multiresolution does not communicate with other layers or rows of data. Due to no sharing of the information across the layer interface, derivative information across the interface is missing throughout the multiresolution process. Accordingly, a certain flow feature across the interface cannot be controlled by thresholding.

## 6. Thresholding

Data compression can be attained by proper thresholding that discards wavelet coefficients which fall below a certain threshold value. In the case of multi-wavelets, a rather complicated treatment is needed since a vector instead of a scalar must be handled.

There are two options in thresholding for multi-wavelets. One is to threshold independently individual coefficient in the vector form by a certain rule. The other is to threshold the whole vector. The former can control the thresholding actively by different weighting factor for individual coefficient. Since individual coefficient can represent average or various orders of derivatives, one can determine which derivative features can be preserved more clearly. Disadvantage of this method is that, in actual data compression, it requires more storage for information data of location where the coefficients are set to zero.

The second method needs a normalization of a vector in order that a multi-wavelet vector can be evaluated in a scalar form. Depending the normed value of wavelet vector, it will be determined whether thresholding will be on or off. For normalization of wavelet vector, we follow a thresholding method suggested by T. Downie and B. Silverman [18], used in noise reduction in signal processing, which is based on the covariance structure of multi-wavelets. In the multi-wavelet analysis, the pre-transformation introduces highly correlated coefficients. Since the wavelet coefficients in the same element (or interval) are highly correlated, a cut-off method based on vectors rather than on each coefficient is preferred [23, 25].

In numerical implementation, both methods (individual coefficient and vector thresholding) are tested.

For the covariance method, by the following transformation,

$$\theta = \mathbf{R}^T \mathbf{V} \mathbf{R} \tag{29}$$

where $\mathbf{V}$ is the covariance matrix, a positive reference value $\theta$ is obtained which will eventually be thresholded.

Once the threshold value $\lambda$ is chosen, the analogous thresholding rule is applied as

$$\hat{\mathbf{R}} = \mathbf{R} \cdot \mathbf{I} \quad (\theta \leq \lambda) \tag{30}$$

which keeps the wavelet vector $\mathbf{R}$ if $\theta$ is greater than $\lambda$ but otherwise sets the vector to zero. As an alternative, soft thresholding (proportional shrinking) may also be applied with the purpose of smooth transition of the values, although reduction of the compression ratio is expected,

$$\hat{\mathbf{R}} = \mathbf{R} \cdot \frac{\max(\theta - \lambda, 0)}{\theta} \tag{31}$$

The covariance matrix is given by

$$\mathbf{V} = \sigma^2 \mathbf{T}^T \mathbf{T} \tag{32}$$

in signal processing, which has noise components with variance $\sigma^2$. Here, $\mathbf{T}$ is post transformation matrix. However, in fluid simulation data compression, variance strength is not an issue, therefore $\sigma$ is simply set to one. By the choice of threshold value $\lambda$, the data compression ratio and accuracy of the method is controlled: a small value of $\lambda$ reduces the data compression ratio but increases the accuracy. In this case, $\lambda$ is determined by the statistical characterization of the flow field. In other words, semi-manual inspection is carried out for the flow field properties including the maximum value of the property.

Multiresolution reduces the magnitude of represented data by a factor of two as the level goes to the next coarser grid ($m$ decreases). Accordingly, the threshold value $\lambda$ on level $m$ is set by an initial error bound value multiplied by $2^m$.

For the hybrid method that combines 3D, 2D and 1D multiresolution, the degenerated version of multiresolution should use smaller cut-off reference value since the norm of the wavelet vector is less than that of 3D multiresolution. It is also due to missing derivative information across the interface. The reduction of the reference value is analyzed in the following by using the transformation matrix. However, there is a possible danger that different cut-off reference values could be applied in the neighboring layers because the maximum value in one layer is different from the other layer. In the implementation, a unified reference value is set throughout the 2D version domain, and also in 1D domain, respectively.

## 7. Data Compression for Overlapping Grid Solution

In practical numerical simulations for the integrated design, the solutions are obtained on the basis of multiple grid sets. Individual grid set is generated for each component of the system and outer grid is also generated for the whole computation domain. However, it is inevitable to have the overlapped regions between two neighboring sub-grid systems. The overlapped part of grids is remained untouched from the beginning of computation in solution algorithm and the values at the region are just garbage values that are completely different from nearby meaningful simulation values. Therefore, most advanced CFD software including OVERFLOW and visualization software has an option to handle the area, so called IJK-Blanking.

In wavelets application, the IJK-Blanking area could adversely affect the decomposition process. Since the multiresolution is dyadic coarsening process, data at one point will be interacted with far away data on coarser level of multiresolution. The garbage values in the IJK-Blanking area, introduces quite big the detail coefficients since there are too much differences between the garbage values and the properly simulated results. This big detail coefficients can hardly be set to zero, leading to reduced data compression ratio and the possibility that, around IJK-Blanking area, the reconstructed data is quite different from the original data set.

For practical data compression application with supercompact multi-wavelets, the method should handle various situations that can arise in actual CFD data set. In this application, a preprocessing procedure for the IJK-Blanking area is proposed. The garbage values in the area are replaced with reasonable values that are computed by relaxation method. For the relaxation treatment, Laplacian operator is used, and at the boundary of the IJK-Blanking, simulated values from the solution solver are used as elliptic boundary condition.

## 8. Results

Data compression based on supercompact wavelets is performed for the incompressible solutions as well as other fidelity level of solutions (e.g. full potential equations solutions and compressible solutions). For practical application for overlapping grid solutions, the relaxation preprocessing technique is also evaluated.

Incompressible flow solution example is a wing tip vortex propagation case [22]. This data set comes from the three dimensional incompressible Navier-Stokes flow solver, **INS3D-UP** developed by kwak, et al [19]. **INS3D** is regarded as an excellent simulation code for aerodynamics and turbomachinery flow [22]. The code is using an artificial compressibility algorithm and modified Baldwin-Barth turbulence model. The data is given based on C-O type topology on curvilinear coordinate and the size is (115 X 189 X 115).

This data, with 2.5 million grid points, can be successfully compressed on a scalar pressure field and on a vector form of the solution. The vector form of incompressible solution consists of **4** components, the velocity vector and pressure. For the pre- and post- processing, decomposition and reconstruction, and thresholding procedure, the evaluation of the coefficients in 3D transformation matrix **T**, wavelet coefficients **L**, and the covariance matrix **V** are required. Since the evaluation of those values is pre-computed once before the actual processes, the computing time is significantly reduced.

Suitability of the problem size to wavelet application is more important than the adaptability of the wavelet to feature characteristics. For the (115 X 189 X 115) grid, several rescaling strategies were explored in an attempt to adapt the problem to the dyadic multiresolution process. For various orders (2-4) of accuracy and levels of multiresolution (2-4), the grid size was rescaled either by expanding or shrinking the size or problem. Efficiency test in data compression ratio indicates that the most favorable multiresolution-applicable grid size is (113 X 177 X 113) for third order accuracy and three stages of multiresolution. For this model problem, hybrid method is not adopted.

Fig. 3 exhibits successful data compression for the pressure field even with curvilinear coordinates. Except around the outer boundary, these two contour plots are indistinguishable. The maximum error (located at the outer boundary) is 7.93 X $10^{-2}$; the L2 error norm normalized by the number of total grid points is 2 X $10^{-6}$. For the grid size of (113 X 177 X 113) and third order of accuracy, the compression ratio ($\mu_1$) on the vector domain is 40. In other words, only 2 percentage of the storage in the original solutions is required to store wavelets values in decomposed domain. However, the actual ratio ($\mu_2$) considering overheads and shrinking effect drops to 8.11.

The vector solution was also compressed, on a component-by-component basis. Frequently a situation arises such that a certain property of the solution should be preserved (e.g. conservation of momentum and energy). However, the burden of preservation is aggravated by the fact that the number of frequently used descriptors is around 50 in the fluid dynamics field. All kinds of combinations of primitive or conservative variables may produce compression errors in an unexpected nonlinear manner. In particular, derivative information such as vorticity can introduce huge difference from the original solution. At this stage, it is not possible to confine errors on all variables to a certain bound. One remedy is when preservation of a certain parameter needs to be maintained, multiresolution for one component can be substituted with a property preservation procedure. In this case, a reduction of the compression ratio is inevitable.

The present research is being conducted in order to assess the vector compression without particular attention to preservation processes. The compression ratio ($\mu_1$) is 45.87 and actual ratio ($\mu_2$) is 8.36. The maximum error, which is located at the outer boundary, is 2.56 X $10^{-1}$ with normalized L2 error of 3.67 X $10^{-6}$. Three dimensional total velocity magnitude contours yield the same pattern of scalar pressure contours in Figure 3 (b).

664

|  | $\mu_1$ | $\mu_2$ | $L_\infty$ | $L^2$ |
|---|---|---|---|---|
| **Scalar (pressure)** | 40 | 8.11 | $7.93 \times 10^{-2}$ | $2 \times 10^{-6}$ |
| **Vector (no density)** | 45.87 | 8.36 | $2.56 \times 10^{-1}$ | $3.67 \times 10^{-6}$ |

Table 2: Data compression ratio

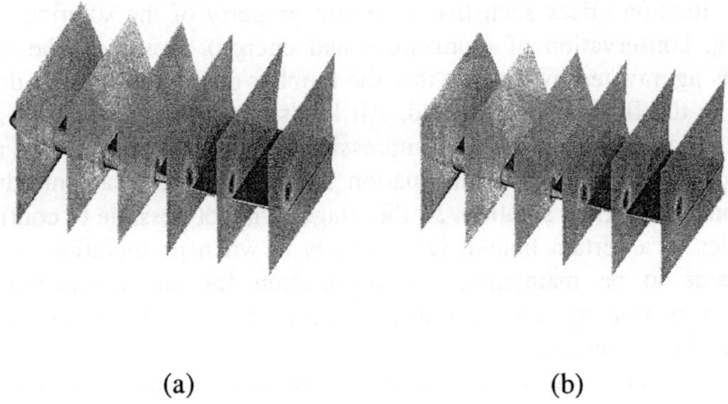

(a) (b)

Fig 3. Pressure contour for wingtip vortex propagation solution ($115 \times 189 \times 115$); compression ratio is 40:1  third order scheme; three level multiresolution; shrinking boundary treatment. (a) Before compression, (b) after compression.

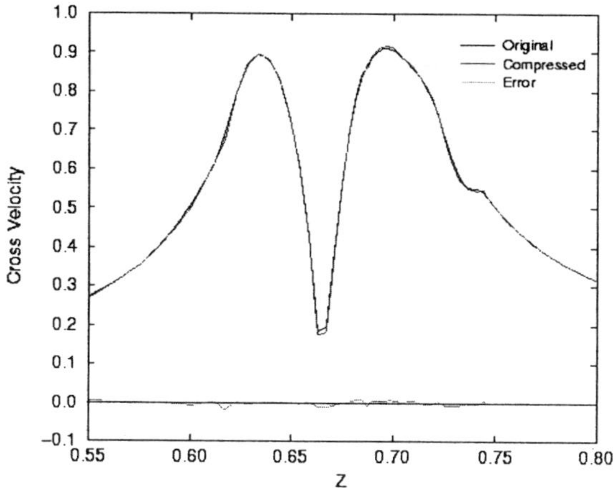

Fig. 4. Cross velocity magnitude across vortex core at x/c=1.5.

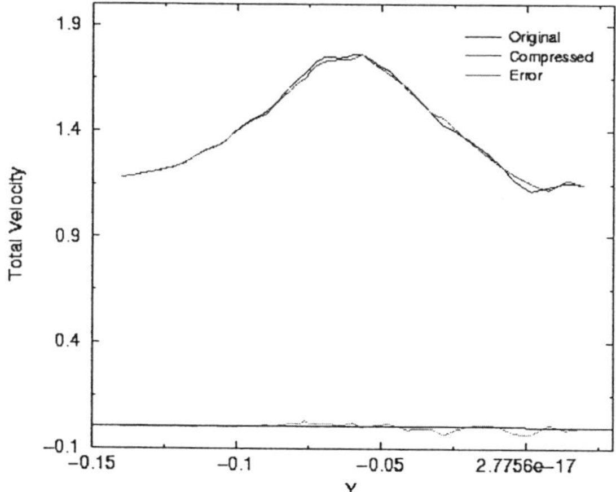

Fig. 5. Total velocity magintude across vortex core at x/c= 1.5.

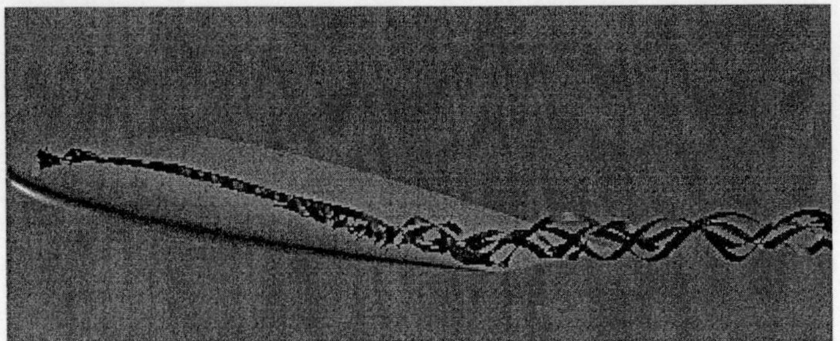

Fig. 6 Streamline ribbons for wingtip vortex propagation solution
before compression

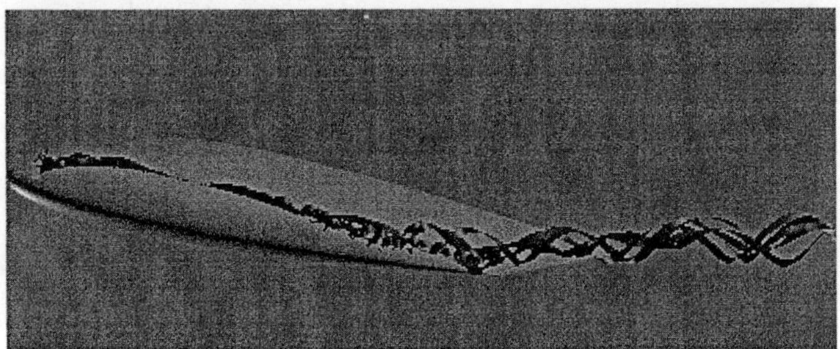

Fig.7 Streamline ribbons for wingtip vortex propagation solution
after reconstruction from compressed data

Fig. 4 and 5 shows cross velocity magnitude and total velocity magnitude across
the vortex core in the cutting plane normal to the **x** (flow) direction, respectively. It
may include some sampling and interpolation errors during data extraction. Figure 4
shows the typical anatomy of a fully developed vortex core: cross velocity magnitude
$\sqrt{(v^2 + w^2)}$ approaches zero in the center of the vortex core and has peak values
axisymmetrically. However, the wingtip vortex is not perfectly axisymmetric. The
figure also shows the initial development of a secondary vortex in the right (outboard
side of wing). The compressed data produces only one weak point on the right side
peak at $z = 0.68$. From Fig. 5, it is evident that total velocity magnitude, three
velocity combination, has a bit less smoothness and more error, compared to cross
velocity magnitude which has two value combination.

Fig. 6 and 7 show streamline ribbons, which indicate particle trajectory and vortex propagation as well as strength of local vorticity. This also shows that the range of influence in compressed data is a little broader than in the original data. While high data compression is achieved, the streamline error, which is sensitive to variation in flow conditions and usually grows exponentially along with the flow path, appears to be small.

The second data set is wing-fuselage simulation results obtained from the full potential equation solver developed by T. Holst [31,32,33]. The data set is composed of three overlapping grid solutions. The sub-grid data size is as follows:

$(97 \times 49 \times 49)$ for overall grid; $(192 \times 49 \times 13)$ for wing grid; $(192 \times 13 \times 49)$ for wing grid. Since the data size is favorable to multiresolution application, there is not much loss of compression ratio due to shrinking or expansion of the application size. On wavelet vector domain, the compression ratio $(\mu_1)$ is 11.5. Considering all overheads that include some location information storage, etc, the compression ratio drops to 5.4. Compression error is still at the same order of the other cases. With third order accurate multiresolution, the L2 error norm is $3.8 \times 10^{-6.}$ (6.26: L2 ratio) Fig. 8 the density contours at some wing stations show that the contour with data compression is identical to the original solutions. Fig. 9 also shows that internal energy contours at fuselage stations are the same in both compressed data and original data.

Since it is overlapping grid solutions, special care needs to be taken at overlapping region. Some area of overlapped region is called IJK-Blanking where garbage values reside that are not updated at all during governing equation solving process. In order to prevent adverse effect of garbage values onto neighboring area during decomposition, the garbage values need to be replaced with more reasonable values. As a preprocessing to multiresolution, smoothing for the garbage values is conducted with 5 point Laplacian operator along with adequate boundary conditions. At the IJK-Blanking boundary, the values of non IJK-Blanking area are used as boundary condition. Fig. 10 shows pressure coefficient curves at the conjunction of wing and fuselage where IJK-Blanking is located. The one without any treatment for IJK-Blanking produces unnecessary kink and oscillations at the fuselage sideline before and after wing.

It is because the completely nonsense garbage values deteriorate the multiresolution process. With smoothing, the kink is completely removed and the oscillation at the trailing edge is substantially reduced. This is nearly identical to uncompressed case. The smoothing can be executed in independent pre-processing program, so it does not affect compression procedure itself.

Before compression      After reconstruction from compressed data

Fig. 8  Density contours at wing stations of wing-body configuration

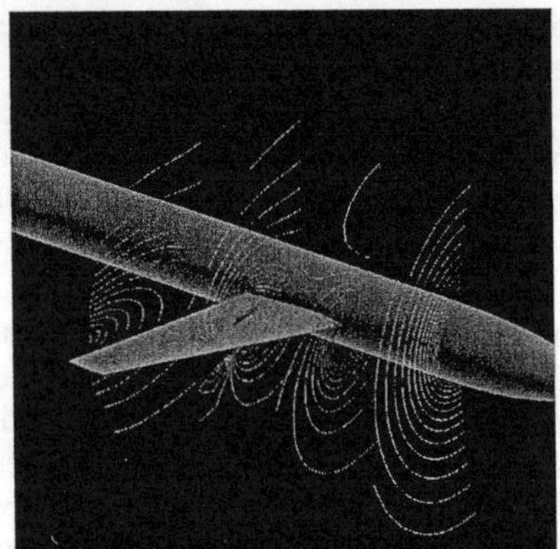 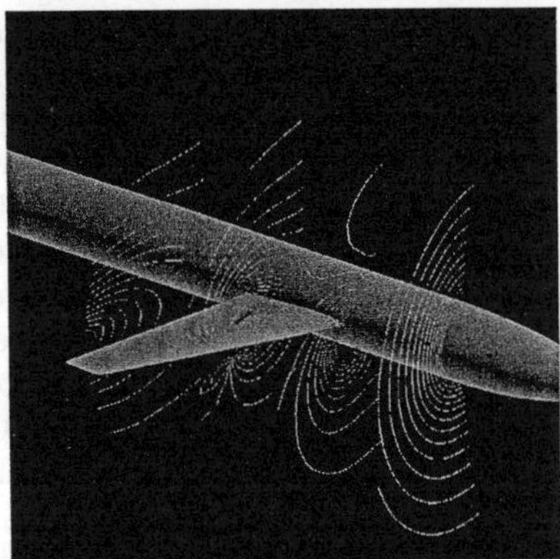

Before compression      After reconstruction from compressed data

Fig. 9  Internal energy contours at fuselage stations  for wing body configuration

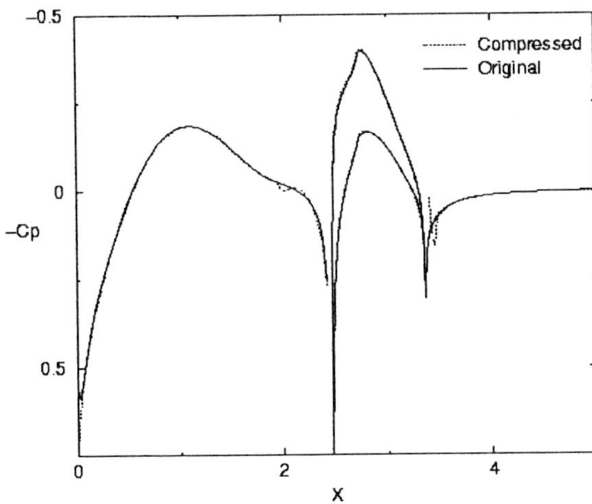

Fig. 10 Cp curve at sideline of fuselage and conjunction line of wing and fuselage.


## 9. Concluding Remarks

Numerical simulation outputs with various fidelity levels are compressed with multiresolution process based on supercompact multi-wavelets. Incompressible Navier-Stokes solutions and potential flow solutions are successfully compressed by the proper wavelet choice and vector thresholding techniques, yielding high compression ratio at around order of two.

The supercompact wavelet is an appropriate choice for data compression of computational fluid dynamics simulations. This can be justified by the fact that it offers strong compact support and accordingly, it can avoid unnecessary interaction with other remotely located data. This compactness becomes more important where strong nonlinearity prevails such as discontinuous shock waves and in wall boundary regions. Numerical evidence also illustrates that these supercompact multi-wavelets can be utilized without modification even near boundaries.

Supercompact multi-wavelets are generalized to three dimensions with the use of multidimensional scaling and wavelet functions rather than with a tensor approach.

Three dimensional data from actual computational simulations is compressed by using the supercompact wavelet and vector thresholding techniques.

The data compression ratio is determined by various factors: wavelet adaptability for flow feature, memory overhead from wavelet usage, the domain size, and the number of possible multiresolution levels. The adaptability may vary depending on individual features such as shocks and vortices. The supercompact wavelet is considered as a proper choice in this regard.

## References


[1]     Engquist B, Osher S, Zhong S. Fast wavelets based algorithms for linear evolutions. ICASE Report, 1992.

[2]     Vasilyev O, Paolucci S. A dynamically adaptive multilevel wavelet collocation method for solving partial differential equations in a finite domain. JCP 1996;125:498-512.

[3]     Harten A. Multiresolution algorithm for the numerical solution of hyperbolic conservation laws. Comm. Pure Appl Math 1995;48:1305-42.

[4]     Gerritsen M. Designing an efficient solution strategy for fluid flows. PhD Thesis, Stanford University, 1996.

[5]     Jameson L, Adachi T, Ukai O, Yuasa A. Wavelet –based numerical methods. Int J Computat Fluid Dyn 1998;10:267-80.

[6]     Sweldens W. The lifting scheme: a new philosophy in biorthogonal wavelet constructions. Proc SPIE 1995;2569:68-79.

[7]     Sweldens W, Schroder P. Building your own wavelets at home. ACM SIGGRAPH Course Note # 13, 1996.

[8]     Strang G. Wavelets and dilation equations: a brief introduction. SIAM Rev 1989;31:614-27.

[9]     Daubechies I. Orthonormal basis of compactly supported wavelets. Comm Pure Appl Math 1988;41:909-96.

[10]   Geronimo J, Hardin D, Massopust P. Fractal functions and wavelet expansions based on several scaling functions. J Approx Theory 1994;78:373-401.

[11]   Strang G, Strela V. Orthogonal multiwavelets with vanishing moments. J Opt Engng 1994;33:2104-7.

[12]   Donovan G, Geronimo J, Hardin D. Interwining multiresolution analyses and the construction of piecewise polynomial wavelets. SIAM J Math Anal 1996;27(6):1791-815.

[13]   Beam RM, Warming RF. Discrete multiresolution analysis using hermite

interpolation: biorthogonal multiwavelets. Technical report, SIAM J Sci Comput 2000: 22(4):1269-317.

[14] Alpert B. A class of bases in $l^2$ for the sparse representation of integral operators. SIAM J Math Anal 1993;24(1):246-62.

[15] Beam RM, Warming RF. Multiresolution analysis and supercompact multiwavelets. SIAM J Sci Comput 2000: 22(4):1238-68.

[16] Harten A. Discrete multi-resolution analysis and generalized wavelets. Appl Numer Math 1993;12:152-92.

[17] Harten A. Multiresolution representation and numerical algorithms: a brief review. Technical Report 94-59, ICASE report, 1994.

[18] Downie TR, Silverman BW. The discrete multiple wavelet transform and thresholding. IEEE Trans Sig Proc, 1996.

[19] Rogers SE, Kwak D, Kiris C. Steady and unsteady solutions of the incompressible Navier-Stokes equations. AIAA J 1991;29(4):603-10.

[20] Buning Pieter G, et al. OVERFLOW user's manual. Version 1.8, NASA Ames Research Center, February 1998.

[21] Rott N. On the viscous core of a line vortex. J Appl Math Phys 1958;9b(5/6):543-53.

[22] Dacles-Mariani J, Zilliac GG, Chow JS, Bradshaw D. Numerical/experimental study of a wingtip vortex in the near field. AIAA J 1995;33(9):1561-8.

[23] Lee D, Beam RM, Warming RF. "Supercompact multiwavelets for flow field simulation", Computers & Fluids 2001: 30: 783-805.

[24] Lee D. "Multiresolution with Supercompact Wavelets", 5th High Performance Computing and Communication/Co- mputational Aerosciences (HPCC/CAS) Workshop, NASA Ames Research Center, February 15-17, 2000.

[25] Lee D. "Supercompact multi-wavelets for three dimensional flow field simulation", AIAA Paper 2000-0396, 38th AIAA Aerosapce Sciences Meeting & Exhibit, Reno, NV, January 10-13,2000.

[26] Cox M, Ellsworth D. Managing big data for scientific visualization. In ACM SIGGRAPH '97 Course # 4, Exploring Gigabyte Datasets in Real-Time: Algorithms, Data Management, and Time-Critical Design, 1997.

[27] Blanford, Ronald P. (TRW Systems Group), Wavelet encoding and variable resolution progressive Transmission, The 1993 Space and Earth Science Data Compression Workshop, NASA TR 19930015359 N(93N24548)

[28] Sung J, Multiresolution Modeling of Curvilinear Grids, PhD thesis, Arizona State University, 1997.

[29] Dahmen W, Gottschlich-Muller B, Muller S, Multiresolution Schemes for

Conservation Laws, IGPM 1998: 157

[30] Cohen A., Daubechies I., Jawerth B., Vial P. Multiresolution analysis, wavelets and fast wavelet transform on an Interval, CRAS Paris, Ser. A 1993: t. 316:417-421

[31] Holst T. L. Multizone chimera algorithm for solving the full-potential equation, AIAA J Aircraft 1998: 35(3): 412-21.

[32] Holst T. L. Transonic flow computations using nonlinear potential methods, Progress in Aerospace Sciences 2000:36:1-61.

[33] Holst T. L. On approximate factorization schemes for solving the full potential equation, 1997, NASA TM 110435

# The Historic Remains 'Drifting Cup in a Meandering Stream' in China, Korea and Japan

**Keun-Shik Chang**

*Department of Mechanical Engineering, Korea Advanced Institute of Science and Technology*
*373-1 Kusong-dong, Yusung-ku, Taejon 305-701, Korea*
*kschang@kaist.ac.kr*

## Abstract

The 'Drifting Cup in a Meandering Stream' was initiated in the ancient China by a famous poet and calligrapher called Wang Xhi Zhi. He summoned 42 social celebrities for a spring party held at Ranting (Orchid Pavilion) where a meandering stream was pulled from a nearby creek. They seated themselves along the stream to compose Chinese poems. Cups filled with rice wine were drifted on the stream and the rule was to finish the poem before the cup arrived at the seat of each poet. If a poet had failed, he had to pardon himself by drinking three cups of rice wine. This witty archaic event has gained popularity afterwards and spread to neighbor countries to become a fabulous culture long enjoyed by the rulers in this region. Many Meandering Streams have been built even up to relatively modern times. The author has looked for a list of the Meandering Streams in China, Korea and Japan, and then visited on leisure time many of those historic remains to examine their designs and functions. The author has investigated by experimental model and computer simulation the flow of some Meandering Streams. The author found that there had been a crucial change in the configuration of the Meandering Streams in the history in order to control motion of the drifting cup. Curiously enough, Posuk-Chung Pavilion located in Kyongju, Korea, is in the midst of such morphological transformation .

## 1. The 'Drifting Cup in a Meandering Stream' in the far-east Asian countries

The first Meandering Stream Feast held on the 3rd of March, A.D. 353, in Siaoxing, Zhejiang Province of China was well documented by the host scholar Wang Xhi Zhi himself in his article 'Narration of the Orchid Pavilion.' The term 'Drifting Cup on a Meandering Stream' was originated from this book. The Meandering Stream Feast has quickly spread afterwards in the far-east Asian countries, China, Korea, and Japan. The Posuk-Chung Pavilion had been used some time in the first millennium of the Silla Dynasty in Korea. Similar feasts had been held in Japan as early as sixth century, according to Nihon Shoki, the history book of Japan.

The Ya Shang Ting in the Forbidden City, Liu Bei Ting in the Tan Zhe Si Buddhist temple, Zuo Shi Yan Liu in the ruins of Yuan Ming Yuan Flower Garden are the Meandering Streams in Beijing used by the Emperor Chien Long in Ching Dynasty. Each of these Beijing canals has a pavilion with its roof casting a shade, except the ruined one in the Yuan Ming Yuan. Japan also has a number of her own historic remains at Jyonangu, Gosho and Kamikamo-Jinsha, all in Kyoto, Mouetsuji in Hiraizumi, Heijiokyo in Nara, Gorakuen in Okayama, Tazaifu-Tenmangu in Fukuoka, and Isoteien in Kagoshima. In many of the palaces, temples and gardens where these Menadering Streams are located, the old-fashioned Meandering Stream Feast is annually reproduced either in the spring or fall before a large eager gallery.

It is not known when the Posuk-Chung Pavilion shown in Fig. 1 was exactly built but Samkuksagi, the Korean history book, recorded that the 49th King Hunkang of Silla Dynasty had made an excursion with his subordinates to the Posuk-Chung Pavilion in the year A.D. 879. Unfortunately, however, the wonderful culture of Meandering Stream Feast has been suddenly eradicated in Korea because of a tragic incident that the 55th King Kyongae was slain on this ill-fated Posuk-Chung Pavilion in A.D. 927 by an infiltrated army led by Gyunwhon, the leader of a rival country called Hoobaekje. This misfortune was followed by the fall of a millennium-old Silla Dynasty in the year A.D. 935.

Posuk-Chung is a meandering canal made of 63 pieces of hammered granite for the side  walls and bottom plates, assembled to have a rectangular cross-section shape. The main loop measures more than 10 meters in the long axis and 5 meters in the short axis. The loop is roughly of an oval shape but

**Fig. 1 Meandering Stream of Posuk-Chung Pavilion**

a couple of highly serpentine curves are introduced to break the monotony of the shape. The canal roughly measures about 0.3m wide and 0.22m deep in its cross section. There is at present no pavilion left above the canal, contrary to the name: no record shows when it was lost. Having been defunct such a long time, the canal has neither water supply nor drainage conduit.

In the present study, Posuk-Chung canal is computationally simulated by solving the Navier-Stokes equations. Experiment was also conducted by fabricating a 1/6.6 scale model using metal plates. To obtain an accurate shape of the irregular canal, the author took an aerial photograph aboard a thermal balloon. To remove branches of the tree that hide the canal in the photograph, local pictures taken from the ground were patched using Adobe Photoshop. The Chinese Meandering Stream, Zuo Shi Yan Liu, in the Yuan Ming Yuan Flower Garden is also simulated to examine the stream characteristics. The morphology of Meandering Streams is investigated in Section 4 to identify the historical change of canal design

## 2. Computer Simulation and Model Experiment

The meandering stream would manifest elements of the river flow, meandering and braiding having three-dimensional characteristics. One thing noteworthy is that surface tension becomes more pronounced in the reduced scale. Flows separated from the main stream will be either stagnant, reversed or recirculating in the canal. The recirculating vortex can capture wine cup drifted into the particular region, if this region is not so small. Release of the wine cup from the vortex trap is also possible since the three-dimensional vortical flow fluctuates with time in its size and strength. An experimental study made earlier [1] revealed that the channel flow was indeed complicated. Measuring flow variables like velocity, pressure, surface waves, surface tension, and passage of drifting objects would be not only exhausting but also impractical.

We consider the three-dimensional incompressible Navier-Stokes equations.

$$\rho \frac{\partial \vec{u}}{\partial t} + \rho \vec{u} \cdot \nabla \vec{u} = -\nabla p + \rho \vec{g} + \mu \nabla^2 \vec{u} \tag{1}$$

$$\nabla \cdot u = 0 \tag{2}$$

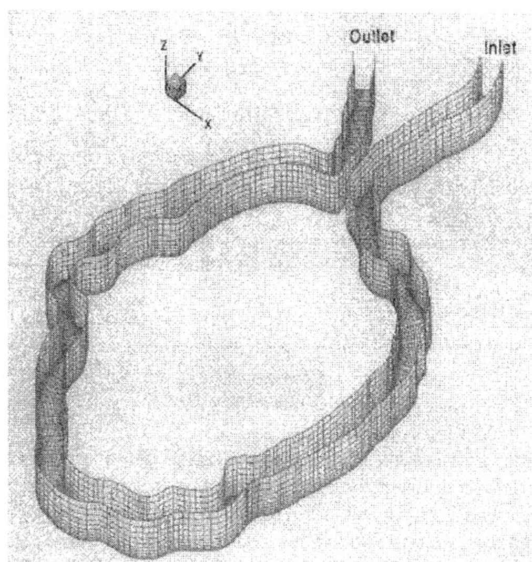

**Fig. 2 Surface grid of the Posuk-Chung numerical model.**

To deal with geometric complexity, the author used the FEM technique. An in-house code called DOLFINS (Drastically Operator-Lightened Finite-element Incompressible Navier-Stokes Solver) was used [2]. To make the problem tractable, the author assumes that there is no shear stress on the free surface and the channel flow is driven by pressure gradient, not by gravitational force.

The experimental scale model was visualized by the powder method. Surface streaklines were obtained in the photographs taken with different exposures. It was difficult to map the passage of drifting cup in the scale model due to lack of dynamic similitude: depending on the hydrophobic or hydrophilic property of the cup material, the surface tension could become unreasonably large relative to other forces like gravitational force, inertia force, shear force and unsteady acceleration force.

### 3. Flow in the Posuk-Chung Meandering Stream

Fig. 2 represents the numerical model of the canal, having 10,780 grid points on the surface among the total 84,780 points used in the whole domain [3]. The flow enters the inlet plane with a specified uniform velocity and leaves the outlet without any flow gradient. The flow is assumed to have constant depth as well as smooth surface on the side and bottom of the canal. The Reynolds number of the prototype is about in the order $10^5$. In the experiment Reynolds number was in the range 500 to 5000. Computation was performed in the unsteady laminar flow regime with Reynolds number 1000.

The velocity vectors obtained computationally on the free surface are plotted in Fig. 3. They suggest location and strength of the vortices distributed along the canal. The three photographs of Fig. 4 show experimental streaklines photographed at the vortex regions labelled B-b, C-c and D-d in Fig.3. The experimental results in Fig. 4 are well compared with the numerical results in Fig. 3.

Now we simulate computationally the trajectory of a particle mass drifting on the surface, which is an ultimate simplification of the 'drifting cup on a meandering stream'. First the Stokes number is defined by

$$St = \frac{m_P}{C_D \cdot \frac{1}{2} \rho A \left| V_f - V_P \right|} \tag{3}$$

The motion of the particle mass relative to the fluid is determined by

676

$$V_P = V_f + \left(V_{P0} - V_f\right)\exp\left(-\frac{\Delta t}{St}\right)$$ (4)

in which $V_{P0}$ is the initial velocity of the particle mass, $V_P$ and $V_f$ are respectively the particle velocity and fluid   velocity, and $m_P$ is the mass of the particle. The particle trajectory is highly dependent on the initial condition and particle properties such as particle mass and drag coefficient. Fig. 5 shows a particle trajectory calculated with Stokes number 0.3.  We can summarize that the particle mass will, in general, stream near the outer bank of the canal due to the centrifugal force and could be trapped in a few recirculating vortex regions.  It can also escape the vortex trap on the reason explained earlier.


## 4. Morphological Evolution of Meandering Streams

**Construction Methods of Meandering Stream**
   The man-made Meandering Stream is ultimately a replica of the natural creek. It is reduced in scale and brought to the garden for the purpose of Meandering Stream Feast and landscape. Initially the shape of the Meandering Stream had to be close to the natural setting. Design features were introduced in the later stage to introduce a certain control function to the canal.  Initially Meandering Stream was built relatively large. The bed of canal was laid with stones and pebbles and the bank was decorated with rocks and flowers.  The Meandering Streams in Ran Ting, Isoteien Garden in Kagoshima, Jyonangu in Kyoto, and Mouetsuji in Hiraizumi are all good examples.   We will call this type of canals as 'the landscape type.'  Fig. 6 shows a modern Meandering Stream Feast being held in the canal of Jyonangu. The cup is drifted on a wood saucer in this picture.

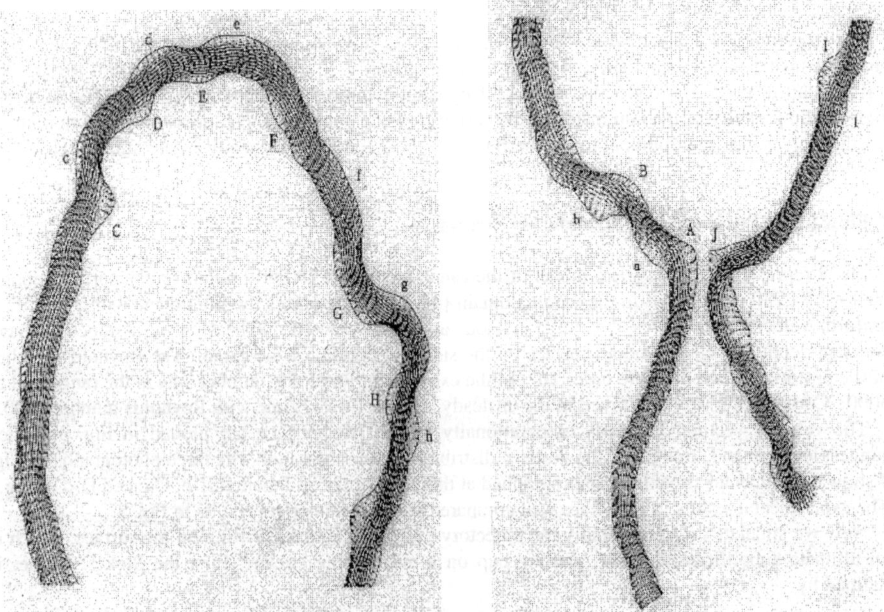

Fig. 3 Velocity vectors of the meandering stream.

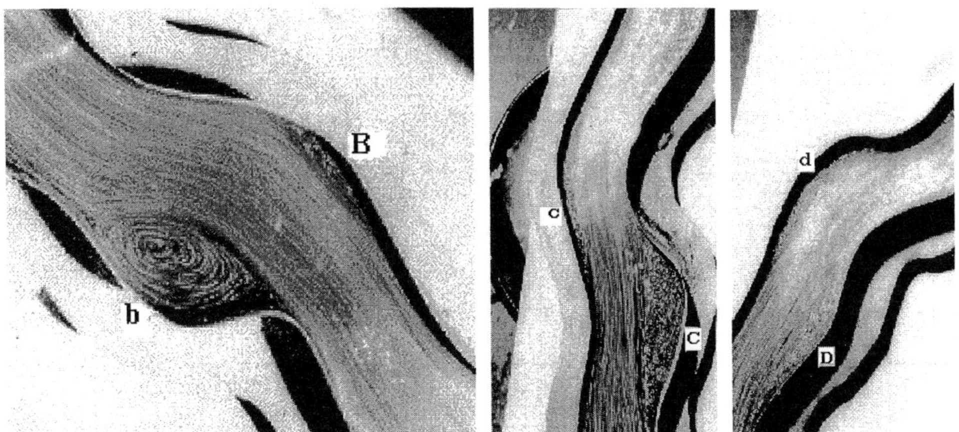



**Fig. 4 Surface streakline in the model experiment.**

To award controllability and durability to the Meandering Stream, the canals were fabricated in later stage with hammered granites or marble stones carved out. The Posuk-Chung Pavilion is an assemblage of finished granites; the Meandering Streams used by Chien Long Emperor in Ching Dynasty are a few pieces of marble stone laid side by side with the meandering narrow grooves carved out on the surface.

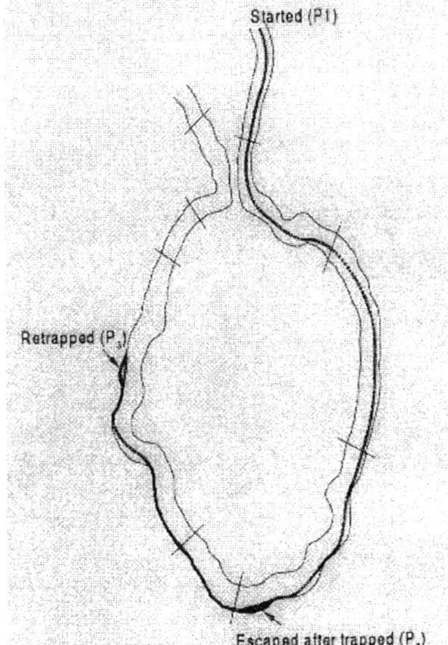



**Fig. 5 Trajectory of the particle mass in the channel.**

## Clock Function of a Meandering Stream

Since the canal is used for the enjoyment of composing Chinese poems in a festival or contest mood, it would be better if the drifting cup had accurate time measuring function as it streams down the canal. The Meandering Streams of Ching Dynasty are definitely equipped with this function since the cup drifting in a narrow groove of width about 0.1m cannot be astray or end up in a vortex trap. Recall that in 'the landscape type' designs like Ranting or most Japanese Meandering Streams, which

**Fig. 6 The modern Meandering Stream Feast in Jyonan-gu canal, Kyoto, Japan.
Courtesy of Professor Y. Nakayama in Future Technology Research Institute, Tokyo.**

has canals of width about 0.5m, the cup would frequently be stranded by the separated vortices lying in ambush along the winding channel. In contrast, the groove of the canals in Beijing is precisely symmetric and extremely serpentine, perhaps in an attempt to allow more time to compose lengthy Chinese poems; see Fig. 7 below as an example. A Meandering Stream designed to have this clock function does not need to be very large. Therefore the canal can be fabricated with stones and consequently a pavilion can be built above. We will call this type of canals as 'the pavilion type.' Fig. 8 represents the streamlines computed on a rough two-dimensional geometrical model of the Chinese canal shown in Fig. 7. Here it is observed that the number and scale of the vortices are very much reduced in the narrow channel.

**Fig. 7 Meandering Stream of Yuan Ming Yuan in Beijing, China.**

## Transition of Morphology

The morphology of Posuk-Chung Meandering Stream curiously lies in between 'the landscape type' and 'the pavilion type.' It may be considered of pavilion type since it entirely consists of hammered stones; there was indeed a pavilion in the beginning as its name suggested. It is not as large as Ran Ting or many Japanese Meandering Streams. The drifting cup in the Posuk-Chung canal of width about 0.3m is surely far better controlled than the one in the Meandering Streams of 'the landscape type.

On the other hand, Posuk-Chung Meandering Stream is still regarded large in comparison with the more compact Chinese canals in Beijing. The Posuk-Chung canal is wider and simpler than the ones in Beijing having grooves very narrow and serpentine. There is spacious room inside and around the Posuk-Chung canal so that, for example, more than ten persons can be seated on the occasion of a festival which is not true for the Beijing canals. The drifting cup can evade into a few vortical shades but far less frequently than in 'the landscape type' Meandering Streams.

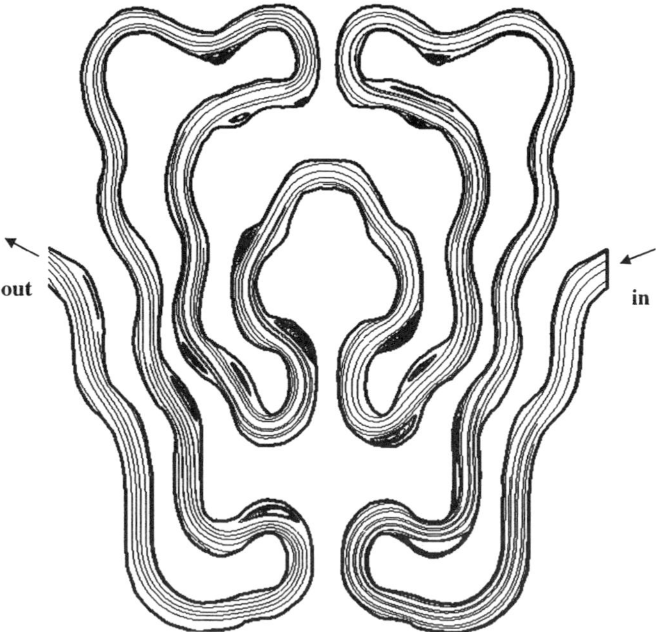


**Fig. 8 The computed streamlines obtained for the Yuan Ming Yuan canal in Beijing, China**

## 5. Conclusion

The Meandering Stream Feast has been a culture long admired by the rulers who understood Chinese poems in the far-east Asian countries which borrowed Chinese characters. The present study has offered an excellent opportunity to reveal how an old culture arisen in one nation could quickly spread to the neighbour countries to be modified. In this paper, the Meandering Streams in China, Korea and Japan are listed and reviewed. Flows in some of the canals are investigated experimentally and computationally. Based on this flow information and by comparing the overall shape or size of the canals, it was possible to classify the Meandering Streams into 'the landscape type' and 'the pavilion type.' There has been definitely a morphological transformation from the earlier 'landscape type' to the later 'pavilion type.' Curiously enough, the Posuk-Chung Pavilion seems to lead this transition, presumably as the first Meandering Stream that has brought the canal under the pavilion roof.

## References

[1] Chang K.S., A study of Vessel-Floating Curved Water Passage of Posuk Pavilion in Kyungju from the Fluid Science View Point, Misulcharyo, Vol. 46, pp. 101-110 and pp. 114-115, December 1990, National Central Museum, Korea (in Korean).

[2] Shim E.B., Chang K.S., Three-Dimensional Vortex Flow past a Tilting Disc Valve Using a Segregated Finite Element Scheme, Computational Fluid Dynamics J., Vol. 3, No. 1, pp. 205-222, April 1994.

[3] Chang K.S., Shim E.B., Numerical Flow Visualization of Posuk-Chung, the Ninth Century Remains of Poetry-Making Curved Water Channel in Korea, Atlas of Visualization III, pp. 241-255, CRC Press, New York, 1997.